## Straight Lines

Slope: $m = \dfrac{y_2 - y_1}{x_2 - x_1}$

Point-slope form: $y - y_1 = m(x - x_1)$

Slope-intercept form: $y = mx + b$

Horizontal line: $y = b$

Vertical line: $x = a$

If $m_1 = m_2$, then $L_1 \parallel L_2$

If $m_1 = -\dfrac{1}{m_2}$, then $L_1 \perp L_2$

Distance: $d = \sqrt{(x_2 - x_1)^2 + (y_2 - y_1)^2}$

Midpoint: $x_m = \dfrac{x_1 + x_2}{2}$; $y_m = \dfrac{y_1 + y_2}{2}$

Quadratic Formula: If $ax^2 + bx + c = 0$ ($a \neq 0$),

then $x = \dfrac{-b \pm \sqrt{b^2 - 4ac}}{2a}$

## Properties of Logarithms

$x = a^y$ is equivalent to $y = \log_a x$, $a > 0$, $a \neq 0$.

1. $\log_a (M \cdot N) = \log_a M + \log_a N$

2. $\log_a \left(\dfrac{M}{N}\right) = \log_a M - \log_a N$

3. $\log_a M^n = n \log_a M$

4. $\log_a \sqrt[n]{M} = \dfrac{1}{n} \cdot \log_a M$

5. $\log_a 1 = 0$

6. $\log_a \dfrac{1}{M} = -\log_a M$

7. $\log_a a^x = x$

8. $a^{\log_a x} = x$

9. $\log x = \log_{10} x$

10. $\ln x = \log_e x$

# Mathematics for Technical Education

## THIRD EDITION

**Dale Ewen**
Parkland Community College

**Joan S. Gary**
Parkland Community College

**James E. Trefzger**
Parkland Community College

**Prentice Hall**
Upper Saddle River, New Jersey     Columbus, Ohio

**Library of Congress Cataloging-in-Publication Data**

Ewen, Dale
    Mathematics for technical education. — 3rd ed. / Dale Ewen,
  Joan S. Gary, James E. Trefzger.
      p.  cm.
    Includes index.
    ISBN 0-13-895517-4
    1. Mathematics.   I. Gary, Joan S.   II. Trefzger, James E.   III. Title.
QA93.2.E94 1998                                             97-35563
512'.13'0246—dc21                                          CIP

Editor: Stephen Helba
Production Editor: Christine M. Harrington
Design Coordinator: Karrie M. Converse
Text Designer: John Edeen
Cover Designer: Proof Positive/Farrowlyne Assoc., Inc.
Production Manager: Patricia A. Tonneman
Illustrations: Jane Lopez
Marketing Manager: Frank Mortimer, Jr.

This book was set in Times Roman and Optima by York Graphic Services and was printed and bound by Quebecor Printing/Book Press. The cover was printed by Phoenix Color Corp.

 © 1998, 1983, 1976 by Prentice-Hall, Inc.
Simon & Schuster/A Viacom Company
Upper Saddle River, New Jersey 07458

Printed in the United States of America

10 9 8 7 6 5

ISBN: 0-13-895517-4

Prentice-Hall International (UK) Limited, *London*
Prentice-Hall of Australia Pty. Limited, *Sydney*
Prentice-Hall of Canada, Inc., *Toronto*
Prentice-Hall Hispanoamericana, S. A., *Mexico*
Prentice-Hall of India Private Limited, *New Delhi*
Prentice-Hall of Japan, Inc., *Tokyo*
Simon & Schuster Asia Pte. Ltd., *Singapore*
Editora Prentice-Hall do Brasil, Ltda., *Rio de Janeiro*

# Preface

*Mathematics for Technical Education,* Third Edition, provides the necessary mathematics skills for students considering a career in a technical or engineering technology program and emphasizes the mathematical concepts that students need to be successful. This third edition has been carefully reviewed, and special efforts have been taken to emphasize clarity and accuracy of presentation.

The text presents the following major areas: fundamental concepts of operations with numbers, the metric system, and measurement; fundamental algebraic concepts; exponential and logarithmic functions; right-triangle trigonometry, the trigonometric functions; and trigonometric formulas and identities; complex numbers; and analytic geometry.

## KEY FEATURES

- Numerous detailed, illustrated examples
- Chapter review summaries
- Chapter review exercises
- Highlighting of important formulas and principles
- Comprehensive development and consistent use of measurement and significant digits throughout the text
- Basic instruction on using a scientific calculator throughout the text
- Basic instruction on using a graphing calculator given in Appendix D that may be integrated throughout the text as a faculty option
- More than 5000 exercises
- Essential geometry reviewed in Appendix A
- Answer Manual for even-numbered exercises

### Illustration of Some Key Features

**Exercises**   To reinforce key concepts for students, we've provided a large variety of well-illustrated exercises. For an example, see page 406 displayed on page iv of this Preface.

**Calculator Flow Charts**   Calculator flow charts are used to show students the sequence of the step-by-step operations and the displays at each step. Page 400, displayed on page iv, shows the scientific calculator flowchart for using the Law of Cosines.

**Chapter End Matter**   A chapter summary and a chapter review are provided at the end of each chapter to reinforce students' understanding of the concepts and to help them review for quizzes and examinations.

the sun and Earth and Venus is 25°. (Assume that Earth and Venus have circular orbits around the sun.)

17. A farmer wants to extend the roof of a barn in Fig. 11.26 to add a storage area for machinery. If 20.0 ft of clearance is needed at the lowest point, how wide can the addition be?

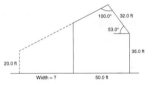

**Figure 11.26**

18. In the framework in Fig. 11.27, we know that $AE = CD$, $AB = BC$, $BD = BE$, and $AC \parallel ED$. Find the measure of each.
   (a) $\angle AEB$     (b) $\angle A$     (c) $BE$     (d) $DE$

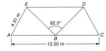

**Figure 11.27**

19. In the roof truss in Fig. 11.28, we know that $AB = CD$, $AG = DE$, $GF = FE$, $BG = CE$, and $BF = CF$. Find the measure of each.
   (a) $BG$    (b) $\angle ABG$    (c) $BF$    (d) $\angle GFB$    (e) $\angle FBC$

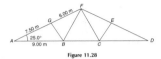

**Figure 11.28**

20. Find the distance between the peaks of two hills across the gorge in Fig. 11.29. Points $A$ and $B$ are trees on the peaks of the two hills. Point $C$ is where you stand. Measure a length of 100 m from point $C$ to point $D$. Then $\angle BCD$ measures 115.0°, $\angle CDA$ measures 120.5°, $\angle BCA$ measures 86.5°, and $\angle ADB$ measures 98.1°.

---

which is the Pythagorean theorem. The Pythagorean theorem is thus a special case of the law of cosines.

**EXAMPLE 1**

If $a = 112$ m, $b = 135$ m, and $C = 104.3°$, solve the triangle.

First, draw a triangle as in Fig. 11.16 and find $c$ by using the law of cosines.

**Figure 11.16**

$$c^2 = a^2 + b^2 - 2ab \cos C$$
$$c^2 = (112 \text{ m})^2 + (135 \text{ m})^2 - 2(112 \text{ m})(135 \text{ m})(\cos 104.3°)$$
$$c = 196 \text{ m}$$

This side may be found using a calculator as follows:

| Flow chart | Buttons pushed | Display |
|---|---|---|
| Enter $112^2$ | [1][1][2][$x^2$] | $112^2$ |
| Push plus | [+] | |
| Enter $135^2$ | [1][3][5][$x^2$] | $135^2$ |
| Push minus | [−] | |
| Enter 2 | [2] | 2 |
| Push times | [×] | |
| Enter 112 | [1][1][2] | 112 |
| Push times | [×] | |
| Enter 135 | [1][3][5] | 135 |
| Push times | [×] | |
| Enter cos 104.3° | [cos][1][0][4][.][3] | cos 104.3 |
| Push equals | [=] | 38238.3 |
| Find square root | [√][ans][=] | 195.546 |

So $c = 196$ m rounded to three significant digits.

# TO THE FACULTY

The topics have been arranged with the assistance of faculty who teach in a variety of technical programs. However, we have also allowed for many other compatible arrangements. The topics are presented in an intuitive manner, with technical applications integrated throughout whenever possible. The large number of detailed examples and exercises are features that students and faculty alike find essential.

The text is written at a language level and a mathematics level that is cognizant of and beneficial to most students in technical programs. We assume that students have a mathematics background that includes one year of high school algebra or its equivalent and some geometry. The introductory chapters are written so that students who are deficient in some topics may also be successful. The material in this book is designed to be completed in two semesters or three quarters, and it serves as a foundation for more advanced work in mathematics. A companion book, *Technical Calculus,* smoothly follows this text for engineering technology and other programs that require a development of practical calculus.

Chapters 1 through 3 provide the basic skills that are needed early in almost any technical program. Chapters 4 through 9 complete the basic algebraic foundation, while Chapters 10 through 14 include the trigonometry necessary for the technologies. Chapter 15 completes a comprehensive mathematics background needed in many programs. Some programs include this chapter at the end of the first year, while other programs include this chapter at the beginning of an introductory calculus course.

A companion Instructor's Manual containing answers for even-numbered exercises is available.

# TO THE STUDENT

Mathematics provides the essential framework for and is the basic language of all the technologies. With this basic understanding of mathematics, you will be able to quickly understand your chosen field of study and will be able to independently guide your own lifelong education. Without this basic understanding, you will likely struggle and often feel frustrated not only in your mathematics and support sciences courses but also in your technical courses.

Technology and the world of work will continue to change rapidly. Your own working career will likely change several times during your lifetime. Mathematical, problem-solving, and critical-thinking skills will be crucial as opportunities develop in your own career path in a rapidly changing world.

# ACKNOWLEDGMENTS

The authors especially thank the many faculty and students who have used the previous editions and those who have offered suggestions. If anyone wishes to correspond with us regarding suggestions, criticisms, questions, or errors, please contact Dale Ewen directly at Parkland Community College, 2400 W. Bradley, Champaign, IL 61821, or through Prentice Hall.

We extend our sincere and special thanks to our Prentice Hall editor, Stephen Helba; to our production editor, Christine Harrington; and to Joyce Ewen for her excellent proofing assistance.

# Contents

# 1
# Fundamental Concepts

## INTRODUCTION

In a steel mill, the temperature of molten steel, the size of the ingots, the time it takes a batch of molten steel to be poured, and the electronic circuitry that keeps the assembly lines moving—all involve measurements and calculations using those measurements.

In this chapter, we introduce many of the basic mathematical terms and methods of calculating that are required in electronics, construction, and manufacturing plant operations.

The International System of Units, often referred to as the metric system, or the SI Metric system, is used throughout the world. Because so many businesses and consumable products (cars, entertainment systems, machinery, etc.) are international, we need to know how to work with the SI metric system. Likewise, units that are used to measure electricity, temperature, and other things that technicians work with are explained, as well as conversion from one unit of measure to another.

### Objectives

- Add, subtract, multiply, and divide signed numbers.
- Work with numbers in scientific notation.
- Know the units of the SI metric system.
- Convert units within the SI metric system and between the English and the SI metric systems.
- Know the difference between the accuracy of a measurement and the precision of a measurement.
- Perform arithmetic operations with measurements.

## 1.1   THE REAL NUMBER SYSTEM

The **positive integers** are the counting numbers; that is, 1, 2, 3, . . . . Note that the positive integers form an infinite set. The **negative integers** may be defined as the set of

1

opposites of the positive integers; that is, $-1, -2, -3, \ldots$. **Zero** is the dividing point between the positive integers and the negative integers and is neither positive nor negative. The set of **integers** consists of the positive integers, the negative integers, and zero.

The **rational numbers** are those numbers that can be represented as the ratio of two integers, such as $\frac{3}{4}$, $\frac{-7}{5}$, and $\frac{5}{1}$. The **irrational numbers** are those numbers that cannot be represented as the ratio of two integers, such as $\sqrt{3}$, $\sqrt[3]{16}$, and $\pi$.

The set of **real numbers** is the set consisting of the rational numbers and the irrational numbers. With respect to the real number line as in Fig. 1.1, we say there is a one-to-one correspondence between the real numbers and the points on the number line; that is, for each real number there is a corresponding point on the number line, and for each point on the number line there is a corresponding real number. As a result, we say the number line is dense, or "filled."

**Figure 1.1**  The real number line

A **prime number** is defined as a positive integer greater than one that is evenly divisible* only by itself and one. The first ten prime numbers are 2, 3, 5, 7, 11, 13, 17, 19, 23, and 29.

The following are some properties of real numbers:

**1.** $a + b = b + a$ — Commutative property of addition

**2.** $ab = ba$ — Commutative property of multiplication

**3.** $(a + b) + c = a + (b + c)$ — Associative property of addition

**4.** $(ab)c = a(bc)$ — Associative property of multiplication

**5.** $a(b + c) = ab + ac$ — Distributive property of multiplication over addition

**6.** $a + (-a) = 0$ — Additive inverse or negative property

**7.** $a \cdot \dfrac{1}{a} = 1 \quad (a \neq 0)$ — Multiplicative inverse or reciprocal property

**8.** $a + 0 = a$ — Identity element of addition

**9.** $a \cdot 1 = a$ — Identity element of multiplication

## Operations with Signed Numbers

The rules for operations with signed numbers are often stated in terms of their absolute values. Therefore, we shall first define absolute value.

The **absolute value** of a real number $n$, written $|n|$, is defined as

$$|n| = n \quad \text{if} \quad n \geq 0 \quad (\geq \text{ means greater than or equal to.})$$
$$|n| = -n \quad \text{if} \quad n < 0 \quad (< \text{ means less than.})$$

*By "evenly divisible" we mean that the remainder is zero after the indicated division is completed.

**EXAMPLE 1**

Find the absolute value of each number.
(a) $|+7| = 7$
(b) $|-3| = -(-3) = 3$
(c) $|0| = 0$

---

**ADDING TWO SIGNED NUMBERS**

1. If the numbers have the *same* signs, add their absolute values and use the common sign before the sum.

2. If the numbers have *different* signs, find the difference of their absolute values. To this result attach the sign of the number whose absolute value is larger.

---

**EXAMPLE 2**

Add each pair of signed numbers.
(a) $(+7) + (+2) = +9$
(b) $(-3) + (-2) = -5$
(c) $(+6) + (-4) = +2$
(d) $(-8) + (+1) = -7$

---

**ADDING THREE OR MORE SIGNED NUMBERS**

1. Find the sum of the positive numbers.

2. Find the sum of the negative numbers.

3. Add the resulting positive sum and negative sum.

---

**EXAMPLE 3**

Add $(+2) + (-3) + (-7) + (+5) + (-9) + (+3) + (-6)$.

**Step 1:** $(+2) + (+5) + (+3) = +10$
**Step 2:** $(-3) + (-7) + (-9) + (-6) = \underline{-25}$
**Step 3:** $\phantom{(-3) + (-7) + (-9) + (-6) = }-15$

---

**SUBTRACTING SIGNED NUMBERS**

To subtract signed numbers, change the sign of the subtrahend (number being subtracted), and add the resulting signed numbers: $a - b = a + (-b)$.

---

**EXAMPLE 4**

Subtract.
(a) $(-6) - (+2) = (-6) + (-2) = -8$
(b) $(+5) - (+9) = (+5) + (-9) = -4$
(c) $(-3) - (-8) = (-3) + (+8) = +5$
(d) $(+9) - (-2) = (+9) + (+2) = +11$
(e) $(-3) - (-7) - (+5) = (-3) + (+7) + (-5) = -1$

## MULTIPLYING (OR DIVIDING) TWO SIGNED NUMBERS

1. If the numbers have the *same* signs, find the product (or quotient) of their absolute values and place a positive sign before the result.

2. If the numbers have *different* signs, find the product (or quotient) of their absolute values and place a negative sign before the result.

**EXAMPLE 5**

Multiply.
(a) $(+3)(+2) = +6$
(b) $(-8)(-6) = +48$
(c) $(+7)(-9) = -63$
(d) $(-5)(+4) = -20$

**EXAMPLE 6**

Divide.
(a) $(+8) \div (+2) = +4$
(b) $(-20) \div (-4) = +5$
(c) $(-36) \div (+6) = -6$
(d) $(+26) \div (-2) = -13$

## MULTIPLYING AND/OR DIVIDING THREE OR MORE SIGNED NUMBERS

Multiply and/or divide the absolute values of the numbers. Then place a negative sign before the result if there are an odd number of negative numbers, or place a positive sign if there are an even number of negative numbers.

**EXAMPLE 7**

(a) Multiply.

$$(+2)(-4)(+5)(-1)(-3) = -120$$

(Note that there are three negative signs.)

(b) Simplify.

$$\frac{(+3)(-4)(+1)}{(-2)(-5)(-6)} = +\frac{1}{5}$$

(Note that there are four negative signs.)

## Exercises 1.1

*Find each absolute value.*

**1.** $|-15|$      **2.** $|+9|$      **3.** $|+11|$      **4.** $|-2|$

**5.** $|-8|$      **6.** $|+1|$      **7.** $|3-5|$      **8.** $|1-(-6)|$

*Perform the indicated operations and simplify.*

**9.** $(-3) + (-6)$             **10.** $(-5) + (+6)$

**11.** $(+7) + (-9)$             **12.** $(-2) + (-4)$

**13.** $(+6) + (+4)$             **14.** $(+4) + (-10)$

**15.** $(-5) + (+8)$

**16.** $(-8) + (-5)$

**17.** $\left(-\dfrac{1}{3}\right) + \left(-\dfrac{3}{4}\right)$

**18.** $\left(+\dfrac{1}{2}\right) + \left(-\dfrac{5}{16}\right)$

**19.** $\left(-3\dfrac{4}{9}\right) + \left(+4\dfrac{5}{12}\right)$

**20.** $\left(-1\dfrac{5}{8}\right) + \left(-2\dfrac{1}{12}\right)$

**21.** $(-13) + (+3) + (-7) + (+6) + (-2)$

**22.** $(+5) + (-1) + (-9) + (-6) + (+4)$

**23.** $(-2) + (-3) + (+6) + (-4) + (+1) + (-5)$

**24.** $(+1) + (-7) + (-6) + (+4) + (+11) + (-3) + (-4)$

**25.** $(+4) - (+9)$

**26.** $(+8) - (-2)$

**27.** $(-2) - (-8)$

**28.** $(-11) - (+6)$

**29.** $(+7) - (-2)$

**30.** $(-2) - (-5)$

**31.** $(-6) - (+8)$

**32.** $(+6) - (+8)$

**33.** $4 - 7$

**34.** $14 - 20$

**35.** $-12.5 - 3.2$

**36.** $-0.75 - (-1.45)$

**37.** $\left(-\dfrac{2}{3}\right) - \left(-\dfrac{1}{4}\right)$

**38.** $\left(+\dfrac{2}{5}\right) - \left(-\dfrac{1}{4}\right)$

**39.** $\left(+\dfrac{1}{12}\right) - \left(+1\dfrac{2}{3}\right)$

**40.** $\left(-1\dfrac{3}{8}\right) - \left(+2\dfrac{5}{12}\right)$

**41.** $(+3) - (-6) - (+9) - (-6)$

**42.** $(-7) - (+6) - (+8) - (-2) - (-4)$

**43.** $(+3) + (-4) - (-7) - (+2) + (+7)$

**44.** $(-2) - (-5) + (+4) - (+6) + (-9)$

**45.** $8 + 9 - 16 + 4 - 5 - 6 + 1$

**46.** $5 - 7 - 6 + 2 - 3 + 19$

**47.** $9 - 7 + 4 + 3 - 8 - 6 - 6 + 2$

**48.** $-4 + 6 - 7 - 5 + 6 - 8 - 1$

**49.** $(-5)(-6)$

**50.** $(-2)(+6)$

**51.** $(+2)(-12)$

**52.** $(-5)(-7)$

**53.** $(-8) \div (-2)$

**54.** $(-16) \div (+8)$

**55.** $\dfrac{+54}{-30}$

**56.** $\dfrac{-8}{-28}$

**57.** $\left(-\dfrac{2}{3}\right)\left(-\dfrac{9}{16}\right)$

**58.** $\left(-1\dfrac{1}{4}\right)\left(+3\dfrac{1}{5}\right)$

**59.** $\left(+1\dfrac{3}{8}\right) \div \left(-1\dfrac{5}{6}\right)$

**60.** $\left(-1\dfrac{1}{8}\right) \div \left(-3\dfrac{3}{8}\right)$

**61.** $\dfrac{-\dfrac{7}{16}}{\dfrac{21}{32}}$

**62.** $\dfrac{2\dfrac{1}{2}}{-1\dfrac{3}{4}}$

**63.** $(+2)(-7)(-6)(-1)$

**64.** $(-6)(+2)(+4)(-5)$

**65.** $(-2)(+3)(-1)(+5)(+2)$

**66.** $(-3)(+5)(-8)(+2)(-1)$

**67.** $\dfrac{(-6)(+4)(-7)}{(-12)(-2)}$

**68.** $\dfrac{(-9)(-4)}{(+6)(-10)}$

**69.** $\dfrac{(+36)(-18)(+5)}{(-4)(-8)(+30)}$

**70.** $\dfrac{(+6)(-15)(-24)}{(-12)(+6)(+20)(-3)}$

**71.** The temperature at 6:00 A.M. is $-15°$. By noon the temperature increases by $35°$. Find the temperature at noon.

**72.** During March, Mary's small business had an income of $1875 and expenses of $2055. Find her monthly profit.

**73.** The temperature at 2:00 P.M. is 18°. The temperature at 11:00 P.M. is $-12°$. Find the difference in temperatures.

**74.** Bill, a diver, is 120 ft below the surface of the Pacific Ocean. Heather is directly above Bill in a balloon that is 260 ft above the Pacific Ocean. Find the distance between Bill and Heather.

**75.** A diver is 65 ft below the surface of the Atlantic Ocean. A second diver is 25 ft below the first diver. A third diver is 15 ft above the second diver. Find the depth of the third diver.

## 1.2   ZERO AND ORDER OF OPERATIONS

The following properties of zero with respect to addition, subtraction, and multiplication are well known and widely used:

**1.** $a + 0 = a$

**2.** $a - 0 = a$

**3.** $a \cdot 0 = 0$

**4.** If $a \cdot b = 0$, then either $a = 0$ or $b = 0$.

Observe that the number zero must be treated as a very special case in the real number system when it comes to division.

From the definition of division, we have

$$\frac{a}{b} = q \quad \text{if and only if} \quad a = b \cdot q$$

For example,

$$\frac{18}{3} = 6 \quad \text{if and only if} \quad 18 = 3 \cdot 6$$

First, if $a = 0$ and $b \neq 0$, what is the nature of the quotient $q$? By substituting in the preceding definition, we have

$$\frac{0}{b} = q \quad \text{only if} \quad 0 = b \cdot q$$

Since the product of $b$ and $q$ is 0 and $b \neq 0$, then $q = 0$. That is, $\frac{0}{b} = 0 \ (b \neq 0)$.

On the other hand, if $a \neq 0$ and $b = 0$, what is the nature of $q$? By substituting, we have

$$\frac{a}{0} = q \quad \text{only if} \quad a = 0 \cdot q$$

The product $0 \cdot q$ equals 0 leads to a contradiction since $a \neq 0$; that is, there is no real number $q$ which, when multiplied by 0, gives a number $a$ that is not 0. Hence, we say that no quotient exists and division by zero is *meaningless*.

Finally, if $a = 0$ and $b = 0$, what is the nature of $q$? Substituting yields

$$\frac{0}{0} = q \quad \text{only if} \quad 0 = 0 \cdot q$$

Note that any real number value of $q$ satisfies $0 = 0 \cdot q$; that is, there is no unique value of $q$ that can be determined. Hence, we say that no unique quotient exists and that the form $\dfrac{0}{0}$ is *indeterminate*.

**EXAMPLE 1**

Perform the indicated operations where possible.

(a) $8 \cdot 0$    (b) $\dfrac{0}{5}$    (c) $\dfrac{12}{0}$    (d) $\dfrac{0}{0}$

(a) $8 \cdot 0 = 0$

(b) $\dfrac{0}{5} = 0$

(c) $\dfrac{12}{0}$ is meaningless.

(d) $\dfrac{0}{0}$ is indeterminate.

**EXAMPLE 2**

Find the values of $x$ that make $\dfrac{4x + 3}{(5x - 1)(2x + 4)}$ meaningless.

This fraction is meaningless for those values of $x$ for which *only* the denominator is zero. Thus, set each factor in the denominator equal to zero and solve for $x$.

$$5x - 1 = 0 \qquad 2x + 4 = 0$$
$$5x = 1 \qquad\quad 2x = -4$$
$$x = \frac{1}{5} \qquad\quad x = -2$$

A fraction is indeterminate for those values of $x$ for which *both* the numerator and the denominator are zero.

The operations with zero are summarized as follows.

> **OPERATIONS WITH ZERO**
>
> **1.** $a + 0 = a$
>
> **2.** $a - 0 = a$
>
> **3.** $a \cdot 0 = 0$
>
> **4.** If $a \cdot b = 0$, then either $a = 0$ or $b = 0$.
>
> **5.** $\dfrac{0}{b} = 0 \quad (b \neq 0)$
>
> **6.** $\dfrac{a}{0}$ is meaningless $(a \neq 0)$.
>
> **7.** $\dfrac{0}{0}$ is indeterminate.

When we treat a group of numbers or quantities as a unit, we use the following grouping symbols: parentheses ( ), brackets [ ], braces { }, and the fraction bar —.

What is the value of $8 - 3 \cdot 2$? Is it 10? Is it 2? Some other number? It is very important that each arithmetic operation have only one result and that we each perform the exact same operations on a given computation or problem. As a result, the following order of operations convention is followed by everyone.

---

**ORDER OF OPERATIONS CONVENTION**

1. Do all operations within any grouping symbols beginning with the innermost pair if grouping symbols are contained within each other.
2. Then do all multiplications and divisions in the order in which they occur from left to right.
3. Finally, do all additions and subtractions in the order in which they occur from left to right.

---

**EXAMPLE 3**

Evaluate.

$$
\begin{aligned}
2 - \{6 + [(2 - 7) + (4 + 6)] - 7\} + 1 &= 2 - \{6 + [-5 + 10] - 7\} + 1 \\
&= 2 - \{6 \;\; + \;\; 5 \;\;\;\; - 7\} + 1 \\
&= 2 - \;\;\;\;\;\;\;\; 4 \;\;\;\;\;\;\;\;\;\;\;\;\; + 1 \\
&= -1
\end{aligned}
$$

**EXAMPLE 4**

Evaluate.

**Step 1:** $\qquad 6(2 + 5) - 33 \div 11 - 2 \cdot 6 = 6(7) - 33 \div 11 - 2 \cdot 6$

**Step 2:** $\qquad\qquad\qquad\qquad\qquad\qquad\quad = 42 - \;\;\; 3 \;\;\;\; - \;\; 12$

**Step 3:** $\qquad\qquad\qquad\qquad\qquad\qquad\quad = 27$

**EXAMPLE 5**

Evaluate.

**Step 1:** $\qquad \dfrac{-3(4 - 6) - 18 \div 3 \cdot 2}{18 \cdot 2 \div 4 + 1} = \dfrac{-3(-2) - 18 \div 3 \cdot 2}{18 \cdot 2 \div 4 + 1}$

**Step 2:** $\qquad\qquad\qquad\qquad\qquad\quad = \dfrac{6 - 12}{9 + 1}$

**Step 3:** $\qquad\qquad\qquad\qquad\qquad\quad = \dfrac{-6}{10}$

$\qquad\qquad\qquad\qquad\qquad\qquad\qquad\quad = -\dfrac{3}{5}$

Now let us return to the earlier problem of finding the value of $8 - 3 \cdot 2$. According to the preceding convention, $8 - 3 \cdot 2 = 8 - 6 = 2$.

# Exercises 1.2

*Perform the indicated operations where possible.*

**1.** $0 - 3$

**2.** $-4 + 0$

**3.** $6 \cdot 0$

**4.** $(0 - 3)0$

**5.** $\dfrac{9 \cdot 0}{3}$

**6.** $\dfrac{8 + (-8)}{5}$

**7.** $\dfrac{13}{6 - 6}$

**8.** $\dfrac{5 \cdot 0}{18 - 18}$

**9.** $\dfrac{7 + (-7)}{6 - 6(3 + 2)}$

**10.** $\dfrac{2 - 2}{2 - |-8| + 6}$

**11.** $\dfrac{5 \cdot 0}{-6 - (-6)}$

**12.** $\dfrac{6 \cdot 15 \cdot 8 \cdot 10}{18 \cdot 0 \cdot 16}$

*Find the values of x that make each fraction meaningless.*

**13.** $\dfrac{5x}{x - 2}$

**14.** $\dfrac{3x - 7}{2x + 1}$

**15.** $\dfrac{2 - x}{x(3x + 4)}$

**16.** $\dfrac{4 + 3x}{2x(x + 1)}$

**17.** $\dfrac{5x + 1}{(x + 1)(2x - 1)}$

**18.** $\dfrac{3}{x(x + 1)(x - 1)}$

*Find the values of x that make each fraction indeterminate.*

**19.** $\dfrac{6x^2}{2x}$

**20.** $\dfrac{2 - x}{(2x - 7)(x - 2)}$

**21.** $\dfrac{12x - 10}{(3 + x)(5 - 6x)}$

**22.** $\dfrac{(1 - 2x)(3x - 7)}{5x(2x - 1)}$

**23.** $\dfrac{(2x + 1)(x - 3)}{(6 - 2x)(2x + 1)}$

**24.** $\dfrac{(3x - 6)(1 - x)}{(x - 1)(2x - 6)}$

*Evaluate.*

**25.** $15 + 2 \cdot 4$

**26.** $20 - 8 \cdot 2$

**27.** $3 - 8(5 - 2)$

**28.** $(6 + 2)3 - 5$

**29.** $12 + 4 \div 2$

**30.** $18 \div 6 - 2$

**31.** $6 - 7(4 + 1)$

**32.** $4 + 2(9 - 6)$

**33.** $5 - [4 + 6(2 - 7) - 3]$

**34.** $19 + \{6 - [5 + 2(8 + 2)] - 8\} + 1$

**35.** $26 \cdot 2 \div 13 + 4 - 7 \cdot 2$

**36.** $36 \div 9 \cdot 3 - 6 \cdot 4 + 1$

**37.** $3(3 - 7) \div 6 - 2$

**38.** $24 \div 6 - 1 + 4 \cdot 3$

**39.** $6 \cdot 8 \div 2 \cdot 72 \div 24 + 4$

**40.** $12 \cdot 9 \div 18 \cdot 64 \div 8 + 2$

**41.** $18 \div 6 \cdot 24 \div 4 \div 6$

**42.** $4 \cdot 9 \div 3 \div 6 \cdot 9 \cdot 2 \div 12$

**43.** $7 + 6(3 + 2) - 6 - 5(4 - 2)$

**44.** $5 - 3(7 - 2) - 5 \cdot 2 - 2(4 - 7)$

**45.** $\dfrac{4 - 7(6 - 2)}{9 \div 3 + 7}$

**46.** $\dfrac{16 \cdot 2 \div 8(2 - 3)}{5 - 2 \cdot 3}$

**47.** $\dfrac{5 \cdot 12 \div 6 \cdot 2 - (-4)}{6 - 2(5 + 4)}$

**48.** $\dfrac{48 \div 4 \div 3 \cdot 6 - 3}{4 + 2(6 - 2)}$

## 1.3 SCIENTIFIC NOTATION AND POWERS OF 10

In technical work it is often necessary to refer to very large and very small numbers. **Scientific notation** is a method of writing such numbers while avoiding the writing of many zeros. When a number is written in scientific notation, it is expressed as a product of a decimal between 1 and 10 and a power of 10.

First, let's review positive powers of 10.

$$10^1 = 10 \qquad\qquad = 10$$
$$10^2 = 10 \cdot 10 \qquad\qquad = 100$$
$$10^3 = 10 \cdot 10 \cdot 10 \qquad\qquad = 1000$$
$$10^4 = 10 \cdot 10 \cdot 10 \cdot 10 \qquad\qquad = 10{,}000$$

$$10^n = \underbrace{10 \cdot 10 \cdot 10 \cdots\cdots 10}_{n \text{ factors}} = \underbrace{1000\cdots 0}_{n \text{ zeros}}$$

For example, if $n = 5$, $10^5 = 100{,}000$; if $n = 7$, $10^7 = 10{,}000{,}000$.

Recall the following properties for negative powers of 10:

$$10^{-1} = \frac{1}{10} \qquad\qquad = \frac{1}{10} = 0.1$$

$$10^{-2} = \frac{1}{10 \cdot 10} \qquad\qquad = \frac{1}{10^2} = 0.01$$

$$10^{-3} = \frac{1}{10 \cdot 10 \cdot 10} \qquad\qquad = \frac{1}{10^3} = 0.001$$

$$10^{-4} = \frac{1}{10 \cdot 10 \cdot 10 \cdot 10} \qquad\qquad = \frac{1}{10^4} = 0.0001$$

$$10^{-n} = \underbrace{\frac{1}{10 \cdot 10 \cdot 10 \cdots\cdots 10}}_{n \text{ factors}} = \frac{1}{10^n} = \underbrace{0.000\cdots 01}_{(n-1) \text{ zeros}}$$

For example, if $n = 6$, $10^{-6} = 0.000001$; if $n = 9$, $10^{-9} = 0.000000001$.

The laws of exponents involving powers of 10 are shown in the following boxes. A more detailed study of the general laws of exponents will be given later.

$$\boxed{\begin{array}{l} \textbf{LAW 1} \\[4pt] 10^m \cdot 10^n = 10^{m+n} \end{array}}$$

**EXAMPLE 1**

Multiply.
(a) $10^3 \cdot 10^2 = 10^{3+2} = 10^5$        (b) $10^6 \cdot 10^{-2} = 10^{6+(-2)} = 10^4$
(c) $10^{-3} \cdot 10^{-4} = 10^{(-3)+(-4)} = 10^{-7}$

$$\boxed{\begin{array}{l} \textbf{LAW 2} \\[4pt] \dfrac{10^m}{10^n} = 10^{m-n} \end{array}}$$

**EXAMPLE 2**

Divide.
(a) $\dfrac{10^8}{10^2} = 10^{8-2} = 10^6$        (b) $10^{-2} \div 10^5 = 10^{(-2)-5} = 10^{-7}$

(c) $\dfrac{10^6}{10^{-3}} = 10^{6-(-3)} = 10^9$

$$\boxed{\begin{array}{l} \textbf{LAW 3} \\[4pt] (10^m)^n = 10^{mn} \end{array}}$$

**EXAMPLE 3**

Raise each power of 10 to the indicated power.
(a) $(10^2)^3 = 10^{(2)(3)} = 10^6$        (b) $(10^{-4})^2 = 10^{(-4)(2)} = 10^{-8}$
(c) $(10^{-3})^{-2} = 10^{(-3)(-2)} = 10^6$

$$\boxed{\begin{array}{l} \textbf{LAW 4} \\[4pt] 10^{-n} = \dfrac{1}{10^n} \quad \text{and} \quad \dfrac{1}{10^{-n}} = 10^n \end{array}}$$

**EXAMPLE 4**

Rewrite each power of 10 using positive exponents.
(a) $10^{-3} = \dfrac{1}{10^3}$        (b) $\dfrac{1}{10^{-4}} = 10^4$

Note that the second part of Law 4 can be shown by using the first part; that is,

$$\frac{1}{10^{-n}} = \frac{1}{\dfrac{1}{10^n}} = 1 \div \frac{1}{10^n} = 1 \times \frac{10^n}{1} = 10^n$$

The zero power of 10 is one; that is, $10^0 = 1$. To show this, we use the substitution principle, which states that if $a = b$ and $a = c$, then $b = c$.

$$\frac{10^n}{10^n} = 1 \qquad \text{(Any number other than zero divided by itself equals one.)}$$

and

$$\frac{10^n}{10^n} = 10^{n-n} \qquad \text{(Law 2)}$$
$$= 10^0$$

Therefore,

$$10^0 = 1 \qquad \text{(Substitution)}$$

---

**LAW 5**

$10^0 = 1$

---

With the preceding properties of powers of 10 in mind, let's now write numbers in scientific notation.

---

**SCIENTIFIC NOTATION**

A number is expressed in scientific notation as a product of a decimal between 1 and 10 and a power of 10; that is, $N \times 10^m$ where $1 \le N < 10$ and $m$ is an integer.

---

A number greater than 10 is expressed as a product of a decimal between 1 and 10 and a positive power of 10.

**EXAMPLE 5**

Write each number greater than 10 in scientific notation.
(a) $\quad 2380 = 2.38 \quad \times 10^3$
(b) $\quad 52{,}600 = 5.26 \quad \times 10^4$
(c) $31{,}000{,}000 = 3.1 \quad \times 10^7$
(d) $\quad 681.4 = 6.814 \times 10^2$
(e) $\quad 32.765 = 3.2765 \times 10^1$

A number between 0 and 1 is expressed as a product of a decimal between 1 and 10 and a negative power of 10.

**EXAMPLE 6**

Write each number less than 1 in scientific notation.
(a) $\quad 0.0617 = 6.17 \times 10^{-2}$
(b) $\quad 0.00024 = 2.4 \quad \times 10^{-4}$
(c) $\quad 0.63 = 6.3 \quad \times 10^{-1}$
(d) $0.0000004 = 4 \quad \times 10^{-7}$

A number between 1 and 10 is expressed as a product of a decimal between 1 and 10 and the zero power of 10.

### EXAMPLE 7

Write each number between 1 and 10 in scientific notation.
(a) $8.23 = 8.23 \times 10^0$
(b) $1.04 = 1.04 \times 10^0$

---

**CHANGING A NUMBER FROM DECIMAL FORM TO SCIENTIFIC NOTATION**

**1.** Move the decimal point to a position immediately after the first nonzero digit reading from left to right.

**2.** If the decimal point is moved to the *left,* the exponent of 10 is the number of places that the decimal point has been moved.

**3.** If the decimal point is moved to the *right,* the exponent of 10 is the negative of the number of places that the decimal point has been moved.

**4.** If the original position of the decimal point is after the first nonzero digit, the exponent of 10 is zero.

**CHANGING A NUMBER FROM SCIENTIFIC NOTATION TO DECIMAL FORM**

**1.** Multiply the decimal part by the power of 10 by moving the decimal point *to the right* the same number of decimal places as indicated by the power of 10 if it is *positive.*

**2.** Multiply the decimal part by the power of 10 by moving the decimal point *to the left* the same number of decimal places as indicated by the power of 10 if it is *negative.*

Supply zeros as needed.

---

### EXAMPLE 8

Write each number in decimal form.
(a) $3.45 \times 10^2 \ = 345$
(b) $1.06 \times 10^5 \ = 106{,}000$
(c) $2.77 \times 10^{-2} = 0.0277$
(d) $8.15 \times 10^{-6} = 0.00000815$
(e) $4.92 \times 10^0 \ = 4.92$

*Note:* In this text we use numerous examples to illustrate how to use a scientific calculator that uses algebraic logic. The display modes of various brands and models of such calculators may vary, but the order in which the buttons are pushed is standard. In Appendix D we give detailed instructions for those students and/or classes that prefer to use a graphing calculator. Sections where a graphing calculator alternative is appropriate are indicated by a calculator symbol and a section number reference in the left margin at the beginning of the section. You may want to refer to Section D.1 for initial instructions.

Numbers expressed in scientific notation can be entered into many calculators. The results may then also be given in scientific notation.

**EXAMPLE 9**

Multiply $(6.5 \times 10^8)(1.4 \times 10^{-15})$ and write the result in scientific notation.

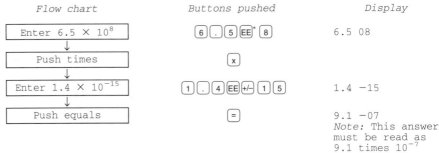

| Flow chart | Buttons pushed | Display |
|---|---|---|
| Enter 6.5 × 10⁸ | 6 . 5 EE* 8 | 6.5  08 |
| Push times | x | |
| Enter 1.4 × 10⁻¹⁵ | 1 . 4 EE +/− 1 5 | 1.4  −15 |
| Push equals | = | 9.1  −07 |

*Note:* This answer must be read as $9.1$ times $10^{-7}$

*Some calculators have a button marked EXP.

Thus, the product is $9.1 \times 10^{-7}$.

**EXAMPLE 10**

Find $\dfrac{3.24 \times 10^{-5}}{7.2 \times 10^{-12}}$ and write the result in scientific notation.

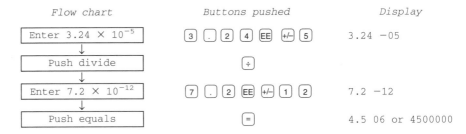

| Flow chart | Buttons pushed | Display |
|---|---|---|
| Enter 3.24 × 10⁻⁵ | 3 . 2 4 EE +/− 5 | 3.24  −05 |
| Push divide | ÷ | |
| Enter 7.2 × 10⁻¹² | 7 . 2 EE +/− 1 2 | 7.2  −12 |
| Push equals | = | 4.5  06 or 4500000 |

Thus, the quotient is $4.5 \times 10^6$.

**EXAMPLE 11**

Find the value of $\dfrac{(-6.3 \times 10^4)(-5.07 \times 10^{-9})(8.11 \times 10^{-6})}{(5.63 \times 10^{12})(-1.84 \times 10^7)}$ and write the result in scientific notation, rounded to three significant digits.

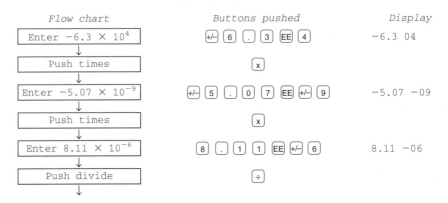

| Flow chart | Buttons pushed | Display |
|---|---|---|
| Enter −6.3 × 10⁴ | +/− 6 . 3 EE 4 | −6.3  04 |
| Push times | x | |
| Enter −5.07 × 10⁻⁹ | +/− 5 . 0 7 EE +/− 9 | −5.07  −09 |
| Push times | x | |
| Enter 8.11 × 10⁻⁶ | 8 . 1 1 EE +/− 6 | 8.11  −06 |
| Push divide | ÷ | |

Thus, the result rounded to three significant digits is $-2.50 \times 10^{-29}$.

Perhaps a quick review of significant digits is in order. All nonzero digits are significant. Zeros between significant digits are significant; for example, 809 m has three significant digits. Zeros to the right of a significant digit *and* a decimal point are significant; for example, 52.30 m has four significant digits. A more detailed discussion is given in Section 1.11.

## Exercises 1.3

*Perform the indicated operations using the laws of exponents. Express each result as a power of 10.*

**1.** $10^3 \cdot 10^8$

**2.** $10^4 \div 10^{-5}$

**3.** $(10^3)^3$

**4.** $10^{-4} \cdot 10^{-6}$

**5.** $(10^{-4})^2$

**6.** $\dfrac{1}{10^{-6}}$

**7.** $10^{-2} \div 10^{-5}$

**8.** $(10^3)^{-2}$

**9.** $\dfrac{1}{10^{-4}}$

**10.** $(10^{-2})^3$

**11.** $\dfrac{10^{-2} \cdot 10^{-7} \cdot 10^{-3} \cdot 10^0}{10^3 \cdot 10^4 \cdot 10^{-2}}$

**12.** $\dfrac{10^3 \cdot 10^{-5} \cdot 10^6}{10^{-2} \cdot 10^{-7} \cdot 10^5}$

**13.** $(10^{-2} \cdot 10^3 \cdot 10^{-5})^{-4}$

**14.** $(10^2 \cdot 10^{-5} \cdot 10^9)^3$

**15.** $\left(\dfrac{10^2 \cdot 10^{-4}}{10^{-3} \cdot 10^6}\right)^3$

**16.** $\left(\dfrac{10^2 \cdot 10^{-5} \cdot 10^{-3}}{10^{-4} \cdot 10^6}\right)^5$

**17.** $\left(\dfrac{10^{-3} \cdot 10^5 \cdot 10^{-7}}{10^2 \cdot 10^{-1} \cdot 10^{-5}}\right)^{-4}$

**18.** $\left(\dfrac{10^5 \cdot 10^{-2}}{10^{-6} \cdot 10^3 \cdot 10^{-4}}\right)^{-3}$

*Write each number in scientific notation.*

**19.** 2070

**20.** 60,000

**21.** 0.091

**22.** 0.00001264

**23.** 5.61

**24.** 370,000

**25.** 8,500,000

**26.** 6700

**27.** 0.000006

**28.** 11.7

**29.** 10,060

**30.** 0.000102

*Write each number in decimal form.*

**31.** $1.27 \times 10^2$

**32.** $1.105 \times 10^{-4}$

**33.** $6.14 \times 10^{-5}$

**34.** $2.96 \times 10^0$

**35.** $9.24 \times 10^6$

**36.** $7.72 \times 10^{-1}$

**37.** $6.96 \times 10^{-9}$

**38.** $3.14 \times 10^7$

**39.** $9.66 \times 10^0$

**40.** $3.18 \times 10^{-3}$

**41.** $5.03 \times 10^4$

**42.** $1.19 \times 10^{-8}$

*Find each value. Round each result to three significant digits and write it in scientific notation.*

**43.** $(6.43 \times 10^8)(5.16 \times 10^{10})$

**44.** $(4.16 \times 10^{-5})(3.45 \times 10^{-7})$

**45.** $(1.456 \times 10^{12})(-4.69 \times 10^{-18})$

**46.** $(-5.93 \times 10^9)(7.055 \times 10^{-12})$

**47.** $(7.46 \times 10^8) \div (8.92 \times 10^{18})$

**48.** $(1.38 \times 10^{-6}) \div (4.324 \times 10^6)$

**49.** $\dfrac{-6.19 \times 10^{12}}{7.755 \times 10^{-8}}$

**50.** $\dfrac{1.685 \times 10^{10}}{1.42 \times 10^{24}}$

**51.** $\dfrac{(5.26 \times 10^{-8})(8.45 \times 10^6)}{(-6.142 \times 10^9)(1.056 \times 10^{-12})}$

**52.** $\dfrac{(-2.35 \times 10^{-9})(1.25 \times 10^{11})(4.65 \times 10^{17})}{(8.75 \times 10^{23})(-5.95 \times 10^{-6})}$

**53.** $\dfrac{(4.68 \times 10^{-15})(5.19 \times 10^{-7})}{(-7.27 \times 10^{-16})(4.045 \times 10^{-8})(1.68 \times 10^{24})}$

**54.** $\dfrac{(3.86 \times 10^5)(5.15 \times 10^{-9})(1.91 \times 10^8)}{(2.34 \times 10^{-4})(1.35 \times 10^6)(9.05 \times 10^{10})}$

## 1.4 INTRODUCTION TO THE SI METRIC SYSTEM

**Measurement** can be defined as the comparison of a quantity with a standard unit. Centuries ago, parts of the human body were used as standards of measurement. However, such standards were neither uniform nor acceptable to all. Later, each of the various standards was defined. A fourteenth-century legal definition of the standard English inch was the following: The length of three barley corns, round and dry, taken from the center of the ear, and laid end to end. To complicate matters, many countries introduced or defined their own standards, which were not related to those of other countries. In 1670 Gabriel Mouton, a Frenchman, recognized the need for a simple, uniform, worldwide measurement system, and he proposed a decimal system that was the basis for the modern metric system. By the 1800s, metric standards were adopted worldwide. In 1960 the modern metric system was adopted; it is identified in all languages by the abbreviation SI ( for Système International d'Unités—the international system of units of measurement, written in French).

The SI metric system has seven **basic units** which are listed in Table 1.1.

### TABLE 1.1

| Basic unit | SI abbreviation | Used for measuring |
|------------|-----------------|--------------------|
| metre*     | m               | length             |
| kilogram   | kg              | mass               |
| second     | s               | time               |
| ampere     | A               | electric current   |
| kelvin     | K               | temperature        |
| candela    | cd              | light intensity    |
| mole       | mol             | molecular substance |

*See note in Table 1.2, p. 18.

All other SI units are called **derived units;** that is, they can be defined in terms of these seven basic units (see Fig. 1.2). For example, the newton (N) is defined as

1 kg · m/s² (kilogram metre per second per second). Other commonly used derived SI units are listed in Table 1.2.

*The chart below shows graphically how the 17 SI derived units with special names are derived in a coherent manner from the base and supplementary units. It was provided by the National Institute of Standards and Technology.*

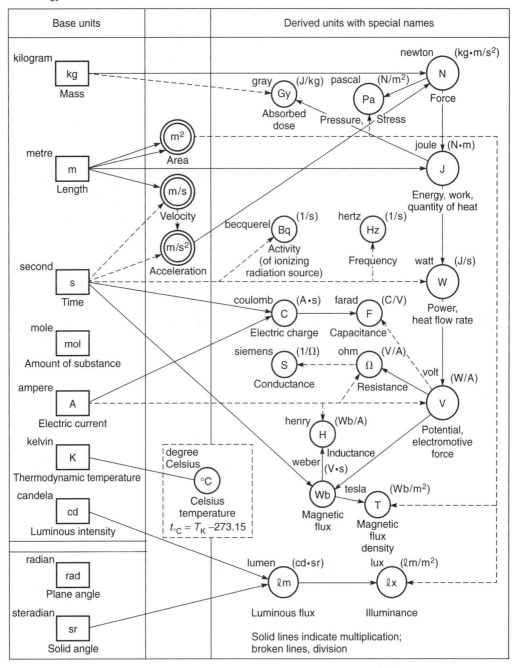

**Figure 1.2** Relationships of SI units with names

**TABLE 1.2**

| Derived unit | SI abbreviation | Used for measuring |
|---|---|---|
| litre* | L | capacity |
| cubic metre | $m^3$ | volume |
| square metre | $m^2$ | area |
| newton | N | force |
| metre per second | m/s | speed |
| joule | J | energy |
| watt | W | power |
| ohm | Ω | resistance |
| volt | V | voltage |
| farad | F | capacitance |
| henry | H | inductance |

*At present, there is some difference of opinion in the United States on the spelling of *metre* and *litre*. We have chosen the *re* spellings for two reasons. First, this is the internationally accepted spelling for all English-speaking countries. Second, the word *meter* already has many different meanings—parking meter, electric meter, odometer, and so on. Many people feel that the metric unit of length should be distinctive and readily recognizable, which the *re* spelling is.

Since the SI metric system is a decimal, or base 10, system, it is very similar to our decimal number system and our decimal money system. It is an easy system to use because calculations are based on the number 10 and its multiples. Special prefixes are used to name these multiples and submultiples which may be used with almost all SI units. Since the same prefixes are repeatedly used, memorization of many conversions has been significantly reduced. Table 1.3 shows these prefixes and their corresponding symbols.

**EXAMPLE 1**

Write the SI abbreviation for 23 centimetres.
 The symbol for the prefix *centi* is c.
 The symbol for the unit *metre* is m.
 The SI abbreviation for 23 centimetres is 23 cm.

**EXAMPLE 2**

Write the SI metric unit for the abbreviation 65 kg.
 The prefix for k is *kilo*.
 The unit for g is *gram*.
 The SI metric unit for 65 kg is 65 kilograms.

## Exercises 1.4

*Give the metric prefix for each value.*

**1.** 1000          **2.** 0.01          **3.** 100

**4.** 0.1          **5.** 0.001          **6.** 10

**TABLE 1.3  Prefixes for SI units**

| Multiple or submultiple[a] decimal form | Power of 10 | Prefix | Prefix symbol | Pronunciation | Meaning |
|---|---|---|---|---|---|
| 1,000,000,000,000 | $10^{12}$ | tera | T | tĕr'ă | one trillion times |
| 1,000,000,000 | $10^{9}$ | giga | G | jĭg'ă | one billion times |
| 1,000,000 | $10^{6}$ | mega | M | mĕg'ă | one million times |
| 1,000 | $10^{3}$ | kilo[b] | k | kĭl'ō | one thousand times |
| 100 | $10^{2}$ | hecto | h | hĕk'tō | one hundred times |
| 10 | $10^{1}$ | deka | da | dĕk'ă | ten times |
| 0.1 | $10^{-1}$ | deci | d | dĕs'ĭ | one-tenth of |
| 0.01 | $10^{-2}$ | centi[b] | c | sĕnt'ĭ | one-hundredth of |
| 0.001 | $10^{-3}$ | milli[b] | m | mĭl'ĭ | one-thousandth of |
| 0.000001 | $10^{-6}$ | micro | $\mu$ | mī'krō | one-millionth of |
| 0.000000001 | $10^{-9}$ | nano | n | nǎn'ō | one-billionth of |
| 0.000000000001 | $10^{-12}$ | pico | p | pē'kō | one-trillionth of |

[a]Factor by which the unit is multiplied.

[b]Most commonly used prefixes.

**7.** 1,000,000          **8.** 0.000001

*Give the metric symbol, or abbreviation, for each prefix.*

**9.** hecto          **10.** kilo          **11.** milli

**12.** deci          **13.** mega          **14.** deka

**15.** centi          **16.** micro

*Write the abbreviation for each quantity.*

**17.** 133 millimetres          **18.** 63 dekagrams          **19.** 18 kilolitres

**20.** 42 centimetres          **21.** 19 centigrams          **22.** 25 milligrams

**23.** 72 hectometres          **24.** 17 decilitres

*Write the SI unit for each abbreviation.*

**25.** 14 m          **26.** 182 L          **27.** 19 g

**28.** 147 kg          **29.** 17 mm          **30.** 23 dL

**31.** 25 dam  **32.** 17 mg  **33.** 16 Mm

**34.** 250 $\mu$g

**35.** The basic metric unit of length is ____.

**36.** The basic metric unit of mass is ____.

**37.** Two common metric units of volume are ____ and ____.

**38.** The basic metric unit for electric current is ____.

**39.** The basic metric unit for time is ____.

**40.** The common metric unit for power is ____.

## 1.5  LENGTH

The basic SI unit of length is the metre (m) (see Fig. 1.3).

Long distances are measured in kilometres (km); 1 km = 1000 m (see Fig. 1.4). We use the centimetre (cm) to measure short distances, such as the length of this book or the width of a board (see Fig. 1.5).

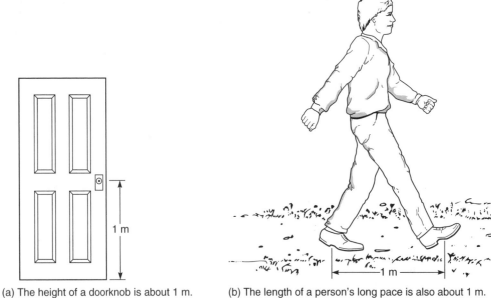

(a) The height of a doorknob is about 1 m.  (b) The length of a person's long pace is also about 1 m.

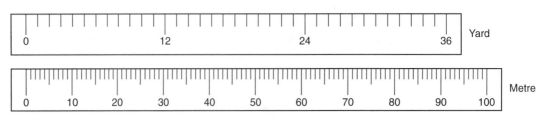

(c) One metre is a little more than 1 yd.

**Figure 1.3**  Length of one metre

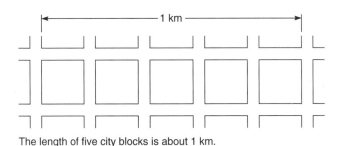

The length of five city blocks is about 1 km.

**Figure 1.4**

The width of your small fingernail is about 1 cm.

**Figure 1.5**

The millimetre (mm) is used to measure very small lengths, such as the thickness of this book or the depth of a tire tread (see Fig. 1.6).

The thickness of a dime is about 1 mm.

**Figure 1.6**

A metric ruler is shown in Fig. 1.7. The large numbered divisions are centimetres. They are divided into 10 equal parts, called millimetres.

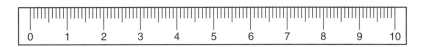

**Figure 1.7** Metric ruler

## Conversion Factors

To change from one unit or set of units to another, we shall use what is commonly called a **conversion factor.** We know that we can multiply any number or quantity by 1 (one) without changing the value of the original quantity. We also know that any fraction whose numerator and denominator are equal is equal to 1. For example,

$$\frac{3}{3} = 1 \qquad \frac{15 \text{ m}}{15 \text{ m}} = 1 \qquad \frac{4.5 \text{ kg}}{4.5 \text{ kg}} = 1$$

In addition, since 1 m = 100 cm, $\frac{1 \text{ m}}{100 \text{ cm}} = 1$. Similarly, $\frac{100 \text{ cm}}{1 \text{ m}} = 1$, because the numerator equals the denominator. We call such names for 1 *conversion factors*. The information necessary for forming a conversion factor is usually found in tables.

As in the case 1 m = 100 cm, there are two conversion factors for each set of data:

$$\frac{1 \text{ m}}{100 \text{ cm}} \quad \text{and} \quad \frac{100 \text{ cm}}{1 \text{ m}}$$

---

**CHOOSING CONVERSION FACTORS**

The correct choice for a particular conversion factor is the one in which the old units are in the numerator of the original expression and in the denominator of the conversion factor, or in the denominator of the original expression and in the numerator of the conversion factor. That is, we want the old units to divide (cancel) each other.

---

**EXAMPLE 1**

Change 245 cm to metres.
    As we saw in the preceding discussion, the two possible conversion factors are

$$\frac{1 \text{ m}}{100 \text{ cm}} \quad \text{and} \quad \frac{100 \text{ cm}}{1 \text{ m}}$$

We choose the conversion factor with centimetres in the *denominator* so that the centimetre units cancel each other.

$$245 \text{ cm} \times \frac{1 \text{ m}}{100 \text{ cm}} = 2.45 \text{ m}$$

*Note:* Conversions *within* the metric system involve only moving the decimal point.

**EXAMPLE 2**

Change 5 m to centimetres.

$$5 \text{ m} \times \frac{100 \text{ cm}}{1 \text{ m}} = 500 \text{ cm}$$

**EXAMPLE 3**

Change 29.5 mm to centimetres.
    We choose the conversion factor with millimetres in the denominator so that the millimetre units cancel each other.

$$29.5 \text{ mm} \times \frac{1 \text{ cm}}{10 \text{ mm}} = 2.95 \text{ cm}$$

**EXAMPLE 4**

Change 0.08 km to centimetres.
    First, change to metres and then to centimetres.

$$0.08 \text{ km} \times \frac{1000 \text{ m}}{1 \text{ km}} = 80 \text{ m}$$

$$80 \text{ m} \times \frac{100 \text{ cm}}{1 \text{ m}} = 8000 \text{ cm}$$

## Exercises 1.5

*Which unit is longer?*

**1.** 1 metre or 1 centimetre

**2.** 1 metre or 1 millimetre

**3.** 1 metre or 1 kilometre

**4.** 1 centimetre or 1 millimetre

**5.** 1 centimetre or 1 kilometre

**6.** 1 millimetre or 1 kilometre

*Which metric unit (km, m, cm, or mm) should you use to measure the following?*

**7.** Length of a pipe wrench.

**8.** Thickness of a saw blade.

**9.** Height of a house.

**10.** Distance around an automobile racing track.

**11.** Diameter of a hypodermic needle.

**12.** Width of a table.

**13.** Distance between Boston and New York.

**14.** Length of a hurdle race.

**15.** Thread size on a spark plug.

**16.** Width of a house lot.

*Fill in each blank with the most reasonable metric unit (km, m, cm, or mm).*

**17.** Your car is about 6 _____ long.

**18.** Your pencil is about 20 _____ long.

**19.** The distance between New York and San Francisco is about 4000 _____.

**20.** Your pencil is about 7 _____ thick.

**21.** The ceiling in my bedroom is about 240 _____ high.

**22.** The length of a football field is about 90 _____.

**23.** A jet plane usually flies about 9 _____ high.

**24.** A standard size film for cameras is 35 _____.

**25.** The diameter of my car tire is about 60 _____.

**26.** The zipper on my jacket is about 70 _____ long.

**27.** Maria drives 8 _____ to college each day.

**28.** Bill, our basketball center, is 203 _____ tall.

**29.** The width of your hand is about 80 _____.

**30.** A hand saw is about 70 _____ long.

**31.** A newborn baby is usually about 45 _____ long.

**32.** The standard metric piece of plywood is 1200 _____ wide and 2400 _____ long.

*Fill in each blank.*

**33.** 1 km = _____ m

**34.** 1 mm = _____ m

**35.** 1 m = _____ cm

**36.** 1 m = _____ hm

**37.** 1 dm = _____ m

**38.** 1 dam = _____ m

**39.** 1 m = _____ mm

**40.** 1 m = _____ dm

**41.** 1 hm = _____ m

**42.** 1 cm = _____ m

**43.** 1 m = _____ km

**44.** 1 m = _____ dam

**45.** 1 cm = _____ mm

**46.** Change 230 m to cm.

**47.** Change 230 m to km.

**48.** Change 576 mm to cm.

**49.** Change 198 km to m.

**50.** Change 25 dm to dam.

**51.** Change 840 cm to m.

**52.** Change 85 hm to km.

**53.** Change 475 cm to mm.

**54.** Change 8.5 mm to $\mu$m.

**55.** Change 4 m to $\mu$m.

**56.** Your height is ____ cm.

**57.** Your height is ____ m.

**58.** Your height is ____ mm.

## 1.6 MASS

The mass of an object is the quantity of material making up the object. One unit of mass in the metric system is the *gram* (g). The gram is defined as the mass of one cubic centimetre ($cm^3$) of water at its maximum density (see Fig. 1.8).

(a) A common paper clip has a mass of about 1 g.        (b) Three aspirin have a mass of about 1 g.

**Figure 1.8**

Since the gram is so small, the **kilogram** (kg) is the basic unit of mass in the metric system. One kilogram is defined as the mass of one cubic decimetre ($dm^3$) of water at its maximum density.

For very, very small masses, such as medicine dosages, we use the *milligram* (mg). One grain of salt has a mass of about 1 milligram.

The metric ton (1000 kg) is used to measure the mass of very large quantities, such as coal on a barge, a trainload of grain, or a shipload of ore.

**EXAMPLE 1**

Change 84 kg to grams.

We choose the conversion factor with kilograms in the denominator so that the kilogram units cancel each other.

$$84 \ \cancel{kg} \times \frac{1000 \ g}{1 \ \cancel{kg}} = 84{,}000 \ g$$

**EXAMPLE 2**

Change 500 mg to grams.

$$500 \ \cancel{mg} \times \frac{1 \ g}{1000 \ \cancel{mg}} = 0.5 \ g$$

## Exercises 1.6

*Which unit is larger?*

**1.** 1 gram or 1 centigram

**2.** 1 gram or 1 milligram

**3.** 1 gram or 1 kilogram

**4.** 1 centigram or 1 milligram

**5.** 1 centigram or 1 kilogram

**6.** 1 milligram or 1 kilogram

*Which metric unit (kg, g, mg, or metric ton) should you use to measure the following?*

**7.** Your mass.

**8.** An aspirin.

**9.** A bag of fertilizer.

**10.** A bar of soap.

**11.** A trainload of grain.

**12.** A sewing needle.

**13.** A small can of corn.

**14.** A channel catfish.

**15.** A vitamin capsule.

**16.** A car.

*Fill in each blank with the most reasonable metric unit (kg, g, mg, or metric ton).*

**17.** A newborn baby's mass is about 3 ____.

**18.** The elevator in a department store has a load limit of 2000 ____.

**19.** Millie's diet calls for 250 ____ of meat.

**20.** A 200-car train can carry 11,000 ____ of soybeans.

**21.** A truckload shipment of copper pipe has a mass of about 900 ____.

**22.** A carrot has a mass of about 75 ____.

**23.** A candy recipe calls for 150 ____ of chocolate.

**24.** By just looking, you can tell your best friend has a mass of 70 ____.

**25.** A pencil weighs about 10 ____.

**26.** Postage rates for letters would be based on the ____.

**27.** A heavyweight boxer weighed in at 93 ____.

**28.** A nickel has a mass of 5 ____.

**29.** A spaghetti recipe calls for 1 ____ of ground beef.

**30.** A spaghetti recipe calls for 150 ____ of tomato paste.

**31.** A grain elevator shipped 10,000 ____ of wheat last year.

**32.** A slice of bread has a mass of about 25 ____.

**33.** I bought a 5-____ bag of potatoes at the store today.

**34.** My grandmother takes 250-____ capsules for her arthritis.

*Fill in each blank.*

**35.** 1 kg = ____ g

**36.** 1 mg = ____ g

**37.** 1 g = ____ cg

**38.** 1 g = ____ hg

**39.** 1 dg = ____ g

**40.** 1 dag = ____ g

**41.** 1 g = ____ mg

**42.** 1 g = ____ dg

**43.** 1 hg = ____ g

**44.** 1 cg = ____ g

**45.** 1 g = ____ kg

**46.** 1 g = ____ dag

**47.** 1 g = ____ $\mu$g

**48.** 1 mg = ____ $\mu$g

**49.** Change 565 g to mg.

**50.** Change 565 g to kg.

**51.** Change 850 mg to g.

**52.** Change 275 kg to g.

**53.** Change 50 dg to g.

**54.** Change 485 dag to dg.

**55.** Change 80 kg to mg.

**56.** Change 5 metric tons to kg.

**57.** Change 15 hg to kg.

**58.** Change 57 $\mu$g to g.

**59.** Change 500 $\mu$g to mg.

**60.** Change 40,000 kg to metric tons.

## 1.7 VOLUME AND AREA

### Volume

A common unit of volume in the metric system is the *litre* (L). The litre is commonly used for liquid volumes (see Fig. 1.9).

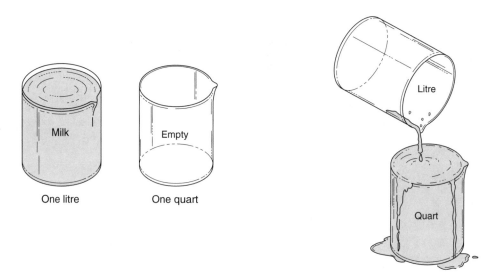

**Figure 1.9**  One litre is a little more than one quart

The cubic metre ($m^3$) (see Fig. 1.10) and the cubic centimetre ($cm^3$) (see Fig. 1.11) are also used to measure volume. The cubic metre is the volume in a cube 1 m on an edge. It is used to measure large volumes. The cubic centimetre is the volume in a cube 1 cm on an edge. For comparison purposes, one teacher's desk and clutter could be boxed into 2 cubic metres side by side, and the eraser on a pencil could fit into a cubic centimetre.

The volume of any solid is the number of cubic units in that solid.

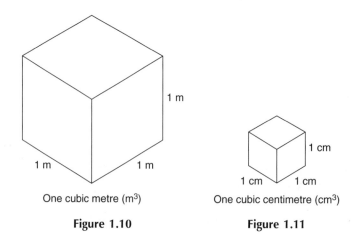

One cubic metre ($m^3$)

One cubic centimetre ($cm^3$)

**Figure 1.10**

**Figure 1.11**

**EXAMPLE 1**

Find the volume of a box 4 cm long, 3 cm wide, and 2 cm high.

Each cube in Fig. 1.12 shows 1 cm$^3$. By simply counting the number of cubes (cubic centimetres), we find that the volume of the box is 24 cm$^3$.

We can also find the volume of the box by using the formula

$$V = lwh$$
$$= (4 \text{ cm})(3 \text{ cm})(2 \text{ cm})$$
$$= 24 \text{ cm}^3 \qquad (\textit{Note: } \text{cm} \times \text{cm} \times \text{cm} = \text{cm}^3)$$

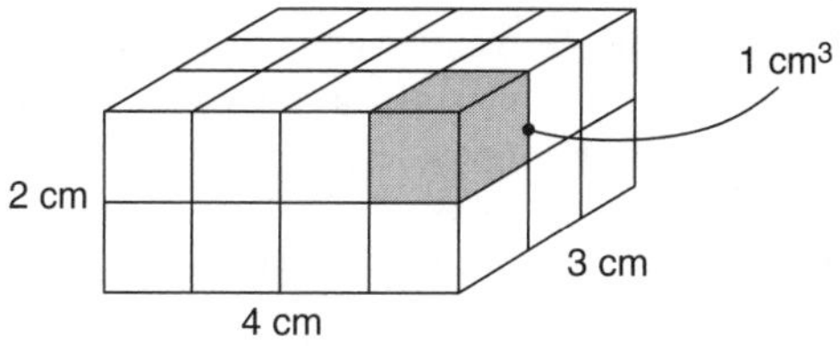

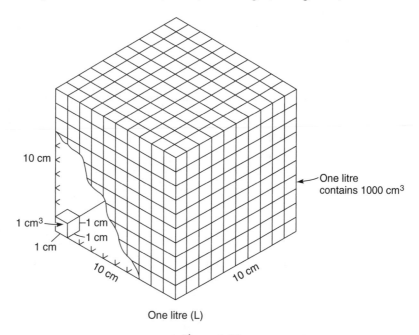

**Figure 1.12**

The relationship between the litre and the cubic centimetre deserves some special mention. The litre is defined as the capacity in a volume of 1 cubic decimetre (dm$^3$). That is, 1 L of liquid fills a cube 1 dm (10 cm) on an edge (see Fig. 1.13).

One litre (L)

**Figure 1.13**

The volume of this cube can also be found by using the formula

$$V = lwh$$
$$= (10 \text{ cm})(10 \text{ cm})(10 \text{ cm})$$
$$= 1000 \text{ cm}^3$$

That is,

$$1 \text{ L} = 1000 \text{ cm}^3$$

Then

$$\frac{1}{1000} \text{ L} = 1 \text{ cm}^3$$

or

$$1 \text{ mL} = 1 \text{ cm}^3$$

Milk, soda, and gasoline are usually sold by the litre in countries using the metric system. Liquid medicine, vanilla extract, and lighter fluid are usually sold by the milli-litre. Many metric cooking recipes are given in millilitres. Very large quantities of oil would be sold by the kilolitre (1000 litres).

Recall that the kilogram was defined as the mass of one cubic decimetre of water. Since $1 \text{ dm}^3 = 1 \text{ L}$, one litre of water has a mass of one kilogram.

**EXAMPLE 2**

Change 0.75 L to millilitres.

$$0.75 \text{ L} \times \frac{1000 \text{ mL}}{1 \text{ L}} = 750 \text{ mL}$$

**EXAMPLE 3**

Change 0.75 $\text{cm}^3$ to cubic millimetres.

$$0.75 \text{ cm}^3 \times \left(\frac{10 \text{ mm}}{1 \text{ cm}}\right)^3 = 750 \text{ mm}^3$$

**EXAMPLE 4**

Change $3.25 \times 10^6 \text{ mm}^3$ to cubic metres.

$$3.25 \times 10^6 \text{ mm}^3 \times \left(\frac{1 \text{ m}}{1000 \text{ mm}}\right)^3 = 3.25 \times 10^{-3} \text{ m}^3$$

## Area

The basic unit of area in the metric system is the square metre ($\text{m}^2$), the area in a square whose sides are each 1 m long (see Fig. 1.14). The square centimetre ($\text{cm}^2$) and the square

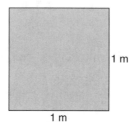

1 m

1 m

One square metre ($\text{m}^2$)        **Figure 1.14**

millimetre (mm$^2$) are smaller units of area. Larger units of area are the square kilometre (km$^2$) and the hectare (ha).

**EXAMPLE 5**

Find the area of a rectangle 4 m long and 3 m wide.

Each square in Fig. 1.15 represents 1 m$^2$. By simply counting the number of squares (square metres), we find that the area of the rectangle is 12 m$^2$.

We can also find the area of the rectangle by using the formula $A = lw$.

$$A = lw$$
$$= (4 \text{ m})(3 \text{ m})$$
$$= 12 \text{ m}^2 \qquad (\textit{Note:} \text{ m} \times \text{m} = \text{m}^2)$$

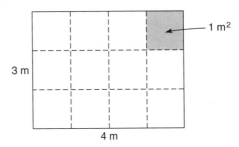

**Figure 1.15**

**EXAMPLE 6**

Change 1500 cm$^2$ to square metres.

$$1500 \text{ cm}^2 \times \left(\frac{1 \text{ m}}{100 \text{ cm}}\right)^2 = 0.15 \text{ m}^2$$

**EXAMPLE 7**

Change 0.6 km$^2$ to square metres.

$$0.6 \text{ km}^2 \times \left(\frac{1000 \text{ m}}{1 \text{ km}}\right)^2 = 600,000 \text{ m}^2$$

The hectare is the fundamental SI unit for land area. An area of one hectare equals the area of a square 100 m on a side (see Fig. 1.16). The hectare is used because it is more convenient to say and use than square hectometre. The metric prefixes are *not* used with the hectare. That is, instead of saying "2 kilohectares," we say "2000 hectares."

1 hectare (ha) = 10,000 m$^2$ = 1 hm$^2$

100 m

100 m

One hectare       **Figure 1.16**

When converting between metric and English land area units, use the relationship

$$1 \text{ hectare} = 2.47 \text{ acres}$$

## Exercises 1.7

*Which unit is larger?*

1. 1 litre or 1 centilitre
2. 1 millilitre or 1 kilolitre
3. 1 cubic millimetre or 1 cubic centimetre
4. 1 cubic centimetre or 1 cubic metre
5. 1 square kilometre or 1 square hectometre
6. 1 square millimetre or 1 square decimetre

*Which metric unit ($m^3$, L, mL, $m^2$, $cm^2$, or ha) would you use to measure the following?*

7. Oil in your car's crankcase.
8. Water in a bathtub.
9. Floor space in a house.
10. Cross section of a piston.
11. Storage space in a miniwarehouse.
12. Coffee in an office coffeepot.
13. Size of a field of corn.
14. Page size of a newspaper.
15. A dose of cough syrup.
16. Size of a ranch.
17. Cargo space in a truck.
18. Gasoline in your car's gas tank.
19. Piston displacement of an engine.
20. Paint needed to paint a house.
21. Drops to put in your eyes.
22. Size of a plot of timber.

*Fill in the blank with the most reasonable metric unit ($m^3$, L, mL, $m^2$, $cm^2$, or ha).*

23. Go to the store and get 4 _____ of root beer for the party.
24. I drank 200 _____ of orange juice for breakfast.
25. Harold bought a 30-_____ tarpaulin for his truck.
26. The cross section of a log is 3200 _____.
27. A farmer buys gasoline for a storage tank in bulk. When filled, the storage tank holds 4000 _____.
28. Our city water tower holds 500 _____ of water.
29. Bill planted 60 _____ of soybeans this year.
30. I need some copper tubing with a cross section of 3 _____.
31. A building contractor ordered 15 _____ of concrete for a driveway.
32. I must heat 420 _____ of living space in my house.
33. Our house has 210 _____ of floor space.
34. My little brother mows 5 _____ of lawn each week.
35. Amy is told by her doctor to drink 2 _____ of water each day.
36. My coffee cup holds about 200 _____ of coffee.

*Fill in each blank.*

37. 1 L = _____ mL
38. 1 kL = _____ L
39. 1 L = _____ daL
40. 1 L = _____ kL
41. 1 L = _____ hL
42. 1 L = _____ dL
43. 1 daL = _____ L
44. 1 mL = _____ L

**45.** 1 mL = ____ cm$^3$

**46.** 1 L = ____ cm$^3$

**47.** 1 m$^3$ = ____ cm$^3$

**48.** 1 cm$^3$ = ____ mL

**49.** 1 cm$^3$ = ____ L

**50.** 1 dm$^3$ = ____ L

**51.** 1 m$^2$ = ____ cm$^2$

**52.** 1 km$^2$ = ____ m$^2$

**53.** 1 cm$^2$ = ____ mm$^2$

**54.** 1 mm$^2$ = ____ m$^2$

**55.** 1 dm$^2$ = ____ m$^2$

**56.** 1 ha = ____ m$^2$

**57.** 1 km$^2$ = ____ ha

**58.** 1 ha = ____ km$^2$

**59.** Change 6500 mL to L.

**60.** Change 0.80 L to mL.

**61.** Change 1.4 L to mL.

**62.** Change 8 mL to L.

**63.** Change 225 cm$^3$ to mm$^3$.

**64.** Change 6 m$^3$ to cm$^3$.

**65.** Change 2 m$^3$ to mm$^3$.

**66.** Change 530 mm$^3$ to cm$^3$.

**67.** Change 175 cm$^3$ to mL.

**68.** Change 145 cm$^3$ to L.

**69.** Change 1 m$^3$ to L.

**70.** Change 160 mm$^3$ to L.

**71.** Change 7.5 L to cm$^3$.

**72.** Change 350 L to m$^3$.

**73.** Change 5000 mm$^2$ to cm$^2$.

**74.** Change 1.25 km$^2$ to m$^2$.

**75.** Change 5 m$^2$ to cm$^2$.

**76.** Change 150 cm$^2$ to mm$^2$.

**77.** Change $4 \times 10^8$ m$^2$ to km$^2$.

**78.** Change $5 \times 10^7$ cm$^2$ to m$^2$.

**79.** What is the mass of 750 mL of water?

**80.** What is the mass of 1 m$^3$ of water?

## 1.8 TEMPERATURE, TIME, CURRENT, AND POWER

### Temperature

The basic SI unit for temperature is the *kelvin* (K), which is used mostly in scientific and engineering work. Everyday temperatures are measured in degrees *Celsius* (°C). The United States has been measuring temperatures in degrees *Fahrenheit* (°F).

The theoretical lowest temperature is called **absolute zero.** At this temperature, there is no heat in an object, and its molecules have stopped moving. The kelvin scale begins with absolute zero as its zero point, so it has no negative temperature readings. The units on the kelvin and Celsius scales are the same size. On the Celsius scale, water freezes at 0° and boils at 100° (see Fig. 1.17). Each degree Celsius is $\frac{1}{100}$ of the difference between the boiling temperature and the freezing temperature of water.

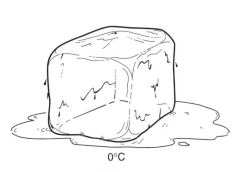

(a) Water freezes at 0°C.

(b) Water boils at 100°C.

**Figure 1.17**

Absolute zero on the Celsius scale is about −273°C (see Fig. 1.18). Water freezes at 273 K and boils at 373 K. Note that kelvin temperatures do *not* use the degree symbol (°). In formula form, we have

$$K = C + 273$$

where $K$ = degrees kelvin and $C$ = degrees Celsius.

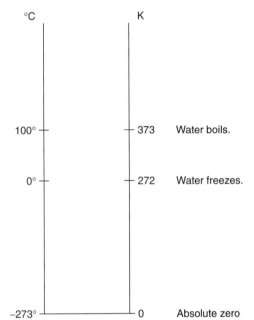

**Figure 1.18** Comparison of Celsius and kelvin scales

Figure 1.19 shows some approximate temperature readings in degrees Celsius and Fahrenheit with some related activity.

The formulas for converting between degrees Celsius ($C$) and degrees Fahrenheit ($F$) are as follows:

$$C = \frac{5}{9}(F - 32)$$

$$F = \frac{9}{5}C + 32$$

**EXAMPLE 1**

Change 50°F to degrees Celsius.

$$C = \frac{5}{9}(F - 32)$$

$$C = \frac{5}{9}(50° - 32°)$$

$$= \frac{5}{9}(18°)$$

$$= 10°$$

That is, 50°F = 10°C.

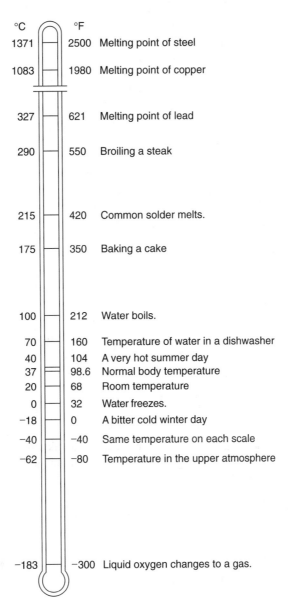

| °C | °F | |
|---|---|---|
| 1371 | 2500 | Melting point of steel |
| 1083 | 1980 | Melting point of copper |
| 327 | 621 | Melting point of lead |
| 290 | 550 | Broiling a steak |
| 215 | 420 | Common solder melts. |
| 175 | 350 | Baking a cake |
| 100 | 212 | Water boils. |
| 70 | 160 | Temperature of water in a dishwasher |
| 40 | 104 | A very hot summer day |
| 37 | 98.6 | Normal body temperature |
| 20 | 68 | Room temperature |
| 0 | 32 | Water freezes. |
| −18 | 0 | A bitter cold winter day |
| −40 | −40 | Same temperature on each scale |
| −62 | −80 | Temperature in the upper atmosphere |
| −183 | −300 | Liquid oxygen changes to a gas. |

**Figure 1.19**

**EXAMPLE 2**

Change 30°C to degrees Fahrenheit.

$$F = \frac{9}{5} C + 32$$

$$F = \frac{9}{5} (30°) + 32°$$

$$= 54° + 32°$$

$$= 86°$$

That is, 30°C = 86°F.

**EXAMPLE 3**

Change 14°F to degrees Celsius.

$$C = \frac{5}{9}(F - 32)$$

$$C = \frac{5}{9}(14° - 32°)$$

$$= \frac{5}{9}(-18°)$$

$$= -10°$$

That is, 14°F = −10°C.

**EXAMPLE 4**

Change −52°C to degrees Fahrenheit.

$$F = \frac{9}{5}C + 32$$

$$F = \frac{9}{5}(-52°) + 32°$$

$$= -93.6° + 32°$$

$$= -61.6°$$

That is, −52°C = −61.6°F.

## Time

The basic metric unit of time is the second (s), which is the same in all systems of measurement. We measure longer periods of time in minutes (min), hours (h), and days.

$$1 \text{ day} = 24 \text{ h}$$
$$1 \text{ h} = 60 \text{ min}$$
$$1 \text{ min} = 60 \text{ s}$$

**EXAMPLE 5**

Change 3 h 20 min to seconds.
    First,

$$3 \,\cancel{h} \times \frac{60 \text{ min}}{1 \,\cancel{h}} = 180 \text{ min}$$

So

$$3 \text{ h } 20 \text{ min} = 180 \text{ min} + 20 \text{ min}$$
$$= 200 \text{ min}$$

Then

$$200 \,\cancel{\text{min}} \times \frac{60 \text{ s}}{1 \,\cancel{\text{min}}} = 12{,}000 \text{ s}$$

Very short periods of time are measured in parts of a second, given with the appropriate metric prefix. These units are commonly used in electronics.

### EXAMPLE 6

What is the meaning of each unit?
(a) 1 ms = 1 millisecond = $10^{-3}$ s and means one-thousandth of a second.
(b) 1 $\mu$s = 1 microsecond = $10^{-6}$ s and means one-millionth of a second.
(c) 1 ns = 1 nanosecond = $10^{-9}$ s and means one-billionth of a second.
(d) 1 ps = 1 picosecond = $10^{-12}$ s and means one-trillionth of a second.

### EXAMPLE 7

Change 15 ms to seconds.
  Since 1 ms = $10^{-3}$ s,

$$15 \text{ ms} \times \frac{10^{-3} \text{ s}}{1 \text{ ms}} = 15 \times 10^{-3} \text{ s} = 0.015 \text{ s}$$

### EXAMPLE 8

Change 0.000000075 s to nanoseconds.
  Since 1 ns = $10^{-9}$ s,

$$0.000000075 \text{ s} \times \frac{1 \text{ ns}}{10^{-9} \text{ s}} = 75 \text{ ns}$$

## Current

The basic metric unit of electric current is the ampere, A, which is the same as in the English system. The ampere is a fairly large unit of current, so smaller currents are measured in parts of an ampere and are shown by the appropriate metric prefix.

### EXAMPLE 9

What is the meaning of each unit?
(a) 1 mA = 1 milliampere = $10^{-3}$ A and means one-thousandth of an ampere.
(b) 1 $\mu$A = 1 microampere = $10^{-6}$ A and means one-millionth of an ampere.

### EXAMPLE 10

Change 150 $\mu$A to amperes.
  Since 1 $\mu$A = $10^{-6}$ A,

$$150 \text{ } \mu\text{A} \times \frac{10^{-6} \text{ A}}{1 \text{ } \mu\text{A}} = 0.00015 \text{ A}$$

### EXAMPLE 11

Change 0.0075 A to milliamperes.
  Since 1 mA = $10^{-3}$ A,

$$0.0075 \text{ A} \times \frac{1 \text{ mA}}{10^{-3} \text{ A}} = 7.5 \text{ mA}$$

## Power

The common metric unit for both mechanical and electrical power is the watt (W).

**EXAMPLE 12**

What is the meaning of each unit?
(a) 1 MW = 1 megawatt = $10^6$ W and means one million watts.
(b) 1 kW = 1 kilowatt = $10^3$ W and means one thousand watts.
(c) 1 mW = 1 milliwatt = $10^{-3}$ W and means one-thousandth of a watt.

**EXAMPLE 13**

Change 5 MW to watts.
Since 1 MW = $10^6$ W,

$$5 \text{ MW} \times \frac{10^6 \text{ W}}{1 \text{ MW}} = 5 \times 10^6 \text{ W} = 5,000,000 \text{ W}$$

**EXAMPLE 14**

Change 0.275 W to milliwatts.
Since 1 mW = $10^{-3}$ W,

$$0.275 \text{ W} \times \frac{1 \text{ mW}}{10^{-3} \text{ W}} = 275 \text{ mW}$$

## Exercises 1.8

*Use Fig. 1.19 to choose the most reasonable answer for each statement.*

1. The freezing temperature of water.
   (a) 32°C      (b) 100°C      (c) 0°C      (d) −32°C

2. The boiling temperature of water.
   (a) 212°C      (b) 100°C      (c) 50°C      (d) 1000°C

3. Normal body temperature.
   (a) 100°C      (b) 50°C      (c) 98.6°C      (d) 37°C

4. The body temperature of a person who is chilled and has a fever.
   (a) 100°C      (b) 101°C      (c) 39°C      (d) 37°C

5. The temperature on a cold winter day in Chicago.
   (a) 15°C      (b) 5°C      (c) 30°C      (d) −15°C

6. The temperature on a hot summer day in the California desert.
   (a) 110°C      (b) 42°C      (c) 60°C      (d) 85°C

7. The temperature on the beach in Florida on a warm winter day.
   (a) 85°C      (b) 30°C      (c) 20°C      (d) 70°C

8. The baking temperature of a cherry pie.
   (a) 85°C      (b) 350°C      (c) 500°C      (d) 215°C

9. The serving temperature of hot soup.
   (a) 20°C      (b) 115°C      (c) 50°C      (d) 150°C

10. The temperature at which freezing rain is most likely to occur.
    (a) 32°C      (b) 25°C      (c) 0°C      (d) −15°C

**11.** The weather forecast calls for a high temperature of 15°C. What should you plan to wear?
    **(a)** A heavy coat.         **(b)** A sweater.         **(c)** A short-sleeve shirt.

**12.** The weather forecast calls for a low temperature of 3°C. What should you plan to do?
    **(a)** Sleep with the air conditioner on.     **(b)** Sleep with an extra blanket on.
    **(c)** Protect your plants from frost.     **(d)** Sleep with the windows open.

**13.** The oven temperature for making pizza.
    **(a)** 100°C         **(b)** 80°C         **(c)** 450°C         **(d)** 225°C

**14.** The temperature of a lake's water is 28°C. What could you do?
    **(a)** Swim.         **(b)** Ice-fish.
    **(c)** Dress warmly and walk along the shore.

**15.** The setting for the thermostat in your home.
    **(a)** 68°C         **(b)** 19°C         **(c)** 30°C         **(d)** 50°C

*Fill in each blank.*

**16.** 59°F = _____ °C         **17.** 25°C = _____ °F

**18.** 375°C = _____ °F         **19.** 140°F = _____ °C

**20.** 5°F = _____ °C         **21.** −12°F = _____ °C

**22.** −12°C = _____ °F         **23.** 550°C = _____ °F

**24.** 50°C = _____ K         **25.** 50 K = _____ °C

**26.** The temperature in a crowded room is 80°F. What is the Celsius reading?

**27.** The temperature of an iced-tea drink is 5°C. What is the Fahrenheit reading?

**28.** The boiling point (temperature at which a liquid changes to a gas) of liquid nitrogen is −196°C. What is the Fahrenheit reading?

**29.** The melting point (temperature at which a solid changes to a liquid) of mercury is −38°F. What is the Celsius reading?

*During the forging and heat-treating of steel, it is important to use the color of heated steel as an indicator of its temperature. Fill in the following chart, which shows the color of heat-treated steel and the corresponding approximate temperatures in Celsius and Fahrenheit.*

| | *Color* | *°C* | *°F* |
|---|---|---|---|
| **30.** | White | | 2200 |
| **31.** | Yellow | 1100 | |
| **32.** | Orange | | 1725 |
| **33.** | Cherry red | 718 | |
| **34.** | Dark red | 635 | |
| **35.** | Faint red | | 900 |
| **36.** | Pale blue | 310 | |

*Fill in each blank.*

**37.** The basic metric unit of time is the _____. Its abbreviation is _____.

**38.** The basic metric unit of electric current is the _____. Its abbreviation is _____.

**39.** The common metric unit of power is the _____. Its abbreviation is _____.

*Which is larger?*

**40.** 1 watt or 1 milliwatt

**41.** 1 milliampere or 1 microampere

**42.** 1 millisecond or 1 nanosecond

**43.** 1 megawatt or 1 milliwatt

**44.** 1 ps or 1 $\mu$s

**45.** 1 mA or 1 A

**46.** 1 mW or 1 $\mu$W

*Write the abbreviation for each unit.*

**47.** 2.7 microamperes

**48.** 3.6 microseconds

**49.** 9.5 kilowatts

**50.** 150 milliamperes

**51.** 15 nanoseconds

**52.** 12 amperes

**53.** 135 megawatts

**54.** 75 picoseconds

*Fill in each blank.*

**55.** 1 kW = _____ W

**56.** 1 A = _____ $\mu$A

**57.** 1 s = _____ ns

**58.** 1 mA = _____ A

**59.** 1 MW = _____ W

**60.** 1 $\mu$s = _____ s

**61.** 1 mA = _____ $\mu$A

**62.** 1 ns = _____ ps

**63.** 1 MW = _____ $\mu$W

**64.** 1 $\mu$s = _____ ns

**65.** Change 6 A to mA.

**66.** Change 2500 W to kW.

**67.** Change 42 mW to $\mu$W.

**68.** Change 245 $\mu$s to s.

**69.** Change 7800 mA to A.

**70.** Change 1 h 25 min to min.

**71.** Change 4 h 25 min 15 s to s.

**72.** Change $7 \times 10^6$ s to h.

**73.** Change 3 s to ns.

**74.** Change $6 \times 10^{10}$ $\mu$W to W.

**75.** Change $4 \times 10^{10}$ $\mu$W to MW.

**76.** Change 1 h to ps.

## 1.9 SPECIALIZED CONVERSIONS FOR ELECTRONICS

Let's first list the metric prefixes and the corresponding powers of 10 that are multiples of three because these are the most commonly used in electronics and physics. The list is given in Table 1.4.

**TABLE 1.4**

| Prefix | Symbol | Power of 10 |
| --- | --- | --- |
| tera | T | $10^{12}$ |
| giga | G | $10^9$ |
| mega | M | $10^6$ |
| kilo | k | $10^3$ |
| unit | | $10^0$ |
| milli | m | $10^{-3}$ |
| micro | $\mu$ | $10^{-6}$ |
| nano | n | $10^{-9}$ |
| pico | p | $10^{-12}$ |

We have converted to and from base units; that is, we have changed from mA to A, from W to MW, from ns to s, and so on. Now we need to consider how to quickly change from one prefix unit to another; that is, we need to change from mA to $\mu$A, from M$\Omega$ to k$\Omega$, from ps to ns, and so on. Such conversions can be accomplished by first changing to the base unit and then to the desired prefix unit.

**EXAMPLE 1**

Change 25,000 $\mu$A to mA.

Since micro means $10^{-6}$, 1 $\mu$A $= 10^{-6}$ A. So we first change $\mu$A to A.

$$25,000 \ \mu A \times \frac{10^{-6} \ A}{1 \ \mu A} = 0.025 \ A$$

Milli means $10^{-3}$, and 1 mA $= 10^{-3}$ A. We change A to mA as follows:

$$0.025 \ A \times \frac{1 \ mA}{10^{-3} \ A} = 25 \ mA$$

Thus, 25,000 $\mu$A $= 25$ mA.

This two-step conversion can be done in one step as follows:

$$25,000 \ \mu A \times \frac{10^{-6} \ A}{1 \ \mu A} \times \frac{1 \ mA}{10^{-3} \ A} = 25,000 \times 10^{-3} \ mA$$
$$= 25 \times 10^3 \times 10^{-3} \ mA$$
$$= 25 \ mA$$

**EXAMPLE 2**

Change 0.5 M$\Omega$ to k$\Omega$.

$$0.5 \ M\Omega \times \frac{10^6 \ \Omega}{1 \ M\Omega} \times \frac{1 \ k\Omega}{10^3 \ \Omega} = 0.5 \times 10^3 \ k\Omega$$
$$= 500 \ k\Omega$$

**EXAMPLE 3**

Change 160 ps to ns.

$$160 \ ps \times \frac{10^{-12} \ s}{1 \ ps} \times \frac{1 \ ns}{10^{-9} \ s} = 160 \times 10^{-3} \ ns$$
$$= 0.16 \ ns$$

**EXAMPLE 4**

Change 0.005 kW to $\mu$W.

$$0.005 \ kW \times \frac{10^3 \ W}{1 \ kW} \times \frac{1 \ \mu W}{10^{-6} \ W} = 0.005 \times 10^9 \ \mu W$$
$$= 5 \times 10^{-3} \times 10^9 \ \mu W$$
$$= 5 \times 10^6 \ \mu W$$

EXAMPLE 5

Change 7500 ps to $\mu$s.

$$7500\ \text{ps} \times \frac{10^{-12}\ \text{s}}{1\ \text{ps}} \times \frac{1\ \mu\text{s}}{10^{-6}\ \text{s}} = 7500 \times 10^{-6}\ \mu\text{s}$$

$$= 0.0075\ \mu\text{s}$$

Note that 7500 ps = 7.5 ns.

## Engineering Notation

Often a unit is expressed in **engineering notation,** where the numerical coefficient is between 1 and 1000 and its prefix represents a power of 10 in the form $10^{3n}$, where $n$ is an integer. To change to engineering notation, move the decimal point in groups of three digits until the number is between 1 and 1000. Write the corresponding power of 10; then change this power of 10 to its corresponding prefix.

### EXAMPLE 6

Change 1,350,000 $\Omega$ to engineering notation.

$$1{,}350{,}000\ \Omega = 1.35 \times 10^{6}\ \Omega = 1.35\ \text{M}\Omega \qquad (\textit{Note: } 10^{6} = \text{M})$$

### EXAMPLE 7

Change 34,000 W to engineering notation.

$$34{,}000\ \text{W} = 34 \times 10^{3}\ \text{W} = 34\ \text{kW} \qquad (\textit{Note: } 10^{3} = \text{k})$$

### EXAMPLE 8

Change 0.0000358 s to engineering notation.

$$0.0000358\ \text{s} = 35.8 \times 10^{-6}\ \text{s} = 35.8\ \mu\text{s} \qquad (\textit{Note: } 10^{-6} = \mu)$$

## Reciprocal Unit Relationships

The **reciprocal** of a nonzero number $n$ is $\dfrac{1}{n}$ or $n^{-1}$. The following list shows some examples.

| The reciprocal of: | Is: |
|---|---|
| 3 | $\dfrac{1}{3}$ |
| 8 | $\dfrac{1}{8}$ |
| $\dfrac{5}{6}$ | $\dfrac{6}{5}$ |
| 10 | $\dfrac{1}{10}$ |

| The reciprocal of: | Is: |
|---|---|
| $0.1$ or $\dfrac{1}{10}$ | $10$ |
| $10^3$ | $10^{-3}$ |
| $10^{-5}$ | $10^5$ |
| $10^a$ | $10^{-a}$ |
| $10^{-a}$ | $10^a$ |

**The Period-Frequency Relationship** The **period** $T$ of a wave equals the reciprocal of its frequency $f$. That is, $T = \dfrac{1}{f}$ or $f = \dfrac{1}{T}$. The unit for $T$ is the second (s), and the unit for $f$ is cycles/second ($\dfrac{1}{s}$) or hertz (Hz). Note that $1 \text{ Hz} = \dfrac{1}{s}$ (1 hertz = 1 cycle/second). The word *cycle* has no physical dimension, so 1 cycle/s = 1/s.

**EXAMPLE 9**

If the period of a wave is 1 ms, find its frequency.

$$f = \frac{1}{T}$$

$$f = \frac{1}{1 \text{ ms}} \qquad (T = 1 \text{ ms})$$

$$= \frac{1}{10^{-3} \text{ s}} \qquad (1 \text{ ms} = 10^{-3} \text{ s})$$

$$= \frac{10^3}{1 \text{ s}} \qquad \left( \frac{1}{10^{-3}} = 10^3 \right)$$

$$= 10^3 \text{ Hz} \qquad \left( \frac{1}{s} = \text{Hz} \right)$$

$$= 1 \text{ kHz} \qquad (Note:\ 10^3 = \text{k})$$

The common reciprocal prefix relationships are listed in Table 1.5.

**EXAMPLE 10**

If the frequency of a wave is 1 MHz, find the period.

$$T = \frac{1}{f}$$

$$T = \frac{1}{1 \text{ MHz}} \qquad (f = 1 \text{ MHz})$$

$$= 1 \text{ } \mu s \qquad \left( \frac{1}{M} = \frac{1}{10^6} = 10^{-6} = \mu \text{ and } \frac{1}{\text{Hz}} = s \right)$$

$$\left( Note:\ \frac{1}{\text{Hz}} = \frac{1}{\dfrac{1}{s}} = 1 \div \frac{1}{s} = 1 \times \frac{s}{1} = s \right)$$

**TABLE 1.5**

| Prefix names | Prefix powers of 10 |
|---|---|
| $\dfrac{1}{\text{tera}} = \text{pico}$ | $\dfrac{1}{10^{12}} = 10^{-12}$ |
| $\dfrac{1}{\text{giga}} = \text{nano}$ | $\dfrac{1}{10^{9}} = 10^{-9}$ |
| $\dfrac{1}{\text{mega}} = \text{micro}$ | $\dfrac{1}{10^{6}} = 10^{-6}$ |
| $\dfrac{1}{\text{kilo}} = \text{milli}$ | $\dfrac{1}{10^{3}} = 10^{-3}$ |
| $\dfrac{1}{\text{milli}} = \text{kilo}$ | $\dfrac{1}{10^{-3}} = 10^{3}$ |
| $\dfrac{1}{\text{micro}} = \text{mega}$ | $\dfrac{1}{10^{-6}} = 10^{6}$ |
| $\dfrac{1}{\text{nano}} = \text{giga}$ | $\dfrac{1}{10^{-9}} = 10^{9}$ |
| $\dfrac{1}{\text{pico}} = \text{tera}$ | $\dfrac{1}{10^{-12}} = 10^{12}$ |

In summary, the reciprocal unit relationships are as follows.

$$\frac{1}{\text{s}} = \text{Hz} \quad \text{and} \quad \frac{1}{\text{Hz}} = \text{s}$$

**EXAMPLE 11**

If the period of a wave is 100 $\mu$s, find the frequency.

$$f = \frac{1}{T}$$

$$f = \frac{1}{100 \ \mu\text{s}}$$

$$= 0.01 \ \text{MHz} \quad \left( Note: \ \frac{1}{100} = 0.01, \ \frac{1}{\mu} = \frac{1}{10^{-6}} = 10^6 = \text{M, and} \ \frac{1}{\text{s}} = \text{Hz} \right)$$

**EXAMPLE 12**

If the frequency of a wave is 10 GHz, find the period.

$$T = \frac{1}{f}$$

$$T = \frac{1}{10 \ \text{GHz}}$$

$$= 0.1 \ \text{ns} \quad \left( Note: \ \frac{1}{10} = 0.1, \ \frac{1}{\text{G}} = \frac{1}{10^{9}} = 10^{-9} = \text{n, and} \ \frac{1}{\text{Hz}} = \text{s} \right)$$

## Exercises 1.9

*Fill in each blank.*

**1.** 24,500 $\mu$F = _____ mF

**2.** 0.0037 $\mu$s = _____ ps

**3.** 0.075 GHz = _____ kHz

**4.** 265,000 ns = _____ ms

**5.** 0.65 mA = _____ $\mu$A

**6.** 75 kV = _____ mV

**7.** 0.00085 M$\Omega$ = _____ m$\Omega$

**8.** 0.008 mW = _____ $\mu$W

**9.** 0.24 ns = _____ ps

**10.** 0.0075 M$\Omega$ = _____ k$\Omega$

**11.** 1450 kV = _____ MV

**12.** 2.5 THz = _____ MHz

**13.** 2750 ps = _____ $\mu$s

**14.** 0.045 k$\Omega$ = _____ m$\Omega$

**15.** 0.018 mF = _____ nF

**16.** 3200 ps = _____ ms

**17.** 1400 kHz = _____ MHz

**18.** 2250 $\mu$H = _____ mH

**19.** 450 ps = _____ $\mu$s

**20.** 0.15 mA = _____ $\mu$A

**21.** $1.3 \times 10^4$ k$\Omega$ = _____ M$\Omega$

**22.** $4.5 \times 10^6$ MV = _____ mV

**23.** $3.5 \times 10^{-6}$ mF = _____ $\mu$F

**24.** $2.1 \times 10^{-4}$ ms = _____ ns

*Change each unit to engineering notation.*

**25.** 145,000 Hz

**26.** 3500 $\Omega$

**27.** 2,100,000 V

**28.** 84,000,000 W

**29.** 118,000,000 Hz

**30.** 1,250,000,000 W

**31.** 0.0085 s

**32.** 0.00025 A

**33.** 0.00008 V

**34.** 0.00000075 s

**35.** 0.0000000872 s

**36.** 0.375 A

**37.** 3250 $\mu$W

**38.** 19,500 ps

**39.** 48,000 mW

**40.** 2700 ns

**41.** 0.0075 M$\Omega$

**42.** 0.00025 $\mu$s

*Use the reciprocal relationship to find f or T, whichever is not given, and write the result in engineering notation.*

**43.** $f = 1$ kHz

**44.** $T = 1$ $\mu$s

**45.** $T = 1$ ns

**46.** $f = 1$ GHz

**47.** $f = 10$ MHz

**48.** $T = 10$ ms

**49.** $T = 100$ ps

**50.** $f = 0.1$ kHz

**51.** $f = 0.01$ GHz

**52.** $f = 100$ kHz

**53.** $T = 10$ $\mu$s

**54.** $T = 10$ ns

**55.** $T = 0.1$ ms

**56.** $f = 0.01$ MHz

**57.** $f = 10$ kHz

**58.** $T = 10$ ps

**59.** $T = 100$ ns

**60.** $f = 1000$ kHz

**61.** A nuclear power plant generated 100 million megawatts of power over its 18 years of service. Convert this to watts in scientific notation.

# 1.10  OTHER CONVERSIONS

One of the most basic and most useful concepts that you can learn is often called **unit analysis.** Unit analysis can be divided into three areas: (1) converting from one set of units to another, as we saw in the previous section; (2) determining and simplifying the units of a physical quantity obtained from the substitution of data into a formula, as in

Section 2.7; and (3) analyzing the derived units in terms of the basic units or other derived units, as introduced in the SI metric system and basic in physics and most technical courses.

Conversion factors can also be used to change units within the English system. The information for forming the conversion factors can be found in the tables in the Appendix.

**EXAMPLE 1**

(a) Change 8 ft to inches

$$8\ \text{ft} \times \frac{12\ \text{in.}}{1\ \text{ft}} = 96\ \text{in.} \qquad \left( Note: \frac{12\ \text{in.}}{1\ \text{ft}} = 1 \right)$$

(b) Change 36 ft to yards.

$$36\ \text{ft} \times \frac{1\ \text{yd}}{3\ \text{ft}} = 12\ \text{yd}$$

(c) Change 2880 in² to square feet.

$$2880\ \text{in}^2 \times \frac{1\ \text{ft}^2}{144\ \text{in}^2} = 20\ \text{ft}^2$$

Converting from English to metric or metric to English also involves a conversion factor in which the numerator equals the denominator. Three-significant-digit accuracy is sufficient.

**EXAMPLE 2**

(a) Change 18 lb to kilograms.

$$18\ \text{lb} \times \frac{1\ \text{kg}}{2.20\ \text{lb}} = 8.18\ \text{kg} \qquad \left( Note: \frac{1\ \text{kg}}{2.20\ \text{lb}} = 1 \right)$$

(b) Change 315 km to miles.

$$315\ \text{km} \times \frac{0.621\ \text{mi}}{1\ \text{km}} = 196\ \text{mi}$$

(c) Change 45 gal to litres.

$$45\ \text{gal} \times \frac{3.79\ \text{L}}{1\ \text{gal}} = 171\ \text{L}$$

**EXAMPLE 3**

Change 90 km/h to metres per second.

This example involves two conversions: kilometres to metres and hours to seconds. To convert km to m (1 km = 1000 m), we have two possible conversion factors:

$$\frac{1\ \text{km}}{1000\ \text{m}} \quad \text{and} \quad \frac{1000\ \text{m}}{1\ \text{km}}$$

We choose the conversion factor with km in the *denominator* so that the km units cancel each other.

To convert h to s (1 h = 3600 s), we again have two possible conversion factors:

$$\frac{1\ \text{h}}{3600\ \text{s}} \quad \text{and} \quad \frac{3600\ \text{s}}{1\ \text{h}}$$

We choose the conversion factor with h in the *numerator* so that the h units cancel each other.

$$90 \, \frac{\cancel{km}}{\cancel{h}} \times \frac{1000 \text{ m}}{1 \, \cancel{km}} \times \frac{1 \, \cancel{h}}{3600 \text{ s}} = 25 \text{ m/s}$$

**EXAMPLE 4**

Change 60 mi/h to ft/s.

$$60 \, \frac{\cancel{mi}}{\cancel{h}} \times \frac{1 \, \cancel{h}}{60 \, \cancel{min}} \times \frac{1 \, \cancel{min}}{60 \text{ s}} \times \frac{5280 \text{ ft}}{1 \, \cancel{mi}} = 88 \text{ ft/s}$$

**EXAMPLE 5**

The weight density of copper is 555 lb/ft$^3$. Find its density in newtons per cubic metre.

$$555 \, \frac{\cancel{lb}}{\cancel{ft^3}} \times \frac{4.45 \text{ N}}{1 \, \cancel{lb}} \times \frac{35.3 \, \cancel{ft^3}}{1 \text{ m}^3} = 87,200 \, \frac{\text{N}}{\text{m}^3}$$

## Exercises 1.10

*Round each result to three significant digits when necessary.*

 1. Change 38,000 ft to **(a)** mi, **(b)** km, and **(c)** yd.
 2. Change 1290 lb to **(a)** oz, **(b)** kg, and **(c)** g.
 3. Change 6.7 km to **(a)** m, **(b)** mi, and **(c)** yd.
 4. Change 250 kg to **(a)** g, **(b)** lb, and **(c)** mg.
 5. Change 168 in. to **(a)** ft, **(b)** cm, **(c)** m, and **(d)** yd.
 6. Change 130 gal to **(a)** qt, **(b)** L, and **(c)** kL.
 7. Change 250,000 ft$^2$ to **(a)** yd$^2$, **(b)** m$^2$, **(c)** acres, and **(d)** ha.
 8. Change 86,000 m$^2$ to **(a)** cm$^2$, **(b)** km$^2$, and **(c)** yd$^2$.
 9. Change 15 ft$^3$ to **(a)** in$^3$, **(b)** m$^3$, and **(c)** L.
10. Change 25 kL to **(a)** L, **(b)** ft$^3$, **(c)** in$^3$, and **(d)** m$^3$.
11. Change 750 mL to **(a)** L, **(b)** pints, and **(c)** qt.
12. Change 150 ha to **(a)** acres, **(b)** m$^2$, **(c)** km$^2$, and **(d)** ft$^2$.
13. Change 20 ft/s to cm/s.
14. Change 55 mi/h to km/h.
15. Change $6 \times 10^{-4}$ °C/min to °C/h.
16. Change 0.03 °C/h to °C/min.
17. Change 0.006 in./h to in./min.
18. Change 0.45 in./min to in./h.
19. Change 30 m/s$^2$ to ft/s$^2$.
20. Change 50 ft/s$^2$ to km/h$^2$.
21. Change 2500 N/m$^2$ to lb/in$^2$.
22. Change 358 lb/ft$^3$ to g/cm$^3$.
23. Change 75 g/cm$^3$ to oz/in$^3$.
24. Change 25 mg/cm$^3$ to g/mm$^3$.

25. Change 560 kcal to Btu.

26. Change 2.5 kW to hp.

27. Change 1600 ft-lb to (a) J (joules) and (b) kcal.

28. Change 200 kg to slugs.

29. A road sign states that it is 125 km to Chicago. How far is it in miles?

30. A camera uses film that is 35 mm wide. What is its width in inches?

31. The diameter of a bolt is 0.625 in. What is its diameter in millimetres?

32. Change $4\frac{15}{32}$ in. to cm.

33. A tank has 75 gal of fuel. How many litres does it contain?

34. A satellite weighs 275 kg. How many pounds does it weigh?

35. A football field is 100 yd long.
    (a) How long is it in ft?
    (b) How long is it in metres?

36. A microwheel weighs 0.065 oz. What is its weight in mg?

37. A hole 0.325 in. in diameter is drilled in a metal plate. What is its diameter in (a) cm and (b) mm?

38. What is your weight in kg and g?

39. A mechanic finds that the fuel consumption of an automobile is 30 mi/gal. How many fl oz does it take to drive the automobile 1 mi?

40. How many $cm^3$ does it take to drive the automobile in Exercise 39 1 km?


## 1.11 MEASUREMENT

### Approximate versus Exact Numbers

Up to this time in your study of mathematics, all numbers and all measurements have probably been treated as exact numbers. An **exact number** is a number that has been determined as a result of counting, such as 24 students enrolled in this class, or that has been defined in some way, such as 1 hour = 60 minutes or 1 inch = 2.54 cm, a conversion definition agreed to by the world governments' bureaus of standards. The treatment of the addition, subtraction, multiplication, and division of exact numbers normally is the emphasis, or main content, of grade-school mathematics.

However, nearly all data of a technical nature involve **approximate numbers;** that is, they have been determined as a result of some measurement process—some direct, as with a ruler, and some indirect, as with a surveying transit. Before studying how to perform the calculations with approximate numbers (measurements), we first must determine the "correctness" of an approximate number. First of all, we realize that no measurement can be found exactly. The length of the cover of this book can be found using many instruments. The better the measuring device used, the better the measurement is.

A measurement may be expressed in terms of its accuracy or its precision.


### Accuracy and Significant Digits

The **accuracy** of a measurement refers to the number of digits, called **significant digits,** which indicate the number of units we are reasonably sure of having counted when making a measurement. The greater the number of significant digits given in a measurement, the better is the accuracy, and vice versa.

### EXAMPLE 1

The average distance between the moon and the earth is 239,000 miles. This measurement indicates measuring 239 thousands of miles, and its accuracy is indicated by three significant digits.

### EXAMPLE 2

A measurement of 0.035 cm indicates measuring 35 thousandths of a centimetre; its accuracy is indicated by two significant digits.

### EXAMPLE 3

A measurement of 0.0200 mg indicates measuring 200 ten-thousandths of a milligram; its accuracy is indicated by three significant digits.

Notice that a zero is sometimes significant and sometimes not. In order to clarify this, we give the following rules for significant digits.

---

**RULES FOR SIGNIFICANT DIGITS**

1. All nonzero digits are significant; for example, 356.4 m has four significant digits (this measurement indicates 3564 tenths of metres).

2. All zeros between significant digits are significant; for example, 406.02 km has five significant digits (this measurement indicates 40,602 hundredths of kilometres).

3. A zero in a number greater than one which is specially tagged, such as by a bar above it, is significant; for example, $13\overline{0},000$ km has three significant digits (this measurement indicates $13\overline{0}$ thousands of kilometres).

4. All zeros to the right of a significant digit *and* a decimal point are significant; for example, 36.10 cm has four significant digits (this measurement indicates $361\overline{0}$ hundredths of centimetres).

5. Zeros to the right in a whole number measurement that are not tagged are *not* significant; for example, 2300 m has two significant digits (23 hundreds of metres).

6. Zeros to the left in a measurement less than one are *not* significant; for example, 0.00252 m has three significant digits (252 hundred-thousandths of a metre).

---

## Precision

The **precision** of a measurement refers to the smallest unit with which a measurement is made; that is, the position of the last significant digit.

### EXAMPLE 4

The precision of the measurement 239,000 mi is 1000 mi. (The position of the last significant digit is in the thousands place.)

**EXAMPLE 5**

The precision of the measurement 0.035 cm is 0.001 cm. (The position of the last significant digit is in the thousandths place.)

**EXAMPLE 6**

The precision of the measurement 0.0200 mg is 0.0001 mg. (The position of the last significant digit is in the ten-thousandths place.)

Unfortunately, many people use *accuracy* and *precision* interchangeably. A measurement of 0.0006 cm has good precision and poor accuracy when compared with the measurement 368.0 cm, which has much better accuracy (one versus four significant digits) and poorer precision (0.0001 cm versus 0.1 cm).

**EXAMPLE 7**

Find the accuracy and precision of each of the following measurements.

|  | *Measurement* | *Accuracy (significant digits)* | *Precision* |
|---|---|---|---|
| (a) | 3463 ft | 4 | 1 ft |
| (b) | 3005 mi | 4 | 1 mi |
| (c) | 10,809 kg | 5 | 1 kg |
| (d) | 36,000 tons | 2 | 1000 tons |
| (e) | 88$\overline{0}$0 mi | 3 | 10 mi |
| (f) | 1,349,000 km | 4 | 1000 km |
| (g) | 600$\overline{0}$ m | 4 | 1 m |
| (h) | 0.00632 kg | 3 | 0.00001 kg |
| (i) | 0.0401 m | 3 | 0.0001 m |
| (j) | 0.0060 g | 2 | 0.0001 g |
| (k) | 14.20 m | 4 | 0.01 m |
| (l) | 30.00 cm | 4 | 0.01 cm |
| (m) | 100.060 g | 6 | 0.001 g |

## Exercises 1.11

*Find the accuracy (the number of significant digits) of each measurement.*

**1.** 205 in.    **2.** 14.7 m    **3.** 60.0 cm

**4.** 35,000 ft    **5.** 6.010 km    **6.** 10,$\overline{0}$00 mi

**7.** 16$\overline{0}$0 Ω    **8.** 120 V    **9.** 0.060 g

**10.** 0.0250 A    **11.** 20$\overline{0}$ mm    **12.** 205,000 Ω

*Find the precision of each measurement.*

**13.** 3.6 cm    **14.** 7.0 m    **15.** 16.00 cm

**16.** 4.100 mi    **17.** 16$\overline{0}$ mm    **18.** 304,000 km

| **19.** 6.00 m | **20.** 360 V | **21.** $30\overline{0}0$ Ω |
| **22.** 0.050 km | **23.** 0.0040 A | **24.** 63.500 g |

*In each set of measurements, find the measurement that is **(a)** the most accurate and **(b)** the most precise.*

**25.** 15.2 m; 0.023 m; 0.06 m          **26.** 256 ft; 400 ft; 270 ft

**27.** 0.642 cm; 0.82 cm; 14.02 cm          **28.** 6.2 m; 4.7 m; 3.0 m

**29.** 0.0270 A; 0.035 A; 0.00060 A; 0.055 A          **30.** 164.00 km; 5.60 km; 4.000 km; 0.05 km

**31.** 305,000 Ω; 38,000 Ω; $4\overline{0}0,000$ Ω; 80,000 Ω

**32.** 1,300,000 V; 35,000 V; $60,\overline{0}00$ V; 20,000 V

*In each set of measurements, find the measurement that is **(a)** the least accurate and **(b)** the least precise.*

**33.** 13.2 m; 0.057 m; 0.08 m          **34.** 372 yd; 300 yd; 560 yd

**35.** 16.8 km; 0.52 km; 15.05 km          **36.** 6.5 kg; 460 kg; 0.075 kg

**37.** 0.0370 A; 0.030 A; 0.00009 A; 0.41 A          **38.** 284.0 mi; 6.35 mi; 7.000 mi; 0.05 mi

**39.** 205,000 Ω; 43,000 Ω; $6\overline{0}0,000$ Ω; 500,000 Ω

**40.** 1,400,000 V; 27,000 V; $50,\overline{0}00$ V; 30,000 V

## 1.12 OPERATIONS WITH MEASUREMENTS

If someone measured the length of one of two parts of a shaft with a micrometer calibrated in 0.01 mm as 12.27 mm and another person measured the second part with a ruler calibrated in mm as 23 mm, would the total length be 35.27 mm? Note that the sum 35.27 mm indicates a precision of 0.01 mm. The precision of the ruler is 1 mm, which means that the measurement 23 mm with the ruler could actually be anywhere between 22.50 mm and 23.50 mm using the micrometer (which has a precision of 0.01 mm). That is, any measurement between 22.50 and 23.50 can be read only as 23 mm using the ruler. Of course, this means that the tenths and hundredths digits in the sum 35.27 mm are actually meaningless. In other words, the sum or difference of measurements can be no more precise than the least precise measurement.

---

**ADDING OR SUBTRACTING MEASUREMENTS**

**1.** Make certain that all of the measurements are expressed in the same units. If they are not, change them all to the same units.

**2.** Add or subtract.

**3.** Then round the result to the same precision as the least precise measurement.

---

**EXAMPLE 1**

Add the measurements 1250 cm, 1562 mm, 2.963 m, and 9.71 m.

First, convert all measurements to the same unit, such as metres.

$$1250 \text{ cm } = 12.5 \text{ m}$$
$$1562 \text{ mm } = 1.562 \text{ m}$$

Next, add.

$$12.5 \ \ \text{m}$$
$$1.562 \ \text{m}$$
$$2.963 \ \text{m}$$
$$\underline{9.71 \ \ \text{m}}$$
$$26.735 \ \text{m} \rightarrow 26.7 \ \text{m}$$

Then round this sum to the same precision as the least precise measurement, which is 12.5 m. Thus, the sum is 26.7 m.

**EXAMPLE 2**

Subtract the measurements: 2567 g − 1.60 kg.

First, convert all measurements to the same units, such as grams.

$$1.60 \ \text{kg} = 16\overline{0}0 \ \text{g}$$

Be careful not to change the number of significant digits. Next, subtract.

$$2567 \ \text{g}$$
$$\underline{16\overline{0}0 \ \text{g}}$$
$$967 \ \text{g} \rightarrow 970 \ \text{g}$$

Then round this difference to the same precision as the least precise measurement, which is $16\overline{0}0$ g. Thus, the difference is 970 g.

Suppose that you need to find the area of a rectangular room that measures 11.4 m by 15.6 m. If you multiply the numbers 11.4 and 15.6, the product 177.84 implies an accuracy of five significant digits. But note that each of the original measurements contains only three significant digits. To rectify this inconsistency, we say that the product or quotient of measurements can be no more accurate than the least accurate measurement.

---

**MULTIPLYING OR DIVIDING MEASUREMENTS**

**1.** First multiply or divide the measurements as given.

**2.** Then round the result to the same number of significant digits as the measurement with the least number of significant digits.

---

**EXAMPLE 3**

Multiply the measurements: 11.4 m × 15.6 m.

$$11.4 \ \text{m} \times 15.6 \ \text{m} = 177.84 \ \text{m}^2$$

Round this product to three significant digits, which is the accuracy of the least accurate measurement (which is the accuracy of each measurement in this example). That is,

$$11.4 \ \text{m} \times 15.6 \ \text{m} = 178 \ \text{m}^2$$

**EXAMPLE 4**

Divide the measurements: 78,000 m² ÷ 654 m.

$$78,000 \ \text{m}^2 \div 654 \ \text{m} = 119.26606 \ \text{m}$$

Round this quotient to two significant digits, which is the accuracy of the least accurate measurement (78,000 m$^2$). That is,

$$78,000 \text{ m}^2 \div 654 \text{ m} = 120 \text{ m}$$

**EXAMPLE 5**

Use the rules for multiplication and division of measurements to evaluate the following:

$$\frac{(25.0 \text{ kg})(14 \text{ m/s})}{0.104 \text{ m}}$$

First, multiply and divide the numbers (3365.384615 . . . ), and round the result to two significant digits, which is the accuracy of the least accurate measurement (14 m/s). Then multiply and divide the units as you did the numbers.

$$\frac{\text{kg (m/s)}}{\text{m}} = \frac{\text{kg}}{\text{s}}$$

Then

$$\frac{(25.0 \text{ kg})(14 \text{ m/s})}{0.104 \text{ m}} = 3400 \text{ kg/s}$$

*Note:* When we multiply or divide measurements, the units do not need to be the same. The units must be the same when we add or subtract measurements. Also, the units are multiplied and/or divided in the same manner as the corresponding numbers.

Any power or root of a measurement should be rounded to the same accuracy as the given measurement.

Obviously, such calculations with measurements should be done with a calculator. When no calculator is available, you may round the original measurements or any intermediate results to one more digit than the accuracy or precision required in the final result.

If both exact numbers and approximate numbers (measurements) occur in the same calculation, only the approximate numbers are used to determine the accuracy or precision of the result.

There are even more sophisticated methods for dealing with the calculations of measurements. The method we use, and indeed if we should even follow any given procedure, depends on the number of measurements and the sophistication needed for a particular situation.

The procedures for operations with measurements shown here are based on methods followed and presented by the American Society for Testing and Materials (ASTM).

## Exercises 1.12

*Use the rules for addition of measurements to find the sum of each set of measurements.*

**1.** 15.7 in.; 6.4 in.

**2.** 178 m; 33.7 m; 61 m

**3.** 45.6 cm; 13.41 cm; 1.407 cm; 24.4 cm

**4.** 406 g; 1648.5 g; 39.74 g; 68.1 g

**5.** 1.0443 g; 0.00134 g; 0.08986 g; 0.001359 g

**6.** 7.639 mi; 14.48 mi; 1.004 mi; 0.68 mi

**7.** 14 V; 1.005 V; 0.018 V; 3.5 V

**8.** 130.5 cm; 14.4 cm; 1.457 m

**9.** 10.505 cm; 9.35 mm; 13.65 cm

**10.** 1850 cm; 1276 mm; 2.816 m; 4.02 m

*Use the rules for subtraction of measurements to subtract the second measurement from the first.*

**11.** 16.3 cm
12.4 cm

**12.** 120.2 cm
13.8 cm

**13.** 15.02 mm
12.6 mm

**14.** 162 mm
15.3 cm

**15.** 16.61 oz
11.372 oz

**16.** 94.1 g
32.74 g

**17.** 6.000 in.
2.004 in.

**18.** 0.54861 in.
0.234 in.

**19.** Four pieces of metal of thickness 0.149 in., 0.407 in., 1.028 in., and 0.77 in. are to be bolted together. What is the total thickness of the four pieces?

**20.** Five pieces of metal of thickness 2.47 mm, 10.4 mm, 3.70 mm, 1.445 mm, and 8.300 mm are clamped together. What is the total thickness of the five pieces?

**21.** Find the current going through $R_5$ in the circuit in Fig. 1.20. *Hint:* $R_1 + R_2 + R_3 = R_4 + R_5$.

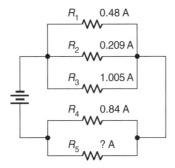

**Figure 1.20**

**22.** Find the sum of the following resistances: 15 $\Omega$, 120 $\Omega$, 6.5 $\Omega$, 0.025 $\Omega$, and 2375 $\Omega$.

*Use the rules for multiplication and division of measurements to evaluate each of the following.*

**23.** (17.7 m)(48.2 m)

**24.** (540 cm)(28.0 cm)

**25.** (4.6 in.)(0.0285 in.)

**26.** (8.2 km)(6.75 km)

**27.** (34.2 cm)(26.1 cm)(28.9 cm)

**28.** (0.065 m)(0.0282 m)(0.0375 m)

**29.** 19.4 m$^3$ ÷ 9.3 m$^2$

**30.** 4300 V ÷ 14.5 A

**31.** $\dfrac{490 \text{ cm}}{6.73 \text{ s}^2}$

**32.** $\dfrac{5.03 \text{ km}}{4.7 \text{ s}}$

**33.** $\dfrac{0.447 \text{ N}}{(1.43 \text{ m})(4.0 \text{ m})}$

**34.** $\dfrac{(120 \text{ V})^2}{50.0 \ \Omega}$

**35.** $\dfrac{(4\overline{0} \text{ kg})(3.0 \text{ m/s})^2}{5.50 \text{ m}}$

**36.** $\dfrac{190 \text{ g}}{(3.4 \text{ cm})(1.6 \text{ cm})(8.4 \text{ cm})}$

**37.** Find the area of a rectangle measured as 6.2 cm by 17.5 cm ($A = lw$).

**38.** The formula for the volume of a rectangular solid is $V = lwh$, where $l$ = length, $w$ = width, and $h$ = height. Find the volume of a rectangular solid when $l = 12.4$ ft, $w = 9.6$ ft, and $h = 5.4$ ft.

**39.** Find the volume of a cube with each edge 8.50 cm long ($V = e^3$, where $e$ is the length of each edge).

**40.** The formula $s = 4.90t^2$ gives the distance $s$ in metres a body falls in a given time $t$. Find the distance a ball falls in 2.6 s.

41. Given K.E. $= \frac{1}{2}mv^2$, where $m = 2.37 \times 10^6$ kg, and $v = 10.4$ m/s. Find K.E.

42. A formula for finding the horsepower of an engine is $p = \dfrac{d^2n}{2.50}$, where $d$ is the diameter of each cylinder in inches and $n$ is the number of cylinders. What is the horsepower of an 8-cylinder engine if each cylinder has a diameter of 3.00 in.? (*Note:* Eight is an exact number. The number of significant digits in an exact number has no bearing on the number of significant digits in the product or quotient.)

43. Six pieces of metal, each of thickness 2.08 mm, are fitted together. What is the total thickness of the six pieces?

44. Find the volume of a cylinder having a radius of 6.1 m and a height of 8.3 m. The formula for the volume of a cylinder is $V = \pi r^2 h$.

45. In 1993 in the United States 6,336,470,000 bushels (bu) of corn were harvested from 62,920,000 acres. In 1995 there were 7,373,880,000 bu harvested from 64,990,000 acres. What was the yield in bushels per acre for each year, and what was the increase in yield?

# CHAPTER SUMMARY

1. *Basic terms:*
   (a) *Positive integers:* 1, 2, 3, . . .
   (b) *Negative integers:* $-1, -2, -3, . . .$
   (c) *Integers:* . . ., $-3, -2, -1, 0, 1, 2, 3, . . .$
   (d) *Rational numbers:* Numbers that can be represented as the ratio of two integers.
   (e) *Irrational numbers:* Numbers that cannot be represented as the ratio of two integers.
   (f) *Real numbers:* Set of numbers consisting of the rational numbers and the irrational numbers.
   (g) *Prime number:* A positive integer greater than one that is evenly divisible only by itself and one.
   (h) The absolute value of a real number $n$, written $|n|$, is defined as $|n| = n$ if $n \geq 0$ or $|n| = -n$ if $n < 0$.

2. *Basic properties of real numbers:*
   (a) $a + b = b + a$ — Commutative property of addition
   (b) $ab = ba$ — Commutative property of multiplication
   (c) $(a + b) + c = a + (b + c)$ — Associative property of addition
   (d) $(ab)c = a(bc)$ — Associative property of multiplication
   (e) $a(b + c) = ab + ac$ — Distributive property of multiplication over addition
   (f) $a + (-a) = 0$ — Additive inverse or negative property
   (g) $a \cdot \dfrac{1}{a} = 1$ $(a \neq 0)$ — Multiplicative inverse or reciprocal property
   (h) $a + 0 = a$ — Identity element of addition
   (i) $a \cdot 1 = a$ — Identity element of multiplication

3. *Operations with signed numbers:* See Section 1.1.

4. *Operations with zero:*
   (a) $a + 0 = a$
   (b) $a - 0 = a$

(c) $a \cdot 0 = 0$

(d) If $a \cdot b = 0$, then either $a = 0$ or $b = 0$.

(e) $\dfrac{0}{b} = 0$  $(b \neq 0)$

(f) $\dfrac{a}{0}$ is meaningless $(a \neq 0)$

(g) $\dfrac{0}{0}$ is indeterminate.

5. *Order of operations:*
   (a) Do all operations within any grouping symbols beginning with the innermost pair if grouping symbols are contained within each other.
   (b) Then do all multiplications and divisions in the order in which they occur from left to right.
   (c) Finally, do all additions and subtractions in the order in which they occur from left to right.

6. *Laws of exponents:*
   (a) $10^m \cdot 10^n = 10^{m+n}$
   (b) $\dfrac{10^m}{10^n} = 10^{m-n}$
   (c) $(10^m)^n = 10^{mn}$
   (d) $10^{-n} = \dfrac{1}{10^n}$  and  $\dfrac{1}{10^{-n}} = 10^n$
   (e) $10^0 = 1$

7. *Scientific notation:* A number is expressed in scientific notation as a product of a decimal between 1 and 10 and a power of 10 in the form $N \times 10^m$ where $1 \leq N < 10$ and $m$ is an integer.

8. *Metric prefixes:* See Table 1.6.

9. *Metric basic units:* See Table 1.7.

10. *Conversion factor:* The correct choice for a particular conversion factor is the one in which the old units are in the numerator of the original expression and in the denominator of the conversion factor, or in the denominator of the original expression and in the numerator of the conversion factor. That is, we want the old units to cancel each other.

11. *Engineering notation:* A unit is expressed in engineering notation as a product of a decimal between 1 and 1000, and its prefix represents a power of 10 in the form $10^{3n}$, where $n$ is an integer.

12. *Period-frequency relationship:* $T = \dfrac{1}{f}$  or  $f = \dfrac{1}{T}$

13. *Accuracy:* The accuracy of a measurement refers to the number of significant digits it contains. The number of significant digits indicates the number of units we are reasonably sure of having counted when making the measurement.

14. *Rules for significant digits:*
   (a) All nonzero digits are significant.
   (b) All zeros between significant digits are significant.

**TABLE 1.6**

| Multiple or submultiple[a] decimal form | Power of 10 | Prefix | Prefix symbol | Pronun-ciation | Meaning |
|---|---|---|---|---|---|
| 1,000,000,000,000 | $10^{12}$ | tera | T | těr′ă | one trillion times |
| 1,000,000,000 | $10^{9}$ | giga | G | jĭg′ă | one billion times |
| 1,000,000 | $10^{6}$ | mega | M | měg′ă | one million times |
| 1,000 | $10^{3}$ | kilo[b] | k | kĭl′ō | one thousand times |
| 100 | $10^{2}$ | hecto | h | hěk′tō | one hundred times |
| 10 | $10^{1}$ | deka | da | děk′ă | ten times |
| 0.1 | $10^{-1}$ | deci | d | děs′ĭ | one-tenth of |
| 0.01 | $10^{-2}$ | centi[b] | c | sěnt′ĭ | one-hundredth of |
| 0.001 | $10^{-3}$ | milli[b] | m | mĭl′ĭ | one-thousandth of |
| 0.000001 | $10^{-6}$ | micro | $\mu$ | mī′krō | one-millionth of |
| 0.000000001 | $10^{-9}$ | nano | n | năn′ō | one-billionth of |
| 0.000000000001 | $10^{-12}$ | pico | p | pē′kō | one-trillionth of |

[a]Factor by which the unit is multiplied.

[b]Most commonly used prefixes.

**TABLE 1.7**

| Basic unit | SI abbreviation | Used for measuring |
|---|---|---|
| metre | m | length |
| kilogram | kg | mass |
| second | s | time |
| ampere | A | electric current |
| kelvin | K | temperature |
| candela | cd | light intensity |
| mole | mol | molecular substance |

(c) A zero in a number greater than one which is specially tagged, such as by a bar above it, is significant.

(d) All zeros to the right of a significant digit *and* a decimal point are significant.

(e) Zeros to the right in a whole number measurement that are not tagged are *not* significant.

(f) Zeros to the left in a measurement less than one are *not* significant.

15. *Precision:* The precision of a measurement refers to the smallest unit with which the measurement is made; that is, the position of the last significant digit.

16. *To add or subtract measurements:*
    (a) Make certain that all of the measurements are expressed in the same units. If they are not, change them all to the same units.
    (b) Add or subtract.
    (c) Then round the result to the same precision as the least precise measurement.

17. *To multiply and/or divide measurements:*
    (a) First multiply and/or divide the measurements as given.
    (b) Then round the result to the same number of significant digits as the measurement with the least number of significant digits.

## CHAPTER 1 REVIEW

*Perform the indicated operations and simplify.*

**1.** $(-3) + (+6) + (-8)$

**2.** $(+4) - (+7) + (-3) - (-6)$

**3.** $(-3)(+5)(-7)(-2)$

**4.** $\dfrac{(+6)(-12)(+4)}{(+9)(-8)(-2)}$

**5.** $8 - 3 \cdot 4$

**6.** $4 + 10 \div 2$

**7.** $6 - (8 + 4) \div 6$

**8.** $5 + 2(6 - 9)$

**9.** $48 \div 8 \cdot 3 + 2 \cdot 5 - 17$

**10.** $3 \cdot 4 \div 6 - 8 + 6(-3)$

**11.** $\dfrac{3 \cdot 5 + 6(4 \div 2)}{8 \div 4 \cdot 8 - 3}$

**12.** $\dfrac{18 \div 9 \cdot 2 - 3}{5 - (2 - 3)}$

**13.** $\dfrac{4 \cdot 0}{8}$

**14.** $\dfrac{-5 - (15 - 20)}{9 \cdot 0 + (-4 - (-4))}$

**15.** $\dfrac{5 - 6(2 + 4)}{-16 - (-16)}$

**16.** $(10^2)^4$

**17.** $\dfrac{10^{-2}}{10^{-6}}$

**18.** $10^3 \cdot 10^4$

**19.** Write 3,420,000 in scientific notation.

**20.** Write $5.61 \times 10^{-4}$ in decimal form.

*Find each value. Round each result to three significant digits and write it in scientific notation.*

**21.** $(8.54 \times 10^7)(4.97 \times 10^{-14})$

**22.** $\dfrac{1.85 \times 10^{12}}{6.17 \times 10^{-18}}$

*Give the metric prefix for each decimal.*

**23.** 0.01

**24.** 1000

*Give the metric abbreviation for each unit.*

**25.** millilitre

**26.** microgram

*Write the SI unit for each abbreviation.*

**27.** 16 km

**28.** 250 mA

**29.** 1.1 hL

**30.** 18 MW

*Which is larger?*

**31.** 1 litre or 1 millilitre

**32.** 1 kilometre or 1 millimetre

**33.** 1 kilogram or 1 gram

**34.** $1 \text{ m}^3$ or 1 L

**35.** $1 \text{ km}^2$ or 1 ha

**36.** 1 ps or 1 ns

*Fill in each blank.*

**37.** 180 m = _____ km

**38.** 250 mg = _____ g

**39.** 5.7 kL = _____ L

**40.** 1.5 km = _____ m

**41.** $650 \text{ cm}^3$ = _____ mL

**42.** 15 $\mu$s = _____ ns

**43.** 15 MW = _____ W

**44.** $750 \text{ cm}^2$ = _____ $\text{mm}^2$

**45.** $0.75 \text{ m}^3$ = _____ $\text{cm}^3$

**46.** $18,000 \text{ m}^2$ = _____ ha

**47.** 70°F = _____ °C

**48.** −5°C = _____ °F

**49.** Water boils at _____ °C.

**50.** Water freezes at _____ °C.

*Choose the most reasonable quantity.*

**51.** A young couple took a short stroll of 85 cm; 1200 m; 35 km; 1600 mm into a park for a picnic.

**52.** They ate 1 kg; 50 g; 5 g; 75 kg; 150 mg of chicken.

**53.** They drank 16 L; 7 mL; 70 mL; 1.5 L; 18 kL of lemonade.

**54.** The man, being of average height, is 67 cm; 5 m; 170 cm; 3.5 m; 0.5 km; 175 mm tall.

**55.** Linda weighs 50 kg; 150 kg; 175 $\mu$g.

**56.** Bob's new car averages 350 km/L; 12 km/L; 0.25 km/L; 40 km/L.

**57.** He plans to drive no faster than 50 km/h; 80 km/h; 250 km/h; 650 km/h until it is "broken in."

*Fill in each blank.*

**58.** 8850 $\mu$A = _____ mA

**59.** 0.0775 ns = _____ ps

*Change each unit to engineering notation.*

**60.** 48,500,000 $\Omega$

**61.** 0.000075 A

*Use the reciprocal relationship to find f or T, whichever is not given, and write the result in engineering notation.*

**62.** $f$ = 100 MHz

**63.** $T$ = 10 ms

*Round each result in Exercises 64–68 to three significant digits when necessary.*

**64.** Change 3600 ft to **(a)** yd and **(b)** m.

**65.** Change 53.5 kg to **(a)** g and **(b)** lb.

**66.** Change 3600 yd$^2$ to **(a)** square feet and **(b)** square centimetres.

**67.** Change 50 km/h to miles per hour.

**68.** Change 250 kg/m$^3$ to pounds per cubic foot.

*For each measurement, find its **(a)** accuracy and **(b)** precision.*

**69.** 307 m          **70.** 0.050 A          **71.** 12$\overline{0}$,000 V

*Use the rules for addition of measurements to find the sum of each set of measurements.*

**72.** 19.80 L; 14.4 L; 6.000 L; 17.431 L

**73.** 12.600 cm; 10.40 mm; 16.75 cm; 7.005 m

*Use the rules for multiplication and division of measurements to find the value of each of the following.*

**74.** (18.5 m)(21.6 m)          **75.** $\dfrac{49.7 \text{ m}^3}{16.0 \text{ m}^2}$          **76.** $\dfrac{680 \text{ lb}}{(14.5 \text{ in.})(18.6 \text{ in.})}$

# APPLICATION
## Precision-Machined Computer Disks

Imagine a 747 airplane flying $\frac{1}{32}$ of an inch above the ground. This is the equivalent of the head of a disk drive flying 60 miles per hour over a computer disk at a height of 1 to 2 microinches. With such a low glide height, the disk must be extremely smooth and flat. Thickness is also an issue since up to 10 disks may be stacked in one hard drive, with a head on the top and bottom of each disk. When disks are stacked, spacer rings are used to separate them. The disks need to be of uniform thickness so that the spacer rings can be uniform for ease of assembly.

Cerion Technologies produces aluminum disk substrates for the magnetic thin-film disks used in computer hard disk drives. The manufacturing process begins with operators measuring the thickness of the unmachined disks and sorting them into 6 to 8 categories, which differ by $\frac{1}{10,000}$ of an inch. The next step is chemical etching which is done for a specified time determined by the initial thickness of the disks. The chemicals remove the tough oxide surface layer, making the disks easier to grind, and they reduce the thickness variation to $\pm\frac{2}{10,000}$ in., so that after the chemical bath the disks will have four categories of thickness.

Disks have a hole in the center and an outside diameter of 95 mm for desktop computers or 65 mm for laptops. After the etching, the diameters are adjusted and the outer and inner edges are beveled at an angle specified by the customer. The beveling aids in the automated handling of the disks during later processing. The beveled edge tolerance is $\pm\frac{20}{10,000}$ in., while the inner and outer diameter tolerances are $\pm\frac{4}{10,000}$ in. and $\pm\frac{10}{10,000}$ in., respectively. The machine that bevels the edges holds a disk in place by a vacuum, which causes the disk to lose its flatness; the disks are then moved through an oven, since heat reestablishes the flatness.

Finally, the disks are ground smooth and flat to a uniform thickness of 0.0307 in. with a tolerance of $\pm\frac{3}{10,000}$ in. The grinding technology is so important to the process that Cerion custom fabricates its own grinding stones.

Flatness is measured by subtracting the lowest point on the disk from the highest point. Usually the specifications require that the maximum difference be 200 or 300 microinches. Smoothness or surface finish has two measurements: roughness average (RA) and roughness total (RT). RA is the average of the many peaks and valleys as measured with a contact stylus. The RA is usually between 0.4 and 0.5 microinch but must be less than or equal to 0.8 microinch. RT is the difference of the highest peak and the lowest valley, i.e., the worst case. It is usually 4 to 5 microinches but must be less than or equal to 8 microinches.

Because thickness, flatness, and smoothness are all critical to the finished disk substrate, Cerion uses advanced test equipment to monitor disks at each production stage.

This description of Cerion's procedure for manufacturing computer disk substrates illustrates the importance of accuracy and precision in measurements. Almost all manufacturing processes and technical applications use these concepts.

A Cerion laboratory technician uses video monitors to conduct a quality control check.

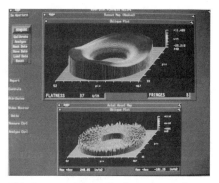

This laser monitor displays the surface finish of a disk substrate.

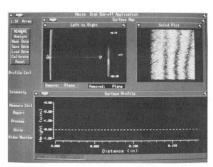

A computer displays a graph of a cross section of a disk surface, where the height is measured in microinches.

*(Photos courtesy of Cerion Technologies, Inc. Champaign, IL.)*

Cerion Technologies, Inc. manufactures substrates used in computer laser printers.

# 2

# Fundamental
# Algebraic Concepts

## INTRODUCTION

A motor causes a gear with 50 teeth to rotate at $40\overline{0}$ rpm (revolutions per minute). This causes a second gear to turn at 125 rpm. How many teeth does the second gear have?

To solve this problem, we need to learn about ratios, proportions, and variation as well as learn how to perform operations on algebraic expressions. In addition to these topics, this chapter lays the foundation for working with formulas, exponents, and radicals. (The preceding problem appears as number 29 in Exercises 2.10.)

### Objectives

- Apply the rules for order of operations.
- Name the number of terms in and the degree of a polynomial.
- Identify the numerical coefficient of a term.
- Use the rules of exponents for positive integral exponents.
- Simplify polynomial expressions.
- Add, subtract, multiply, and divide polynomials.
- Evaluate formulas.

## 2.1 ALGEBRAIC EXPRESSIONS

An **algebraic expression** is a combination of finite sums, differences, products, quotients, roots, and powers of numbers and of letters representing numbers. Some examples of algebraic expressions are

$$6x - 7y + 3 \qquad \frac{5a^2bc}{9a - 1} \qquad (6x + 1)^2 \qquad 3a\sqrt{6a + 4}$$

In an expression a **variable** quantity may be represented by a letter. This letter may be replaced by any number from a given replacement set, such as the set of real numbers.

A **constant** may also be represented by a letter. This letter may be replaced by only one number in a given situation.

A **term** is an expression or part of an expression involving only the product of numbers or letters. Terms may be connected by plus or minus signs, which indicate addition or subtraction of the terms.

Terms may have two or more **factors** connected by signs indicating multiplication. The term $7xyz$ has four factors, namely, 7, $x$, $y$, and $z$. The **coefficient** of a factor (or factors) is the product of the remaining factors. In the term $7xyz$, the coefficient of $xyz$ is 7; the coefficient of $7z$ is $xy$; the coefficient of $7y$ is $xz$. The **numerical coefficient** of $12xy^2$ is 12.

The algebraic expression $a^4$ indicates that the letter $a$ is to be used as a factor four times. We say that $a^4$ is the *fourth power* of $a$; it may also be written $a \cdot a \cdot a \cdot a$. The factor that is expressed as a power is called the **base.** The number that indicates the number of times the base is to be used as a factor is the **exponent.** For example, in the seventh power of $x$, written $x^7$, $x$ is the base and 7 is the exponent.

A *polynomial in one variable, x,* is a special type of algebraic expression defined as follows:

$$a_n x^n + a_{n-1} x^{n-1} + a_{n-2} x^{n-2} + \cdots + a_2 x^2 + a_1 x + a_0$$

where $n$ is a positive integer and the coefficients $a_n, a_{n-1}, a_{n-2}, \ldots, a_2, a_1, a_0$ are real numbers.

When an algebraic expression contains only one term, it is called a **monomial.** A **binomial** is an algebraic expression containing exactly two terms. A **trinomial** is an algebraic expression containing exactly three terms. More generally, an algebraic expression with two or more terms is called a **multinomial.**

### EXAMPLE 1

The following polynomials in one variable are examples of monomials, binomials, and trinomials.

| Monomials (one term) | Binomials (two terms) | Trinomials (three terms) |
|---|---|---|
| $3a$ | $3y^2 + 4y$ | $8x^2 + 4x + 9$ |
| $7z^2$ | $5a^3 - 6$ | $6b^3 - 4b^2 + 7b$ |
| $6m^3$ | $8w^2 + 5w^3$ | $5 + a - a^4$ |

The **degree of a monomial in one variable** is the same as the exponent of the variable.

### EXAMPLE 2

Find the degree of each monomial.
(a)  $4x^2$ has degree 2.  (b)  $-16x^5$ has degree 5.
(c)  $3y^4$ has degree 4.  (d)  8 has degree 0.
*Note:* A constant has degree 0.

The **degree of a polynomial** is the same as the highest-degree monomial in the polynomial.

**EXAMPLE 3**

Find the degree of each polynomial.
(a) $3x^2 - 4x + 7$ has degree 2.
(b) $2y^6 - 7y^4 + 4y^3 - 8y + 10$ has degree 6.
(c) $5y^8 - 9y^7 - y^5 + 4y^3 - y$ has degree 8.

A polynomial is in **decreasing order** if each term is of some degree less than the preceding term. For example, the following polynomial is in decreasing order:

$$5x^6 - 4x^4 + 5x^3 - 9x^2 - 6x + 1$$

A polynomial is in **increasing order** if each term is of some degree larger than the preceding term. For example, the following polynomial is in increasing order:

$$4 - 7x + 3x^2 - x^3 + 5x^6$$

The **degree of a monomial in more than one variable** equals the sum of the exponents of its variables.

**EXAMPLE 4**

Find the degree of each monomial.
(a) $3x^2y^2$ has degree 4.     (b) $-2ab^4$ has degree 5.
(c) $12x^2y^3z$ has degree 6.

**EXAMPLE 5**

Find the degree of each polynomial.
(a) $3x^2y^3 - 4xy^2 + xy$ has degree 5.
(b) $4abc - 6a^2bc^3 - 10bd$ has degree 6.

To add and subtract algebraic expressions, combine like terms. **Like terms** have identical letters and powers of letters.

**EXAMPLE 6**

Add $5x^2 + 7x - 4$ and $3x - x^2 - 5$.

$$(5x^2 + 7x - 4) + (3x - x^2 - 5) = (5x^2 - x^2) + (7x + 3x) + (-4 - 5)$$
$$= 4x^2 + 10x - 9$$

You may prefer to arrange the expressions so that the like terms appear in the same vertical column and then add.

$$\begin{array}{r} 5x^2 + 7x - 4 \\ - x^2 + 3x - 5 \\ \hline 4x^2 + 10x - 9 \end{array}$$

To subtract one multinomial from another, use the subtraction principle.

$$a - b = a + (-b)$$

That is, to subtract, add the opposite of each quantity being subtracted.

**EXAMPLE 7**

Subtract $6x^2 - 5x + 4$ from $-3x^2 + 4x + 7$.

$$(-3x^2 + 4x + 7) - (6x^2 - 5x + 4) = -3x^2 + 4x + 7 - 6x^2 + 5x - 4$$
$$= -9x^2 + 9x + 3$$

---

**REMOVING GROUPING SYMBOLS**

1. To remove grouping symbols preceded by a plus sign, remove the grouping symbols and leave the sign of each term unchanged within the grouping symbols.

2. To remove grouping symbols preceded by a minus sign, remove the grouping symbols and change the sign of each term within the grouping symbols.

3. When sets of grouping symbols are contained within each other, remove each set starting with the innermost set and finishing with the outermost set.

---

**EXAMPLE 8**

Perform the indicated operations and simplify.

$$(3x^2 - 5x + 4) - (6x^2 - 6x + 1) + (2x^2 - 4x)$$
$$= 3x^2 - 5x + 4 - 6x^2 + 6x - 1 + 2x^2 - 4x$$
$$= -x^2 - 3x + 3$$

**EXAMPLE 9**

Perform the indicated operations and simplify.

$$-[3x + (x - y) - (3y + 2x)] - [-(x - 5y) + (y - x)]$$
$$= -[3x + x - y - 3y - 2x] - [-x + 5y + y - x]$$
$$= -[2x - 4y] - [-2x + 6y]$$
$$= -2x + 4y + 2x - 6y$$
$$= -2y$$

How do powers affect the order of operations that we discussed in Section 1.2? Since raising a number to a power is a form of multiplication, we simplify all powers before multiplications and divisions. Therefore, when exponent operations are included, the order of operations is as follows.

---

**ORDER OF OPERATIONS**

1. Do all operations within any grouping symbols beginning with the innermost pair if grouping symbols are contained within each other.

2. Simplify all powers.

3. Do all multiplications and divisions in the order in which they occur from left to right.

4. Do all additions and subtractions in the order in which they occur from left to right.

---

To evaluate an algebraic expression, substitute the given numerical values in place of the letters, and follow the order of operations as given in the preceding box.

**EXAMPLE 10**

Evaluate

$$4a^3b + 5b + 2[6 - b]^2$$

when $a = -2$ and $b = -3$.
First, substitute as follows:

$$4(-2)^3(-3) + 5(-3) + 2[6 - (-3)]^2$$
$$= 4(-2)^3(-3) + 5(-3) + 2[9]^2 \qquad \text{(Do the subtraction within the brackets first.)}$$

$$= 4(-8)(-3) + 5(-3) + 2(81) \qquad \text{(Then simplify all powers.)}$$
$$= \quad 96 \quad - \quad 15 \quad + 162 \qquad \text{(Do all multiplications from left to right.)}$$

$$= 243 \qquad \text{(Do all additions and subtractions from left to right.)}$$

**EXAMPLE 11**

Evaluate

$$\frac{3x^2 + 8y}{4z^3}$$

when $x = 4$, $y = -5$, and $z = 2$.
First, substitute as follows:

$$\frac{3(4)^2 + 8(-5)}{4(2)^3}$$

$$= \frac{3(16) + 8(-5)}{4(8)} \qquad \text{(Simplify the powers.)}$$

$$= \frac{48 - 40}{32} \qquad \text{(Do all multiplications.)}$$

$$= \frac{8}{32} \qquad \text{(Subtract. \textit{Note:} Treat the fraction bar here as grouping symbols.)}$$

$$= \frac{1}{4} \qquad \text{(Divide.)}$$

## Exercises 2.1

*Classify each expression as a monomial, binomial, or trinomial.*

**1.** $3x + 4$

**2.** $5a^2$

**3.** $8x + 2x^2 - x^3$

**4.** $5x^2 + 6x$

**5.** $-3xy$

**6.** $7x^3 + 4x^2 + x$

**7.** $x^2 + x$

**8.** $1 - x^2$

**9.** $3ab^2 - 4a^2b - 5a^2b^2$

**10.** $5x^2 - 5y^2 + 5z^2$

**11.** $8ab - 3a^2$

**12.** $4 - x^3y^3$

*Find the degree of each monomial.*

**13.** $5a^2$

**14.** $-6x^3$

**15.** $-7x^4$

**16.** $10x^5$

**17.** $3y^{10}$

**18.** 4

**19.** $4x^2y^3$

**20.** $6xy$

**21.** $-2a^2b^3c$

**22.** $19p^2q^3r^5$

**23.** $10ab^2c$

**24.** $-16xy^3z^4$

*Write each polynomial in decreasing order and find its degree.*

**25.** $3x^2 + 2 + 5x$

**26.** $1 - 8x^2 + x^3 - x^5$

**27.** $5x^2 + 9x^8 - 5x^4 + 6x^3$

**28.** $8x^5 - 5x^4 - 2x^6 - x$

**29.** $3y^3 + 5 - 3y + 4y^5$

**30.** $7 - 3z + 4z^3 - 6z^2$

*Write each polynomial in increasing order and find its degree.*

**31.** $2x^3 - 3x^4 + 4x$

**32.** $a^3 - 1 + a^2 - a$

**33.** $5c^3 - 8c^5 + c - 7 + 3c^4$

**34.** $6x^3 - 4x^5 - x + 7x^4 + 2$

**35.** $5y^3 - 6y^6 + 2 - 8y^4 + 2y$

**36.** $7y^2 + 4 - 2y^5 - 6y + y^8$

*Find the degree of each polynomial.*

**37.** $8a^2b - 4ab^2 + 6ab$

**38.** $5xy^4 - 14x^2y + 7$

**39.** $6x^3y - 5x^2y^2 + 6x^5 - 2$

**40.** $6ab^4 - 4a^2b^2 + 6b^3 - 4$

**41.** $3x^2y^4z - 4xy^3z^4 + 9x^3y^2z^3$

**42.** $16x^3yz + 12x^4yz - 4x^5y^2 + 3$

**43.** $a^3b^4c + 4a^2b^2c^2 - 6a^9 - 3b^2$

**44.** $2x^5y^5 + 3x^2y^3z - 5y^6 + z^8$

*Perform the indicated operations and simplify.*

**45.** $(3x^2 - 4x + 8) + (6x^2 - x - 3)$

**46.** $(-4x^2 + 3x - 2) + (x^2 - x + 4)$

**47.** $(5x^2 + 7x - 9) + (-x^2 - 6x + 4)$

**48.** $(3x^2 + 2x - 8) + (9x^2 - 8x + 2)$

**49.** $(-3x^2 - 5x - 3) - (5x^2 - 2x - 7)$

**50.** $(-6x^2 + 2x - 4) - (-8x^2 - 7)$

**51.** $(-6x^2 + 7) - (4x^2 + x - 3)$

**52.** $(5x^2 + 2x) - (3x^2 - 5x + 2)$

**53.** $(3x^2 + 4x - 4) + (-x^2 - x + 2) - (-2x^2 + 2x + 8)$

**54.** $(-x^2 + 7x - 9) + (10x^2 - 11x + 4) - (-12x^2 - 15x + 3)$

**55.** $(-4x^2 + 6x - 2) - (5x^2 + 7x - 4) + (5x^2 - 6x + 1)$

**56.** $(5x^2 - 3x + 9) - (-2x^2 - 3x + 4) - (4x^2 - 2x - 1)$

**57.** $(3x^2 - 1 + 2x) - (9x^2 + 3 - 9x) - (3x - 4 - 2x^2)$

**58.** $(4 - 3x - x^2) - (3x^2 - 1 - 4x) + (4x^2 + 3 - x)$

**59.** $(5x^2 - 12x - 1) - (11x^2 + 4) + (4x + 7) - (3x - 2)$

**60.** $(3x^2 + 5) - (6x - 7) - (3x^2 + 6x - 13) + (5x - 9)$

**61.** $(3x^3 + 5x - 2) + (6x^2 - 10x + 1) - (4x^2 - 1) - (-5x^3 - 3)$

**62.** $(x^2 + 1) - (x^3 - 1) - (x^2 - x + 1) + (1 - x^3)$

**63.** $(3x^2 + 2x - 1) - (1 - 5x) - (3x^2 + x) + (3x - 4x^2) - (6x^2 + x^3)$

**64.** $(5x + 4) - (6x^2 + 4x - 7) - (6 - 5x + x^2) + (3x - 7x^2) + (5x^2 - 14) - (-3x)$

**65.** $-(x - 3y) - [(x + 2y) + (3y - 2x)]$

**66.** $[(5x + 3y) + (-2x - 2y)] - [-(x + 6y) - (-3x + 2y)]$

**67.** $-\{(5x + 3y) - (2x + 5y) - [3x + (4y + x)]\}$

**68.** $-[-(5x - 6y) - (-3x + 2y) + 6x - (4y - x)]$

*Evaluate each expression when $a = -2$, $b = 3$, $c = -1$, and $d = 1$.*

**69.** $a - b$
**70.** $a + 2c$
**71.** $3a - 2c$

**72.** $4d - 3bc$
**73.** $3a^2b^3c^2$
**74.** $4a^3b^4c^2d^3$

**75.** $(-b + 3cd)^3$
**76.** $(a + 6c)^2$
**77.** $b - a(c + d)$

**78.** $\dfrac{2a + 5b}{3b - 2c}$
**79.** $\dfrac{12a + 6b^2}{9c + 10a}$
**80.** $\dfrac{4a^3 - 5b}{3c^2 + d}$

**81.** $(4a^2bc)^3$
**82.** $(-a^2cd)^4$
**83.** $\dfrac{5}{ab} - \dfrac{6a}{5b} - \dfrac{cd}{b}$

**84.** $\left(\dfrac{4a^2 + 3b}{b - a}\right)^2$

## 2.2  EXPONENTS AND RADICALS

D.1, D.2

In Section 1.3 we studied the laws of exponents involving powers of 10. There are similar, but more general, laws of exponents for any base, given in the following box.

---

**LAWS OF EXPONENTS**

In each of the following, $a$ and $b$ are real numbers and $m$ and $n$ are positive integers.

**1.** $a^m \cdot a^n = a^{m+n}$

**2.** **(a)** $\dfrac{a^m}{a^n} = a^{m-n}$    $m > n$   and   $a \neq 0$

  **(b)** $\dfrac{a^m}{a^n} = \dfrac{1}{a^{n-m}}$   $m < n$   and   $a \neq 0$

**3.** $(a^m)^n = a^{mn}$

**4.** $(ab)^n = a^n b^n$

**5.** $\left(\dfrac{a}{b}\right)^n = \dfrac{a^n}{b^n}$   $b \neq 0$

---

The following examples illustrate the laws of exponents.

**EXAMPLE 1**

$$x^2 \cdot x^3 = x^{2+3} = x^5 \qquad \text{Law 1}$$

This result can also be shown as follows:

$$x^2 \cdot x^3 = (x \cdot x)(x \cdot x \cdot x) = x^5$$

**EXAMPLE 2**

$$\frac{x^7}{x^4} = x^{7-4} = x^3 \qquad \text{Law 2(a)}$$

This result can also be shown as follows:

$$\frac{x^7}{x^4} = \frac{\cancel{x} \cdot \cancel{x} \cdot \cancel{x} \cdot \cancel{x} \cdot x \cdot x \cdot x}{\cancel{x} \cdot \cancel{x} \cdot \cancel{x} \cdot \cancel{x}} = x^3$$

**EXAMPLE 3**

$$\frac{x^2}{x^5} = \frac{1}{x^{5-2}} = \frac{1}{x^3} \qquad \text{Law 2(b)}$$

This result can also be shown as follows:

$$\frac{x^2}{x^5} = \frac{\cancel{x} \cdot \cancel{x}}{\cancel{x} \cdot \cancel{x} \cdot x \cdot x \cdot x} = \frac{1}{x^3}$$

**EXAMPLE 4**

$$(x^3)^4 = x^{(3)(4)} = x^{12} \qquad \text{Law 3}$$

This result can also be shown as follows:

$$(x^3)^4 = x^3 \cdot x^3 \cdot x^3 \cdot x^3 = x^{12}$$

**EXAMPLE 5**

$$(3y^3)^4 = (3)^4(y^3)^4 \qquad \text{Law 4}$$
$$= 81y^{12} \qquad \text{Law 3}$$

This result can also be shown as follows:

$$(3y^3)^4 = (3y^3)(3y^3)(3y^3)(3y^3) = 81y^{12}$$

**EXAMPLE 6**

$$\left(\frac{x^4}{y^3}\right)^2 = \frac{x^8}{y^6} \qquad \text{Laws 3 and 5}$$

This result can also be shown as follows:

$$\left(\frac{x^4}{y^3}\right)^2 = \left(\frac{x^4}{y^3}\right)\left(\frac{x^4}{y^3}\right) = \frac{x^8}{y^6}$$

The next general law of exponents is given in the following box.

$$\boxed{a^0 = 1 \quad a \neq 0}$$

To show that this law is valid, we use the same reasoning as in Section 1.3 by using the substitution principle: If $a = b$ and $a = c$, then $b = c$.

$$\frac{a^n}{a^n} = 1 \qquad \text{(Any number other than zero divided by itself equals one.)}$$

and

$$\frac{a^n}{a^n} = a^{n-n} \quad \text{(Law 2, where } m = n\text{)}$$

$$= a^0$$

Therefore,

$$a^0 = 1 \qquad \text{(Substitution)}$$

The inverse process of raising a number to a power is called **finding the root of a number.** Square roots and cube roots are the roots most often used in technical problems. The $n$th root of a number $a$ is written $\sqrt[n]{a}$, where $n$ is the **index,** $a$ is the **radicand,** and the symbol $\sqrt{\phantom{a}}$ is called a **radical sign.**

The square root of a number $a$, written $\sqrt{a}$, is that number which, when multiplied by itself, is $a$. The index for square root is 2, but it is not usually written. For example, $\sqrt{36} = 6$ because $(6)(6) = 36$; however, $(-6)(-6) = 36$ is also true. Therefore, it seems that

$$\sqrt{36} = 6 \quad \text{and} \quad \sqrt{36} = -6$$

From the principle of substitution we know that

$$\text{if} \quad p = q \quad \text{and} \quad p = r$$
$$\text{then} \quad q = r$$

Applying this principle to $\sqrt{36} = 6$ and $\sqrt{36} = -6$, then we must accept the statement that $6 = -6$, which we know is false. Therefore, one of the assumptions must be false. The quantity $\sqrt{36}$ is a real number, and each real number has only one value. Mathematicians have agreed that the quantity $\sqrt{36}$ is a positive number and must have a positive value. As a result, we say that $\sqrt{36} = 6$ is true and $\sqrt{36} = -6$ is false.

What is the square root of a negative number, such as $-49$? That is, what real number squared equals $-49$? The product $(7)(7) = +49$ and the product $(-7)(-7) = +49$; hence, there is no real number whose square is $-49$. In fact, there is no real number whose square is negative. For this reason, the square root of a negative number is undefined within the set of real numbers.

As a result of the preceding discussions, we can now define the square root of a *nonnegative* number $a$, written $\sqrt{a}$, as that nonnegative number which, when multiplied by itself, is $a$.

**EXAMPLE 7**

Simplify each square root.
(a) $\sqrt{81} = 9$        (b) $\sqrt{100} = 10$
(c) $\sqrt{0} = 0$        (d) $\sqrt{25} = 5$
(e) $\sqrt{8^2} = 8$        (f) $\sqrt{10^8} = 10^4$
(g) $\sqrt{-4}$ is not a real number.        (h) $-\sqrt{16} = -(+4) = -4$

The cube root of a number $a$, written $\sqrt[3]{a}$, is that number which, when multiplied by itself three times, is $a$. For example, $\sqrt[3]{8} = 2$ because $2 \cdot 2 \cdot 2 = 8$.

**EXAMPLE 8**

Simplify each cube root.
(a)   $\sqrt[3]{27} = 3$      because          $(3)(3)(3) = 27$
(b)   $\sqrt[3]{64} = 4$      because          $(4)(4)(4) = 64$
(c)   $\sqrt[3]{8} = 2$      because          $(2)(2)(2) = 8$
(d)   $\sqrt[3]{-125} = -5$   because   $(-5)(-5)(-5) = -125$
(e)   $\sqrt[3]{-27} = -3$   because   $(-3)(-3)(-3) = -27$   (*Note:* A negative quantity under the radical sign does not present a problem for cube roots.)

(f)  $-\sqrt[3]{-8} = -(-2) = 2$

Roots of numbers, or radicals, are discussed in more detail in Chapter 8. One property of radicals that needs some discussion now is as follows.

$$\sqrt{ab} = \sqrt{a}\,\sqrt{b} \quad \text{where } a > 0 \quad \text{and} \quad b > 0$$

That is, *the square root of a product of positive numbers equals the product of their square roots.* This property is used to simplify radicals when either $a$ or $b$ is a perfect square. The **perfect square** of a number is the square of a rational number. The first ten positive integral perfect squares are 1, 4, 9, 16, 25, 36, 49, 64, 81, and 100.
Examples 9–11 illustrate how to simplify square roots.

**EXAMPLE 9**

Simplify $\sqrt{18}$.

$$\sqrt{18} = \sqrt{9 \cdot 2} \qquad \text{( Find the largest perfect square factor of 18.)}$$
$$= \sqrt{9}\,\sqrt{2}$$
$$= 3\sqrt{2}$$

**EXAMPLE 10**

Simplify $\sqrt{48}$.

$$\sqrt{48} = \sqrt{16 \cdot 3} \qquad \text{( Find the largest perfect square factor of 48.)}$$
$$= \sqrt{16}\,\sqrt{3}$$
$$= 4\sqrt{3}$$

**EXAMPLE 11**

Simplify $\sqrt{360}$.

$$\sqrt{360} = \sqrt{36 \cdot 10} \qquad \text{( Find the largest perfect square factor of 360.)}$$
$$= \sqrt{36}\,\sqrt{10}$$
$$= 6\sqrt{10}$$

The square root of a number is simplified when the number under the radical contains no perfect square factors.
To find the decimal value of the square root of a number, use the square root key as shown in the following example.

**EXAMPLE 12**

Find $\sqrt{268}$ rounded to three significant digits.

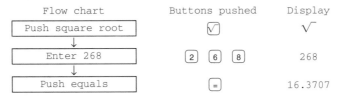

| Flow chart | Buttons pushed | Display |
|---|---|---|
| Push square root | ☑ | √ |
| Enter 268 | 2  6  8 | 268 |
| Push equals | = | 16.3707 |

The result is 16.4 rounded to three significant digits.

For order of operation purposes, treat a radical as a grouping symbol. That is, do all operations under the radical before finding the root, as in the following example.

**EXAMPLE 13**

Simplify $\sqrt{6^2 + 18}$.

$$\sqrt{6^2 + 18} = \sqrt{36 + 18} \qquad \text{(Find the power.)}$$
$$= \sqrt{54} \qquad \text{(Add.)}$$
$$= \sqrt{9 \cdot 6} \qquad \text{(Find the largest perfect square factor of 54.)}$$
$$= \sqrt{9}\,\sqrt{6}$$
$$= 3\sqrt{6}$$

# Exercises 2.2

*Using the laws of exponents, perform the indicated operations and simplify.*

**1.** $x^5 \cdot x^7$ 

**2.** $y^3 y^6$ 

**3.** $(3a^2)(4a^3)$

**4.** $5y^3 \cdot 6y^4$ 

**5.** $\dfrac{m^9}{m^3}$ 

**6.** $\dfrac{c^{12}}{c^3}$

**7.** $\dfrac{x^2}{x^6}$ 

**8.** $\dfrac{y^2}{y^8}$ 

**9.** $\dfrac{12x^8}{4x^4}$

**10.** $\dfrac{36x^{10}}{2x^2}$ 

**11.** $\dfrac{15x^2}{3x^5}$ 

**12.** $\dfrac{5y^6}{35y^8}$

**13.** $(a^2)^3$ 

**14.** $(x^5)^2$ 

**15.** $(c^4)^4$

**16.** $(b^7)^6$ 

**17.** $(9a)^2$ 

**18.** $(4m^2)^3$

**19.** $(2x^2)^5$ 

**20.** $(3c^6)^4$ 

**21.** $\left(\dfrac{3}{4}\right)^2$

**22.** $\left(\dfrac{a}{2}\right)^3$ 

**23.** $\left(\dfrac{2}{a^3}\right)^4$ 

**24.** $\left(\dfrac{x^2}{y}\right)^5$

**25.** $4^0$ 

**26.** $(3x)^0$ 

**27.** $3x^0$

**28.** $7(x^2)^0$ 

**29.** $(-3x)^2$ 

**30.** $(-2x^2)^5$

**31.** $(-t^3)^4$ 

**32.** $(-s^2)^5$ 

**33.** $(-a^2)^3$

**34.** $(-c^3)^6$ 

**35.** $(-2a^2b)^3$ 

**36.** $(-a^2b^3)^4$

**37.** $(3x^2y^3)^2$ 

**38.** $(-5a^2b^4)^2$ 

**39.** $(-3x^3y^4z)^3$

**40.** $(2a^2b^3c^4)^3$  |  **41.** $\left(\dfrac{2x^2}{3y^3}\right)^2$  |  **42.** $\left(\dfrac{-3x^4}{4y^2}\right)^3$

**43.** $\left(\dfrac{-4x}{3y^2}\right)^2$  |  **44.** $\left(\dfrac{5xy^2}{7z^3}\right)^2$  |  **45.** $\left(\dfrac{-1}{6y^3}\right)^2$

**46.** $\left(\dfrac{-2}{3x^3}\right)^3$

*Simplify each root. In Exercises 47–60, the result is an integer.*

**47.** $\sqrt{4}$  |  **48.** $\sqrt{16}$  |  **49.** $\sqrt{64}$

**50.** $\sqrt{49}$  |  **51.** $\sqrt{121}$  |  **52.** $\sqrt{144}$

**53.** $\sqrt{5^{16}}$  |  **54.** $\sqrt{10^6}$  |  **55.** $\sqrt[3]{125}$

**56.** $\sqrt[3]{343}$  |  **57.** $\sqrt[3]{-216}$  |  **58.** $\sqrt[3]{-64}$

**59.** $\sqrt[3]{512}$  |  **60.** $\sqrt[3]{1000}$  |  **61.** $\sqrt{45}$

**62.** $\sqrt{12}$  |  **63.** $\sqrt{50}$  |  **64.** $\sqrt{80}$

**65.** $\sqrt{72}$  |  **66.** $\sqrt{75}$  |  **67.** $\sqrt{4^2 + 32}$

**68.** $\sqrt{12^2 - 44}$  |  **69.** $\sqrt{3 \cdot 4^2 - 4 \cdot 2^2}$  |  **70.** $\sqrt{3^4 + 5 \cdot 2^3}$

*Evaluate and round to three significant digits.*

**71.** $\sqrt{329}$  |  **72.** $\sqrt{492}$  |  **73.** $\sqrt{2596}$

**74.** $\sqrt{87,500}$  |  **75.** $\sqrt{0.00472}$  |  **76.** $\sqrt{0.924}$

**77.** $\sqrt{16 + 36}$  |  **78.** $\sqrt{81 - 49}$  |  **79.** $\sqrt{5^2 + 8^2}$

**80.** $\sqrt{9^2 - 9 \cdot 2^2}$

**81.** $\sqrt{(2.73 \times 10^4)^2 + (1.00 \times 10^5)^2}$

**82.** $\sqrt{(3.45 \times 10^{-3})^2 + (6.85 \times 10^{-4})^2}$

**83.** $\sqrt{(115)^2 + (15.5 - 84.6)^2}$

**84.** $\dfrac{1}{2\pi \sqrt{(1.23 \times 10^{-5})(4.45 \times 10^{-12})}}$

## 2.3 MULTIPLICATION OF ALGEBRAIC EXPRESSIONS

To multiply monomials, multiply their numerical coefficients and multiply each set of like letters using the laws of exponents.

**EXAMPLE 1**

Multiply: $(-3a^2b^3)(5ab^2)$

$$(-3a^2b^3)(5ab^2) = (-3 \cdot 5)(a^2 \cdot a)(b^3 \cdot b^2) = -15a^3b^5$$

**EXAMPLE 2**

Multiply: $(2x^2yz^2)(4x^3z)(6xy^2z)$

$$(2x^2yz^2)(4x^3z)(6xy^2z) = (2 \cdot 4 \cdot 6)(x^2 \cdot x^3 \cdot x)(y \cdot y^2)(z^2 \cdot z \cdot z) = 48x^6y^3z^4$$

To multiply a multinomial by a monomial, multiply each term of the multinomial by the monomial using the distributive property:

$$a(b + c) = ab + ac$$

**EXAMPLE 3**

Multiply:   $5a(6b + 3c)$

$$5a(6b + 3c) = (5a)(6b) + (5a)(3c) = 30ab + 15ac$$

**EXAMPLE 4**

Multiply:   $4x(3x^2 - 2x + 5)$

$$4x(3x^2 - 2x + 5) = (4x)(3x^2) + (4x)(-2x) + (4x)(5)$$
$$= 12x^3 - 8x^2 + 20x$$

**EXAMPLE 5**

Multiply:   $6a^2b(-2a^3b^2 + 3a^2b - b + 1)$

$$6a^2b(-2a^3b^2 + 3a^2b - b + 1)$$
$$= (6a^2b)(-2a^3b^2) + (6a^2b)(3a^2b) + (6a^2b)(-b) + (6a^2b)(1)$$
$$= -12a^5b^3 + 18a^4b^2 - 6a^2b^2 + 6a^2b$$

In general, to multiply two multinomials, multiply each term of the first multinomial by each term of the second and simplify. *Note:* The method is most similar to the method of multiplying whole numbers.

**EXAMPLE 6**

Multiply:   $(5x + 4)\,(3x - 6)$

$$
\begin{array}{r}
5x + \phantom{0}4 \\
3x - \phantom{0}6 \\
\hline
-30x - 24 \\
15x^2 + 12x \phantom{- 24} \\
\hline
15x^2 - 18x - 24
\end{array}
$$

**EXAMPLE 7**

Multiply:   $(3x^2 + 4x - 7)\,(2x^3 - x^2 - 2)$

$$
\begin{array}{r}
3x^2 + 4x \phantom{0} - \phantom{0}7 \\
2x^3 - \phantom{0} x^2 - \phantom{0}2 \\
\hline
-6x^2 - 8x \phantom{0} + 14 \\
-3x^4 - \phantom{0}4x^3 + 7x^2 \phantom{00000} \\
6x^5 + 8x^4 - 14x^3 \phantom{000000000} \\
\hline
6x^5 + 5x^4 - 18x^3 + \phantom{0} x^2 - 8x \phantom{0} + 14
\end{array}
$$

Next, we discuss a special method to mentally find the product of two binomials, a special product that occurs again and again in our work. This method is called the **FOIL**

**method.** The initials FOIL (first, outer, inner, last) are used to help keep track of the order of multiplying the terms of the two binomials as illustrated in the following examples.

**EXAMPLE 8**

Multiply:   $(2x + 3)(4x - 5)$

Outer product

$(2x + 3)(4x - 5)$

Inner product

F:   product of *First* terms of the binomials:   $(2x)(4x) = 8x^2$
O:   *Outer* product:   $(2x)(-5) = -10x$ ⎫
I:   *Inner* product:   $(3)(4x) \ = \ 12x$ ⎬   sum = $\quad 2x$
L:   product of *Last* terms of the binomials:   $(3)(-5) = \underline{\qquad -15}$
$(2x + 3)(4x - 5) = 8x^2 + 2x - 15$

**EXAMPLE 9**

Multiply:   $(4x - 2)(5x - 6)$

Outer product

$(4x - 2)(5x - 6)$

Inner product

F:   $(4x)(5x) = \qquad\qquad 20x^2$
O:   $(4x)(-6) = -24x$ ⎫
I:   $(-2)(5x) = -10x$ ⎬ sum = $\quad -34x$
L:   $(-2)(-6) = \underline{\qquad\qquad 12}$
$(4x - 2)(5x - 6) = 20x^2 - 34x + 12$

You should do these steps mentally and write only the final result.

**EXAMPLE 10**

Multiply:   $(2x + 3) (5x + 1)$

$$(2x + 3)(5x + 1) = 10x^2 + 17x + 3$$

**EXAMPLE 11**

Multiply:   $(6x + 4) (2x - 3)$
$$(6x + 4)(2x - 3) = 12x^2 - 10x - 12$$

**EXAMPLE 12**

Multiply:   $(5x^2 - 6)(4x^2 - 1)$
$$(5x^2 - 6)(4x^2 - 1) = 20x^4 - 29x^2 + 6$$

**EXAMPLE 13**

Multiply:   $(3x + 2)(3x - 2)$
$$(3x + 2)(3x - 2) = 9x^2 - 4$$

To find the power of a binomial, first rewrite the binomial as a product, and then use the most appropriate multiplication method.

**EXAMPLE 14**

Multiply: $(x - 3)^2$

$$(x - 3)^2 = (x - 3)(x - 3) = x^2 - 6x + 9$$

**EXAMPLE 15**

Multiply: $(3x - 2)^3$

$$(3x - 2)^3 = (3x - 2)(3x - 2)(3x - 2)$$
$$= (9x^2 - 12x + 4)(3x - 2)$$

Use FOIL to find the product of the first two binomials. Then find the final product as follows:

$$
\begin{array}{r}
9x^2 - 12x + 4 \\
3x - 2 \\
\hline
-18x^2 + 24x - 8 \\
27x^3 - 36x^2 + 12x \\
\hline
27x^3 - 54x^2 + 36x - 8
\end{array}
$$

# Exercises 2.3

*Find each product.*

**1.** $(4x^2)(8x^3)$

**2.** $(-6x^3)(12x^7)$

**3.** $(-4a^2b)(6a^3b^2)$

**4.** $(-4x^2y)(-7x^4y^2)$

**5.** $(12a^2bc^3)(-4ac^2)$

**6.** $(-2x^2y^2z)(7x^2yz^4)$

**7.** $(-3a^2b^4c)(2ab^2c^5)(-4ab^3)$

**8.** $(3x^2yz^5)(5x^3y^2)(-2y^3z^2)$

**9.** $3a(4a - 7b)$

**10.** $-5c(a + 2b - 3c)$

**11.** $3x(2x^2 + 4x - 5)$

**12.** $5x(-3x^2 - x + 7)$

**13.** $-5x^2(3x^2 - 5x + 8)$

**14.** $-3x^3(4x^2 + 6x - 2)$

**15.** $6ab^3(4a^2b - 8a^3b^4)$

**16.** $4a^2b(-2a^2b + 6ab^3)$

**17.** $-3a^2b^4(-a^4b^3 + 3ab - b^2)$

**18.** $-8a^3b^2(-2ab^3 - 3ab + a^2)$

**19.** $(3x - 7)(2x + 5)$

**20.** $(6x - 1)(5x - 3)$

**21.** $(6x + 3)(8x + 5)$

**22.** $(5x + 7)(3x - 2)$

**23.** $(3x + 4)(3x - 4)$

**24.** $(5x + 9)(5x - 9)$

**25.** $(4x + 1)(6x - 1)$

**26.** $(5x + 2)(3x + 4)$

**27.** $(3x - 7)(2x - 3)$

**28.** $(6x - 3)(x + 6)$

**29.** $(3x + 8y)(6x + 4y)$

**30.** $(2x + 7y)(9x - y)$

**31.** $(5s - 9t)(8s + 2t)$

**32.** $(4a - 5b)(7a - 10b)$

**33.** $(-3x + 4)(5x + 6)$

**34.** $(-11x + 2)(-10x - 3)$

**35.** $(3x^2 - 1)(2x^2 + 7)$

**36.** $(5x^2 + 4)(6x^2 + 9)$

**37.** $(5x^2 - 6)(6x^2 - 5)$

**38.** $(3x^2 + 4)(4x^2 - 5)$

**39.** $(x^2 + x + 2)(x^2 - x + 3)$

**40.** $(3x^2 + 8x - 2)(5x^2 + x - 7)$

**41.** $(x + y - 7)(x - y + 4)$

**42.** $(x - 3y + 7)(2x - y - 3)$

**43.** $(3x^2 + 2x - 6)(5x^2 - 4x - 1)$

**44.** $(4x^2 - 6x + 4)(-2x^2 - x + 5)$

**45.** $(3x^2 + 5x + 2)(4x^2 - 3)$

**46.** $(3x^2 - 5x - 4)(2x^2 - 6x)$

**47.** $(2x - 5)^2$

**48.** $(4x + 3)^2$

**49.** $(3x + 8)^2$

**50.** $(6x - 4)^2$

**51.** $(-5x + 2)^2$

**52.** $(-3x^2 - 4)^2$

**53.** $(3x - 4)^2(x^2 - 2x + 1)$

**54.** $(2x + 5)^2(x^2 + 3x - 1)$

**55.** $(2x - 1)^3$

**56.** $(4x + 3)^3$

**57.** $(2a + 5b)^3$

**58.** $(2 - 4x)^3$

## 2.4  DIVISION OF ALGEBRAIC EXPRESSIONS

To divide monomials, divide their numerical coefficients and divide each set of like letters using the laws of exponents.

**EXAMPLE 1**

Divide:  $\dfrac{24x^2y^5}{3xy^3}$

$$\frac{24x^2y^5}{3xy^3} = \frac{24}{3} \cdot \frac{x^2}{x} \cdot \frac{y^5}{y^3}$$

$$= 8xy^2 \qquad \text{(Subtract exponents.)}$$

**EXAMPLE 2**

Divide:  $\dfrac{18a^3b^6}{-8a^5b^2}$

$$\frac{18a^3b^6}{-8a^5b^2} = \frac{18}{-8} \cdot \frac{a^3}{a^5} \cdot \frac{b^6}{b^2}$$

$$= -\frac{9}{4} \cdot \frac{1}{a^2} \cdot b^4 \qquad \text{(Subtract exponents.)}$$

$$= -\frac{9b^4}{4a^2}$$

An alternate method for this division of monomials involves cancellation as follows:

$$\frac{18a^3b^6}{-8a^5b^2} = \frac{\overset{9}{\cancel{18}}\,\overset{1}{\cancel{a^3}}\,\overset{b^4}{\cancel{b^6}}}{\underset{-4}{\cancel{-8}}\,\underset{a^2}{\cancel{a^5}}\,\underset{1}{\cancel{b^2}}} = -\frac{9b^4}{4a^2}$$

To divide a multinomial by a monomial, divide each term in the multinomial by the monomial and simplify.

**EXAMPLE 3**

Divide: $\dfrac{15x^2 - 6x}{3x}$

$$\frac{15x^2 - 6x}{3x} = \frac{15x^2}{3x} - \frac{6x}{3x} = 5x - 2$$

**EXAMPLE 4**

Divide: $\dfrac{6a^3 - 10a^2 + 4a}{2a^2}$

$$\frac{6a^3 - 10a^2 + 4a}{2a^2} = \frac{6a^3}{2a^2} - \frac{10a^2}{2a^2} + \frac{4a}{2a^2}$$

$$= 3a - 5 + \frac{2}{a}$$

---

**DIVIDING ONE MULTINOMIAL (DIVIDEND) BY A SECOND MULTINOMIAL (DIVISOR)**

1. Arrange each multinomial in descending powers of one of the variables.

2. To find the first term of the quotient, divide the first term of the dividend by the first term of the divisor.

3. Multiply the divisor by the first term of the quotient, and subtract the product from the dividend.

4. Continue this procedure of dividing the first term of each remainder by the first term of the divisor until the final remainder is zero or until the remainder is of degree less than the divisor.

---

**EXAMPLE 5**

Divide $3x^3 + 2x^2 - 7x + 2$ by $x + 2$.

$$
\begin{array}{r}
3x^2 - 4x + 1 \quad \text{(Quotient)} \\
x + 2 \overline{\smash{)}3x^3 + 2x^2 - 7x + 2} \quad \text{(Dividend)} \\
\underline{3x^3 + 6x^2} \\
-4x^2 - 7x \\
\underline{-4x^2 - 8x} \\
x + 2 \\
\underline{x + 2} \\
0 \quad \text{(Remainder)}
\end{array}
$$

(Divisor) $x + 2$

That is,

$$\frac{3x^3 + 2x^2 - 7x + 2}{x + 2} = 3x^2 - 4x + 1$$

**EXAMPLE 6**

Divide: $\dfrac{2x^3 + 11x^2 + 10}{2x + 3}$

$$\begin{array}{r} x^2 + 4x - 6 \phantom{+10} \\ 2x + 3{\overline{\smash{\big)}\,2x^3 + 11x^2 \phantom{+00} + 10}} \\ \underline{2x^3 + \phantom{1}3x^2} \phantom{+ 10} \\ 8x^2 \phantom{+12x+10} \\ \underline{8x^2 + 12x} \phantom{+10} \\ -12x + 10 \\ \underline{-12x - 18} \\ 28 \end{array}$$

[Since 28 is the remainder (which is of degree less than the divisor, $2x + 3$), it is usually written in the form of a quotient, $\dfrac{28}{2x + 3}$.]

That is,

$$\frac{2x^3 + 11x^2 + 10}{2x + 3} = x^2 + 4x - 6 + \frac{28}{2x + 3}$$

**EXAMPLE 7**

Divide: $\dfrac{10x^4 - 11x^3 - 15x^2 + 28x - 12}{2x^2 + x - 3}$

$$\begin{array}{r} 5x^2 - \phantom{1}8x + \phantom{1}4 \phantom{00} \\ 2x^2 + x - 3{\overline{\smash{\big)}\,10x^4 - 11x^3 - 15x^2 + 28x - 12}} \\ \underline{10x^4 + \phantom{1}5x^3 - 15x^2} \phantom{+28x - 12} \\ -16x^3 \phantom{- 15x^2} + 28x \phantom{- 12} \\ \underline{-16x^3 - \phantom{1}8x^2 + 24x} \phantom{-12} \\ 8x^2 + \phantom{1}4x - 12 \\ \underline{8x^2 + \phantom{1}4x - 12} \end{array}$$

Thus,

$$\frac{10x^4 - 11x^3 - 15x^2 + 28x - 12}{2x^2 + x - 3} = 5x^2 - 8x + 4$$

## Exercises 2.4

*Find each quotient.*

**1.** $\dfrac{24x^2}{4x}$

**2.** $\dfrac{-18x^2y^3}{3xy}$

**3.** $\dfrac{36a^3b^5}{-9a^2b^2}$

**4.** $\dfrac{-48a^5b^3}{-12a^2b}$

**5.** $\dfrac{45x^6y}{72x^3y^2}$

**6.** $\dfrac{32a^5b^2}{28a^8b^4}$

**7.** $\dfrac{-25x^3}{20x^5y^2}$

**8.** $\dfrac{40a^2b^4}{-15b^5}$

**9.** $\dfrac{3a(2b^2)^2}{(12ab)^2}$

**10.** $\dfrac{(3x^2y)(4x^2)^2}{xy(2x^3)^3}$

11. $\dfrac{(4st^2)^2(3t^2)^3}{t^2(9st)^2}$

12. $\dfrac{(3a^2)^2(ab^2)^2(2b)}{4a(3ab)^2(b^2)^3}$

13. $\dfrac{24x^2 - 16x + 8}{8}$

14. $\dfrac{36x^2 + 18x + 12}{6}$

15. $\dfrac{15x^5 - 20x^4 + 10x^2}{5x}$

16. $\dfrac{27x^3 - 33x^2 - 21x}{3x}$

17. $\dfrac{-28x^5 + 35x^4 - 49x^3}{7x^3}$

18. $\dfrac{20x^4 - 25x^3 + 10x^2 - 15x}{5x^2}$

19. $\dfrac{-64x^7 + 48x^5 + 36x^3 + 24x}{8x^4}$

20. $\dfrac{27x^8 - 18x^5 + 36x^4 - 12x}{9x^3}$

21. $\dfrac{4a^2b - 6a^2b^2 + 8ab}{2ab}$

22. $\dfrac{8a^2b^3 + 12ab^4 - 16a^3b^2}{8a^2b}$

23. $\dfrac{9m^2n^2 + 12mn^3 - 15m^3}{-3mn}$

24. $\dfrac{25p^2q^3 - 40p^4 + 20pq^2}{-10p^2q^2}$

25. $\dfrac{224x^4y^2z^5 - 168x^3y^3z^4 - 112xy^4z^2}{28xy^2z^2}$

26. $\dfrac{175m^2n^3p + 125mnp^4 - 225n^2p^3}{-75mn^2p^3}$

27. $\dfrac{2x^2 + x - 15}{x + 3}$

28. $\dfrac{x^2 - 2x - 3}{x - 3}$

29. $\dfrac{x^3 - 3x^2 + 5x - 6}{x - 2}$

30. $\dfrac{2x^3 + 5x^2 + 14}{x + 3}$

31. $\dfrac{2x^3 - 5x^2 + 8x + 1}{2x - 1}$

32. $\dfrac{x^3 - x^2 - 2x + 8}{x + 2}$

33. $\dfrac{6x^3 + 31x^2 + 26x - 15}{3x + 5}$

34. $\dfrac{20x^3 - 54x^2 + 44x - 12}{4x - 6}$

35. $\dfrac{-16x^3 + 37x - 24}{4x - 3}$

36. $\dfrac{25x^3 - 56x + 52}{5x - 4}$

37. $\dfrac{9x^4 - 3x^3 - 2x^2 + 12x + 4}{3x + 1}$

38. $\dfrac{8x^4 - 12x^3 + 10x^2 - 19x}{2x - 3}$

39. $\dfrac{2x^3 + x^4 - 10 + 11x - 6x^2}{x^2 - x + 2}$

40. $\dfrac{x^6 + x^4 - x^3 + x^2 + 1}{x^2 + x + 1}$

41. $\dfrac{x^3 - 64}{x + 4}$

42. $\dfrac{x^4 + 4}{x^2 - 2x + 2}$

43. $\dfrac{8x^3 + 1}{2x + 1}$

44. $\dfrac{16a^4 + 1}{2a - 1}$

## 2.5  LINEAR EQUATIONS

An **equation** is a statement that two algebraic expressions are equal. In particular, a **linear equation** in one variable is an equation in which each term containing that variable is of the first degree. To **solve** an equation means to find what number or numbers can replace the variable to make the equation a true statement.

**EXAMPLE 1**

Solve $2x + 4 = 6x - 8$.

If we let $x = 2$, then

$$2(2) + 4 = 6(2) - 8$$

which is a *false* statement. Hence, 2 *is not* a solution. But if we let $x = 3$, then

$$2(3) + 4 = 6(3) - 8$$

which is a *true* statement. Hence, 3 *is* a solution (sometimes called a **root**).

---

### FOUR BASIC PROPERTIES USED TO SOLVE EQUATIONS

1. If the same quantity is added to each side of an equation, the resulting equation is equivalent to the original equation.[*]
2. If the same quantity is subtracted from each side of an equation, the resulting equation is equivalent to the original equation.
3. If each side of an equation is multiplied by the same (nonzero) quantity, the resulting equation is equivalent to the original equation.
4. If each side of an equation is divided by the same (nonzero) quantity, the resulting equation is equivalent to the original equation.

---

**EXAMPLE 2**

Solve $x - 6 = 11$.

$$x - 6 = 11$$
$$x - 6 + 6 = 11 + 6 \qquad \text{Property 1}$$
$$x = 17$$

**EXAMPLE 3**

Solve $x + 4 = 1$.

$$x + 4 = 1$$
$$x + 4 - 4 = 1 - 4 \qquad \text{Property 2}$$
$$x = -3$$

**EXAMPLE 4**

Solve $\dfrac{x}{4} = 9$.

$$\frac{x}{4} = 9$$
$$\left(\frac{x}{4}\right)4 = (9)4 \qquad \text{Property 3}$$
$$x = 36$$

---

[*] Two equations are equivalent when they have the same solutions.

**EXAMPLE 5**

Solve $3x = 15$.

$$3x = 15$$
$$\frac{3x}{3} = \frac{15}{3} \qquad \text{Property 4}$$
$$x = 5$$

---

### SOLVING A FIRST-DEGREE EQUATION IN ONE UNKNOWN OR VARIABLE

1. Eliminate any fractions by multiplying each side by the lowest common denominator of all fractions in the equation.
2. Remove any grouping symbols.
3. Combine like terms on each side of the equation.
4. Isolate all the unknown terms on one side of the equation and all other terms on the other side of the equation.
5. Again combine like terms where possible.
6. Divide each side by the coefficient of the unknown.
7. Check your solution by substituting it in the original equation.

---

**EXAMPLE 6**

Solve $3x + 4 = 28$.

$$3x + 4 = 28$$
$$3x = 24 \qquad \text{(Subtract 4 from each side.)}$$
$$x = 8 \qquad \text{(Divide each side by 3.)}$$

*Check*:

$$3(8) + 4 = 28 \qquad \text{(Substitute } x = 8.\text{)}$$
$$28 = 28$$

**EXAMPLE 7**

Solve $\frac{1}{3}x = \frac{1}{4}x - 2$.

$$\frac{1}{3}x = \frac{1}{4}x - 2$$
$$4x = 3x - 24 \qquad \text{[Multiply each side by the L.C.D. (lowest common denominator), 12.]}$$
$$x = -24 \qquad \text{(Subtract } 3x \text{ from each side.)}$$

*Check:*

$$\frac{1}{3}(-24) = \frac{1}{4}(-24) - 2 \qquad \text{(Substitute } x = -24.\text{)}$$
$$-8 = -8$$

**EXAMPLE 8**

Solve $4(x + 1) = 6 - (3x - 12)$.

$$4(x + 1) = 6 - (3x - 12)$$
$$4x + 4 = 6 - 3x + 12 \qquad \text{(Remove parentheses.)}$$
$$4x + 4 = -3x + 18 \qquad \text{(Combine like terms.)}$$
$$7x + 4 = 18 \qquad \text{(Add } 3x \text{ to each side.)}$$
$$7x = 14 \qquad \text{(Subtract 4 from each side.)}$$
$$x = 2 \qquad \text{(Divide each side by 7.)}$$

*Check:*

$$4(2 + 1) = 6 - (3 \cdot 2 - 12) \qquad \text{(Substitute } x = 2.\text{)}$$
$$12 = 12$$

**EXAMPLE 9**

Solve the following:

$$\frac{2(6y + 1)}{5} - 3(y + 1) = \frac{3 - y}{4} - 3$$
$$8(6y + 1) - 60(y + 1) = 5(3 - y) - 60 \qquad \text{(Multiply each side by the L.C.D., 20.)}$$
$$48y + 8 - 60y - 60 = 15 - 5y - 60 \qquad \text{(Remove parentheses.)}$$
$$-12y - 52 = -5y - 45 \qquad \text{(Combine like terms.)}$$
$$-7y - 52 = -45 \qquad \text{(Add } 5y \text{ to each side.)}$$
$$-7y = 7 \qquad \text{(Add 52 to each side.)}$$
$$y = -1 \qquad \text{(Divide each side by } -7.\text{)}$$

*Check:*

$$\frac{2[6(-1) + 1]}{5} - 3(-1 + 1) = \frac{3 - (-1)}{4} - 3 \qquad \text{(Substitute } y = -1.\text{)}$$
$$-2 - 0 = 1 - 3$$
$$-2 = -2$$

**EXAMPLE 10**

Solve $8.24(3.6x + 18.6) = 246.7$ using a calculator.

$$8.24(3.6x + 18.6) = 246.7$$
$$(8.24)(3.6x) + (8.24)(18.6) = 246.7 \qquad \text{(Remove parentheses.)}$$
$$(8.24)(3.6x) = 246.7 - (8.24)(18.6) \qquad \text{[Subtract } (8.24)(18.6) \text{ from each side.]}$$
$$x = \frac{246.7 - (8.24)(18.6)}{(8.24)(3.6)} \qquad \text{[Divide each side by } (8.24)(3.6).\text{]}$$
$$x = 3.15 \qquad \text{(Rounded to three significant digits)}$$

*Note:* When you are checking a root that has been rounded, there will often be a slight difference on each side of the equation.

## Exercises 2.5

*Solve each equation.*

**1.** $x + 9 = 7$

**2.** $x - 4 = 12$

**3.** $x - 6 = 10$

**4.** $x + 3 = -4$

**5.** $5x = 35$

**6.** $\dfrac{x}{4} = 5$

**7.** $\dfrac{x}{8} = -9$

**8.** $6x = 72$

**9.** $4x - 6 = 14$

**10.** $7x + 8 = 36$

**11.** $5y + 6 = 16$

**12.** $2x - 4 = 10$

**13.** $8 - 3x = 14$

**14.** $7 - 2y = 3$

**15.** $3x - 2 = 5x + 8$

**16.** $4x + 9 = 7x - 15$

**17.** $8 - 6x = 5x + 25$

**18.** $1 - 7x = 16 - 2x$

**19.** $3(x - 5) = 27$

**20.** $-4(3x + 2) = 14$

**21.** $4(y + 2) = 30 - (y - 3)$

**22.** $2(x - 3) = 4 + (x - 14)$

**23.** $6(3x + 1) + (2x + 2) = 5x$

**24.** $4(2y - 3) - (3y + 7) = 6$

**25.** $4(2x - 1) - 2x = 3 - 5(2x - 5)$

**26.** $6x - 2(4x - 2) - (x - 4) = x + 4$

**27.** $-[x - (4x + 5)] = 2 + 3(2x + 3)$

**28.** $10y - (4 - y) = 2[-3 - (2 + y) - 12]$

**29.** $\dfrac{2}{3}y - 4 = \dfrac{3}{4}y - 5$

**30.** $\dfrac{3}{5}x + 2 = \dfrac{1}{3}x + 10$

**31.** $2 - \dfrac{3}{7}x = \dfrac{x}{4} - \dfrac{15}{2}$

**32.** $\dfrac{1}{2}y + \dfrac{1}{6} = \dfrac{2}{3}y + \dfrac{1}{9}$

**33.** $3\left(\dfrac{2}{5}x - 7\right) = 4\left(\dfrac{1}{3}x - 5\right)$

**34.** $-2\left(\dfrac{1}{2}x + 6\right) = 7\left(5 - \dfrac{1}{2}x\right) + 8$

**35.** $\dfrac{3(y + 2)}{4} - \dfrac{y - 3}{2} = \dfrac{3}{4}y$

**36.** $\dfrac{5(x - 4)}{6} + \dfrac{x - 2}{12} = \dfrac{2x}{3} + 1$

**37.** $\dfrac{4(x - 3)}{9} - \dfrac{2x}{3} = 1$

**38.** $\dfrac{2}{3}\left(x - \dfrac{1}{2}\right) - \left(x + \dfrac{1}{4}\right) = \dfrac{3}{2}x$

**39.** $\dfrac{3}{5}\left(\dfrac{2}{3}y + \dfrac{1}{5}\right) = \dfrac{2}{3}\left(\dfrac{3}{4}y - \dfrac{1}{3}\right)$

**40.** $\dfrac{1}{4}\left(\dfrac{4}{5}x - 3\right) = -\dfrac{1}{2}\left(\dfrac{2}{5}x + \dfrac{3}{2}\right)$

*Use a calculator to solve each equation. Round each result to three significant digits.*

**41.** $18.7x + 253 = 28.6x$

**42.** $6.19x - 39.4 = 82.4$

**43.** $4.12x + 6.18 = 12.6x - 3.6$

**44.** $0.96 + 4.08x = 6.21 - 1.33x$

**45.** $6.3 - 0.4(9.36 - 2x) = 1.2x$

**46.** $4.5x + 3.2 = 0.6(5x + 14.5)$

**47.** $2.76(1.81x + 59.2) + 16.7 = 763$

**48.** $29.2(2.4x - 6.5) = 10(3.4 - x)$

**49.** $2\pi(2.50 \times 10^6)L = 2590$

**50.** $\dfrac{V_1}{2.62 \times 10^4} = \dfrac{5.63 \times 10^5}{4.16 \times 10^7}$

**51.** $2.50 \times 10^{-3} = \dfrac{(9.00 \times 10^9)(8.50 \times 10^{-7})q}{(0.15)^2}$

**52.** $1.59 \times 10^{-3} = \dfrac{1}{2\pi f(1.55 \times 10^{-6})}$

## 2.6  FORMULAS

A **formula** is an equation, usually expressed in letters, that shows the relationship between quantities.

### EXAMPLE 1

The formula $P = VI$ states that the electrical power, $P$, equals the product of the voltage drop, $V$, and the current, $I$.

### EXAMPLE 2

The formula $p = \dfrac{F}{A}$ states that the pressure, $p$, equals the quotient of the force, $F$, and the area, $A$.

### EXAMPLE 3

The formula $F = \dfrac{mv^2}{r}$ states that the centripetal force, $F$, of a rotating body equals the product of the mass, $m$, and the square of the velocity, $v$, divided by $r$, the radius of the circular path.

---

**SOLVING A FORMULA**

To **solve a formula for a given letter** means to isolate the given letter on one side of the equation and express it in terms of all the remaining letters by having all the other letters and numbers appear on the opposite side of the equation. We solve a formula using the same principles that we use in solving any equation.

---

### EXAMPLE 4

Solve $P = VI$ for $V$.

$$P = VI$$

$$\frac{P}{I} = V \qquad \text{(Divide each side by } I.)$$

Note that $\dfrac{P}{I} = V$ and $V = \dfrac{P}{I}$ are equivalent equations.

### EXAMPLE 5

Solve $p = \dfrac{F}{A}$ for $F$ and then for $A$.

First solve for $F$.

$$p = \frac{F}{A}$$

$$pA = F \qquad \text{(Multiply each side by } A.)$$

Now solve for $A$.

$$p = \frac{F}{A}$$

$$pA = F \qquad \text{(Multiply each side by } A.)$$

$$A = \frac{F}{p} \qquad \text{(Divide each side by } p.)$$

**EXAMPLE 6**

Solve $F = \dfrac{mv^2}{r}$ for $v^2$.

$$F = \frac{mv^2}{r}$$

$$Fr = mv^2 \qquad \text{(Multiply each side by } r.)$$

$$\frac{Fr}{m} = v^2 \qquad \text{(Divide each side by } m.)$$

**EXAMPLE 7**

Solve $V = E - Ir$ for $r$.
   One way:

$$V = E - Ir$$

$$V - E = -Ir \qquad \text{(Subtract } E \text{ from each side.)}$$

$$\frac{V - E}{-I} = r \qquad \text{(Divide each side by } -I.)$$

Alternate way:

$$V = E - Ir$$

$$V + Ir = E \qquad \text{(Add } Ir \text{ to each side.)}$$

$$Ir = E - V \qquad \text{(Subtract } V \text{ from each side.)}$$

$$r = \frac{E - V}{I} \qquad \text{(Divide each side by } I.)$$

Note that the two results are equivalent. Take the first result, $\dfrac{V - E}{-I}$, and multiply numerator and denominator by $-1$.

$$\left(\frac{V - E}{-I}\right)\left(\frac{-1}{-1}\right) = \frac{-V + E}{I} = \frac{E - V}{I}$$

## Exercises 2.6

*Solve each formula for the given letter. These formulas are in common use in various technical and scientific areas.*

**1.** $W = JQ$   for $J$

**2.** $v = f\lambda$   for $\lambda$

**3.** $R_T = R_1 + R_2 + R_3$   for $R_2$

**4.** $V = IZ$   for $I$

**5.** $E = IR$   for $R$

**6.** $C_T = C_1 + C_2 + C_3 + C_4$   for $C_3$

**7.** $C = \dfrac{Q}{V}$   for $Q$

**8.** $I = \dfrac{Q}{t}$   for $t$

**9.** $V = \dfrac{W}{Q}$   for $Q$

**10.** $E = mc^2$   for $m$

**11.** $Q = \dfrac{I^2 Rt}{J}$   for $R$

**12.** $R = \rho \dfrac{L}{A}$   for $A$

**13.** $P = \dfrac{(\text{O.D.})}{N + 2}$   for $N$

**14.** $\text{hp} = \dfrac{D^2 N}{2.5}$   for $N$

**15.** $R = \dfrac{kL}{D^2}$   for $L$

**16.** $F = \dfrac{wa}{g}$   for $g$

**17.** $R = \dfrac{\pi}{2P}$   for $P$

**18.** $X_L = 2\pi f L$   for $f$

**19.** $\dfrac{V}{V'} = \dfrac{T}{T'}$   for $T$

**20.** $\dfrac{V}{V'} = \dfrac{p'}{p}$   for $V'$

**21.** $C = \dfrac{5}{9}(F - 32)$   for $F$

**22.** $F = \dfrac{9}{5}C + 32$   for $C$

**23.** $\dfrac{I_s}{I_p} = \dfrac{N_p}{N_s}$   for $N_s$

**24.** $\dfrac{V_1 P_1}{T_1} = \dfrac{V_2 P_2}{T_2}$   for $T_1$

**25.** $\dfrac{\Delta d}{d} = 1.22\dfrac{\lambda}{a}$   for $a$

**26.** $\dfrac{I_1}{I_2} = \dfrac{r_2^2}{r_1^2}$   for $I_2$

**27.** $l = n\dfrac{\lambda}{2}$   for $\lambda$

**28.** $l = (2n + 1)\dfrac{\lambda}{4}$   for $\lambda$

**29.** $\Delta L = \alpha L(T - T_0)$   for $T$

**30.** $A = ab + \dfrac{d}{2}(a + c)$   for $d$

**31.** $f' = f\left(\dfrac{v + v_0}{v - v_s}\right)$   for $v$

**32.** $f' = f\left(\dfrac{v + v_0}{v - v_s}\right)$   for $v_s$

**33.** $F = \dfrac{q_1 q_2}{4\pi\epsilon_0 r^2}$   for $q_1$

**34.** $\alpha = \dfrac{\Delta\rho}{\rho\,\Delta T}$   for $\Delta T$

**35.** $E - IR - \dfrac{q}{C} = 0$   for $R$

**36.** $X_C = \dfrac{1}{2\pi f C}$   for $f$

**37.** $\dfrac{1}{R_T} = \dfrac{1}{R_1} + \dfrac{1}{R_2}$   for $R_2$

**38.** $\dfrac{1}{C_T} = \dfrac{1}{C_1} + \dfrac{1}{C_2}$   for $C_T$

**39.** $\dfrac{1}{R_T} = \dfrac{1}{R_1} + \dfrac{1}{R_2} + \dfrac{1}{R_3}$   for $R_T$

**40.** $\dfrac{1}{C_T} = \dfrac{1}{C_1} + \dfrac{1}{C_2} + \dfrac{1}{C_3}$   for $C_2$

**41.** $\dfrac{1}{f} = \dfrac{1}{s_0} + \dfrac{1}{s_i}$   for $s_0$

**42.** $\dfrac{1}{f} = \dfrac{1}{s_0} + \dfrac{1}{s_i}$   for $f$

**43.** $\dfrac{1}{f} = (n - 1)\left(\dfrac{1}{R'} - \dfrac{1}{R''}\right)$   for $n$

**44.** $\dfrac{1}{f} = (n - 1)\left(\dfrac{1}{R'} - \dfrac{1}{R''}\right)$   for $R''$

**45.** $eV = hf - \phi$   for $f$

**46.** $hf = -13.6\left(\dfrac{1}{m^2} - \dfrac{1}{n^2}\right)$   for $m$

**47.** $I_2 = \dfrac{Z_3}{Z_2 + Z_3} I_T$   for $Z_2$

**48.** $I_3 = \dfrac{Z_2}{Z_2 + Z_3} I_T$   for $Z_2$

**49.** $R_A = \dfrac{R_1 R_3}{R_1 + R_2 + R_3}$   for $R_1$

**50.** $R_B = \dfrac{R_1 R_2}{R_1 + R_2 + R_3}$   for $R_3$

## 2.7 SUBSTITUTION OF DATA INTO FORMULAS

Problem solving is basic to all technical fields. An important part of problem solving involves analyzing the given data, finding a formula that relates the given quantities with the unknown quantity, substituting the given data into this formula, and solving for the unknown quantity. Working with formulas in this way is one of the most important skills that you will learn.

---

**USING A FORMULA TO SOLVE A PROBLEM WHERE ALL BUT THE UNKNOWN QUANTITY IS GIVEN**

1. Solve the formula for the unknown quantity.
2. Substitute each known quantity with its units.
3. Use the order of operations procedures to find the numerical quantity and to simplify the units.

---

This is the most useful method when you use a calculator, which we assume you will. We shall use this method most often in this text.

### EXAMPLE 1

Given the formula $A = bh$, $A = 150$ m$^2$, and $b = 25$ m. Find $h$.
First solve for $h$.

$$A = bh$$

$$\frac{A}{b} = \frac{bh}{b} \qquad \text{(Divide each side by } b.)$$

$$\frac{A}{b} = h$$

Then substitute the data.

$$h = \frac{A}{b}$$

$$h = \frac{150 \text{ m}^2}{25 \text{ m}}$$

$$= 6.0 \text{ m}$$

*Note:* Follow the rules for calculations with measurements in this section.

### EXAMPLE 2

Given the formula $P = 2l + 2w$, $P = 712$ cm, and $w = 128$ cm. Find $l$.
First solve for $l$.

$$P = 2l + 2w$$

$$P - 2w = 2l + 2w - 2w \qquad \text{(Subtract } 2w \text{ from each side.)}$$

$$P - 2w = 2l$$

$$\frac{P - 2w}{2} = \frac{2l}{2} \qquad\qquad\qquad \text{(Divide each side by 2.)}$$

$$\frac{P - 2w}{2} = l \qquad\qquad\qquad \left(\text{or } l = \frac{P}{2} - w\right)$$

Then substitute the data.

$$l = \frac{P - 2w}{2}$$

$$l = \frac{712 \text{ cm} - 2(128 \text{ cm})}{2}$$

$$= \frac{712 \text{ cm} - 256 \text{ cm}}{2}$$

$$= \frac{456 \text{ cm}}{2} = 228 \text{ cm}$$

**EXAMPLE 3**

Given the formula $V = \frac{1}{2}lw(D + d)$, $V = 54.85$ in³, $l = 4.50$ in., $w = 3.65$ in., and $d = 3.25$ in. Find $D$.

First solve for $D$.

$$V = \frac{1}{2}lw(D + d)$$

$$2V = lw(D + d) \qquad \text{(Multiply each side by 2.)}$$

$$2V = lwD + lwd \qquad \text{(Remove parentheses.)}$$

$$2V - lwd = lwD \qquad \text{(Subtract } lwd \text{ from each side.)}$$

$$\frac{2V - lwd}{lw} = D \qquad \text{(Divide each side by } lw.)$$

$$D = \frac{2V}{lw} - d$$

Then substitute the data.

$$D = \frac{2(54.85 \text{ in}^3)}{(4.50 \text{ in.})(3.65 \text{ in.})} - 3.25 \text{ in.}$$

$$= 6.68 \text{ in.} - 3.25 \text{ in.}$$

$$= 3.43 \text{ in.}$$

**EXAMPLE 4**

Given $R = \frac{\rho L}{A}$, $R = 3.25 \ \Omega$, $\rho = 1.72 \times 10^{-6} \ \Omega$ cm, and $A = 0.0250$ cm². Find $L$.

$$R = \frac{\rho L}{A}$$

$$RA = \rho L$$

$$L = \frac{RA}{\rho}$$

$$= \frac{(3.25 \ \Omega)(0.0250 \text{ cm}^2)}{1.72 \times 10^{-6} \ \Omega \text{ cm}}$$

$$= 47,200 \text{ cm} \quad \text{or} \quad 472 \text{ m}$$

**EXAMPLE 5**

If a satellite orbits the earth in a circular path $50\overline{0}$ mi above the earth's surface, find the acceleration due to the pull of gravity at that height.

The formula to use is $g = \dfrac{GM}{s^2}$, where $g$ is the acceleration due to the force or pull of gravity; $G$ is a constant, $6.67 \times 10^{-11} \dfrac{m^3}{kg\, s^2}$; $M$ is the mass of the earth, $5.96 \times 10^{24}$ kg; and $s$ is the distance from the center of the earth to the satellite.

The average radius of the earth is 3960 mi; therefore, the average radius of the orbit is 3960 mi + 500 mi = 4460 mi or $7.18 \times 10^6$ m.

Substituting, we have

$$g = \frac{GM}{s^2}$$

$$g = \frac{\left(6.67 \times 10^{-11} \dfrac{m^3}{kg\, s^2}\right)(5.96 \times 10^{24}\ kg)}{(7.18 \times 10^6\ m)^2}$$

$$= 7.71\ m/s^2 \text{ or } 25.3\ ft/s^2$$

( For comparison purposes, the value of $g$ on the earth's surface is 9.80 m/s$^2$, or 32.2 ft/s$^2$.)

Formulas involving reciprocals are often used in electronics and physics. We next consider an alternate method for substituting data into such formulas and solving for a specified letter using a calculator. First, solve for the reciprocal of the specified variable. Then substitute the given data and follow the calculator flow chart as shown in the following example.

**EXAMPLE 6**

Given the formula $\dfrac{1}{R} = \dfrac{1}{R_1} + \dfrac{1}{R_2} + \dfrac{1}{R_3}$, $R = 1.5\ \Omega$, $R_1 = 3.0\ \Omega$, $R_3 = 12\ \Omega$. Find $R_2$.

First, solve the formula for the reciprocal of $R_2$.

$$\frac{1}{R} = \frac{1}{R_1} + \frac{1}{R_2} + \frac{1}{R_3}$$

$$\frac{1}{R_2} = \frac{1}{R} - \frac{1}{R_1} - \frac{1}{R_3}$$

Then substitute the data.

$$\frac{1}{R_2} = \frac{1}{1.5\ \Omega} - \frac{1}{3.0\ \Omega} - \frac{1}{12\ \Omega}$$

Next, use your calculator as follows:

| Flow chart | Buttons pushed | Display |
|---|---|---|
| Enter 1.5 | $\boxed{1}$ $\boxed{.}$ $\boxed{5}$ | 1.5 |
| Find reciprocal | $\boxed{1/x}$ or $\boxed{x^{-1}}$ | $1.5^{-1}$ |
| Push minus | $\boxed{-}$ | |
| Enter 3 | $\boxed{3}$ | 3 |
| Find reciprocal | $\boxed{1/x}$ or $\boxed{x^{-1}}$ | $3^{-1}$ |
| Push minus | $\boxed{-}$ | |

Thus $R_2 = 4.0\ \Omega$.

## Exercises 2.7

*Solve for the given letter. Then substitute the data to find the value of the given letter. Follow the rules for calculating with measurements.*

| Formula | Data | Find |
|---|---|---|
| **1.** $A = bh$ | $A = 24.0\ \text{in}^2$, $h = 6.00$ in. | $b$ |
| **2.** $C = \pi d$ | $C = 358$ cm | $d$ |
| **3.** $A = \dfrac{1}{2}bh$ | $A = 144\ \text{m}^2$, $b = 8.00$ m | $h$ |
| **4.** $A = 2\pi rh$ | $A = 1450\ \text{ft}^2$, $r = 16.0$ ft | $h$ |
| **5.** $V = \pi r^2 h$ | $V = 1950\ \text{m}^3$, $r = 12.6$ m | $h$ |
| **6.** $P = 2(a + b)$ | $P = 248$ in., $b = 45$ in. | $a$ |
| **7.** $V = lwh$ | $V = 2.50 \times 10^4\ \text{ft}^3$, $l = 44.5$ ft, $h = 19.7$ ft | $w$ |
| **8.** $A = \left(\dfrac{a + b}{2}\right)h$ | $A = 205.2\ \text{m}^2$, $a = 16.50$ m, $b = 19.50$ m | $h$ |
| **9.** $C = \dfrac{5}{9}(F - 32°)$ | $C = 55°$ | $F$ |
| **10.** $(\Delta L) = \alpha L(T - T_0)$ | $\Delta L = 0.025$ m, $\alpha = 1.3 \times 10^{-5}/°\text{C}$ $L = 25.000$ m, $T_0 = 25°\text{C}$ | $T$ |

**11.** Given $Q = mc\Delta T$, where $m = 50\overline{0}$ g, $c = 0.214$ cal/g°C, and $\Delta T = 40.0°$C. Find $Q$.

**12.** Given $E = \dfrac{wv^2}{2g}$, where $w = 2.8 \times 10^4$ N, $v = 2.5 \times 10^4$ m/s, and $g = 9.80\ \text{m/s}^2$. Find $E$.

**13.** Given $X_L = 2\pi fL$, where $X_L = 75.0\ \Omega$ and $f = 60.0$ Hz. Find $L$.

**14.** Given $I = \dfrac{E}{R_L + r}$, where $I = 2.00$ A, $R_L = 28.0\ \Omega$, and $r = 0.500\ \Omega$. Find $E$.

**15.** Given $P$ (in hp) $= \dfrac{(\text{force in lb})(\text{distance in ft})}{\left(550\ \dfrac{\text{ft-lb}}{\text{s}}\right)(\text{time in s})}$ hp. What horsepower engine is required to

hoist $8\overline{0}$ tons of coal per hour from a mine shaft $20\overline{0}$ ft deep?

**16.** Given $\dfrac{V_p}{V_s} = \dfrac{N_p}{N_s}$, where $V_p = 12\overline{0}$ V, $V_s = 3{,}50\overline{0}$ V, and $N_s = 12{,}\overline{0}00$ turns. Find $N_p$.

---

\* For some calculators, you do not need to push the equals key after the reciprocal button.

17. Given $\dfrac{P_1}{P_2} = \dfrac{V_2}{V_1}$, where $P_1 = 270$ kPa, $V_1 = 750$ cm$^3$, and $P_2 = 210$ kPa. Find $V_2$.

18. Given $X_C = \dfrac{1}{2\pi fC}$, where $X_C = 10\bar{0}0$ $\Omega$ and $C = 2.65 \times 10^{-6}$ F. Find $f$.

19. Given $\dfrac{\Delta d}{d} = 1.22\dfrac{\lambda}{a}$, where $\Delta d = 1.75$ mm, $\lambda = 5.50 \times 10^{-7}$ m, and $a = 2.00 \times 10^{-3}$ m. Find $d$.

20. Given $F = k\dfrac{q_1 q_2}{r^2}$, where $F = -2.5 \times 10^{-2}$ N, $k = 9.00 \times 10^9$ N $\cdot$ m$^2$/C$^2$, $q_1 = 8.0 \times 10^{-7}$ C, and $r = 0.15$ m. Find $q_2$.

21. Given $\dfrac{V_1 P_1}{T_1} = \dfrac{V_2 P_2}{T_2}$, where $V_1 = 125$ m$^3$, $P_1 = 13.5$ MPa, $T_1 = 295$ K, $P_2 = 18.4$ MPa, and $T_2 = 305$ K. Find $V_2$.

22. Given $\dfrac{F_1}{F_2} = \dfrac{r_1^2}{r_2^2}$, where $F_2 = 60\bar{0}0$ N, $r_2 = 12.0$ cm, and $r_1 = 2.50$ cm. Find $F_1$.

23. Given $f = \dfrac{nv}{2l}$, where $f = 384$ Hz $= 384$/s, $n = 1$, and $v = 348$ m/s. Find $l$ (in cm).

24. Given $f = \dfrac{(2n + 1)v}{4l}$, where $f = 384$ Hz $= 384$/s, $n = 2$, and $l = 112$ cm. Find $v$ (in m/s).

25. Given $f' = f\left(\dfrac{v - v_0}{v + v_s}\right)$, where $f = 425$ Hz $= 425$/s, $v = 343$ m/s, $v_0 = 0$, and $v_s = 25.0$ m/s. Find $f'$.

26. Given $f' = f\left(\dfrac{v + v_0}{v + v_s}\right)$, where $f' = 438$ Hz $= 438$/s, $v = 343$ m/s, $v_0 = 35.0$ m/s, and $v_s = 15.0$ m/s. Find $f$.

27. Given $\dfrac{1}{R} = \dfrac{1}{R_1} + \dfrac{1}{R_2}$, where $R = 60.0$ $\Omega$ and $R_2 = 24\bar{0}$ $\Omega$. Find $R_1$.

28. Given $\dfrac{1}{Z} = \dfrac{1}{Z_1} + \dfrac{1}{Z_2} + \dfrac{1}{Z_3}$, where $Z_1 = 25.0$ $\Omega$, $Z_2 = 15.0$ $\Omega$, and $Z_3 = 50.0$ $\Omega$. Find $Z$.

29. Given $\dfrac{1}{R} = \dfrac{1}{R_1} + \dfrac{1}{R_2} + \dfrac{1}{R_3}$, where $R_1 = 3\bar{0}$ $\Omega$, $R_2 = 4\bar{0}$ $\Omega$, and $R_3 = 5\bar{0}$ $\Omega$. Find $R$.

30. Given $\dfrac{1}{R} = \dfrac{1}{R_1} + \dfrac{1}{R_2} + \dfrac{1}{R_3}$, where $R = 15$ $\Omega$, $R_1 = 3\bar{0}$ $\Omega$, and $R_3 = 6\bar{0}$ $\Omega$. Find $R_2$.

31. Given $\dfrac{1}{f} = \dfrac{1}{s_0} + \dfrac{1}{s_i}$, where $f = 2\bar{0}$ cm and $s_0 = 15$ cm. Find $s_i$.

32. Given $\dfrac{1}{f} = (n - 1)\left(\dfrac{1}{R'} - \dfrac{1}{R''}\right)$, where $R' = 2\bar{0}$ cm, $R'' = 4\bar{0}$ cm, and $n = 1.50$. Find $f$.

33. Given $s = v_i t + \dfrac{1}{2}at^2$, where $s = 40.0$ m, $t = 3.70$ s, and $a = -6.00$ m/s$^2$. Find $v_i$.

34. Given $v_f^2 = v_i^2 + 2as$, where $v_f = 0$, $v_i = 25.0$ m/s, and $a = -7.50$ m/s$^2$. Find $s$.

35. Given $I_1 = \dfrac{R_2}{R_1 + R_2}I_2$, where $R_1 = 5.0$ $\Omega$, $R_2 = 25.0$ $\Omega$, and $I_1 = 0.125$ mA. Find $I_2$.

36. Given $R_A = \dfrac{R_1 R_3}{R_1 + R_2 + R_3}$, where $R_A = 2.00$ $\Omega$, $R_1 = 4.00$ $\Omega$, and $R_2 = 2.00$ $\Omega$. Find $R_3$.

## 2.8 APPLICATIONS INVOLVING LINEAR EQUATIONS

Many technical problems can be expressed mathematically as linear equations. Solving such a problem then becomes a problem of solving a linear equation. What follows is a suggested outline for the solution to this type of problem. The most difficult task is that of "translating" the stated problem into mathematical terms. Because of the numerous ways of describing any given problem, there is no simple process that can be given for making the mathematical translation. The ability to make an accurate translation will come from experience. We cannot overemphasize the need to work out numerous problems to gain this experience.

---

**SOLVING APPLICATION PROBLEMS**

**Step 1:** Read the problem carefully at least two times.
   **(a)** The first time you should read straight through from beginning to end. Do not stop this time to think about setting up an equation. You are only after a general impression at this point.
   **(b)** Now read through *completely* for the second time. This time begin to think ahead to the next steps.

**Step 2:** If possible, draw a picture or diagram. This will often help you to visualize the possible mathematical relationships needed in order to write the equation.

**Step 3:** Choose a symbol to represent the unknown quantity (the quantity not given as a specific value) in the problem. Be sure to label this symbol, the unknown variable, to indicate what it represents.

**Step 4:** Write an equation that you think expresses the information given in the problem. To obtain this equation, you will be looking for information that will express two quantities that can be set equal to each other. Do not rush through this step. Try expressing your equation in words to see if it corresponds to the original problem.

**Step 5:** Solve for the unknown variable.

**Step 6:** Check your solution in the equation from Step 4. If it does not check, you have made a mathematical error. You will need to repeat Step 5. Also check your solution in the original verbal problem for reasonableness.

---

*Note:* Do not use rules for calculating with measurements in this section.

**EXAMPLE 1**

A piece of lumber 7 ft 7 in. long is to be sawed into 9 equal pieces. If the loss per cut is $\frac{1}{8}$ in., how long will each piece be?

After reading the problem, draw a diagram (see Fig. 2.1).
Let

$$x = \text{the length of each sawed piece}$$

$$\frac{1}{8} = \text{the length of each cut}$$

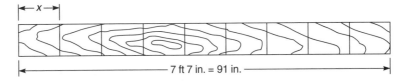

**Figure 2.1**

Then

$$9x = \text{the total length of the 9 pieces}$$

$$8\left(\frac{1}{8}\right) = 1 = \text{the total length of the 8 cuts} \qquad (\textit{Note:} \ 8 \text{ cuts will be needed.})$$

Since the total of the lengths of the pieces and the cuts equals the total length of the original board, we have the equation

$$9x + 1 = 91$$
$$9x = 90 \qquad \text{(Subtract 1 from each side.)}$$
$$x = 10 \qquad \text{(Divide each side by 9.)}$$

*Check:*

$$9 \text{ pieces 10 in. each} = 90 \text{ in.}$$

$$8 \text{ cuts } \frac{1}{8} \text{ in. each} = \underline{\ \ 1 \text{ in.}}$$

$$\text{Total} = 91 \text{ in.} = 7 \text{ ft 7 in.}$$

**EXAMPLE 2**

Bill wants to use 160 m of fence to enclose a rectangular portion of a lot. He wants the length to be 20 m longer than the width. Find the dimensions of the fenced portion.

Let

$$x = \text{width}$$
$$x + 20 = \text{length}$$
$$160 = \text{perimeter}$$

Draw a diagram (see Fig. 2.2).

The perimeter of a rectangle is given by the formula

$$2l + 2w = P$$

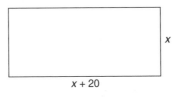

$x + 20$

**Figure 2.2**

Substituting, we have

$$2(x + 20) + 2x = 160$$
$$2x + 40 + 2x = 160 \quad \text{(Remove parentheses.)}$$
$$4x + 40 = 160 \quad \text{(Combine like terms.)}$$
$$4x = 120 \quad \text{(Subtract 40 from each side.)}$$
$$x = 30 \quad \text{(Divide each side by 4.)}$$
$$x + 20 = 50$$

Thus, the width is 30 m and the length is 50 m.

*Check:*

$$2l + 2w = 2(50 \text{ m}) + 2(30 \text{ m})$$
$$= 100 \text{ m} + 60 \text{ m}$$
$$= 160 \text{ m}$$

which is the amount of fence used.

### EXAMPLE 3

Helen bought 50 acres of land for $111,000. Part of the land cost $1500 per acre and the rest cost $3500 per acre. How much land was sold at each rate?

Let

$$x = \text{the amount of land sold at \$1500 per acre}$$
$$50 - x = \text{the amount of land sold at \$3500 per acre}$$

Therefore,

$$1500x + 3500(50 - x) = 111{,}000$$
$$1500x + 175{,}000 - 3500x = 111{,}000$$
$$-2000x + 175{,}000 = 111{,}000$$
$$-2000x = -64{,}000$$
$$x = 32$$
$$50 - x = 18$$

Thus,

$$32 \text{ acres sold at \$1500 per acre}$$
$$18 \text{ acres sold at \$3500 per acre}$$

*Check:*

$$32 \text{ acres at \$1500 per acre} = \$\ 48{,}000$$
$$18 \text{ acres at \$3500 per acre} = \underline{\$\ 63{,}000}$$
$$\text{Total} = \$111{,}000$$

### EXAMPLE 4

The cruising air speed of a small plane is 200 km/h. The wind is blowing from the north at 60 km/h. How far north can the pilot expect to fly and still return within 3 h?

Let

$$200 \text{ km/h} = \text{speed of plane in still air}$$
$$60 \text{ km/h} = \text{speed of wind}$$
$$140 \text{ km/h} = \text{speed of plane traveling north}$$
$$260 \text{ km/h} = \text{speed of plane traveling south}$$
$$x = \text{elapsed time traveling north in hours}$$
$$3 - x = \text{elapsed time traveling south in hours}$$

The basic formula is $d = rt$ (distance = rate $\times$ time). We can make a chart of the information.

|  | $d$ | $r$ | $t$ |
|---|---|---|---|
| Traveling north | $140x$ | $140$ | $x$ |
| Traveling south | $260(3 - x)$ | $260$ | $3 - x$ |

$$\uparrow$$
$$d = rt$$

Here the distances are the same; that is,

$$140x = 260(3 - x)$$
$$140x = 780 - 260x$$
$$400x = 780$$
$$x = 1.95 \text{ h}$$

which is the time traveling north.

The distance traveled north is then

$$d = 140x$$
$$= (140 \text{ km/h})(1.95 \text{ h})$$
$$= 273 \text{ km}$$

We leave the check to you.

**EXAMPLE 5**

A 12-qt cooling system is checked and found to be filled with a solution that is 40% antifreeze. The desired strength of the solution is 60% antifreeze. How many quarts of solution need to be drained and replaced with pure antifreeze to reach the desired strength?

The amount of 40% antifreeze solution that needs to be drained equals the amount of pure antifreeze that must be added in order to increase the concentration of the mixture to 60% antifreeze. Let us denote this amount by $x$. This relationship can be shown by the diagram in Fig. 2.3.

Write an equation in terms of antifreeze.

$$(0.40)(12) - (0.40)(x) + (1.00)(x) = (0.60)(12)$$
$$4.8 \quad - \quad 0.4x \quad + \quad x \quad = \quad 7.2$$
$$4.8 + 0.60x = 7.2$$
$$0.60x = 2.4$$
$$x = 4 \text{ qt}$$

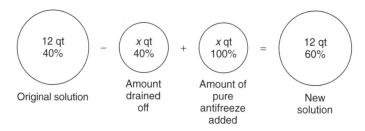

**Figure 2.3**

That is, 4 qt of the old solution needs to be drained off and replaced with 4 qt of pure antifreeze. The check is left to you.

## Exercises 2.8

*Solve the following application problems. (Do not use the rules for calculating with measurements.)*

1. Separate $216 into two parts so that one part is 3 times the other.

2. The difference between two numbers is 6. Their sum is 30. Find the numbers.

3. A set of 8 built-in bookshelves is to be constructed in a room with a floor-to-ceiling clearance of 9 ft 6 in. If each shelf is $\frac{3}{4}$ in. thick and equal spacing is desired between shelves, what is the space between the shelves? (Assume there is no shelf against the ceiling and no shelf on the floor.)

4. A deck railing 23 ft 8 in. long is to be made by 8 evenly spaced redwood posts 4 in. by 4 in. (Posts are located on the ends.) What is the distance between each post.

5. The length of a rectangle is 4 m longer than its width. The perimeter is 56 m. Find the length and the width of the rectangle.

6. The perimeter of an isosceles triangle is 122 cm. Its base is 4 cm shorter than one of its equal sides. Find lengths of the sides of the triangle.

7. Forty acres of land were purchased for $20,400. The land facing the highway cost $650 per acre. The remainder cost $450 per acre. How much land was sold at $650 per acre? At $450 per acre?

8. A man deposited $6800 into two different savings accounts. One account earns interest at 8%, and the other at $10\frac{1}{2}$%. The total interest earned from both accounts at the end of one year was $669. Find the amount deposited at each rate.

9. A man earned $360 per week for Smith Contracting Co. He quit this job to work for Jones Contracting Co. for $415 per week. It cost him $1210 in moving expenses to relocate near his new place of employment. How many weeks will it take his increase in salary to cover the cost of his moving expenses?

10. A woman wishes to enclose a rectangular yard with a fence. She plans to use the back of her house (70 ft long) as one side of the enclosed yard. If she has 50 yd of fencing, find the dimensions of the largest yard that can be enclosed.

11. A freight train leaves a station and travels at 45 mi/h. Three hours later a passenger train leaves and travels at 75 mi/h.
    (a) How long does it take the passenger train to overtake the freight train?
    (b) How far will each train travel?

12. A plane has a cruising speed of 270 mi/h. The wind velocity is 30 mi/h.
    (a) What is the speed of the plane flying with the wind?
    (b) What is the speed of the plane flying against the wind?
    (c) How far can the plane fly with the wind and return in 9 h?

13. Two planes are 760 mi apart, leave at the same time, fly toward each other, and meet in 4 h. If their speeds differ by 20 mi/h, find the speed of each plane.

14. A plane averaging 120 mi/h and an express train averaging 60 mi/h depart at the same time from the same city headed for the same destination. If the train arrives 45 min later than the plane, how far did they each travel?

15. A man has two piles of solder (a mixture of tin and lead) available. One pile is 30% tin and the other is 70% tin. How much of each must be used to make a 20-lb mixture that is 45% tin?

16. A bar of metal contains 10% silver. Another bar contains 15% silver. How many kilograms of each must be taken to make a 10-kg bar containing 12% silver?

17. A 12-qt cooling system is checked and found to be filled with 25% antifreeze. The desired strength is 50% antifreeze. How many quarts need to be drained and replaced with pure antifreeze to reach the desired strength?

18. In testing a hybrid engine, engineers are trying various mixtures of gasoline and methanol. How much of a 90% gasoline mixture and a 75% gasoline mixture would be needed for 600 L of an 85% gasoline mixture?

19. The sum of three electric currents is 4.65 A. The greatest current is 4 times the least. The third current is 0.75 A greater than the least. Find the three currents.

20. A test solution of gasohol contains 60 L of gasoline and 15 L of alcohol. How much pure alcohol must be added so that the resulting solution is 25% alcohol?

21. Enclose a rectangular plot with 280 m of fencing so that the length is twice the width and the area is divided into two equal parts. Find the length and the width of the lot.

22. Lee is in charge of a fireworks display for a Fourth of July celebration. She notices that she hears the explosion 2.40 seconds after she sees the display in the sky. If sound travels at 331 m/s, how far is she from the aerial display?

23. The sum of three resistances connected in series in a circuit is 540 Ω. The second resistance is 15 Ω greater than the first, and the third is 60 Ω greater than the second. Find the size of each resistance.

24. One of three angles is 5° more than the smallest angle. The third angle is three times the middle angle. The sum of the three angles is 120°. Find the measure of each of the three angles.

# 2.9   RATIO AND PROPORTION

A **ratio** is the quotient of two numbers or quantities. The ratio $\dfrac{30 \text{ m}}{40 \text{ m}}$ compares two linear (length) measurements. This ratio may also be represented by a division sign (30 m ÷ 40 m), by a colon (30 m : 40 m), or by the word *to* (30 m to 40 m). Ratios should be expressed in simplest form; that is,

$$\frac{30 \text{ m}}{40 \text{ m}} = \frac{3}{4}$$

Note that the ratio of two like quantities has no units.

In technical applications ratios are sometimes used to compare unlike quantities (a ratio of unlike quantities is often called a **rate**). For example, if we travel 300 mi in 6 h, our average speed can be expressed by the ratio $\dfrac{300 \text{ mi}}{6 \text{ h}}$, or $50 \dfrac{\text{mi}}{\text{h}}$. Other examples of ratios include pressure, which may be defined as the ratio of force per unit area; mass density, which may be defined as the ratio of mass per unit volume; and power, which may be defined as the ratio of work per unit time.

**EXAMPLE 1**

Find the ratio of 90 cm to 72 cm.

$$\frac{90 \text{ cm}}{72 \text{ cm}} = \frac{5}{4}$$

**EXAMPLE 2**

Find the ratio of 24 yd : 12 ft.
 First, change to a common unit.

$$\frac{24 \text{ yd}}{12 \text{ ft}} = \frac{24 \text{ yd}}{4 \text{ yd}} = \frac{6}{1}$$

or

$$\frac{24 \text{ yd}}{12 \text{ ft}} = \frac{72 \text{ ft}}{12 \text{ ft}} = \frac{6}{1}$$

**EXAMPLE 3**

The mechanical advantage (*MA*) of a simple machine such as a pulley system may be defined as the following ratio:

$$MA = \frac{\text{resistance force}}{\text{effort force}}$$

If 130 N of effort lift a 6500 N load, the *MA* is

$$\frac{6500 \text{ N}}{130 \text{ N}} = \frac{50}{1}$$

**EXAMPLE 4**

The mechanical advantage (*MA*) of a hydraulic press may be calculated by the ratio

$$MA = \frac{r_l^2}{r_s^2}$$

where

$$r_l = \text{radius of the larger or resistance piston}$$
$$r_s = \text{radius of the smaller or effort piston}$$

Find the *MA* of a hydraulic press that has a large piston of radius 10 in. and a small piston of radius 2 in.

$$MA = \frac{r_l^2}{r_s^2} = \frac{(10 \text{ in.})^2}{(2 \text{ in.})^2} = \frac{100 \text{ in}^2}{4 \text{ in}^2} = \frac{25}{1}$$

**EXAMPLE 5**

Mass density is defined as the ratio of mass per unit volume. Find the mass density of a piece of rock that has a volume of 320 cm$^3$ and a mass of 880 g.

$$\text{Mass density} = \frac{\text{mass}}{\text{volume}} = \frac{880 \text{ g}}{320 \text{ cm}^3} = 2.75 \text{ g/cm}^3$$

A **proportion** is a statement that two ratios are equal. The proportion $\frac{a}{b} = \frac{c}{d}$ may also be written as $a : b : : c : d$ or as $a : b = c : d$. The first and fourth terms, $a$ and $d$, are called the **extremes;** the second and third terms, $b$ and $c$, are called the **means.**

---

In any proportion the product of the means equals the product of the extremes; that is, if

$$\frac{a}{b} = \frac{c}{d}$$

then

$$bc = ad$$

---

We can use this principle to solve proportions.

**EXAMPLE 6**

Solve the proportion $\frac{x}{12} = \frac{5}{6}$.

$$\frac{x}{12} = \frac{5}{6}$$
$$6x = 60$$
$$x = 10$$

**EXAMPLE 7**

Solve the proportion $\frac{x}{21 - x} = \frac{3}{4}$.

$$\frac{x}{21 - x} = \frac{3}{4}$$
$$4x = 3(21 - x)$$
$$4x = 63 - 3x$$
$$7x = 63$$
$$x = 9$$

**EXAMPLE 8**

Cut a board 14 ft long into two pieces whose lengths are in the ratio 2 : 5. Find the lengths of the two pieces.

Let

$$x = \text{length of one piece}$$
$$14 - x = \text{length of the other piece}$$
$$\frac{x}{14 - x} = \frac{2}{5}$$
$$2(14 - x) = 5x$$
$$28 - 2x = 5x$$
$$28 = 7x$$
$$4 = x$$

Therefore, $x = 4$ ft is the length of one piece, and $14 - x = 10$ ft is the length of the other.

**EXAMPLE 9**

A car travels 275 mi using 11 gal of gasoline. How far can the car travel on 35 gal?
This problem can be solved using a proportion as follows:

$$\frac{275 \text{ mi}}{11 \text{ gal}} = \frac{x \text{ mi}}{35 \text{ gal}}$$
$$11x = (275)(35)$$
$$x = \frac{(275)(35)}{11}$$
$$x = 875 \text{ mi}$$

## Exercises 2.9

*Express each ratio as a fraction in simplest form.*

1. 36 mi : 8 mi

2. $\dfrac{240 \text{ ft}}{12 \text{ yd}}$

3. 2500 m to 25 km

4. $\dfrac{36 \text{ h}}{45 \text{ min}}$

5. $12 \text{ ft}^2 : 24 \text{ in}^2$

6. 18 kg to 360 g

7. $\dfrac{1.75 \text{ m}^2}{35 \text{ cm}^2}$

8. $12 \text{ ft}^3 : 24 \text{ in}^3$

9. $\dfrac{(28 \text{ in.})^2}{(7 \text{ in.})^2}$

10. $\dfrac{(8 \text{ cm})^3}{(2 \text{ cm})^3}$

11. Find the mechanical advantage of a pulley system in which it takes 250 lb of effort to lift 7500 lb of load.

12. Find the mechanical advantage of a hydraulic press that has pistons of radii 36 cm and 4 cm (see Example 4).

13. The scale of a given map is $\frac{3}{4}$ in. to 2500 ft. Express this ratio in simplest form.

14. Pressure may be defined as the ratio of force to area.
    (a) Find the pressure when a force of 170 lb is exerted on an area of $4\frac{1}{4}$ in.$^2$.
    (b) Find the pressure (in kPa) when a force of 72 N is exerted on an area of 4 cm$^2$
        [1 pascal (Pa) = 1 N/m$^2$].

**15.** A flywheel has 72 teeth, and a starter drive gear has 18 teeth. Find the ratio of flywheel teeth to drive-gear teeth.

**16.** A transformer has 200 turns in the primary coil and 18,000 turns in the secondary coil. Find the ratio of secondary turns to primary turns.

**17.** A 1850-ft$^2$ house sells for $80,475. Find the ratio of cost to area (price per square foot).

**18.** If 27 ft$^3$ of cement are needed to make 144 ft$^3$ of concrete, find the ratio of concrete to cement.

**19.** A 350-gal spray tank can cover 14 acres. Find the rate of application in gallons per acre.

**20.** A bearing bronze mix includes 192 lb of copper and 30 lb of lead. What is the ratio of copper to lead?

**21.** What is the alternator-to-engine-drive ratio if the alternator turns at 1125 rpm when the engine is idling at 500 rpm?

**22.** If 35 gal of oil flow through a feeder pipe in 10 min, express the rate of flow in gallons per minute.

**23.** A flywheel has 72 teeth and a starter drive gear has 15 teeth. Find the ratio of flywheel teeth to drive-gear teeth.

**24.** A certain transformer has a voltage of 18 V in the primary circuit and 5850 V in the secondary circuit. Find the ratio of the primary voltage to the secondary voltage.

**25.** If the total yield from a 45-acre field is 6075 bu, express the yield in bushels per acre.

**26.** The ratio of the voltage drops across two resistors connected in series equals the ratio of their resistances. Find the ratio of the voltage drops across a 960-Ω resistor to those across a 400-Ω resistor.

**27.** Steel can be worked into a lathe at a cutting speed of 25 ft/min. Stainless steel can be worked at 15 ft/min. What is the ratio of the cutting speed of steel to the cutting speed of stainless steel?

**28.** The power gain of an amplifier is the ratio $\dfrac{\text{power output}}{\text{power input}}$. Find the power gain of an amplifier whose output is 18 W and input is 0.72 W.

*Solve each proportion for x.*

**29.** $\dfrac{x}{9} = \dfrac{81}{27}$

**30.** $\dfrac{25}{96} = \dfrac{75}{x}$

**31.** $\dfrac{64}{2x} = \dfrac{16}{84}$

**32.** $\dfrac{1}{75} = \dfrac{5x}{125}$

**33.** $\dfrac{x}{5-x} = \dfrac{2}{3}$

**34.** $\dfrac{x}{8-x} = \dfrac{1}{3}$

**35.** $\dfrac{x}{28-x} = \dfrac{3}{4}$

**36.** $\dfrac{a}{b} = \dfrac{c}{x}$

**37.** $\dfrac{m}{x} = \dfrac{n}{p}$

**38.** $\dfrac{x}{m} = \dfrac{p}{q}$

**39.** $\dfrac{x}{2a} = \dfrac{4}{a}$

**40.** $\dfrac{x}{-a} = \dfrac{a}{b}$

*Solve each proportion using a calculator. Round each result to three significant digits.*

**41.** $\dfrac{30.2}{276} = \dfrac{85.6}{x}$

**42.** $\dfrac{284}{7.8} = \dfrac{x}{13.1}$

**43.** $\dfrac{32.5}{115} = \dfrac{2450}{x}$

**44.** $\dfrac{x}{563.2} = \dfrac{34.5}{244}$

**45.** $\dfrac{x}{0.477} = \dfrac{2.75}{16.1}$

**46.** $\dfrac{47.9}{x} = \dfrac{0.355}{0.0115}$

**47.** The sum of the length and the width of a proposed feed lot is 1260 ft. Find the length and the width if their ratio must be 5 : 2.

**48.** Divide $450 in wages between two people so that the money is in the ratio of 4 : 5.

**49.** On an assembly line 68,340 parts need to be divided into two groups in the ratio of $\frac{5}{12}$. How many parts are in each group?

**50.** The ratio of the length and the width of a rectangular field is 5 : 6. Find the dimensions of the field if its perimeter is 4400 m.

**51.** Brass is an alloy of copper and zinc in the ratio of 3 : 2. How much copper and how much zinc are contained in 2500 lb of brass?

**52.** If 12 machines can produce 27,000 bolts in 6 h, how many bolts can be produced in 10 h? How many machines would it take to increase the production to 45,000 bolts in 6 h?

**53.** If 2000 bolts cost $240, what is the cost of 750 bolts? Of 2800 bolts?

**54.** If $\frac{1}{4}$ in. on a map represents 25 mi, what distance is represented by $2\frac{1}{8}$ in.?

**55.** If a man receives $182.40 for 30 h of work, how much will he receive for 36 h of work?

**56.** A car travels 54 km on 4.5 L of gasoline.
   **(a)** How far can the car travel on 40 L?
   **(b)** How much gasoline would be used on a 3600-km trip?
   **(c)** Find the car's gas "mileage" in kilometres per litre.

**57.** A fuel pump delivers 45 mL of fuel in 540 strokes. How many strokes are needed to use 75 mL of fuel?

**58.** If $2\frac{3}{4}$ ft$^3$ of sand make 8 ft$^3$ of concrete, how much sand is needed to make 120 ft$^3$ of concrete?

**59.** If a builder sells a 1500-ft$^2$ home for $63,000, find the price of a 2100-ft$^2$ home of similar quality. *Note:* Assume that the price per square foot remains constant.

**60.** If 885 bricks are used in constructing a wall 15 ft long, how many bricks are needed for a similar wall 35 ft long?

**61.** The owner of a building assessed at $45,000 is billed $638.40 for property taxes. How much should the owner of a similar building next door, assessed at $78,000, be billed for property taxes?

**62.** How much wettable powder do you put in a 300-gal spray tank if 20 gal of water and 3 lb of chemical pesticide are applied per acre?

**63.** A farmer uses 150 lb of insecticide on a 40-acre field. How many pounds are needed for a 180-acre field at the same rate of application?

**64.** If 100 bu of corn per acre remove 90 lb of nitrogen, phosphorus, and potassium (N, P, and K), how many pounds of N, P, and K, are removed by a yield of 135 bu per acre?

**65.** If a farmer has a total yield of 38,400 bu of corn from a 320-acre farm, what total yield should be expected from a similar 520-acre farm?

**66.** A copper wire 750 ft long has a resistance of 1.563 Ω. How long is a copper wire of the same area whose resistance is 2.605 Ω? *Note:* The resistance of these wires is proportional to their length.

**67.** If the voltage drop across a 28-Ω resistor is 52 V, what is the voltage drop across a 63-Ω resistor that is in series with the first one? *Note:* Resistors in series have voltage drops proportional to their resistances.

**68.** If the ratio of secondary turns to primary turns in a given transformer is 35 to 4, how many secondary turns are there if the primary coil has 136 turns?

69. An 8-V automotive coil has 250 turns of wire in the primary circuit. If the secondary voltage is 15,000 V, how many secondary turns are in the coil? *Note:* The ratio of secondary voltage to primary voltage equals the ratio of secondary turns to primary turns.

70. The clutch linkage on a vehicle has an overall advantage of 24 : 1. If the pressure plate applies a force of 504 lb, how much force must the driver apply to release the clutch?

71. The clutch linkage on a vehicle has an overall advantage of 30 : 1. If the driver applies a force of 15 N to the clutch, how much force is applied to the pressure plate?

72. A certain concrete dry mix is composed of 1 part cement, $2\frac{1}{2}$ parts sand, and 4 parts gravel by volume. Assume that 60 ft$^3$ of gravel are used.
    (a) How much cement is needed?
    (b) How much sand is needed?

73. A triangle whose sides are in the ratio of 3 : 4 : 5 is a right triangle, as shown in Fig. 2.4. If the length of the hypotenuse is 85 m, what are the lengths of the legs?

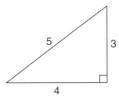

**Figure 2.4**

74. Assume that 75 m$^3$ of the concrete mix in Exercise 72 are needed for a job.
    (a) How much cement is needed?
    (b) How much sand is needed?
    (c) How much gravel is needed?

75. You have 360 ft of fence. What size lot in the shape of a 3 : 4 : 5 right triangle can be fenced using all the fence?

# 2.10 VARIATION

Often scientific laws and technical principles are given in terms of **variation.** To help you understand and use this idea and terminology effectively in your other courses, we now develop the basic concepts of variation.

> **DIRECT VARIATION**
>
> If two quantities, $y$ and $x$, change and their ratio remains constant $\left(\dfrac{y}{x} = k\right)$, the quantities **vary directly,** or $y$ is **directly proportional** to $x$. In general, this relationship is written in the form $y = kx$, where $k$ is the **proportionality constant.**

**EXAMPLE 1**

Consider the following data.

| y | 4 | 10 | 6 | 20 | 240 |
|---|---|----|---|----|-----|
| x | 6 | 15 | 9 | 30 | 360 |

Note that $y$ varies directly with $x$ because the ratio $\frac{y}{x}$ is always $\frac{2}{3}$. Also note that $y = \frac{2}{3}x$, where $k = \frac{2}{3}$.

In general, if $y$ varies directly as $x$ (or $y = kx$) and $x$ *increases,* then $y$ also *increases.* Similarly, if $x$ *decreases,* then $y$ also *decreases.*

### EXAMPLE 2

The circumference, $C$, of a circle varies directly with the radius, $r$. This relation is written $C = kr$ where $k = 2\pi$, the proportionality constant.

### EXAMPLE 3

The weight, $w$, of a body varies directly with its mass, $m$. This relation is written $w = km$ where $k = g$, the acceleration of gravity which is the proportionality constant.

### EXAMPLE 4

In a given wire the electrical resistance, $R$, varies directly with the wire's length, $L$. This relation is written $R = kL$, where $k$ is the proportionality constant.

### EXAMPLE 5

Charles's law states that if the pressure on a gas is constant, its volume, $V$, varies directly with its absolute temperature, $T$. This relation is written $V = kT$, where $k$ is the proportionality constant.

---

**INVERSE VARIATION**

If two quantities, $y$ and $x$, change and their product remains constant ($yx = k$), the quantities **vary inversely,** or $y$ is **inversely proportional** to $x$. In general, this relation is written $y = \frac{k}{x}$, where $k$ is again called the proportionality constant.

---

### EXAMPLE 6

Observe the following data:

| y | 6 | 2 | 12 | 3 | 96 |
|---|---|---|----|---|----|
| x | 8 | 24 | 4 | 16 | $\frac{1}{2}$ |

Note that $y$ varies inversely with $x$ because the product is always 48. Also note that $y = \frac{48}{x}$, where $k = 48$.

In general, if $y$ varies inversely as $x$ (or $y = k/x$) and $x$ *increases,* then $y$ *decreases.* Similarly, if $x$ *decreases,* then $y$ *increases.*

### EXAMPLE 7

Boyle's law states that if the temperature of a gas is constant, its volume, $V$, varies inversely with its pressure, $p$. This relation is written $V = \dfrac{k}{p}$, where $k$ is the proportionality constant.

### EXAMPLE 8

The rate, $r$, at which an automobile covers a distance of 200 mi varies inversely with the time, $t$. This relation is written $r = \dfrac{k}{t}$ or $r = \dfrac{200}{t}$, where $k = 200$ is the proportionality constant.

For many relationships, one quantity varies directly or inversely with a power of the other.

### EXAMPLE 9

The area, $A$, of a circle varies directly with the square of its radius, $r$. This relation is written $A = kr^2$, where $k = \pi$, which is the proportionality constant.

### EXAMPLE 10

In wire of a given length, the electrical resistance, $R$, varies inversely with the square of its diameter, $D$. This relation is written $R = \dfrac{k}{D^2}$, where $k$ is the proportionality constant.

---

**JOINT VARIATION**

One quantity **varies jointly** with two or more quantities when it varies directly with the product of these quantities. In general, this relation is written $y = kxz$, where $k$ is the proportionality constant.

---

### EXAMPLE 11

Coulomb's law for magnetism states that the force, $F$, between two magnetic poles varies jointly with the strengths, $s_1$ and $s_2$, of the poles and inversely with the square of their distance apart, $d$. This relation is written $F = \dfrac{ks_1s_2}{d^2}$, where $k$ is the proportionality constant.

Once we are able to express a given variation sentence as an equation, the next step is to find the value of $k$, the proportionality constant. This value of $k$ can then be used for all the different sets of data in a given problem or situation.

### EXAMPLE 12

Suppose that $y$ varies directly with the square of $x$ and that $x = 2$ when $y = 16$. Find $y$ when $x = 3$ and when $x = \frac{1}{2}$.

First write the variation equation:

$$y = kx^2$$

Then, to find $k$, substitute the set of data which includes a value for each variable.

$$16 = k2^2$$
$$16 = 4k$$
$$4 = k$$

Therefore,

$$y = 4x^2$$

For $x = 3$,

$$y = 4(3)^2 = 36$$

For $x = \dfrac{1}{2}$,

$$y = 4\left(\dfrac{1}{2}\right)^2 = 1$$

**EXAMPLE 13**

At a given temperature the electrical resistance, $R$, of a wire varies directly with its length, $L$, and inversely with the square of its diameter, $D$. The resistance of 20.0 m of copper wire of diameter 0.81 mm is 0.67 $\Omega$. What is the resistance of 40.0 m of copper wire 1.20 mm in diameter?

First write the variation equation:

$$R = \dfrac{kL}{D^2}$$

Then find $k$.

$$0.67\ \Omega = \dfrac{k(20.0\ \text{m})}{(0.81\ \text{mm})^2}$$

$$k = 0.022\ \dfrac{\Omega\ \text{mm}^2}{\text{m}}$$

Therefore,

$$R = \dfrac{0.022L}{D^2}$$

For $L = 40.0$ m and $D = 1.20$ mm,

$$R = \dfrac{\left(0.022\ \dfrac{\Omega\ \text{mm}^2}{\text{m}}\right)(40.0\ \text{m})}{(1.20\ \text{mm})^2} = 0.61\ \Omega$$

## Exercises 2.10

*For each set of data, determine whether y varies directly with x or inversely with x. Also find k.*

**1.**

| y | 6 | 15 | 54 | 1.5 |
|---|---|----|----|-----|
| x | 8 | 20 | 72 | 2 |

**2.**

| y | 4 | 2 | $\dfrac{1}{2}$ | 5 |
|---|---|---|-----|---|
| x | 5 | 10 | 40 | 4 |

**3.**

| y | $\dfrac{3}{4}$ | $\dfrac{1}{4}$ | $\dfrac{13}{20}$ | $\dfrac{9}{16}$ |
|---|---|---|---|---|
| x | 2 | 6 | $\dfrac{60}{26}$ | $\dfrac{8}{3}$ |

**4.**

| y | 14 | 21 | $\dfrac{7}{3}$ | $\dfrac{14}{5}$ |
|---|---|---|---|---|
| x | 16 | 28 | 4 | $\dfrac{16}{5}$ |

**5.**

| y | 12 | 18 | 0.75 | 1.5 |
|---|---|---|---|---|
| x | 20 | 30 | 2.5 | 5 |

**6.**

| y | 2 | $\dfrac{1}{3}$ | 18 | $\dfrac{2}{11}$ |
|---|---|---|---|---|
| x | 7 | $\dfrac{7}{6}$ | 63 | $\dfrac{7}{11}$ |

*Write a variation equation for each.*

**7.** $y$ varies directly with $z$.

**8.** $p$ varies inversely with $q$.

**9.** $a$ varies jointly with $b$ and $c$.

**10.** $m$ varies directly with the square of $n$.

**11.** $r$ varies directly with $s$ and inversely with the square root of $t$.

**12.** $d$ varies jointly with $e$ and the cube of $f$.

**13.** $f$ varies jointly with $g$ and $h$ and inversely with the square of $j$.

**14.** $m$ varies directly with the square root of $n$ and inversely with the cube of $p$.

*First find k; then find the given quantity for Exercises 15–28.*

**15.** $y$ varies directly with $x$; $y = 8$ when $x = 24$. Find $y$ when $x = 36$.

**16.** $m$ varies directly with $n$; $m = 198$ when $n = 22$. Find $m$ when $n = 35$.

**17.** $y$ varies inversely with $x$; $y = 9$ when $x = 6$. Find $y$ when $x = 18$.

**18.** $d$ varies inversely with $e$; $d = \dfrac{4}{5}$ when $e = \dfrac{9}{16}$. Find $d$ when $e = \dfrac{5}{3}$.

**19.** $y$ varies directly with the square root of $x$; $y = 24$ when $x = 16$. Find $y$ when $x = 36$.

**20.** $y$ varies jointly with $x$ and the square of $z$; $y = 150$ when $x = 3$ and $z = 5$. Find $y$ when $x = 12$ and $z = 8$.

**21.** $p$ varies directly with $q$ and inversely with the square of $r$; $p = 40$ when $q = 20$ and $r = 4$. Find $p$ when $q = 24$ and $r = 6$.

**22.** $m$ varies inversely with $n$ and the square root of $p$; $m = 18$ when $n = 2$ and $p = 36$. Find $m$ when $n = 9$ and $p = 64$.

**23.** Use Charles's law in Example 5 to find the volume of oxygen when the absolute temperature is $30\overline{0}$ K if the volume is $150\overline{0}$ cm$^3$ when the temperature is $25\overline{0}$ K.

**24.** Use Boyle's law in Example 7 to find the volume of nitrogen when the pressure is 25.0 lb/in$^2$ if the volume is $350\overline{0}$ ft$^3$ when the pressure is 15.0 lb/in$^2$.

**25.** Charles's law and Boyle's law combined state that the volume, $V$, of a gas varies directly with its absolute temperature, $T$, and inversely with its pressure, $p$. If the volume of acetylene is $15{,}00\overline{0}$ ft$^3$ when the absolute temperature is $40\overline{0}$°R and its pressure is 20.0 lb/in$^2$, find its volume when its temperature is 575°R and its pressure is 50.0 lb/in$^2$.

**26.** The electric power, $P$, generated by current varies jointly with the voltage drop, $V$, and the current, $I$. If 6.00 A of current in a $22\overline{0}$-V circuit generate 1320 W of power, how much power do 8.00 A of current in a 115-V circuit generate?

27. The electric power, $P$, used in a circuit varies directly with the square of the voltage drop, $V$, and inversely with the resistance, $R$. If 180 W of power in a circuit are used by a $9\overline{0}$-V drop with a resistance of 45 Ω, find the power used when the voltage drop is 120 V and the resistance is $3\overline{0}$ Ω.

28. The power, $P$, of an engine varies directly with the square of the radius, $r$, of its piston. If a $4\overline{0}$-hp engine has a piston of radius 2.0 in., what power does an engine with a piston of radius 4.0 in. have?

29. For two rotating gears as in Fig. 2.5, the number of teeth, $t$, for either gear is inversely proportional to the number of revolutions, $n$, that the gear makes per unit of time. A gear with 50 teeth rotating at $40\overline{0}$ rpm (revolutions per minute) turns a second gear at 125 rpm. How many teeth does the second gear have?

30. A gear with 75 teeth rotates at 32 rpm and turns a second gear with 25 teeth. How fast does the second gear rotate?

31. For two rotating pulleys connected with a belt as in Fig. 2.6, the diameter, $d$, of either pulley varies inversely with the number of revolutions, $n$, that the pulley makes per unit of time. A small pulley, 25 cm in diameter and rotating at 72 rpm, rotates a large pulley that is 75 cm in diameter. What is the speed of the large pulley?

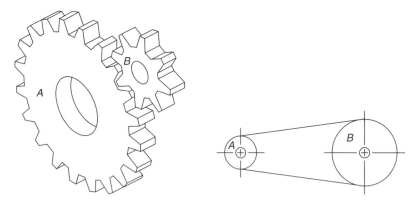

Figure 2.5                                    Figure 2.6

32. One pulley is 22 cm and revolves at 680 rpm. A second pulley revolves at 440 rpm. What is its diameter?

33. One pulley is 15 in. in diameter and revolves at 150 rpm. A second pulley is 45 in. in diameter.

    (a) What is the speed of the second pulley?

    (b) If the speed of the first pulley doubles, what is the speed of the second pulley?

34. The number of workers needed to complete a particular job is inversely proportional to the number of hours that they work. If 12 electricians can complete a job in 72 h, how long will it take 8 electricians to complete the same job? Of course, we are assuming that each person works at the same rate no matter how many persons are assigned to the job.

## CHAPTER SUMMARY

1. An **algebraic expression** is a combination of finite sums, differences, products, quotients, roots, and powers of numbers and of letters representing numbers.

**2.** A **polynomial in one variable** is defined as follows:

$$a_n x^n + a_{n-1}x^{n-1} + a_{n-2}x^{n-2} + \cdots + a_2x^2 + a_1x + a_0$$

where $n$ is a positive integer and the coefficients $a_n, a_{n-1}, a_{n-2}, \ldots, a_2, a_1, a_0$ are real numbers.

**3.** *Basic terms:*
   (a) *Monomial*: algebraic expression with one term.
   (b) *Binomial*: algebraic expression with two terms.
   (c) *Trinomial*: algebraic expression with three terms.
   (d) *Multinomial*: algebraic expression with two or more terms.

**4.** *Degree:*
   (a) The degree of a monomial in one variable is the same as the exponent of the variable.
   (b) The degree of a monomial in more than one variable equals the sum of the exponents of its variables.
   (c) The degree of a polynomial is the same as the highest-degree monomial in the polynomial.
   (d) A polynomial is in decreasing order if each term is of some degree less than the preceding term.
   (e) A polynomial is in increasing order if each term is of some degree larger than the preceding term.

**5.** *Removing grouping symbols:*
   (a) To remove grouping symbols preceded by a plus sign, remove the grouping symbols and leave the sign of each term unchanged within the grouping symbols.
   (b) To remove grouping symbols preceded by a minus sign, remove the grouping symbols and change the sign of each term within the grouping symbols.
   (c) When sets of grouping symbols are contained within each other, remove each set starting with the innermost set and finishing with the outermost set.

**6.** *Order of operations:*
   (a) Do all operations within any grouping symbols beginning with the innermost pair if grouping symbols are contained within each other.
   (b) Simplify all powers.
   (c) Do all multiplications and divisions in the order in which they occur from left to right.
   (d) Do all additions and subtractions in the order in which they occur from left to right.

**7.** To evaluate an algebraic expression, substitute the given numerical values in place of the letters, and follow the order of operations.

**8.** *Laws of exponents:* In each of the following, $a$ and $b$ are real numbers and $m$ and $n$ are positive integers.
   (a) $a^m \cdot a^n = a^{m+n}$
   (b) (i) $\dfrac{a^m}{a^n} = a^{m-n}$   $m > n$   and   $a \neq 0$
   (ii) $\dfrac{a^m}{a^n} = \dfrac{1}{a^{n-m}}$   $m < n$   and   $a \neq 0$
   (c) $(a^m)^n = a^{mn}$

(d) $(ab)^n = a^n b^n$

(e) $\left(\dfrac{a}{b}\right)^n = \dfrac{a^n}{b^n}$   $b \neq 0$

(f) $a^0 = 1$   $a \neq 0$

9. *Roots:*
   (a) The square root of a number, $\sqrt{a}$, is that nonnegative number which, when multiplied by itself, is $a$.
   (b) The cube root of a number, $\sqrt[3]{a}$, is that number which, when multiplied by itself three times, is $a$.
   (c) The $n$th root of a number $a$ is written $\sqrt[n]{a}$, where $n$ is the index, $a$ is the radicand, and $\sqrt{\phantom{x}}$ is called the radical sign.
   (d) The square root property states that $\sqrt{ab} = \sqrt{a}\,\sqrt{b}$, where $a > 0$ and $b > 0$.
   (e) The square root of a number is simplified when the number under the radical contains no perfect square factors.

10. *Multiplication of algebraic expressions:*
   (a) To multiply monomials, multiply their numerical coefficients and multiply each set of like letters using the laws of exponents.
   (b) To multiply a multinomial by a monomial, multiply each term of the multinomial by the monomial.
   (c) To multiply two multinomials, multiply each term of the first multinomial by each term of the second and simplify.
   (d) Use the FOIL method to multiply two binomials mentally.

11. *Division of algebraic expressions:*
   (a) To divide monomials, divide their numerical coefficients and divide each set of like letters using the laws of exponents.
   (b) To divide a multinomial by a monomial, divide each term of the multinomial by the monomial.
   (c) To divide one multinomial by a second multinominal:

   Arrange each multinomial in descending powers of one of the variables.

   To find the first term of the quotient, divide the first term of the dividend by the first term of the divisor.

   Multiply the divisor by the first term of the quotient and subtract the product from the dividend.

   Continue this procedure of dividing the first term of each remainder by the first term of the divisor until the final remainder is zero or until the remainder is of degree less than the divisor.

12. *Equations:*
   (a) An equation is a statement that two algebraic expressions are equal.
   (b) A linear equation in one variable is an equation in which each term containing that variable is of first degree.
   (c) *Four basic properties used to solve equations:*

   If the same quantity is added to each side of an equation, the resulting equation is equivalent to the original equation.

   If the same quantity is subtracted from each side of an equation, the resulting equation is equivalent to the original equation.

If each side of an equation is multiplied by the same (nonzero) quantity, the resulting equation is equivalent to the original equation.

If each side of an equation is divided by the same (nonzero) quantity, the resulting equation is equivalent to the original equation.

(d) To solve a first-degree equation in one variable:

Eliminate any fractions by multiplying each side by the lowest common denominator of all fractions in the equation.

Remove any grouping symbols.

Combine like terms on each side of the equation.

Isolate all the unknown terms on one side of the equation and all other terms on the other side of the equation.

Again combine like terms where possible.

Divide each side by the coefficient of the unknown.

Check your solution by substituting it in the original equation.

13. A **formula** is an equation, usually expressed in letters, that shows the relationship between quantities.

14. To **solve a formula for a given letter** means to isolate the given letter on one side of the equation and express it in terms of all the remaining letters by having all the other letters and numbers appear on the opposite side of the equation. We solve a formula using the same principles that we use in solving any equation.

15. *To use a formula to solve a problem where all but the unknown quantity is given:*

Solve the formula for the unknown quantity.

Substitute each known quantity with its units.

Use the order of operations procedures to find the numerical quantity and to simplify the units.

16. *To solve application problems:*

**Step 1:** Read the problem carefully at least two times.

(a) The first time you should read straight through from beginning to end. Do not stop this time to think about setting up an equation. You are only after a general impression at this point.

(b) Now read through *completely* for the second time. This time begin to think ahead to the next steps.

**Step 2:** If possible, draw a picture or diagram. This will often help you to visualize the possible mathematical relationships needed in order to write the equation.

**Step 3:** Choose a symbol to represent the unknown quantity (the quantity not given as a specific value) in the problem. Be sure to label this symbol, the unknown variable, to indicate what it represents.

**Step 4:** Write an equation that you think expresses the information given in the problem. To obtain this equation, you will be looking for information that will express two quantities that can be set equal to each other. Do not rush through this step. Try expressing your equation in words to see if it corresponds to the original problem.

**Step 5:** Solve for the unknown variable.

**Step 6:** Check your solution in the equation from Step 4. If it does not check, you have made a mathematical error. You will need to repeat Step 5. Also check your solution in the original verbal problem for reasonableness.

**17.** A **ratio** is the quotient of two numbers or quantities.

**18.** A **proportion** is a statement that two ratios are equal.

**19.** In any proportion, the product of the means equals the product of the extremes.

**20.** *Variation:*

(a) Direct: If two quantities, $y$ and $x$, change and their ratio remains constant $\left(\dfrac{y}{x} = k\right)$, the quantities **vary directly,** or $y$ is **directly proportional** to $x$. In general, this relation is written in the form $y = kx$, where $k$ is the **proportionality constant.**

(b) Inverse: If two quantities, $y$ and $x$, change and their product remains constant $(yx = k)$, the quantities **vary inversely,** or $y$ is **inversely proportional** to $x$. In general, this relation is written $y = \dfrac{k}{x}$, where $k$ is again called the proportionality constant.

(c) Joint: One quantity **varies jointly** with two or more quantities when it varies directly with the product of these quantities.

# CHAPTER 2 REVIEW

**1.** Classify $4x^3 - 6x^2 + 5x$ as a monomial, binomial, or trinomial, and give its degree.

*Perform the indicated operations and simplify.*

**2.** $(3x^2 + 5x - 7) + (13x^2 - x - 14)$

**3.** $(-5x^2 - 2x - 3) + (12x^2 + x - 8)$

**4.** $(-4x^2 - 3x + 1) - (-5x^2 + 2x - 1)$

**5.** $(7x^2 + x - 5) - (2x^2 - 3x - 5)$

**6.** $-[(2a - 3b) - (-4a + b) + (-a - 4b)]$

**7.** $-[-(2a - b) - (a - b) + (6a + 5b)]$

**8.** Evaluate $\dfrac{3a^2b - 6b}{2b + 1}$ for $a = -2$ and $b = -3$.

*Perform the indicated operations and simplify.*

**9.** $y^5 \cdot y^4$

**10.** $\dfrac{b^{12}}{b^3}$

**11.** $(3x^2)(-5x^4)$

**12.** $\dfrac{18m^6}{3m^2}$

**13.** $(a^3)^4$

**14.** $(5a^3)^2$

**15.** $\left(\dfrac{y}{x^2}\right)^3$

**16.** $(5x^2)^0$

**17.** $(-s^3)^3$

**18.** $(x^5 \cdot x^4)^2$

**19.** $(2a^3b^2)^3$

**20.** $\left(\dfrac{-3x^6}{6x^2}\right)^2$

**21.** $\left(\dfrac{-4x^8}{2x^2}\right)^3$

*Simplify each root.*

**22.** $\sqrt{49}$      **23.** $\sqrt[3]{27}$      **24.** $\sqrt{63}$      **25.** $\sqrt{108}$

*Find each square root rounded to three significant digits.*

**26.** $\sqrt{3147}$          **27.** $\sqrt{0.0205}$

*Find each product.*

**28.** $(-3a^2bc^3)(8a^3bc)$      **29.** $5x(2x - 6y)$

**30.** $-3x^2(4x^3 - 3x^2 - x + 4)$      **31.** $3a^2b(-5a^2b^3 + 3a^3 - 5b^2)$

**32.** $(2x + 7)(3x - 4)$      **33.** $(5x - 3)(6x - 9)$

**34.** $(4x + 6)(5x + 2)$      **35.** $(2x - 3)(3x + 2)$

**36.** $(8x - 5)^2$      **37.** $(x + y - 7)(2x - y + 3)$

*Find each quotient.*

**38.** $\dfrac{81x^2y^3z}{27x^2yz^2}$      **39.** $\dfrac{(4a)^2(5a^2b)^2}{50a^3b^2}$

**40.** $\dfrac{48x^2 - 24x + 15}{3x}$      **41.** $\dfrac{15x^3 - 25x + 35}{5x^2}$

**42.** $\dfrac{18m^2n^5 + 24mn - 8m^6n^2}{6m^2n^3}$      **43.** $\dfrac{6x^2 - x - 12}{2x - 3}$

**44.** $\dfrac{3x^3 + 5x - 3}{x + 1}$      **45.** $\dfrac{2x^3 - 5x^2 + 8x + 7}{2x - 1}$

*Solve each equation.*

**46.** $6x + 7 = -17$      **47.** $8 - 3x = 12 - 2x$

**48.** $3(x + 8) = 2(x - 3)$      **49.** $3(2x - 1) - 2(3x - 1) = 3x + 1$

**50.** $\dfrac{1}{4}x + 8 = \dfrac{3}{4}x - \dfrac{7}{2}$      **51.** $\dfrac{2}{3}x - \dfrac{4}{5} = \dfrac{1}{5}x + \dfrac{4}{3}$

**52.** $\dfrac{5(x - 2)}{6} - \dfrac{3x}{2} = \dfrac{5}{3}x$      **53.** $\dfrac{3}{4}(2x - 3) - \dfrac{2}{3}(6x - 2) = x$

**54.** Solve $24.6x - 45.2 = 0.4(39.2x + 82.5)$ and round the result to three significant digits.

*Solve each formula for the indicated letter.*

**55.** $S = \dfrac{\pi D}{12}$    for $D$      **56.** $C = \dfrac{Q}{V}$    for $V$

**57.** $v = v_0 - gt$    for $t$      **58.** $(\Delta V) = \beta V(T - T_0)$    for $T_0$

**59.** Given $v = v_0 + at$, $v = 18\overline{0}$ m/s, $a = 9.80$ m/s$^2$, and $t = 10.0$ s. Find $v_0$.

**60.** Given $Z = \sqrt{R^2 + X_L^2}$, $R = 40.0$ Ω, and $X_L = 37.7$ Ω. Find Z.

**61.** Given $\dfrac{1}{R} = \dfrac{1}{R_1} + \dfrac{1}{R_2} + \dfrac{1}{R_3}$, $R = 4.5$ Ω, $R_1 = 12.5$ Ω, and $R_2 = 8.5$ Ω. Find $R_3$ rounded to three significant digits.

**62.** A mechanic needs to insert shims for a total thickness of 0.090 in. If there is already an 0.018-in. shim in place, how many 0.004-in. shims must be inserted?

**63.** A rectangle is 5 m longer than it is wide. Its perimeter is 90 m. Find its dimensions.

**64.** A bar of metal contains 15% silver. Another bar contains 20% silver. How many ounces of each must be taken to make a 30-oz bar containing 18% silver?

**65.** A car is traveling at 55 mi/h. A policeman is 10 mi behind traveling at 75 mi/h and is dispatched to intercept this car. How long will it take the policeman to overtake the car?

*Express each ratio in simplest form.*

**66.** $\dfrac{360 \text{ ft}}{441 \text{ ft}}$

**67.** $\dfrac{600 \text{ mm}}{1.5 \text{ m}}$

**68.** $\dfrac{(2.8 \text{ in.})^2}{(1.2 \text{ in.})^2}$

*Solve each proportion for x.*

**69.** $\dfrac{13}{24} = \dfrac{x}{96}$

**70.** $\dfrac{39}{12} = \dfrac{3x}{48}$

**71.** $\dfrac{x}{60 - x} = \dfrac{2}{3}$

**72.** $\dfrac{a}{6 + x} = \dfrac{b}{a}$

**73.** $\dfrac{x}{c} = \dfrac{b}{a}$

**74.** $\dfrac{270}{15} = \dfrac{x}{3.15}$

**75.** If 8000 people use 40,000 gal of water per day, how much water will 20,000 people use?

**76.** Separate 2370 parts into two groups in the ratio of 3 : 7. How many parts are in each group?

*In Exercises 77–80, write the variation equation.*

**77.** *y* varies directly with the square root of *z*.

**78.** *y* varies jointly with *v* and the square of *u*.

**79.** *y* varies directly with *p* and inversely with *q*.

**80.** *y* varies jointly with *m* and *n* and inversely with the square of *p*.

**81.** *y* varies jointly with *x* and *z*; *y* = 576 when *x* = 24 and *z* = 8. Find *y* when *x* = 36 and *z* = 48.

**82.** *y* varies directly with the square of *p* and inversely with the square root of *q*; *y* = 6 when *p* = 4 and *q* = 16. Find *y* when *p* = 8 and *q* = 4.

**83.** Hooke's law states that within the limits of perfect elasticity, the force, *F*, needed to stretch a spring varies directly with the distance, *d*, the spring is stretched. If 40.0 N stretch a spring 1.5 m, what force is needed to stretch the spring 6.0 m?

**84.** The centrifugal force, *F*, of a body moving in a circular path varies jointly with its weight, *W*, and the square of its velocity, *v*. If 19.2 lb of centrifugal force are produced by a 20.0-lb weight moving at 16.0 ft/s, what force is produced by a 50.0-lb weight moving at 10.0 ft/s?

**85.** Coulomb's law for electric charges states that the force, *F*, between two point charges varies jointly with the charges, $q_1$ and $q_2$, and inversely with the square of the distance, *d*, between them. Two charges of +20 $\mu$C (microcoulombs) and +12 $\mu$C 6.0 cm apart produce a repulsive force of 600 N. What attractive force is produced by two charges of −36 $\mu$C and +12 $\mu$C 12 cm apart?

# 3

# Right-Triangle
# Trigonometry

## INTRODUCTION

Two roads on opposite sides of a river are to be connected by a bridge. Because of rock formations, the roads are not directly opposite each other, and the least expensive way to connect them is to build the bridge at an angle of 75.0° with the near side of the river. If the width of the river is 89.1 m, what must be the length of the bridge?

To solve this problem we need to use a trigonometric function. In this chapter we learn how to find a side or an angle of a right triangle given other pieces of information. We use the given information and our knowledge of trigonometric functions to solve problems that are not solvable using only algebra. (The preceding problem is number 7 in Exercises 3.4.)

### Objectives

- Understand the degree/minute/second and radian measures of an angle.
- Know the Pythagorean theorem.
- Know the ratio definitions of the trigonometric functions.
- Know the values of the trigonometric functions for key angles.
- Use a calculator to evaluate trigonometric functions.
- Solve right triangles.

## 3.1   THE TRIGONOMETRIC RATIOS

Trigonometry is concerned with the measurement of the parts of a triangle—sides and angles.* It is a basic tool used in the development of mathematics and various fields of tech-

---

*You may find it helpful to refer to Appendix A, Review of Geometry, to solve some of the exercises and to help you to follow some of the mathematical developments. Note that Section A.1 relates to angles.

nology. Hipparchus and Ptolemy created this branch of mathematics in the second century B.C. Earliest applications were in astronomy, surveying, and navigation. Trigonometry continues to play a crucial role in today's problems of engineering, science, and technology. It is used, for example, to explain wave phenomena in physics, such as with sound and electricity. As we shall soon see, trigonometry is based on working with certain ratios.

The generation or formation of an **angle** is the result of rotating a line segment about a fixed point from one position to another. To be more specific, consider a fixed point $O$ and the line segment $OP$ (see Fig. 3.1). Before the line segment is rotated about point $O$, its initial (beginning) position, $OP$, is called the **initial side** of the angle $\theta$ (theta) as shown. The final position of the line segment, $OQ$, is called the **terminal side** of the angle. The fixed point $O$ is called the **vertex** of the angle.

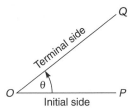

**Figure 3.1** Angle $\theta$

A question now arises: How much is the rotation; that is, how large is the angle? Angles can be measured using any of three basic units of measure: revolutions, degrees, and radians.

One revolution (rev) [see Fig. 3.2(a)] corresponds to one complete rotation of the initial side. When an object, such as a gear, makes several rotations, its angle displacement is usually measured in revolutions. This is a common unit in industry.

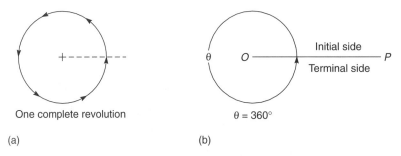

One complete revolution

(a)

$\theta = 360°$

(b)

**Figure 3.2** One complete revolution = 360°

A **degree** is $\frac{1}{360}$ of one complete rotation of a line about a point. That is, if we rotate the line $OP$ until it reaches its original position, we will have rotated an angle of 360 degrees, as shown in Fig. 3.2(b).

$$1 \text{ complete rotation} = 360 \text{ degrees}$$
$$1 \text{ degree} = \frac{1}{360} \text{ of a complete rotation}$$

Thus, we measure angles by measuring how many degrees the terminal side of the angle has been rotated from the initial side. The symbol ° is used to denote degrees.

**EXAMPLE 1**

Construct an angle of $\frac{1}{6}$ rotation.

An angle generated by $\frac{1}{6}$ of a rotation has a measure of $\frac{1}{6} \times 360° = 60°$ as shown in Fig. 3.3.

The protractor (see Fig. 3.4) is an instrument marked in degrees that is commonly used to measure angles.

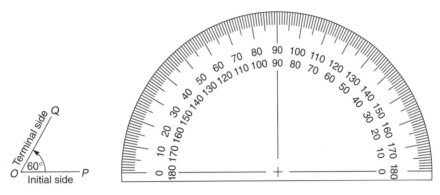

**Figure 3.3**

**Figure 3.4** Protractor

An **acute angle** is an angle whose measure is less than 90° [see Fig. 3.5(a)]. An **obtuse angle** is an angle whose measure is more than 90° but less than 180° [see Fig. 3.5(b)].

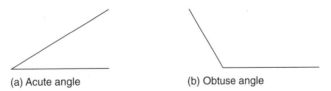

(a) Acute angle

(b) Obtuse angle

**Figure 3.5**

Just as the linear unit yards can be broken down into smaller units of feet and inches, degrees can be broken down into smaller units. These are the minute and the second, and they are not to be confused with time measurement. A **minute** in trigonometry is $\frac{1}{60}$ of a degree. A degree is divided into 60 equal parts called minutes. The symbol ′ is used to denote minutes.

$$1' = \frac{1}{60}°$$

$$1° = 60'$$

A **second** is defined to be $\frac{1}{60}$ of a minute. The symbol $''$ is used to denote seconds.

$$1'' = \frac{1}{60}'$$

$$1' = 60''$$

**EXAMPLE 2**

Change $63°12'$ to degrees.

Since $12' = \frac{12}{60}°$,

$$63°12' = 63\frac{12}{60}°$$

$$= 63\frac{1}{5}°$$

$$= 63.2°$$

**EXAMPLE 3**

Change $147.3°$ to degrees and minutes.

$$147.3° = 147\frac{3}{10}°$$

$$= 147\frac{18}{60}°$$

$$= 147°18'$$

**EXAMPLE 4**

Change $78°15'45''$ to degrees.

First

$$15' = \frac{15}{60}° = 0.25°$$

and

$$45'' = \frac{45'}{60} = 0.75'$$

$$= \frac{0.75}{60}° = 0.0125°$$

Then

$$78°15'45'' = 78° + 0.25° + 0.0125°$$

$$= 78.2625°$$

Calculators are often used to do these conversions. Consult Appendix B or your calculator manual.

The radian unit is discussed in Chapter 10.

A right triangle has one right angle, two acute angles, a hypotenuse, and two legs. As shown in Fig. 3.6, the right angle is usually labeled with the capital letter $C$. The ver-

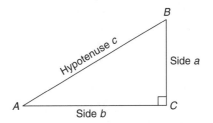

**Figure 3.6** Common labels of angles and sides of a right triangle.

tices of the two acute angles are labeled with the capital letters $A$ and $B$. The hypotenuse is usually labeled with the lowercase letter $c$. The legs are the sides opposite the acute angles. The leg (side) opposite angle $A$ is labeled $a$, and the leg opposite angle $B$ is labeled $b$. Note that each side of the triangle is labeled with the lowercase of the letter of the angle opposite it.

The **Pythagorean theorem** gives the relationship among the sides of a right triangle.

---

**PYTHAGOREAN THEOREM**

$$c^2 = a^2 + b^2 \quad \text{or} \quad c = \sqrt{a^2 + b^2}$$

---

In words, in a right triangle, the square of the length of the hypotenuse is equal to the sum of the squares of the lengths of the two legs.

**EXAMPLE 5**

Find the length of the hypotenuse in the right triangle in Fig. 3.7.
Using the Pythagorean theorem, we obtain

$$c = \sqrt{a^2 + b^2}$$
$$c = \sqrt{(115 \text{ m})^2 + (184 \text{ m})^2}$$
$$= 217 \text{ m} \qquad \text{(to three significant digits)}$$

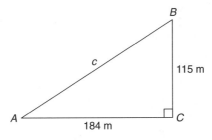

**Figure 3.7**

**EXAMPLE 6**

Find the length of side $a$ in the right triangle in Fig. 3.8.
Use the Pythagorean theorem,

$$c^2 = a^2 + b^2$$

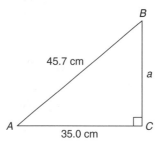

Figure 3.8

and solve for $a$.

$$a^2 = c^2 - b^2$$
$$a = \sqrt{c^2 - b^2}$$
$$a = \sqrt{(45.7 \text{ cm})^2 - (35.0 \text{ cm})^2}$$
$$= 29.4 \text{ cm} \quad \text{(to three significant digits)}$$

The following six trigonometric ratios express the relationships between an acute angle of a right triangle and the lengths of two of its sides (see Fig. 3.9):

The **sine** of angle $A$, abbreviated sin $A$, is equal to the ratio of the length of the side opposite angle $A$, $a$, to the length of the hypotenuse, $c$.

The **cosine** of angle $A$, abbreviated cos $A$, is the ratio of the length of the side adjacent to angle $A$, $b$, to the length of the hypotenuse, $c$.

The **tangent** of angle $A$, abbreviated tan $A$, is equal to the ratio of the length of the side opposite angle $A$, $a$, to the length of the side adjacent to angle $A$, $b$.

The **cotangent** of angle $A$, abbreviated cot $A$, is equal to the ratio of the length of the side adjacent to angle $A$, $b$, to the length of the side opposite angle $A$, $a$.

The **secant** of angle $A$, abbreviated sec $A$, is equal to the ratio of the length of the hypotenuse, $c$, to the length of the side adjacent to angle $A$, $b$.

The **cosecant** of angle $A$, abbreviated csc $A$, is equal to the ratio of the length of the hypotenuse, $c$, to the length of the side opposite angle $A$, $a$.

$$\sin A = \frac{\text{side opposite } A}{\text{hypotenuse}} = \frac{a}{c} \quad \text{or} \quad \sin B = \frac{b}{c}$$

$$\cos A = \frac{\text{side adjacent to } A}{\text{hypotenuse}} = \frac{b}{c} \quad \text{or} \quad \cos B = \frac{a}{c}$$

$$\tan A = \frac{\text{side opposite } A}{\text{side adjacent to } A} = \frac{a}{b} \quad \text{or} \quad \tan B = \frac{b}{a}$$

$$\cot A = \frac{\text{side adjacent to } A}{\text{side opposite } A} = \frac{b}{a} \quad \text{or} \quad \cot B = \frac{a}{b}$$

$$\sec A = \frac{\text{hypotenuse}}{\text{side adjacent to } A} = \frac{c}{b} \quad \text{or} \quad \sec B = \frac{c}{a}$$

$$\csc A = \frac{\text{hypotenuse}}{\text{side opposite } A} = \frac{c}{a} \quad \text{or} \quad \csc B = \frac{c}{b}$$

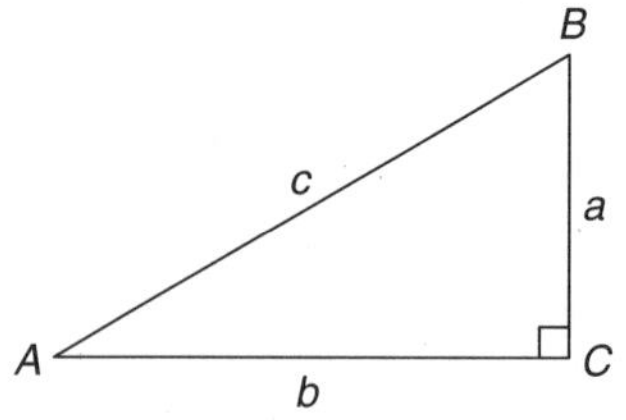

Figure 3.9

We shall discuss the trigonometric ratios of nonacute angles in Chapter 10.

**EXAMPLE 7**

Find the six trigonometric ratios of angle $A$ in the right triangle in Fig. 3.10.

Using the Pythagorean theorem, we have

$$c = \sqrt{(3.00 \text{ m})^2 + (4.00 \text{ m})^2}$$

$$= 5.00 \text{ m}$$

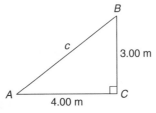

**Figure 3.10**

Use the definitions as follows:

$$\sin A = \frac{\text{side opposite } A}{\text{hypotenuse}} = \frac{3.00 \text{ m}}{5.00 \text{ m}} = 0.600$$

$$\cos A = \frac{\text{side adjacent to } A}{\text{hypotenuse}} = \frac{4.00 \text{ m}}{5.00 \text{ m}} = 0.800$$

$$\tan A = \frac{\text{side opposite } A}{\text{side adjacent to } A} = \frac{3.00 \text{ m}}{4.00 \text{ m}} = 0.750$$

$$\cot A = \frac{\text{side adjacent to } A}{\text{side opposite } A} = \frac{4.00 \text{ m}}{3.00 \text{ m}} = 1.33$$

$$\sec A = \frac{\text{hypotenuse}}{\text{side adjacent to } A} = \frac{5.00 \text{ m}}{4.00 \text{ m}} = 1.25$$

$$\csc A = \frac{\text{hypotenuse}}{\text{side opposite } A} = \frac{5.00 \text{ m}}{3.00 \text{ m}} = 1.67$$

Let us further examine this relationship between an acute angle of a right triangle and the lengths of two of its sides. Draw any two right triangles with a common angle $A$, as shown in Fig. 3.11. Note that $\triangle ABC$ and $\triangle ADE$ are similar. Two triangles are similar when two angles of one triangle have the same measure as two corresponding angles of the second. In Fig. 3.11, angle $A$ is a common angle to both $\triangle ABC$ and $\triangle ADE$; angle

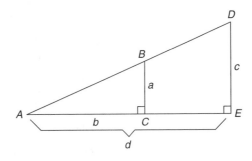

**Figure 3.11**

*ACB* and angle *AED* are both right angles and therefore are of equal measure. From geometry, we know that the corresponding sides of similar triangles are proportional. This means that $\frac{a}{b} = \frac{c}{d}$. Thus, no matter how points *B* and *D* are chosen, the numerical ratios $\frac{a}{b}$ and $\frac{c}{d}$, which are used to define tan *A*, are the same.

Similar arguments using Fig. 3.11 can be given to show that the other trigonometric ratios do not depend upon the choice of the points in determining the defining ratios. Angle *A* is the only determining factor.

**EXAMPLE 8**

Find the six trigonometric ratios of angle *A* (a) using △*ABC* and (b) using △*ADE* in Fig. 3.12.

(a) First find the length of hypotenuse *AB* in △*ABC*.

$$AB = \sqrt{3^2 + 4^2} = \sqrt{9 + 16} = \sqrt{25} = 5$$

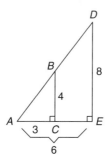

**Figure 3.12**

Then, using the definitions, we have

$$\sin A = \frac{\text{side opposite } A}{\text{hypotenuse}} = \frac{4}{5}$$

$$\cos A = \frac{\text{side adjacent to } A}{\text{hypotenuse}} = \frac{3}{5}$$

$$\tan A = \frac{\text{side opposite } A}{\text{side adjacent to } A} = \frac{4}{3}$$

$$\cot A = \frac{\text{side adjacent to } A}{\text{side opposite } A} = \frac{3}{4}$$

$$\sec A = \frac{\text{hypotenuse}}{\text{side adjacent to } A} = \frac{5}{3}$$

$$\csc A = \frac{\text{hypotenuse}}{\text{side opposite } A} = \frac{5}{4}$$

(b) Find the length of the hypotenuse *AD* in △*ADE*.

$$AD = \sqrt{6^2 + 8^2} = \sqrt{36 + 64} = \sqrt{100} = 10$$

Then

$$\sin A = \frac{\text{side opposite } A}{\text{hypotenuse}} = \frac{8}{10} = \frac{4}{5}$$

$$\cos A = \frac{\text{side adjacent to } A}{\text{hypotenuse}} = \frac{6}{10} = \frac{3}{5}$$

$$\tan A = \frac{\text{side opposite } A}{\text{side adjacent to } A} = \frac{8}{6} = \frac{4}{3}$$

$$\cot A = \frac{\text{side adjacent to } A}{\text{side opposite } A} = \frac{6}{8} = \frac{3}{4}$$

$$\sec A = \frac{\text{hypotenuse}}{\text{side adjacent to } A} = \frac{10}{6} = \frac{5}{3}$$

$$\csc A = \frac{\text{hypotenuse}}{\text{side opposite } A} = \frac{10}{8} = \frac{5}{4}$$

Note that any other right triangle with the same angle $A$ would give the same six ratios.

Look closely at each of the six ratios in Examples 7 and 8. Do you see each of the following relationships?

**1.** The values of $\sin A$ and $\csc A$ are reciprocals of each other.

**2.** The values of $\cos A$ and $\sec A$ are reciprocals of each other.

**3.** The values of $\tan A$ and $\cot A$ are reciprocals of each other.

Now, look closely at the six definitions. Note that this same reciprocal relationship exists there, too. The corresponding pairs of reciprocals are called **reciprocal trigonometric functions.** They are summarized in the following box.

$$\sin \theta = \frac{1}{\csc \theta} \qquad \csc \theta = \frac{1}{\sin \theta}$$

$$\cos \theta = \frac{1}{\sec \theta} \qquad \sec \theta = \frac{1}{\cos \theta}$$

$$\tan \theta = \frac{1}{\cot \theta} \qquad \cot \theta = \frac{1}{\tan \theta}$$

## Exercises 3.1

*Use a protractor to draw each angle.*

**1.** 35°  **2.** 126°  **3.** 240°  **4.** 333°

*Change each angle to degrees.*

**5.** 15′  **6.** 6′  **7.** 120′  **8.** 47′

*Change each angle to minutes.*

**9.** $\frac{1}{2}^\circ$  **10.** $\frac{2}{3}^\circ$  **11.** 0.4°  **12.** 0.7°

*Change each angle to degrees.*

**13.** 37°12′  **14.** 142°30′  **15.** 75°47′  **16.** 120°11′

*Change each angle to degrees and minutes.*

**17.** $69\frac{1}{3}^{\circ}$  **18.** 183.5°  **19.** 23.3°  **20.** $7\frac{2}{5}^{\circ}$

*Change each angle to degrees.*

**21.** 34°24′15″  **22.** 65°27′36″  **23.** 19°18′27″  **24.** 135°48′9″

*Change each angle to degrees, minutes, and seconds.*

**25.** 18.21°  **26.** 35.84°  **27.** 8.925°  **28.** 29.608°

*Use △ABC in Fig. 3.13 for Exercises 29–60.*

**29.** The side opposite angle $A$ is ____.  **30.** The side opposite angle $B$ is ____.

**31.** The hypotenuse is ____.  **32.** The side adjacent to angle $A$ is ____.

**33.** The side adjacent to angle $B$ is ____.  **34.** The angle opposite side $a$ is ____.

**35.** The angle opposite side $b$ is ____.  **36.** The angle opposite side $c$ is ____.

**37.** The angle adjacent to side $a$ is ____.  **38.** The angle adjacent to side $b$ is ____.

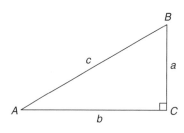

**Figure 3.13**

*Find the length of the third side of each right triangle, rounded to three significant digits.*

**39.** $a = 5.00$ cm, $b = 12.0$ cm  **40.** $a = 13.5$ ft, $b = 18.5$ ft

**41.** $a = 115$ mi, $c = 208$ mi  **42.** $a = 46.7$ km, $c = 75.6$ km

**43.** $b = 377$ yd, $c = 506$ yd  **44.** $b = 1450$ mi, $c = 2960$ mi

**45.** $a = 35.7$ m, $b = 16.8$ m  **46.** $a = 105$ m, $c = 537$ m

**47.** $a = 2.25$ cm, $c = 3.75$ cm  **48.** $b = 155$ mi, $c = 208$ mi

*Find the six trigonometric ratios of angle A rounded to three significant digits.*

**49.** $a = 5.00$ cm, $b = 12.0$ cm  **50.** $a = 13.5$ ft, $b = 18.5$ ft

**51.** $a = 335$ m, $c = 685$ m  **52.** $a = 19.8$ km, $c = 40.5$ km

**53.** $b = 3.00$ km, $c = 6.00$ km  **54.** $b = 239$ mi, $c = 307$ mi

*Find the six trigonometric ratios of angle B rounded to three significant digits.*

**55.** $a = 5.00$ cm, $b = 12.0$ cm  **56.** $a = 13.5$ ft, $b = 18.5$ ft

**57.** $a = 4.60$ m, $c = 9.25$ m  **58.** $a = 1.62$ km, $c = 4.05$ km

**59.** $b = 4.50$ ft, $c = 9.00$ ft  **60.** $b = 27.5$ in., $c = 51.2$ in.

**61.** Find the six trigonometric ratios of angle $A$ **(a)** using $\triangle ABC$ and **(b)** using $\triangle ADE$ in Fig. 3.14. Round each result to three significant digits.

**62.** Find the six trigonometric ratios of angle $B$ **(a)** using $\triangle BAC$ and **(b)** using $\triangle BFG$ in Fig. 3.15. Round each result to three significant digits.

**63.** Show that $\tan A = \dfrac{\sin A}{\cos A}$ by using the defining ratios. (*Hint:* The right-hand expression is a complex fraction that reduces to $\tan A$.)

**64.** Given that $\sin A = 0.8387$ and $\cos A = 0.5446$, compute the remaining four trigonometric functions of $A$. Round each result to four significant digits.

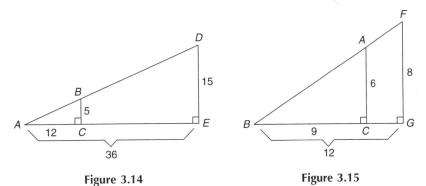

**Figure 3.14**          **Figure 3.15**

## 3.2   VALUES OF THE TRIGONOMETRIC RATIOS

D.1, D.5

Our work with trigonometry would be quite limited if we always had to rely on computing trigonometric ratios of an angle by the methods used thus far. In practice we are often given the measurement of the angle in degrees or radians and need the value of one of the trigonometric ratios of this angle. Until about 1980, the only practical, inexpensive source that provided the values of the trigonometric ratios was tables. But now with the universal use and acceptance of the inexpensive hand calculator, such tables, being cumbersome to use and read, are not necessary. In this text we shall present trigonometry using a calculator. Before learning to use a calculator we shall look at how these trigonometric values have been obtained for certain angles.

Let us first consider the angle of measure $60°$. For this purpose we construct an equilateral triangle as shown in Fig. 3.16. From geometry we know that the sum of the angles of any triangle equals $180°$. Each angle of triangle $OPQ$ must then be $\frac{180}{3}°$, or $60°$.

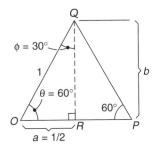

**Figure 3.16**   Equilateral triangle

We shall let each side be of length 1. By construction, $OR = a = \frac{1}{2}$ which is the side adjacent to angle $\theta$ in right triangle $ORQ$. The opposite side, $QR = b$, can be computed by using the Pythagorean theorem.

$$1^2 = a^2 + b^2$$

$$1 = \left(\frac{1}{2}\right)^2 + b^2$$

$$1 - \frac{1}{4} = b^2$$

$$b^2 = \frac{3}{4}$$

$$b = \sqrt{\frac{3}{4}} = \frac{\sqrt{3}}{2}$$

Using the trigonometric definitions and $\triangle ORQ$, we have

$$\sin 60° = \frac{\dfrac{\sqrt{3}}{2}}{1} = \frac{\sqrt{3}}{2}$$

$$\cos 60° = \frac{\dfrac{1}{2}}{1} = \frac{1}{2}$$

$$\tan 60° = \frac{\dfrac{\sqrt{3}}{2}}{\dfrac{1}{2}} = \sqrt{3}$$

$$\cot 60° = \frac{\dfrac{1}{2}}{\dfrac{\sqrt{3}}{2}} = \frac{1}{\sqrt{3}} = \frac{\sqrt{3}}{3}$$

$$\sec 60° = \frac{1}{\dfrac{1}{2}} = 2$$

$$\csc 60° = \frac{1}{\dfrac{\sqrt{3}}{2}} = \frac{2}{\sqrt{3}} = \frac{2\sqrt{3}}{3}$$

If we let $\phi = 30°$ and use the fact that in Fig. 3.16 side $b$ is $\dfrac{\sqrt{3}}{2}$, we can compute the trigonometric ratios of 30°. Side $a$ is considered to be the side opposite angle $\phi$, and side $b$ is the adjacent side. The same right triangle, $\triangle ORQ$, has been redrawn in Fig. 3.17 for convenience in viewing angle $\phi$.

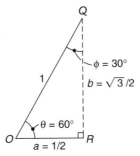

**Figure 3.17**

Again using the trigonometric definitions and $\triangle ORQ$, we have

$$\sin 30° = \dfrac{\dfrac{1}{2}}{1} = \dfrac{1}{2}$$

$$\cos 30° = \dfrac{\dfrac{\sqrt{3}}{2}}{1} = \dfrac{\sqrt{3}}{2}$$

$$\tan 30° = \dfrac{\dfrac{1}{2}}{\dfrac{\sqrt{3}}{2}} = \dfrac{1}{\sqrt{3}} = \dfrac{\sqrt{3}}{3}$$

$$\cot 30° = \dfrac{\dfrac{\sqrt{3}}{2}}{\dfrac{1}{2}} = \sqrt{3}$$

$$\sec 30° = \dfrac{1}{\dfrac{\sqrt{3}}{2}} = \dfrac{2}{\sqrt{3}} = \dfrac{2\sqrt{3}}{3}$$

$$\csc 30° = \dfrac{1}{\dfrac{1}{2}} = 2$$

Two acute angles are **complementary** if their sum is 90°. Thus, angles of 30° and 60° are complementary angles. In fact, it is always true that the two acute angles of any right triangle are complementary since their sum must be 90°.

Observe that we have found

$$\sin 30° = \dfrac{1}{2} = \cos 60°$$

$$\sin 60° = \dfrac{\sqrt{3}}{2} = \cos 30°$$

In fact, this is true of any two complementary angles $\theta$ and $\phi$.

$$\sin \theta = \cos \phi$$

From this relationship we can interpret the "co" in cosine to mean that it is the sine of the *co*mplementary angle. The sine and cosine are said to be **cofunctions.** Similar statements can be made for tangent and cotangent as well as for secant and cosecant.

We shall use Fig. 3.18 to evaluate the trigonometric ratios of 45°.

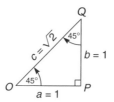

**Figure 3.18**  Isosceles right triangle

Since $\triangle OPQ$ is isosceles, the adjacent side and opposite side are equal. If we let $a = b = 1$, then $c = \sqrt{1^2 + 1^2} = \sqrt{2}$. Therefore,

$$\sin 45° = \frac{1}{\sqrt{2}} = \frac{\sqrt{2}}{2}$$

$$\cos 45° = \frac{1}{\sqrt{2}} = \frac{\sqrt{2}}{2}$$

$$\tan 45° = \frac{1}{1} = 1$$

$$\cot 45° = \frac{1}{1} = 1$$

$$\sec 45° = \frac{\sqrt{2}}{1} = \sqrt{2}$$

$$\csc 45° = \frac{\sqrt{2}}{1} = \sqrt{2}$$

However, we shall need to know the values of trigonometric ratios of angles besides 30°, 45°, and 60°. For this purpose we shall use calculators. Most hand calculators have buttons to evaluate the sine, cosine, and tangent of an angle. The calculator examples shown in this text are designed for use with calculators using algebraic logic. For some calculators, you may have to use the "2nd" button. Be certain that your calculator is in the degree mode.

**EXAMPLE 1**

Find sin 26° rounded to four significant digits.

That is, sin 26° = 0.4384 rounded to four significant digits.

**EXAMPLE 2**

Find cos 36.75° rounded to four significant digits.

| Flow chart | Buttons pushed | Display |
|---|---|---|
| Enter cos | [cos] | cos |
| ↓ | | |
| Enter 36.75 | [3] [6] [·] [7] [5] | 36.75 |
| ↓ | | |
| Push equals | [=] | 0.801254 |

That is, cos 36.75° = 0.8013 rounded to four significant digits.

If a calculator does not have buttons for cotangent, secant, and cosecant, how can you evaluate these functions? Recall that

$$\cot \theta = \frac{1}{\tan \theta} \qquad \sec \theta = \frac{1}{\cos \theta} \quad \text{and} \quad \csc \theta = \frac{1}{\sin \theta}$$

These are called the *reciprocal trigonometric functions*.

**EXAMPLE 3**

Find cot 27.5° rounded to four significant digits.

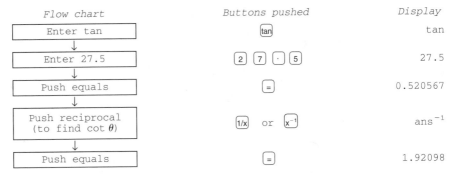

| Flow chart | Buttons pushed | Display |
|---|---|---|
| Enter tan | [tan] | tan |
| ↓ | | |
| Enter 27.5 | [2] [7] [·] [5] | 27.5 |
| ↓ | | |
| Push equals | [=] | 0.520567 |
| ↓ | | |
| Push reciprocal (to find cot θ) | [1/x] or [x⁻¹] | ans⁻¹ |
| ↓ | | |
| Push equals | [=] | 1.92098 |

That is, cot 27.5° = 1.921 rounded to four significant digits.

A calculator may also be used to find the angle when the value of a trigonometric ratio is given. The procedure is shown by the following examples. We shall first limit our angle $\theta$ for $0° \leq \theta \leq 90°$ in this chapter.

**EXAMPLE 4**

Find $\theta$ to the nearest tenth of a degree when $\sin \theta = 0.4321$.

| Flow chart | Buttons pushed | Display |
|---|---|---|
| Enter sin⁻¹ | [2nd F] → [sin] | sin⁻¹ |
| ↓ | | |
| Enter 0.4321 | [·] [4] [3] [2] [1] | 0.4321 |
| ↓ | | |
| Find θ | [=] | 25.6009 |

Thus, $\theta = 25.6°$ to the nearest tenth of a degree.

### EXAMPLE 5

Find $\theta$ to the nearest tenth of a degree when $\cos \theta = 0.6046$.

| *Flow chart* | *Buttons pushed* | *Display* |
|---|---|---|
| Enter cos⁻¹ | [2nd F] → [cos] | cos⁻¹ |
| ↓ | | |
| Enter 0.6046 | [·] [6] [0] [4] [6] | 0.6046 |
| ↓ | | |
| Find $\theta$ | [=] | 52.7999 |

Thus, $\theta = 52.8°$ to the nearest tenth of a degree.

### EXAMPLE 6

Find $\theta$ to the nearest tenth of a degree when $\tan \theta = 2.584$.

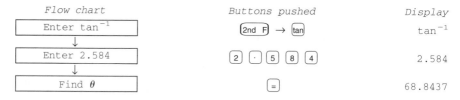

| *Flow chart* | *Buttons pushed* | *Display* |
|---|---|---|
| Enter tan⁻¹ | [2nd F] → [tan] | tan⁻¹ |
| ↓ | | |
| Enter 2.584 | [2] [·] [5] [8] [4] | 2.584 |
| ↓ | | |
| Find $\theta$ | [=] | 68.8437 |

Thus, $\theta = 68.8°$ to the nearest tenth of a degree.

### EXAMPLE 7

Find $\theta$ to the nearest tenth of a degree when $\sec \theta = 1.365$.

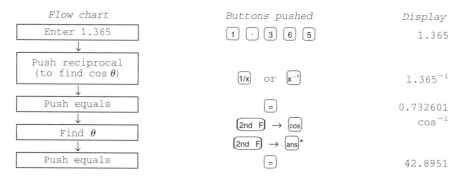

| *Flow chart* | *Buttons pushed* | *Display* |
|---|---|---|
| Enter 1.365 | [1] [·] [3] [6] [5] | 1.365 |
| ↓ | | |
| Push reciprocal (to find cos $\theta$) | [1/x] or [x⁻¹] | 1.365⁻¹ |
| ↓ | | |
| Push equals | [=] | 0.732601 |
| ↓ | | |
| Find $\theta$ | [2nd F] → [cos] [2nd F] → [ans]* | cos⁻¹ |
| ↓ | | |
| Push equals | [=] | 42.8951 |

Thus, $\theta = 42.9°$ to the nearest tenth of a degree.

### EXAMPLE 8

Find $\tan 58°16'24''$ rounded to four significant digits.

The DMS key will be used here, as illustrated in Appendix B. If your calculator does not have this key, you should consult your calculator manual or consider using the method in Appendix B to be used when your calculator does not have such a special key.

---

*Your calculator may not require this step.

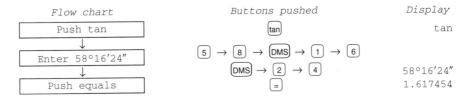

| Flow chart | Buttons pushed | Display |
|---|---|---|
| Push tan | tan | tan |
| Enter 58°16′24″ | $5 \to 8 \to$ DMS $\to 1 \to 6$<br>DMS $\to 2 \to 4$ | 58°16′24″ |
| Push equals | = | 1.617454 |

Thus, tan 58°16′24″ = 1.6175 rounded to five significant digits.

**EXAMPLE 9**

Find $\theta$ to the nearest second when sin $\theta$ = 0.2587.

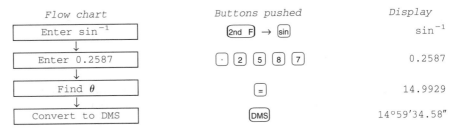

| Flow chart | Buttons pushed | Display |
|---|---|---|
| Enter $\sin^{-1}$ | 2nd F $\to$ sin | $\sin^{-1}$ |
| Enter 0.2587 | · 2 5 8 7 | 0.2587 |
| Find $\theta$ | = | 14.9929 |
| Convert to DMS | DMS | 14°59′34.58″ |

Thus, $\theta$ = 14°59′35″ rounded to the nearest second.

## Exercises 3.2

*Use a calculator to find the value of each rounded to four significant digits.*

**1.** sin 18.5°          **2.** cos 27.6°          **3.** tan 41.4°

**4.** sin 13.6°          **5.** cos 77.2°          **6.** tan 87.1°

**7.** sec 34.7°          **8.** csc 80.5°          **9.** cot 34.0°

**10.** sec 19.0°          **11.** csc 49.8°          **12.** cot 74.1°

**13.** sin 46.72°          **14.** cos 19.51°          **15.** tan 73.8035°

**16.** csc 34.9625°          **17.** sec 8.3751°          **18.** cot 16.3795°

*Use a calculator to find each angle rounded to the nearest tenth of a degree.*

**19.** sin $\theta$ = 0.4305          **20.** cos $\theta$ = 0.7771          **21.** tan $\theta$ = 0.4684

**22.** sin $\theta$ = 0.2096          **23.** cos $\theta$ = 0.1463          **24.** tan $\theta$ = 1.357

**25.** tan $\theta$ = 3.214          **26.** cos $\theta$ = 0.5402          **27.** sin $\theta$ = 0.1986

**28.** cot $\theta$ = 3.270          **29.** sec $\theta$ = 2.363          **30.** csc $\theta$ = 5.662

**31.** cot $\theta$ = 0.5862          **32.** sec $\theta$ = 3.341          **33.** csc $\theta$ = 2.221

**34.** csc $\theta$ = 1.333          **35.** sec $\theta$ = 6.005          **36.** cot $\theta$ = 8.307

*Use a calculator to find each angle rounded to the nearest hundredth of a degree.*

**37.** cos $\theta$ = 0.4836          **38.** sin $\theta$ = 0.1920          **39.** cot $\theta$ = 1.5392

**40.** tan $\theta$ = 2.5575          **41.** csc $\theta$ = 2.4075          **42.** sec $\theta$ = 1.2566

*Use a calculator to find the value of each rounded to four significant digits.*

**43.** sin 36°24′

**44.** cos 48°18′

**45.** tan 52°43′38″

**46.** tan 17°35′52″

**47.** cos 9°56′21″

**48.** sin 21°8′46″

**49.** cot 36°15′44″

**50.** sec 31°8′27″

**51.** csc 84°35′53″

**52.** cot 51°40′11″

**53.** sec 72°27″

**54.** csc 7°52″

*Use a calculator to find each angle rounded to the nearest second.*

**55.** sin $\theta$ = 0.8556

**56.** cos $\theta$ = 0.2749

**57.** tan $\theta$ = 6.2662

**58.** tan $\theta$ = 0.3254

**59.** cos $\theta$ = 0.5966

**60.** sin $\theta$ = 0.2694

**61.** cot $\theta$ = 0.8678

**62.** sec $\theta$ = 3.3424

**63.** csc $\theta$ = 2.3770

**64.** cot $\theta$ = 4.5065

**65.** sec $\theta$ = 1.1678

**66.** csc $\theta$ = 8.1407

## 3.3 SOLVING RIGHT TRIANGLES

D.1, D.5

Many technical problems involve finding unknown values of sides or angles of triangles. In mathematics we call this process **solving a triangle.** Every triangle consists of six parts: three sides and three angles. If at least one side and two other parts are known, it is possible to find the values of the other three parts.

In this chapter we are concerned only with right triangles. Since one angle is 90°, we need to know only a side and one other part to find the other three parts. When you are solving triangles, it is often very helpful to draw a diagram of the triangle to be solved. For this purpose we shall follow customary practice to label the angles and sides as shown in Fig. 3.19.

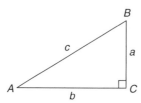

**Figure 3.19**

From the Pythagorean theorem, we know that

$$c^2 = a^2 + b^2 \quad \text{or} \quad c = \sqrt{a^2 + b^2}$$

Recall that the two acute angles of a right triangle are complementary. That is,

$$A + B = 90°$$
$$A = 90° - B$$
$$B = 90° - A$$

This tells us that once we know the value of one acute angle, we then can find the value of the other.

We now look at solving right triangles. Using examples, we shall see how to find the unknown parts.

When calculations with measurements involve a trigonometric ratio, we use the following rule of thumb:

| Angles expressed to the nearest: | Lengths of sides of a triangle will contain: |
|---|---|
| 1° | Two significant digits |
| 0.1°  or  1′ | Three significant digits |
| 0.01°  or  1″ | Four significant digits |

**EXAMPLE 1**

Find side $c$ given side $a = 6.00$ cm and side $b = 8.00$ cm.

First, draw a triangle as in Fig. 3.20. Then

$$c = \sqrt{a^2 + b^2}$$
$$c = \sqrt{(6.00 \text{ cm})^2 + (8.00 \text{ cm})^2}$$
$$= \sqrt{36.0 \text{ cm}^2 + 64.0 \text{ cm}^2}$$
$$= \sqrt{100.00 \text{ cm}^2}$$
$$= 10.0 \text{ cm}$$

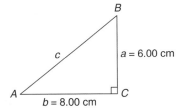

**Figure 3.20**

**EXAMPLE 2**

Find side $a$ given side $b = 6.00$ m and side $c = 11.0$ m.

First, draw a triangle as in Fig. 3.21. Since

$$c^2 = a^2 + b^2$$

we have

$$a^2 = c^2 - b^2$$
$$a = \sqrt{c^2 - b^2}$$
$$a = \sqrt{(11.0 \text{ m})^2 - (6.00 \text{ m})^2}$$
$$= 9.22 \text{ m}$$

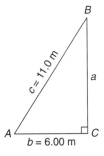

**Figure 3.21**

These two examples did not require the use of a trigonometric ratio. However, such use will be necessary in the following examples.

**EXAMPLE 3**

Find angle $B$ to the nearest tenth of a degree given $b = 7.00$ m and $c = 9.00$ m.
Draw a triangle as in Fig. 3.22. From the definition of $\sin B$,

$$\sin B = \frac{\text{side opposite } B}{\text{hypotenuse}}$$

$$= \frac{7.00 \text{ m}}{9.00 \text{ m}} = 0.7778$$

So

$$B = 51.1° \qquad \text{(to the nearest tenth of a degree)}$$

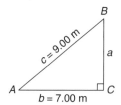

**Figure 3.22**

This angle is found using a calculator as follows:

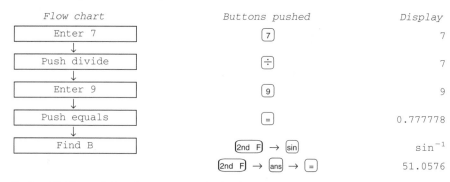

So $B = 51.1°$ rounded to the nearest tenth of a degree.

**EXAMPLE 4**

Find angle $A$ to the nearest hundredth of a degree given $b = 4.250$ km and $c = 9.750$ km.
Draw a triangle as in Fig. 3.23. In this case we have

$$\cos A = \frac{\text{side adjacent to } A}{\text{hypotenuse}}$$

$$= \frac{4.250 \text{ km}}{9.750 \text{ km}} = 0.4359$$

So

$$A = 64.16° \qquad \text{(to the nearest hundredth of a degree)}$$

**Figure 3.23**

**EXAMPLE 5**

Find side $a$ given $c = 12.00$ mi and $B = 24.00°$.
　　Draw a triangle as in Fig. 3.24.

$$\cos B = \frac{\text{side adjacent to } B}{\text{hypotenuse}}$$

or

$$\cos 24.00° = \frac{a}{12.00 \text{ mi}}$$

$$a = (12.00 \text{ mi})(\cos 24.00°)$$

$$= (12.00 \text{ mi})(0.9135)$$

$$= 10.96 \text{ mi}$$

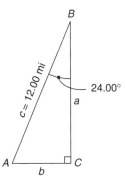

**Figure 3.24**

　　In the next example we solve a right triangle completely; that is, we find all sides and angles not given or known.

**EXAMPLE 6**

Solve the right triangle given $a = 6.00$ ft and $b = 4.00$ ft.
　　Draw a triangle as in Fig. 3.25.

**Method 1:**

$$\tan A = \frac{\text{side opposite } A}{\text{side adjacent to } A}$$

$$= \frac{6.00 \text{ ft}}{4.00 \text{ ft}} = 1.500$$

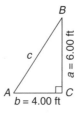

**Figure 3.25**

So

$$A = 56.3° \qquad \text{(to the nearest tenth of a degree)}$$

To find $B$ we let

$$B = 90° - A$$
$$B = 90° - 56.3°$$
$$= 33.7°$$

Finally,

$$c = \sqrt{a^2 + b^2}$$
$$c = \sqrt{(4.00 \text{ ft})^2 + (6.00 \text{ ft})^2}$$
$$= \sqrt{52.0 \text{ ft}^2} = 7.21 \text{ ft}$$

**Method 2:** We could have found $c = 7.21$ ft first. Then

$$\sin A = \frac{\text{side opposite } A}{\text{hypotenuse}}$$

$$= \frac{6.00 \text{ ft}}{7.21 \text{ ft}} = 0.8322$$

So

$$A = 56.3°$$

Angle $B$ would be found in the same way as in Method 1.

A triangle is not determined if only the angles are known. If $A = 60°$ and $B = 30°$, then we know that the triangle with sides $a = \dfrac{1}{2}$, $b = \dfrac{\sqrt{3}}{2}$, and $c = 1$ is a possible solution. But so is the triangle with sides $a = 1$, $b = \sqrt{3}$, and $c = 2$. In fact, there are infinitely many other solutions as well. Thus, unless the value of a side is given, we are unable to determine a specific triangle.

*Note:* While all six trigonometric ratios may be used to solve a right triangle, we usually restrict our choices here to sine, cosine, and tangent because these buttons appear on calculators. In later chapters the other three trigonometric ratios will become more important and will receive more attention.

## Exercises 3.3

*Solve each right triangle. Assume the standard labeling, as shown in Fig. 3.26.*

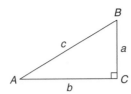

**Figure 3.26**

In Exercises 1–10, find each side to three significant digits and each angle to the nearest tenth of a degree.

**1.** $a = 4.00$ ft, $b = 8.00$ ft

**2.** $a = 7.00$ m, $B = 43.2°$

**3.** $c = 21.0$ cm, $A = 27.3°$

**4.** $a = 5.00$ in., $c = 9.00$ in.

**5.** $b = 7.50$ m, $c = 13.4$ m

**6.** $a = 9.20$ cm, $A = 72.4°$

**7.** $a = 12.4$ mi, $b = 7.70$ mi

**8.** $c = 9.40$ km, $B = 17.3°$

**9.** $b = 25\bar{0}$ km, $B = 37.0°$

**10.** $a = 12\bar{0}$ yd, $b = 95\bar{0}$ yd

In Exercises 11–18, find each side to four significant digits and each angle to the nearest hundredth of a degree.

**11.** $a = 14.21$ cm, $c = 37.42$ cm

**12.** $c = 7.300$ m, $A = 49.35°$

**13.** $a = 6755$ mi, $A = 68.75°$

**14.** $b = 13{,}530$ km, $c = 25{,}550$ km

**15.** $c = 45.32$ m, $B = 15.80°$

**16.** $a = 500\bar{0}$ ft, $B = 25.00°$

**17.** $b = 2572$ ft, $c = 4615$ ft

**18.** $a = 3.512$ mi, $c = 5.205$ mi

In Exercises 19–26, find each side to two significant digits and each angle to the nearest degree.

**19.** $b = 1500$ mi, $c = 3500$ mi

**20.** $a = 15$ ft, $A = 3\bar{0}°$

**21.** $c = 45$ m, $B = 5\bar{0}°$

**22.** $a = 36$ ft, $b = 16$ ft

**23.** $a = 140$ ft, $A = 37°$

**24.** $b = 3700$ m, $A = 59°$

**25.** $a = 3.5$ mi, $B = 22°$

**26.** $a = 0.36$ mi, $c = 0.44$ mi

In Exercises 27–34, find each side to four significant digits and each angle to the nearest second.

**27.** $a = 1753$ m, $B = 37°41'30''$

**28.** $a = 28{,}570$ ft, $b = 37{,}550$ ft

**29.** $a = 495.5$ ft, $c = 617.0$ ft

**30.** $c = 5.632$ km, $A = 18°6'45''$

**31.** $a = 37.52$ m, $A = 58°11'25''$

**32.** $b = 7753$ ft, $c = 8455$ ft

**33.** $c = 6752$ ft, $B = 27°5'16''$

**34.** $b = 55.60$ km, $B = 75°7'8''$

## 3.4 APPLICATIONS OF THE RIGHT TRIANGLE

Many applications, both technical and nontechnical, are solved using the trigonometric definitions and right triangles. The following examples illustrate some of these basic applications.

### EXAMPLE 1

A mine shaft extends down at an angle of 5°. How long will the shaft have to be to reach a vein of ore which is 65 ft directly below the surface?

**Figure 3.27**

First, draw a triangle as in Fig. 3.27.

$$\sin 5° = \frac{65 \text{ ft}}{s}$$

$$s(\sin 5°) = 65 \text{ ft} \qquad \text{(Multiply each side by } s.)$$

$$s = \frac{65 \text{ ft}}{\sin 5°} \qquad \text{(Divide each side by } \sin 5°.)$$

$$= 750 \text{ ft} \qquad \text{(Round to two significant digits.)}$$

**EXAMPLE 2**

On one grade a railroad track rises 1.00 m for each 42.0 m of track. What angle does the track make with the level ground?

From Fig. 3.28,

$$\sin \theta = \frac{1.00 \text{ m}}{42.0 \text{ m}}$$

$$\sin \theta = 0.0238$$

$$= 1.4° \qquad \text{(to the nearest tenth of a degree)}$$

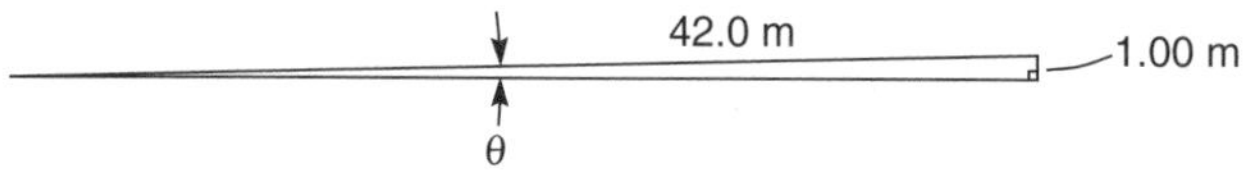

42.0 m     1.00 m

$\theta$

**Figure 3.28**

The **angle of depression** is the angle between the horizontal and the line of sight to an object that is *below* the horizontal. The **angle of elevation** is the angle between the horizontal and the line of sight to an object that is *above* the horizontal. In Fig. 3.29, $\alpha$ is the angle of depression for an observer in the helicopter sighting down to the building on the ground, and $\beta$ is the angle of elevation for an observer in the building sighting up to the helicopter. (Note: $\alpha = \beta$.)

**EXAMPLE 3**

A horizontal distance of 225 ft is measured from the base of a vertical cliff. The angle of elevation from this distance measures 67°. Find the height of the cliff.

First, draw a sketch as in Fig. 3.30. Then

$$\tan 67° = \frac{h}{225 \text{ ft}}$$

$$h = (\tan 67°)(225 \text{ ft}) \qquad \text{(Multiply each side by 225 ft.)}$$

$$= 530 \text{ ft}$$

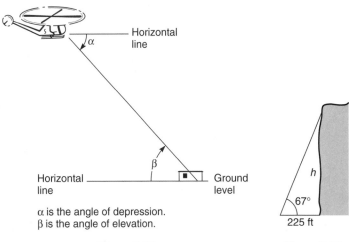

$\alpha$ is the angle of depression.
$\beta$ is the angle of elevation.

**Figure 3.29**                    **Figure 3.30**

**EXAMPLE 4**

Carla is piloting a helicopter in the wilderness at 1200 ft above the ground searching for a downed plane. As Carla spots it, she measures its angle of depression as 53°. She also spots a road whose angle of depression is 15°. Find the distance of the downed plane to the road (see Fig. 3.31).

$$\tan 37° = \frac{x}{1200 \text{ ft}}$$

$$x = (\tan 37°)(1200 \text{ ft}) \qquad \text{(Multiply each side by 1200 ft.)}$$

$$= 9\overline{0}0 \text{ ft}$$

$$\tan 75° = \frac{y}{1200 \text{ ft}}$$

$$y = (\tan 75°)(1200 \text{ ft})$$

$$= 4500 \text{ ft}$$

The distance from the downed plane to the road is $4500 \text{ ft} - 9\overline{0}0 \text{ ft} = 3600 \text{ ft}$.

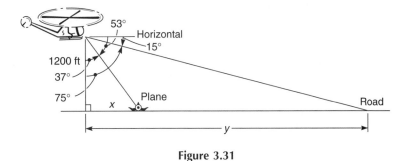

**Figure 3.31**

**EXAMPLE 5**

In ac (alternating current) circuits the relationship between the impedance, $Z$, the resistance, $R$, and the phase angle, $\phi$, is shown by the right triangle in Fig. 3.32.

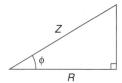

**Figure 3.32**

(a) If the resistance is 94 Ω and the phase angle is 36°, find the impedance.

$$\cos \phi = \frac{R}{Z}$$

$$\cos 36° = \frac{94 \ \Omega}{Z}$$

$$Z(\cos 36°) = 94 \ \Omega \qquad \text{(Multiply each side by Z.)}$$

$$Z = \frac{94 \ \Omega}{\cos 36°} \qquad \text{(Divide each side by cos 36°.)}$$

$$= 120 \ \Omega \qquad \text{(Round to two significant digits.)}$$

(b) If the resistance is 145 Ω and the impedance is 210 Ω, what is the phase angle?

$$\cos \phi = \frac{R}{Z}$$

$$\cos \phi = \frac{145 \ \Omega}{210 \ \Omega} = 0.6905$$

$$\phi = 46° \qquad \text{(Round to the nearest degree.)}$$

## Exercises 3.4

**1.** A tower casts a shadow 42 m long when the angle of elevation of the sun is 51°. Find the height of the tower.

**2.** Find the width of the river in Fig. 3.33.

**3.** A roadbed rises 250 ft for each 3600 ft of road. Find the angle of elevation of the roadbed.

**4.** When directly over one town, the pilot of a plane noticed that the angle of depression of another town is 11°. If the altimeter registered 8400 ft, what is the ground distance (in mi) between the two towns?

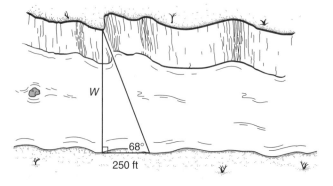

**Figure 3.33**

5. A smokestack is $19\overline{0}$ ft high. Find the length of a guy wire which must be fastened to the stack 25 ft from the top and which makes an angle of 40.0° with the ground.

6. A railroad track has an angle of elevation of 1.0°. What is the difference in altitudes of two points on the track which are (a) 1.00 mi apart and (b) 1.00 km apart?

7. A bridge is to be built across a river at an angle of 75.0° with the near side of the river. If the width of the river is 89.1 m, what must be the length of the bridge?

8. If the span of the bridge in Exercise 7 is to be 16.0 m above the roadway and the angle of elevation of the approach is to be 5.0°, how long will the approach be?

9. The cliff shown in Fig. 3.34 has a dangerous boulder on its face which may fall on the roadway below. At a point 275 ft from the base of the cliff, the angle of elevation to the boulder is 42.0° and the angle of elevation to the top of the cliff is 62.0°. How far down from the top will a scaffold have to be lowered to reach the boulder?

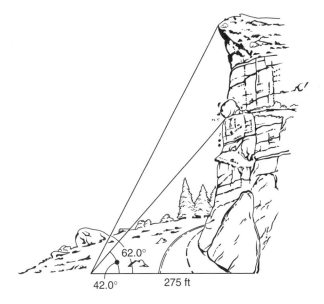

62.0°

42.0°          275 ft

**Figure 3.34**

10. The pathway for a mineshaft is described as follows: The first shaft extends north and down at an angle of 3.5° for 225 ft to a vertical shaft which is 125 ft long. A third shaft then extends north and down at an angle of 6.8° for 175 ft. What is the total depth below ground level at the end of the third shaft? Also, what is the net horizontal distance from the ground opening to the end of the third shaft?

11. The right triangle in Fig. 3.35 shows the relationship among impedance, resistance, and reactance in an ac circuit, where

$$Z = \text{impedance (in } \Omega)$$
$$R = \text{resistance (in } \Omega)$$
$$X = \text{reactance (in } \Omega)$$
$$\phi = \text{phase angle}$$

(a) If the reactance is 82.6 $\Omega$ and the resistance is 112 $\Omega$, find the impedance and the phase angle to the nearest tenth of a degree.

(b) If the resistance is 250 $\Omega$ and the phase angle is 23°, find the impedance and the reactance.

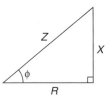

**Figure 3.35** Impedance-resistance relationship

12. A machinist needs to drill 5 holes in a circular plate. The centers of the holes must be 8.00 cm from the center of the plate and be spaced in a circular pattern equidistant from each other. How far apart (straight-line distance) will the centers of the adjacent holes be placed?

13. The corner of the metal plate in Fig. 3.36 is cut off. Find length $x$ and angle $\alpha$.

14. A piece of electrical conduit 14.8 m long cuts across the corner of a building in Fig. 3.37. Find length $x$ and angle $\beta$.

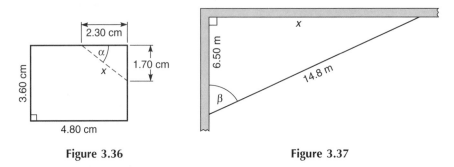

**Figure 3.36**                    **Figure 3.37**

15. Find lengths $x$, $y$, and $z$ and angles $\alpha$, $\beta$, $\theta$, and $\phi$ in Fig. 3.38.

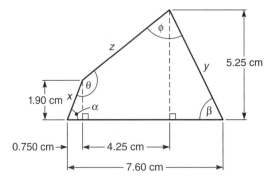

**Figure 3.38**

16. In the trapezoid in Fig. 3.39, find lengths $x$ and $y$ and angles $\alpha$, $\beta$, $\theta$, and $\phi$.

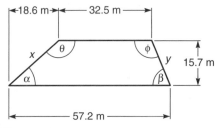

**Figure 3.39**

**17.** Figure 3.40 is a schematic for a thread. Find length *x*.

**18.** Find length *x* on the bolt in Fig. 3.41.

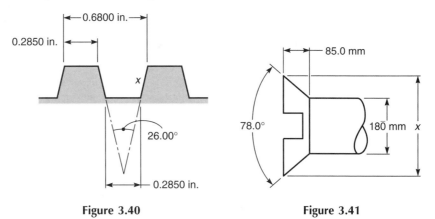

**Figure 3.40**                                   **Figure 3.41**

**19.** Find angle $\theta$ of the taper in Fig. 3.42.

**20.** Find the width, *x*, of the dovetail wedge in Fig. 3.43.

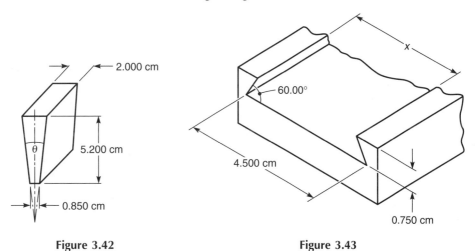

**Figure 3.42**                                   **Figure 3.43**

**21.** Find length *x* and angles $\alpha$ and $\beta$ for the retaining wall in Fig. 3.44.

**22.** Find length *x* and angle $\theta$ from the footings for a foundation plan in Fig. 3.45.

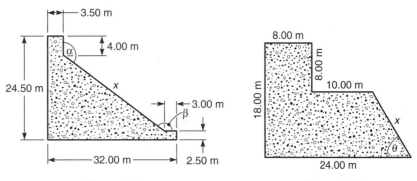

**Figure 3.44**                                   **Figure 3.45**

**23.** From the sketch of a rolling piping offset in Fig. 3.46, find the travel if the roll is 3.00 ft, the offset is 5.00 ft, and the advance is 10.00 ft.

**24.** From Fig. 3.46, find the length of the advance if the roll is 20.0 in., the offset is 36.0 in., and the travel is 72.0 in.

**25.** Using the data in Exercise 23, find angle *EBA*.

**26.** Using the data in Exercise 24, find angle *CBA*.

**27.** A cylindrical tank of diameter 16.0 ft is placed in a corner as in Fig. 3.47. Using 45° fittings and assuming 1.0 ft clearance between walls, pipe, and tank, what is the length, *x*, of the pipe that cuts across the corner?

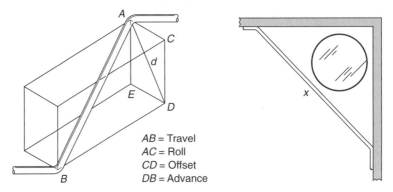

AB = Travel
AC = Roll
CD = Offset
DB = Advance

**Figure 3.46**          **Figure 3.47**

**28.** Find the radius of the circle in Fig. 3.48.

**29.** Find the radius of the circle in Fig. 3.49.

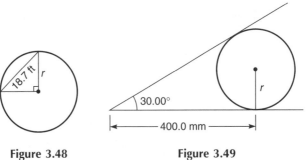

**Figure 3.48**          **Figure 3.49**

**30.** Find the radius of the circle in Fig. 3.50.

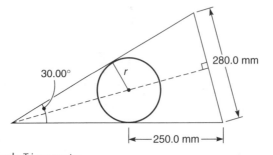

**Figure 3.50**

**31.** Find the radius of the circle in Fig. 3.51.

**32.** Find the radius of the smaller circle in Fig. 3.52.

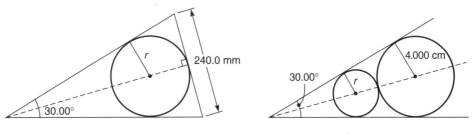

Figure 3.51                                  Figure 3.52

**33.** At a horizontal distance of 125.5 ft from the base of a tower, the angle of elevation to its top is 71°24′30″. Find the height of the tower.

**34.** Given $AB = 18.34$ in. and $BC = 10.15$ in. in Fig. 3.53. Find angle $A$ to the nearest second and length $AC$.

**35.** A line joins a vertex of a square to the midpoint of an opposite side and forms a right triangle. Find the two acute angles rounded to the nearest second.

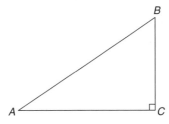

Figure 3.53

## CHAPTER SUMMARY

**1.** 1 complete rotation $= 360°$
$$1° = 60'$$
$$1' = 60''$$

**2.** *Pythagorean theorem:* $c^2 = a^2 + b^2$ · or $c = \sqrt{a^2 + b^2}$ (See Fig. 3.54.)

**3.** *The trigonometric ratios (see Fig. 3.54):*

(a) $\sin A = \dfrac{\text{side opposite } A}{\text{hypotenuse}} = \dfrac{a}{c}$ or $\sin B = \dfrac{b}{c}$

(b) $\cos A = \dfrac{\text{side adjacent to } A}{\text{hypotenuse}} = \dfrac{b}{c}$ or $\cos B = \dfrac{a}{c}$

(c) $\tan A = \dfrac{\text{side opposite } A}{\text{side adjacent to } A} = \dfrac{a}{b}$ or $\tan B = \dfrac{b}{a}$

(d) $\cot A = \dfrac{\text{side adjacent to } A}{\text{side opposite } A} = \dfrac{b}{a}$ or $\cot B = \dfrac{a}{b}$

(e) $\sec A = \dfrac{\text{hypotenuse}}{\text{side adjacent to } A} = \dfrac{c}{b}$ or $\sec B = \dfrac{c}{a}$

(f) $\csc A = \dfrac{\text{hypotenuse}}{\text{side opposite } A} = \dfrac{c}{a}$ or $\csc B = \dfrac{c}{b}$

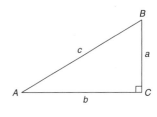

Figure 3.54

**4.** *Reciprocal trigonometric functions:*

(a) $\sin \theta = \dfrac{1}{\csc \theta}$      (d) $\csc \theta = \dfrac{1}{\sin \theta}$

(b) $\cos \theta = \dfrac{1}{\sec \theta}$      (e) $\sec \theta = \dfrac{1}{\cos \theta}$

(c) $\tan \theta = \dfrac{1}{\cot \theta}$      (f) $\cot \theta = \dfrac{1}{\tan \theta}$

**5.** The two acute angles of a triangle are complementary; that is,

$$A + B = 90°$$

**6.** *General rounding procedure:*

| Angles expressed to the nearest: | Lengths of sides of a triangle will contain: |
|---|---|
| 1° | Two significant digits |
| 0.1° or 1′ | Three significant digits |
| 0.01° or 1″ | Four significant digits |

## CHAPTER 3 REVIEW

*Change each angle to degrees.*

**1.** 129°30′

**2.** 76°12′

*Change each angle to degrees and minutes.*

**3.** $35\dfrac{2}{3}°$

**4.** 314.3°

**5.** Change 16°27′45″ to degrees.

**6.** Change 38.405° to degrees, minutes, and seconds.

*Find the length of the third side of each right triangle rounded to three significant digits.*

**7.** $a = 16.0$ m, $c = 36.0$ m

**8.** $a = 18.7$ mi, $b = 25.5$ mi

**9.** Find the six trigonometric ratios of angle $A$ rounded to three significant digits when $b = 127$ cm and $c = 235$ cm.

*Use a calculator to find the value of each rounded to four significant digits.*

**10.** cos 14.6°

**11.** sin 51.7°

**12.** tan 29.5°

**13.** sec 16.7°

**14.** cot 29.1°

**15.** csc 79.2°

*Use a calculator to find each angle rounded to the nearest tenth of a degree.*

**16.** $\sin \theta = 0.6075$

**17.** $\cos \theta = 0.3522$

**18.** $\tan \theta = 1.2345$

**19.** $\sec \theta = 1.3290$

**20.** $\cot \theta = 0.9220$

**21.** $\csc \theta = 1.2222$

*Use a calculator to find the value of each rounded to four significant digits.*

**22.** sin 41°37′55″

**23.** tan 75°9′27″

**24.** sec 34°14′35″

*Use a calculator to find each angle rounded to the nearest second.*

**25.** $\cos \theta = 0.4470$        **26.** $\tan \theta = 0.2408$        **27.** $\csc \theta = 3.4525$

*Solve each right triangle.*

**28.** $a = 7.00$ m, $b = 9.50$ m     **29.** $b = 15.75$ cm, $B = 36.50°$

**30.** $c = 1700$ km, $A = 20\overline{0}°$     **31.** $a = 245.7$ m, $A = 35°14'32''$

**32.** A ranger at the top of a fire tower observes the angle of depression to a fire to be 3°. If the tower is 250 ft tall, what is the ground distance to the fire from the tower?

**33.** A tower which is known to be 175 feet high is sighted from the ground. The angle of elevation is found to be 2°. How far away is the tower?

**34.** A roadbed rises 175 ft for each $41\overline{0}0$ ft of road. Find the angle of elevation of the roadbed.

**35.** Find lengths $x$ and $y$ and angles $\alpha$ and $\beta$ in the trapezoid in Fig. 3.55.

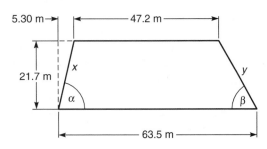

**Figure 3.55**

**36.** A roof is to be built to cover a building 56.0 ft wide as in Fig. 3.56. If the slope of the roof is to be 20.0°, how high, $x$, should the roof rise at the center?

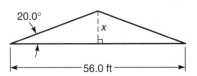

**Figure 3.56**

**37. (a)** If the reactance in an ac circuit is 75 $\Omega$ and the resistance is 42 $\Omega$, find the impedance and the phase angle. (See Exercise 11 of Section 3.4.)

    **(b)** If the reactance is 94 $\Omega$ and the phase angle is 47°, find the resistance and the impedance.

# APPLICATION
## The Global Positioning System

Emerging from the dense forest into a clearing, the hikers looked around and then at each other—they did not know the way back to their vehicle. But because they were using GPS, the Global Positioning System, they were not lost. How does GPS work?

A handheld instrument records signals from satellites orbiting the earth in a fixed pattern. It determines the time of the transmission and reception of a signal and uses that to calculate the distance between the receiver and a satellite. This distance determines an imaginary sphere with the transmitting satellite at the center. The intersection of several such spheres with the surface of the earth determines the location of the GPS receiver. Synchronizing the clocks to get an accurate time to calculate the distance from a satellite to a point on earth is crucial. Because the clocks cannot be perfectly synchronized, the initial distance may not be precise. A few mathematical calculations determine how to compensate for the errors, and thus determine the location.

The GPS has 24 satellites orbiting the earth at an altitude of about 20,000 kilometres. This altitude was a compromise between a much higher altitude, which would require fewer satellites but more powerful transmitters, and a lower altitude, which would allow the use of lower-powered transmitters but would necessitate hundreds of satellites. Reception from at least four satellites is needed to fix a position. The position is described using a coordinate system on the earth. There are several systems, but the most common uses circles of longitude and latitude.

The equator is the initial circle of latitude and is defined as 0°. Others are parallel to the equator going up to 90° at the north pole and to 90° at the south pole. A position in Champaign, Illinois, has a latitude of 40°06.52′ North. The initial circle of longitude, called the Prime Meridian, passes through the north and south poles and is perpendicular to the equator. It is defined to be 0°. The Prime Meridian also passes through Greenwich, England. The meridians, or great circles passing through the north and south poles, are positioned in degrees going up to 180° east or west of the Prime Meridian. A position in Champaign, Illinois, has a longitude of 88°16.20′ West. Thus, knowing the longitude and latitude of a position identifies it.

As the hikers explored the area, they used their GPS receiver to record the *location* of landmarks along the way. When they were ready to return to their vehicle, the receiver displayed the bearing or compass direction to the vehicle (or landmark) and the heading, i.e., the direction in which they were moving.

The global positioning system is also used for tracking delivery trucks, for mapping agricultural fields to determine fertilizer applications, and as a component in the electronic maps now available in some vehicles.

## Bibliography

Herring, Thomas A. *"The Global Positioning System." Scientific American* (February 1996).
*Magellan GPS 2000 Satellite Navigator User Guide.* Magellan Systems Corporation, 960 Overland
Cont., San Dimas, CA 91773.

Hand held Global Positioning System (GPS)
satellite navigation receiver. *(Photo courtesy of
Photo Researchers, Inc.)*

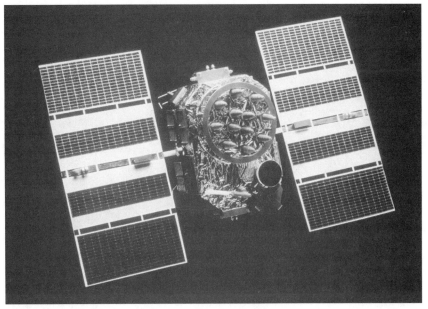

GPS satellite. *(Photo courtesy of Tom Stack & Associates.)*

# 4

# Equations and Their Graphs

## INTRODUCTION

The path of a golf ball in flight, the relationship between the strength of a beam and its length, and the cost of manufacturing a particular item can all be described using the concept of a function. In this chapter we study functions and relate their equations to a graph. This graphic representation which connects algebra and geometry is extremely helpful in solving problems.

### Objectives

- Identify relations that are functions.
- Determine the domain and the range of a function.
- Work with function notation.
- Solve equations graphically.
- Find the slope of a line.
- Find the equation of a line given appropriate information.
- Know the characteristics of parallel and perpendicular lines.
- Know the distance and midpoint formulas.

## 4.1 FUNCTIONS

In common usage, a relation means that two or more things have something in common. We say that a brother and a sister are related because they have the same parents or that a person's career potential is related to his or her education and work experience.

In mathematics a **relation** is defined as a set of ordered pairs of numbers in the form $(x, y)$. Sometimes an equation, a rule, a data chart, or some other type of description is given that states the relationship between $x$ and $y$. In an ordered pair the first element or variable, called the **independent variable,** may be represented by any letter, but $x$ is normally used. The second element or variable is normally represented by the letter $y$ and is called the **dependent variable** because its value depends on the particular choice of the independent variable.

All of the numbers that can be used as the first element of an ordered pair or as replacements for the independent variable of a given relation form a set of numbers called the **domain.** The domain is often referred to as the set of all $x$'s. We can think of these $x$-values as "inputs." The **range** of a relation is the set of numbers that can be used as the second element of an ordered pair or as replacements for the dependent variable. The range is often referred to as the set of all $y$'s. We can think of these $y$-values as "outputs."

### EXAMPLE 1

Given the relation described in ordered pair form $A = \{(1, 2), (3, 5), (7, 9), (6, 3)\}$, find its domain and its range.

The domain is the set of first elements: $\{1, 3, 6, 7\}$. The range is the set of second elements: $\{2, 3, 5, 9\}$.

*Note:* Braces $\{ \quad \}$ are normally used to group elements of sets.

### EXAMPLE 2

Given the relation in equation form $y = x^2$, find its domain and its range.

The domain is the set of possible replacements for the independent variable $x$. Note that there are no restrictions on the numbers that you may substitute for $x$. That is, we may replace $x$ by any real number. We say that the domain is the set of real numbers.

After each replacement of $x$, there is no possible way that we can obtain a negative value for $y$ because the square of any real number is always positive or zero. Thus, the range is the set of nonnegative real numbers, or $y \geq 0$.

### EXAMPLE 3

Find the domain and the range of the relation $y = \sqrt{x - 4}$.

Note that no value of $x$ less than 4 may be used because the square root of any negative number is not a real number. Thus, the domain is the set of real numbers greater than or equal to 4, or $x \geq 4$.

After each possible $x$-replacement, the square root of the resulting value is never negative, so the range is $y \geq 0$.

---

### FUNCTION

A **function** is a special relation: a set of ordered pairs in which no two distinct ordered pairs have the same first element.

---

In equation form, a relation is a function when for each possible value of the first or independent variable, there is only one corresponding value of the second or dependent variable. In brief, for a relation to be a function, each value of $x$ must correspond to one, and only one, value of $y$.

### EXAMPLE 4

Is the relation $B = \{(3, 2), (6, 7), (5, 3), (1, 1), (3, 7)\}$ a function? Find its domain and its range.

*B* is not a function because it contains two different ordered pairs that have the same first element: (3, 2) and (3, 7). In other words, the fact that both 2 and 7 correspond to 3 causes the relation *B* not to be a function. The domain of *B* is {1, 3, 5, 6}. The range of *B* is {1, 2, 3, 7}.

Does the set $A = \{(1, 2), (3, 5), (7, 9), (6, 3)\}$ from Example 1 describe a function? Yes, because no two ordered pairs have the same first element.

### EXAMPLE 5

Is the relation $x = y^2$ a function? Find its domain and its range.

Can we find two ordered pairs that have the same first element? Yes, for example, (9, 3) and (9, −3) as well as (16, 4) and (16, −4) and many others. Therefore, $x = y^2$ is not a function because for at least one *x*-value, there corresponds more than one *y*-value.

To find the domain, note that each *x*-value is the square of a real number and can never be negative. Thus, the domain is $x \geq 0$.

There are no restrictions on replacements for *y*; therefore, the range is the set of all real numbers.

Consider the relations in Examples 2 and 3. Are they functions? Note that in the relation $y = x^2$, for each value of *x* there is only one corresponding value of *y*. For example, (2, 4), (−2, 4), (3, 9), (−3, 9), (4, 16), (−4, 16), and so forth. Therefore, $y = x^2$ is a function.

Example 3 was the relation $y = \sqrt{x - 4}$. Here we find that for each *x*-value there corresponds only one *y*-value [for example, (5, 1), (8, 2), (10, $\sqrt{6}$), . . .]. Therefore, $y = \sqrt{x - 4}$ is a function.

In summary, a function is a relationship between two sets of numbers, the domain and the range, which relates each number, *x*, in the domain to one and only one number, *y*, in the range.

Next let's consider the following, more intuitive function: On a summer vacation trip, you are driving 65 mi/h using cruise control on an interstate highway. You want to relate the distance and the time you are traveling. First, we know that distance equals rate times time, or $d = rt$. Also, since $r = 65$, we have the relation $d = 65t$. As you drive along, you begin to think how far you can drive in one hour:

$$d = 65t = 65(1) = 65 \text{ mi}$$

How far can you drive in three hours?

$$d = 65t = 65(3) = 195 \text{ mi}$$

Is this relation a function? Yes, because for each value of *t*, there is one and only one value of *d*; that is, during each driving time period, there is one and only one distance traveled. What are the domain and the range? First, note that $t \geq 0$ and $d \geq 0$. While there are no theoretical upper limits on *t* and *d*, the practical limits depend on the amount of time and the distance that you want to travel.

## Functional Notation

To say that *y* is a function of *x* means that for each value of *x* from the domain of the function, we can find exactly one value of *y* from the range. This statement is said so often that we have developed the following notation, called **functional notation,** to write that $y$ is a function of $x$:

$$y = f(x)$$

with $f(x)$ read "$f$ of $x$." *Note:* $f(x)$ does **not** mean $f$ times $x$.

In each of the following equations, $y$ can be replaced by $f(x)$, and the resulting equation is written in functional notation.

| *Equation* | *Functional notation form* |
|---|---|
| $y = 3x - 4$ | $f(x) = 3x - 4$ |
| $y = 5x^2 - 8x + 7$ | $f(x) = 5x^2 - 8x + 7$ |
| $y = \sqrt{6 - 2x}$ | $f(x) = \sqrt{6 - 2x}$ |

Functional notation can be used to simplify statements. For example, find the value of $y = 3x^2 + 5x - 6$ for $x = 2$. Using substitution, we replace $x$ with 2 as follows:

$$y = 3x^2 + 5x - 6$$
$$y = 3(2)^2 + 5(2) - 6 = 16$$

The statement "Find the value of $y = 3x^2 + 5x - 6$ for $x = 2$" may be abbreviated using functional notation as follows:

$$\text{Given } f(x) = 3x^2 + 5x - 6, \text{ find } f(2).$$

### EXAMPLE 6

Given the function $f(x) = 5x - 4$, find each of the following:
(a) $f(0)$
    Replace $x$ with 0 as follows:

$$f(0) = 5(0) - 4 = 0 - 4 = -4$$

(b) $f(7)$
    Replace $x$ with 7 as follows:

$$f(7) = 5(7) - 4 = 35 - 4 = 31$$

A function is usually named by a specific letter, such as $f(x)$, where $f$ names the function. Other letters, such as $g$ in $g(x)$ and $h$ in $h(x)$, are often used to represent or name functions.

### EXAMPLE 7

Given the function $g(x) = \sqrt{x + 4} + 3x^2$, find each of the following:
(a) $g(5)$
    Replace $x$ with 5 as follows

$$g(5) = \sqrt{5 + 4} + 3(5)^2 = 3 + 75 = 78$$

(b) $g(-3)$
    Replace $x$ with $-3$ as follows:

$$g(-3) = \sqrt{-3 + 4} + 3(-3)^2 = 1 + 27 = 28$$

(c) $g(-10)$

Replace $x$ with $-10$ as follows:

$$g(-10) = \sqrt{-10 + 4} + 3(-10)^2 = \sqrt{-6} + 300$$

which is not a real number because $\sqrt{-6}$ is not a real number. Another way of responding to Part (c) is to say, "Since $-10$ is not in the domain of $g(x)$, $g(-10)$ has no real value."

Letters may also be used with functional notation as illustrated by the following example.

**EXAMPLE 8**

Given the function $f(x) = x^2 - 4x$, find each of the following:

(a) $f(a)$

Replace $x$ with $a$ as follows:

$$f(a) = a^2 - 4a$$

(b) $f(3c^2)$

Replace $x$ with $3c^2$ as follows:

$$f(3c^2) = (3c^2)^2 - 4(3c^2) = 9c^4 - 12c^2$$

(c) $f(a + 5)$

Replace $x$ with $a + 5$ as follows:

$$\begin{aligned}
f(a + 5) &= (a + 5)^2 - 4(a + 5) \\
&= a^2 + 10a + 25 - 4a - 20 \\
&= a^2 + 6a + 5
\end{aligned}$$

Other letters, such as $t$ in $f(t)$ and $r$ in $f(r)$, are used in applications to name independent variables.

**EXAMPLE 9**

Given the function $f(t) = 0.50t + 5.4$, find each of the following:

(a) $f(3.2)$

Replace $t$ with 3.2 as follows:

$$\begin{aligned}
f(t) &= 0.50t + 5.4 \\
f(3.2) &= 0.50(3.2) + 5.4 \\
&= 1.6 + 5.4 \\
&= 7.0
\end{aligned}$$

(b) $f(t_0)$

Replace $t$ with $t_0$ as follows:

$$\begin{aligned}
f(t) &= 0.50t + 5.4 \\
f(t_0) &= 0.50t_0 + 5.4
\end{aligned}$$

## Exercises 4.1

*Determine whether or not each relation is a function. Write its domain and its range.*

1. $A = \{(2, 4), (3, 7), (9, 2)\}$

2. $B = \{(5, 2), (3, 3), (1, 2)\}$

**3.** $C = \{(2, 5), (7, 3), (2, 1), (1, 3)\}$

**4.** $D = \{(0, 2), (5, -1), (2, 7), (5, 1)\}$

**5.** $E = \{(3, 2), (5, 2), (2, 2), (-2, 2)\}$

**6.** $F = \{(3, 4), (3, -4), (-3, -4), (-3, 4)\}$

**7.** $y = 2x + 5$

**8.** $y = -3x$

**9.** $y = x^2 + 1$

**10.** $y = 2x^2 - 3$

**11.** $x = y^2 - 2$

**12.** $x = 3y^2 + 4$

**13.** $y = \sqrt{x + 3}$

**14.** $y = \sqrt{3 - 6x}$

**15.** $y = 6 + \sqrt{2x - 8}$

**16.** $y = 16 - \sqrt{x + 5}$

**17.** Given the function $f(x) = 8x - 12$, find

   **(a)** $f(4)$       **(b)** $f(0)$       **(c)** $f(-2)$

**18.** Given the function $g(x) = 20 - 4x$, find

   **(a)** $g(6)$       **(b)** $g(0)$       **(c)** $g(-3)$

**19.** Given $g(x) = 10x + 15$, find

   **(a)** $g(2)$       **(b)** $g(0)$       **(c)** $g(-4)$

**20.** Given $f(x) = x^2 - 4$, find

   **(a)** $f(6)$       **(b)** $f(0)$       **(c)** $f(-6)$

**21.** Given $h(x) = 3x^2 + 4x$, find

   **(a)** $h(5)$       **(b)** $h(0)$       **(c)** $h(-2)$

**22.** Given $f(x) = -2x^2 + 6x - 7$, find

   **(a)** $f(3)$       **(b)** $f(0)$       **(c)** $f(-1)$

**23.** Given $f(t) = \dfrac{5 - t^2}{2t}$, find

   **(a)** $f(1)$       **(b)** $f(-3)$       **(c)** $f(0)$

**24.** Given $g(t) = \sqrt{21 - 5t}$, find

   **(a)** $g(1)$   **(b)** $g(-3)$   **(c)** $g(2)$   **(d)** $g(8)$

**25.** Given $f(x) = 6x + 8$, find

   **(a)** $f(a)$       **(b)** $f(4a)$       **(c)** $f(c^2)$

**26.** Given $g(x) = 8x^2 - 7x$, find

   **(a)** $g(z)$       **(b)** $g(2y)$       **(c)** $g(3t^2)$

**27.** Given $h(x) = 4x^2 - 12x$, find

   **(a)** $h(x + 2)$    **(b)** $h(x - 3)$    **(c)** $h(2x + 1)$

**28.** Given $f(y) = y^2 - 3y + 6$, find

   **(a)** $f(y - 1)$    **(b)** $f(y^2 + 1)$    **(c)** $f(1 - 4y)$

**29.** Given $f(x) = 3x - 1$ and $g(x) = x^2 - 6x + 1$, find

   **(a)** $f(x) + g(x)$   **(b)** $f(x) - g(x)$   **(c)** $[f(x)][g(x)]$   **(d)** $f(x + h)$

**30.** Given $f(t) = 5 - 2t + t^2$ and $g(t) = t^2 - 4t + 4$, find

   **(a)** $f(t) + g(t)$   **(b)** $g(t) - f(t)$   **(c)** $[f(t)][g(t)]$   **(d)** $g(t + h)$

*Find the domain of each function.*

**31.** $f(x) = \dfrac{3x + 4}{x - 2}$

**32.** $f(t) = \dfrac{8}{6t + 3}$

**33.** $g(t) = \dfrac{2t + 4t^2}{(t - 6)(t + 3)}$

**34.** $g(x) = \dfrac{3x - 10}{x^2 + 4}$

**35.** $f(x) = \dfrac{12}{\sqrt{15 - 3x}}$

**36.** $g(t) = \dfrac{9}{\sqrt{5t + 20}}$

## 4.2 GRAPHING EQUATIONS

D.3, D.4

Consider a plane in which two number lines intersect at right angles. Let the point of intersection be the zero point of each line and call it the **origin.** Each line is called an **axis.** The horizontal number line is usually called the ***x*-axis,** and the vertical line is usually called the ***y*-axis.** On each axis the same scale (unit length) is preferred but not always possible in all applications. Such a system is called the **rectangular coordinate system,** or the **Cartesian coordinate system.** (The name *Cartesian* is after Descartes, a seventeenth-century French mathematician. He first conceived this idea of combining algebra and geometry together in such a way that each could aid the study of the other.) The plane is divided by the axes into four regions called **quadrants.** The quadrants are numbered as shown in Fig. 4.1.

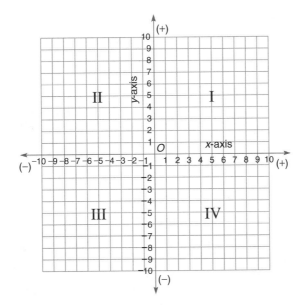

**Figure 4.1** The rectangular coordinate system

In the plane there is a point that corresponds to each ordered pair of real numbers $(x, y)$. Likewise, there is an ordered pair $(x, y)$ that corresponds to each point in the plane. Together $x$ and $y$ are called the **coordinates** of the point; $x$ is called the **abscissa,** and $y$ is called the **ordinate.** This relationship is called a **one-to-one correspondence.** The location, or position, of a point in the plane corresponding to a given ordered pair is found by first counting right or left from $O$ (origin) the number of units along the $x$-axis indicated by the first number of the ordered pair (right if positive, left if negative). Then from this point reached on the $x$-axis, count up or down the number of units indicated by the second number of the ordered pair (up if positive, down if negative).

**EXAMPLE 1**

Plot the point corresponding to each ordered pair in the number plane:

$A(3, 1)$   $B(2, -3)$   $C(-4, -2)$   $D(-3, 0)$   $E(-6, 2)$   $F(0, 2)$   (See Fig. 4.2.)

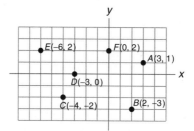

**Figure 4.2**

To graph equations we plot a sample of ordered pairs and connect them with a smooth curve. To obtain the sample, we need to generate ordered pairs from a given equation. One way to generate these ordered pairs is by randomly choosing a value for $x$, replacing this value for $x$ in the equation, and solving for $y$.

**EXAMPLE 2**

Graph $y = 2x - 3$.

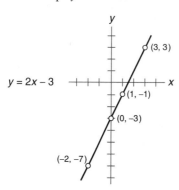

| $x$ | $y$ | $y = 2x - 3$    or    $f(x) = 2x - 3$ |
|-----|-----|---------------------------------------|
| 1   | $-1$ | $y = 2(1) - 3 = -1$ |
| 3   | 3   | $y = 2(3) - 3 = 3$ |
| $-2$ | $-7$ | $y = 2(-2) - 3 = -7$ |
| 0   | $-3$ | $y = 2(0) - 3 = -3$ |

**Figure 4.3**

Plot the ordered pairs and connect them with a smooth line as in Fig. 4.3.

A **linear equation** with two unknowns is an equation of degree one in the form $ax + by = c$ with $a$ and $b$ not both 0. Its graph is always a straight line. Therefore, two ordered pairs are sufficient to graph a linear function, since two points determine a straight line. However, finding a third point provides good insurance against a careless error.

**EXAMPLE 3**

Graph $y = -3x + 5$.

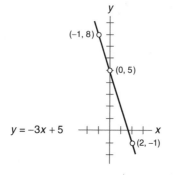

| $x$ | $y$ | $y = -3x + 5$    or    $g(x) = -3x + 5$ |
|-----|-----|-----------------------------------------|
| 0   | 5   | $y = -3(0) + 5 = 5$ |
| 2   | $-1$ | $y = -3(2) + 5 = -1$ |
| $-1$ | 8   | $y = -3(-1) + 5 = 8$ |

See Fig. 4.4.

**Figure 4.4**

The graph of an equation that is not linear is usually a curve of some kind and hence requires several points to sketch a smooth curve.

**EXAMPLE 4**

Graph $y = x^2 - 4$.

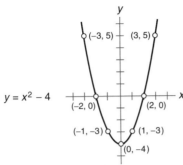

**Figure 4.5**

| $x$ | $y$ | $y = x^2 - 4$ |
|---|---|---|
| 0 | $-4$ | $y = (0)^2 - 4 = -4$ |
| 1 | $-3$ | $y = (1)^2 - 4 = -3$ |
| 2 | 0 | $y = (2)^2 - 4 = 0$ |
| 3 | 5 | $y = (3)^2 - 4 = 5$ |
| $-1$ | $-3$ | $y = (-1)^2 - 4 = -3$ |
| $-2$ | 0 | $y = (-2)^2 - 4 = 0$ |
| $-3$ | 5 | $y = (-3)^2 - 4 = 5$ |

See Fig. 4.5.

**EXAMPLE 5**

Graph $y = 2x^2 + x - 5$.

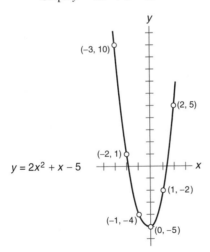

**Figure 4.6**

| $x$ | $y$ | $y = 2x^2 + x - 5$ |
|---|---|---|
| 0 | $-5$ | $y = 2(0)^2 + (0) - 5 = -5$ |
| 1 | $-2$ | $y = 2(1)^2 + (1) - 5 = -2$ |
| 2 | 5 | $y = 2(2)^2 + (2) - 5 = 5$ |
| $-1$ | $-4$ | $y = 2(-1)^2 + (-1) - 5 = -4$ |
| $-2$ | 1 | $y = 2(-2)^2 + (-2) - 5 = 1$ |
| $-3$ | 10 | $y = 2(-3)^2 + (-3) - 5 = 10$ |

See Fig. 4.6.

For a more complicated function, more ordered pairs are usually required to obtain a smooth curve. It may also be necessary to change the scale of the graph in order to plot enough ordered pairs to obtain a smooth curve. To change the scale means to enlarge or reduce the unit length on the axes according to a specified ratio. This ratio is chosen on the basis of fitting the necessary values in a given space allowed for the graph.

**EXAMPLE 6**

Graph $y = x^3 + 4x^2 - x - 4$.

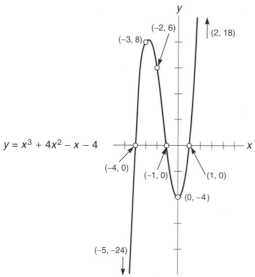

**Figure 4.7**

| $x$ | $y$ | $y = x^3 + 4x^2 - x - 4$ |
|---|---|---|
| 0 | $-4$ | $y = (0)^3 + 4(0)^2 - (0) - 4 = -4$ |
| 1 | 0 | $y = (1)^3 + 4(1)^2 - (1) - 4 = 0$ |
| 2 | 18 | $y = (2)^3 + 4(2)^2 - (2) - 4 = 18$ |
| 3 | 56 | $y = (3)^3 + 4(3)^2 - (3) - 4 = 56$ |
| $-1$ | 0 | $y = (-1)^3 + 4(-1)^2 - (-1) - 4 = 0$ |
| $-2$ | 6 | $y = (-2)^3 + 4(-2)^2 - (-2) - 4 = 6$ |
| $-3$ | 8 | $y = (-3)^3 + 4(-3)^2 - (-3) - 4 = 8$ |
| $-4$ | 0 | $y = (-4)^3 + 4(-4)^2 - (-4) - 4 = 0$ |
| $-5$ | $-24$ | $y = (-5)^3 + 4(-5)^2 - (-5) - 4 = -24$ |

See Fig. 4.7.

**EXAMPLE 7**

Graph $y = \sqrt{2x - 6}$.

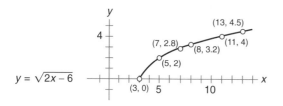

**Figure 4.8**

| $x$ | $y$ | $y = \sqrt{2x - 6}$ | |
|---|---|---|---|
| 3 | 0 | $y = \sqrt{2(3) - 6} = \sqrt{0} = 0$ | |
| 5 | 2 | $y = \sqrt{2(5) - 6} = \sqrt{4} = 2$ | |
| 7 | 2.8 | $y = \sqrt{2(7) - 6} = \sqrt{8} = 2.8$ | (approx.) |
| 8 | 3.2 | $y = \sqrt{2(8) - 6} = \sqrt{10} = 3.2$ | (approx.) |
| 11 | 4 | $y = \sqrt{2(11) - 6} = \sqrt{16} = 4$ | |
| 13 | 4.5 | $y = \sqrt{2(13) - 6} = \sqrt{20} = 4.5$ | (approx.) |

See Fig. 4.8.

## Solving Equations by Graphing

Equations may be solved graphically. This method is particularly useful when an algebraic method is very cumbersome, cannot be recalled, or does not exist; it is especially useful in technical applications.

***Solving for $y = 0$***   To solve the equation $y = x^2 - x - 6$ for $y = 0$ graphically means to find the point or points, if any, where the graph crosses the line $y = 0$ (the $x$-axis).

**EXAMPLE 8**

Solve $y = x^2 - x - 6$ for $y = 0$ graphically.

First, graph the equation $y = x^2 - x - 6$ (see Fig. 4.9).

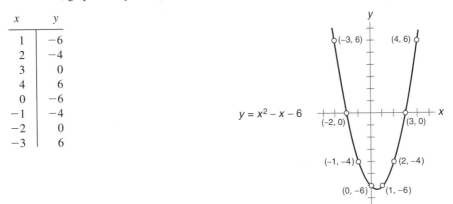

| $x$ | $y$ |
|---|---|
| 1 | $-6$ |
| 2 | $-4$ |
| 3 | 0 |
| 4 | 6 |
| 0 | $-6$ |
| $-1$ | $-4$ |
| $-2$ | 0 |
| $-3$ | 6 |

**Figure 4.9**

Then note the values of $x$ where the curve crosses the $x$-axis: $x = -2$ and $x = 3$. Therefore, from the graph the solutions of $y = x^2 - x - 6$ for $y = 0$ are $x = -2$ and $x = 3$.

Sometimes the curve crosses the $x$-axis between the unit marks on the $x$-axis. In this case we must estimate as closely as possible the point of intersection of the curve and the $x$-axis. If a particular problem requires greater accuracy, we can scale the graph to allow a more accurate estimation.

**EXAMPLE 9**

Solve $y = x^2 + 2x - 4$ for $y = 0$ graphically.

First, graph the equation $y = x^2 + 2x - 4$ (see Fig. 4.10).

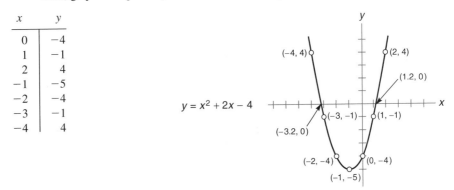

| $x$ | $y$ |
|---|---|
| 0 | $-4$ |
| 1 | $-1$ |
| 2 | 4 |
| $-1$ | $-5$ |
| $-2$ | $-4$ |
| $-3$ | $-1$ |
| $-4$ | 4 |

**Figure 4.10**

The values of $x$ where the curve crosses the $x$-axis are approximately 1.2 and $-3.2$. Therefore, the approximate solutions of $y = x^2 + 2x - 4$ for $y = 0$ are $x = 1.2$ and $x = -3.2$.

*Solving for y = k*

**EXAMPLE 10**

Solve the equation from Example 9, which was $y = x^2 + 2x - 4$, for $y = 4$ and $y = -3$.

First find the values of $x$ where the curve crosses the line $y = 4$. From the graph in Fig. 4.11, the $x$-values are 2 and $-4$. Therefore, solving for $y = 4$, we find that the solutions of $y = x^2 + 2x - 4$ are $x = 2$ and $x = -4$.

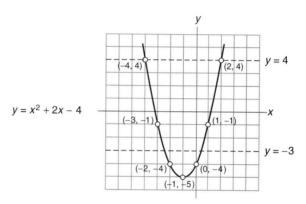

**Figure 4.11**

Next find the values of $x$ where the curve crosses the line $y = -3$. From the graph in Fig. 4.11, the approximate $x$-values are 0.4 and $-2.4$. That is, for $y = -3$, the solutions of $y = x^2 + 2x - 4$ are approximately $x = 0.4$ and $x = -2.4$.

Note that $y = x^2 + 2x - 4$ has no solutions for $y = -7$.

**EXAMPLE 11**

The voltage, $V$ in volts, in a given circuit varies with time, $t$ in ms, according to the equation $V = 6t^2 + t$. Solve for $t$ when $V = 0$, 70, and 120.

| $t$ | $V$ |
|---|---|
| 0 | 0 |
| 1 | 7 |
| 2 | 26 |
| 3 | 57 |
| 4 | 100 |
| 5 | 155 |

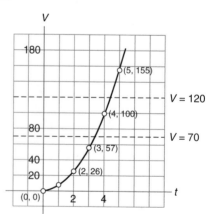

$V = 6t^2 + t$

**Figure 4.12**

In Fig. 4.12, we used the following scales.

$$t: \quad 1 \text{ square} = 1 \text{ ms}$$

$$V: \quad 1 \text{ square} = 20 \text{ V}$$

From the graph,

$$\text{at } V = 0 \text{ V}, \quad t = 0 \text{ ms}$$
$$\text{at } V = 70 \text{ V}, \quad t = 3.3 \text{ ms}$$
$$\text{at } V = 120 \text{ V}, \quad t = 4.4 \text{ ms}$$

Negative values of time, $t$, are not meaningful in this example.

**EXAMPLE 12**

The work, $w$, done in a circuit varies with time, $t$, according to the equation $w = 8t^2 + 4t$. Solve for $t$ when $w = 60$, 120, and 250.

| $t$ | $w$ |
|---|---|
| 0 | 0 |
| 1 | 12 |
| 2 | 40 |
| 3 | 84 |
| 4 | 144 |
| 5 | 220 |
| 6 | 312 |

$w = 8t^2 + 4t$

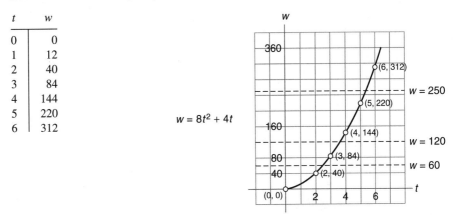

**Figure 4.13**

In Fig. 4.13, we used these scales

$$t: \quad 1 \text{ square} = 1 \text{ unit}$$
$$w: \quad 1 \text{ square} = 40 \text{ units}$$

From the graph,

$$\text{at } w = 60, \quad t = 2.5$$
$$\text{at } w = 120, \quad t = 3.6$$
$$\text{at } w = 250, \quad t = 5.4$$

## Exercises 4.2

**1.** Plot the point corresponding to each ordered pair in the number plane.

| | | |
|---|---|---|
| $A(2, 5)$ | $B(-3, -6)$ | $C(-4, 0)$ |
| $D(2, -1)$ | $E(3, 4)$ | $F(0, 3)$ |
| $G(0, -3)$ | $H(5, 0)$ | $I(-4, -1)$ |
| $J(-2, 2)$ | $K(3, -5)$ | $L(-3, 4)$ |

**2.** Plot the point corresponding to each ordered pair in the number plane.

| | | |
|---|---|---|
| $P_1(3, 2)$ | $P_2(-2, 5)$ | $P_3(-1, -6)$ |
| $P_4(4, -7)$ | $P_5(0, 4)$ | $P_6(-5, 0)$ |
| $P_7(-4, 4)$ | $P_8(6, 0)$ | $P_9(-8, 5)$ |
| $P_{10}(7, -2)$ | $P_{11}(-5, -3)$ | $P_{12}(0, -8)$ |

*Graph each equation.*

**3.** $y = 2x + 1$

**4.** $y = 3x - 4$

**5.** $-2x - 3y = 6$

**6.** $2y = -4x - 3$

**7.** $y = x^2 - 9$

**8.** $y = x^2 + x - 6$

**9.** $y = x^2 - 5x + 4$

**10.** $y = x^2 + 3$

**11.** $y = 2x^2 + 3x - 2$

**12.** $y = -x^2 + 2x + 4$

**13.** $y = x^2 + 2x$

**14.** $y = x^2 - 4x$

**15.** $y = -2x^2 + 4x$

**16.** $y = -\dfrac{1}{4}x^2 - \dfrac{3}{2}x + 2$

**17.** $y = x^3 - x^2 - 10x + 8$

**18.** $y = x^3 - 4x^2 + x + 6$

**19.** $y = x^3 + 2x^2 - 7x + 4$

**20.** $y = x^3 - 8x - 3$

**21.** $y = \sqrt{x + 4}$

**22.** $y = \sqrt{3x - 12}$

**23.** $y = \sqrt{12 - 6x}$

**24.** $y = \sqrt{3 - x}$

*Solve each equation graphically for the given values.*

**25.** Exercise 7 for $y = 0$, $-5$, and 2.

**26.** Exercise 8 for $y = 0$, 6, and $-3$.

**27.** Exercise 9 for $y = 0$, 2, and $-4$.

**28.** Exercise 10 for $y = 0$, 4, and 6.

**29.** Exercise 11 for $y = 0$, 3, and 5.

**30.** Exercise 12 for $y = 0$, 4, and $-2$.

**31.** Exercise 13 for $y = 0$, 3, and 6.

**32.** Exercise 14 for $y = 0$, $-2$, and 3.

**33.** Exercise 15 for $y = 0$, 5, $-4$, and $-1\frac{1}{2}$.

**34.** Exercise 16 for $y = 0$, $-1$, and 1.5.

**35.** Exercise 17 for $y = 0$, 2, and $-2$.

**36.** Exercise 18 for $y = 0$, 2, and 8.

**37.** Exercise 19 for $y = 0$, 4, and 8.

**38.** Exercise 20 for $y = 0$, 2, and $-3$.

*Solve each equation graphically for the given values.*

**39.** $y = x^2 + 3x - 4$ for $y = 0$, 6, and $-2$.

**40.** $y = 2x^2 - 5x - 3$ for $y = 2$, 0, and $-3$.

**41.** $y = -\dfrac{1}{2}x^2 + 2$ for $y = 0$, 4, and $-4$.

**42.** $y = -\dfrac{1}{4}x^2 + x$ for $y = 0$, $\dfrac{1}{2}$, and $-4$.

**43.** $y = x^3 - 3x^2 + 1$ for $y = 0$, $-2$, and $-0.5$.

**44.** $y = -x^3 + 3x + 2$ for $y = 2$, 0, and 3.

**45.** The resistance, $r$, of a resistor in a circuit of constant current varies with time, $t$ in ms, according to the equation

$$r = 10t^2 + 20$$

Solve for $t$ when $r = 90\ \Omega$, $180\ \Omega$, and $320\ \Omega$.

**46.** An object dropped from an airplane 2500 m above the ground falls according to the equation $h = 2500 - 4.95t^2$, where $h$ is the height in metres above the ground and $t$ is the time in seconds. Find the times for the object to fall to a height of 2000 m, 1200 m, and 600 m above the ground. Also find the time it takes to hit the ground.

**47.** The energy dissipated (work lost), $w$, by a resistor varies with the time, $t$ in ms, according to the equation $w = 5t^2 + 6t$. Solve for $t$ when $w = 2$, 4, and 10.

48. The resistance $r$ in ohms, in a given circuit varies with time, $t$ in ms, according to the equation $r = 10 + \sqrt{t}$. Find $t$ when $r = 14.1\ \Omega$, $14.3\ \Omega$, and $14.7\ \Omega$. (*Hint:* Choose a suitable scale for the graph and graph only the part you need.)

49. A given inductor carries a current expressed by the equation $i = t^3 - 15$ where $i$ is the current in amperes and $t$ is the time in seconds. Find $t$ when $i$ is 5 A and 15 A.

50. A charge, $q$ in coulombs, flowing in a given circuit varies with the time, $t$ in $\mu$s, according to the equation $q = t^2 - \dfrac{t^3}{3}$. Find $t$ when $q$ is 5 coulombs (C) and 10 C.

51. A machinist needs to drill four holes 2.00 in. apart in a straight line in a metal plate as shown below. The first hole is placed at the origin, and the line forms an angle of 36.0° with the vertical axis. Find the coordinates of the other three holes.

52. A machinist often uses a coordinate system to drill holes by placing the origin at the most convenient location. A bolt circle is the circle formed by completing an arc through the centers of the bolt holes in a piece of metal. Find the coordinates of the centers of eight equally spaced, $\frac{1}{4}$-in. holes on a bolt circle of radius 4.00 in. as shown below.

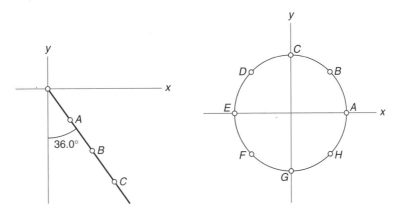

## 4.3 THE STRAIGHT LINE

D.3, D.4

**Analytic geometry** is the study of the relationships between algebra and geometry. The concepts of analytic geometry provide us with ways of algebraically analyzing a geometrical problem. Likewise, with these concepts we can often solve an algebraic problem by viewing it geometrically.

We now develop several basic relations between equations and their graphs. The **slope** of a nonvertical line is the ratio of the difference of the $y$-coordinates of any two points on the line to the difference of their $x$-coordinates when the differences are taken in the same order (see Fig. 4.14).

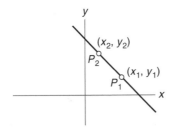

**Figure 4.14**

**EXAMPLE 1**

Find the slope of the line passing through $(-2, 1)$ and $(3, 5)$.

If we let $x_1 = -2$, $y_1 = 1$, $x_2 = 3$, and $y_2 = 5$ as in Fig. 4.15, then

$$m = \frac{y_2 - y_1}{x_2 - x_1} = \frac{5 - 1}{3 - (-2)} = \frac{4}{5}$$

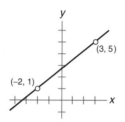

**Figure 4.15**

Note that if we reverse the order of taking the differences of the coordinates, the result is the same.

$$\frac{y_1 - y_2}{x_1 - x_2} = \frac{1 - 5}{-2 - 3} = \frac{-4}{-5} = \frac{4}{5} = m$$

**EXAMPLE 2**

Find the slope of the line passing through $(-2, 4)$ and $(6, -6)$.

If we let $x_1 = -2$, $y_1 = 4$, $x_2 = 6$, and $y_2 = -6$ as in Fig. 4.16, then

$$m = \frac{y_2 - y_1}{x_2 - x_1} = \frac{-6 - 4}{6 - (-2)} = \frac{-10}{8} = -\frac{5}{4}$$

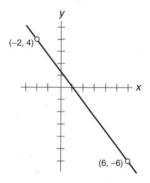

**Figure 4.16**

Note that in Example 1, the line slopes upward from left to right, while in Example 2 the line slopes downward. In general, we have the following:

1. If a line has positive slope, then the line slopes upward from left to right ("rises").

2. If a line has negative slope, then the line slopes downward from left to right ("falls").

3. If the line has zero slope, then the line is horizontal ("flat").

4. If the line is vertical, then the line has no slope as $x_1 = x_2$, or $x_2 - x_1 = 0$. In this case, the ratio $\dfrac{y_2 - y_1}{x_2 - x_1}$ is undefined because division by zero is undefined.

We can use these facts to assist us in graphing a line if we know the slope of the line and one point $P$ on the line. The line can be sketched by drawing a line through the given point $P$ and a point $Q$ which is plotted by moving one unit to the right of $P$, then moving vertically $m$ units. That is, a point moving along a line will move vertically an amount equal to $m$, the slope, for every unit move to the right as in Fig. 4.17.

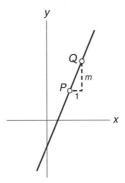

**Figure 4.17**

**EXAMPLE 3**

Graph a line with slope $-2$ that passes through the point (1, 3).

Since the slope is $-2$, points on the line drop 2 units for every unit move to the right. The line passes through (1, 3) and (2, 1) as in Fig. 4.18.

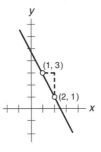

**Figure 4.18**

Knowing the slope and one point on the line will also determine the equation of the straight line. Let $m$ be the slope of a given nonvertical straight line, and let $(x_1, y_1)$ be the coordinates of a point on this line. If $(x, y)$ is any other point on the line as in Fig. 4.19, then we have

$$\frac{y - y_1}{x - x_1} = m$$

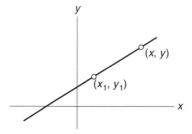

**Figure 4.19**

By multiplying each side of the equation by $(x - x_1)$, we obtain the following.

---

**POINT-SLOPE FORM OF A STRAIGHT (OBLIQUE) LINE**

If $m$ is the slope and $(x_1, y_1)$ is any point on a nonvertical straight line, its equation is

$$y - y_1 = m(x - x_1)$$

---

**EXAMPLE 4**

Find the equation of the line with slope 3 that passes through the point $(-1, 2)$.

Here $m = 3$, $x_1 = -1$, and $y_1 = 2$. Using the point-slope form, we have the equation

$$y - y_1 = m(x - x_1)$$
$$y - 2 = 3(x - (-1))$$

Simplifying, we have

$$y - 2 = 3x + 3$$
$$y = 3x + 5$$

The point-slope form can also be used to find the equation of a straight line that passes through two points.

**EXAMPLE 5**

Find the equation of the line passing through the points $(2, -3)$ and $(-2, 5)$.

First, find the slope.

$$m = \frac{y_2 - y_1}{x_2 - x_1} = \frac{5 - (-3)}{-2 - 2} = \frac{5 + 3}{-4} = \frac{8}{-4} = -2$$

Substitute $m = -2$ and the point $(2, -3)$ in the point-slope form.

$$y - y_1 = m(x - x_1)$$
$$y - (-3) = -2(x - 2)$$
$$y + 3 = -2x + 4$$
$$2x + y - 1 = 0$$

*Note:* We could have used the other point $(-2, 5)$ in the point-slope form to obtain the equation

$$y - 5 = -2(x - (-2))$$

which also simplifies to

$$2x + y - 1 = 0$$

A nonvertical line will intersect the $y$-axis at some point in the form $(0, b)$ as in Fig. 4.20. This ordinate ($y$-value) $b$ is called the **$y$-intercept** of the line. If the slope of the line is $m$, then

$$y - y_1 = m(x - x_1)$$
$$y - b = m(x - 0)$$
$$y - b = mx$$
$$y = mx + b$$

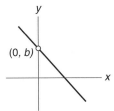

Figure 4.20

---

**SLOPE-INTERCEPT FORM OF A STRAIGHT (OBLIQUE) LINE**

If $m$ is the slope and $(0, b)$ is the $y$-intercept of a nonvertical straight line, its equation is

$$y = mx + b$$

---

**EXAMPLE 6**

Find the equation of the line with slope $\frac{1}{2}$ that crosses the $y$-axis at $b = -3$.
   Using the slope-intercept form, we have

$$y = mx + b$$
$$y = \frac{1}{2}x + (-3)$$
$$y = \frac{1}{2}x - 3$$

or

$$x - 2y - 6 = 0$$

A line parallel to the $x$-axis has slope $m = 0$ (see Fig. 4.21). Its equation is

$$y = mx + b$$
$$y = (0)x + b$$
$$y = b$$

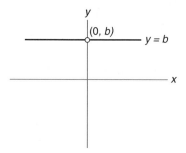

**Figure 4.21**  Horizontal line

---

**EQUATION OF A HORIZONTAL LINE**

If a horizontal line passes through the point $(a, b)$, its equation is

$$y = b$$

---

**EXAMPLE 7**

Find the equation of the line parallel to and 3 units above the $x$-axis.
  The equation is $y = 3$.

By writing the equation of a nonvertical straight line in the slope-intercept form, we can quickly determine the line's slope and a point on the line (the point where it crosses the $y$-axis).

**EXAMPLE 8**

Find the slope and the $y$-intercept of $3y - x + 6 = 0$. Graph the line.
  Write the equation in slope-intercept form; that is, solve for $y$.

$$3y - x + 6 = 0$$

$$3y = x - 6$$

$$y = \frac{x}{3} - 2$$

$$y = \left(\frac{1}{3}\right)x + (-2)$$

So $m = \dfrac{1}{3}$ and $b = -2$ (see Fig. 4.22).

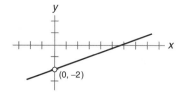

**Figure 4.22**

**EXAMPLE 9**

Describe and graph the line whose equation is $y = -5$.

This is a line parallel to and 5 units below the $x$-axis (see Fig. 4.23).

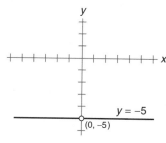

$y = -5$

$(0, -5)$

**Figure 4.23**

If a line is vertical, then we cannot use any of these equations since the line has no slope. However, note that in this case, as shown in Fig. 4.24, the line crosses the $x$-axis at some point in the form $(a, 0)$. All points on the line have the same abscissa as the point $(a, 0)$. This characterizes the line, giving us the following equation.

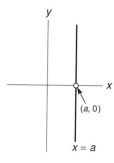

$(a, 0)$

$x = a$

**Figure 4.24**   Vertical Line

---

**EQUATION OF A VERTICAL LINE**

If a vertical line passes through the point $(a, b)$, its equation is

$$x = a$$

---

**EXAMPLE 10**

Describe and graph the line whose equation is $x = 2$.

This is a line perpendicular to the $x$-axis that crosses the $x$-axis at the point $(2, 0)$ (see Fig. 4.25).

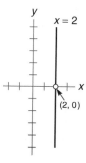

$x = 2$

$(2, 0)$

**Figure 4.25**

**EXAMPLE 11**

Write the equation of the line perpendicular to the $x$-axis that crosses the $x$-axis at the point $(-3, 0)$.

The equation is $x = -3$.

*Note:* All the equations presented in this section can be put in the form

$$Ax + By + C = 0 \quad \text{with } A \text{ and } B \text{ not both } 0$$

This is known as the **general form** of the equation of the line and agrees with our definition of a linear equation whose graph is a straight line.

## Exercises 4.3

*Find the slope of each line passing through the given points.*

**1.** $(4, 2), (3, 1)$        **2.** $(-3, 2), (-1, -2)$        **3.** $(4, -5), (2, 3)$

**4.** $(-6, -4), (5, -3)$        **5.** $(-3, 2), (6, 2)$        **6.** $(4, -7), (4, 3)$

**7.** $(5, 7), (-3, 2)$        **8.** $(-3, 6), (-1, 3)$

*Graph each line passing through the given point with the given slope.*

**9.** $(2, -1), m = 2$        **10.** $(0, 1), m = -3$        **11.** $(-3, -2), m = \dfrac{1}{2}$

**12.** $(4, 4), m = -\dfrac{1}{3}$        **13.** $(4, 0), m = -2$        **14.** $(-3, 1), m = 4$

**15.** $(0, -3), m = -\dfrac{3}{4}$        **16.** $(5, -2), m = \dfrac{3}{2}$

*Find the equation of the line with the given properties.*

**17.** Passes through $(-2, 8)$ with slope of $-3$.

**18.** Passes through $(3, -5)$ with slope of $2$.

**19.** Passes through $(-3, -4)$ with slope of $\frac{1}{2}$.

**20.** Passes through $(6, -7)$ with slope of $-\frac{3}{4}$.

**21.** Passes through $(-2, 7)$ and $(1, 4)$.

**22.** Passes through $(1, 6)$ and $(4, -3)$.

**23.** Passes through $(6, -8)$ and $(-4, -3)$.

**24.** Passes through $(-2, 2)$ and $(7, -1)$.

**25.** Crosses the $y$-axis at $-2$ with slope of $-5$.

**26.** Crosses the $y$-axis at $8$ with slope of $\frac{1}{3}$.

**27.** Has $y$-intercept of $7$ and slope of $2$.

**28.** Has $y$-intercept of $-4$ and slope of $-\frac{3}{4}$.

**29.** Parallel to and $5$ units above the $x$-axis.

**30.** Parallel to and $2$ units below the $x$-axis.

**31.** Perpendicular to the $x$-axis and crosses the $x$-axis at $(-2, 0)$.

**32.** Perpendicular to the $x$-axis and crosses the $x$-axis at $(5, 0)$.

**33.** Parallel to the $x$-axis containing the point $(2, -3)$.

**34.** Parallel to the $y$-axis containing the point $(-5, -4)$.

**35.** Perpendicular to the $x$-axis containing the point $(-7, 9)$.

**36.** Perpendicular to the $y$-axis containing the point $(4, 6)$.

*Find the slope and the y-intercept of each straight line.*

**37.** $x + 4y = 12$      **38.** $-2x + 3y + 9 = 0$      **39.** $4x - 2y + 14 = 0$

**40.** $3x - 6y = 0$      **41.** $y = 6$      **42.** $x = -4$

*Graph each equation.*

**43.** $y = 3x - 2$      **44.** $y = -2x + 5$      **45.** $5x - 2y + 4 = 0$

**46.** $4x + 3y + 6 = 0$      **47.** $x = 7$      **48.** $x = -2$

**49.** $y = -3$      **50.** $y = 2$      **51.** $6x + 8y = 24$

**52.** $3x - 5y = 30$      **53.** $x - 3y = -12$      **54.** $x + 6y = 8$

**55.** A certain metal rod with temperature $-15.0°C$ is 43.0 cm long and at $55.0°C$ is 43.2 cm long. These data can be listed in $(x, y)$ form as $(-15.0, 43.0)$ and $(55.0, 43.2)$. Find the slope (as a simplified fraction) of the straight line passing through these two points.

## 4.4 PARALLEL AND PERPENDICULAR LINES

**PARALLEL LINES**

Two lines are parallel if either one of the following conditions holds:

**1.** They are both perpendicular to the $x$-axis [see Fig. 4.26(a)].

**2.** They both have the same slope [see Fig. 4.26(b)]. That is, if the equations of the lines are

$$L_1: \quad y = m_1x + b_1 \quad \text{and} \quad L_2: \quad y = m_2x + b_2$$

then

$$m_1 = m_2$$

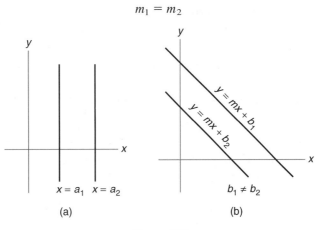

(a)          (b)

**Figure 4.26**

PERPENDICULAR LINES

Two lines are perpendicular if either one of the following conditions holds:

1. One line is vertical with equation $x = a$, and the other is horizontal with equation $y = b$.

2. Neither is vertical and the slope of one line is the negative reciprocal of the other. That is, if the equations of the lines are

$$L_1: \quad y = m_1 x + b_1 \quad \text{and} \quad L_2: \quad y = m_2 x + b_2$$

then

$$m_1 = -\frac{1}{m_2}$$

To show this last relationship, consider the triangle in Fig. 4.27, where $L_1$ is perpendicular to $L_2$. Let

$(c, 0)$ represent the point $P$

$(d, 0)$ represent the point $R$

$(e, f)$ represent the point $Q$

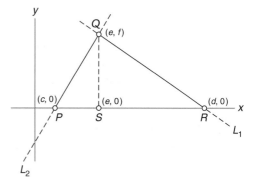

**Figure 4.27**

Draw $QS$ perpendicular to the $x$-axis. Then $S$ must be represented by $(e, 0)$.

Triangles $PSQ$ and $QSR$ in Fig. 4.27 are similar. (Note that angle $PQS$ equals angle $QRS$.) From geometry we know that

$$\frac{PS}{QS} = \frac{QS}{SR} \tag{4.1}$$

In this case,

$$PS = e - c \quad \text{(the distance from } c \text{ to } e \text{ on the } x\text{-axis)}$$
$$QS = f \quad \text{(the distance from 0 to } f \text{ on the } y\text{-axis)}$$
$$SR = d - e \quad \text{(the distance from } e \text{ to } d \text{ on the } x\text{-axis)}$$

Substituting these values in Equation (4.1), we have

$$\frac{e - c}{f} = \frac{f}{d - e}$$

Multiplying each side of the equation by $(d - e)f$ gives

$$f^2 = (d - e)(e - c) \tag{4.2}$$

Compute slopes $m_1$ and $m_2$ as follows:

$$m_1 = \frac{f - 0}{e - d} = \frac{f}{e - d}$$

$$m_2 = \frac{f - 0}{e - c} = \frac{f}{e - c}$$

$$(m_1)(m_2) = \frac{f}{e - d} \cdot \frac{f}{e - c} = \frac{f^2}{(e - d)(e - c)}$$

Substituting from Equation 4.2 we have

$$(m_1)(m_2) = \frac{(d - e)(e - c)}{(e - d)(e - c)} = \frac{d - e}{e - d} = -\left(\frac{e - d}{e - d}\right) = -1$$

or

$$(m_1)(m_2) = -1$$

Dividing each side of this equation by $m_2$ we have

$$m_1 = \frac{-1}{m_2}$$

### EXAMPLE 1

Determine whether the lines given by the equations $3y + 6x - 5 = 0$ and $2y - x + 7 = 0$ are perpendicular.

Change each equation into slope-intercept form; that is, solve for $y$.

$$y = -2x + \frac{5}{3} \qquad \text{(Slope is } -2\text{.)}$$

and

$$y = \frac{1}{2}x - \frac{7}{2} \qquad \text{(Slope is } \frac{1}{2}\text{.)}$$

Since

$$-2 = \frac{-1}{\frac{1}{2}} \qquad \left(-2 \text{ is the negative reciprocal of } \frac{1}{2}.\right)$$

the lines are perpendicular.

### EXAMPLE 2

Find the equation of the line through $(-3, 2)$ and perpendicular to $2y - 3x + 5 = 0$.

We can find the slope of the desired line by finding the negative reciprocal of the slope of the given line. First find the slope of the line $2y - 3x + 5 = 0$. Writing this equation in slope-intercept form, we have

$$y = \frac{3}{2}x - \frac{5}{2}$$

The slope of this line is $m = \frac{3}{2}$. The slope of the line perpendicular to this line is then equal to $-\frac{2}{3}$, the negative reciprocal of $\frac{3}{2}$. Now using the point-slope form, we have

$$y - y_1 = m(x - x_1)$$

$$y - 2 = -\frac{2}{3}(x - (-3))$$

$$y - 2 = -\frac{2}{3}(x + 3)$$

or

$$2x + 3y = 0$$

**EXAMPLE 3**

Find the equation of the line through $(2, -5)$ and parallel to $3x + y = 7$.
First, find the slope of the given line by solving its equation for $y$.

$$y = -3x + 7$$

Its slope is $-3$. The slope of any line parallel to this line has the same slope. Now, write the equation of the line with slope $-3$ passing through $(2, -5)$.

$$y - y_1 = m(x - x_1)$$
$$y - (-5) = -3(x - 2)$$
$$y + 5 = -3x + 6$$
$$y = -3x + 1 \quad \text{or} \quad 3x + y = 1$$

## Exercises 4.4

*Determine whether each given pair of equations represents lines that are parallel, perpendicular, or neither.*

1. $x + 3y - 7 = 0$; $-3x + y + 2 = 0$
2. $x + 2y - 11 = 0$; $x + 2y + 4 = 0$
3. $-x + 4y + 7 = 0$; $x + 4y - 5 = 0$
4. $2x + 7y + 4 = 0$; $7x - 2y - 5 = 0$
5. $y - 5x + 13 = 0$; $y - 5x + 9 = 0$
6. $-3x + 9y + 22 = 0$; $x + 3y - 17 = 0$

*Find the equation of the line that satisfies each set of conditions.*

7. Passes through $(-1, 5)$ and is parallel to $-2x + y + 13 = 0$.
8. Passes through $(2, -2)$ and is perpendicular to $3x - 2y - 14 = 0$.
9. Passes through $(-7, 4)$ and is perpendicular to $5y = x$.
10. Passes through $(2, -10)$ and is parallel to $2x + 3y - 7 = 0$.
11. Passes through the origin and is parallel to $3x - 4y = 12$.
12. Passes through the origin and is perpendicular to $4x + 5y = 17$.
13. Has an $x$-intercept of 6 and is perpendicular to $4x + 6y = 9$.
14. Has a $y$-intercept of $-2$ and is parallel to $6x - 4y = 11$.
15. Has a $y$-intercept of 8 and is parallel to $y = 2$.
16. Has an $x$-intercept of $-4$ and is perpendicular to $y = 6$.
17. Has an $x$-intercept of 7 and is parallel to $x = -4$.
18. Has a $y$-intercept of $-9$ and is perpendicular to $x = 5$.

**19.** The vertices of a quadrilateral are $A(-2, 3)$, $B(2, 2)$, $C(9, 6)$, and $D(5, 7)$.
   **(a)** Is the quadrilateral a parallelogram? Why or why not?
   **(b)** Is the quadrilateral a rectangle? Why or why not?

**20.** The vertices of a quadrilateral are $A(-4, 1)$, $B(0, -2)$, $C(6, 6)$, and $D(2, 9)$.
   **(a)** Is the quadrilateral a parallelogram? Why or why not?
   **(b)** Is the quadrilateral a rectangle? Why or why not?

## 4.5 THE DISTANCE AND MIDPOINT FORMULAS

D.1, D.2

We now wish to find the distance between two points on a straight line. Suppose $P$ has the coordinates $(x_1, y_1)$ and $Q$ has the coordinates $(x_2, y_2)$. Then a triangle similar to that in Fig. 4.28 can be constructed. Note that $R$ must have the coordinates $(x_2, y_1)$. (Point $R$ has the same $x$-coordinate as $Q$ and the same $y$-coordinate as $P$.)

Using the Pythagorean theorem, we have

$$PQ^2 = PR^2 + QR^2 \qquad (4.3)$$

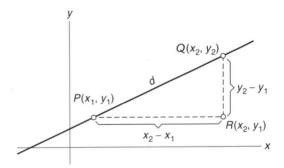

**Figure 4.28**   $d$ is the distance between points $P$ and $Q$.

Observe that

$$PR = x_2 - x_1 \qquad \text{(the horizontal distance between } x_1 \text{ and } x_2 \text{ on the } x\text{-axis)}$$
$$QR = y_2 - y_1 \qquad \text{(the vertical distance between } y_1 \text{ and } y_2 \text{ on the } y\text{-axis)}$$

Substituting these values for $PR$ and $QR$ in Equation (4.3) gives

$$PQ^2 = (x_2 - x_1)^2 + (y_2 - y_1)^2$$

---

**DISTANCE FORMULA**

The distance between two points $P(x_1, y_1)$ and $Q(x_2, y_2)$ is given by the formula

$$d = PQ = \sqrt{(x_2 - x_1)^2 + (y_2 - y_1)^2}$$

---

**EXAMPLE 1**

Find the distance, $d$, between $(3, 4)$ and $(-2, 7)$.

$$d = \sqrt{(x_2 - x_1)^2 + (y_2 - y_1)^2}$$
$$d = \sqrt{(-2 - 3)^2 + (7 - 4)^2}$$
$$= \sqrt{(-5)^2 + (3)^2} = \sqrt{25 + 9} = \sqrt{34}$$

Note that we can reverse the order of (3, 4) and (−2, 7) in the formula for computing $d$ without affecting the result.

$$d = \sqrt{(3 - (-2))^2 + (4 - 7)^2}$$
$$= \sqrt{(5)^2 + (-3)^2} = \sqrt{25 + 9} = \sqrt{34}$$

---

**MIDPOINT FORMULA**

The coordinates of point $Q(x_m, y_m)$ which is midway between two points $P(x_1, y_1)$ and $R(x_2, y_2)$ are given by

$$x_m = \frac{x_1 + x_2}{2} \qquad y_m = \frac{y_1 + y_2}{2}$$

---

Figure 4.29 illustrates the midpoint formula. First look at points $P$, $Q$, and $R$. Triangles $PSQ$ and $QTR$ are congruent. This means that

$$PS = QT$$

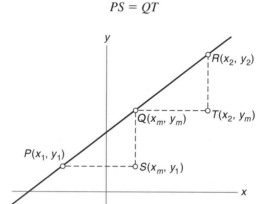

**Figure 4.29** Point $Q$ is the midpoint between points $P$ and $R$.

Since

$$PS = x_m - x_1$$

and

$$QT = x_2 - x_m$$

then

$$x_m - x_1 = x_2 - x_m$$

or

$$2x_m - x_1 = x_2$$
$$2x_m = x_1 + x_2$$
$$x_m = \frac{x_1 + x_2}{2}$$

The formula for $y_m$ is found in the same manner.

**EXAMPLE 2**

Find the point midway between $(2, -3)$ and $(-4, 6)$.

$$x_m = \frac{x_1 + x_2}{2} = \frac{2 + (-4)}{2} = \frac{-2}{2} = -1$$

$$y_m = \frac{y_1 + y_2}{2} = \frac{-3 + 6}{2} = \frac{3}{2}$$

The midpoint is $(-1, \frac{3}{2})$.

## Exercises 4.5

*Find the distance between each pair of points.*

**1.** $(4, -7); (-5, 5)$      **2.** $(4, 3); (-2, -1)$      **3.** $(3, -2); (10, -2)$

**4.** $(6, -2); (6, 4)$      **5.** $(5, -2); (1, 2)$      **6.** $(2, -3); (-1, 1)$

**7.** $(3, -5); (3, 2)$      **8.** $(2, -4); (6, -4)$

*Find the coordinates of the point midway between each pair of points.*

**9.** $(2, 3); (5, 7)$      **10.** $(0, 5); (2, -4)$      **11.** $(3, -2); (0, 0)$

**12.** $(2, -3); (4, -3)$      **13.** $(11, 4); (-11, -9)$      **14.** $(4, 10); (-6, -8)$

*The vertices of each $\triangle ABC$ are given below. For each triangle, find **(a)** the perimeter, **(b)** whether it is a right triangle, **(c)** whether it is isosceles, and **(d)** its area if it is a right triangle.*

**15.** $A(2, 8); B(10, 2); C(10, 8)$      **16.** $A(0, 0); B(3, 3); C(3, -3)$

**17.** $A(-3, 6); B(5, 0); C(4, 9)$      **18.** $A(-6, 3); B(-3, 7); C(1, 4)$

**19.** Given $\triangle ABC$ with vertices $A(7, -1)$, $B(9, 1)$, and $C(-3, 5)$, find the distance from $A$ to the midpoint of side $BC$.

**20.** Find the distance from $B$ to the midpoint of side $AC$ in Exercise 19.

**21.** Find the equation of the line parallel to the line $3x - 6y = 10$ and through the midpoint of $AB$, where $A(4, 2)$ and $B(8, -6)$.

**22.** Find the equation of the line perpendicular to the line $2x + 5y = 12$ and through the midpoint of $AB$, where $A(-3, -4)$ and $B(7, -8)$.

**23.** Find the equation of the line perpendicular to the line $4x + 8y = 16$ and through the midpoint of $AB$, where $A(-8, 12)$ and $B(6, 10)$.

**24.** Find the equation of the line parallel to the line $5x - 6y = 30$ and through the midpoint of $AB$, where $A(3, 11)$ and $B(7, -5)$.

*In Exercises 25–28, start with a graph and then use the distance formula and the slopes to confirm the given geometric figure.*

**25.** Show that $ABCD$ is a rectangle, where $A(-2, 2); B(1, 3); C(2, 0);$ and $D(-1, -1)$.

**26.** Show that $ABCD$ is a parallelogram, where $A(2, 6); B(7, 2); C(8, 5);$ and $D(3, 9)$.

**27.** Show that $ABCD$ is a trapezoid with one right angle, where $A(-12, 8); B(3, 2); C(5, 7); D(-5, 11)$.

**28.** Show that if the coordinates of the vertices of a triangle are $(a, b)$, $(a + c, b)$, and $(a + c, b + c)$, the triangle is a right triangle.

# CHAPTER SUMMARY

1. *Basic terms:*
   (a) *Relation:* Set of ordered pairs, usually in the form $(x, y)$.
   (b) *Independent variable:* First element of an ordered pair, usually $x$.
   (c) *Dependent variable:* Second element of an ordered pair, usually $y$.
   (d) *Domain:* Set of all first elements of the ordered pairs in a relation, or set of all $x$'s.
   (e) *Range:* Set of all second elements of the ordered pairs in a relation, or set of all $y$'s.
   (f) *Function:* Set of ordered pairs in which no two distinct ordered pairs have the same first element.
   (g) *Functional notation:* To write an equation in variables $x$ and $y$ in functional notation, solve for $y$ and replace $y$ with $f(x)$.
   (h) *Linear equation with two unknowns:* An equation of degree one in the form $ax + by = c$ where $a$ and $b$ are not both 0.

2. *Slope of a line:* If $P_1(x_1, y_1)$ and $P_2(x_2, y_2)$ represent any two points on a line, then the slope $m$ of the line is

$$m = \frac{y_2 - y_1}{x_2 - x_1}$$

   (a) If a line has positive slope, the line slopes upward from left to right.
   (b) If a line has negative slope, the line slopes downward from left to right.
   (c) If a line has zero slope, the line is horizontal.
   (d) If a line has no slope, the line is vertical.

3. *Point-slope form of a line:* If $m$ is the slope and $P_1(x_1, y_1)$ is any point on a nonvertical line, its equation is

$$y - y_1 = m(x - x_1)$$

4. *Slope-intercept form of a line:* If $m$ is the slope and $(0, b)$ is the $y$-intercept of a nonvertical line, its equation is

$$y = mx + b$$

5. *Equation of a horizontal line:* If a horizontal line passes through the point $(a, b)$, its equation is

$$y = b$$

6. *Equation of a vertical line:* If a vertical line passes through the point $(a, b)$, its equation is

$$x = a$$

7. *General form of the equation of a straight line:*

$$Ax + By + C = 0 \quad \text{where } A \text{ and } B \text{ are not both } 0$$

8. *Parallel lines:* Two lines are parallel if either one of the following conditions holds:
   (a) They are both perpendicular to the $x$-axis.
   (b) They both have the same slope. That is, if the equations of the lines are

$$L_1: \quad y = m_1 x + b_1 \quad \text{and} \quad L_2: \quad y = m_2 x + b_2$$

then

$$m_1 = m_2$$

9. *Perpendicular lines:* Two lines are perpendicular if either one of the following conditions holds:
   (a) One line is vertical with equation $x = a$, and the other is horizontal with equation $y = b$.
   (b) Neither is vertical and the slope of one line is the negative reciprocal of the other. That is, if the equations of the lines are

$$L_1: \quad y = m_1x + b_1 \quad \text{and} \quad L_2: \quad y = m_2x + b_2$$

then

$$m_1 = -\frac{1}{m_2}$$

10. *Distance formula:* The distance between two points $P(x_1, y_1)$ and $Q(x_2, y_2)$ is given by the formula

$$d = PQ = \sqrt{(x_2 - x_1)^2 + (y_2 - y_1)^2}$$

11. *Midpoint formula:* The coordinates of the point $Q(x_m, y_m)$ which is midway between two points $P(x_1, y_1)$ and $R(x_2, y_2)$ are given by

$$x_m = \frac{x_1 + x_2}{2} \quad \text{and} \quad y_m = \frac{y_1 + y_2}{2}$$

# CHAPTER 4 REVIEW

*Determine whether or not each relation is a function. Write its domain and its range.*

1. $A = \{(2, 3), (3, 4), (4, 5), (5, 6)\}$
2. $B = \{(2, 6), (6, 4), (2, 1), (4, 3)\}$
3. $y = -4x + 3$
4. $y = x^2 - 5$
5. $x = y^2 + 4$
6. $y = \sqrt{4 - 8x}$

7. Given $f(x) = 5x + 14$, find
   (a) $f(2)$      (b) $f(0)$      (c) $f(-4)$

8. Given $g(t) = 3t^2 + 5t - 12$, find
   (a) $g(2)$      (b) $g(0)$      (c) $g(-5)$

9. Given $h(x) = \dfrac{4x^2 - 3x}{2\sqrt{x} - 1}$, find
   (a) $h(2)$      (b) $h(5)$      (c) $h(-15)$      (d) $h(1)$

10. Given $g(x) = x^2 - 6x + 4$, find
    (a) $g(a)$      (b) $g(2x)$      (c) $g(z - 2)$

*Graph each equation.*

11. $y = 4x + 5$
12. $y = x^2 + 4$
13. $y = x^2 + 2x - 8$
14. $y = 2x^2 + x - 6$
15. $y = -x^2 - x + 4$
16. $y = \sqrt{2x}$
17. $y = \sqrt{-2 - 4x}$
18. $y = x^3 - 6x$

*Solve each graphically.*

**19.** Exercise 12 for $y = 5, 7$, and 2.     **20.** Exercise 13 for $y = 0, -2$, and 3.

**21.** Exercise 15 for $y = 2, 0$, and $-2$.     **22.** Exercise 18 for $y = 0, 2$, and $-3$.

**23.** The current, $i$, in a given circuit varies with the time, $t$, according to $i = 2t^2$. Find $t$ when $i = 2, 6$, and 8.

**24.** A given capacitor receives between its terminals a voltage, $V$, where $V = 4t^3 + t$ when $t$ is in s. Find $t$ when $V$ is 40 and 60.

*Use the points $(3, -4)$ and $(-6, -2)$ in Exercises 25–27.*

**25.** Find the slope of the line through the two points.

**26.** Find the distance between the two points.

**27.** Find the coordinates of the point midway between the two points.

*Find the equation of the line that satisfies each condition in Exercises 28–31.*

**28.** Passes through $(4, 7)$ and $(6, -4)$.

**29.** Passes through $(-3, 1)$ with a slope of $\frac{2}{3}$.

**30.** Crosses the $y$-axis at $-3$ with a slope of $-\frac{1}{3}$.

**31.** Is parallel to and 3 units to the left of the $y$-axis.

**32.** Find the slope and the $y$-intercept of $3x - 2y - 6 = 0$.

**33.** Graph $3x - 4y = 12$.

*Using the slope and the intercept of each line, determine whether each given pair of equations represents lines that are parallel, perpendicular, or neither.*

**34.** $x - 2y + 3 = 0$; $8x + 4y - 9 = 0$

**35.** $2x - 3y + 4 = 0$; $-8x + 12y = 16$

**36.** $3x - 2y + 5 = 0$; $2x - 3y + 9 = 0$

**37.** $x = 2$; $y = -3$

**38.** $x = 4$; $x = 7$

**39.** Find the equation of the line parallel to the line $2x - y + 4 = 0$ that passes through the point $(5, 2)$.

**40.** Find the equation of the line perpendicular to the line $3x + 5y - 6 = 0$ that passes through the point $(-4, 0)$.

# 5

# Factoring and Algebraic Fractions

## INTRODUCTION

The properties that allow us to change the way algebraic expressions look without changing their values or their characteristics are very important. Expressions often need a different format so that we can simplify them or enter them into a calculator or a computer.

In this chapter we will study factoring, simplifying algebraic fractions, and solving equations with algebraic fractions.

### Objectives

- In an algebraic expression, find the greatest common factor.
- Identify and factor the difference of two squares.
- Identify and factor the square of a binomial.
- Factor trinomials.
- Factor using the method of grouping.
- Identify and factor the sum or difference of two cubes.
- Add, subtract, multiply, and divide algebraic fractions.
- Solve equations containing algebraic fractions.

## 5.1 SPECIAL PRODUCTS

In Chapter 2 we introduced certain fundamental algebraic concepts and operations. Before proceeding, we must develop some additional algebraic techniques that are necessary in the development of later topics.

Certain types of algebraic products occur so often that we must be able to recognize them on sight. We must also be able to do the multiplications quickly and mentally.

To illustrate the first two types of special algebraic products, we give the general form of each followed by two examples.

$$a(x + y + z) = ax + ay + az$$

**EXAMPLE 1**

$$3x(x^2 + 2x - 4) = 3x^3 + 6x^2 - 12x$$

**EXAMPLE 2**

$$4a^2b^3(7a^4b^3 - 5a^2b^3 - ab + 3a^2) = 28a^6b^6 - 20a^4b^6 - 4a^3b^4 + 12a^4b^3$$

$$(x + y)(x - y) = x^2 - y^2$$

**EXAMPLE 3**

$$(3a + 4)(3a - 4) = 9a^2 - 16$$

**EXAMPLE 4**

$$(6a^2 + 5xy^3)(6a^2 - 5xy^3) = 36a^4 - 25x^2y^6$$

There are two general forms of the *square of a binomial.*

$$(x + y)^2 = x^2 + 2xy + y^2$$

**EXAMPLE 5**

$$(2x + 3y)^2 = 4x^2 + 12xy + 9y^2$$

**EXAMPLE 6**

$$(5ab^2 + 4c)^2 = 25a^2b^4 + 40ab^2c + 16c^2$$

$$(x - y)^2 = x^2 - 2xy + y^2$$

**EXAMPLE 7**

$$(3x - 4)^2 = 9x^2 - 24x + 16$$

**EXAMPLE 8**

$$(2a^2 - 5y^3)^2 = 4a^4 - 20a^2y^3 + 25y^6$$

The *product of two binomials* may be found mentally using the FOIL method discussed in Section 2.3.

**EXAMPLE 9**

$$(x + 4)(x + 6) = x^2 + 10x + 24$$

**EXAMPLE 10**

$$(x + 3)(x - 5) = x^2 - 2x - 15$$

**EXAMPLE 11**

$$(2x + 3)(5x + 4) = 10x^2 + 23x + 12$$

**EXAMPLE 12**

$$(3x - 4)(-2x + 7) = -6x^2 + 29x - 28$$

The general form of the *square of a trinomial* is as follows.

$$(x + y + z)^2 = x^2 + y^2 + z^2 + 2xy + 2xz + 2yz$$

**EXAMPLE 13**

$$(4x + y + 3z)^2 = 16x^2 + y^2 + 9z^2 + 8xy + 24xz + 6yz$$

The *square of a trinomial* may also be found using grouping and the square of a binomial as follows for Example 13:

$$(4x + y + 3z)^2 = [(4x + y) + 3z]^2$$
$$= (4x + y)^2 + 2(4x + y)(3z) + (3z)^2$$
$$= 16x^2 + 8xy + y^2 + 24xz + 6yz + 9z^2$$

**EXAMPLE 14**

$$(3a - b^2 - 2c^3)^2 = 9a^2 + b^4 + 4c^6 - 6ab^2 - 12ac^3 + 4b^2c^3$$

Or

$$(3a - b^2 - 2c^3)^2 = [(3a - b^2) - 2c^3]^2$$
$$= (3a - b^2)^2 - 2(3a - b^2)(2c^3) + (2c^3)^2$$
$$= 9a^2 - 6ab^2 + b^4 - 12ac^3 + 4b^2c^3 + 4c^6$$

There are two general forms of the *cube of a binomial*.

$$(x + y)^3 = x^3 + 3x^2y + 3xy^2 + y^3$$

**EXAMPLE 15**

$$(2a + 5b)^3 = (2a)^3 + 3(2a)^2(5b) + 3(2a)(5b)^2 + (5b)^3$$
$$= 8a^3 + 60a^2b + 150ab^2 + 125b^3$$

**EXAMPLE 16**

$$(4x + y)^3 = (4x)^3 + 3(4x)^2(y) + 3(4x)(y)^2 + (y)^3$$
$$= 64x^3 + 48x^2y + 12xy^2 + y^3$$

$$(x - y)^3 = x^3 - 3x^2y + 3xy^2 - y^3$$

**EXAMPLE 17**

$$(4a - b)^3 = (4a)^3 - 3(4a)^2(b) + 3(4a)(b)^2 - (b)^3$$
$$= 64a^3 - 48a^2b + 12ab^2 - b^3$$

**EXAMPLE 18**

$$(3x - 2y)^3 = (3x)^3 - 3(3x)^2(2y) + 3(3x)(2y)^2 - (2y)^3$$
$$= 27x^3 - 54x^2y + 36xy^2 - 8y^3$$

## Exercises 5.1

*Find each product mentally.*

1. $-8x^2(5x^3 - 9x^2 + 10x)$
2. $5x^2(3x^5 - 4x^3 + 2x^2 - 6x + 7)$
3. $6x^3y^5(4xy - 7x^3y^2 + 9x^4)$
4. $6a^2b^4(-2a^2b^2 + 5ab^2 - 9b^3)$
5. $(2x + 7)(2x - 7)$
6. $(3x^2 + 8yz^2)(3x^2 - 8yz^2)$
7. $(7x^2 + 2y)^2$
8. $(8a^2b + 5c)^2$
9. $(2a^2 - 3b)^2$
10. $(9x - 4y^3)^2$
11. $(x + 4)(x + 12)$
12. $(x - 9)(x + 8)$
13. $(2x + 5)(-3x - 8)$
14. $(4x - 7)(5x - 6)$
15. $(2a - b + 3c)^2$
16. $(-3a + 2b - c)^2$
17. $(2a + b)^3$
18. $(4x + 5y)^3$
19. $(5 - 2x^2)^3$
20. $(2x - 4y^3)^3$
21. $(6a + 5b)(7a - 3b)$
22. $(5x - y)(4x - 8y)$
23. $(4x + 7y)^2$
24. $(2x - 10y)^2$
25. $(2a + 9b)(6a - 11b)$
26. $(2x^2y + 7z^3)^2$
27. $(3a^3 + 10b^2)^3$
28. $(8 - a^2b^3)^2$
29. $-4a^2b^3(6a^3b^4 - 9a^5b^7 + 20a^5)$
30. $(2 + a)(4 - 2a)$
31. $(9a^3 - 12b^2)^2$
32. $(12a - 5b)(15a + 2b)$
33. $(x - 3y - 5z)^2$
34. $(-3x + 7)(7x + 3)$
35. $(9a - 4b)(6a - 10b)$
36. $(80 - a^5)(80 + a^5)$
37. $(6x^3 + 4)(5x^3 - 4)$
38. $(4x^2 - 5)(3x^2 + 7)$
39. $(3xy + 9yz)(5xy - 8yz)$
40. $(-5y + z)^2$
41. $(5abc + 6)(5abc - 6)$
42. $(1 - a^2)^2$
43. $(3x^2 - \frac{1}{2}y)(x^2 + \frac{2}{3}y)$
44. $(2x^2 + \frac{1}{4}y^3)(2x^2 + \frac{1}{2}y^3)$
45. $(x - \frac{2}{3}y)^2$
46. $(4x^2 - \frac{1}{2}y)^2$
47. $x^2(1 + x^2)^2$
48. $x^2(1 - x^2)^3$

## 5.2 FACTORING ALGEBRAIC EXPRESSIONS

Factoring is the process of writing an algebraic expression as a product of two or more factors*—usually prime factors. The process is the opposite of multiplying algebraic expressions; that is, the reverse of the product forms in the previous section.

*Recall that in the product $ab$, $a$ and $b$ are called **factors**.

The first type of factoring involves finding the largest monomial factor that divides each factor "evenly," †dividing each term by this factor, and then writing the result as a product.

> **GREATEST COMMON FACTOR**
>
> $$ax + ay + az = a(x + y + z)$$

**EXAMPLE 1**

Factor $4x + 8y$.

$$4x + 8y = 4(x + 2y)$$

**EXAMPLE 2**

Factor $21x^3 - 14x^2 + 28x$.

$$21x^3 - 14x^2 + 28x = 7x(3x^2 - 2x + 4)$$

**EXAMPLE 3**

Factor $50xy^2z - 80xyz - 32xz$.

$$50xy^2z - 80xyz - 32xz = 2xz(25y^2 - 40y - 16)$$

*Note:* You should always try this type of factoring first, because many times it simplifies the remaining factor into something that factors more easily.

The difference of two perfect squares is the product of the square root of the first *plus* the square root of the second times the square root of the first *minus* the square root of the second.

> **DIFFERENCE OF TWO PERFECT SQUARES**
>
> $$x^2 - y^2 = (x + y)(x - y)$$

**EXAMPLE 4**

Factor $9a^2 - b^2$.

$$9a^2 - b^2 = (3a + b)(3a - b)$$

**EXAMPLE 5**

Factor $36x^2 - 49y^4$.

$$36x^2 - 49y^4 = (6x + 7y^2)(6x - 7y^2)$$

**EXAMPLE 6**

Factor $x^2 - 4a^2b^2$.

$$x^2 - 4a^2b^2 = (x + 2ab)(x - 2ab)$$

†The remainder is zero after dividing.

*Note:* The sum of two perfect squares, $x^2 + y^2$, cannot be factored.

**EXAMPLE 7**

Factor $81x^4 - y^8$.

$$81x^4 - y^8 = (9x^2 + y^4)(9x^2 - y^4)$$
$$= (9x^2 + y^4)(3x + y^2)(3x - y^2)$$

*Note:* To factor this expression completely, we must also factor $9x^2 - y^4$.

The factors of a general trinomial are often binomial factors. To find these binomial factors, you "undo" the FOIL multiplication.

**EXAMPLE 8**

Factor $x^2 + 7x + 10$.

This type of factoring involves some trial and error. Here we need two numbers whose sum is 7 and whose product is 10. They are $+2$ and $+5$.

$$x^2 + 7x + 10 = (x + 2)(x + 5)$$

**EXAMPLE 9**

Factor $x^2 - 9x + 18$.

We need two numbers whose sum is $-9$ and whose product is 18: $-3$ and $-6$.

$$x^2 - 9x + 18 = (x - 3)(x - 6)$$

**EXAMPLE 10**

Factor $x^2 - 3x - 28$.

We need two numbers whose sum is $-3$ and whose product is $-28$: $+4$ and $-7$.

$$x^2 - 3x - 28 = (x + 4)(x - 7)$$

A short summary about the signs in trinomials might be helpful. If the trinomial to be factored is one of the following forms, use the corresponding sign patterns.

**1.** $x^2 + px + q = (x + \quad)(x + \quad)$
**2.** $x^2 - px + q = (x - \quad)(x - \quad)$
**3.** $x^2 + px - q = (x + \quad)(x - \quad)$
**4.** $x^2 - px - q = (x + \quad)(x - \quad)$

**EXAMPLE 11**

Factor $x^2 + 10x + 24$.

We need two numbers whose sum is 10 and whose product is 24: $+4$ and $+6$.

$$x^2 + 10x + 24 = (x + 4)(x + 6)$$

**EXAMPLE 12**

Factor $x^2 - 15x + 56$.

We need two numbers whose sum is $-15$ and whose product is 56: $-7$ and $-8$.

$$x^2 - 15x + 56 = (x - 7)(x - 8)$$

**EXAMPLE 13**

Factor $x^2 + 3x - 40$.

We need two numbers whose sum is 3 and whose product is $-40$: $+8$ and $-5$.

$$x^2 + 3x - 40 = (x + 8)(x - 5)$$

**EXAMPLE 14**

Factor $x^2 - 7x - 18$.

We need two numbers whose sum is $-7$ and whose product is $-18$: $+2$ and $-9$.

$$x^2 - 7x - 18 = (x + 2)(x - 9)$$

**EXAMPLE 15**

Factor $8x^2 + 14x + 3$.

When the coefficient of the squared term is not 1, the factoring involves more trial and error. Here, we need the following:

**1.** Pairs of factors whose product is $8x^2$. There are two possible pairs:

$$(8x \quad )(x \quad ) \quad \text{and} \quad (2x \quad )(4x \quad )$$

**2.** Pairs of factors whose product is 3. There is only one pair: $3 \cdot 1$.

**3.** To arrange these factors so that the middle term is $14x$. If the coefficient of the squared term is positive, the signs in the factors take the same forms as discussed in the summary after Example 10. The possibilities are as follows.

| *Factors* | *Middle term* | |
|---|---|---|
| $(8x + 1)(x + 3)$ | $25x$ | |
| $(8x + 3)(x + 1)$ | $11x$ | |
| $(2x + 1)(4x + 3)$ | $10x$ | |
| $(2x + 3)(4x + 1)$ | $14x$ | (the correct combination) |

Therefore, $8x^2 + 14x + 3 = (2x + 3)(4x + 1)$.

**EXAMPLE 16**

Factor $12x^2 - 28x + 15$.

**Step 1:** The pairs of factors of $12x^2$ are

$$(12x \quad )(x \quad ), \quad (2x \quad )(6x \quad ), \quad \text{and} \quad (3x \quad )(4x \quad )$$

**Step 2:** The pairs of factors of 15 are $1 \cdot 15$ and $3 \cdot 5$.

**Step 3:** Arrange these factors so that the middle term is $-28x$. The possibilities are as follows:

| *Factors* | *Middle term* | |
|---|---|---|
| $(12x - 1)(x - 15)$ | $-181x$ | |
| $(12x - 15)(x - 1)$ | $-27x$ | |
| $(12x - 3)(x - 5)$ | $-63x$ | |
| $(12x - 5)(x - 3)$ | $-41x$ | |
| $(2x - 1)(6x - 15)$ | $-36x$ | |
| $(2x - 15)(6x - 1)$ | $-92x$ | |
| $(2x - 3)(6x - 5)$ | $-28x$ | (the correct combination) |
| $(2x - 5)(6x - 3)$ | $-36x$ | |

|  Factors | Middle term |
| :---: | :---: |
| $(3x - 1)(4x - 15)$ | $-49x$ |
| $(3x - 15)(4x - 1)$ | $-63x$ |
| $(3x - 3)(4x - 5)$ | $-27x$ |
| $(3x - 5)(4x - 3)$ | $-29x$ |

*Note:* Both signs inside the factors must be negative. Therefore,

$$12x^2 - 28x + 15 = (2x - 3)(6x - 5)$$

Of course, we do not go through this long listing process for every factoring. The work is often done mentally, and we stop when we have found the combination of factors that gives the desired middle term.

**EXAMPLE 17**

Factor $12x^2 - 7x - 12$.

$$12x^2 - 7x - 12 = (4x + 3)(3x - 4)$$

**EXAMPLE 18**

Factor $30x^2 - 59x + 9$.

$$30x^2 - 59x + 9 = (5x - 9)(6x - 1)$$

**EXAMPLE 19**

Factor $10x^2 + 27xy - 9y^2$.

$$10x^2 + 27xy - 9y^2 = (10x - 3y)(x + 3y)$$

**EXAMPLE 20**

Factor $4x^2 + 28xy^2 + 45y^4$.

$$4x^2 + 28xy^2 + 45y^4 = (2x + 5y^2)(2x + 9y^2)$$

**EXAMPLE 21**

Factor $-6x^2 - 11x + 10$.

$$-6x^2 - 11x + 10 = (-3x + 2)(2x + 5)$$

An alternate approach to factoring a trinomial whose first term is negative involves dividing each term first by $-1$ and then proceeding as usual. Following this approach in this example leads to the following:

$$-6x^2 - 11x + 10 = -(6x^2 + 11x - 10)$$
$$= -(3x - 2)(2x + 5)$$

Do you see that this result is equivalent to our previous result?

**EXAMPLE 22**

Factor $60x^2 + 20x - 105$.

$$60x^2 + 20x - 105 = 5(12x^2 + 4x - 21)$$
$$= 5(2x + 3)(6x - 7)$$

*Note:* Dividing each term by the common monomial factor, 5, simplified the remaining trinomial factor.

**EXAMPLE 23**

Factor $180x^2 + 279x + 108$.

$$180x^2 + 279x + 108 = 9(20x^2 + 31x + 12)$$
$$= 9(5x + 4)(4x + 3)$$

**EXAMPLE 24**

Factor $9x^2 + 24xy + 16y^2$.

$$9x^2 + 24xy + 16y^2 = (3x + 4y)(3x + 4y) = (3x + 4y)^2$$

**EXAMPLE 25**

Factor $16x^2 - 8xy + y^2$.

$$16x^2 - 8xy + y^2 = (4x - y)(4x - y) = (4x - y)^2$$

When the factors of a trinomial are the same two binomial factors, the trinomial is called a **perfect square trinomial.** The general forms are as follows.

> **PERFECT SQUARE TRINOMIALS**
> $$x^2 + 2xy + y^2 = (x + y)^2$$
> $$x^2 - 2xy + y^2 = (x - y)^2$$

## Exercises 5.2

*Factor each expression completely.*

**1.** $6x + 9y$

**2.** $15x - 20$

**3.** $10x + 25y - 45z$

**4.** $8x^2 - 12xy + 10xz$

**5.** $12x^2 - 30xy + 6xz$

**6.** $10x^2 + 25x$

**7.** $8x^3y^2 - 6xy^3 + 12xy^2z$

**8.** $15x^7y^3 - 6x^6y^2 + 12x^3yz$

**9.** $x^2 - 16$

**10.** $x^2 - 64$

**11.** $9x^2 - 25y^2$

**12.** $49x^2 - 100y^2$

**13.** $2e^2 - 72$

**14.** $20x^2 - 45$

**15.** $16d^2 - 100$

**16.** $64x^2 - 4y^2$

**17.** $4R^2 - 4r^2$

**18.** $49 - a^4$

**19.** $x^2 + 6x + 8$

**20.** $x^2 + 8x + 15$

**21.** $b^2 + 11b + 24$

**22.** $m^2 + 11m + 30$

**23.** $x^2 - 9x + 18$

**24.** $x^2 - 9x + 14$

**25.** $a^2 - 18a + 32$

**26.** $c^2 + 22c + 40$

**27.** $x^2 - 2x - 35$

**28.** $x^2 + 4x - 12$

**29.** $a^2 + 3a - 4$

**30.** $q^2 - 3q - 28$

**31.** $2x^2 - x - 6$

**32.** $5x^2 + 32x + 12$

**33.** $15x^2 - 31x + 14$

**34.** $8x^2 + 26x - 45$

**35.** $45y^2 + 59y + 6$

**36.** $8m^2 - 14m + 3$

**37.** $35a^2 + 2a - 1$

**38.** $16g^2 + 8g + 1$

**39.** $9c^2 - 24c + 16$

**40.** $6m^2 - 13m + 5$

**41.** $25b^2 + 60b + 20$

**42.** $15t^2 + 69t - 30$

**43.** $35t^2 - 4ts - 15s^2$

**44.** $12a^2 + 8ab - 15b^2$

**45.** $30k^2 - 95kt + 50t^2$

**46.** $24m^2 + 34m + 10$

**47.** $54x^2 + 27x - 42$

**48.** $8x^2 - 24xy - 1440y^2$

**49.** $4a^2 + 20a + 25$

**50.** $9x^2 + 42x + 49$

**51.** $9x^2 - 48xy^2 + 64y^4$

**52.** $4a^4 + 12a^2b + 9b^2$

**53.** $25x^2 + 10xy^3 + y^6$

**54.** $9c^4 - 12c^2y^2 + 4y^4$

**55.** $x^4 - 81$

**56.** $x^4 - 16$

**57.** $a^4 - 3a^2 - 40$

**58.** $b^4 + 21b^2 - 100$

**59.** $t^4 - 13t^2 + 36$

**60.** $x^4 - x^2 - 12$

## 5.3 OTHER FORMS OF FACTORING

Some algebraic expressions may be factored by grouping their terms so that they are of the types we have already studied. We shall consider those which can be grouped in one of the following ways:

**1.** With terms having a common binomial factor.

**2.** As the difference of two squares.

**3.** In the form of a trinomial.

Factoring out a common binomial factor is similar to factoring out a common monomial factor.

**EXAMPLE 1**

Factor $ax + bx$.

$$ax + bx = x(a + b)$$

**EXAMPLE 2**

Factor $a(x - y) + b(x - y)$.

$$a(x - y) + b(x - y) = (x - y)(a + b)$$

**EXAMPLE 3**

Factor $h(x - 1) - 2(x - 1)$.

$$h(x - 1) - 2(x - 1) = (x - 1)(h - 2)$$

In most cases, however, the algebraic expressions are not grouped so that the common binomial factors are so obvious.

**EXAMPLE 4**

Factor $3x - 3y + ax - ay$.

Group the first two terms and the last two terms, factor out common monomial factors, and then factor out the resulting binomial factors.

$$3x - 3y + ax - ay = 3(x - y) + a(x - y)$$
$$= (x - y)(3 + a)$$

**EXAMPLE 5**

Factor $a^2x - x - 3a^2 + 3$.

$$a^2x - x - 3a^2 + 3 = x(a^2 - 1) - 3(a^2 - 1)$$
$$= (a^2 - 1)(x - 3)$$
$$= (a + 1)(a - 1)(x - 3)$$

**EXAMPLE 6**

Factor $x^3 + x^2 + x + 1$.

$$x^3 + x^2 + x + 1 = x^2(x + 1) + (x + 1)$$
$$= (x + 1)(x^2 + 1)$$

If an algebraic expression can be expressed as the difference of two squares, it can be factored similarly to the form $x^2 - y^2 = (x + y)(x - y)$.

**EXAMPLE 7**

Factor $(a + b)^2 - 16$.

$$(a + b)^2 - 16 = (a + b + 4)(a + b - 4)$$

**EXAMPLE 8**

Factor $25a^2 - (a + 2b)^2$.

$$25a^2 - (a + 2b)^2 = (5a)^2 - (a + 2b)^2$$
$$= [5a + (a + 2b)][5a - (a + 2b)]$$
$$= (6a + 2b)(4a - 2b)$$
$$= 2(3a + b)(2)(2a - b)$$
$$= 4(3a + b)(2a - b)$$

**EXAMPLE 9**

Factor $y^2 - 4y + 4 - 25x^2$.

Group the first three terms which form a perfect square trinomial, factor into a perfect square, and then factor the resulting difference of two squares.

$$(y^2 - 4y + 4) - 25x^2 = (y - 2)^2 - 25x^2$$
$$= (y - 2 + 5x)(y - 2 - 5x)$$

The last type of group factoring involves algebraic expressions that can be grouped in the form $ax^2 + bx + c$.

**EXAMPLE 10**

Factor $(a + b)^2 + 8(a + b) + 15$.

Note that $(a + b)^2 + 8(a + b) + 15$ is in the form $x^2 + 8x + 15$ which factors $(x + 3)(x + 5)$. Therefore,

$$(a + b)^2 + 8(a + b) + 15 = (a + b + 3)(a + b + 5)$$

The last type of factoring we shall study here is factoring the sum and difference of two perfect cubes. The general forms and three examples of each follow.

---

**SUM OF TWO PERFECT CUBES**

$$x^3 + y^3 = (x + y)(x^2 - xy + y^2)$$

---

**EXAMPLE 11**

Factor $a^3 + 8$.

$$a^3 + 8 = a^3 + 2^3 = (a + 2)(a^2 - 2a + 4)$$

**EXAMPLE 12**

Factor $y^3 + 27$.

$$y^3 + 27 = y^3 + 3^3 = (y + 3)(y^2 - 3y + 9)$$

**EXAMPLE 13**

Factor $27x^3 + 64y^6$.

$$27x^3 + 64y^6 = (3x)^3 + (4y^2)^3 = (3x + 4y^2)(9x^2 - 12xy^2 + 16y^4)$$

---

**DIFFERENCE OF TWO PERFECT CUBES**

$$x^3 - y^3 = (x - y)(x^2 + xy + y^2)$$

---

Note the similarities in these two forms.

**EXAMPLE 14**

Factor $a^3 - 1$.

$$a^3 - 1 = a^3 - 1^3 = (a - 1)(a^2 + a + 1)$$

**EXAMPLE 15**

Factor $125x^3 - 8y^3$.

$$125x^3 - 8y^3 = (5x)^3 - (2y)^3 = (5x - 2y)(25x^2 + 10xy + 4y^2)$$

**EXAMPLE 16**

Factor $x^6 - 64$.

$$x^6 - 64 = (x^3 + 8)(x^3 - 8)$$
$$= (x + 2)(x^2 - 2x + 4)(x - 2)(x^2 + 2x + 4)$$

## Exercises 5.3

*Factor each expression completely.*

1. $a(m + n) - b(m + n)$
2. $m(x^2 - 16) - 3(x^2 - 16)$
3. $mx + my + nx + ny$
4. $ab + 3a - bc - 3c$
5. $3x + y - 6x^2 - 2xy$
6. $x^2 + ax + xy + ay$
7. $6x^3 - 4x^2 + 3x - 2$
8. $x^4 - 8x + x^3y - 8y$
9. $(x + y)^2 - 4z^2$
10. $(x - y)^2 - 16a^4$
11. $100 - 49(x - y)^2$
12. $9(a - 4)^2 - 64b^4$
13. $x^2 - 6x + 9 - 4y^2$
14. $x^2 + 10x + 25 - 36y^2$
15. $4x^2 - 4y^2 + 4y - 1$
16. $4 - a^2 - 2ab - b^2$
17. $(x + y)^2 + 13(x + y) + 36$
18. $3(x - y)^2 - 14(x - y) + 8$
19. $24(a + b)^2 - 14(a + b) - 5$
20. $x^2 + 2x(y + z) + (y + z)^2$
21. $a^3 + b^3$
22. $27x^3 + 8y^3$
23. $x^3 - 64$
24. $27x^3 - 8y^3$
25. $a^3b^3 + c^3$
26. $x^3 - y^3z^3$
27. $a^3 - 27b^3$
28. $8a^3 + b^3$
29. $27a^3 + 64b^3$
30. $64x^6 - 125y^3$
31. $27a^6 - 8b^9$
32. $8a^3b^6 + 125c^{12}$
33. $x^6 - y^6$
34. $a^6b^6 - c^6$

## 5.4  EQUIVALENT FRACTIONS

In order to solve many problems that occur in the technical and applied sciences, it is necessary to use fractions. A **fraction** may be defined as the quotient of two numbers or algebraic expressions. Both the numerator and denominator of a fraction may be multiplied or divided by the same nonzero number without changing the value of the fraction.

Two fractions are **equivalent** when both the numerator and denominator of one fraction can be multiplied or divided by the same nonzero number in order to change one fraction to the other. For example, $\frac{2}{3}$ is equivalent to $\frac{6}{9}$, and $\frac{12}{16}$ is equivalent to $\frac{3}{4}$, because

$$\frac{2}{3} = \frac{2 \cdot 3}{3 \cdot 3} = \frac{6}{9} \quad \text{and} \quad \frac{12}{16} = \frac{12 \div 4}{16 \div 4} = \frac{3}{4}$$

A fraction has three signs associated with it:

1. The sign of the fraction.
2. The sign of the numerator.
3. The sign of the denominator.

Any two of these three signs may be changed without changing the value of the fraction. For example,

$$-\frac{-3}{+4} = -\frac{+3}{-4} = +\frac{+3}{+4} = +\frac{-3}{-4}$$

When the numerator or the denominator is an algebraic expression, its sign is changed by placing a negative sign before the numerator or denominator in parentheses or by changing the signs of *all* the terms. For example,

$$-\frac{a+b-c}{d-3} = \frac{-(a+b-c)}{d-3} = \frac{-a-b+c}{d-3}$$

or

$$-\frac{a+b-c}{d-3} = \frac{a+b-c}{-(d-3)} = \frac{a+b-c}{-d+3} \quad \text{or} \quad \frac{a+b-c}{3-d}$$

A fraction is in **lowest terms** when its numerator and denominator have no common factors except 1. To reduce a fraction to lowest terms, divide the numerator and the denominator by their common factors.

### EXAMPLE 1

Reduce $\dfrac{8xy^3}{12x^4y^2}$ to lowest terms.

The common factors of the numerator and the denominator are $4xy^2$. Dividing both numerator and denominator by $4xy^2$, we have

$$\frac{8xy^3}{12x^4y^2} = \frac{2y}{3x^3}$$

Some prefer to use "cancellation marks" to indicate this division; that is,

$$\frac{\overset{2}{\cancel{8x}}\,\overset{y}{\cancel{y^3}}}{\underset{3}{\cancel{12x^4}}\,\underset{x^3}{\cancel{y^2}}} = \frac{2y}{3x^3}$$

Remember, *cancellation* is a nontechnical word for the mathematical operation of division. Think *division* whenever these marks are used.

### EXAMPLE 2

Reduce $\dfrac{x^2+5x+6}{x^2+6x+9}$ to lowest terms.

Before finding common factors, we must first factor the numerator and the denominator.

$$\frac{x^2+5x+6}{x^2+6x+9} = \frac{(x+2)(x+3)}{(x+3)(x+3)} = \frac{x+2}{x+3}$$

### EXAMPLE 3

Reduce $\dfrac{4-9x^2}{9x^2+3x-2}$ to lowest terms.

$$\frac{4-9x^2}{9x^2+3x-2} = \frac{(2-3x)(2+3x)}{(3x+2)(3x-1)} = \frac{2-3x}{3x-1}$$

(*Note:* $2+3x = 3x+2$.)

When factors *differ only in sign,* such as $x - y$ and $y - x$, one may be replaced by the opposite of the other; that is,

$$x - y = -(y - x) \quad \text{because} \quad -(y - x) = -y + x = x - y$$

or

$$y - x = -(x - y) \quad \text{because} \quad -(x - y) = -x + y = y - x$$

### EXAMPLE 4

Reduce $\dfrac{x^2 - 2x - 8}{16 - x^2}$ to lowest terms.

$$\frac{x^2 - 2x - 8}{16 - x^2} = \frac{(x - 4)(x + 2)}{(4 + x)(4 - x)}$$

$$= \frac{-(4 - x)(x + 2)}{(4 + x)(4 - x)} \qquad [\textit{Note: } x - 4 = -(4 - x)]$$

$$= -\frac{x + 2}{4 + x} \quad \text{or} \quad \frac{-x - 2}{4 + x}$$

### EXAMPLE 5

Reduce $\dfrac{30 - 19x - 5x^2}{5x^2 + 14x - 24}$ to lowest terms.

$$\frac{30 - 19x - 5x^2}{5x^2 + 14x - 24} = \frac{(5 + x)(6 - 5x)}{(5x - 6)(x + 4)}$$

$$= \frac{(5 + x)(6 - 5x)}{-(6 - 5x)(x + 4)} \qquad [\textit{Note: } 5x - 6 = -(6 - 5x)]$$

$$= -\frac{5 + x}{x + 4} \quad \text{or} \quad \frac{-x - 5}{x + 4}$$

## Exercises 5.4

*Reduce each fraction to lowest terms.*

1. $\dfrac{3x}{18x}$

2. $\dfrac{18xy}{24y}$

3. $\dfrac{30x^2y^4}{54x^2y^5}$

4. $\dfrac{27x^2y^3z^4}{57x^5y^7}$

5. $\dfrac{8(x + 4)^3}{2(x + 4)^5}$

6. $\dfrac{6x^2(2x - 1)^6}{2x(2x - 1)^2}$

7. $\dfrac{18(x - 3)^3(1 - 5x)^2}{9(1 - 5x)^2(x - 3)}$

8. $\dfrac{12(3x - 4)^3(2x + 5)^2}{3(2x + 5)^3(3x - 4)^5}$

9. $\dfrac{5m + 15}{6m + 18}$

10. $\dfrac{6x^2 + 3x}{12x + 6}$

11. $\dfrac{6m + 6}{m^2 - 1}$

12. $\dfrac{2t - t^2}{4t - t^2}$

13. $\dfrac{x^2 + x}{x + 1}$

14. $\dfrac{3x^2 + 21x}{6x^2 - 12x}$

15. $\dfrac{x^2 + 5x + 4}{x^2 - 6x - 7}$

16. $\dfrac{x^2 - 2x - 24}{2x^2 + 7x - 4}$

17. $\dfrac{6x^2 + x - 12}{6x^2 + 19x - 36}$

18. $\dfrac{6x^2 + 19x + 10}{6x^2 - 5x - 6}$

19. $\dfrac{18 - 12x}{6x - 9}$

20. $\dfrac{15x - 25}{50 - 30x}$

21. $\dfrac{x^2 - 1}{1 - x}$

22. $\dfrac{6x^3 - 5x^2 - 4x}{16 - 9x^2}$

23. $\dfrac{a^2 - 4}{a^2 + 4a + 4}$

24. $\dfrac{4c^2 + 12c + 9}{4c^2 - 9}$

25. $\dfrac{m^2 - 16m}{16 - m}$

26. $\dfrac{1 - y}{y^2 - y}$

27. $\dfrac{t^2 - t^3}{t^2 - 1}$

28. $\dfrac{1 - 4x + 4x^2}{4x^2 - 1}$

29. $\dfrac{x^2 + xy}{x^2 + xy - 2x - 2y}$

30. $\dfrac{a^2 - ab + 3a - 3b}{a^2 - ab}$

31. $\dfrac{y^3 - 8}{y^2 + 2y + 4}$

32. $\dfrac{a^2b^2 - 16b^2}{a^2b + 9ab + 20b}$

33. $\dfrac{xy - 3x - 2y + 6}{y^3 - 27}$

34. $\dfrac{x^3 - y^3}{x^2 - y^2}$

35. $\dfrac{x^2 + 4x + 16}{x^3 - 64}$

36. $\dfrac{x^4 - 16}{x^4 - 2x^2 - 8}$

37. $\dfrac{3 - 4x - 4x^2}{4x^2 - 8x + 3}$

38. $\dfrac{y^2 - x^2}{x^2 - 2xy + y^2}$

39. $\dfrac{3a^3 + 3b^3}{3a^2 + 6ab + 3b^2}$

40. $\dfrac{x^2 - 2xy + y^2 - z^2}{z^2 + y^2 - x^2 + 2yz}$

# 5.5 MULTIPLICATION AND DIVISION OF ALGEBRAIC FRACTIONS

The product of two or more fractions is a fraction whose numerator is the product of the numerators and whose denominator is the product of the denominators; that is,

$$\frac{a}{b} \cdot \frac{c}{d} = \frac{ac}{bd}$$

It is usually helpful to first factor each of the terms of all numerators and denominators; divide by those common factors, if any; and then multiply the numerators and multiply the denominators. Cancellation marks are useful for showing each completed division.

**EXAMPLE 1**

Multiply $\dfrac{3xy^2}{5ab} \cdot \dfrac{20a^2}{6x^2y}$

$$\dfrac{\overset{1}{\cancel{3}}\,\overset{1}{\cancel{x}}\,\overset{y}{\cancel{y^2}}}{\underset{11}{\cancel{5ab}}} \cdot \dfrac{\overset{\overset{2}{\cancel{4}}\,a}{\cancel{20}\,a^2}}{\underset{2x\ 1}{\cancel{6x^2y}}} = \dfrac{2ay}{bx}$$

**EXAMPLE 2**

Multiply $\dfrac{x^2 - 4x + 4}{x^2 + x - 6} \cdot \dfrac{x^2 + 3x}{x^2 - x}$.

Factor each numerator and denominator.

$$\dfrac{\overset{x-2}{\cancel{(x-2)^2}}}{\underset{1}{\cancel{(x+3)}}\underset{1}{\cancel{(x-2)}}} \cdot \dfrac{\overset{1}{\cancel{x}}\overset{1}{\cancel{(x+3)}}}{\underset{1}{\cancel{x}}(x-1)} = \dfrac{x-2}{x-1}$$

To divide one fraction by another, "invert the divisor" and multiply, or multiply the first fraction by the reciprocal of the second.

$$\boxed{\dfrac{a}{b} \div \dfrac{c}{d} = \dfrac{a}{b} \cdot \dfrac{d}{c} = \dfrac{ad}{bc}}$$

**EXAMPLE 3**

Divide $\dfrac{xy}{a^2b} \div \dfrac{x^2y}{ab^3}$.

$$\dfrac{xy}{a^2b} \div \dfrac{x^2y}{ab^3} = \dfrac{\overset{1\ 1}{\cancel{xy}}}{\underset{a\ 1}{\cancel{a^2b}}} \cdot \dfrac{\overset{1\ b^2}{\cancel{ab^3}}}{\underset{x\ 1}{\cancel{x^2y}}} = \dfrac{b^2}{ax}$$

**EXAMPLE 4**

Divide $\dfrac{x^2 - 9}{4x^2 - 9} \div \dfrac{2x^2 - x - 15}{2x^2 + x - 6}$.

$$\dfrac{x^2 - 9}{4x^2 - 9} \div \dfrac{2x^2 - x - 15}{2x^2 + x - 6} = \dfrac{x^2 - 9}{4x^2 - 9} \cdot \dfrac{2x^2 + x - 6}{2x^2 - x - 15}$$

$$= \dfrac{(x + 3)\overset{1}{\cancel{(x-3)}}}{(2x + 3)\underset{1}{\cancel{(2x-3)}}} \cdot \dfrac{\overset{1}{\cancel{(2x-3)}}(x + 2)}{\underset{1}{\cancel{(x-3)}}(2x + 5)}$$

$$= \dfrac{(x + 3)(x + 2)}{(2x + 3)(2x + 5)}$$

*Note:* Results left in factored form are usually preferred.

**EXAMPLE 5**

Divide $\dfrac{18 + 3t - t^2}{1 + 2t + t^2} \div \dfrac{2t^2 - 14t + 12}{t^2 - 1}$

$$\frac{18 + 3t - t^2}{1 + 2t + t^2} \div \frac{2t^2 - 14t + 12}{t^2 - 1} = \frac{18 + 3t - t^2}{1 + 2t + t^2} \cdot \frac{t^2 - 1}{2t^2 - 14t + 12}$$

$$= \frac{(3 + t)(6 - t)}{(1 + t)^2} \cdot \frac{(t + 1)(t - 1)}{2(t - 6)(t - 1)}$$

[Replace $6 - t$ by $(-1)(t - 6)$ in the first numerator.]

$$= \frac{(3 + t)(-1)\cancel{(t - 6)}}{\cancelto{1+t}{(1 + t)^2}} \cdot \frac{\cancel{(t + 1)}\cancel{(t - 1)}}{2\cancel{(t - 6)}\cancel{(t - 1)}}$$

$$= \frac{-t - 3}{2(t + 1)}$$

**EXAMPLE 6**

Perform the indicated operations.

$$\frac{x^2 + xy}{x^2 + 3xy + 2y^2} \cdot \frac{x^2 - xy + y^2}{x^2 - 2xy + y^2} \div \frac{x^4 + xy^3}{(x^2 - y^2)^2}$$

$$= \frac{x^2 + xy}{x^2 + 3xy + 2y^2} \cdot \frac{x^2 - xy + y^2}{x^2 - 2xy + y^2} \cdot \frac{(x^2 - y^2)^2}{x^4 + xy^3}$$

$$= \frac{\cancel{x}\cancel{(x + y)}}{\cancel{(x + y)}(x + 2y)} \cdot \frac{\cancel{x^2 - xy + y^2}}{\cancel{(x - y)^2}} \cdot \frac{(x + y)^2(x - y)^2}{\cancel{x}\cancel{(x + y)}\cancel{(x^2 - xy + y^2)}}$$

$$= \frac{x + y}{x + 2y}$$

# Exercises 5.5

*Perform the indicated operations and simplify.*

**1.** $\dfrac{9}{16} \cdot \dfrac{4}{3}$

**2.** $\dfrac{16}{25} \cdot \dfrac{5}{32}$

**3.** $\dfrac{15}{32} \div \dfrac{3}{16}$

**4.** $\dfrac{5}{9} \div \dfrac{5}{18}$

**5.** $\dfrac{x^4}{3} \cdot \dfrac{6x}{x^5}$

**6.** $\dfrac{4t^4}{6t} \cdot \dfrac{12t^2}{9t^3}$

**7.** $\dfrac{6a^4}{a^2} \div \dfrac{18a^2}{a^5}$

**8.** $\dfrac{8c^2}{4c^3} \div \dfrac{6c^4}{9c}$

9. $\dfrac{3ab}{4x^2y} \cdot \dfrac{6x^2}{5ab^3}$

10. $\dfrac{15pq^2}{13m^5n^3} \cdot \dfrac{39mn^4}{5p^4q^3}$

11. $\dfrac{4ax^2}{9b^2y} \div \dfrac{5(ax)^2}{18(by^2)^2}$

12. $\dfrac{(5a)^2}{6x} \div \dfrac{5a^3}{-6x^2}$

13. $\dfrac{(2a^2b^3)^2}{(4ab^2)^3} \cdot \dfrac{(8a^2b)^2}{12a^3b}$

14. $\dfrac{(3x^2y)^3}{(6x^2y^2)^2} \div \dfrac{(12x^3y^4)^2}{(9x^2)^3}$

15. $\dfrac{4x + 8}{8xy} \cdot \dfrac{16x}{32x + 64}$

16. $\dfrac{2x}{8x + 4} \cdot \dfrac{16x + 8}{6}$

17. $\dfrac{5k + 5}{12} \div \dfrac{9k + 9}{4}$

18. $(8c - 16) \div \dfrac{3c - 6}{12}$

19. $\dfrac{(y + 2)^2}{y} \cdot \dfrac{y^2}{y^2 - 4}$

20. $\dfrac{9a}{(a - 1)^2} \cdot \dfrac{a^2 - 1}{a^2}$

21. $(m^2 - 1) \cdot \dfrac{1 + m}{1 - m}$

22. $(s^2 - 9) \div \dfrac{3 - s}{3s}$

23. $\dfrac{x^2 - 64}{x + 2} \div \dfrac{8 - x}{x}$

24. $\dfrac{y^2 - 25}{5y} \cdot \dfrac{10}{5 - y}$

25. $\dfrac{x - 2}{x - y} \cdot \dfrac{x^2 - y^2}{x^2 - 4x + 4}$

26. $\dfrac{x + 1}{x + y} \cdot \dfrac{x^2 - y^2}{x^2 - 1}$

27. $\dfrac{x}{(x - 3)^2} \cdot \dfrac{9 - x^2}{x^4}$

28. $\dfrac{x^2 - y^2}{z^2 - y^2} \cdot \dfrac{xy - xz}{y - x}$

29. $\dfrac{x^2 + 6x + 9}{x^2 - 9} \cdot \dfrac{x^2 - 6x + 9}{x - 3}$

30. $\dfrac{t^2 - 10t + 25}{t^2 - 25} \div \dfrac{t^2 + 10t + 25}{t + 5}$

31. $\dfrac{x^2 - x - 12}{x^2 + 2x + 1} \cdot \dfrac{x^2 - 4x - 5}{x^2 - 9x - 20}$

32. $\dfrac{x^2 - x - 2}{x^2 + 7x + 6} \cdot \dfrac{x^2 + 3x - 18}{x^2 - 4x + 4}$

33. $\dfrac{9x^2 - 4}{6x^2 - 5x - 6} \div \dfrac{9x^2 - 12x + 4}{6x^2 - 13x + 6}$

34. $\dfrac{10x^2 + 11x + 3}{3x^2 - 19x + 28} \div \dfrac{4x^2 - 8x - 5}{6x^2 - 29x + 35}$

35. $\dfrac{15a^2 + 7ab - 2b^2}{6a^2 - 11ab - 10b^2} \cdot \dfrac{2a^2 - 13ab + 20b^2}{25a^2 - b^2}$

36. $\dfrac{18m^2 + 27mn + 10n^2}{6m^2 + 19mn + 10n^2} \div \dfrac{12m^2 - 8mn - 15n^2}{16m^2 + 34mn - 15n^2}$

37. $\dfrac{x^2 - y^2}{x^2 + 2xy + y^2} \div \dfrac{x^2 - 2xy + y^2}{xy - y^2}$

38. $\dfrac{x^2y - 16y^3}{xy^2 - 4y^3} \div \dfrac{x^2 + 3xy - 4y^2}{y - x}$

39. $\dfrac{x^2 - 16}{x^2 - 9} \div \dfrac{3x^2 + 13x + 4}{2x^2 + x - 21}$

40. $\dfrac{25a^2b^2 - 49}{4ab + 1} \div \dfrac{5ab + 7}{16a^2b^2 + 16ab + 3}$

41. $\dfrac{x^3 - 8}{x^2 - 2x + 4} \cdot \dfrac{x^3 + 8}{x^2 + 2x + 4}$

42. $\dfrac{(x - y)^3}{(x + y)^2} \cdot \dfrac{x^2 + 2xy + y^2}{x^2 - 2xy + y^2}$

43. $\dfrac{xy + y - x^2 - x}{12xy} \div \dfrac{x^2 + x}{4x^2}$

44. $\dfrac{2x^2 + 10x + xy + 5y}{x^2 - 25} \div \dfrac{x^2 + 5x + 25}{x^3 - 125}$

45. $\dfrac{x^3 + 27}{x^2 - 9} \cdot \dfrac{5x^2 + x - 18}{5x^2 - 9x} \div \dfrac{x^2 - 3x + 9}{5x^2 - 6x - 27}$

46. $\dfrac{x^2 - 16}{x^2 - x - 12} \cdot \left(\dfrac{x + 3}{x + 4}\right)^2 \div \dfrac{x^3 + 27}{x^3 + 64}$

**47.** $\dfrac{18x^2 + 9x - 20}{6x^2 - 5x - 4} \div \left( \dfrac{2x^2 - 9x - 5}{9x^2 - 16} \cdot \dfrac{12x^2 - 10x}{4x^2 + 4x + 1} \right)$

**48.** $\dfrac{2x^2 - 9x + 7}{x^2 - 1} \div \left( \dfrac{2x^2 + 9x + 4}{x^2 + 4x + 3} \cdot \dfrac{x + 3}{x + 4} \right)^2$

# 5.6 ADDITION AND SUBTRACTION OF ALGEBRAIC FRACTIONS

Fractions may be added or subtracted if they have a common denominator. If they do not have a common denominator, the lowest common denominator is the most convenient to use because it involves the least amount of computation.

---

**FINDING THE LOWEST COMMON DENOMINATOR (L.C.D.)**

**1.** Factor each denominator into its prime factors; that is, factor each denominator completely.

**2.** Then the L.C.D. is the product formed by using each of the different factors the greatest number of times that it occurs in any *one* of the given denominators.

---

**EXAMPLE 1**

Find the L.C.D. for $\dfrac{3}{18}, \dfrac{15}{60},$ and $\dfrac{19}{27}$.

The prime factors of the denominators are

$$18 = 2 \cdot 3 \cdot 3$$
$$60 = 2 \cdot 2 \cdot 3 \cdot 5$$
$$27 = 3 \cdot 3 \cdot 3$$

Thus,

$$\text{L.C.D.} = 2 \cdot 2 \cdot 3 \cdot 3 \cdot 3 \cdot 5 = 540$$

because 2 occurs at most twice, 3 occurs at most three times, and 5 occurs at most one time in any *one* given denominator.

**EXAMPLE 2**

Find the L.C.D. for $\dfrac{1}{4x}, \dfrac{5}{6x^2y},$ and $\dfrac{-7}{15y^3}$.

The prime factors of the denominators are

$$4x = 2 \cdot 2 \cdot x$$
$$6x^2y = 2 \cdot 3 \cdot x \cdot x \cdot y$$
$$15y^3 = 3 \cdot 5 \cdot y \cdot y \cdot y$$

So

$$\text{L.C.D.} = 2 \cdot 2 \cdot 3 \cdot 5 \cdot x \cdot x \cdot y \cdot y \cdot y = 60x^2y^3$$

because 2 occurs at most twice, 3 occurs at most once, 5 occurs at most once, $x$ occurs at most twice, and $y$ occurs at most three times in any *one* given denominator.

**EXAMPLE 3**

Find the L.C.D. for

$$\frac{3x}{x^2 - 2xy + y^2} \qquad \frac{4y}{x^2 - y^2} \quad \text{and} \quad \frac{6x + 2y}{x^2 - xy - 2y^2}$$

First find the prime factors of the denominators.

$$x^2 - 2xy + y^2 = (x - y)(x - y)$$
$$x^2 - y^2 = (x + y)(x - y)$$
$$x^2 - xy - 2y^2 = (x + y)(x - 2y)$$
$$\text{L.C.D.} = (x - y)(x - y)(x + y)(x - 2y)$$

because in any *one* given denominator the factor $x - y$ occurs at most twice, the factor $x + y$ occurs at most once, and the factor $x - 2y$ occurs at most once.

---

### ADDING OR SUBTRACTING FRACTIONS

**1.** If the fractions do not have a common denominator, change each fraction to an equivalent fraction having the L.C.D. (For each fraction, divide the L.C.D. by the denominator. Then multiply both numerator and denominator by this result in order to obtain an equivalent fraction having the L.C.D.)

**2.** Add or subtract the numerators in the order they occur, and place this result over the L.C.D.

**3.** Reduce the resulting fraction to lowest terms, if possible.

---

**EXAMPLE 4**

Perform the indicated operations.

$$\frac{5}{6a^2b} - \frac{3}{ab^2} + \frac{1}{4a^3}$$

The L.C.D. is $12a^3b^2$. Now divide $12a^3b^2$ by each denominator.

$$\frac{12a^3b^2}{6a^2b} = 2ab \qquad \frac{12a^3b^2}{ab^2} = 12a^2 \qquad \frac{12a^3b^2}{4a^3} = 3b^2$$

Multiply each numerator and denominator by the corresponding result.

$$\frac{5}{6a^2b} - \frac{3}{ab^2} + \frac{1}{4a^3} = \frac{5}{6a^2b} \cdot \frac{2ab}{2ab} - \frac{3}{ab^2} \cdot \frac{12a^2}{12a^2} + \frac{1}{4a^3} \cdot \frac{3b^2}{3b^2}$$

$$= \frac{10ab - 36a^2 + 3b^2}{12a^3b^2}$$

**EXAMPLE 5**

Perform the indicated operations and simplify.

$$\frac{x - 4}{2x} + \frac{x + 2}{x + 1} - \frac{3x + 2}{x^2 + x}$$

The L.C.D. is $2x(x + 1)$.

$$\frac{x-4}{2x}+\frac{x+2}{x+1}-\frac{3x+2}{x^2+x} = \frac{x-4}{2x}\cdot\frac{x+1}{x+1}+\frac{x+2}{x+1}\cdot\frac{2x}{2x}-\frac{3x+2}{x(x+1)}\cdot\frac{2}{2}$$

$$x(x+1)$$

$$=\frac{(x^2-3x-4)+(2x^2+4x)-(6x+4)}{2x(x+1)}$$

$$=\frac{3x^2-5x-8}{2x(x+1)}$$

$$=\frac{(3x-8)(x+1)}{2x(x+1)}$$

$$=\frac{3x-8}{2x}$$

**EXAMPLE 6**

Perform the indicated operations and simplify.

$$\frac{4x}{4x^2-1}-\frac{3x+1}{1-2x}+\frac{2x}{1+2x}$$

$$(2x+1)(2x-1)\qquad -(2x-1)$$

*Note:* If we write $1-2x$ as $-(2x-1)$, the L.C.D. is $(2x+1)(2x-1)$.

$$\frac{4x}{(2x+1)(2x-1)}+\frac{3x+1}{2x-1}\cdot\frac{2x+1}{2x+1}+\frac{2x}{1+2x}\cdot\frac{2x-1}{2x-1}$$

$$=\frac{4x+(6x^2+5x+1)+(4x^2-2x)}{(2x+1)(2x-1)}$$

$$=\frac{10x^2+7x+1}{(2x+1)(2x-1)}$$

$$=\frac{(2x+1)(5x+1)}{(2x+1)(2x-1)}$$

$$=\frac{5x+1}{2x-1}$$

# Exercises 5.6

*Find the lowest common denominator for each set of denominators.*

1. $6x, 8$
2. $15, 20m$
3. $2k, 4k^2$
4. $3t, 12t^3$
5. $6ab^2, 5a^2b$
6. $4ab^3c^4, 12a^2bc, 16ac^2$
7. $5x, x-1$
8. $x+3, 3x$
9. $3x+6, 6x+12$
10. $4a-12, 6a-18$
11. $x^2-25, (x-5)^2$
12. $b^2-16, (b+4)^2$
13. $x^2+6x+8, x^2-x-6$
14. $6x^2+3x-1, 6x^2-5x-4$
15. $3x^3-12x^2, 6x^2+12x, 2x^3-4x^2-16x$
16. $18a(a+5)^2, 6(a^2-25), 5a-25$
17. $6(c+2)(c-4)^2, 12c(c+2)(c-4), 3c(c-4)^2$
18. $t^2(t+3)^2, t^2(t-3)^2, t(t^2-9)$

*Perform the indicated operations and simplify.*

**19.** $\dfrac{3}{14} + \dfrac{5}{14}$

**20.** $\dfrac{9}{20} - \dfrac{3}{20}$

**21.** $\dfrac{16}{a} - \dfrac{9}{a}$

**22.** $\dfrac{x}{3a} + \dfrac{5x}{3a}$

**23.** $\dfrac{6}{x+1} + \dfrac{9}{x+1}$

**24.** $\dfrac{6a}{2x+3} - \dfrac{8a}{2x+3}$

**25.** $\dfrac{7}{36} + \dfrac{7}{45}$

**26.** $\dfrac{19}{24} - \dfrac{31}{60}$

**27.** $\dfrac{2x-1}{12} + \dfrac{3x+5}{20} - \dfrac{x}{4}$

**28.** $\dfrac{3a+b}{6} - \dfrac{2a-b}{8} + \dfrac{5a}{12}$

**29.** $\dfrac{5}{6y} - \dfrac{7}{2y}$

**30.** $\dfrac{2}{3y} + \dfrac{1}{y^2}$

**31.** $\dfrac{4}{3p} + \dfrac{2}{p^2}$

**32.** $\dfrac{3}{4t} - \dfrac{5}{3t}$

**33.** $\dfrac{5}{4x^2} + \dfrac{1}{3x^3} - \dfrac{3}{2x}$

**34.** $\dfrac{5}{3a^2b} - \dfrac{3}{2a} - \dfrac{1}{4ab^3}$

**35.** $\dfrac{1}{a} + \dfrac{1}{b} + \dfrac{1}{c}$

**36.** $\dfrac{a}{b} + \dfrac{c}{a} + 1$

**37.** $3x + \dfrac{4-x}{2x}$

**38.** $9 - \dfrac{2x-1}{3y}$

**39.** $\dfrac{2}{c-1} + \dfrac{1}{c}$

**40.** $\dfrac{4}{s-3} - \dfrac{1}{s}$

**41.** $\dfrac{6}{t+2} - \dfrac{3}{t-2}$

**42.** $\dfrac{5}{x-5} + \dfrac{2}{x+5}$

**43.** $\dfrac{2}{a} + \dfrac{1}{a+1}$

**44.** $\dfrac{2}{a+3} - \dfrac{4}{a+2}$

**45.** $\dfrac{d}{d+4} - \dfrac{2d}{d-5}$

**46.** $\dfrac{3t}{t-2} + \dfrac{4t}{t+4}$

**47.** $\dfrac{8}{x-4} + \dfrac{2}{4-x}$

**48.** $\dfrac{c}{c-6} + \dfrac{2c}{6-c}$

**49.** $\dfrac{4}{3a+9} - \dfrac{6}{4a+12}$

**50.** $\dfrac{12}{5r-25} - \dfrac{5}{3r-15}$

**51.** $\dfrac{5a+1}{a^2-1} - \dfrac{2a-3}{a+1}$

**52.** $\dfrac{2x+5}{x+6} + \dfrac{14x}{x^2-36}$

**53.** $\dfrac{6}{x-3} - \dfrac{5}{3-x}$

**54.** $\dfrac{2x}{x^2-1} + \dfrac{2-x}{1-x}$

**55.** $\dfrac{2}{r^2-4r+4} + \dfrac{1}{r^2+r-6}$

**56.** $\dfrac{4}{y^2-5y+6} - \dfrac{3}{y^2-y-2}$

**57.** $\dfrac{1}{x-2} + \dfrac{1}{x^2-5x+6}$

**58.** $\dfrac{1}{6x} + \dfrac{1}{3x-6} - \dfrac{1}{2x+4}$

**59.** $\dfrac{t+4}{2t-4} - \dfrac{2t+5}{t^2-t-2} + \dfrac{3}{4}$

**60.** $\dfrac{2}{x^2-y^2} - \dfrac{1}{x^2-3xy+2y^2} - \dfrac{3}{x^2-xy-2y^2}$

**61.** $\dfrac{3x-2}{x^2-3x-4} + \dfrac{2x-3}{x^2-x-12}$

**62.** $\dfrac{3x+7}{6x^2+x-35} - \dfrac{3x-7}{6x^2+29x+35}$

**63.** $\dfrac{a^2+b^2}{a^2-b^2} - \dfrac{a}{a+b} + \dfrac{b}{b-a}$

**64.** $\dfrac{2x}{x^2-4} - \dfrac{2+x}{2-x} - \dfrac{x-2}{x+2}$

**65.** $\dfrac{x+3}{x^2-7x+12} + \dfrac{x+3}{16-x^2}$

**66.** $\dfrac{3x^2-18x+5}{2x^2+x-1} - \dfrac{2x+1}{1-2x} - \dfrac{3x-7}{x+1}$

**67.** $\dfrac{5x}{8x+2} - \dfrac{3}{4x^2+x} + \dfrac{1}{2x}$

**68.** $\dfrac{1}{x^2+2x+1} + \dfrac{3}{2x+2} - \dfrac{5}{4x+4}$

**69.** $\dfrac{1}{t-2} - \dfrac{6t}{t^3-8}$

**70.** $\dfrac{c}{1-c^3} + \dfrac{1}{c^2-1} - \dfrac{2}{c^2+c+1}$

**71.** $\dfrac{x^2+x}{x^3+1} + 1$

**72.** $\dfrac{x^2-1}{x^3-1} + \dfrac{x-1}{x^2+x+1}$

## 5.7 COMPLEX FRACTIONS

A **complex fraction** is a fraction that contains a fraction in the numerator, the denominator, or both. There are basically two ways to simplify a complex fraction.

**Method 1:** Multiply both the numerator and denominator of the complex fraction by the L.C.D. of all the fractions that appear in the numerator and denominator.

**Method 2:** Simplify the numerator and the denominator separately. Then divide the simplified numerator by the simplified denominator. Finally simplify that result, if possible.

### EXAMPLE 1

Simplify.

$$\dfrac{1 - \dfrac{1}{x}}{x - \dfrac{1}{x^2}}$$

**Method 1:** The L.C.D. of all the denominators of the fractions in the complex fraction is $x^2$.

$$\dfrac{\left(1 - \dfrac{1}{x}\right)x^2}{\left(x - \dfrac{1}{x^2}\right)x^2} = \dfrac{x^2 - x}{x^3 - 1}$$

$$= \dfrac{x(x-1)}{(x-1)(x^2+x+1)} = \dfrac{x}{x^2+x+1}$$

**Method 2:**

$$\dfrac{1 - \dfrac{1}{x}}{x - \dfrac{1}{x^2}} = \dfrac{\dfrac{x-1}{x}}{\dfrac{x^3-1}{x^2}} = \dfrac{x-1}{x} \div \dfrac{x^3-1}{x^2}$$

$$= \dfrac{x-1}{x} \cdot \dfrac{x^2}{(x-1)(x^2+x+1)} = \dfrac{x}{x^2+x+1}$$

To simplify an expression containing a complex fraction, first simplify the complex fraction and then perform the remaining indicated operations.

**EXAMPLE 2**

Simplify.

$$x - \frac{x}{1 - \dfrac{x}{1 - x}} = x - \frac{x}{\dfrac{1 - x - x}{1 - x}}$$

$$= x - \frac{x}{\dfrac{1 - 2x}{1 - x}} \cdot \frac{(1 - x)}{(1 - x)}$$

$$= x - \frac{x(1 - x)}{1 - 2x}$$

$$= \frac{x(1 - 2x)}{1 - 2x} - \frac{x(1 - x)}{1 - 2x}$$

$$= \frac{(x - 2x^2) - (x - x^2)}{1 - 2x} = \frac{-x^2}{1 - 2x}$$

## Exercises 5.7

*Perform the indicated operations and simplify.*

**1.** $\dfrac{\dfrac{5}{x}}{\dfrac{15}{x + 2}}$

**2.** $\dfrac{\dfrac{12}{r - 3}}{\dfrac{3}{2r}}$

**3.** $\dfrac{\dfrac{t - 1}{6t}}{\dfrac{t + 1}{9t}}$

**4.** $\dfrac{\dfrac{3x^2 y}{9z}}{\dfrac{12xy}{3z}}$

**5.** $\dfrac{\dfrac{s + t}{4k}}{\dfrac{2s + 2t}{8k}}$

**6.** $\dfrac{\dfrac{x - 4y}{6x}}{\dfrac{8x - 32y}{12}}$

**7.** $\dfrac{2 + \dfrac{7}{16}}{2 - \dfrac{1}{4}}$

**8.** $\dfrac{\dfrac{2}{9} - 6}{\dfrac{1}{3} + 1}$

**9.** $\dfrac{a - 4}{2 - \dfrac{8}{a}}$

**10.** $\dfrac{a - 12}{1 - \dfrac{12}{a}}$

**11.** $\dfrac{a - \dfrac{25}{a}}{a + 5}$

**12.** $\dfrac{a + \dfrac{27}{a^2}}{a + 3}$

**13.** $\dfrac{\dfrac{3}{x} + \dfrac{5}{y}}{\dfrac{x}{3} + \dfrac{y}{5}}$

**14.** $\dfrac{\dfrac{x}{y} - \dfrac{y}{z}}{\dfrac{y}{x} - \dfrac{z}{y}}$

**15.** $\dfrac{1 - y}{1 - \dfrac{1}{y}}$

**16.** $\dfrac{\dfrac{1}{y} - 1}{y - \dfrac{1}{y}}$

**17.** $\dfrac{1 + \dfrac{1}{a}}{1 - \dfrac{1}{a}}$

**18.** $\dfrac{\dfrac{2}{c} - 1}{\dfrac{2}{c} + 1}$

**19.** $\dfrac{\dfrac{1}{m} + \dfrac{1}{n}}{\dfrac{1}{m} - \dfrac{1}{n}}$

**20.** $\dfrac{\dfrac{x}{y} + 2}{\dfrac{x}{y} - \dfrac{4y}{x}}$

**21.** $\dfrac{x - 3 - \dfrac{28}{x}}{x + 10 + \dfrac{24}{x}}$

**22.** $\dfrac{12x + 16 + \dfrac{5}{x}}{18x - 27 - \dfrac{35}{x}}$

**23.** $\dfrac{\dfrac{x^2 - y^2}{x^3 + y^3}}{\dfrac{x^2 + 2xy + y^2}{x^2 - xy + y^2}}$

**24.** $\dfrac{\dfrac{8x^2 - 2x - 21}{6x^2 + 11x + 3}}{\dfrac{4x^2 + 20x + 25}{12x^3 + 19x^2 + 5x}}$

**25.** $\dfrac{\dfrac{x}{x + y} + \dfrac{y}{x - y}}{\dfrac{x^2 + y^2}{x^2 - y^2}}$

**26.** $\dfrac{\dfrac{x - y}{x + y} - \dfrac{x + y}{x - y}}{\dfrac{x + y}{x - y} - \dfrac{x - y}{x + y}}$

**27.** $\dfrac{\dfrac{6a^2 - 5ab - 4b^2}{a - b}}{\dfrac{3a^2 - 7ab + 4b^2}{2a + b}}$

**28.** $\dfrac{x + \dfrac{y^2}{x + y}}{y + \dfrac{x^2}{x + y}}$

**29.** $\dfrac{1 + \dfrac{1}{x^2 - 1}}{1 + \dfrac{1}{x - 1}}$

**30.** $\dfrac{3 + \dfrac{5x}{x^2 - 4}}{3 + \dfrac{2}{x - 2}}$

**31.** $1 - \dfrac{1}{2 - \dfrac{1}{x}}$

**32.** $3 + \dfrac{1}{4 - \dfrac{2}{y}}$

**33.** $1 + \dfrac{1}{x + \dfrac{1}{x - 1}}$

**34.** $1 + \dfrac{x + 1}{x + \dfrac{1}{x + 2}}$

**35.** $\dfrac{\dfrac{1}{x} - \dfrac{2}{x^2} - \dfrac{3}{x^3}}{x + 6 + \dfrac{5}{x}} - \dfrac{\dfrac{1}{x^2} - \dfrac{9}{x^4}}{1 + \dfrac{2}{x} - \dfrac{15}{x^2}}$

**36.** $\dfrac{\dfrac{x^2 + 7x + 10}{x + 2}}{x - 6} + \dfrac{x - 2}{\dfrac{x^2 - 2x}{x - 3}}$

**37.** $1 - \dfrac{1}{1 + \dfrac{1}{1 - \dfrac{1}{1 + \dfrac{1}{1 - \dfrac{1}{1 + 1}}}}}$

**38.** $1 + \dfrac{1}{1 + \dfrac{1}{1 + \dfrac{1}{1 + \dfrac{1}{1 + \dfrac{1}{1 + 1}}}}}$

# 5.8 EQUATIONS WITH FRACTIONS

To solve an equation with fractions, multiply each side by the L.C.D. The resulting equation will not contain any fractions and thus may be solved as those equations in Chapter 2.

**EXAMPLE 1**

Solve.

$$\frac{x}{12} + \frac{1}{6} = \frac{x+1}{8}$$

The L.C.D. of 12, 6, and 8 is 24; therefore, multiply each side of the preceding equation by 24.

$$24\left(\frac{x}{12} + \frac{1}{6}\right) = 24\left(\frac{x+1}{8}\right)$$
$$2x + 4 = 3(x + 1)$$
$$2x + 4 = 3x + 3$$
$$1 = x$$

*Check:* Substitute 1 for $x$ each time it occurs in the *original equation.*

$$\frac{(1)}{12} + \frac{1}{6} = \frac{(1)+1}{8}$$
$$\frac{1}{12} + \frac{2}{12} = \frac{2}{8}$$
$$\frac{3}{12} = \frac{2}{8}$$
$$\frac{1}{4} = \frac{1}{4}$$

It is common practice in science and technology to solve a given formula for the unknown letter before the substitution of the data is made. To solve a formula involving fractions, follow the same procedure as above.

**EXAMPLE 2**

Solve for $r$.

$$\frac{E}{e} = \frac{R+r}{r}$$

The L.C.D. is $er$; therefore, multiply each side of the preceding equation by $er$.

$$er\left(\frac{E}{e}\right) = er\left(\frac{R+r}{r}\right)$$
$$Er = eR + er$$
$$Er - er = eR \qquad \text{(Subtract } er \text{ from each side.)}$$
$$(E - e)r = eR \qquad \text{( Factor.)}$$
$$r = \frac{eR}{E - e} \qquad \text{(Divide each side by } E - e.\text{)}$$

**EXAMPLE 3**

Solve.

$$\frac{x}{x-5} = \frac{5}{x-5} + 2$$

The L.C.D. is $x - 5$; therefore, multiply each side by $x - 5$.

$$(x-5)\left[\frac{x}{x-5}\right] = (x-5)\left[\frac{5}{x-5} + 2\right]$$

$$x = 5 + 2(x - 5)$$
$$x = 5 + 2x - 10$$
$$x = 2x - 5$$
$$5 = x$$

*Check:*

$$\frac{5}{5-5} = \frac{5}{5-5} + 2$$

$$\frac{5}{0} \neq \frac{5}{0} + 2$$

because $\dfrac{5}{0}$ is meaningless.

The only apparent solution, $x = 5$, does not satisfy the *original* equation. Therefore, the original equation has no solution; that is, there is no real number that can replace $x$ in the original equation and produce a true statement.

---

The technique of multiplying each side of an equation by a **variable** quantity gives the possibility of obtaining *extraneous,* or false, solutions.

---

For example, the simple equation

$$x = 3$$

obviously has only one solution. If we multiply each side by $x$, we have

$$x^2 = 3x$$

But this equation has two solutions: 3 and 0. Here, 0 is an extraneous solution.

---

In solving an equation using the technique of multiplying each side by a variable quantity, the only way to determine whether or not an apparent solution is extraneous is to check each apparent solution in the *original* equation.

---

**EXAMPLE 4**

Solve.

$$\frac{x}{x+2} + \frac{4}{x-2} = \frac{x^2}{\underset{(x+2)(x-2)}{x^2 - 4}}$$

The L.C.D. is $(x + 2)(x - 2)$.

$$x(x - 2) + 4(x + 2) = x^2$$
$$x^2 - 2x + 4x + 8 = x^2$$
$$2x + 8 = 0$$
$$2x = -8$$
$$x = -4$$

*Check:* Substitute the apparent solution, $x = -4$, in the original equation.

$$\frac{(-4)}{(-4) + 2} + \frac{4}{(-4) - 2} = \frac{(-4)^2}{(-4)^2 - 4}$$

$$\frac{-4}{-2} + \frac{4}{-6} = \frac{16}{12}$$

$$2 - \frac{2}{3} = \frac{4}{3}$$

$$\frac{4}{3} = \frac{4}{3}$$

Since $x = -4$ satisfies the original equation, $x = -4$ is the solution, or root.

As we have seen, multiplying each side of an equation by a variable quantity can lead to extraneous solutions. This operation requires checking the solutions. It is also important *not* to divide each side by a variable quantity, because this operation may result in the loss of solutions. For example, the equation

$$x^2 \doteq 5x$$

has two solutions: 5 and 0. If we divide each side of this equation by $x$, we have

$$x = 5$$

and we have lost a solution.

You should note that there is a great difference between a fractional expression involving addition, subtraction, multiplication, and division of algebraic fractions and an algebraic fractional equation. The only time that you multiply by the L.C.D. to eliminate fractions is when an equal sign is present, which indicates a fractional equation.

## Exercises 5.8

*Solve and check.*

**1.** $\dfrac{x}{3} - \dfrac{x}{6} = 2$

**2.** $\dfrac{y}{4} + \dfrac{y}{12} = 5$

**3.** $\dfrac{3m}{4} + \dfrac{2m}{5} = \dfrac{23}{2}$

**4.** $\dfrac{5t}{4} - \dfrac{4t}{9} = \dfrac{29}{4}$

**5.** $\dfrac{x}{6} + 2 = \dfrac{x + 3}{4}$

**6.** $\dfrac{t - 6}{6} + \dfrac{1}{10} = \dfrac{t}{15}$

**7.** $\dfrac{s + 4}{8} - 1\dfrac{1}{8} = \dfrac{3s}{4}$

**8.** $\dfrac{4m}{5} + 7\dfrac{2}{3} = \dfrac{m + 5}{15}$

**9.** $\dfrac{x + 5}{2} = \dfrac{x + 4}{3}$

**10.** $\dfrac{x + 3}{5} = \dfrac{x - 7}{4}$

**11.** $\dfrac{x}{20} + \dfrac{1}{4} = \dfrac{x + 2}{5}$

**12.** $\dfrac{x}{15} + \dfrac{1}{3} = \dfrac{x - 3}{5}$

**13.** $\dfrac{2x + 6}{4} + 1 = \dfrac{3x + 1}{8}$

**14.** $\dfrac{4x-3}{9} - 2 = \dfrac{x-4}{3}$

**15.** $\dfrac{5-x}{7} - \dfrac{2x-1}{14} = 1$

**16.** $\dfrac{2-x}{24} - \dfrac{3x+4}{16} = 2$

*In Exercises 17–28, solve for the indicated letter.*

**17.** $F = \dfrac{mv^2}{r}$  for $r$

**18.** $R = \dfrac{\rho L}{A}$  for $\rho$

**19.** $I_L = \dfrac{V}{R + R_L}$  for $R$

**20.** $\dfrac{1}{f} = \dfrac{1}{p} + \dfrac{1}{q}$  for $f$

**21.** $\dfrac{1}{R} = \dfrac{1}{R_1} + \dfrac{1}{R_2}$  for $R_1$

**22.** $\dfrac{1}{R} = \dfrac{1}{R_1} + \dfrac{1}{R_2} + \dfrac{1}{R_3}$  for $R_2$

**23.** $y = \dfrac{x+y}{x}$  for $y$

**24.** $y = \dfrac{x-y}{x}$  for $x$

**25.** $\dfrac{x}{a+b} = \dfrac{x+a}{b}$  for $x$

**26.** $\dfrac{1}{x+a} - \dfrac{a}{x-a} = \dfrac{ax}{x^2 - a^2}$  for $x$

**27.** $V = \dfrac{Q}{R_1} - \dfrac{Q}{R_2}$  for $Q$

**28.** $V = \dfrac{Q}{R_1} - \dfrac{Q}{R_2}$  for $R_2$

*Solve and check.*

**29.** $\dfrac{3}{2x} = \dfrac{5}{x} - 7$

**30.** $\dfrac{7}{2x} - \dfrac{5}{3} = \dfrac{1}{x}$

**31.** $\dfrac{x-3}{x} = \dfrac{1}{x} + \dfrac{2}{3}$

**32.** $\dfrac{2x+5}{x} = \dfrac{2}{x} + \dfrac{1}{4}$

**33.** $\dfrac{1}{x-2} = \dfrac{3}{x+4}$

**34.** $\dfrac{6}{3-2x} = \dfrac{2}{5-3x}$

**35.** $\dfrac{x+2}{x-3} = \dfrac{x+1}{x-1}$

**36.** $\dfrac{6t-3}{2t+1} = \dfrac{9t}{3t+4}$

**37.** $\dfrac{2x^2+8}{x^2-1} - \dfrac{x-4}{x+1} = \dfrac{x}{x-1}$

**38.** $\dfrac{x-3}{x+3} = \dfrac{2x+3}{x-3} - \dfrac{x^2}{x^2-9}$

**39.** $\dfrac{x+2}{x+7} = \dfrac{2}{x-2} + \dfrac{x^2+3x-28}{x^2+5x-14}$

**40.** $\dfrac{1}{x-4} + \dfrac{1}{x-5} + \dfrac{1}{x^2-9x+20} = 0$

**41.** $\dfrac{x}{x+4} + \dfrac{x+1}{x-5} = 2$

**42.** $\dfrac{-3x^2+5x}{x^2+6x+9} + 4 = \dfrac{x+5}{x+3}$

**43.** $\dfrac{x^2+3x+7}{x^2-x-12} + \dfrac{x+6}{x+3} = \dfrac{2x+1}{x-4}$

**44.** $\dfrac{7x^2+x-17}{2x^2-5x+3} - \dfrac{3x-4}{2x-3} = \dfrac{2x+1}{x-1}$

# CHAPTER SUMMARY

1. Special products:
   (a) *Product of a monomial and a polynomial:*
   $$a(x + y + z) = ax + ay + az$$
   (b) $(x + y)(x - y) = x^2 - y^2$
   (c) *Square of a binomial:*
   $$(x + y)^2 = x^2 + 2xy + y^2$$
   $$(x - y)^2 = x^2 - 2xy + y^2$$
   (d) *Square of a trinomial:*
   $$(x + y + z)^2 = x^2 + y^2 + z^2 + 2xy + 2xz + 2yz$$
   (e) *Cube of a binomial:*
   $$(x + y)^3 = x^3 + 3x^2y + 3xy^2 + y^3$$
   $$(x - y)^3 = x^3 - 3x^2y + 3xy^2 - y^3$$

2. *Factoring:*
   (a) *Greatest common factor:*
   $$ax + ay + az = a(x + y + z)$$
   (b) *Difference of two perfect squares:*
   $$x^2 - y^2 = (x + y)(x - y)$$
   (c) *Trinomials by trial and error.*
   (d) *Perfect square trinomials:*
   $$x^2 + 2xy + y^2 = (x + y)^2$$
   $$x^2 - 2xy + y^2 = (x - y)^2$$
   (e) *Sum and difference of two perfect cubes:*
   $$x^3 + y^3 = (x + y)(x^2 - xy + y^2)$$
   $$x^3 - y^3 = (x - y)(x^2 + xy + y^2)$$

3. *Equivalent fractions:* Two fractions are equivalent when both numerator and denominator of one fraction can be multiplied or divided by the same nonzero number in order to change one fraction to the other.

4. *Signs of fractions:* A fraction has three signs—the sign of the fraction, the sign of the numerator, and the sign of the denominator. Any two of these three signs may be changed without changing the value of the fraction. For example,
$$-\frac{-a}{+b} = -\frac{+a}{-b} = +\frac{+a}{+b} = +\frac{-a}{-b}$$

5. *Lowest terms:* A fraction is in lowest terms when its numerator and denominator have no common factors except 1.

6. *Factors that differ in sign:* When factors differ in sign, one may be replaced by the opposite of the other; that is,

$$x - y = -(y - x) \quad \text{or} \quad y - x = -(x - y)$$

7. *To multiply fractions:*

$$\frac{a}{b} \cdot \frac{c}{d} = \frac{ac}{bd}$$

8. *To divide fractions:*

$$\frac{a}{b} \div \frac{c}{d} = \frac{a}{b} \cdot \frac{d}{c} = \frac{ad}{bc}$$

9. *To find the lowest common denominator (L.C.D.):*
   (a) Factor each denominator completely.
   (b) Then the L.C.D. is the product formed by using each of the different factors the greatest number of times that it occurs in any *one* of the given denominators.

10. *Adding or subtracting fractions:*
    (a) If the fractions do not have a common denominator, change each fraction to an equivalent fraction having the L.C.D. (For each fraction, divide the L.C.D. by the denominator. Then multiply both numerator and denominator by this result in order to obtain an equivalent fraction having the L.C.D.)
    (b) Add or subtract the numerators in the order they occur, and place this result over the L.C.D.
    (c) Reduce the resulting fraction to lowest terms, if possible.

11. *Complex fraction:* A fraction that contains a fraction in the numerator, the denominator, or both. There are basically two ways to simplify a complex fraction.

**Method 1:** Multiply both numerator and denominator of the complex fraction by the L.C.D. of all the fractions that appear in the numerator and denominator.

**Method 2:** Simplify the numerator and the denominator separately. Then divide the simplified numerator by the simplified denominator. Finally simplify that result, if possible.

12. *To solve an equation with fractions:* Multiply each side by the lowest common denominator. When you multiply each side by a variable quantity you *must check* each apparent solution in the original equation.

# CHAPTER 5 REVIEW

*Find each product mentally.*

1. $-6a^2b^4(a^3 - 12ab^2 - 9b^5)$

2. $(3x + 13)(5x - 4)$

3. $(4x - 7)(4x + 7)$

4. $(5a^2 - 9b^2)^2$

5. $(4a - b^2)^3$

6. $(6x + 7yz^2)(5x - 2yz^2)$

*Factor each expression **completely**.*

7. $18a^3b - 9a^2b^4 + 9a^3b^2$

8. $16x^2 - 9y^2$

**9.** $5x^2 + 28x + 32$

**10.** $14x^2 + 59x - 18$

**11.** $4a^2 - 20a + 25$

**12.** $m^3 - n^3$

**13.** $3x^2 - 18x - 21$

**14.** $x^4 - 81$

**15.** $15x^2 + 23x - 90$

**16.** $10x^2 + 25xy - 60y^2$

**17.** $x^2 + 6x + 9 - 25y^2$

**18.** $ax + 2bx - 4ay - 8by$

*Reduce each fraction to lowest terms.*

**19.** $\dfrac{90x^4y^9}{48x^8y^3}$

**20.** $\dfrac{16x^2 - 40x}{8x^2 - 20x}$

**21.** $\dfrac{x^2 - 4x - 32}{2x^2 + 11x + 12}$

**22.** $\dfrac{-2x^2 + 7x + 4}{x^2 - 16}$

*Perform the indicated operations and simplify.*

**23.** $\dfrac{2x^2 + 5x - 25}{4x^2 + 21x + 5} \cdot \dfrac{4x^2 + x}{4x^2 - 25}$

**24.** $\dfrac{6x^2 + 11x - 7}{4x^2 + 4x + 1} \div \dfrac{3x^2 + 4x - 7}{-2x^2 + x + 1}$

**25.** $\dfrac{y^6 + 125}{4y^2 + 20} \div \dfrac{y^4 - 5y^2 + 25}{12y}$

**26.** $\dfrac{2 - x}{x^2 + 10x + 21} \cdot \dfrac{3x^2 + 21x}{18x + 36} \div \dfrac{x^3 - 4x}{x + 3}$

**27.** $\dfrac{4}{3a} + \dfrac{9}{4a^2} - \dfrac{5}{2}$

**28.** $4 - \dfrac{3x + 4}{2x + 1}$

**29.** $\dfrac{x}{x^2 - 7x + 10} + \dfrac{x}{2x^2 - 5x + 2}$

**30.** $\dfrac{x - 1}{2x^2 + 3x + 1} + \dfrac{x - 3}{3x^2 - x - 4}$

**31.** $\dfrac{\dfrac{1}{x} + \dfrac{1}{x^2}}{x + \dfrac{1}{x^2}}$

**32.** $4 + \dfrac{2}{3 + \dfrac{1}{y}}$

*Solve for x and check.*

**33.** $\dfrac{x + 2}{5} + \dfrac{x + 4}{30} = \dfrac{x}{6}$

**34.** $\dfrac{ax + b}{b} = \dfrac{bx}{a} + \dfrac{a}{b}$

**35.** $\dfrac{x + 1}{x - 3} + 1 = \dfrac{4}{x - 3}$

**36.** $\dfrac{4}{x - 4} - \dfrac{x^2}{16 - x^2} = 1$

# 6

# Systems of Linear Equations

## INTRODUCTION

Joyce owns a field with an irrigation system consisting of two networks of pipes. During the month of July she contracted to have the water turned on the first and third Fridays. On the first Friday the larger network operated 9 hours and the other 8 hours, delivering 730 gallons of water. On the third Friday, the larger network operated 5 hours and the other 7 hours, delivering 495 gallons. What is the rate of flow (gallons per hour) in each network?

To solve this problem, we need to use two equations with two unknowns to represent all the information. This very powerful problem-solving method is called a *system of equations*. It allows separate conditions or limitations to be expressed in a relatively simple manner while tying all the conditions together in a system.

In this chapter we describe methods of solving systems of linear equations. In the preceding problem the larger network delivers 50 gal/h, while the smaller delivers 35 gal/h.

## Objectives

- Solve systems of equations by substitution.
- Solve systems of equations by the addition-subtraction method.
- Evaluate determinants using determinant properties.
- Use Cramer's rule.
- Use the method of partial fractions to rewrite rational expressions as the sum or difference of simpler expressions.

## 6.1   SOLVING A SYSTEM OF TWO LINEAR EQUATIONS

Systems of linear equations in the form

$$a_1x + b_1y = c_1$$
$$a_2x + b_2y = c_2$$

D.3, D.4

may be solved in a number of ways. In this section we shall study solutions by graphing, by the addition-subtraction method, and by the method of substitution. Any ordered pair $(x, y)$ that satisfies both equations is called a **solution,** or root, of the system.

We saw in Chapter 4 that the graph of a linear equation with two variables is a straight line. The solution of a system of two linear equations with two variables can be discussed in general terms of their graphs as follows.

---

**GRAPHS OF LINEAR SYSTEMS OF EQUATIONS WITH TWO VARIABLES**

1. The two lines may intersect at a common, single point. This point, in ordered pair form $(x, y)$, is the solution of the system.

2. The two lines may be parallel with no points in common; hence, the system has no solution.

3. The two lines may coincide; the solution of the system is the set of all points on the common line.

---

These results are illustrated in Fig. 6.1.

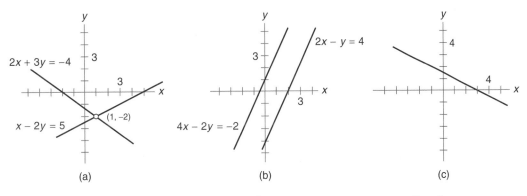

|  (a)  |  (b)  |  (c)  |

$2x + 3y = -4$
$x - 2y = 5$

The lines intersect at the point $(1, -2)$, which is the common solution.

$2x - y = 4$
$4x - 2y = -2$

The lines are parallel, and hence there is no common solution.

$x + 2y = 3$
$2x + 4y = 6$

The lines coincide, and as a result any solution of one equation is a solution of the other.

**Figure 6.1**

When the two lines intersect, the system of equations is called **independent and consistent.** When the two lines are parallel, the system of equations is called **inconsistent.** When the two lines coincide, the system of equations is called **dependent.**

Graphical methods of solving systems of linear equations usually give only approximate solutions. Algebraic methods give exact solutions.* The first algebraic method we shall study is called the **addition-subtraction method.**

---

*Do not use the rules for calculating with measurements in this chapter.

SOLVING A PAIR OF LINEAR EQUATIONS BY THE
ADDITION-SUBTRACTION METHOD

1. If necessary, multiply each side of one or both equations by some number so that the numerical coefficients of one of the variables are of equal absolute value.

2. If these coefficients of equal absolute value have like signs, subtract one equation from the other. If they have unlike signs, add the equations. That is, do whatever is necessary to eliminate that variable.

3. Solve the resulting equation for the remaining variable.

4. Substitute the solution for the variable found in Step 3 in either of the original equations, and solve this resulting equation for the second variable.

5. Check by substituting the solution in both original equations.

**EXAMPLE 1**

Solve

$$2x + 3y = -4$$
$$x - 2y = 5$$

using the addition-subtraction method.

$$2x + 3y = -4$$
$$x - 2y = 5 \qquad \text{(Multiply each side of this equation by 2, and the numerical coefficients of } x \text{ will be of equal absolute value.)}$$

$$\begin{aligned} 2x + 3y &= -4 \\ \underline{2x - 4y} &= \underline{10} \qquad \text{(Subtract the equations.)} \\ 7y &= -14 \qquad \text{(Solve for } y.) \\ y &= -2 \end{aligned}$$

Substitute $y = -2$ in either original equation.

$$x - 2y = 5$$
$$x - 2(-2) = 5$$
$$x = 1$$

The solution is $(1, -2)$.

*Check:* Substitute the solution in both original equations.

$$\begin{array}{cc} 2x + 3y = -4 & x - 2y = 5 \\ 2(1) + 3(-2) = -4 & (1) - 2(-2) = 5 \\ -4 = -4 & 5 = 5 \end{array}$$

**EXAMPLE 2**

Solve

$$3x - 4y = -36$$
$$10x + 3y = -22$$

using the addition-subtraction method.

$$3x - 4y = -36$$
$$10x + 3y = -22$$

(Multiply each side of the first equation by 3 and each side of the second equation by 4; the numerical coefficients of $y$ will be of equal absolute value.)

$$9x - 12y = -108$$
$$\underline{40x + 12y = -88}$$

(Add the equations.)

$$49x \qquad = -196$$

(Solve for $x$.)

$$x = -4$$

Substitute $x = -4$ in either original equation.

$$3x - 4y = -36$$
$$3(-4) - 4y = -36$$
$$-4y = -24$$
$$y = 6$$

The solution is $(-4, 6)$.

*Check:* Substitute the solution in both original equations as before.

The second algebraic method of solving systems of linear equations is called the **method of substitution.**

---

**SOLVING A PAIR OF LINEAR EQUATIONS BY THE METHOD OF SUBSTITUTION**

1. From either of the two given equations, solve for one variable in terms of the other.
2. Substitute this result from Step 1 in the *other* equation. Note that this step eliminates one variable.
3. Solve the equation obtained from Step 2 for the remaining variable.
4. From the equation obtained in Step 1, substitute the solution for the variable found in Step 3, and solve this resulting equation for the second variable.
5. Check by substituting the solution in both original equations.

---

**EXAMPLE 3**

Solve

$$3x + 4y = -15$$
$$y = -2x$$

using the method of substitution.

Since the second equation is already solved for $y$, substitute $y = -2x$ in the first equation.

$$3x + 4y = -15$$
$$3x + 4(-2x) = -15 \qquad \text{(Substitute for } y.\text{)}$$
$$3x - 8x = -15 \qquad \text{(Solve for } x.\text{)}$$
$$-5x = -15$$
$$x = 3$$

Substitute $x = 3$ in the second equation.

$$y = -2x$$
$$y = -2(3)$$
$$y = -6$$

The solution is $(3, -6)$.

    *Check:* Substitute in both original equations.

**EXAMPLE 4**

Solve

$$3x + y = 3$$
$$2x - 4y = 16$$

by the method of substitution.

    Solve the first equation for $y$.

$$3x + y = 3$$
$$y = -3x + 3$$

Substitute $y = -3x + 3$ in the second equation.

$$2x - 4y = 16$$
$$2x - 4(-3x + 3) = 16 \qquad \text{(Substitute for } y\text{.)}$$
$$2x + 12x - 12 = 16 \qquad \text{(Solve for } x\text{.)}$$
$$14x = 28$$
$$x = 2$$

Substitute $x = 2$ in the equation $y = -3x + 3$.

$$y = -3x + 3$$
$$y = -3(2) + 3$$
$$y = -3$$

The solution is $(2, -3)$.

    *Check:* Substitute the solution in both original equations.

**EXAMPLE 5**

Solve

$$8x + 5y = -16$$
$$2x - 3y = 13$$

by the method of substitution.

    Solve the second equation for $x$.

$$2x - 3y = 13$$
$$2x = 3y + 13$$
$$x = \frac{3y + 13}{2}$$

Substitute $x = \dfrac{3y + 13}{2}$ in the first equation.

$$8x + 5y = -16$$

$$8\left(\frac{3y + 13}{2}\right) + 5y = -16 \qquad \text{(Substitute for } x\text{.)}$$

$$4(3y + 13) + 5y = -16 \qquad \text{(Solve for } y\text{.)}$$

$$12y + 52 + 5y = -16$$

$$17y = -68$$

$$y = -4$$

Substitute $y = -4$ in the equation $x = \dfrac{3y + 13}{2}$.

$$x = \frac{3y + 13}{2}$$

$$x = \frac{3(-4) + 13}{2}$$

$$x = \frac{1}{2}$$

The solution is $\left(\dfrac{1}{2}, -4\right)$.

*Check:* Substitute the solution in both original equations.

A special case of the substitution method is the **comparison method.** That is,

$$\text{if} \quad a = c \quad \text{and} \quad b = c$$

$$\text{then} \quad a = b$$

**EXAMPLE 6**

Solve

$$3x - 4 = 5y$$

$$6 - 2x = 5y$$

Since the left side of each equation equals the same quantity, we have

$$3x - 4 = 6 - 2x$$

which eliminates a variable. Then

$$5x = 10$$

$$x = 2$$

Substitute $x = 2$ in the first equation.

$$3x - 4 = 5y$$

$$3(2) - 4 = 5y$$

$$2 = 5y$$

$$\frac{2}{5} = y$$

The solution is $\left(2, \dfrac{2}{5}\right)$.

*Check:* Substitute the solution in both original equations.

*Note:* The basic idea of solving any system of equations is to choose the method that eliminates a variable most easily. This is determined by the form of the given equations.

**EXAMPLE 7**

Solve

$$2x - y = 4$$
$$4x - 2y = -2$$

Using the addition-subtraction method, multiply each side of the first equation by 2.

$$
\begin{array}{l}
4x - 2y = 8 \\
\underline{4x - 2y = -2} \quad \text{(Subtract the equations.)} \\
\phantom{4x - 2y = {}}0 = 10
\end{array}
$$

When we obtain the result $0 = a$ $(a \neq 0)$, this means that there is no common solution, the lines when graphed are parallel, and the system of equations is inconsistent.

**EXAMPLE 8**

Solve

$$x + 2y = 3$$
$$2x + 4y = 6$$

Using the addition-subtraction method, multiply each side of the first equation by 2.

$$
\begin{array}{l}
2x + 4y = 6 \\
\underline{2x + 4y = 6} \quad \text{(Subtract the equations.)} \\
\phantom{2x + 4y = {}}0 = 0
\end{array}
$$

When we obtain the result $0 = 0$, this means that the system is dependent and the lines coincide when graphed. Since one equation is a multiple of the other, the equations are equivalent and therefore form the same straight line when graphed.

Many technical applications can be expressed mathematically as a system of linear equations. Examples will be given in this section and in the following sections to demonstrate how to set up an appropriate system of equations for a given problem. Except for Steps 4, 5, and 6, the steps in approaching the solution of a problem stated verbally are the same as for problems involving only one linear equation. You may wish to review Section 2.8 at this time.

---

**STEPS FOR PROBLEM SOLVING**

1. Read the problem carefully at least two times.
2. If possible, draw a picture or a diagram.
3. Write what facts are given and what unknown quantities are to be found.
4. Choose a symbol to represent each quantity to be found (there will be more than one). Be sure to label each symbol to indicate what it represents.

---

**5.** Write appropriate equations relating these variables from the information given in the problem.* Watch for information that is not stated but which should be assumed. For example, rate × time = distance.

**6.** Solve for the unknown variables by the methods presented in this chapter.

**7.** Check your solution in the original equations.

**8.** Check your solution in the original verbal problem.

**EXAMPLE 9**

The sum of two currents is 200 mA. The larger current is three times the smaller. What are the two currents?

Let

$$x = \text{the smaller current}$$

$$y = \text{the larger current}$$

The first sentence gives the first equation that follows; the second sentence gives the second equation.

$$x + y = 200$$
$$y = 3x$$

Substitute the second equation into the first.

$$x + 3x = 200$$
$$4x = 200$$
$$x = 50$$

Substitute $x = 50$ into $y = 3x$.

$$y = 3(50) = 150$$

The solution is (50, 150). That is, the smaller current is 50 mA and the larger current is 150 mA.

The solution checks in both original equations and in the problem as it is originally stated.

**EXAMPLE 10**

A boat can travel 24 mi upstream in 6 h. It takes 4 h to make the return trip downstream. Find the rate of the current and the speed of the boat (in still water).

Let

$$x = \text{the speed of the boat}$$

$$y = \text{the rate of the current}$$

Then

$$x + y = \text{the speed of the boat going downstream (the boat is traveling with the current)}$$

$$x - y = \text{the speed of the boat going upstream (the boat is traveling against the current)}$$

*There should be as many equations as there are unknown variables.

Let's make a chart of the information.

|  | $d$ | $r$ | $t$ |
|---|---|---|---|
| Traveling downstream | 24 | $x + y$ | 4 |
| Traveling upstream | 24 | $x - y$ | 6 |

We now have

$$4(x + y) = 24 \qquad \text{(time} \times \text{rate} = \text{distance)}$$
$$6(x - y) = 24$$

This gives us the system

$$4x + 4y = 24$$
$$6x - 6y = 24$$

Use the addition-subtraction method.

$$\begin{aligned} 12x + 12y &= 72 \\ \underline{12x - 12y} &= \underline{48} \\ 24x \phantom{+12y} &= 120 \\ x \phantom{+12y} &= 5 \end{aligned}$$

Substitute $x = 5$ in one of the original equations.

$$\begin{aligned} 6x - 6y &= 24 \\ 6(5) - 6y &= 24 \\ 30 - 6y &= 24 \\ -6y &= -6 \\ y &= 1 \end{aligned}$$

The solution is (5, 1). The rate of the current is 1 mi/h and the speed of the boat is 5 mi/h. The solution should be checked in both original equations and in the original problem.

## Exercises 6.1

*Solve each system of equations graphically by graphing each equation and finding the point of intersection. Check by substituting the solution ordered pair in both original equations.*

**1.** $2x + 3y = 5$
$x - 2y = 6$

**2.** $3x - 2y = 0$
$3x + y = -9$

**3.** $5x + 2y = -5$
$2x + 3y = 9$

**4.** $-2x + 4y = -6$
$3x - 2y = 9$

**5.** $9x - 6y = 15$
$-3x + 2y = -5$

**6.** $-3x + y = 4$
$y = 3x$

*Solve each system of equations by the addition-subtraction method. Check by substituting the solution in both original equations.*

**7.** $4x + y = 15$
$3x + y = 13$

**8.** $-2x + 5y = -30$
$2x - 3y = 22$

**9.** $-4x + 3y = -22$
$4x - 5y = 34$

**10.** $7x + 3y = 1$
$2x + 3y = 11$

**11.** $4x - 5y = -34$
$2x + 3y = 16$

**12.** $3x - 2y = -3$
$7x - 10y = 1$

**13.** $3x + 2y = -15$
$x + 5y = -5$

**14.** $x + 3y = 6$
$5x - 2y = -4$

**15.** $4x - 5y = 7$
$-2x + 3y = -3$

**16.** $6x + 7y = -8$
$18x - 4y = 26$

**17.** $4x + 3y = -3$
$12x + 9y = -12$

**18.** $2x - y = 6$
$10x - 5y = 30$

**19.** $12x + 5y = -18$
$8x - 7y = -74$

**20.** $7x - 10y = -6$
$24x - 16y = 16$

**21.** $-12x + 15y = -43$
$9x - 12y = 34$

**22.** $24x + 84y = 185$
$42x + 12y = 155$

*Solve each system of equations by the method of substitution and check.*

**23.** $2x - 5y = -36$
$y = 4x$

**24.** $6x - 11y = -3$
$x = 2y$

**25.** $8x + 9y = -5$
$x = -3y$

**26.** $12x + 3y = -6$
$y = -5x$

**27.** $3x - 5y = 64$
$y = 6x - 2$

**28.** $6x + 8y = 154$
$x = 3y + 4$

**29.** $3x + 4y = -6$
$2x - y = -15$

**30.** $2x + 3y = -29$
$x + 4y = -32$

**31.** $5x - 7y = 23$
$3x + 2y = -11$

**32.** $6x + 7y = 16$
$-5x - 6y = -13$

*Solve each system of equations by any method and check.*

**33.** $\dfrac{2}{3}x + \dfrac{1}{2}y = \dfrac{4}{9}$

$\dfrac{3}{4}x - \dfrac{2}{3}y = -\dfrac{23}{72}$

**34.** $\dfrac{5}{4}x - \dfrac{2}{5}y = -1$

$\dfrac{1}{4}x + \dfrac{2}{3}y = 11$

**35.** $\dfrac{3}{5}x + 2y = \dfrac{5}{2}$

$\dfrac{2}{5}x - 4y = 3$

**36.** $\dfrac{7}{3}x + \dfrac{2}{5}y = -17$

$\dfrac{2}{5}x + \dfrac{4}{3}y = 54$

**37.** $1.4x - 2.7y = 5.66$
$0.5x + 2y = 3.8$

**38.** $x + y = 40.3$
$0.02x + 0.05y = 1.46$

**39.** $0.002x + 0.008y = 2.28$
$0.04x + 0.09y = 28.8$

**40.** $1.57x + 2.04y = 20.262$
$2.16x - 8.42y = -47.342$

**41.** The sum of two capacitors is 55 microfarads ($\mu$F). The difference between them is 25 $\mu$F. What is the size of each capacitor?

**42.** Find the two acute angles in a right triangle if the difference of their measures is 20°.

**43.** In one concrete mix there is four times as much gravel as cement. If the total volume of the mix is 11.5 m$^3$, how much of each ingredient is in the mix?

**44.** The sum of two inductors is 90 millihenries (mH). The larger is 3.5 times the smaller. What is the size of each inductor?

**45.** A farmer has a 3% solution and an 8% solution of a pesticide. How much of each must he mix to get 2000 L of 4% solution for his sprayer?

**46.** In testing gasohol mixtures, two mixtures are on hand: mixture A (90% gasoline, 10% alcohol) and mixture B (80% gasoline, 20% alcohol). How much of each mixture must be combined to get 100 gal of an 84% gasoline, 16% alcohol mixture?

**47.** A boat can travel 20 km upstream in 5 h. It takes 4 h to make the return trip downstream. Find the rate of current and the speed of the boat (in still water).

48. Two planes are 60 km apart. Flying toward each other, they meet in 10 min. When flying in the same direction (at the same speeds as before), the faster plane overtakes the slower plane in 30 min. What is the speed of each plane in kilometres per hour?

49. A rectangular yard has been fenced using 96 m of fence. Find the length and the width of the yard if the length of the yard is 12 m longer than the width.

50. The perimeter of an isosceles triangle is 87 cm. The shortest side is 12 cm less than the longer sides. Find the length of the three sides of the triangle.

51. In a parallel electrical circuit, the products of the current and the resistance are equal in all branches. If the total current is 900 mA through any two branches, find the current flowing through branches having resistances of 50 $\Omega$ and 300 $\Omega$.

52. The total current in a parallel circuit equals the sum of the currents in the branches. If the total current in six parallel branches is 1.90 A, and if some of the branches have currents of 0.25 A and others have 0.45 A, how many of each type of branch are in the circuit?

53. The circuit in Fig. 6.2 gives the following system of linear equations:

$$4.3I_1 + 2.3(I_1 + I_2) = 27$$
$$2.8I_2 + 2.3(I_1 + I_2) = 35.2$$

Find $I_1$ and $I_2$ in mA.

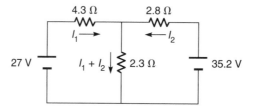

**Figure 6.2**

54. Harold needs a loan for $40,000. He gets part at 11% at a credit union. He gets the rest at 14.5% at a bank. If the total annual interest he must pay is $5520, how much did he receive from each lending institution?

55. Find the distance between the parallel lines $3x + 4y = 25$ and $6x + 8y = 15$.

56. Find the distance between the parallel lines $y = 3x - 4$ and $y = 3x + 5$.

## 6.2 OTHER SYSTEMS OF EQUATIONS

A **literal equation** is one in which letter coefficients are used in place of numerical coefficients. For example, in the following equations, $a$ and $b$ represent known quantities or coefficients, and $x$ and $y$ are the variables or unknown quantities. To solve such systems of literal equations, use either of the algebraic methods discussed in Section 6.1.

**EXAMPLE 1**

Solve

$$ax + by = ab$$
$$bx - ay = b^2$$

using the addition-subtraction method.

$$a^2x + aby = a^2b$$ (Multiply each side of the first equation by $a$.)

$$b^2x - aby = b^3$$ (Multiply each side of the second equation by $b$.)

$$\overline{a^2x + b^2x = a^2b + b^3}$$ (Add the two equations.)

$$(a^2 + b^2)x = b(a^2 + b^2)$$ (Factor.)

$$x = \frac{b(a^2 + b^2)}{a^2 + b^2}$$ (Solve for $x$.)

$$x = b$$

Substitute $x = b$ in the first equation.

$$ax + by = ab$$

$$a(b) + by = ab$$

$$by = 0$$

$$y = 0$$

The solution is $(b, 0)$.

The equations in the system

$$\frac{6}{x} + \frac{4}{y} = -2$$

$$\frac{9}{x} - \frac{7}{y} = 10$$

are not linear, or first-degree, equations. When each is multiplied by the L.C.D., $xy$, they become

$$6y + 4x = -2xy$$

$$9y - 7x = 10xy$$

each of which is second degree. However, we must assume that $x \neq 0$ and $y \neq 0$. Such a pair of equations can be solved using the algebraic methods from Section 6.1.

**EXAMPLE 2**

Solve

$$\frac{6}{x} + \frac{4}{y} = -2$$

$$\frac{9}{x} - \frac{7}{y} = 10$$

using the addition-subtraction method.

$$\frac{18}{x} + \frac{12}{y} = -6$$ (Multiply each side of the first equation by 3.)

$$\frac{18}{x} - \frac{14}{y} = 20$$ (Multiply each side of the second equation by 2.)

$$\overline{\frac{26}{y} = -26}$$ (Subtract.)

$$26 = -26y$$ (Solve for $y$.)

$$-1 = y$$

Substitute $y = -1$ in the first equation.

$$\frac{6}{x} + \frac{4}{y} = -2$$

$$\frac{6}{x} + \frac{4}{(-1)} = -2$$

$$\frac{6}{x} = 2$$

$$6 = 2x$$

$$3 = x$$

The solution is $(3, -1)$.

*Check:* Substitute the solution in both original equations.

*Note:* A system of equations as the preceding one can also be solved by first letting $A = \dfrac{1}{x}$ and $B = \dfrac{1}{y}$. Then solve for $A$ and $B$, and finally solve for $x$ and $y$.

## Exercises 6.2

*Solve each system of equations.*

**1.** $ax + y = b$
$bx - y = a$

**2.** $x + y = a - b$
$x - y = a + b$

**3.** $x + y = a^2 + ab$
$ax = by$

**4.** $2ax + b^2y = a^2b^4 + 2ab^2$
$y = a^2x$

**5.** $ax - by = 0$
$bx + ay = 1$

**6.** $ax + by = 1$
$bx + ay = 1$

**7.** $ax + by = c$
$bx + ay = c$

**8.** $ax + by = b - a$
$bx - ay = b - a$

**9.** $(a + 3b)x - by = a$
$(a - b)x - ay = b$

**10.** $(a + b)x - (a - b)y = 4ab$
$(a - b)x - (a + b)y = 0$

**11.** $\dfrac{3}{x} + \dfrac{4}{y} = 2$
$\dfrac{6}{x} + \dfrac{12}{y} = 5$

**12.** $\dfrac{12}{x} - \dfrac{15}{y} = 7$
$\dfrac{9}{x} + \dfrac{30}{y} = -3$

**13.** $\dfrac{3}{x} + \dfrac{2}{y} = 0$
$\dfrac{2}{x} - \dfrac{5}{y} = 19$

**14.** $\dfrac{6}{x} - \dfrac{4}{y} = 12$
$\dfrac{5}{x} - \dfrac{3}{y} = 11$

**15.** $\dfrac{1}{x} + \dfrac{1}{y} = \dfrac{13}{6}$
$\dfrac{1}{x} - \dfrac{1}{y} = \dfrac{5}{6}$

**16.** $\dfrac{3}{x} + \dfrac{12}{y} = 32$
$\dfrac{4}{x} + \dfrac{9}{y} = 21\dfrac{2}{3}$

**17.** $\dfrac{6}{s} + \dfrac{4}{t} = 16$
$\dfrac{9}{s} - \dfrac{5}{t} = 2$

**18.** $\dfrac{24}{m} + \dfrac{18}{n} = 72$
$\dfrac{36}{m} + \dfrac{45}{n} = 116$

**19.** $\dfrac{1}{x} + \dfrac{1}{y} = a$
$\dfrac{1}{x} - \dfrac{1}{y} = b$

**20.** $\dfrac{a}{x} + \dfrac{b}{y} = 2$
$\dfrac{b}{x} + \dfrac{a}{y} = 1$

## 6.3  SOLVING A SYSTEM OF THREE LINEAR EQUATIONS

The graph of a linear equation with three variables in the form

$$ax + by + cz = d$$

is a **plane.** The solution of a system of three linear equations with three variables can be discussed in general terms of their graphs as follows:

---

**GRAPHS OF LINEAR SYSTEMS OF EQUATIONS WITH THREE VARIABLES**

1. The three planes may intersect at a common, single point. This point, in ordered triple form $(x, y, z)$, is then the solution of the system.

2. The three planes may intersect along a common line. The infinite set of points that satisfy the equation of the line is the solution of the system.

3. The three planes may not have any points in common; the system has no solution. For example, the planes may be parallel, or they may intersect triangularly with no points common to all three planes.

4. The three planes may coincide; the solution of the system is the set of all points in the common plane.

---

Graphical solutions of three linear equations with three unknowns are not used because three-dimensional graphing is required and is not practical by hand. However, three-dimensional graphing is useful to illustrate the general cases (see Fig. 6.3).

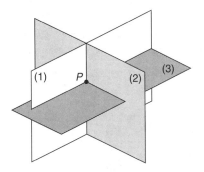

(a) One common solution at point $P$

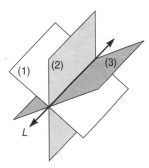

(b) Common solutions at line $L$

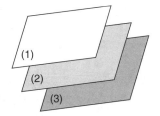

(c) No points in common

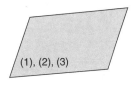

(d) All points in common

**Figure 6.3**

A linear equation with four or more variables cannot be represented in three-dimensional space. However, systems of linear equations with four or more variables have many applications, and the methods presented in this chapter will be extended to solve such systems.

The first method presented here for solving a system of linear equations with three variables is the **addition-subtraction method.**

---

**THE ADDITION-SUBTRACTION METHOD FOR A SYSTEM OF THREE LINEAR EQUATIONS WITH THREE VARIABLES**

**1.** Choose a variable to be eliminated. Eliminate it from any pair of equations by using the techniques of the addition-subtraction method from Section 6.1.

**2.** Eliminate this same variable from *any other pair* of equations.

**3.** The result of Steps 1 and 2 is a pair of linear equations in two unknowns. Solve this pair for the two variables.

**4.** Solve for the third variable by substituting the results from Step 3 in any one of the original equations.

**5.** Check by substituting the solution in all three original equations.

---

**EXAMPLE 1**

Solve

$$x + 2y - 6z = -17 \tag{1}$$
$$2x - 5y + z = 28 \tag{2}$$
$$-3x + 4y + 2z = -21 \tag{3}$$

Let's choose to eliminate $x$ first. To eliminate $x$ from any pair of equations, such as (1) and (2), multiply each side of Equation (1) by 2 and subtract.

$$
\begin{array}{r}
2x + 4y - 12z = -34 \\
\underline{2x - 5y + \phantom{0}z = 28} \\
9y - 13z = -62
\end{array}
\tag{4}
$$

To eliminate $x$ from any other pair of equations, such as (1) and (3), multiply each side of Equation (1) by 3 and add.

$$
\begin{array}{r}
3x + 6y - 18z = -51 \\
\underline{-3x + 4y + \phantom{0}2z = -21} \\
10y - 16z = -72
\end{array}
\tag{5}
$$

We have now reduced the system of three equations in three variables to a system of two equations in two variables, namely, (4) and (5).

$$9y - 13z = -62 \tag{4}$$
$$10y - 16z = -72 \tag{5}$$

$$
\begin{array}{ll}
90y - 130z = -620 & \text{[Multiply (4) by 10.]} \\
\underline{90y - 144z = -648} & \text{[Multiply (5) by 9.]} \\
14z = 28 & \text{(Subtract.)} \\
z = 2
\end{array}
$$

Substitute $z = 2$ in (5).

$$10y - 16z = -72$$
$$10y - 16(2) = -72$$
$$10y = -40$$
$$y = -4$$

Substitute $z = 2$ and $y = -4$ in any original equation; we shall use Equation (1).

$$x + 2y - 6z = -17$$
$$x + 2(-4) - 6(2) = -17$$
$$x = 3$$

The solution is $(3, -4, 2)$.

     *Check:* Substitute the solution in all three original equations.

**EXAMPLE 2**

Solve

$$4x - 2y + z = 12 \qquad \textbf{(1)}$$
$$-y + 3z = -11 \qquad \textbf{(2)}$$
$$5x + 3y \phantom{{}= {}} = 45 \qquad \textbf{(3)}$$

     Let's choose to eliminate $z$ first. To eliminate $z$ from Equations (1) and (2), multiply each side of the first equation by 3 and subtract.

$$12x - 6y + 3z = 36$$
$$\underline{\phantom{12x - }{-y + 3z} = -11}$$
$$12x - 5y \phantom{{}= {}} = 47 \qquad \textbf{(4)}$$

We now have reduced the system to Equations (3) and (4).

$$5x + \phantom{1}3y = 45 \qquad \textbf{(3)}$$
$$12x - \phantom{1}5y = 47 \qquad \textbf{(4)}$$
$$25x + 15y = 225 \qquad \text{[Multiply (3) by 5.]}$$
$$\underline{36x - 15y = 141} \qquad \text{[Multiply (4) by 3.]}$$
$$61x \phantom{{}+ 15y} = 366 \qquad \text{(Add.)}$$
$$x = 6$$

Substitute $x = 6$ in (3).

$$5x + 3y = 45$$
$$5(6) + 3y = 45$$
$$3y = 15$$
$$y = 5$$

Substitute $x = 6$ and $y = 5$ in any original equation; we shall use (1).

$$4x - 2y + z = 12$$
$$4(6) - 2(5) + z = 12$$
$$z = -2$$

The solution is $(6, 5, -2)$.

*Check:* Substitute the solution in all three original equations.

**EXAMPLE 3**

A person deposits $6000 into three different savings accounts. One account earns interest at 5%, another at $5\frac{1}{2}$%, and another at 6%. The total interest earned at the end of one year was $340. How much was deposited into each account if the 6% account earned $135 more than the 5% account?

Let

$$x = \text{the amount deposited at } 6\%$$
$$y = \text{the amount deposited at } 5\tfrac{1}{2}\%$$
$$z = \text{the amount deposited at } 5\%$$

Then

| | |
|---|---|
| $0.06x + 0.055y + 0.05z = 340$ | (the sum of the interest earned from each account) |
| $0.06x \qquad\quad - 0.05z = 135$ | (the difference in earnings between the 6% and 5% accounts) |
| $x + \qquad y + \qquad z = 6000$ | (the total sum deposited) |

This gives us a system of three linear equations in three unknowns. Multiply each equation by some quantity which will result in removing the decimal points.

$$60x + 55y + 50z = 340{,}000 \qquad (1)$$
$$6x \qquad\qquad - 5z = 13{,}500 \qquad (2)$$
$$x + \quad y + \quad z = 6{,}000 \qquad (3)$$

Let us choose to eliminate $y$ first. To eliminate $y$ from Equations (1) and (3), multiply each side of (3) by 55 and subtract.

$$
\begin{array}{r}
60x + 55y + 50z = 340{,}000 \\
\underline{55x + 55y + 55z = 330{,}000} \\
5x \qquad\quad - 5z = 10{,}000 \qquad (4)
\end{array}
$$

We have now reduced the system to Equations (2) and (4).

$$
\begin{array}{r}
6x - 5z = 13{,}500 \qquad (2) \\
\underline{5x - 5z = 10{,}000} \qquad (4) \\
x \qquad\quad = 3500 \qquad \text{(Subtract.)}
\end{array}
$$

Substitute $x = 3500$ in (2).

$$6x - 5z = 13{,}500$$
$$6(3500) - 5z = 13{,}500$$
$$21{,}000 - 5z = 13{,}500$$
$$-5z = -7500$$
$$z = 1500$$

Substitute $x = 3500$ and $z = 1500$ in (3).

$$x + y + \qquad z = 6000$$
$$(3500) + y + (1500) = 6000$$
$$y \qquad\qquad = 1000$$

The solution is (3500, 1000, 1500), which means that

$3500 was deposited at 6%

$1000 was deposited at $5\frac{1}{2}\%$

$1500 was deposited at 5%

*Check:* Substitute the solution in all three equations and in the original verbal problem.

## Exercises 6.3

*Solve each system of equations and check.*

**1.** $x + y + z = 4$
$x - y + z = 0$
$x - y - z = 2$

**2.** $x + y + z = 9$
$x - 2y + 3z = 36$
$2x - y + z = 27$

**3.** $3x - 5y - 6z = -19$
$3y + 6z = 15$
$4x - 2y - 5z = -2$

**4.** $x + 2z = -8$
$y - 3z = 2$
$x - y + z = 2$

**5.** $2x + 3y - 5z = 56$
$6x - 4y + 7z = -42$
$x - 2y + 3z = -26$

**6.** $5x - y + 3z = 28$
$8x + 2y - 4z = 24$
$9x - 4y + 3z = 32$

**7.** $x + y + z = 16$
$y + z = 3$
$x - z = 11$

**8.** $9x - 7y + 3z = -48$
$2x + 3y + 4z = -4$
$-3x + 2y - 5z = 8$

**9.** $2x - 4y + z = 17$
$4x + 5y - z = -8$
$x - 3y + 5z = 16$

**10.** $x + 3y - 6z = -1$
$2x - y + z = 10$
$5x - 2y + 3z = 27$

**11.** $\dfrac{1}{x} + \dfrac{1}{y} + \dfrac{1}{z} = 3\dfrac{5}{6}$
$\dfrac{3}{x} - \dfrac{5}{y} - \dfrac{2}{z} = \dfrac{10}{3}$
$\dfrac{2}{x} + \dfrac{3}{y} + \dfrac{6}{z} = 7$

**12.** $\dfrac{3}{x} + \dfrac{2}{y} - \dfrac{3}{z} = -\dfrac{7}{2}$
$\dfrac{5}{x} + \dfrac{3}{y} - \dfrac{4}{z} = -\dfrac{53}{12}$
$-\dfrac{2}{y} + \dfrac{8}{z} = 11\dfrac{1}{6}$

**13.** $2x + 3y + 4z - w = 7$
$-x - 2y + 3z + 2w = -3$
$3x + y - 3w = -5$
$4x + 2z - 5w = -19$

**14.** $5x - 2y - z + 2w = -44$
$3x + 3y - 4z + w = -27$
$-2x + 4y + 5z - 2w = 48$
$4x - 5y + 6z + 4w = -23$

**15.** The sum of three resistors, $R_1$, $R_2$, and $R_3$, connected in series is 1950 ohms ($\Omega$). The sum of $R_1$ and $R_3$ is 1800 $\Omega$. If $R_1$ is eight times $R_2$, what is the size of each resistor?

**16.** The following equations are derived from a circuit diagram. Solve for the currents $I_1$, $I_2$, and $I_3$ in amperes (A).

$$5I_1 + 9I_2 = 37$$
$$15I_2 + 5I_3 = 70$$
$$I_1 + I_2 - I_3 = 0$$

**17.** The perimeter of a triangle is 65 cm. The longest side is 5 cm longer than the medium side. The medium side is twice the length of the shortest side. Find the length of each side of the triangle.

**18.** Seventy-five acres of land were purchased for \$142,500. The land facing the highway cost \$2700 per acre. The land facing the railroad cost \$2200 per acre, and the remainder cost \$1450 per acre. There were 5 acres more facing the railroad than the highway. How much land was sold at each price?

**19.** Three cylindrical rods are welded together as shown in Fig. 6.4. Find the diameter of each rod.

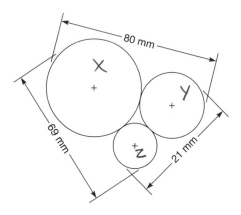

**Figure 6.4**

**20.** A chemist has three acid solutions. The first contains 20% acid, the second contains 30% acid, and the third contains 60% acid. She uses all three solutions to obtain a mixture of 80 L containing 40% acid. She uses twice as much of the 60% solution as the 20% solution. How many litres of each solution should she use?

**21.** If $f(x) = ax^2 + bx + c$, find $a$, $b$, and $c$ such that the graph passes through the points $P_1(1, 4)$, $P_2(3, 10)$, and $P_3(-1, 14)$.

## 6.4  DETERMINANTS

A **determinant** is a square array of numbers such as

$$\begin{vmatrix} a_1 & b_1 \\ a_2 & b_2 \end{vmatrix}$$

where the array has as many rows as columns. The number of rows or columns determines the order of the determinant. The preceding one is a *determinant of second order.* The numbers $a_1$, $b_1$, $a_2$, and $b_2$ are called **elements.** The elements $a_1$ and $b_1$ form the first row; the elements $a_2$ and $b_2$ form the second row; the elements $a_1$ and $a_2$ form the first column; and the elements $b_1$ and $b_2$ form the second column. The value of a second-order determinant is given by the following.

---

**VALUE OF SECOND-ORDER DETERMINANT**

$$\begin{vmatrix} a_1 & b_1 \\ a_2 & b_2 \end{vmatrix} = a_1 b_2 - a_2 b_1$$

---

The elements $a_1$ and $b_2$ form what is called the **principal diagonal.** The value of a second-order determinant is the product of the elements of the principal diagonal minus the product of the elements of the other diagonal. This can also be shown using the following diagram.

**EXAMPLE 1**

Evaluate $\begin{vmatrix} 2 & 4 \\ 6 & 3 \end{vmatrix}$.

$$\begin{vmatrix} 2 & 4 \\ 6 & 3 \end{vmatrix} = (2)(3) - (6)(4) = 6 - 24 = -18$$

**EXAMPLE 2**

Evaluate $\begin{vmatrix} -3 & 9 \\ -2 & 4 \end{vmatrix}$.

$$\begin{vmatrix} -3 & 9 \\ -2 & 4 \end{vmatrix} = (-3)(4) - (-2)(9) = -12 + 18 = 6$$

**EXAMPLE 3**

Evaluate $\begin{vmatrix} 0 & -4 \\ -2 & 7 \end{vmatrix}$.

$$\begin{vmatrix} 0 & -4 \\ -2 & 7 \end{vmatrix} = (0)(7) - (-2)(-4) = 0 - 8 = -8$$

A determinant of third order has three rows and three columns. A third-order determinant is in the form

$$\begin{vmatrix} a_1 & b_1 & c_1 \\ a_2 & b_2 & c_2 \\ a_3 & b_3 & c_3 \end{vmatrix}$$

The elements, rows, and columns of a determinant of third order are defined in the same way as a determinant of second order. The value of a third-order determinant is defined as

$$\begin{vmatrix} a_1 & b_1 & c_1 \\ a_2 & b_2 & c_2 \\ a_3 & b_3 & c_3 \end{vmatrix} = a_1b_2c_3 + a_3b_1c_2 + a_2b_3c_1 - a_3b_2c_1 - a_1b_3c_2 - a_2b_1c_3$$

Let's rearrange the terms on the right side of the equation as

$$a_1(b_2c_3 - b_3c_2) - a_2(b_1c_3 - b_3c_1) + a_3(b_1c_2 - b_2c_1)$$

These terms can then be expressed as the following $2 \times 2$ determinants:

$$a_1 \begin{vmatrix} b_2 & c_2 \\ b_3 & c_3 \end{vmatrix} - a_2 \begin{vmatrix} b_1 & c_1 \\ b_3 & c_3 \end{vmatrix} + a_3 \begin{vmatrix} b_1 & c_1 \\ b_2 & c_2 \end{vmatrix}$$

Each $2 \times 2$ determinant is called a **minor** of an element in the $3 \times 3$ determinant. In general, the minor of a given element of a determinant is the resulting determinant after the row and the column that contain the element have been deleted, as shown in Table 6.1.

**TABLE 6.1**

| Element | Minor | Original Determinant |
|---|---|---|
| $a_1$ | $\begin{vmatrix} b_2 & c_2 \\ b_3 & c_3 \end{vmatrix}$ | $\begin{vmatrix} a_1 & b_1 & c_1 \\ a_2 & b_2 & c_2 \\ a_3 & b_3 & c_3 \end{vmatrix}$ |
| $a_2$ | $\begin{vmatrix} b_1 & c_1 \\ b_3 & c_3 \end{vmatrix}$ | $\begin{vmatrix} a_1 & b_1 & c_1 \\ a_2 & b_2 & c_2 \\ a_3 & b_3 & c_3 \end{vmatrix}$ |
| $a_3$ | $\begin{vmatrix} b_1 & c_1 \\ b_2 & c_2 \end{vmatrix}$ | $\begin{vmatrix} a_1 & b_1 & c_1 \\ a_2 & b_2 & c_2 \\ a_3 & b_3 & c_3 \end{vmatrix}$ |

The value of any determinant of any order may be found by finding the sums and differences of *any* row or column of the products of the elements and the corresponding minors. The following diagram should help you remember how to determine the signs (sum or difference) of the various products of elements and minors:

$$\begin{vmatrix} + & - & + & - & \cdots \\ - & + & - & + & \cdots \\ + & - & + & - & \cdots \\ - & + & - & + & \cdots \\ . & . & . & . \\ . & . & . & . \\ . & . & . & . \end{vmatrix}$$

*Note:* Begin with a $+$ sign in the upper-left position. Then alternate the signs across each row and down each column.

**EXAMPLE 4**

Evaluate the determinant

$$\begin{vmatrix} 1 & 3 & -5 \\ 6 & -7 & -2 \\ -4 & 8 & 2 \end{vmatrix}$$

using expansion by minors (a) down the first column and (b) across the second row.

(a) The elements and the corresponding minors down the first column are determined as shown in Table 6.2.

**TABLE 6.2**

| Element | Minor | Original Determinant |
|---|---|---|
| 1 | $\begin{vmatrix} -7 & -2 \\ 8 & 2 \end{vmatrix}$ | $\begin{vmatrix} 1 & 3 & -5 \\ 6 & -7 & -2 \\ -4 & 8 & 2 \end{vmatrix}$ |
| 6 | $\begin{vmatrix} 3 & -5 \\ 8 & 2 \end{vmatrix}$ | $\begin{vmatrix} 1 & 3 & -5 \\ 6 & -7 & -2 \\ -4 & 8 & 2 \end{vmatrix}$ |
| $-4$ | $\begin{vmatrix} 3 & -5 \\ -7 & -2 \end{vmatrix}$ | $\begin{vmatrix} 1 & 3 & -5 \\ 6 & -7 & -2 \\ -4 & 8 & 2 \end{vmatrix}$ |

The signs down the first column are $+ \ - \ +$. Therefore,

$$
\begin{vmatrix} 1 & 3 & -5 \\ 6 & -7 & -2 \\ -4 & 8 & 2 \end{vmatrix} = \ +1 \begin{vmatrix} -7 & -2 \\ 8 & 2 \end{vmatrix} \ - \ 6 \begin{vmatrix} 3 & -5 \\ 8 & 2 \end{vmatrix} \ + \ (-4) \begin{vmatrix} 3 & -5 \\ -7 & -2 \end{vmatrix}
$$

$$
\begin{aligned}
&= 1[(-7)(2) - (8)(-2)] - 6[(3)(2) - (8)(-5)] + (-4)[(3)(-2) - (-7)(-5)] \\
&= \quad 1[-14 + 16] \quad - \quad 6[6 + 40] \quad + \quad (-4)[-6 - 35] \\
&= \qquad\qquad 2 \qquad\quad - \qquad\quad 276 \qquad + \qquad\quad 164 \\
&= \qquad\quad -110
\end{aligned}
$$

(b) The elements and the corresponding minors across the second row are determined as shown in Table 6.3.

**TABLE 6.3**

| Element | Minor | Original Determinant |
|---|---|---|
| 6 | $\begin{vmatrix} 3 & -5 \\ 8 & 2 \end{vmatrix}$ | $\begin{vmatrix} 1 & 3 & -5 \\ 6 & -7 & -2 \\ -4 & 8 & 2 \end{vmatrix}$ |
| $-7$ | $\begin{vmatrix} 1 & -5 \\ -4 & 2 \end{vmatrix}$ | $\begin{vmatrix} 1 & 3 & -5 \\ 6 & -7 & -2 \\ -4 & 8 & 2 \end{vmatrix}$ |
| $-2$ | $\begin{vmatrix} 1 & 3 \\ -4 & 8 \end{vmatrix}$ | $\begin{vmatrix} 1 & 3 & -5 \\ 6 & -7 & -2 \\ -4 & 8 & 2 \end{vmatrix}$ |

The signs across the second row are $- + -$. Therefore,

$$\begin{vmatrix} 1 & 3 & -5 \\ 6 & -7 & -2 \\ -4 & 8 & 2 \end{vmatrix} = -6\begin{vmatrix} 3 & -5 \\ 8 & 2 \end{vmatrix} \quad + \quad (-7)\begin{vmatrix} 1 & -5 \\ -4 & 2 \end{vmatrix} \quad - \quad (-2)\begin{vmatrix} 1 & 3 \\ -4 & 8 \end{vmatrix}$$

$$= -6[(3)(2) - (8)(-5)] + (-7)[(1)(2) - (-4)(-5)] - (-2)[(1)(8) - (-4)(3)]$$

$$\begin{array}{ccccc} = & -6[6 + 40] & - & 7[2 - 20] & + & 2[8 + 12] \\ = & -276 & + & 126 & + & 40 \\ = & -110 & & & & \end{array}$$

**EXAMPLE 5**

Evaluate the determinant

$$\begin{vmatrix} 3 & 0 & 5 & 4 \\ 2 & 1 & 3 & 4 \\ 0 & -2 & -1 & -3 \\ 1 & 0 & 2 & 1 \end{vmatrix}$$

Since the value of a determinant may be found by expansion by minors across any row or down any column, let's expand down the second column; it has the most zero elements and, therefore, involves the least work.

The signs down the second column are $- + - +$. Therefore,

$$\begin{vmatrix} 3 & 0 & 5 & 4 \\ 2 & 1 & 3 & 4 \\ 0 & -2 & -1 & -3 \\ 1 & 0 & 2 & 1 \end{vmatrix} = -0\begin{vmatrix} 2 & 3 & 4 \\ 0 & -1 & -3 \\ 1 & 2 & 1 \end{vmatrix} + 1\begin{vmatrix} 3 & 5 & 4 \\ 0 & -1 & -3 \\ 1 & 2 & 1 \end{vmatrix} - (-2)\begin{vmatrix} 3 & 5 & 4 \\ 2 & 3 & 4 \\ 1 & 2 & 1 \end{vmatrix}$$

$$+ 0\begin{vmatrix} 3 & 5 & 4 \\ 2 & 3 & 4 \\ 0 & -1 & -3 \end{vmatrix}$$

$$= 0 + 1(4) + 2(-1) + 0$$

$$= 2$$

*Note:* The second and third $3 \times 3$ determinants may be evaluated by expansion of three $2 \times 2$ minors.

The method for evaluating determinants by expansion by minors works for *any* order determinant. In Appendix C, an optional method for evaluating *only* third-order determinants is shown.

## Exercises 6.4

*Evaluate each determinant.*

**1.** $\begin{vmatrix} 4 & 3 \\ 2 & 5 \end{vmatrix}$          **2.** $\begin{vmatrix} 6 & -1 \\ 4 & 5 \end{vmatrix}$          **3.** $\begin{vmatrix} 2 & -3 \\ -4 & 5 \end{vmatrix}$

**4.** $\begin{vmatrix} 5 & 1 \\ 3 & -4 \end{vmatrix}$

**5.** $\begin{vmatrix} 3 & -7 \\ 6 & -1 \end{vmatrix}$

**6.** $\begin{vmatrix} -5 & -2 \\ 7 & 3 \end{vmatrix}$

**7.** $\begin{vmatrix} -5 & 8 \\ 9 & -6 \end{vmatrix}$

**8.** $\begin{vmatrix} 4 & 12 \\ 5 & 8 \end{vmatrix}$

**9.** $\begin{vmatrix} 4 & 0 \\ 6 & -3 \end{vmatrix}$

**10.** $\begin{vmatrix} 2 & -5 \\ -7 & 0 \end{vmatrix}$

**11.** $\begin{vmatrix} -4 & -2 \\ -6 & -7 \end{vmatrix}$

**12.** $\begin{vmatrix} -1 & 5 \\ -4 & 2 \end{vmatrix}$

**13.** $\begin{vmatrix} -7 & -2 \\ 4 & 1 \end{vmatrix}$

**14.** $\begin{vmatrix} 5 & -7 \\ -2 & -6 \end{vmatrix}$

**15.** $\begin{vmatrix} 8 & -9 \\ 7 & 4 \end{vmatrix}$

**16.** $\begin{vmatrix} 6 & -9 \\ -8 & 7 \end{vmatrix}$

**17.** $\begin{vmatrix} m & -n \\ n^2 & n \end{vmatrix}$

**18.** $\begin{vmatrix} -mn & -n \\ m & -mn \end{vmatrix}$

**19.** $\begin{vmatrix} 1 & 1 & -4 \\ -3 & 7 & 11 \\ 2 & 1 & -5 \end{vmatrix}$

**20.** $\begin{vmatrix} 2 & 0 & -6 \\ 4 & -1 & -7 \\ -3 & 3 & 2 \end{vmatrix}$

**21.** $\begin{vmatrix} -1 & 3 & 8 \\ 0 & 0 & -6 \\ -5 & 2 & -3 \end{vmatrix}$

**22.** $\begin{vmatrix} 2 & -5 & 7 \\ 0 & -8 & 1 \\ 3 & -2 & -1 \end{vmatrix}$

**23.** $\begin{vmatrix} 1 & 5 & -3 \\ 2 & -2 & 2 \\ -5 & -6 & 1 \end{vmatrix}$

**24.** $\begin{vmatrix} 2 & 3 & -6 \\ -5 & 1 & -2 \\ 0 & 3 & 0 \end{vmatrix}$

**25.** $\begin{vmatrix} 1 & -3 & 4 \\ 4 & 6 & -2 \\ 1 & -3 & 4 \end{vmatrix}$

**26.** $\begin{vmatrix} -2 & -5 & -2 \\ 1 & -7 & 1 \\ 3 & 2 & 3 \end{vmatrix}$

**27.** $\begin{vmatrix} 3 & 6 & -9 \\ 1 & -5 & 2 \\ 0 & 0 & 0 \end{vmatrix}$

**28.** $\begin{vmatrix} -6 & 0 & 5 \\ 3 & 0 & -2 \\ 7 & 0 & 4 \end{vmatrix}$

**29.** $\begin{vmatrix} 1 & 3 & -7 & 2 \\ -5 & 3 & 0 & -2 \\ 1 & -2 & 5 & 1 \\ 3 & 1 & 0 & 2 \end{vmatrix}$

**30.** $\begin{vmatrix} 2 & -1 & 7 & -5 \\ -3 & -2 & 7 & 4 \\ -1 & 0 & 3 & 2 \\ -2 & 0 & 5 & 1 \end{vmatrix}$

**31.** $\begin{vmatrix} 3 & -1 & 6 & 2 \\ -5 & 3 & -8 & 7 \\ 1 & 0 & -5 & 0 \\ 2 & -6 & 3 & 1 \end{vmatrix}$

**32.** $\begin{vmatrix} 1 & 1 & 0 & 1 \\ 1 & 0 & 1 & 1 \\ 1 & 1 & 1 & 0 \\ 0 & 1 & 1 & 1 \end{vmatrix}$

**33.** $\begin{vmatrix} 1 & 3 & 6 & -2 \\ 0 & 2 & -5 & 7 \\ 0 & 0 & 3 & 1 \\ 0 & 0 & 0 & 4 \end{vmatrix}$

**34.** $\begin{vmatrix} 1 & 1 & 1 & 4 \\ 0 & 1 & 1 & 1 \\ 0 & 0 & 1 & 1 \\ 0 & 0 & 0 & 1 \end{vmatrix}$

**35.** $\begin{vmatrix} 3 & -2 & 0 & 2 & -1 \\ 1 & 2 & -3 & 1 & 2 \\ 0 & -1 & 0 & 6 & 1 \\ -3 & 2 & 0 & 6 & -7 \\ 0 & 0 & 0 & 5 & 2 \end{vmatrix}$

**36.** $\begin{vmatrix} -1 & 1 & 5 & 6 & 2 \\ 3 & 2 & 0 & 0 & -1 \\ -2 & 0 & -3 & 1 & 1 \\ 0 & 2 & 0 & 0 & 0 \\ 6 & -1 & 0 & 1 & 2 \end{vmatrix}$

## 6.5 PROPERTIES OF DETERMINANTS

Evaluation of determinants using expansion by minors allows us to evaluate a determinant of any order. As you saw in Section 6.4, the amount of work is considerable. However, a few basic properties of determinants significantly lessen the effort and time needed to evaluate a determinant.

**EXAMPLE 1**

(a) $\begin{vmatrix} 3 & -2 & 6 \\ 0 & 0 & 0 \\ 1 & 5 & -8 \end{vmatrix} = 0$

(b) $\begin{vmatrix} 5 & 3 & 0 \\ -7 & 2 & 0 \\ 8 & -6 & 0 \end{vmatrix} = 0$

**EXAMPLE 2**

(a) $\begin{vmatrix} 1 & -3 & 5 \\ -2 & -6 & 3 \\ 1 & -3 & 5 \end{vmatrix} = 0$

(b) $\begin{vmatrix} 6 & 2 & 2 \\ 0 & 5 & 5 \\ 2 & -3 & -3 \end{vmatrix} = 0$

**EXAMPLE 3**

(a) $\begin{vmatrix} 3 & -2 & 7 \\ 5 & 8 & 4 \\ 0 & 6 & -9 \end{vmatrix} = - \begin{vmatrix} 5 & 8 & 4 \\ 3 & -2 & 7 \\ 0 & 6 & -9 \end{vmatrix}$

(b) $\begin{vmatrix} 0 & 3 & -5 \\ -6 & 2 & 3 \\ 9 & 7 & 1 \end{vmatrix} = - \begin{vmatrix} -5 & 3 & 0 \\ 3 & 2 & -6 \\ 1 & 7 & 9 \end{vmatrix}$

**EXAMPLE 4**

(a) $\begin{vmatrix} 3 & 5 & -6 \\ 4 & -8 & 12 \\ 3 & 0 & 7 \end{vmatrix} = 4 \begin{vmatrix} 3 & 5 & -6 \\ 1 & -2 & 3 \\ 3 & 0 & 7 \end{vmatrix}$

(b) $3 \begin{vmatrix} 1 & -4 & 6 \\ 3 & 0 & 8 \\ 2 & 4 & -3 \end{vmatrix} = \begin{vmatrix} 1 & -4 & 18 \\ 3 & 0 & 24 \\ 2 & 4 & -9 \end{vmatrix}$

The next property is probably the most useful. As you saw in the previous section, it was nice to have zero elements. This last property allows us to make more zeros before we expand by minors.

If every element of a row (or a column) is multiplied by the same real number $k$, and if the resulting products are added to another row (or another column), the value of the determinant remains the same.

**EXAMPLE 5**

Evaluate

$$\begin{vmatrix} 3 & -2 & 6 \\ -6 & 0 & 4 \\ 5 & 1 & -1 \end{vmatrix}$$

Let's multiply the third row by 2 and add the resulting products to the first row to make another zero in the second column.

$$\begin{vmatrix} (2)(5)+3 & (2)(1)+(-2) & (2)(-1)+6 \\ -6 & 0 & 4 \\ 5 & 1 & -1 \end{vmatrix} = \begin{vmatrix} 13 & 0 & 4 \\ -6 & 0 & 4 \\ 5 & 1 & -1 \end{vmatrix}$$

Expanding down the second column, we find that the first two minors result in zero. Thus,

$$\begin{vmatrix} 13 & 0 & 4 \\ -6 & 0 & 4 \\ 5 & 1 & -1 \end{vmatrix} = -1 \begin{vmatrix} 13 & 4 \\ -6 & 4 \end{vmatrix} = -1[52-(-24)] = -76$$

Note that if we evaluate this determinant by expansion by minors, we obtain the same result:

$$\begin{vmatrix} 3 & -2 & 6 \\ -6 & 0 & 4 \\ 5 & 1 & -1 \end{vmatrix}$$

Let's expand by minors down the second column as follows:

$$-(-2)\begin{vmatrix} -6 & 4 \\ 5 & -1 \end{vmatrix} + 0\begin{vmatrix} 3 & 6 \\ 5 & -1 \end{vmatrix} - 1\begin{vmatrix} 3 & 6 \\ -6 & 4 \end{vmatrix}$$

$$= 2(6-20) \qquad + 0(-3-30) - 1(12+36)$$

$$= -28 \qquad\qquad + 0 \qquad\qquad - 48$$

$$= -76$$

**EXAMPLE 6**

Evaluate

$$\begin{vmatrix} -7 & -6 & 5 & 1 \\ 5 & 3 & 0 & 7 \\ 0 & -1 & 4 & -2 \\ -8 & -3 & 8 & -10 \end{vmatrix}$$

Let's make two more zeros in the third row. First multiply the second column by 4 and add the resulting products to the third column. Then multiply the second column by $-2$ and add the resulting products to the fourth column.

$$\begin{vmatrix} -7 & -6 & (4)(-6)+5 & (-2)(-6)+1 \\ 5 & 3 & (4)(3)+0 & (-2)(3)+7 \\ 0 & -1 & (4)(-1)+4 & (-2)(-1)+(-2) \\ -8 & -3 & (4)(-3)+8 & (-2)(-3)+(-10) \end{vmatrix} = \begin{vmatrix} -7 & -6 & -19 & 13 \\ 5 & 3 & 12 & 1 \\ 0 & -1 & 0 & 0 \\ -8 & -3 & -4 & -4 \end{vmatrix}$$

Expanding across the third row, we find that the first, third, and fourth minors result in zero. Therefore,

$$\begin{vmatrix} -7 & -6 & -19 & 13 \\ 5 & 3 & 12 & 1 \\ 0 & -1 & 0 & 0 \\ -8 & -3 & -4 & -4 \end{vmatrix} = -(-1)\begin{vmatrix} -7 & -19 & 13 \\ 5 & 12 & 1 \\ -8 & -4 & -4 \end{vmatrix} = (+1)(-4)\begin{vmatrix} -7 & -19 & 13 \\ 5 & 12 & 1 \\ 2 & 1 & 1 \end{vmatrix}$$

Let's make two zeros in the third row. First multiply the third column by $-2$ and add the resulting products to the first column. Then multiply the third column by $-1$ and add the resulting products to the second column.

$$(-4)\begin{vmatrix} (-2)(13)+(-7) & (-1)(13)+(-19) & 13 \\ (-2)(1)+5 & (-1)(1)+12 & 1 \\ (-2)(1)+2 & (-1)(1)+1 & 1 \end{vmatrix} = (-4)\begin{vmatrix} -33 & -32 & 13 \\ 3 & 11 & 1 \\ 0 & 0 & 1 \end{vmatrix}$$

Expanding across the third row, we find that the first two minors result in zero. Thus,

$$(-4)\begin{vmatrix} -33 & -32 & 13 \\ 3 & 11 & 1 \\ 0 & 0 & 1 \end{vmatrix} = (-4)(+1)\begin{vmatrix} -33 & -32 \\ 3 & 11 \end{vmatrix} = (-4)[-363+96] = 1068$$

*Note:* You could have chosen to make zeros in any row or column. In any case, the result would be 1068.

## Exercises 6.5

*Evaluate each determinant using Properties 1–5.*

1. $\begin{vmatrix} 4 & 3 & -8 \\ 2 & -7 & 9 \\ 0 & 0 & 0 \end{vmatrix}$

2. $\begin{vmatrix} -3 & 6 & -3 \\ 1 & 5 & 1 \\ 7 & 0 & 7 \end{vmatrix}$

3. $\begin{vmatrix} 3 & 6 & 0 & -8 \\ 2 & 7 & 5 & 4 \\ 0 & -5 & 8 & 1 \\ 2 & 7 & 5 & 4 \end{vmatrix}$

4. $\begin{vmatrix} 6 & 5 & 0 & -9 \\ -3 & -4 & 0 & 2 \\ 2 & 3 & 0 & -7 \\ 7 & -1 & 0 & 12 \end{vmatrix}$

5. $\begin{vmatrix} 4 & 6 & -6 \\ 0 & 0 & -9 \\ 1 & -2 & 7 \end{vmatrix}$

6. $\begin{vmatrix} 2 & 6 & -5 \\ -3 & -5 & 0 \\ 7 & 2 & 0 \end{vmatrix}$

7. $\begin{vmatrix} 3 & 0 & 8 & 0 \\ 0 & 7 & 0 & 0 \\ 0 & 0 & -7 & 5 \\ 0 & 0 & 1 & 0 \end{vmatrix}$

8. $\begin{vmatrix} 0 & 0 & 5 & 1 \\ 1 & 0 & 3 & 0 \\ 2 & 6 & 8 & 0 \\ 3 & 0 & 0 & 0 \end{vmatrix}$

9. $\begin{vmatrix} 3 & 0 & 0 & 0 \\ 0 & 1 & 0 & 0 \\ 0 & 0 & -2 & 0 \\ 0 & 0 & 0 & 5 \end{vmatrix}$

**10.** $\begin{vmatrix} 0 & 0 & 0 & -5 \\ 0 & 0 & -7 & 0 \\ 0 & 2 & 0 & 0 \\ 3 & 0 & 0 & 0 \end{vmatrix}$

**11.** $\begin{vmatrix} 1 & 0 & 4 \\ 6 & 3 & 2 \\ 5 & 4 & 1 \end{vmatrix}$

**12.** $\begin{vmatrix} 2 & 7 & 4 \\ 3 & 0 & 1 \\ 6 & 3 & 2 \end{vmatrix}$

**13.** $\begin{vmatrix} 3 & 1 & -6 \\ -4 & 0 & 7 \\ 5 & 6 & 2 \end{vmatrix}$

**14.** $\begin{vmatrix} -3 & 2 & 5 \\ 4 & 7 & 0 \\ 6 & -3 & -1 \end{vmatrix}$

**15.** $\begin{vmatrix} 5 & 3 & 7 \\ 1 & 2 & 3 \\ 6 & 2 & 5 \end{vmatrix}$

**16.** $\begin{vmatrix} 2 & -4 & 6 \\ 4 & -2 & 1 \\ -8 & 2 & 1 \end{vmatrix}$

**17.** $\begin{vmatrix} 3 & -7 & 5 \\ 2 & 4 & 6 \\ -9 & 7 & -2 \end{vmatrix}$

**18.** $\begin{vmatrix} 4 & 3 & -9 \\ 5 & -6 & -4 \\ -7 & 9 & 2 \end{vmatrix}$

**19.** $\begin{vmatrix} 1 & 0 & 5 & 0 \\ 6 & 2 & -3 & 7 \\ -1 & 2 & 3 & -4 \\ -3 & 2 & -7 & 1 \end{vmatrix}$

**20.** $\begin{vmatrix} -5 & -6 & 3 & 7 \\ 3 & -1 & 0 & 0 \\ 2 & 1 & 0 & 5 \\ 4 & -4 & -7 & 0 \end{vmatrix}$

**21.** $\begin{vmatrix} 1 & 3 & -7 & 2 \\ 5 & 6 & 1 & 5 \\ 4 & -8 & 9 & -2 \\ -6 & 7 & -4 & 3 \end{vmatrix}$

**22.** $\begin{vmatrix} 2 & 5 & 2 & -3 \\ 3 & 4 & -4 & 4 \\ 2 & -8 & 1 & 9 \\ -2 & 6 & 3 & -7 \end{vmatrix}$

**23.** $\begin{vmatrix} 1 & 0 & 3 & 2 & 5 \\ 3 & 2 & 3 & -2 & 3 \\ -4 & -2 & 4 & 9 & -6 \\ -3 & 7 & 6 & 2 & 3 \\ 0 & 2 & -2 & -4 & 5 \end{vmatrix}$

**24.** $\begin{vmatrix} 2 & 2 & 3 & 5 & -7 \\ 3 & -6 & 9 & 12 & 15 \\ 4 & -5 & 3 & 7 & 2 \\ 2 & 0 & -4 & 2 & 7 \\ 3 & 3 & 4 & -3 & 3 \end{vmatrix}$

# 6.6  SOLVING A SYSTEM OF LINEAR EQUATIONS USING DETERMINANTS

Suppose we were to solve the general system of two linear equations in two variables:

$$a_1 x + b_1 y = c_1 \qquad \text{(1)}$$
$$a_2 x + b_2 y = c_2 \qquad \text{(2)}$$

By the addition-subtraction method we would proceed as follows:

$$
\begin{aligned}
a_1 b_2 x + b_1 b_2 y &= b_2 c_1 && \text{[Multiply (1) by } b_2.] \\
\underline{a_2 b_1 x + b_1 b_2 y = b_1 c_2} && \text{[Multiply (2) by } b_1.] \\
a_1 b_2 x - a_2 b_1 x &= b_2 c_1 - b_1 c_2 && \text{(Subtract.)} \\
(a_1 b_2 - a_2 b_1) x &= b_2 c_1 - b_1 c_2 && \text{( Factor.)} \\
x &= \frac{b_2 c_1 - b_1 c_2}{a_1 b_2 - a_2 b_1}
\end{aligned}
$$

By similarly eliminating $x$, we can show that

$$y = \frac{a_1c_2 - a_2c_1}{a_1b_2 - a_2b_1}$$

Note the following:

**1.** $a_1b_2 - a_2b_1$ is the value of and may be represented by the determinant

$$\begin{vmatrix} a_1 & b_1 \\ a_2 & b_2 \end{vmatrix}$$

**2.** $b_2c_1 - b_1c_2$ is the value of

$$\begin{vmatrix} c_1 & b_1 \\ c_2 & b_2 \end{vmatrix}$$

**3.** $a_1c_2 - a_2c_1$ is the value of

$$\begin{vmatrix} a_1 & c_1 \\ a_2 & c_2 \end{vmatrix}$$

Therefore, the solution of

$$a_1x + b_1y = c_1$$
$$a_2x + b_2y = c_2$$

may be written in determinant form as follows.

---

**DETERMINANT SOLUTION
OF TWO LINEAR EQUATIONS**

$$x = \frac{\begin{vmatrix} c_1 & b_1 \\ c_2 & b_2 \end{vmatrix}}{\begin{vmatrix} a_1 & b_1 \\ a_2 & b_2 \end{vmatrix}} \qquad y = \frac{\begin{vmatrix} a_1 & c_1 \\ a_2 & c_2 \end{vmatrix}}{\begin{vmatrix} a_1 & b_1 \\ a_2 & b_2 \end{vmatrix}}$$

---

If the determinant of the numerator is not zero and the determinant of the denominator is zero, the system is *inconsistent*. If the determinants of numerator and denominator are both zero, the system is *dependent*. If the determinant of the denominator is not zero, there is a unique solution and the system is *independent and consistent*.

Note that the determinant of each denominator is made up of the coefficients of $x$ and $y$. To find the determinant of the numerator, use these steps:

**1.** For $x$, take the determinant of the denominator and replace the coefficients of $x$, $a$'s, by the corresponding constants, $c$'s.

**2.** For $y$, take the determinant of the denominator and replace the coefficients of $y$, $b$'s, by the corresponding constants, $c$'s.

This method of solution is called **Cramer's rule** for solving a system of two linear equations in two variables. To use this method, the equations must be in the general form shown in the preceding Equations (1) and (2).

**EXAMPLE 1**

Solve using determinants.

$$2x - 3y = 22$$
$$5x + 4y = -14$$

$$x = \frac{\begin{vmatrix} 22 & -3 \\ -14 & 4 \end{vmatrix}}{\begin{vmatrix} 2 & -3 \\ 5 & 4 \end{vmatrix}} = \frac{(22)(4) - (-14)(-3)}{(2)(4) - 5(-3)} = \frac{46}{23} = 2$$

$$y = \frac{\begin{vmatrix} 2 & 22 \\ 5 & -14 \end{vmatrix}}{\begin{vmatrix} 2 & -3 \\ 5 & 4 \end{vmatrix}} = \frac{(2)(-14) - (5)(22)}{23} = \frac{-138}{23} = -6$$

The solution is $(2, -6)$.

*Check:* Substitute the solution in both original equations.

**EXAMPLE 2**

Solve using determinants.

$$3x + 4y = 10$$
$$6x + 8y = -5$$

$$x = \frac{\begin{vmatrix} 10 & 4 \\ -5 & 8 \end{vmatrix}}{\begin{vmatrix} 3 & 4 \\ 6 & 8 \end{vmatrix}} = \frac{(10)(8) - (-5)(4)}{(3)(8) - (6)(4)} = \frac{100}{0}$$

Since the numerator of the determinant is not zero and the denominator is zero, the system is inconsistent (the lines are parallel). There are no solutions.

**EXAMPLE 3**

Solve using determinants.

$$mx + ny = mn$$
$$nx - my = n^2$$

$$x = \frac{\begin{vmatrix} mn & n \\ n^2 & -m \end{vmatrix}}{\begin{vmatrix} m & n \\ n & -m \end{vmatrix}} = \frac{(mn)(-m) - (n^2)(n)}{m(-m) - (n)(n)}$$

$$= \frac{-m^2n - n^3}{-m^2 - n^2} = \frac{-n(m^2 + n^2)}{-(m^2 + n^2)} = n$$

$$y = \frac{\begin{vmatrix} m & mn \\ n & n^2 \end{vmatrix}}{\begin{vmatrix} m & n \\ n & -m \end{vmatrix}} = \frac{(m)(n^2) - (n)(mn)}{-m^2 - n^2}$$

$$= \frac{mn^2 - mn^2}{-m^2 - n^2} = \frac{0}{-m^2 - n^2} = 0$$

The solution is $(n, 0)$.

The general system of three linear equations in three variables is given as

$$a_1x + b_1y + c_1z = d_1$$
$$a_2x + b_2y + c_2z = d_2$$
$$a_3x + b_3y + c_3z = d_3$$

If we used the addition-subtraction method, we would find the following solutions for $x$, $y$, and $z$:

$$x = \frac{d_1b_2c_3 + d_3b_1c_2 + d_2b_3c_1 - d_3b_2c_1 - d_1b_3c_2 - d_2b_1c_3}{a_1b_2c_3 + a_3b_1c_2 + a_2b_3c_1 - a_3b_2c_1 - a_1b_3c_2 - a_2b_1c_3}$$

$$y = \frac{a_1d_2c_3 + a_3d_1c_2 + a_2d_3c_1 - a_3d_2c_1 - a_1d_3c_2 - a_2d_1c_3}{a_1b_2c_3 + a_3b_1c_2 + a_2b_3c_1 - a_3b_2c_1 - a_1b_3c_2 - a_2b_1c_3}$$

$$z = \frac{a_1b_2d_3 + a_3b_1d_2 + a_2b_3d_1 - a_3b_2d_1 - a_1b_3d_2 - a_2b_1d_3}{a_1b_2c_3 + a_3b_1c_2 + a_2b_3c_1 - a_3b_2c_1 - a_1b_3c_2 - a_2b_1c_3}$$

This general solution may be written in terms of determinants as follows.

---

**DETERMINANT SOLUTION OF THREE LINEAR EQUATIONS**

$$x = \frac{\begin{vmatrix} d_1 & b_1 & c_1 \\ d_2 & b_2 & c_2 \\ d_3 & b_3 & c_3 \end{vmatrix}}{\begin{vmatrix} a_1 & b_1 & c_1 \\ a_2 & b_2 & c_2 \\ a_3 & b_3 & c_3 \end{vmatrix}} \qquad y = \frac{\begin{vmatrix} a_1 & d_1 & c_1 \\ a_2 & d_2 & c_2 \\ a_3 & d_3 & c_3 \end{vmatrix}}{\begin{vmatrix} a_1 & b_1 & c_1 \\ a_2 & b_2 & c_2 \\ a_3 & b_3 & c_3 \end{vmatrix}} \qquad z = \frac{\begin{vmatrix} a_1 & b_1 & d_1 \\ a_2 & b_2 & d_2 \\ a_3 & b_3 & d_3 \end{vmatrix}}{\begin{vmatrix} a_1 & b_1 & c_1 \\ a_2 & b_2 & c_2 \\ a_3 & b_3 & c_3 \end{vmatrix}}$$

---

If the determinant of the numerator is not zero and the determinant of the denominator is zero, the system is *inconsistent.* If the determinants of numerator and denominator are both zero, the system may be *dependent* or *inconsistent.* If the determinant of the denominator is not zero, there is a unique solution, and the system is *independent and consistent.*

Note that the determinant of each denominator is made up of the coefficients of $x$, $y$, and $z$. To find the determinant of the numerator, use these steps:

1. For $x$, take the determinant of the denominator and replace the coefficients for $x$, $a$'s, by the corresponding constants, $d$'s.

2. For $y$, take the determinant of the denominator and replace the coefficients of $y$, $b$'s, by the corresponding constants, $d$'s.

3. For $z$, take the determinant of the denominator and replace the coefficients of $z$, $c$'s, by the corresponding constants, $d$'s.

This method is called **Cramer's rule** for solving a system of three linear equations in three variables.

**EXAMPLE 4**

Solve using determinants.

$$3x - y + 4z = -15$$
$$2x + 5y \quad\quad = 29$$
$$x - 6y - z = -24$$

Using Cramer's rule, we have

$$x = \frac{\begin{vmatrix} -15 & -1 & 4 \\ 29 & 5 & 0 \\ -24 & -6 & -1 \end{vmatrix}}{\begin{vmatrix} 3 & -1 & 4 \\ 2 & 5 & 0 \\ 1 & -6 & -1 \end{vmatrix}} = \frac{\begin{vmatrix} -111 & -25 & 0 \\ 29 & 5 & 0 \\ -24 & -6 & -1 \end{vmatrix}}{\begin{vmatrix} 7 & -25 & 0 \\ 2 & 5 & 0 \\ 1 & -6 & -1 \end{vmatrix}}$$

(In both numerator and denominator, multiply the third row by 4, and add the result to the first row.)

$$= \frac{(-1)\begin{vmatrix} -111 & -25 \\ 29 & 5 \end{vmatrix}}{(-1)\begin{vmatrix} 7 & -25 \\ 2 & 5 \end{vmatrix}}$$

(Expand down the third column.)

$$= \frac{(-1)(-555 + 725)}{(-1)(35 + 50)}$$

$$= \frac{-170}{-85}$$

$$= 2$$

Next, find $y$ as follows:

$$y = \frac{\begin{vmatrix} 3 & -15 & 4 \\ 2 & 29 & 0 \\ 1 & -24 & -1 \end{vmatrix}}{\begin{vmatrix} 3 & -1 & 4 \\ 2 & 5 & 0 \\ 1 & -6 & -1 \end{vmatrix}} = \frac{\begin{vmatrix} 7 & -111 & 0 \\ 2 & 29 & 0 \\ 1 & -24 & -1 \end{vmatrix}}{-85}$$

(Multiply the third row by 4 and add the result to the first row.)

$$= \frac{(-1)\begin{vmatrix} 7 & -111 \\ 2 & 29 \end{vmatrix}}{-85}$$

(Expand down the third column.)

$$= \frac{(-1)(203 + 222)}{-85}$$

$$= \frac{-425}{-85} = 5$$

Then find $z$ as follows:

$$z = \frac{\begin{vmatrix} 3 & -1 & -15 \\ 2 & 5 & 29 \\ 1 & -6 & -24 \end{vmatrix}}{\begin{vmatrix} 3 & -1 & 4 \\ 2 & 5 & 0 \\ 1 & -6 & -1 \end{vmatrix}} = \frac{\begin{vmatrix} 0 & 17 & 57 \\ 0 & 17 & 77 \\ 1 & -6 & -24 \end{vmatrix}}{-85}$$

(Multiply the third row by $-3$ and add the result to the first row. Then multiply the third row by $-2$ and add the result to the second row.)

$$= \frac{(1)\begin{vmatrix} 17 & 57 \\ 17 & 77 \end{vmatrix}}{-85}$$

(Expand down the first column.)

$$= \frac{(1)(17)\begin{vmatrix} 1 & 57 \\ 1 & 77 \end{vmatrix}}{-85}$$

(Property 4)

$$= \frac{17(77 - 57)}{-85} = \frac{340}{-85} = -4$$

The solution is $(2, 5, -4)$.

*Check:* Substitute the solution in all three original equations.

You may also find the value of the third variable by substituting the first two values into any equation and then solving for the third variable. For example, substitute $x = 2$ and $y = 5$ into the third equation and solve for $z$ as follows:

$$x - 6y - z = -24$$
$$2 - 6(5) - z = -24$$
$$2 - 30 - z = -24$$
$$-z = 4$$
$$z = -4$$

## Exercises 6.6

*Solve each system of equations using determinants.*

1. $3x + 5y = -1$
   $2x - 3y = 12$

2. $6x - 2y = 36$
   $5x + 4y = 47$

3. $8x - 3y = -43$
   $5x - 7y = -73$

4. $4x - 2y = 6$
   $-3x + 4y = -17$

5. $6x - 7y = 28$
   $-4x + 5y = -20$

6. $3x - 8y = 9$
   $-9x - 6y = -27$

7. $3x + 4y = 5$
   $6x + 8y = 10$

8. $-x + 4y = 7$
   $3x - 12y = -21$

9. $12x - 16y = 24$
   $15x - 20y = 36$

10. $-2x + 3y = 6$
    $8x - 12y = 24$

11. $15x - 6y = -15$
    $9x + 12y = 4$

12. $3x - 5y = 3$
    $-5x + 7y = -2$

13. $8x + 7y = 18$
    $y = 4x$

14. $20x - 5y = -13$
    $x = -3y$

15. $ax + by = 2$
    $bx + ay = 4$

16. $ax - 3y = b$
    $bx - 2y = a$

17. $5x + ay = b$
    $2x - by = a$

18. $4x - by = b$
    $6x - ay = a$

19. $ax + by = c$
    $y = bx$

20. $ax - by = a$
    $x = ay$

**21.** The sum of two voltages is 210 volts (V). The larger voltage is 15 V less than twice the smaller voltage. Find the voltages.

**22.** A man has available two different mixtures of solder. One mixture is 25% tin and the other is 65% tin. How much of each must he use to make a 56-kg mixture of 40% tin?

*Solve each system of equations using determinants.*

**23.**
$$3x - 4y + 7z = -26$$
$$-2x + y - 3z = 9$$
$$12x + 15z = -36$$

**24.**
$$5x - 2y - 3z = -3$$
$$4y + 3z = -2$$
$$x - y + 9z = 60$$

**25.**
$$3x + 2y + 5z = -7$$
$$8x - 3y + 2z = 10$$
$$7x - 2y + 4z = 1$$

**26.**
$$3x - 7y - 2z = -38$$
$$6x + 5y - z = 63$$
$$-2x - 4y + 5z = -28$$

**27.**
$$7x - 5y - 7z = 8$$
$$9x + 3y - 6z = 33$$
$$4x - 2y - 8z = 28$$

**28.**
$$x - 6y - 4z = 0$$
$$2x - 3y + 5z = 46$$
$$9x + 7y + 8z = 3$$

**29.**
$$3x - 4y - 5z = 0$$
$$9x + 6y + 10z = 11$$
$$12x + 2y - 20z = 36$$

**30.**
$$4x - 6y - 8z = 18$$
$$12x + 15y + 16z = -13$$
$$20x - 12y - 24z = 60$$

**31.**
$$2x + 5z = 29$$
$$5y - 7z = -40$$
$$8x + y = 15$$

**32.**
$$-4x + 7y = 37$$
$$6x - 2z = 24$$
$$3y - 5z = 36$$

**33.**
$$4x + 6y + 8z = -8$$
$$-x + 5z = -19$$
$$5y + 7z = -21$$

**34.**
$$3x - 5y = 20$$
$$9x + 8y - 2z = -25$$
$$5y - 8z = 8$$

**35.** The perimeter of a triangle is 21 cm. The longest side is 2 cm longer than the shortest side. These two sides together are twice the size of the remaining side. Find the lengths of the sides of the triangle.

**36.** As a result of Kirchhoff's laws, the following system of equations was obtained:

$$I_1 - I_2 + I_3 = 0$$
$$2.2I_1 + 0.5I_2 = 12.6$$
$$3.4I_1 - 3.8I_2 = -10.25$$

Determine the indicated currents in amperes (A).

*Cramer's rule may be extended to solve a system of n linear equations in n unknowns. Use Cramer's rule to solve each system of equations.*

**37.**
$$3x - 2y + z - 3w = -20$$
$$2x + 5z + w = -3$$
$$5x + y - z + 4w = 30$$
$$6x - 3y + 4z = -11$$

**38.**
$$8x - 2y + z - w = 38$$
$$4y - 7z + w = -28$$
$$3x - 5z = -3$$
$$3x - 2y + 7w = 23$$

**39.**
$$2x + 2y - z + 3w - 4v = 13$$
$$3x + 7y - z - 3v = -8$$
$$3y + 2z + w - v = 5$$
$$-2x + 3z - w + v = 0$$
$$5x + 7w = 38$$

**40.**
$$3x - 4y + 2z + 2w + v = -2$$
$$x + 3z + 2v = -1$$
$$3y + 4z - 8w + 3v = -13$$
$$7z + 2w - v = 5$$
$$5x + z + 3w = 11$$

## 6.7 PARTIAL FRACTIONS

We add two fractions such as

$$\frac{5}{x+1} + \frac{6}{x-2} = \frac{5(x-2)}{(x+1)(x-2)} + \frac{6(x+1)}{(x+1)(x-2)}$$

$$= \frac{5x - 10 + 6x + 6}{(x+1)(x-2)}$$

$$= \frac{11x - 4}{(x+1)(x-2)}$$

At times, we need to express a fraction as the sum of two or more fractions that are each simpler than the original; that is, we reverse the operation. Such simpler fractions whose numerators are of lower degree than their denominators are called **partial fractions.**

We separate our study of partial fractions into four cases. In each case we assume that the given fraction is expressed in lowest terms and the degree of each numerator is less than the degree of its denominator.

---

**CASE 1: NONREPEATED LINEAR DENOMINATOR FACTORS**

For every nonrepeated factor $ax + b$ of the denominator of a given fraction, there corresponds the partial fraction $\dfrac{A}{ax+b}$, where $A$ is a constant.

---

**EXAMPLE 1**

Find the partial fractions of $\dfrac{11x-4}{(x+1)(x-2)}$.

The possible partial fractions are $\dfrac{A}{x+1}$ and $\dfrac{B}{x-2}$. So we have

$$\frac{11x-4}{(x+1)(x-2)} = \frac{A}{x+1} + \frac{B}{x-2}$$

Then multiply each side of this equation by the L.C.D.: $(x+1)(x-2)$.

$$11x - 4 = A(x-2) + B(x+1)$$

Removing parentheses and rearranging terms, we have

$$11x - 4 = Ax - 2A + Bx + B$$
$$11x - 4 = Ax + Bx - 2A + B$$
$$11x - 4 = (A+B)x - 2A + B$$

Next, the coefficients of $x$ must be equal and the constant terms must be equal. This gives the following system of linear equations:

$$A + B = 11$$
$$-2A + B = -4$$

Subtracting the two equations gives

$$3A = 15$$
$$A = 5$$

Substituting $A = 5$ into either of the preceding equations gives

$$B = 6$$

Then

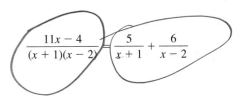

$$\frac{11x - 4}{(x + 1)(x - 2)} = \frac{5}{x + 1} + \frac{6}{x - 2}$$

**EXAMPLE 2**

Find the partial fractions of $\dfrac{3x^2 - 27x - 12}{x(2x + 1)(x - 4)}$.

The possible partial fractions are

$$\frac{A}{x} \qquad \frac{B}{2x + 1} \qquad \text{and} \qquad \frac{C}{x - 4}$$

So we have

$$\frac{3x^2 - 27x - 12}{x(2x + 1)(x - 4)} = \frac{A}{x} + \frac{B}{2x + 1} + \frac{C}{x - 4}$$

Now multiply each side of this equation by the L.C.D.: $x(2x + 1)(x - 4)$.

$$3x^2 - 27x - 12 = A(2x + 1)(x - 4) + Bx(x - 4) + Cx(2x + 1)$$

Removing parentheses and rearranging terms, we have

$$3x^2 - 27x - 12 = 2Ax^2 - 7Ax - 4A + Bx^2 - 4Bx + 2Cx^2 + Cx$$
$$3x^2 - 27x - 12 = (2A + B + 2C)x^2 + (-7A - 4B + C)x - 4A$$

Then the coefficients of $x^2$ must be equal, the coefficients of $x$ must be equal, and the constant terms must be equal. This gives the following system of linear equations:

$$\begin{aligned}
2A + B + 2C &= 3 \\
-7A - 4B + C &= -27 \\
-4A &= -12
\end{aligned}$$

Note that $A = 3$ from the third equation. Substituting $A = 3$ into the first two equations gives

$$\begin{aligned}
6 + B + 2C &= 3 \\
-21 - 4B + C &= -27
\end{aligned}$$

or

$$\begin{aligned}
B + 2C &= -3 \\
-4B + C &= -6
\end{aligned}$$

Multiplying the second equation by 2 gives

$$\begin{aligned}
B + 2C &= -3 \\
-8B + 2C &= -12
\end{aligned}$$

Subtracting these two equations gives

$$9B = 9$$

$$B = 1$$

Then $C = -2$ and

$$\frac{3x^2 - 27x - 12}{x(2x + 1)(x - 4)} = \frac{3}{x} + \frac{1}{2x + 1} - \frac{2}{x - 4}$$

---

**CASE 2: REPEATED LINEAR DENOMINATOR FACTORS**

For every factor $(ax + b)^k$ of the denominator of a given fraction, there correspond the possible partial fractions

$$\frac{A_1}{ax + b}, \frac{A_2}{(ax + b)^2}, \frac{A_3}{(ax + b)^3}, \ldots, \frac{A_k}{(ax + b)^k}$$

where $A_1, A_2, A_3, \ldots, A_k$ are constants.

---

**EXAMPLE 3**

Find the partial fractions of $\dfrac{-x^2 - 8x + 27}{x(x - 3)^2}$.

The possible partial fractions are

$$\frac{A}{x} \quad \frac{B}{x - 3} \quad \text{and} \quad \frac{C}{(x - 3)^2}$$

So we have

$$\frac{-x^2 - 8x + 27}{x(x - 3)^2} = \frac{A}{x} + \frac{B}{x - 3} + \frac{C}{(x - 3)^2}$$

Then multiply each side of this equation by the L.C.D.: $x(x - 3)^2$.

$$-x^2 - 8x + 27 = A(x - 3)^2 + Bx(x - 3) + Cx$$

Removing parentheses and rearranging terms, we have

$$-x^2 - 8x + 27 = Ax^2 - 6Ax + 9A + Bx^2 - 3Bx + Cx$$

$$-x^2 - 8x + 27 = (A + B)x^2 + (-6A - 3B + C)x + 9A$$

Equating coefficients, we have

$$A + B = -1$$

$$-6A - 3B + C = -8$$

$$9A = 27$$

From the third equation, we have $A = 3$. Substituting $A = 3$ into the first equation gives $B = -4$. Then substituting $A = 3$ and $B = -4$ into the second equation gives $C = -2$. Thus,

$$\frac{-x^2 - 8x + 27}{x(x - 3)^2} = \frac{3}{x} - \frac{4}{x - 3} - \frac{2}{(x - 3)^2}$$

**EXAMPLE 4**

Find the partial fractions of $\dfrac{3x^2 - 12x + 17}{(x - 2)^3}$.

Since $x - 2$ is repeated as a linear factor three times, the possible partial fractions are

$$\frac{A}{x - 2} \quad \frac{B}{(x - 2)^2} \quad \text{and} \quad \frac{C}{(x - 2)^3}$$

So we have

$$\frac{3x^2 - 12x + 17}{(x - 2)^3} = \frac{A}{x - 2} + \frac{B}{(x - 2)^2} + \frac{C}{(x - 2)^3}$$

Then multiply each side of this equation by the L.C.D.: $(x - 2)^3$.

$$3x^2 - 12x + 17 = A(x - 2)^2 + B(x - 2) + C$$

Removing parentheses and rearranging terms, we have

$$3x^2 - 12x + 17 = Ax^2 - 4Ax + 4A + Bx - 2B + C$$
$$3x^2 - 12x + 17 = Ax^2 + (-4A + B)x + 4A - 2B + C$$

Equating coefficients, we have the system

$$A = 3$$
$$-4A + B = -12$$
$$4A - 2B + C = 17$$

Substituting $A = 3$ into the second equation, we have $B = 0$. Then substituting $A = 3$ and $B = 0$ into the third equation, we have $C = 5$. Thus,

$$\frac{3x^2 - 12x + 17}{(x - 2)^3} = \frac{3}{x - 2} + \frac{5}{(x - 2)^3}$$

---

**CASE 3: NONREPEATED QUADRATIC DENOMINATOR FACTORS**

For every nonrepeated factor $ax^2 + bx + c$ of the denominator of a given fraction, there corresponds the partial fraction

$$\frac{Ax + B}{ax^2 + bx + c} \qquad \text{where } A \text{ and } B \text{ are constants.}$$

---

**EXAMPLE 5**

Find the partial fractions of $\dfrac{11x^2 + 8x - 12}{(2x^2 + x - 4)(x + 1)}$.

The possible partial fractions are

$$\frac{Ax + B}{2x^2 + x - 4} \quad \text{and} \quad \frac{C}{x + 1}$$

So we have

$$\frac{11x^2 + 8x - 12}{(2x^2 + x - 4)(x + 1)} = \frac{Ax + B}{2x^2 + x - 4} + \frac{C}{x + 1}$$

Then multiply each side of this equation by the L.C.D.: $(2x^2 + x - 4)(x + 1)$.

$$11x^2 + 8x - 12 = (Ax + B)(x + 1) + C(2x^2 + x - 4)$$

Removing parentheses and rearranging terms, we have

$$11x^2 + 8x - 12 = Ax^2 + Ax + Bx + B + 2Cx^2 + Cx - 4C$$
$$11x^2 + 8x - 12 = (A + 2C)x^2 + (A + B + C)x + B - 4C$$

Equating coefficients, we have

$$A + 2C = 11$$
$$A + B + C = 8$$
$$B - 4C = -12$$

The solution of this system of linear equations is $A = 5$, $B = 0$, $C = 3$. Then

$$\frac{11x^2 + 8x - 12}{(2x^2 + x - 4)(x + 1)} = \frac{5x}{2x^2 + x - 4} + \frac{3}{x + 1}$$

**EXAMPLE 6**

Find the partial fractions of $\dfrac{2x^3 - 2x^2 + 8x + 7}{(x^2 + 1)(x^2 + 4)}$.

The possible partial fractions are

$$\frac{Ax + B}{x^2 + 1} \quad \text{and} \quad \frac{Cx + D}{x^2 + 4}$$

So we have

$$\frac{2x^3 - 2x^2 + 8x + 7}{(x^2 + 1)(x^2 + 4)} = \frac{Ax + B}{x^2 + 1} + \frac{Cx + D}{x^2 + 4}$$

Then multiply each side of this equation by the L.C.D.: $(x^2 + 1)(x^2 + 4)$.

$$2x^3 - 2x^2 + 8x + 7 = (Ax + B)(x^2 + 4) + (Cx + D)(x^2 + 1)$$

Removing parentheses and rearranging terms, we have

$$2x^3 - 2x^2 + 8x + 7 = Ax^3 + Bx^2 + 4Ax + 4B + Cx^3 + Dx^2 + Cx + D$$
$$2x^3 - 2x^2 + 8x + 7 = (A + C)x^3 + (B + D)x^2 + (4A + C)x + 4B + D$$

Equating coefficients, we have

$$A + C = 2$$
$$B + D = -2$$
$$4A + C = 8$$
$$4B + D = 7$$

The solution of this system of linear equations is $A = 2$, $B = 3$, $C = 0$, $D = -5$. Then

$$\frac{2x^3 - 2x^2 + 8x + 7}{(x^2 + 1)(x^2 + 4)} = \frac{2x + 3}{x^2 + 1} - \frac{5}{x^2 + 4}$$

---

**CASE 4: REPEATED QUADRATIC DENOMINATOR FACTORS**

For every factor $(ax^2 + bx + c)^k$ of the denominator of a given fraction, there correspond the possible partial fractions

$$\frac{A_1x + B_1}{ax^2 + bx + c}, \frac{A_2x + B_2}{(ax^2 + bx + c)^2}, \frac{A_3x + B_3}{(ax^2 + bx + c)^3}, \cdots, \frac{A_kx + B_k}{(ax^2 + bx + c)^k}$$

where $A_1, A_2, A_3, \ldots, A_k, B_1, B_2, B_3, \ldots, B_k$ are constants.

---

**EXAMPLE 7**

Find the partial fractions of $\dfrac{5x^4 - x^3 + 44x^2 - 5x + 75}{x(x^2 + 5)^2}$.

The possible partial fractions are

$$\frac{A}{x} \qquad \frac{Bx + C}{x^2 + 5} \quad \text{and} \quad \frac{Dx + E}{(x^2 + 5)^2}$$

So we have

$$\frac{5x^4 - x^3 + 44x^2 - 5x + 75}{x(x^2 + 5)^2} = \frac{A}{x} + \frac{Bx + C}{x^2 + 5} + \frac{Dx + E}{(x^2 + 5)^2}$$

Then multiply each side of this equation by the L.C.D.: $x(x^2 + 5)^2$.

$$5x^4 - x^3 + 44x^2 - 5x + 75 = A(x^2 + 5)^2 + (Bx + C)(x^2 + 5)(x) + (Dx + E)x$$

Removing parentheses and rearranging terms, we have

$$5x^4 - x^3 + 44x^2 - 5x + 75 = Ax^4 + 10Ax^2 + 25A + Bx^4 + Cx^3$$
$$+ 5Bx^2 + 5Cx + Dx^2 + Ex$$
$$5x^4 - x^3 + 44x^2 - 5x + 75 = (A + B)x^4 + Cx^3 + (10A + 5B + D)x^2$$
$$+ (5C + E)x + 25A$$

Equating coefficients, we have

$$A + B = 5$$
$$C = -1$$
$$10A + 5B + D = 44$$
$$5C + E = -5$$
$$25A = 75$$

The solution of this system of linear equations is $A = 3$, $B = 2$, $C = -1$, $D = 4$, $E = 0$. Then

$$\frac{5x^4 - x^3 + 44x^2 - 5x + 75}{x(x^2 + 5)^2} = \frac{3}{x} + \frac{2x - 1}{x^2 + 5} + \frac{4x}{(x^2 + 5)^2}$$

If the degree of the numerator is greater than or equal to the degree of the denominator of the original fraction, you must first divide the numerator by the denominator using long division. Then find the partial fractions of the resulting remainder.

**EXAMPLE 8**

Find the partial fractions of $\dfrac{x^3 + 3x^2 + 7x + 4}{x^2 + 2x}$.

Since the degree of the numerator is greater than the degree of the denominator, divide as follows:

$$
\begin{array}{r}
x + 1 \phantom{00000000} \\
x^2 + 2x\overline{)x^3 + 3x^2 + 7x + 4} \\
\underline{x^3 + 2x^2\phantom{0000000}} \\
x^2 + 7x\phantom{000} \\
\underline{x^2 + 2x\phantom{000}} \\
5x + 4
\end{array}
$$

or

$$\frac{x^3 + 3x^2 + 7x + 4}{x^2 + 2x} = x + 1 + \frac{5x + 4}{x^2 + 2x}$$

Now factor the denominator and find the partial fractions of the remainder.

$$\frac{5x + 4}{x(x + 2)} = \frac{A}{x} + \frac{B}{x + 2}$$

$$5x + 4 = A(x + 2) + Bx$$

$$5x + 4 = Ax + 2A + Bx$$

$$5x + 4 = (A + B)x + 2A$$

Then

$$A + B = 5$$

$$2A = 4$$

So $A = 2$ and $B = 3$ and

$$\frac{x^3 + 3x^2 + 7x + 4}{x^2 + 2x} = x + 1 + \frac{2}{x} + \frac{3}{x + 2}$$

## Exercises 6.7

*Find the partial fractions of each expression.*

**1.** $\dfrac{8x - 29}{(x + 2)(x - 7)}$

**2.** $\dfrac{10x - 34}{(x - 4)(x - 2)}$

**3.** $\dfrac{-x - 18}{2x^2 - 5x - 12}$

**4.** $\dfrac{17x - 18}{3x^2 + x - 2}$

**5.** $\dfrac{61x^2 - 53x - 28}{x(3x - 4)(2x + 1)}$

**6.** $\dfrac{11x^2 - 7x - 42}{(2x + 3)(x^2 - 2x - 3)}$

**7.** $\dfrac{x^2 + 7x + 10}{(x + 1)(x + 3)^2}$

**8.** $\dfrac{3x^2 - 18x + 9}{(2x - 1)(x - 1)^2}$

**9.** $\dfrac{48x^2 - 20x - 5}{(4x - 1)^3}$

**10.** $\dfrac{x^2 + 8x}{(x + 4)^3}$

**11.** $\dfrac{11x^2 - 18x + 3}{x(x - 1)^2}$

**12.** $\dfrac{6x^2 + 4x + 4}{x^3 + 2x^2}$

**13.** $\dfrac{-x^2 - 4x + 3}{(x^2 + 1)(x^2 - 3)}$

**14.** $\dfrac{-6x^3 + 2x^2 - 3x + 10}{(2x^2 + 1)(x^2 + 5)}$

**15.** $\dfrac{4x^3 - 21x - 6}{(x^2 + x + 1)(x^2 - 5)}$

**16.** $\dfrac{x^3 + 6x^2 + 2x - 2}{(3x^2 - x - 1)(x^2 + 4)}$

**17.** $\dfrac{4x^3 - 16x^2 - 93x - 9}{(x^2 + 5x + 3)(x^2 - 9)}$

**18.** $\dfrac{12x^2 + 8x - 72}{(x^2 + x - 1)(x^2 - 16)}$

**19.** $\dfrac{8x^4 - x^3 + 13x^2 - 6x + 5}{x(x^2 + 1)^2}$

**20.** $\dfrac{-4x^4 + 6x^3 + 8x^2 - 19x + 17}{(x - 1)(x^2 - 3)^2}$

**21.** $\dfrac{x^5 - 2x^4 - 8x^2 + 4x - 8}{x^2(x^2 + 2)^2}$

**22.** $\dfrac{3x^5 + x^4 + 24x^3 + 10x^2 + 48x + 16}{x^2(x^2 + 4)^2}$

**23.** $\dfrac{6x^2 + 108x + 54}{x^4 - 81}$

**24.** $\dfrac{x^6 + 2x^4 + 3x^2 + 1}{x^2(x^2 + 1)^3}$

**25.** $\dfrac{x^3}{x^2 - 1}$

**26.** $\dfrac{x^4 + x^2}{(x + 1)(x - 2)}$

**27.** $\dfrac{x^3 - x^2 + 8}{x^2 - 4}$

**28.** $\dfrac{2x^3 - 2x^2 + 8x - 3}{x(x - 1)}$

**29.** $\dfrac{3x^4 - 2x^3 - 2x + 5}{x(x^2 + 1)}$

**30.** $\dfrac{x^5 - x^4 - 3x^3 + 7x^2 + 3x + 20}{(x + 2)(x^2 + 2)}$

# CHAPTER SUMMARY

1. *Graphs of linear systems of equations with two variables:*
   (a) The two lines may intersect at a common, single point. This point, in ordered pair form $(x, y)$, is the solution of the system.
   (b) The two lines may be parallel with no points in common; hence, the system has no common solution.
   (c) The two lines may coincide; the solution of the system is the set of all points on the common line.

2. *Solving a pair of linear equations by the addition-subtraction method:*
   (a) If necessary, multiply each side of one or both equations by some number so that the numerical coefficients of one of the variables are of equal absolute value.
   (b) If these coefficients of equal absolute value have like signs, subtract one equation from the other. If they have unlike signs, add the equations. That is, do whatever is necessary to eliminate that variable.
   (c) Solve the resulting equation for the remaining variable.
   (d) Substitute the solution for the variable found in Step (c) in either of the original equations, and solve this resulting equation for the second variable.
   (e) Check by substituting the solution in both original equations.

3. *Solving a pair of linear equations by the method of substitution:*
   (a) From either of the two given equations, solve for one variable in terms of the other.
   (b) Substitute this result from Step (a) in the *other* equation. Note that this step eliminates a variable.
   (c) Solve the equation obtained from Step (b) for the remaining variable.
   (d) From the equation obtained in Step (a), substitute the solution for the variable found in Step (c), and solve this resulting equation for the second variable.
   (e) Check by substituting the solution in both original equations.

4. *Steps for problem solving:*
   (a) Read the problem carefully at least two times.
   (b) If possible, draw a picture or a diagram.
   (c) Write what facts are given and what unknown quantities are to be found.
   (d) Choose a symbol to represent each quantity to be found (there will be more than one). Be sure to label each symbol to indicate what it represents.
   (e) Write appropriate equations relating these variables from the information given in the problem. Watch for information that is not stated but which should be assumed.
   (f) Solve for the unknown variables by the methods presented in this chapter.
   (g) Check your solution in the original equations.
   (h) Check your solution in the original verbal problem.

5. *Graphs of linear systems of equations with three variables:*
   (a) The three planes may intersect at a common, single point. This point, in ordered triple form $(x, y, z)$, is then the solution of the system.
   (b) The three planes may intersect along a common line. The infinite set of points that satisfy the equation of the line is the solution of the system.
   (c) The three planes may not have any points in common; the system has no solution. For example, the planes may be parallel, or they may intersect triangularly with no points common to all three planes.

(d) The three planes may coincide; the solution of the system is the set of all points in the common plane.

6. *The addition-subtraction method for a system of three linear equations with three variables:*
   (a) Choose a variable to be eliminated. Eliminate it from any pair of equations by using the techniques of the addition-subtraction method.
   (b) Eliminate this same variable from *any other pair* of equations.
   (c) The result of Steps (a) and (b) is a pair of linear equations in two unknowns. Solve this pair for the two variables.
   (d) Solve for the third variable by substituting the results from Step (c) in any one of the original equations.
   (e) Check by substituting the solution in all three original equations.

7. A **determinant** is a square array of numbers such as

$$\begin{vmatrix} a_1 & b_1 \\ a_2 & b_2 \end{vmatrix}$$

8. *Value of second-order determinant:*

$$\begin{vmatrix} a_1 & b_1 \\ a_2 & b_2 \end{vmatrix} = a_1 b_2 - a_2 b_1$$

9. The value of a third-order determinant is defined as

$$\begin{vmatrix} a_1 & b_1 & c_1 \\ a_2 & b_2 & c_2 \\ a_3 & b_3 & c_3 \end{vmatrix} = a_1 b_2 c_3 + a_3 b_1 c_2 + a_2 b_3 c_1 - a_3 b_2 c_1 - a_1 b_3 c_2 - a_2 b_1 c_3$$

$$= a_1(b_2 c_3 - b_3 c_2) - a_2(b_1 c_3 - b_3 c_1) + a_3(b_1 c_2 - b_2 c_1)$$

$$= a_1 \begin{vmatrix} b_2 & c_2 \\ b_3 & c_3 \end{vmatrix} - a_2 \begin{vmatrix} b_1 & c_1 \\ b_3 & c_3 \end{vmatrix} + a_3 \begin{vmatrix} b_1 & c_1 \\ b_2 & c_2 \end{vmatrix}$$

Each $2 \times 2$ determinant is called a **minor** of an element in the $3 \times 3$ determinant. The minor of a given element is the resulting determinant after the row and the column which contain the element have been deleted, as shown in Table 6.4.

The value of any determinant of any order may be found by evaluating the sums and differences of *any* row or column of the products of the elements and the corresponding minors. The following diagram shows the signs (sum or difference) of the various products of elements and minors:

$$\begin{vmatrix} + & - & + & - & \cdots \\ - & + & - & + & \cdots \\ + & - & + & - & \cdots \\ - & + & - & + & \cdots \\ \cdot & \cdot & \cdot & \cdot \\ \cdot & \cdot & \cdot & \cdot \\ \cdot & \cdot & \cdot & \cdot \end{vmatrix}$$

**TABLE 6.4**

| Element | Minor | Original Determinant |
|---|---|---|
| $a_1$ | $\begin{vmatrix} b_2 & c_2 \\ b_3 & c_3 \end{vmatrix}$ | $\begin{vmatrix} a_1 & b_1 & c_1 \\ a_2 & b_2 & c_2 \\ a_3 & b_3 & c_3 \end{vmatrix}$ |
| $a_2$ | $\begin{vmatrix} b_1 & c_1 \\ b_3 & c_3 \end{vmatrix}$ | $\begin{vmatrix} a_1 & b_1 & c_1 \\ a_2 & b_2 & c_2 \\ a_3 & b_3 & c_3 \end{vmatrix}$ |
| $a_3$ | $\begin{vmatrix} b_1 & c_1 \\ b_2 & c_2 \end{vmatrix}$ | $\begin{vmatrix} a_1 & b_1 & c_1 \\ a_2 & b_2 & c_2 \\ a_3 & b_3 & c_3 \end{vmatrix}$ |

10. *Properties of determinants:*
    (a) *Property 1:* If every element in a row (or a column) of a determinant is zero, the value of the determinant is zero.
    (b) *Property 2:* If two rows (or two columns) of a determinant are identical, the value of the determinant is zero.
    (c) *Property 3:* If any two rows (or two columns) of a determinant are interchanged, the sign of the value of the determinant is changed.
    (d) *Property 4:* If every element of a row (or a column) is multiplied by the same real number $k$, the value of the determinant is multiplied by $k$.
    (e) *Property 5:* If every element of a row (or a column) is multiplied by the same real number $k$, and if the resulting products are added to another row (or another column), the value of the determinant remains the same.

11. *Cramer's rule:*
    (a) The solution of two linear equations in two variables

$$a_1x + b_1y = c_1$$
$$a_2x + b_2y = c_2$$

may be written in determinant form as follows:

$$x = \frac{\begin{vmatrix} c_1 & b_1 \\ c_2 & b_2 \end{vmatrix}}{\begin{vmatrix} a_1 & b_1 \\ a_2 & b_2 \end{vmatrix}} \qquad y = \frac{\begin{vmatrix} a_1 & c_1 \\ a_2 & c_2 \end{vmatrix}}{\begin{vmatrix} a_1 & b_1 \\ a_2 & b_2 \end{vmatrix}}$$

    (b) The solution of three linear equations in three variables

$$a_1x + b_1y + c_1z = d_1$$
$$a_2x + b_2y + c_2z = d_2$$
$$a_3x + b_3y + c_3z = d_3$$

may be written in determinant form as follows:

$$x = \frac{\begin{vmatrix} d_1 & b_1 & c_1 \\ d_2 & b_2 & c_2 \\ d_3 & b_3 & c_3 \end{vmatrix}}{\begin{vmatrix} a_1 & b_1 & c_1 \\ a_2 & b_2 & c_2 \\ a_3 & b_3 & c_3 \end{vmatrix}} \qquad y = \frac{\begin{vmatrix} a_1 & d_1 & c_1 \\ a_2 & d_2 & c_2 \\ a_3 & d_3 & c_3 \end{vmatrix}}{\begin{vmatrix} a_1 & b_1 & c_1 \\ a_2 & b_2 & c_2 \\ a_3 & b_3 & c_3 \end{vmatrix}} \qquad z = \frac{\begin{vmatrix} a_1 & b_1 & d_1 \\ a_2 & b_2 & d_2 \\ a_3 & b_3 & d_3 \end{vmatrix}}{\begin{vmatrix} a_1 & b_1 & c_1 \\ a_2 & b_2 & c_2 \\ a_3 & b_3 & c_3 \end{vmatrix}}$$

12. *Partial fractions:*

(a) *Case 1: Nonrepeated linear denominator factors.* For every nonrepeated factor $ax + b$ of the denominator of a given fraction, there corresponds the partial fraction $\dfrac{A}{ax + b}$, where $A$ is a constant.

(b) *Case 2: Repeated linear denominator factors.* For every factor $(ax + b)^k$ of the denominator of the given fraction, there correspond the possible partial fractions

$$\frac{A_1}{ax + b}, \frac{A_2}{(ax + b)^2}, \frac{A_3}{(ax + b)^3}, \ldots, \frac{A_k}{(ax + b)^k}$$

where $A_1, A_2, A_3, \ldots, A_k$ are constants.

(c) *Case 3: Nonrepeated quadratic denominator factors.* For every nonrepeated factor $ax^2 + bx + c$ of the denominator of the given fraction, there corresponds the partial fraction

$$\frac{Ax + B}{ax^2 + bx + c} \qquad \text{where } A \text{ and } B \text{ are constants.}$$

(d) *Case 4: Repeated quadratic denominator factors.* For every factor $(ax^2 + bx + c)^k$ of the denominator of the given fraction, there correspond the possible partial fractions

$$\frac{A_1 x + B_1}{ax^2 + bx + c}, \frac{A_2 x + B_2}{(ax^2 + bx + c)^2}, \frac{A_3 x + B_3}{(ax^2 + bx + c)^3}, \ldots, \frac{A_k x + B_k}{(ax^2 + bx + c)^k}$$

where $A_1, A_2, A_3, \ldots, A_k, B_1, B_2, B_3, \ldots, B_k$ are constants.

# CHAPTER 6 REVIEW

*Solve each system of equations graphically and check by substituting the solution ordered pair in both original equations.*

**1.** $x + y = 3$
 $x - y = -1$

**2.** $y - 3x = 2$
 $y - x = 2$

**3.** $2x - y = -10$
 $4x + y = 4$

**4.** $2x - 3y = 6$
 $-4x + 6y = 8$

*Solve each system of equations by the addition-subtraction method and check.*

**5.** $3x - 4y = 5$
 $x + 7y = 10$

**6.** $2x + 3y = 7$
 $4x - 6y = 11$

**7.** $x - 2y = 6$
$\phantom{}4x + \phantom{2}y = 6$

**8.** $3x - 2y = 5$
$\phantom{}2x + 2y = 15$

*Solve each system by the method of substitution and check.*

**9.** $\phantom{3x-5}x = -4y$
$\phantom{}3x - 5y = 17$

**10.** $\phantom{2x-15}y = 3x$
$\phantom{}2x - 15y = 86$

**11.** $5x - \phantom{3}y = 7$
$\phantom{}2x + 3y = 13$

**12.** $\phantom{-}3x + \phantom{1}4y = -7$
$-5x + 12y = 14$

*Solve each system of equations.*

**13.** $8x + by = 4$
$\phantom{}16x - ay = 12$

**14.** $\dfrac{3}{x} - \dfrac{4}{y} = -5$

$\dfrac{6}{x} + \dfrac{5}{y} = 16$

**15.** $x + \phantom{2}y + \phantom{2}z = 2$
$x - \phantom{2}y - 2z = 3$
$x + 2y - \phantom{2}z = \frac{3}{2}$

**16.** $2x + 2y - \phantom{2}z = 4$
$2x - \phantom{2}y + 2z = -2$
$\phantom{2}x - 5y + \phantom{2}z = 8$

*Evaluate each determinant.*

**17.** $\begin{vmatrix} 2 & -3 \\ 6 & 2 \end{vmatrix}$

**18.** $\begin{vmatrix} -1 & 7 \\ 4 & -3 \end{vmatrix}$

**19.** $\begin{vmatrix} 1 & 3 & 5 \\ 7 & 2 & -1 \\ -4 & 2 & -8 \end{vmatrix}$

**20.** $\begin{vmatrix} 2 & 0 & -3 \\ 1 & 4 & 2 \\ 5 & 1 & -1 \end{vmatrix}$

*Solve each system using determinants and check.*

**21.** $4x + 3y = 46$
$\phantom{}2x - 3y = 14$

**22.** $\phantom{-}3x - 2y = 13$
$-6x + 4y = -26$

**23.** $2x + 9y = 4$
$\phantom{}5x - 9y = 10$

**24.** $\phantom{-}5x - 2y = 2$
$-3x + 3y = 1$

**25.** $2x + 2y - \phantom{3}z = 11$
$3x + 4y + \phantom{3}z = -16$
$4x - 8y + 3z = -113$

**26.** $\phantom{2}x + \phantom{3}y + z = 9$
$2x - \phantom{3}y + z = 3$
$\phantom{2}x + 3y - z = 3$

*Solve each system by any method and check.*

**27.** $\phantom{1}2x + 7y = 3$
$10x + 3y = -1$

**28.** $-5x + 2y = 5$
$-3x + 7y = 32$

**29.** $2x - 3y = 10$
$2x + 5y = 2$

**30.** $\phantom{2}x + \phantom{3}y = -25$
$2x - 3y = 70$

**31.** $x + y + z = 7$
$x \phantom{+y} + z = 2$
$\phantom{x} y - z = 4$

**32.** $\phantom{1}7x - 2y + 5z = 9$
$\phantom{1}3x + 4y - 9z = 27$
$10x - 3y + 7z = 7$

**33.** An alloy contains 10% lead. Another alloy contains 20% lead. How many kg of each must be used to make a 30-kg alloy containing 12% lead?

**34.** A person deposits $3600 into two different savings accounts. One account earns interest at $5\frac{1}{2}\%$ and the other at $6\frac{1}{2}\%$. The total interest earned from both accounts at the end of one year is $220.00. Find the amount deposited at $5\frac{1}{2}\%$ and the amount deposited at $6\frac{1}{2}\%$.

**35.** It took a plane 1 h 15 min to make a trip of 100 km flying into a head wind. On the return trip with the same wind velocity, it took only 1 h. Find the wind velocity and the speed of the plane (in still air).

**36.** Jack, John, and Bob work on an assembly line producing hand-cut aluminum parts. Jack and John together can process, on average, three times as many pieces per hour as Bob. Jack can process 2 more pieces per h than John, and together all three can process 32 pieces per h. How many pieces per h does each man process, on average?

*Evaluate each determinant.*

**37.**
$$\begin{vmatrix} 1 & 0 & 3 & 6 \\ 4 & -2 & 1 & -5 \\ -3 & 2 & 5 & 2 \\ 1 & 0 & 7 & -9 \end{vmatrix}$$

**38.**
$$\begin{vmatrix} 5 & 1 & 6 & -3 \\ -2 & 2 & 5 & 1 \\ 3 & -1 & 7 & 5 \\ 3 & 2 & 3 & -2 \end{vmatrix}$$

*Find the partial fractions of each expression.*

**39.** $\dfrac{6x + 14}{(x - 3)(x + 5)}$

**40.** $\dfrac{-x^2 - 6x - 1}{x(x^2 - 1)}$

**41.** $\dfrac{3x^2 + 5x + 1}{x^2(x + 1)}$

**42.** $\dfrac{10x + 4}{(x + 1)^2}$

**43.** $\dfrac{7x^2 - x + 2}{(x^2 + 1)(x - 1)}$

**44.** $\dfrac{5x^2 + 2x + 21}{(x^2 + 4)^2}$

# APPLICATION
## Solar Cell Technology

With the world population increasing exponentially and third world countries striving to improve their quality of life, it is natural to turn to renewable sources of energy to supplement the dwindling supply of fossil fuels. Approximately 1000 watts of energy from the sun reach each square metre of the earth on a clear day at noon. We can harvest this energy with photovoltaic (PV) cells or solar cells. The photovoltaic phenomenon was discovered (but not fully understood) in the nineteenth century. In 1955 Bell Laboratories introduced a single-crystal silicon solar cell. Since then, scientists and manufacturers have made great strides in producing practical PV cells.

Most PV cells consist of two thin layers of pure silicon. One layer is coated with elements that produce a surplus of electrons; the other is coated with elements that produce a deficit of electrons. When sunlight strikes the cells, some of the electrons flow to metallic conductors on the silicon and then through an electrical circuit. A module is a collection of PV cells wired in series and parallel to yield the desired voltage and amperage. Usually a module is sandwiched between pieces of glass and framed in aluminum. Once the electricity has been produced, it needs to be stored. Normally, 12-volt batteries are used, which means that a module should produce 14 to 18 volts of output, since electron flow goes from high to low voltage.

Three types of PV cells are being manufactured. Single-crystal and multicrystalline are named for the type of silicon that is used. The crystals are grown in large blocks and then sliced thin. Their color is deep blue since silicon crystals absorb all other colors. They have an efficiency of approximately 13%. In amorphous PV cells, the silicon is vaporized and deposited on a surface, usually stainless steel or glass. This manufacturing process costs less than growing and slicing crystals and produces an efficiency rate of about 12%.

Other PV thin-film materials that researchers are working with include copper indium diselenide (CIS), cadmium telluride (CdTe), and gallium arsenide (GaAs). One goal is to reduce the manufacturing costs and improve the efficiency so that the cost of PV-generated electricity drops to between 10 and 15 cents per kWh by the year 2000.

PV cells are being built into roofing materials and glass siding for homes. They are the source of electricity in satellites, spacecraft, and calculators. Electric cars already exist; a commuter car powered by solar electricity would help to solve environmental and transportation issues.

### Bibliography

Ashley, Steven. "High-Volume Photovoltaic Plant." *Mechanical Engineering* 118 (April 1996): 12.

Bisio, Attilio, and Boots, Sharon. *The Wiley Encyclopedia of Energy and the Environment.* Vol. 2. New York: John Wiley & Sons, 1997.

"PV Breakthrough Will Make Solar Competitive." *Electric Light and Power* 74 (Sept. 1996): 38.

Real Goods Trading Corporation. *Photovoltaic—A Brief Technical Explanation.* Ukiah, CA: Real Goods Trading Corporation, 1996. www.realgoods.com.

Real Goods Trading Corporation. *Photovoltaic—Construction Types.* Ukiah, CA: Real Goods Trading Corporation, 1996. www.realgoods.com.

Real Goods Trading Corporation. *Photovoltaic—Efficiency.* Ukiah, CA: Real Goods Trading Corporation, 1996. www.realgoods.com.

"Solar Energy Captured in Film." *Science News* 149 (May 4, 1996): 282.

"Thin-Film Photovoltaic Cell Achieves Efficiency." *Laser Focus World* 32 (July 1996): 11.

A photovoltaic cell panel supplies electricity to an island in Lake of the Woods, Canada.

Photovoltaic modules on roof of Georgetown University Intercultural Center in Washington, D.C., providing a power output of 300 kilowatts. *(Photo courtesy of U.S. Department of Energy, Energy Technology Visuals.)*

# 7
# Quadratic Equations

## INTRODUCTION

A baseball player at bat pops a ball straight up. The ball's height, $h$, in feet, after $t$ seconds may be expressed by $h = 64t - 16t^2$. Does the ball ever reach a height of 80 feet? How long is the ball in the air?

The formula that relates the height of the ball in $t$ seconds is called a *quadratic equation*. Variable electric current, specification of box sizes, and the path of a projectile shot or thrown in the air can all be represented by a quadratic equation.

In this chapter we learn how to solve quadratic equations and how to interpret some of their characteristics. (The preceding problem is number 25 in the chapter review exercises.)

### Objectives

- Solve quadratic equations by factoring.
- Solve quadratic equations by completing the square.
- Solve quadratic equations by using the quadratic formula.

## 7.1 SOLVING A QUADRATIC EQUATION BY FACTORING

A **quadratic equation** in one variable is an equation with at least one term of second degree and no other term of higher degree. A quadratic equation is usually expressed in the form $ax^2 + bx + c = 0$ $(a \neq 0)$.

In Chapter 4 equations, including quadratic equations, were solved graphically, which was an approximation method. Methods for obtaining exact solutions for quadratic equations are now presented.

One way to solve quadratic equations is by factoring. When using this method, we shall use the following principle.

> If $ab = 0$, then either $a = 0$ or $b = 0$, or both.

That is, if the product of two factors is zero, one or both of the factors must also be zero.

**EXAMPLE 1**

Solve $(x + 4)(x - 3) = 0$.

By the preceding principle,

$$x + 4 = 0 \quad \text{or} \quad x - 3 = 0$$

so

$$x = -4 \quad \text{or} \quad x = 3$$

---

**SOLVING A QUADRATIC EQUATION BY FACTORING**

1. If necessary, write the equation in the form $ax^2 + bx + c = 0$; that is, "set the equation equal to zero."
2. Factor the nonzero side of the equation.
3. Using the preceding principle, set each factor that contains a variable equal to zero.
4. Solve each resulting linear equation.
5. Check.

---

**EXAMPLE 2**

Solve $2x^2 = 7x + 15$.

**Step 1:** $\qquad\qquad 2x^2 - 7x - 15 = 0$

**Step 2:** $\qquad\qquad (2x + 3)(x - 5) = 0$

**Step 3:** $\qquad\quad 2x + 3 = 0 \quad \text{or} \quad x - 5 = 0$

**Step 4:** $\qquad\qquad\quad 2x = -3$

$$x = -\frac{3}{2} \quad \text{or} \quad x = 5$$

*Check:*

**Step 5:**

$$2\left(-\frac{3}{2}\right)^2 = 7\left(-\frac{3}{2}\right) + 15 \qquad \bigg| \qquad 2(5)^2 = 7(5) + 15$$

$$\frac{9}{2} = -\frac{21}{2} + 15 \qquad\qquad \bigg| \qquad 50 = 35 + 15$$

$$\frac{9}{2} = \frac{9}{2} \qquad\qquad\qquad \bigg| \qquad 50 = 50$$

**EXAMPLE 3**

Solve $6x^2 - 13x - 5 = 0$.

$$(3x + 1)(2x - 5) = 0$$
$$3x + 1 = 0 \quad \text{or} \quad 2x - 5 = 0$$
$$3x = -1 \qquad\qquad 2x = 5$$
$$x = -\frac{1}{3} \quad \text{or} \qquad x = \frac{5}{2}$$

*Check:* Use the same procedure as in Example 2.

**EXAMPLE 4**

Solve $16x^2 + 4x = 0$.

$$4x(4x + 1) = 0$$
$$4x = 0 \quad \text{or} \quad 4x + 1 = 0$$
$$\qquad\qquad\qquad 4x = -1$$
$$x = 0 \quad \text{or} \qquad x = -\frac{1}{4}$$

**EXAMPLE 5**

Solve $4x^2 = 49$.

$$4x^2 - 49 = 0$$
$$(2x + 7)(2x - 7) = 0$$
$$2x + 7 = 0 \quad \text{or} \quad 2x - 7 = 0$$
$$2x = -7 \qquad\qquad 2x = 7$$
$$x = -\frac{7}{2} \quad \text{or} \qquad x = \frac{7}{2}$$

---

**SOLVING $ax^2 = c$**

**1.** Divide each side by $a$.

**2.** Take the square root of each side.

**3.** Simplify the result, if possible.

---

**EXAMPLE 6**

Solve $2x^2 = 32$.

$$2x^2 = 32$$

**Step 1:** $\qquad\qquad\qquad x^2 = 16$

**Step 2:** $\qquad\qquad\qquad x = \pm\sqrt{16}$

**Step 3:** $\qquad\qquad\qquad x = \pm 4$

Note that in Step 2 we introduced the $\pm$ sign because there are **two** numbers whose square is 16, namely, 4 and $-4$.

**EXAMPLE 7**

Solve $9y^2 - 13 = 7$.

$$9y^2 - 13 = 7$$
$$9y^2 = 20$$
$$y^2 = \frac{20}{9}$$
$$y = \pm\sqrt{\frac{20}{9}}$$
$$y = \pm\frac{2\sqrt{5}}{3}$$

You may wish to review simplifying radicals in Section 2.2.

## Exercises 7.1

*Solve each equation.*

**1.** $(x + 4)(x - 7) = 0$

**2.** $(3x + 1)(x + 5) = 0$

**3.** $x^2 - 6x + 8 = 0$

**4.** $x^2 - 9x + 18 = 0$

**5.** $x^2 + 3x = 10$

**6.** $x^2 = x + 42$

**7.** $2x^2 = 3x + 9$

**8.** $2x^2 + 9x = 5$

**9.** $14x^2 + 17x + 5 = 0$

**10.** $36x^2 + 4 = 25x$

**11.** $18x^2 + 56 = 69x$

**12.** $14x^2 + 5x = 24$

**13.** $8x^2 + x = 0$

**14.** $9x^2 - 6x = 0$

**15.** $-7x^2 + 21x = 0$

**16.** $x - 2x^2 = 0$

**17.** $x^2 - 36 = 0$

**18.** $16 - x^2 = 0$

**19.** $16x^2 - 25 = 0$

**20.** $4x^2 = 9$

**21.** $5x^2 - 12 = 0$

**22.** $3x^2 - 8 = 0$

**23.** $4x^2 = 21$

**24.** $7x^2 = 4$

**25.** $5a^2 - 40 = 0$

**26.** $2c^2 - 36 = 0$

**27.** $2 = \dfrac{1}{3}x^2$

**28.** $\dfrac{1}{4} = 3x^2$

**29.** $30x^2 + 16x = 24$

**30.** $24x^2 + 24 = 50x$

**31.** $40x^2 + 100x + 40 = 0$

**32.** $27x^2 + 45x = 18$

**33.** $\dfrac{x - 1}{2x} = \dfrac{5}{x + 12}$

**34.** $\dfrac{3x - 1}{\dfrac{11}{2}} = \dfrac{2}{5 - x}$

**35.** $\dfrac{3x}{x - 4} = x + 2$

**36.** $\dfrac{3x + 16}{9x} = x + \dfrac{1}{3}$

**37.** $\dfrac{x + 1}{x - 2} + \dfrac{x - 1}{x + 1} = \dfrac{9}{2}$

**38.** $\dfrac{4}{x^2 - 1} + \dfrac{1}{x - 1} = \dfrac{7}{3}$

## 7.2 SOLVING QUADRATIC EQUATIONS BY COMPLETING THE SQUARE

Factoring is usually the easiest and fastest method of solving a quadratic equation *if* the equation can be factored. If the equation cannot be factored, then one of two methods is normally used. We shall study one of these methods, **completing the square,** in this section. The other method, the *quadratic formula,* is developed using the method of completing the square. This formula and its development are included in the next section.

---

### SOLVING A QUADRATIC EQUATION BY COMPLETING THE SQUARE

1. The coefficient of the second-degree term *must* equal (positive) 1. If not, divide each side of the equation by its coefficient.

2. Write an equivalent equation with the variable terms on the left side and the constant term on the right side; that is, in the form $x^2 + px = q$.

3. Add the square of one-half of the coefficient of the first-degree term to each side; that is, $\left(\frac{1}{2}p\right)^2$.

$$x^2 + px + \frac{p^2}{4} = q + \frac{p^2}{4}$$

4. The left side is now a perfect square trinomial—as the method's name implies. Rewrite the left side as a square.

$$\left(x + \frac{p}{2}\right)^2 = q + \frac{p^2}{4}$$

5. Take the square root of each side.

$$x + \frac{p}{2} = \pm\sqrt{q + \frac{p^2}{4}}$$

6. Solve for the variable $x$ and simplify, if possible.

$$x = -\frac{p}{2} \pm \sqrt{q + \frac{p^2}{4}}$$

7. Check.

---

**EXAMPLE 1**

Solve $x^2 + 3x - 10 = 0$ by completing the square.

**Step 1:** The coefficient of $x^2$ is 1, so add 10 to each side.

**Step 2:**
$$x^2 + 3x = 10$$

Add the square of one-half of the coefficient of $x$, $\left(\frac{3}{2}\right)^2$ or $\frac{9}{4}$, to each side.

**Step 3:**
$$x^2 + 3x + \frac{9}{4} = 10 + \frac{9}{4}$$

Rewrite the left side as a square and simplify the right side.

$$\left(x + \frac{3}{2}\right)\left(x + \frac{3}{2}\right) = \frac{40}{4} + \frac{9}{4}$$

**Step 4:**
$$\left(x + \frac{3}{2}\right)^2 = \frac{49}{4}$$

Take the square root of each side.

**Step 5:**
$$x + \frac{3}{2} = \pm\frac{7}{2}$$

Solve for $x$ and simplify.

**Step 6:**
$$x = -\frac{3}{2} \pm \frac{7}{2}$$

*Note:* $x = -\dfrac{3}{2} \pm \dfrac{7}{2}$ is a short way of writing

$$x = -\frac{3}{2} + \frac{7}{2} \quad \text{or} \quad x = -\frac{3}{2} - \frac{7}{2}$$

That is, read the equation all the way through a first time by using only the $+$ sign of the $\pm$ sign; then read it through a second time using only the $-$ sign. Therefore,

$$x = \frac{4}{2} = 2 \quad \text{or} \quad x = \frac{-10}{2} = -5$$

*Check:*

**Step 7:**

| $(2)^2 + 3(2) - 10 = 0$ | $(-5)^2 + 3(-5) - 10 = 0$ |
|---|---|
| $4 + 6 \ \ - 10 = 0$ | $25 - 15 \ \ - 10 = 0$ |
| $0 = 0$ | $0 = 0$ |

**EXAMPLE 2**

Solve $4x^2 - 24x + 7 = 0$ by completing the square.

First, divide each side of the equation by the coefficient of $x^2$, which is 4.

$$x^2 - 6x + \frac{7}{4} = 0$$

Add $-\dfrac{7}{4}$ to each side.

$$x^2 - 6x = -\frac{7}{4}$$

Add the square of one-half of the coefficient of $x$, $\left(\dfrac{-6}{2}\right)^2$ or 9, to each side.

$$x^2 - 6x + 9 = -\frac{7}{4} + 9$$

Rewrite the left side as a perfect square and simplify the right side.

$$(x - 3)^2 = \frac{29}{4}$$

Take the square root of each side.

$$x - 3 = \pm \frac{\sqrt{29}}{2}$$

Solve for $x$ and simplify.

$$x = 3 \pm \frac{\sqrt{29}}{2}$$

Or if we combine the terms by using the lowest common denominator,

$$x = \frac{6 \pm \sqrt{29}}{2}$$

**EXAMPLE 3**

Solve $-2x^2 + 3x + 1 = 0$ by completing the square.

$$x^2 - \frac{3}{2}x - \frac{1}{2} = 0 \qquad \text{(Divide each side by } -2.\text{)}$$

$$x^2 - \frac{3}{2}x = \frac{1}{2}$$

$$x^2 - \frac{3}{2}x + \frac{9}{16} = \frac{1}{2} + \frac{9}{16} \qquad \left[ Note: \left( \frac{1}{2} \cdot -\frac{3}{2} \right)^2 = \frac{9}{16} \right]$$

$$\left( x - \frac{3}{4} \right)^2 = \frac{17}{16}$$

$$x - \frac{3}{4} = \pm \frac{\sqrt{17}}{4}$$

$$x = \frac{3}{4} \pm \frac{\sqrt{17}}{4} \quad \text{or} \quad x = \frac{3 \pm \sqrt{17}}{4}$$

## Exercises 7.2

*Solve each equation by completing the square.*

**1.** $x^2 + 6x - 16 = 0$  

**2.** $x^2 + 15x + 54 = 0$

**3.** $x^2 + 35 = 12x$  

**4.** $x^2 = 6x + 27$

**5.** $2x^2 + x = 1$  

**6.** $2x^2 + 3 = 7x$

**7.** $25x^2 = 15x + 18$  

**8.** $3x^2 + 17x + 20 = 0$

**9.** $x^2 + 4x - 7 = 0$  

**10.** $x^2 - 5x + 3 = 0$

**11.** $2x^2 + 2x - 9 = 0$  

**12.** $5x^2 + 4x = 1$

**13.** $3x^2 + 9x + 5 = 0$  

**14.** $-3x^2 - 5x + 4 = 0$

**15.** $-2x^2 + 3x + 9 = 0$  

**16.** $-5x^2 + 8x - 2 = 0$

**17.** $4x^2 + 11x = 3$  

**18.** $3a^2 + 5a = 2$

**19.** $6m^2 + 15m + 3 = 0$  

**20.** $8y^2 - 7y - 10 = 0$

## 7.3  THE QUADRATIC FORMULA

D.6

Instead of using the method of completing the square for each quadratic equation, we may use the method to solve the general quadratic equation ($ax^2 + bx + c = 0$) and obtain a formula for solving any quadratic equation in that form. Let's now solve

$$ax^2 + bx + c = 0$$

by the method of completing the square.

First, divide each side by $a$, the coefficient of $x^2$.

$$x^2 + \frac{b}{a}x + \frac{c}{a} = 0$$

Add $-\dfrac{c}{a}$ to each side.

$$x^2 + \frac{b}{a}x = -\frac{c}{a}$$

Add the square of one-half of the coefficient of $x$, $\left(\dfrac{1}{2} \cdot \dfrac{b}{a}\right)^2$ or $\dfrac{b^2}{4a^2}$, to each side.

$$x^2 + \frac{b}{a}x + \frac{b^2}{4a^2} = \frac{b^2}{4a^2} - \frac{c}{a}$$

Rewrite the left side as a perfect square and simplify the right side.

$$\left(x + \frac{b}{2a}\right)^2 = \frac{b^2 - 4ac}{4a^2}$$

Take the square root of each side.

$$x + \frac{b}{2a} = \pm\frac{\sqrt{b^2 - 4ac}}{2a}$$

Solve for $x$ by adding $-\dfrac{b}{2a}$ to each side, and simplify.

---

**SOLVING A QUADRATIC EQUATION USING THE QUADRATIC FORMULA**

$$x = \frac{-b \pm \sqrt{b^2 - 4ac}}{2a}$$

---

To solve a quadratic equation by using the quadratic formula, we need only to identify $a$, $b$, and $c$ from $ax^2 + bx + c = 0$ and substitute those values in the preceding quadratic formula.

### EXAMPLE 1

Solve $x^2 + 2x - 35 = 0$ using the quadratic formula.

First, identify $a$, $b$, and $c$.

$$a = 1$$
$$b = 2$$
$$c = -35$$

Then substitute these values in the quadratic formula.

$$x = \frac{-b \pm \sqrt{b^2 - 4ac}}{2a}$$

$$x = \frac{-2 \pm \sqrt{(2)^2 - 4(1)(-35)}}{2(1)}$$

$$= \frac{-2 \pm \sqrt{4 + 140}}{2}$$

$$= \frac{-2 \pm \sqrt{144}}{2}$$

$$= \frac{-2 \pm 12}{2}$$

$$x = \frac{-2 + 12}{2} \quad \text{or} \quad x = \frac{-2 - 12}{2}$$

$$x = 5 \quad \quad \text{or} \quad x = -7$$

*Check:*

| | |
|---|---|
| $(5)^2 + 2(5) - 35 = 0$ | $(-7)^2 + 2(-7) - 35 = 0$ |
| $25 + 10 - 35 = 0$ | $49 - 14 - 35 = 0$ |
| $0 = 0$ | $0 = 0$ |

**EXAMPLE 2**

Solve $3x^2 = 4x + 8$ using the quadratic formula.

First, identify $a$, $b$, and $c$. To find $a$, $b$, and $c$, the equation *must* be in the form

$$ax^2 + bx + c = 0$$

That is,

$$3x^2 - 4x - 8 = 0$$

Therefore, $a = 3$, $b = -4$, and $c = -8$. Second, substitute these values in the quadratic formula,

$$x = \frac{-b \pm \sqrt{b^2 - 4ac}}{2a}$$

$$x = \frac{-(-4) \pm \sqrt{(-4)^2 - 4(3)(-8)}}{2(3)}$$

$$= \frac{4 \pm \sqrt{16 + 96}}{6}$$

$$= \frac{4 \pm \sqrt{112}}{6}$$

$$= \frac{4 \pm 4\sqrt{7}}{6} \qquad\qquad (\sqrt{112} = \sqrt{16 \cdot 7} = 4\sqrt{7})$$

$$= \frac{2 \pm 2\sqrt{7}}{3}$$

An examination of the quadratic formula $x = \dfrac{-b \pm \sqrt{b^2 - 4ac}}{2a}$ gives an indication about the type of roots, or solutions, we can expect from the equation $ax^2 + bx + c = 0$, where $a$, $b$, and $c$ are integers.

The quantity under the radical sign $(b^2 - 4ac)$ is called the **discriminant** because it discriminates, or determines, whether the roots are real or imaginary, rational or irrational, or one or two in number.

1. If $b^2 - 4ac > 0$, there are two real solutions.

   (a) If $b^2 - 4ac$ is also a perfect square, then the two solutions are rational.

   (b) If $b^2 - 4ac$ is not a perfect square, then both solutions are irrational.

2. If $b^2 - 4ac = 0$, there is only one real, rational solution.

3. If $b^2 - 4ac < 0$, there are two imaginary solutions (see Chapter 14).

In Section 4.2, we solved equations graphically. The equation $y = x^2 + 2x - 4$ was solved graphically for $y = 0$, $y = 4$, $y = -3$, and $y = -7$; the graph is repeated in Fig. 7.1. Let's now find each of these solutions either by using the quadratic formula or by factoring.

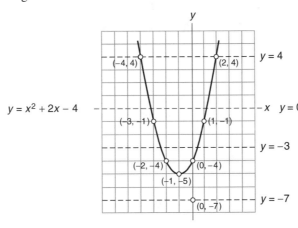

**Figure 7.1**

<table>
<tr><td>Quadratic formula</td><td>Factoring</td></tr>
</table>

| Quadratic formula | Factoring |
|---|---|
| $y = x^2 + 2x - 4$  for $y = 0$ | $y = x^2 + 2x - 4$  for $y = 4$ |
| $0 = x^2 + 2x - 4$ | $4 = x^2 + 2x - 4$ |
| $x = \dfrac{-b \pm \sqrt{b^2 - 4ac}}{2a}$ | $0 = x^2 + 2x - 8$ |
| $x = \dfrac{-2 \pm \sqrt{(2)^2 - 4(1)(-4)}}{2(1)}$ | $0 = (x + 4)(x - 2)$ |
| $= \dfrac{-2 \pm \sqrt{20}}{2}$ | $x + 4 = 0$ or $x - 2 = 0$ |
| $= \dfrac{-2 \pm 2\sqrt{5}}{2}$ | $x = -4$ or $x = 2$ |
| $= -1 \pm \sqrt{5}$ or $1.24, -3.24$ | |

|    *Quadratic formula*    |    *Factoring*    |
|---|---|

*Quadratic formula*

$$y = x^2 + 2x - 4 \quad \text{for } y = -3$$
$$-3 = x^2 + 2x - 4$$
$$0 = x^2 + 2x - 1$$

$$x = \frac{-2 \pm \sqrt{(2)^2 - 4(1)(-1)}}{2(1)}$$

$$= \frac{-2 \pm \sqrt{8}}{2}$$

$$= \frac{-2 \pm 2\sqrt{2}}{2}$$

$$= -1 \pm \sqrt{2} \quad \text{or} \quad 0.414, -2.41$$

*Factoring*

$$y = x^2 + 2x - 4 \quad \text{for } y = -7$$
$$-7 = x^2 + 2x - 4$$
$$0 = x^2 + 2x + 3$$

$$x = \frac{-2 \pm \sqrt{(2)^2 - 4(1)(3)}}{2(1)}$$

$$= \frac{-2 \pm \sqrt{-8}}{2}$$

The discriminant is negative; therefore, there are no real solutions. *Note:* In Fig. 7.1, the curve does not intersect the line $y = -7$.

The quadratic formula may also be used to solve literal quadratic equations; that is, quadratic equations involving letters. Assume that the letters represent positive real numbers.

**EXAMPLE 3**

Solve $m^2x^2 - mnx - n^2 = 0$ for $x$.
      Here $a = m^2$, $b = -mn$, and $c = -n^2$.

$$x = \frac{-b \pm \sqrt{b^2 - 4ac}}{2a}$$

$$x = \frac{-(-mn) \pm \sqrt{(-mn)^2 - 4(m^2)(-n^2)}}{2(m^2)}$$

$$= \frac{mn \pm \sqrt{m^2n^2 + 4m^2n^2}}{2m^2}$$

$$= \frac{mn \pm \sqrt{5m^2n^2}}{2m^2}$$

$$= \frac{mn \pm mn\sqrt{5}}{2m^2}$$

$$= \frac{n \pm n\sqrt{5}}{2m}$$

## Exercises 7.3

*Solve each equation using the quadratic formula.*

**1.** $x^2 - 4x - 32 = 0$      **2.** $x^2 + 10x + 21 = 0$      **3.** $2x^2 = 3x + 9$

**4.** $2x^2 + 9x = 5$      **5.** $12x^2 + 5x = 28$      **6.** $6x^2 = 5x + 25$

**7.** $15x^2 + 75x + 90 = 0$      **8.** $20x^2 - 30x = 200$      **9.** $8x^2 + 27 = 30x$

**10.** $12x^2 - 72x = 135$    **11.** $6x^2 + 15x = 0$    **12.** $8x^2 - 24 = 0$

**13.** $x^2 + 5x = 7$    **14.** $4x^2 = 4x + 11$    **15.** $x^2 = 4x + 14$

**16.** $x^2 = 6x + 19$    **17.** $4x^2 = 8x + 23$    **18.** $20x + 3 = 4x^2$

**19.** $9x^2 + 25 = 42x$    **20.** $9x^2 = 15x + 59$    **21.** $x^2 - 4x + 1 = 0$

**22.** $2x^2 + 7x = -1$    **23.** $-3x^2 + 7x = 2$    **24.** $-6x^2 - 4x + 3 = 0$

**25.** $x^2 + 10x = 24$    **26.** $x^2 + 7 = 8x$    **27.** $4x^2 + 12x + 9 = 0$

**28.** $2x^2 + 5x = 4$    **29.** $2x^2 - 5x + 1 = 0$    **30.** $9x^2 + 25 = 30x$

**31.** $5x^2 + 6x + 4 = 0$    **32.** $3x^2 - 5x + 8 = 0$

*Solve each equation for the indicated values using any method.*

**33.** $y = -2x^2 + 2x$
for $y = 0, -12$, and 7

**34.** $y = x^2 + x - 12$
for $y = 0, 8$, and 6

**35.** $y = 3x^2 + x$
for $y = 0, 4$, and 1

**36.** $y = 3x^2 + 2x$
for $y = 0, 40, 3$, and $-3$

*Solve for x.*

**37.** $m^2x^2 - 2mx + 1 = 0$

**38.** $4m^2x^2 + 6mnx - n^2 = 0$

**39.** $n^2x^2 + 4nx + 4 = 0$

**40.** $3x^2 + 4x - mn = 0$

**41.** $x^2 + 5x = 16a^2 - 20a$

**42.** $x^2 + 6nx + 9n^2 = 0$

*Solve for r.*

**43.** $A = \pi r^2$

**44.** $S = 2\pi r^2 + 2\pi rh$

# 7.4 APPLICATIONS

Let us now present some applications that involve quadratic equations. For consistency, all final results are rounded to three significant digits in this section.

**EXAMPLE 1**

Design a rectangular metal plate to meet the following specifications: (a) the length is 4.00 cm less than twice its width, and (b) its area is 96.0 cm$^2$.

First, draw a diagram as in Fig. 7.2 and let

$$x = \text{width}$$
$$2x - 4 = \text{length}$$
$$A = lw$$
$$96 = (2x - 4)x$$
$$96 = 2x^2 - 4x$$
$$0 = 2x^2 - 4x - 96$$
$$0 = x^2 - 2x - 48$$
$$0 = (x + 6)(x - 8)$$
$$x + 6 = 0 \quad \text{or} \quad x - 8 = 0$$
$$x = -6 \quad \text{or} \quad x = 8$$

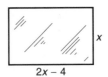

2x − 4        **Figure 7.2**

The solution $x = -6$ is not meaningful, as $x$ refers to a length measurement which must be a positive quantity. Therefore,

$$x = \text{width} = 8.00 \text{ cm}$$
$$2x - 4 = \text{length} = 12.0 \text{ cm}$$

Example 1 represents a situation where we must be very cautious of our interpretation of algebraic solutions. We must be aware that in some cases a numerical solution may not be meaningful to the given physical problem. Such a solution is algebraically correct, but makes no sense when it is applied to the problem.

### EXAMPLE 2

A variable voltage in a given electrical circuit is given by the formula $V = t^2 - 12t + 40$. At what values of $t$ (in seconds) is the voltage $V$ equal to 8.00 volts? To 25.0 volts?

If $V = 8.00$,

$$8 = t^2 - 12t + 40$$
$$0 = t^2 - 12t + 32$$
$$0 = (t - 4)(t - 8)$$
$$t - 4 = 0 \quad \text{or} \quad t - 8 = 0$$
$$t = 4.00 \text{ s} \quad \text{or} \quad t = 8.00 \text{ s}$$

If $V = 25.0$,

$$25 = t^2 - 12t + 40$$
$$0 = t^2 - 12t + 15 \qquad \text{(Does not factor)}$$
$$t = \frac{-b \pm \sqrt{b^2 - 4ac}}{2a}$$
$$t = \frac{-(-12) \pm \sqrt{(-12)^2 - 4(1)(15)}}{2(1)}$$
$$= \frac{12 \pm \sqrt{144 - 60}}{2}$$
$$= \frac{12 \pm \sqrt{84}}{2} \qquad (\sqrt{84} = \sqrt{4 \cdot 21} = 2\sqrt{21})$$
$$= \frac{12 \pm 2\sqrt{21}}{2}$$
$$= 6 \pm \sqrt{21}$$

If decimal solutions are needed ($\sqrt{21}$ is approximately 4.58),

$$t = 6 + \sqrt{21} \quad \text{or} \quad t = 6 - \sqrt{21}$$
$$t = 6 + 4.58 \qquad\qquad t = 6 - 4.58$$
$$t = 10.6 \text{ s} \quad \text{or} \quad t = 1.42 \text{ s}$$

**EXAMPLE 3**

The perimeter of a rectangle is 24.0 m and the area is 20.0 m². Find the dimensions (the length and the width).

First, draw a diagram as in Fig. 7.3 and let

$$x = \text{width}$$
$$12 - x = \text{length}$$

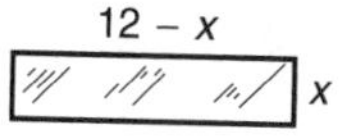

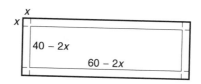

**Figure 7.3**

(*Note:* The perimeter of a rectangle is the sum of the measurements of all four sides. Therefore, if the perimeter is 24.0 m, one width plus one length is 12.0 m.)

$$A = lw$$
$$20 = (12 - x)x$$
$$20 = 12x - x^2$$
$$x^2 - 12x + 20 = 0$$
$$(x - 2)(x - 10) = 0$$
$$x - 2 = 0 \quad \text{or} \quad x - 10 = 0$$
$$x = 2 \qquad\qquad x = 10$$

If $\quad x = 2$ $\qquad\qquad$ If $\quad x = 10$

$x = \text{width} = 2.00 \text{ m}$ $\qquad$ $x = \text{width} = 10.0 \text{ m}$

$12 - x = \text{length} = 10.0 \text{ m}$ $\qquad$ $12 - x = \text{length} = 2.00 \text{ m}$

Since the length is greater than the width, the width is 2.00 m and the length is 10.0 m.

**EXAMPLE 4**

A square is cut out of each corner of a rectangular sheet of metal 40.0 cm × 60.0 cm. The sides are folded up to form a rectangular container. What are the dimensions of the square if the area of the bottom of the container is 1500 cm²? Also, find the volume of the container.

Draw a diagram as in Fig. 7.4 and let

$$x = \text{side of square cut out}$$
$$40 - 2x = \text{width of rectangular container}$$
$$60 - 2x = \text{length of rectangular container}$$

**Figure 7.4**

Then

$$A = lw$$
$$1500 = (60 - 2x)(40 - 2x)$$
$$1500 = 2400 - 200x + 4x^2$$
$$0 = 4x^2 - 200x + 900$$
$$0 = x^2 - 50x + 225$$
$$0 = (x - 5)(x - 45)$$
$$x - 5 = 0 \quad \text{or} \quad x - 45 = 0$$
$$x = 5 \quad \text{or} \quad x = 45 \quad \text{(\textit{Note:} } x = 45 \text{ is physically impossible!)}$$

Therefore, the length of the side of square cut out is 5.00 cm, and

$$V = lwh$$
$$= (50.0 \text{ cm})(30.0 \text{ cm})(5.00 \text{ cm}) = 7500 \text{ cm}^3$$

## Exercises 7.4

*Express the final results using three significant digits.*

1. Separate 13 into two parts whose product is 40.

2. Separate 10 into two parts whose product is 24.

3. The length of a rectangular metal plate is 3.00 cm greater than twice the width. Find the dimensions if its area is 35.0 cm$^2$.

4. A rectangular plot of ground contains 5400 m$^2$. If the length exceeds the width by 30.0 m, what are the dimensions?

5. The area of a triangle is 66.0 m$^2$. Find the lengths of the base and the height if the base is one metre longer than the height $(A = \frac{1}{2}bh)$.

6. A rectangle is 2.00 ft longer than it is wide. Find its dimensions if the area is 48.0 ft$^2$.

7. A variable electric current is given by the formula $i = t^2 - 7t + 12$. If $t$ is given in s, at what times is the current equal to 2.00 amperes (A)? To 0 A? To 4.00 A?

8. A variable voltage is given by the formula $V = t^2 - 14t + 48$. At what times, $t$ in s, is the voltage equal to 3.00 volts (V)? To 35.0 V? To 0 V?

9. A charge in coulombs flows in a given circuit according to the formula $q = 2t^2 - 4t + 4$. Find $t$ in $\mu$s when $q = 2.00$ and when $q = 3.00$.

10. The work done in a circuit varies with time according to the formula $w = 8t^2 - 12t + 20$. Find $t$ in ms when $w = 16.0$, $w = 18.0$, and $w = 0$.

11. A winding links a magnetic field that varies according to the formula $\phi = 0.4t - 2t^2$. Find $t$ in ms when $\phi$ is 0.0200 weber.

12. The perimeter of a rectangle is 22.0 cm and the area is 24.0 cm$^2$. Find the dimensions.

13. The perimeter of a rectangle is 80.0 m and the area is 375 m$^2$. Find the dimensions.

14. A rectangular yard is made using fencing for three sides of the yard and the house for one of the widths. The area of the yard is 288 m$^2$. Find its length and its width if the length is twice the width. How much fencing is needed?

15. A square 4.00 in. on a side is cut out of each corner of a square sheet of aluminum. The sides are folded up to form a rectangular container. If the volume is 400 in$^3$, what was the size of the original sheet of aluminum?

**16.** A square is cut out of each corner of a rectangular sheet of aluminum 20.0 cm × 30.0 cm. The sides are folded up to form a rectangular container. What are the dimensions of the square if the area of the bottom of the container is 416 cm²? Also find the volume of the container.

**17.** A rectangular field is fenced in by using a river as one side. If 2500 m of fencing are used for the 720,000-m² field, what are its dimensions?

**18.** A projectile is shot vertically upward. Its height, $h$ in feet, after time $t$ in s may be expressed by the formula $h = 96t - 16t^2$.
(a) Find $t$ when $h = 80.0$ ft.
(b) Find $t$ when $h = 128$ ft.
(c) Find $t$ when $h = 0$.
(d) Find the maximum height reached by the projectile.

**19.** How wide a strip must be mowed around a rectangular grass plot 40.0 m by 60.0 m for one-half of the grass to be mowed?

**20.** A rectangular sheet of metal 24.0 in. wide is formed into a rectangular trough with an open top and no ends. If the cross-sectional area is 70.0 in², find the depth of the trough.

**21.** A rectangular sheet of metal 48.0 in. wide is formed into a rectangular closed tube. If the cross-sectional area is 108 in², find the length and the width of the cross-section of the tube.

**22.** The cross-sectional area of the L-shaped beam in Fig. 7.5 is 24.0 in². What is the thickness, $x$, of the metal?

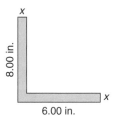

6.00 in.          **Figure 7.5**

**23.** The area of a rectangular lot, 80.0 m × 100 m, is increased by 4000 m². If the length and the width are increased by the same amount, what are the dimensions of the larger lot?

**24.** The area of a rectangular lot, 240 ft × 400 ft, is increased by 15,800 ft². If the width and the length are increased at a ratio of 2:3, what are the dimensions of the larger lot?

**25.** The equivalent resistance of two resistors connected in parallel is 45.0 ohms (Ω). If the resistance of one resistor is three times the other, what is the value of each? The formula for two resistors in parallel is $R = \dfrac{R_1 R_2}{R_1 + R_2}$, where $R$ is the equivalent resistance and $R_1$ and $R_2$ are the individual resistors.

**26.** A 100-W lamp and a 20.0-Ω resistor are connected in series with a 120-V power supply. By Kirchhoff's law, we have

$$\varepsilon = IR + \frac{P}{I}$$

After substituting the known values, we have

$$120 = 20I + \frac{100}{I}$$

Solve for $I$ (in amperes).

**27.** A parking lot $20\bar{0}$ m $\times$ $10\bar{0}$ m is to be tripled in area by adding the region as shown in Fig. 7.6. Find $x$.

**28.** A parking lot $20\bar{0}$ m $\times$ $10\bar{0}$ m is to be tripled in area by adding the region as shown in Fig. 7.7. Find $x$.

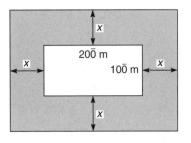

**Figure 7.6**

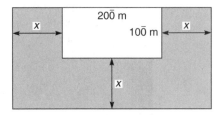

**Figure 7.7**

## CHAPTER SUMMARY

**1.** A quadratic equation in one variable is an equation in the form

$$ax^2 + bx + c = 0 \quad (a \neq 0)$$

**2.** *To solve a quadratic equation by factoring:*
   (a) If necessary, write the equation in the form $ax^2 + bx + c = 0$.
   (b) Factor.
   (c) Set each factor that contains a variable equal to zero.
   (d) Solve each resulting linear equation.
   (e) Check.

**3.** *To solve an equation in the form $ax^2 = c$:*
   (a) Divide each side by $a$.
   (b) Take the square root of each side.
   (c) Simplify the result, if possible.

**4.** *To solve a quadratic equation by completing the square:*
   (a) The coefficient of the second-degree term *must* equal 1. If not, divide each side by its coefficient.
   (b) Write the equation with the variable terms on one side and the constant term on the other.
   (c) Add the square of one-half of the coefficient of the first-degree term to each side.
   (d) Rewrite the variable side, now a perfect square, as the square of a binomial.
   (e) Take the square root of each side.
   (f) Solve for the variable and simplify, if possible.
   (g) Check.

**5.** *The quadratic formula:* To solve a quadratic equation in the form $ax^2 + bx + c = 0$, substitute the values for $a$, $b$, and $c$ into the formula

$$x = \frac{-b \pm \sqrt{b^2 - 4ac}}{2a}$$

6. The **discriminant** is $b^2 - 4ac$, the quantity under the radical in the quadratic formula.
 (a) If $b^2 - 4ac > 0$, there are two real solutions.
  (i) If $b^2 - 4ac$ is also a perfect square, then the two solutions are rational.
  (ii) If $b^2 - 4ac$ is not a perfect square, then both solutions are irrational.
 (b) If $b^2 - 4ac = 0$, there is only one real, rational solution.
 (c) If $b^2 - 4ac < 0$, there are two imaginary solutions.

## CHAPTER 7 REVIEW

*Solve each equation.*

1. $x^2 + 4x = 21$
2. $6x^2 + 40 = 31x$
3. $20x^2 + 21x + 4 = 0$
4. $36x^2 - 1 = 0$
5. $18x^2 + 45x + 18 = 0$
6. $6x^2 - 36x = 0$
7. $-8x^2 + 6x + 9 = 0$
8. $36 - 4x^2 = 0$
9. $x^2 + 10x + 25 = 0$
10. $9x^2 + 16 = 24x$

*Solve each equation by completing the square.*

11. $2x^2 + 13x + 20 = 0$
12. $3x^2 = 4x + 7$
13. $x^2 - 8x + 9 = 0$
14. $-5x^2 + 6x + 1 = 0$

*Solve each equation for x using the quadratic formula.*

15. $3x^2 - 16x + 20 = 0$
16. $4x^2 = 6x + 3$
17. $-2x^2 + 3x + 1 = 0$
18. $6x^2 + 8x - 5 = 0$
19. $mnx^2 - mx + n = 0$
20. $n^4x^2 - 2m^2n^2x + m^4 = 0$

21. Solve $s = vt + \dfrac{1}{2} at^2$ for $t$.

22. Divide 15 into two parts whose product is 36.

23. Three thousand feet of fence are used to enclose a rectangular plot of ground. Find the dimensions if the area is $54\overline{0},000 \text{ ft}^2$.

24. A square is cut out of each corner of a square sheet of aluminum 50.0 cm on a side. The sides are folded up to form a rectangular container. What are the dimensions of the square cut out if the area of the bottom of the container is $90\overline{0} \text{ cm}^2$? Also, find the volume of the container.

25. A ball is thrown vertically upward. Its height, $h$ in feet, after time $t$ in s may be expressed by $h = 64t - 16t^2$.
 (a) At what times is the ball 48.0 ft above the ground?
 (b) Does the ball ever reach a height of 80.0 ft?
 (c) Find the time that the ball is in the air.
 (d) What is the maximum height reached?

26. Find the resistance $R$ (in ohms).

$$\frac{1}{R} + \frac{1}{55\overline{0} - R} = \frac{1}{12\overline{0}}$$

# 8

# Exponents and Radicals

## INTRODUCTION

Radicals are contained in many formulas, such as trigonometric functions of half angles, e.g., $\cos \dfrac{x}{2} = \pm \sqrt{\dfrac{1 + \cos x}{2}}$; the area of a triangle when the sides are known, $A = \sqrt{s(s-a)(s-b)(s-c)}$; magnitude of impedance, $|Z| = \sqrt{R^2 + X^2}$; and angular velocity, $\omega_n = \sqrt{\dfrac{kg}{W}}$.

In this chapter we will simplify radicals, rewrite them using fractional exponents, and solve equations involving them.

### Objectives

- Simplify expressions with integral and rational exponents.
- Simplify radical expressions.
- Add, subtract, and multiply radical expressions.
- Rationalize denominators of algebraic expressions.
- Solve equations containing radical expressions.
- Solve equations that are quadratic in form.

## 8.1 INTEGRAL EXPONENTS

In Chapter 2, the laws of exponents were discussed in terms of positive integral exponents. Before proceeding, we must develop additional properties and further uses for exponents. The laws are repeated here for convenient reference.

$$\boxed{\begin{array}{l} \textbf{LAWS OF EXPONENTS} \\[6pt] \textbf{1. } a^m \cdot a^n = a^{m+n} \\[6pt] \textbf{2. } \dfrac{a^m}{a^n} = a^{m-n} \quad (a \neq 0) \\[6pt] \textbf{3. } (a^m)^n = a^{mn} \\[4pt] \textbf{4. } (ab)^n = a^n b^n \\[6pt] \textbf{5. } \left(\dfrac{a}{b}\right)^n = \dfrac{a^n}{b^n} \qquad (b \neq 0) \\[6pt] \textbf{6. } a^0 = 1 \qquad\qquad (a \neq 0) \end{array}}$$

Next, we discuss negative exponents. If Law 1 is to hold for negative exponents, we have

$$a^m \cdot a^{-m} = a^0 = 1 \quad (a \neq 0)$$

If we divide each side by $a^m$, we have the following equation.

$$\boxed{a^{-m} = \frac{1}{a^m}}$$

Also, if we divide each side by $a^{-m}$, we have

$$a^m = \frac{1}{a^{-m}}$$

Another way of showing $\dfrac{1}{a^{-m}} = a^m$ is

$$\frac{1}{a^{-m}} = \frac{1}{\dfrac{1}{a^m}} = 1 \div \frac{1}{a^m} = 1 \times \frac{a^m}{1} = a^m$$

### EXAMPLE 1

Find each product and write the result using positive exponents.

(a) $a^2 \cdot a^4 = a^{2+4} = a^6$

(b) $3^{-4} \cdot 3^7 = 3^{-4+7} = 3^3$

(c) $5^3 \cdot 5^{-6} = 5^{3+(-6)} = 5^{-3} = \dfrac{1}{5^3}$

(d) $x^{-5} \cdot x^{-4} = x^{(-5)+(-4)} = x^{-9} = \dfrac{1}{x^9}$

### EXAMPLE 2

Find each quotient and write the result using positive exponents.

(a) $\dfrac{4^8}{4^6} = 4^{8-6} = 4^2$

(b) $\dfrac{a^{-2}}{a^3} = a^{(-2)-3} = a^{-5} = \dfrac{1}{a^5}$

(c) $\dfrac{c^4}{c^{-3}} = c^{4-(-3)} = c^7$

(d) $\dfrac{8^{-5}}{8^{-2}} = 8^{(-5)-(-2)} = 8^{-3} = \dfrac{1}{8^3}$

### EXAMPLE 3

Find each power and write the result using positive exponents.

(a) $(3^2)^4 = 3^{(2)(4)} = 3^8$

(b) $(x^{-3})^2 = x^{(-3)(2)} = x^{-6} = \dfrac{1}{x^6}$

(c) $(c^5)^{-3} = c^{(5)(-3)} = c^{-15} = \dfrac{1}{c^{15}}$

(d) $(m^{-2})^{-3} = m^{(-2)(-3)} = m^6$

### EXAMPLE 4

Simplify.

(a) $\left(\dfrac{2}{3}\right)^{-1} = \dfrac{1}{\dfrac{2}{3}} = 1 \div \dfrac{2}{3} = 1 \cdot \dfrac{3}{2} = \dfrac{3}{2}$

(b) $\left(\dfrac{3}{4}\right)^{-2} = \dfrac{1}{\left(\dfrac{3}{4}\right)^2} = \dfrac{1}{\dfrac{9}{16}} = 1 \div \dfrac{9}{16} = 1 \cdot \dfrac{16}{9} = \dfrac{16}{9}$

(c) $2^{-1} + 4^{-2} = \dfrac{1}{2^1} + \dfrac{1}{4^2} = \dfrac{1}{2} + \dfrac{1}{16} = \dfrac{8}{16} + \dfrac{1}{16} = \dfrac{9}{16}$

### EXAMPLE 5

Simplify and write the result using positive exponents.

(a) $(3a^{-2}b^4c^{-5})(6a^6b^{-4}c^{-1}) = (3 \cdot 6)a^{(-2)+6}b^{4+(-4)}c^{(-5)+(-1)}$

$$= 18a^4b^0c^{-6}$$

$$= \dfrac{18a^4}{c^6} \quad \text{(Note: } b^0 = 1.\text{)}$$

(b) $\dfrac{2^2x^5y^{-4}z^{-2}}{4^{-2}x^{-2}y^7z^{-6}} = 2^2 \cdot 4^2 x^{5-(-2)}y^{(-4)-7}z^{(-2)-(-6)}$

$$= 4 \cdot 16x^7y^{-11}z^4$$

$$= \dfrac{64x^7z^4}{y^{11}}$$

(c) $(2x^{-3}y^2z^0)^{-3} = 2^{-3}x^{(-3)(-3)}y^{(2)(-3)}z^{(0)(-3)}$

$$= 2^{-3}x^9y^{-6}z^0$$

$$= \dfrac{x^9}{2^3y^6}$$

$$= \dfrac{x^9}{8y^6}$$

**EXAMPLE 6**

Simplify $xy^{-1} + x^{-1}y$ and write the result using positive exponents.

$$xy^{-1} + x^{-1}y = \frac{x}{y} + \frac{y}{x}$$
$$= \frac{x^2}{xy} + \frac{y^2}{xy}$$
$$= \frac{x^2 + y^2}{xy}$$

# Exercises 8.1

*Simplify and write the result using positive exponents.*

**1.** $x^2 \cdot x^3$

**2.** $c^4 \cdot c^6$

**3.** $\dfrac{5^6}{5^3}$

**4.** $\dfrac{y^2}{y^5}$

**5.** $(7^2)^3$

**6.** $(z^3)^3$

**7.** $m^{-2} \cdot m^5$

**8.** $c^4 \cdot c^{-6}$

**9.** $\dfrac{d^{-3}}{d^4}$

**10.** $\dfrac{p^5}{p^{-2}}$

**11.** $(2^{-3})^2$

**12.** $(4^{-3})^{-3}$

**13.** $y^{-3} \cdot y^{-5}$

**14.** $\dfrac{r^{-4}}{r^{-2}}$

**15.** $(s^{-2})^{-5}$

**16.** $7^{-2} \cdot 7^{-1}$

**17.** $\dfrac{t^{-2}}{t^{-5}}$

**18.** $(2^4)^{-2}$

**19.** $\left(\dfrac{4}{7}\right)^2$

**20.** $\left(\dfrac{5}{2}\right)^3$

**21.** $\left(\dfrac{1}{3}\right)^{-1}$

**22.** $\left(\dfrac{1}{4}\right)^{-1}$

**23.** $\left(\dfrac{5}{6}\right)^{-1}$

**24.** $\left(\dfrac{5}{4}\right)^{-1}$

**25.** $\left(\dfrac{1}{2}\right)^{-3}$

**26.** $\left(\dfrac{1}{5}\right)^{-2}$

**27.** $\left(\dfrac{2}{3}\right)^{-2}$

**28.** $\left(\dfrac{3}{5}\right)^{-3}$

**29.** $\dfrac{a^5}{a^{-5}}$

**30.** $\dfrac{a^{-7}}{a^{-2}}$

**31.** $a^6 \cdot a^{-4}$

**32.** $a^{-2} \cdot a^{-4}$

**33.** $(a^3)^{-4}$

**34.** $(a^{-2})^{-3}$

**35.** $(3a^2)^2$

**36.** $(4a^2b)^2$

**37.** $(2a^{-2}b)^{-2}$

**38.** $(4x^2y^{-1})^{-3}$

**39.** $6k^2(-2k)(4k^{-5})$

**40.** $(3b^2)(-4b^{-6})(-2b)$

**41.** $(3a^{-2}b)(5a^3b^{-3})$

**42.** $(6s^2t^{-4})(3^{-1}st^{-2})$

**43.** $w^{-4} \cdot w^2 \cdot w^{-3}$

**44.** $p^5 \cdot p^{-3} \cdot p^4$

**45.** $\dfrac{x^3 \cdot x^{-4}}{x^{-2} \cdot x^{-5}}$

**46.** $\dfrac{t^{-4} \cdot t^{-3}}{t^3 \cdot t^{-1}}$

**47.** $\dfrac{x^{-2}y^3}{x^3y^{-2}}$

**48.** $\dfrac{a^{-4}b}{a^{-1}b^{-3}}$

**49.** $\left(\dfrac{1}{t^4}\right)^{-2}$

**50.** $\left(\dfrac{1}{s^4}\right)^{-3}$

**51.** $\left(\dfrac{1}{b^{-2}}\right)^{-1}$

**52.** $\left(\dfrac{1}{s^{-1}}\right)^{-1}$

**53.** $\dfrac{ab^{-4}c^5}{a^2b^{-2}c}$

**54.** $\dfrac{a^4b^2c^{-4}}{a^6b^{-3}c^{-4}}$

**55.** $\left(\dfrac{a^2}{a^{-4}}\right)^3$

**56.** $\left(\dfrac{x^{-2}}{x^{-4}}\right)^{-2}$

**57.** $\dfrac{(3a^{-2})^3}{(2a^{-4})^{-2}}$

**58.** $\dfrac{(2a^2)^{-2}}{(4a^{-3})^2}$

**59.** $(2a^4b^0c^{-2})^3$

**60.** $(3a^4b^{-5}c^{-2})^4$

**61.** $\left(\dfrac{14a^3b^{-8}}{2a^{-2}b^{-4}}\right)^2$

**62.** $\left(\dfrac{12a^{-2}b^{-3}}{4a^{-6}b^0}\right)^{-3}$

**63.** $a^{-1} + b^{-1}$

**64.** $(a + b)^{-1}$

**65.** $a^{-2}b + a^{-1}b$

**66.** $ab^{-2} + a^{-1}b$

**67.** $3^{-2} + 9^{-1}$

**68.** $2^{-1} - 4^{-1}$

**69.** $(2^{-1} + 3^{-1})^{-1}$

**70.** $(x^{-1} + y^{-1})^{-1}$

**71.** $\dfrac{1}{a^0 + b^0}$

**72.** $(a^{-1} + b^{-1})^0$

## 8.2 FRACTIONAL EXPONENTS

We have now used the laws of exponents for cases when the exponents were positive integers, zero, and negative integers. Let's next extend their use to rational or fractional exponents.

---

**FRACTIONAL EXPONENTS**

**Definition 1:** $a^{1/n} = \sqrt[n]{a}$

where $n$ is called the index and $a$ is called the radicand.

**Definition 2:** $a^{m/n} = \sqrt[n]{a^m} = (\sqrt[n]{a})^m$

*Note:* $m$ is an integer, $n$ is a positive integer, and $a$ is a real number. If $n$ is even, $a \geq 0$.

---

The examples in Table 8.1 have been worked both in fractional exponential form and in radical form. These examples show that the laws of exponents can be extended to fractional exponents with equivalent results.

For the two preceding definitions, note the restrictions on $n$. First, let us investigate the definitions when $n$ is even. For example,

$$25^{1/2} = \sqrt{25} = 5$$

but

$$(-25)^{1/2} = \sqrt{-25}$$

which is not a real number. Also,

$$16^{1/4} = \sqrt[4]{16} = 2$$

but

$$(-16)^{1/4} = \sqrt[4]{-16}$$

**TABLE 8.1  Examples**

| Fractional exponential form | Equivalent expressions | Radical form |
|---|---|---|
| **1.** $3^{1/2} \cdot 3^{1/2} = 3^{1/2+1/2}$ $= 3^1 = 3$ | $3^{1/2} = \sqrt{3}$ | **1.** $\sqrt{3} \cdot \sqrt{3} = \sqrt{9}$ $= 3$ |
| **2.** $4^{1/3} \cdot 4^{1/3} \cdot 4^{1/3}$ $= 4^{1/3 + 1/3 + 1/3}$ $= 4^1 = 4$ | $4^{1/3} = \sqrt[3]{4}$ | **2.** $\sqrt[3]{4} \cdot \sqrt[3]{4} \cdot \sqrt[3]{4}$ $= \sqrt[3]{64}$ $= 4$ |
| **3.** $(12^{1/2})^2 = 12^{(1/2)(2)}$ $= 12^1 = 12$ | $12^{1/2} = \sqrt{12}$ | **3.** $(\sqrt{12})^2 = \sqrt{12} \cdot \sqrt{12}$ $= \sqrt{144} = 12$ |
| **4.** $(5^{1/3})^3 = 5^{(1/3)(3)}$ $= 5^1 = 5$ | $5^{1/3} = \sqrt[3]{5}$ | **4.** $(\sqrt[3]{5})^3 = \sqrt[3]{5} \cdot \sqrt[3]{5} \cdot \sqrt[3]{5}$ $= \sqrt[3]{125} = 5$ |
| **5.** $\dfrac{8^{2/3}}{8^{1/3}} = 8^{2/3 - 1/3} = 8^{1/3}$ $= (2^3)^{1/3}$ $= 2^{(3)(1/3)} = 2^1 = 2$ | $8^{2/3} = \sqrt[3]{8^2}$ $8^{1/3} = \sqrt[3]{8}$ | **5.** $\dfrac{\sqrt[3]{8^2}}{\sqrt[3]{8}} = \sqrt[3]{\dfrac{8^2}{8}}$ $= \sqrt[3]{8}$ $= 2$ |

which is not a real number. As a third example,

$$4^{3/2} = \sqrt{4^3} = \sqrt{64} = 8$$

but

$$(-4)^{3/2} = \sqrt{(-4)^3} = \sqrt{-64}$$

which is not a real number. However, there is no problem if $n$ is odd. For example,

$$(27)^{1/3} = \sqrt[3]{27} = 3 \quad \text{and} \quad (-27)^{1/3} = \sqrt[3]{-27} = -3$$
$$(32)^{1/5} = \sqrt[5]{32} = 2 \quad \text{and} \quad (-32)^{1/5} = \sqrt[5]{-32} = -2$$
$$8^{2/3} = \sqrt[3]{8^2} = \sqrt[3]{64} = 4 \quad \text{and} \quad (-8)^{2/3} = \sqrt[3]{(-8)^2} = \sqrt[3]{64} = 4$$

**EXAMPLE 1**

Evaluate $9^{3/2}$.

$$9^{3/2} = (9^{1/2})^3$$
$$= (3)^3 \qquad (9^{1/2} = 3)$$
$$= 27$$

**EXAMPLE 2**

Evaluate $16^{3/4}$.

$$16^{3/4} = (16^{1/4})^3$$
$$= (2)^3 \qquad (16^{1/4} = 2)$$
$$= 8$$

**EXAMPLE 3**

Evaluate $27^{-2/3}$.

$$27^{-2/3} = (27^{1/3})^{-2}$$
$$= (3)^{-2} \qquad (27^{1/3} = 3)$$
$$= \frac{1}{3^2}$$
$$= \frac{1}{9}$$

**EXAMPLE 4**

Simplify (a) $\dfrac{x^{5/6}}{x^{2/3}}$ and (b) $(x^{5/6})(x^{2/3})$.

(a) $\dfrac{x^{5/6}}{x^{2/3}} = x^{5/6-2/3}$     (b) $(x^{5/6})(x^{2/3}) = x^{5/6+2/3}$

$$= x^{5/6-4/6} \qquad\qquad\qquad = x^{5/6+4/6}$$
$$= x^{1/6} \qquad\qquad\qquad\quad = x^{9/6}$$
$$\qquad\qquad\qquad\qquad\quad = x^{3/2}$$

**EXAMPLE 5**

Simplify $(x^{3/4})^{2/3}$.

$$(x^{3/4})^{2/3} = x^{(3/4)(2/3)}$$
$$= x^{1/2}$$

**EXAMPLE 6**

Simplify $\left(\dfrac{a^3 b^{-6}}{c^{12}}\right)^{-2/3}$.

$$\left(\frac{a^3 b^{-6}}{c^{12}}\right)^{-2/3} = \frac{a^{(3)(-2/3)} b^{(-6)(-2/3)}}{c^{(12)(-2/3)}}$$
$$= \frac{a^{-2} b^4}{c^{-8}}$$
$$= \frac{b^4 c^8}{a^2}$$

**EXAMPLE 7**

Simplify $a^{2/3}(a^{-1/3} + 2a^{1/3})$.

$$a^{2/3}(a^{-1/3} + 2a^{1/3}) = a^{2/3} \cdot a^{-1/3} + 2a^{2/3} \cdot a^{1/3}$$
$$= a^{1/3} + 2a$$

## Exercises 8.2

*Evaluate each expression.*

**1.** $36^{1/2}$             **2.** $49^{1/2}$             **3.** $64^{1/3}$

**4.** $125^{1/3}$       **5.** $64^{2/3}$       **6.** $8^{2/3}$

**7.** $8^{5/3}$       **8.** $32^{2/5}$       **9.** $16^{-1/2}$

**10.** $25^{-1/2}$       **11.** $27^{-1/3}$       **12.** $8^{-1/3}$

**13.** $27^{2/3}$       **14.** $64^{4/3}$       **15.** $9^{-3/2}$

**16.** $16^{-3/4}$       **17.** $\left(\dfrac{16}{25}\right)^{1/2}$       **18.** $\left(\dfrac{4}{9}\right)^{1/2}$

**19.** $\left(\dfrac{4}{9}\right)^{3/2}$       **20.** $\left(\dfrac{16}{81}\right)^{3/4}$       **21.** $\left(\dfrac{25}{16}\right)^{-3/2}$

**22.** $\left(\dfrac{8}{27}\right)^{-2/3}$       **23.** $(-8)^{-1/3}$       **24.** $(-27)^{-2/3}$

**25.** $3^{3/5} \cdot 3^{7/5}$       **26.** $5^{9/4} \cdot 5^{3/4}$       **27.** $\dfrac{9^{7/4}}{9^{5/4}}$

**28.** $\dfrac{8^{2/3}}{8^{1/3}}$       **29.** $(16^{1/2})^{1/2}$       **30.** $(32^{1/3})^{3/5}$

*Perform the indicated operations. Simplify and express with positive exponents.*

**31.** $x^{4/3} \cdot x^{2/3}$       **32.** $m^{-5/4} \cdot m^{1/4}$

**33.** $x^{3/4} \cdot x^{-1/2}$       **34.** $a^{-2/3} \cdot a^{-1/2}$

**35.** $\dfrac{x^{2/3}}{x^{1/6}}$       **36.** $\dfrac{x^{1/2}}{x^{3/4}}$

**37.** $(x^{2/3})^{-1/2}$       **38.** $(x^{-3/2})^{-2/3}$

**39.** $(x^{2/5} \cdot x^{-4/5})^{5/6}$       **40.** $(x^{-1/2} \cdot x^{-1/3})^{-1/5}$

**41.** $\left(\dfrac{x^{3/2}}{x^{3/4}}\right)^{1/3}$       **42.** $\left(\dfrac{x^{1/4}y^{-3/4}}{x^{1/2}y^{3/2}}\right)^{-1/2}$

**43.** $\left(\dfrac{a^4 b^{-8}}{c^{16}}\right)^{3/4}$       **44.** $\left(\dfrac{a^{-6}}{b^9 c^{-15}}\right)^{-5/3}$

**45.** $(a^{-12} b^9 c^{-6})^{-4/3}$       **46.** $(a^5 b^{10} c^{-20})^{1/5}$

**47.** $x^{2/3}(x^{1/3} + 4x^{4/3})$       **48.** $a^{1/4}(a^{4/3} - a^{1/2})$

**49.** $4t^{1/2}(2t^{1/2} + \frac{1}{2}t^{-1/2})$       **50.** $2s^{3/4}(s^{1/2} - 3s^{-3/4})$

**51.** $2c^{-1/3}(3c^{2/3} + 4c^{-2/3})$       **52.** $5p^{-1/4}(3p^{-3/4} - 6p^{1/4})$

**53.** $(x^{1/2} + y^{1/2})(x^{1/2} - y^{1/2})$       **54.** $(x^{1/2} + y^{-1/2})(x^{1/2} - y^{-1/2})$

**55.** $(x^{1/2} + y^{1/2})^2$       **56.** $(2x^{1/2} + 3y^{1/2})(3x^{1/2} - 4y^{1/2})$

*The following formulas involving fractional exponents are taken from various technical fields. Evaluate as indicated.*

**57.** $v = 14t^{2/5}$    Find $v$ when $t = 32$.

**58.** $w = 3t^{4/3}$    Find $w$ when $t = 8$.

**59.** $y = 4^{2t}$    Find $y$ when $t = \dfrac{5}{4}$.

**60.** $\phi = 0.3t^{5/4}$    Find $\phi$ when $t = 0.0001$.

**61.** $Q = \dfrac{bH^{3/2}}{3}$    Find $Q$ when $b = 12$ and $H = 16$.

**62.** $V = \dfrac{3R^{2/3}S^{1/2}}{2n}$    Find $V$ when $R = 8$, $S = 9$, and $n = 0.01$.

**63.** $H = \dfrac{4M^{3/4}N^{1/3}}{5Q}$   Find $H$ when $M = 16$, $N = 64$, and $Q = 0.04$.

**64.** $V = \dfrac{5D^{1/2}G^{2/5}}{2R^{3/4}}$   Find $V$ when $D = 36$, $G = 32$, and $R = 81$.

**65.** $f = \dfrac{2}{\pi}\left(\dfrac{3EIg}{w}\right)^{1/2} l^{-3/2}$   Find $f$ when $E = 4$, $I = 1$, $g = 32$, $w = 6$, and $l = 4$.

**66.** $J = (2 \times 10^{-6})\dfrac{E^{3/2}}{d^2}$   Find $J$ when $E = 9$ and $d = 2$.

## 8.3   SIMPLEST RADICAL FORM

In Section 2.2 we introduced the *n*th root of a number *a*, written $\sqrt[n]{a}$, but we limited our discussion mostly to square roots ($n = 2$). In Section 8.2 we saw the relationship between radicals and fractional exponents; that is, $\sqrt[n]{a} = a^{1/n}$, where *n* is a positive integer greater than one. Now we need to extend and build on our work with radicals.

Sometimes expressions involving rational or fractional exponents may be worked more easily in radical form. The rules in this section for radical expressions define the operations with expressions in radical form. These rules are extensions of and consistent with the laws of exponents as discussed in the previous section.

Let's begin our discussion of simplifying radicals with square roots. First, recall that the square root of a nonnegative real number is a nonnegative real number. Whenever variables are used with square roots, we must assume that they represent positive real numbers.

---

**SIMPLIFYING SQUARE ROOT QUANTITIES**

A quantity involving a square root is simplified when the following hold:

**1.** The quantity under the radical contains no perfect square factors.

**2.** The radicand contains no fractions.

**3.** The denominator of a fraction contains no radical expression.

---

The following properties are used to simplify square roots.

---

**1.** $\sqrt{ab} = \sqrt{a}\,\sqrt{b}$

**2.** $\sqrt{\dfrac{a}{b}} = \dfrac{\sqrt{a}}{\sqrt{b}}$   $(b \neq 0)$

*Note:* $a$ and $b$ are nonnegative real numbers.

---

**EXAMPLE 1**

Simplify $\sqrt{160}$.

$$\sqrt{160} = \sqrt{16 \cdot 10} \qquad \text{(Find the largest perfect square factor of 160.)}$$
$$= \sqrt{16}\,\sqrt{10}$$
$$= 4\sqrt{10}$$

**EXAMPLE 2**

Simplify $\sqrt{9a^3b^4c^7}$.

$$\sqrt{9a^3b^4c^7} = \sqrt{9 \cdot a^2 \cdot a \cdot b^4 \cdot c^6 \cdot c} \qquad \text{(Find each largest perfect square factor.)}$$
$$= \sqrt{9}\sqrt{a^2}\sqrt{b^4}\sqrt{c^6}\sqrt{ac}$$
$$= 3ab^2c^3\sqrt{ac}$$

**EXAMPLE 3**

Simplify $\sqrt{\dfrac{2}{5}}$.

$$\sqrt{\frac{2}{5}} = \sqrt{\frac{2}{5} \cdot \frac{5}{5}}$$

(When a square root appears in the denominator of a fractional quantity, multiply numerator and denominator by a quantity that makes the denominator a perfect square.)

$$= \sqrt{\frac{10}{25}}$$
$$= \frac{\sqrt{10}}{\sqrt{25}}$$
$$= \frac{\sqrt{10}}{5}$$

*Note:* The procedure of changing a fraction with a radical in the denominator to an equivalent one having no radical in the denominator is called **rationalizing the denominator.**

**EXAMPLE 4**

Simplify $\dfrac{\sqrt{14a^2c}}{\sqrt{24b^2}}$.

$$\frac{\sqrt{14a^2c}}{\sqrt{24b^2}} = \sqrt{\frac{14a^2c}{24b^2}}$$

$$= \sqrt{\frac{7a^2c}{12b^2} \cdot \frac{3}{3}}$$

(Multiply numerator and denominator of the radical by 3 to make the denominator a perfect square.)

$$= \sqrt{\frac{21a^2c}{36b^2}}$$
$$= \frac{a\sqrt{21c}}{6b}$$

**EXAMPLE 5**

Simplify $\dfrac{3}{\sqrt{24}}$.

$$\frac{3}{\sqrt{24}} = \frac{3}{\sqrt{2^3 \cdot 3}}$$

$$= \frac{3}{\sqrt{2^3 \cdot 3}} \cdot \frac{\sqrt{2 \cdot 3}}{\sqrt{2 \cdot 3}}$$

(Multiply numerator and denominator by $\sqrt{2 \cdot 3}$ to make the denominator a perfect square.)

$$= \frac{3\sqrt{6}}{\sqrt{2^4 \cdot 3^2}}$$

$$= \frac{3\sqrt{6}}{2^2 \cdot 3}$$

$$= \frac{\sqrt{6}}{4}$$

Next, recall that the cube root of a positive real number is positive, and the cube root of a negative real number is negative. Whenever variables are used with cube roots, we do not have to assume that they represent positive real numbers as we did with square roots.

---

**SIMPLIFYING CUBE ROOT QUANTITIES**

A quantity involving a cube root is simplified when the following hold:

**1.** The quantity under the radical contains no perfect cube factors.

**2.** The radicand contains no fractions.

**3.** The denominator of a fraction contains no radical expression.

---

The following properties are used to simplify cube roots.

---

**1.** $\sqrt[3]{ab} = \sqrt[3]{a}\,\sqrt[3]{b}$

**2.** $\sqrt[3]{\dfrac{a}{b}} = \dfrac{\sqrt[3]{a}}{\sqrt[3]{b}}$   $(b \neq 0)$

*Note:* $a$ and $b$ are any real numbers.

---

**EXAMPLE 6**

Simplify $\sqrt[3]{54}$.

$$\sqrt[3]{54} = \sqrt[3]{27 \cdot 2} \qquad \text{(Find the largest perfect cube factor.)}$$
$$= \sqrt[3]{27}\,\sqrt[3]{2}$$
$$= 3\sqrt[3]{2}$$

**EXAMPLE 7**

Simplify $\sqrt[3]{32}$.

$$\sqrt[3]{32} = \sqrt[3]{8 \cdot 4} \qquad \text{(Find the largest perfect cube factor.)}$$
$$= \sqrt[3]{8}\,\sqrt[3]{4}$$
$$= 2\sqrt[3]{4}$$

**EXAMPLE 8**

Simplify $\sqrt[3]{128a^5b^7c^9}$.

$$\sqrt[3]{128a^5b^7c^9} = \sqrt[3]{64 \cdot 2 \cdot a^3 \cdot a^2 \cdot b^6 \cdot b \cdot c^9} \qquad \text{(Find each largest perfect cube factor.)}$$
$$= \sqrt[3]{64}\,\sqrt[3]{a^3}\,\sqrt[3]{b^6}\,\sqrt[3]{c^9}\,\sqrt[3]{2a^2b}$$
$$= 4ab^2c^3\sqrt[3]{2a^2b}$$

**EXAMPLE 9**

Simplify $\dfrac{\sqrt[3]{2}}{\sqrt[3]{3}}$.

$$\dfrac{\sqrt[3]{2}}{\sqrt[3]{3}} = \sqrt[3]{\dfrac{2}{3}}$$

$$= \sqrt[3]{\dfrac{2}{3} \cdot \dfrac{9}{9}}$$

(When a cube root appears in the denominator of a fractional quantity, multiply numerator and denominator by a quantity that makes the denominator a perfect cube.)

$$= \sqrt[3]{\dfrac{18}{27}}$$

$$= \dfrac{\sqrt[3]{18}}{\sqrt[3]{27}}$$

$$= \dfrac{\sqrt[3]{18}}{3}$$

**EXAMPLE 10**

Simplify $\sqrt[3]{\dfrac{5a}{18b}}$.

$$\sqrt[3]{\dfrac{5a}{18b}} = \sqrt[3]{\dfrac{5a}{2 \cdot 3^2 \cdot b} \cdot \dfrac{2^2 \cdot 3 \cdot b^2}{2^2 \cdot 3 \cdot b^2}}$$

(Multiply numerator and denominator of the radical by $2^2 \cdot 3 \cdot b^2$ to make the denominator a perfect cube.)

$$= \sqrt[3]{\dfrac{60ab^2}{2^3 \cdot 3^3 \cdot b^3}}$$

$$= \dfrac{\sqrt[3]{60ab^2}}{6b}$$

**EXAMPLE 11**

Simplify $\dfrac{4}{\sqrt[3]{4}}$.

$$\dfrac{4}{\sqrt[3]{4}} = \dfrac{4}{\sqrt[3]{4}} \cdot \dfrac{\sqrt[3]{2}}{\sqrt[3]{2}}$$

(Multiply numerator and denominator by $\sqrt[3]{2}$ to make the denominator a perfect cube.)

$$= \dfrac{4\sqrt[3]{2}}{\sqrt[3]{8}}$$

$$= \dfrac{4\sqrt[3]{2}}{2}$$

$$= 2\sqrt[3]{2}$$

Next, let's extend our rules for radical expressions to any root as follows.

---

**OPERATIONS WITH RADICAL EXPRESSIONS**

**1.** $\sqrt[n]{a} \cdot \sqrt[n]{b} = \sqrt[n]{ab}$

**2.** $\dfrac{\sqrt[n]{a}}{\sqrt[n]{b}} = \sqrt[n]{\dfrac{a}{b}} \quad (b \neq 0)$

**3.** $\sqrt[m]{\sqrt[n]{a}} = \sqrt[mn]{a}$

**4.** $\sqrt[cn]{a^{cm}} = \sqrt[n]{a^m}$

---

where $n$, $m$, and $c$ are positive integers. It is important to remember that $a$ and $b$ must be positive real numbers if the index is even. If the index is odd, the rules are valid for all real values of $a$ and $b$.

Rules 3 and 4 may be shown using fractional exponents as follows:

**3.** $\sqrt[m]{\sqrt[n]{a}} = (a^{1/n})^{1/m} = a^{1/mn} = \sqrt[mn]{a}$

**4.** $\sqrt[cn]{a^{cm}} = a^{cm/cn} = a^{m/n} = \sqrt[n]{a^m}$

Solutions involving radicals are usually expressed in simplest form so that easy comparison of results can be made.

---

**SIMPLEST RADICAL FORM**

**1.** In a radical with index $n$, the radicand contains no factor with exponent greater than or equal to $n$.

**2.** No radical appears in the denominator of any fraction.

**3.** No fractions are under a radical sign.

**4.** The index of a radical is as small as possible.

---

**EXAMPLE 12**

Simplify $\sqrt[4]{32x^5}$.

$$\sqrt[4]{32x^5} = \sqrt[4]{16 \cdot 2 \cdot x^4 \cdot x} \qquad \text{(Find each largest perfect fourth power factor.)}$$
$$= \sqrt[4]{16 \cdot x^4} \sqrt[4]{2x}$$
$$= 2x\sqrt[4]{2x}$$

**EXAMPLE 13**

Simplify $\sqrt[4]{\dfrac{3b}{8a^6}}$.

$$\sqrt[4]{\frac{3b}{8a^6}} = \sqrt[4]{\frac{3b}{2^3 a^6} \cdot \frac{2a^2}{2a^2}} \qquad \text{(Multiply numerator and denominator by } 2a^2 \text{ to make the denominator a perfect fourth power.)}$$
$$= \sqrt[4]{\frac{6a^2 b}{2^4 a^8}}$$
$$= \frac{\sqrt[4]{6a^2 b}}{2a^2}$$

**EXAMPLE 14**

Change each to simplest radical form.

(a) $\sqrt[6]{8}$      (b) $\sqrt[4]{a^4 b^2}$      (c) $\sqrt{\sqrt[3]{5a}}$      (d) $\sqrt[3]{\sqrt[4]{64}}$

Such radicals may be simplified more easily if they are expressed with fractional exponents.

(a)    $\sqrt[6]{8} = 8^{1/6} = (2^3)^{1/6} = 2^{3/6} = 2^{1/2} = \sqrt{2}$

(b)    $\sqrt[4]{a^4 b^2} = (a^4 b^2)^{1/4} = a^{4/4} \cdot b^{2/4} = ab^{1/2} = a\sqrt{b}$

(c)    $\sqrt{\sqrt[3]{5a}} = ((5a)^{1/3})^{1/2} = (5a)^{1/6} = \sqrt[6]{5a}$

(d)    $\sqrt[3]{\sqrt[4]{64}} = ((64)^{1/4})^{1/3} = ((2^6)^{1/4})^{1/3} = 2^{6/12} = 2^{1/2} = \sqrt{2}$

Let's summarize the relationship between odd and even roots as follows.

Let $a$ be a real number and $n$ be a positive integer greater than one. In particular,

**1.** $\sqrt{a^2} = |a|$ and $\sqrt[3]{a^3} = a$.

In general,

**2.** if $n$ is an **even** positive integer,
$$\sqrt[n]{a^n} = |a|$$

and

**3.** if $n$ is an **odd** positive integer,
$$\sqrt[n]{a^n} = a.$$

## Exercises 8.3

*Simplify each radical expression. Assume that all variables represent positive real numbers.*

**1.** $\sqrt{49}$      **2.** $\sqrt{81}$      **3.** $\sqrt{75}$      **4.** $\sqrt{72}$

**5.** $\sqrt{180}$      **6.** $\sqrt{96}$      **7.** $\sqrt{8a^2}$      **8.** $\sqrt{63x^2}$

**9.** $\sqrt{72b^2}$      **10.** $\sqrt{45x^3}$      **11.** $\sqrt{80a^5b^2}$      **12.** $\sqrt{60a^3b^4}$

**13.** $\sqrt{32a^2b^4c^9}$      **14.** $\sqrt{108a^5b^{12}c^7}$      **15.** $\sqrt{\dfrac{3}{4}}$      **16.** $\sqrt{\dfrac{5}{16}}$

**17.** $\sqrt{\dfrac{5}{8}}$      **18.** $\sqrt{\dfrac{7}{12}}$      **19.** $\dfrac{\sqrt{6}}{\sqrt{10}}$      **20.** $\dfrac{\sqrt{8}}{\sqrt{12}}$

**21.** $\dfrac{5}{\sqrt{24}}$      **22.** $\dfrac{8}{\sqrt{50}}$      **23.** $\sqrt{\dfrac{4a}{15b^2}}$      **24.** $\sqrt{\dfrac{12a}{50b}}$

**25.** $\dfrac{2}{\sqrt{8b}}$      **26.** $\dfrac{2y}{\sqrt{20y}}$      **27.** $\dfrac{\sqrt{5a^4b}}{\sqrt{20a^2b^3}}$      **28.** $\dfrac{\sqrt{6x^2y^5}}{\sqrt{8x^5}}$

**29.** $\sqrt[3]{125}$      **30.** $\sqrt[3]{343}$      **31.** $\sqrt[3]{16a^4}$      **32.** $\sqrt[3]{135c^5}$

**33.** $\sqrt[3]{40a^8}$      **34.** $\sqrt[3]{250c^{10}}$      **35.** $\sqrt[3]{54x^5}$      **36.** $\sqrt[3]{64x^7}$

**37.** $\sqrt[3]{56x^7y^5z^3}$      **38.** $\sqrt[3]{72x^4y^6z^2}$      **39.** $\sqrt[3]{\dfrac{3}{8}}$      **40.** $\sqrt[3]{\dfrac{5}{27}}$

**41.** $\sqrt[3]{\dfrac{5}{4}}$      **42.** $\sqrt[3]{\dfrac{5}{9}}$      **43.** $\sqrt[3]{\dfrac{5}{12}}$      **44.** $\sqrt[3]{\dfrac{7}{24}}$

**45.** $\dfrac{1}{\sqrt[3]{2}}$      **46.** $\dfrac{4}{\sqrt[3]{18}}$      **47.** $\sqrt[3]{\dfrac{4a}{50b^2}}$      **48.** $\sqrt[3]{\dfrac{6a}{10a}}$

**49.** $\sqrt[3]{\dfrac{8a}{63a^2b^3}}$      **50.** $\sqrt[3]{\dfrac{5xy^4}{10x^2z}}$      **51.** $\dfrac{\sqrt[3]{5a^4b}}{\sqrt[3]{20a^2b^3}}$      **52.** $\dfrac{\sqrt[3]{6x^4y^5}}{\sqrt[3]{18x^2y^4}}$

**53.** $\sqrt[4]{80x^6}$      **54.** $\sqrt[4]{162a^8}$      **55.** $\sqrt[4]{25a^5b^4}$      **56.** $\sqrt[5]{64a^2b^7}$

**57.** $\sqrt[5]{\dfrac{32a^8b^{12}}{c^6}}$ **58.** $\sqrt[4]{\dfrac{3bc^5}{4a^3}}$ **59.** $\sqrt[4]{a^2}$ **60.** $\sqrt[6]{a^3b^6}$

**61.** $\sqrt[6]{27a^3b^9}$ **62.** $\sqrt[5]{32a^5b^{15}}$ **63.** $\sqrt{\sqrt{3}}$ **64.** $\sqrt{\sqrt[3]{5}}$

**65.** $\sqrt[3]{\sqrt{64}}$ **66.** $\sqrt{\sqrt{32}}$ **67.** $\sqrt[4]{64x^6}$ **68.** $\sqrt[9]{27a^6b^3c^{12}}$

**69.** $\sqrt{4a^2 - 4b^2}$ **70.** $\sqrt[3]{27a^6 + 27b^9}$ **71.** $\sqrt{4(a - b)^2}$ **72.** $\sqrt[3]{27(a^2 + b^3)^3}$

## 8.4 ADDITION AND SUBTRACTION OF RADICALS

Radical quantities, like algebraic terms, may be added or subtracted only if they are like terms. Two radical quantities are **like terms** when their radicands and indices are the same (written in simplest form).

**EXAMPLE 1**

Combine $2\sqrt{5} - 7\sqrt{5} + 3\sqrt{5}$.

$$2\sqrt{5} - 7\sqrt{5} + 3\sqrt{5} = -2\sqrt{5}$$

**EXAMPLE 2**

Combine $\sqrt{48} - \sqrt{27} + 5\sqrt{3}$.

$$\sqrt{48} - \sqrt{27} + 5\sqrt{3} = 4\sqrt{3} - 3\sqrt{3} + 5\sqrt{3} \qquad \left( \begin{array}{l} \sqrt{48} = \sqrt{16 \cdot 3} = 4\sqrt{3} \\ \sqrt{27} = \sqrt{9 \cdot 3} = 3\sqrt{3} \end{array} \right)$$
$$= 6\sqrt{3}$$

**EXAMPLE 3**

Combine $\sqrt{50} + 3\sqrt{\dfrac{1}{2}}$.

$$\sqrt{50} + 3\sqrt{\dfrac{1}{2}} = 5\sqrt{2} + \dfrac{3\sqrt{2}}{2} \qquad \left( \begin{array}{l} \sqrt{50} = \sqrt{25 \cdot 2} = 5\sqrt{2} \\ \sqrt{\dfrac{1}{2}} = \sqrt{\dfrac{1}{2}} \cdot \sqrt{\dfrac{2}{2}} = \dfrac{\sqrt{2}}{2} \end{array} \right)$$
$$= \dfrac{10\sqrt{2} + 3\sqrt{2}}{2}$$
$$= \dfrac{13\sqrt{2}}{2}$$

**EXAMPLE 4**

Combine $4\sqrt[3]{24} - \dfrac{3}{\sqrt[3]{9}} - \sqrt[4]{9}$.

$$= 4 \cdot 2\sqrt[3]{3} - \dfrac{3\sqrt[3]{3}}{3} - \sqrt{3} \qquad \left( \begin{array}{l} \sqrt[3]{24} = \sqrt[3]{8 \cdot 3} = 2\sqrt[3]{3} \\ \dfrac{1}{\sqrt[3]{9}} = \dfrac{1}{\sqrt[3]{9}} \cdot \sqrt[3]{\dfrac{3}{3}} = \dfrac{\sqrt[3]{3}}{\sqrt[3]{27}} = \dfrac{\sqrt[3]{3}}{3} \\ \sqrt[4]{9} = (3^2)^{1/4} = 3^{2/4} = 3^{1/2} = \sqrt{3} \end{array} \right)$$
$$= 8\sqrt[3]{3} - \sqrt[3]{3} - \sqrt{3}$$
$$= 7\sqrt[3]{3} - \sqrt{3}$$

*Note:* $\sqrt[3]{3}$ and $\sqrt{3}$ are not like terms and therefore may not be combined.

# Exercises 8.4

*Perform the indicated operations and simplify when possible.*

1. $3\sqrt{2} - 5\sqrt{2} + \sqrt{2}$
2. $6\sqrt{5} - \sqrt{5} + 7\sqrt{5}$
3. $\sqrt{8} - \sqrt{2}$
4. $\sqrt{3} - \sqrt{27}$
5. $3\sqrt{48} + 4\sqrt{75}$
6. $4\sqrt{32} - 2\sqrt{8}$
7. $5\sqrt{80} - 6\sqrt{45}$
8. $2\sqrt{63} + 4\sqrt{28}$
9. $4\sqrt{12} - \sqrt{27} - 3\sqrt{48}$
10. $2\sqrt{18} - \sqrt{72} + 3\sqrt{75}$
11. $\sqrt{32} + 3\sqrt{50} - 6\sqrt{18}$
12. $4\sqrt{50} - 5\sqrt{72} + 3\sqrt{8}$
13. $3\sqrt{27} - 9\sqrt{48} + \sqrt{75}$
14. $3\sqrt{5} - 6\sqrt{45} - 2\sqrt{20}$
15. $2\sqrt{54} + 3\sqrt{24} - 3\sqrt{96}$
16. $2\sqrt{90} + 3\sqrt{40} - \sqrt{160}$
17. $5\sqrt{2x} - \sqrt{18x} + \sqrt{8x}$
18. $2\sqrt{18c} - \sqrt{72c} + 3\sqrt{50c}$
19. $4\sqrt{32x^2} - \sqrt{218x^2} + 3\sqrt{50x^2}$
20. $6\sqrt{27t^2} - 5\sqrt{108t^2} - 4\sqrt{48t^2}$
21. $3\sqrt[3]{6} - 7\sqrt[3]{6} + \sqrt[3]{6}$
22. $4\sqrt[3]{5} - 6\sqrt[3]{5} + 10\sqrt[3]{5}$
23. $2\sqrt[3]{54} - 2\sqrt[3]{16}$
24. $2\sqrt[3]{81} + 3\sqrt[3]{24}$
25. $\dfrac{2\sqrt[3]{40} + 8\sqrt[3]{5}}{4}$
26. $\dfrac{3\sqrt[3]{32} + 4\sqrt[3]{108}}{6}$
27. $\sqrt[3]{81} + 3\sqrt[3]{3} - \sqrt{3}$
28. $\sqrt[3]{16} + 3\sqrt[3]{2} - \sqrt{2}$
29. $\sqrt[3]{54} - 4\sqrt[3]{16} + \sqrt[3]{128}$
30. $\sqrt[3]{48} - 2\sqrt[3]{162} + \sqrt[3]{750}$
31. $\dfrac{5}{\sqrt{2}} - \dfrac{\sqrt{2}}{2}$
32. $\dfrac{\sqrt{5}}{10} + \dfrac{3}{\sqrt{5}}$
33. $\dfrac{3\sqrt{3}}{2} + \dfrac{2}{\sqrt{3}}$
34. $\dfrac{6\sqrt{2}}{5} + \dfrac{2}{\sqrt{2}}$
35. $\sqrt{\dfrac{2}{3}} + \sqrt{\dfrac{1}{6}} - \sqrt{24}$
36. $\sqrt{\dfrac{5}{8}} + \sqrt{40} - \sqrt{\dfrac{2}{5}}$
37. $\sqrt{3} - \dfrac{1}{\sqrt{3}}$
38. $\dfrac{\sqrt{6}}{2} + \dfrac{1}{\sqrt{6}}$
39. $\dfrac{\sqrt{18}}{6} + \dfrac{\sqrt{2}}{2} + \sqrt{\dfrac{1}{2}}$
40. $\sqrt{\dfrac{1}{3}} + \dfrac{\sqrt{3}}{3} - \dfrac{\sqrt{12}}{6}$
41. $\sqrt{10} - \sqrt{\dfrac{2}{5}}$
42. $\sqrt{\dfrac{7}{3}} + \sqrt{21}$
43. $4\sqrt[3]{3} - \dfrac{6}{\sqrt[3]{9}}$
44. $-2\sqrt[3]{2} + \dfrac{5}{\sqrt[3]{4}}$
45. $\sqrt[3]{\dfrac{3}{4}} - \sqrt[3]{\dfrac{2}{9}} - \dfrac{1}{\sqrt[3]{6}}$
46. $\sqrt[3]{\dfrac{3}{2}} - \sqrt[3]{\dfrac{4}{9}} - \sqrt[6]{144}$
47. $\sqrt{\dfrac{1}{8}} - \sqrt{50} + \sqrt[4]{\dfrac{1}{4}}$
48. $\sqrt[4]{36} - \dfrac{1}{\sqrt[4]{36}} + \sqrt{54}$
49. $\sqrt{50ax^3} + \sqrt{72a^3x^3} - \sqrt{8a^5x}$
50. $\sqrt{27x^3} - \sqrt{48x^3} - \sqrt{3x^5}$
51. $\sqrt[3]{16x^4} + \sqrt[3]{54x^7} + \sqrt[3]{250x}$
52. $\sqrt[3]{a^4b^2} - \sqrt[3]{ab^5} + \sqrt[3]{ab^2}$
53. $\sqrt{\dfrac{a}{6b}} + \sqrt{\dfrac{6b}{a}} - \sqrt{\dfrac{2}{3ab}}$
54. $\sqrt{\dfrac{a}{2}} + \dfrac{1}{\sqrt{2a}} - \sqrt{32a^3}$

**55.** $\sqrt[3]{\dfrac{a}{b^2}} + \sqrt[3]{\dfrac{b}{a^2}} - \dfrac{1}{\sqrt[3]{a^2 b^2}}$ 

**56.** $\sqrt[3]{\dfrac{a^2 b}{9}} - \sqrt[3]{\dfrac{3a^2}{b^2}} + \sqrt[3]{\dfrac{3b}{a}} - \sqrt[3]{24 a^5 b^7}$

## 8.5  MULTIPLICATION AND DIVISION OF RADICALS

When two radicals have the same index, the radicands may be multiplied or divided by using the laws of radicals given in Section 8.3.

$$\sqrt[n]{a} \cdot \sqrt[n]{b} = \sqrt[n]{ab}$$
$$\frac{\sqrt[n]{a}}{\sqrt[n]{b}} = \sqrt[n]{\frac{a}{b}}$$

Do you recall the restrictions on $a$, $b$, and $n$?

**EXAMPLE 1**

Multiply and simplify $\sqrt[4]{4} \cdot \sqrt[4]{8}$.

$$\begin{aligned}
\sqrt[4]{4} \cdot \sqrt[4]{8} &= \sqrt[4]{4 \cdot 8} \\
&= \sqrt[4]{32} \\
&= \sqrt[4]{16 \cdot 2} \\
&= 2\sqrt[4]{2}
\end{aligned}$$

**EXAMPLE 2**

Divide and simplify $\dfrac{\sqrt[3]{54}}{\sqrt[3]{2}}$.

$$\begin{aligned}
\frac{\sqrt[3]{54}}{\sqrt[3]{2}} &= \sqrt[3]{\frac{54}{2}} \\
&= \sqrt[3]{27} \\
&= 3
\end{aligned}$$

When two radicals have different indices, it is necessary to change them so that they have a common index before the product or the quotient may be written as a single radical. Expressing the radicals in fractional exponent form is especially useful here. We then can express the exponents with a common denominator.

**EXAMPLE 3**

Multiply and simplify $\sqrt[3]{4} \cdot \sqrt{6}$.

$$\begin{aligned}
\sqrt[3]{4} \cdot \sqrt{6} &= 4^{1/3} \cdot 6^{1/2} \\
&= 4^{2/6} \cdot 6^{3/6} \\
&= \sqrt[6]{4^2} \cdot \sqrt[6]{6^3} \\
&= \sqrt[6]{4^2 \cdot 6^3} \\
&= \sqrt[6]{(2^2)^2 (2 \cdot 3)^3}
\end{aligned}$$

$$= \sqrt[6]{2^4 \cdot 2^3 \cdot 3^3}$$
$$= \sqrt[6]{2^6 \cdot 2 \cdot 3^3}$$
$$= 2\sqrt[6]{2 \cdot 3^3}$$
$$= 2\sqrt[6]{54}$$

**EXAMPLE 4**

Divide and simplify $\dfrac{\sqrt{12}}{\sqrt[4]{3}}$.

$$\frac{\sqrt{12}}{\sqrt[4]{3}} = \frac{12^{1/2}}{3^{1/4}}$$
$$= \frac{12^{2/4}}{3^{1/4}}$$
$$= \frac{\sqrt[4]{12^2}}{\sqrt[4]{3}}$$
$$= \sqrt[4]{\frac{144}{3}}$$
$$= \sqrt[4]{48}$$
$$= \sqrt[4]{16 \cdot 3}$$
$$= 2\sqrt[4]{3}$$

To multiply expressions containing radicals, use the same procedures that we used to multiply algebraic expressions in Chapter 2.

**EXAMPLE 5**

Expand and simplify $(5 + 2\sqrt{2})(1 - 3\sqrt{2})$.

$$(5 + 2\sqrt{2})(1 - 3\sqrt{2}) = 5 - 13\sqrt{2} - 6(\sqrt{2})^2$$
$$= 5 - 13\sqrt{2} - 12$$
$$= -7 - 13\sqrt{2}$$

**EXAMPLE 6**

Expand and simplify $(2 + 3\sqrt{6})^2$.

$$(2 + 3\sqrt{6})^2 = 4 + 12\sqrt{6} + (3\sqrt{6})^2$$
$$= 4 + 12\sqrt{6} + 54$$
$$= 58 + 12\sqrt{6}$$

When the denominator is in the form $a + \sqrt{b}$ or $a - \sqrt{b}$, we need to find an expression which, when multiplied by the denominator, will give a product free of radicals. For example, by what do we multiply numerator and denominator in $\dfrac{4}{1 + \sqrt{3}}$ to rationalize the denominator? How about $\sqrt{3}$?

$$\frac{4}{1 + \sqrt{3}} \cdot \frac{\sqrt{3}}{\sqrt{3}} = \frac{4\sqrt{3}}{\sqrt{3} + 3} \qquad \text{No!}$$

How about $1 + \sqrt{3}$?

$$\frac{4}{1 + \sqrt{3}} \cdot \frac{1 + \sqrt{3}}{1 + \sqrt{3}} = \frac{4 + 4\sqrt{3}}{1 + 2\sqrt{3} + (\sqrt{3})^2} = \frac{4 + 4\sqrt{3}}{4 + 2\sqrt{3}} \quad \text{No!}$$

How about $1 - \sqrt{3}$?

$$\frac{4}{1 + \sqrt{3}} \cdot \frac{1 - \sqrt{3}}{1 - \sqrt{3}} = \frac{4 - 4\sqrt{3}}{1 - (\sqrt{3})^2} = \frac{4 - 4\sqrt{3}}{-2} = -2 + 2\sqrt{3} \quad \text{Yes!}$$

In general, to rationalize a denominator in the form $a + \sqrt{b}$, multiply numerator and denominator by $a - \sqrt{b}$. To rationalize a denominator in the form $a - \sqrt{b}$, multiply numerator and denominator by $a + \sqrt{b}$.

The expressions $a + \sqrt{b}$ and $a - \sqrt{b}$ are called **conjugates.** Specific examples of conjugates are $1 + \sqrt{3}$ and $1 - \sqrt{3}$, $2 - 3\sqrt{5}$ and $2 + 3\sqrt{5}$, $4 + 3\sqrt{a}$ and $4 - 3\sqrt{a}$, and $2\sqrt{5} - \sqrt{3}$ and $2\sqrt{5} + \sqrt{3}$.

**EXAMPLE 7**

Simplify $\dfrac{4 + \sqrt{3}}{2 - \sqrt{3}}$.

$$\frac{4 + \sqrt{3}}{2 - \sqrt{3}} \cdot \frac{2 + \sqrt{3}}{2 + \sqrt{3}} = \frac{8 + 6\sqrt{3} + 3}{4 - 3} \quad \text{(The conjugate of } 2 - \sqrt{3} \text{ is } 2 + \sqrt{3}.)$$

$$= 11 + 6\sqrt{3}$$

**EXAMPLE 8**

Simplify $\dfrac{3\sqrt{5} - \sqrt{2}}{2\sqrt{5} + 3\sqrt{2}}$.

$$\frac{3\sqrt{5} - \sqrt{2}}{2\sqrt{5} + 3\sqrt{2}} \cdot \frac{2\sqrt{5} - 3\sqrt{2}}{2\sqrt{5} - 3\sqrt{2}} = \frac{30 - 11\sqrt{10} + 6}{20 - 18}$$

$$= \frac{36 - 11\sqrt{10}}{2}$$

## Exercises 8.5

*Perform the indicated operations and simplify when possible.*

**1.** $\sqrt{3} \cdot \sqrt{6}$

**2.** $\sqrt{12} \cdot \sqrt{15}$

**3.** $\sqrt[3]{4} \cdot \sqrt[3]{18}$

**4.** $\sqrt[3]{16} \cdot \sqrt[3]{54}$

**5.** $\dfrac{\sqrt[4]{36}}{\sqrt[4]{9}}$

**6.** $\dfrac{\sqrt[3]{32}}{\sqrt[3]{4}}$

**7.** $\sqrt[3]{2} \cdot \sqrt{5}$

**8.** $\sqrt{2} \cdot \sqrt[4]{2}$

**9.** $\sqrt{2} \cdot \sqrt[3]{4}$

**10.** $\sqrt[3]{4} \cdot \sqrt[6]{8}$

**11.** $\sqrt{27} \cdot \sqrt[3]{9}$

**12.** $\sqrt{12} \cdot \sqrt[4]{9}$

**13.** $\dfrac{\sqrt{3}}{\sqrt[3]{3}}$

**14.** $\dfrac{\sqrt[4]{64}}{\sqrt[3]{4}}$

**15.** $\dfrac{2\sqrt[4]{3}}{3\sqrt{2}}$

**16.** $\dfrac{2\sqrt{3}}{6\sqrt[4]{2}}$

**17.** $(2 + \sqrt{5})(1 - \sqrt{5})$

**18.** $(6 - \sqrt{3})(5 - \sqrt{3})$

**19.** $(5 - \sqrt{3})(5 + \sqrt{3})$

**20.** $(6 + 2\sqrt{5})(6 - 2\sqrt{5})$

**21.** $(3\sqrt{3} + 2)(4\sqrt{3} + 3)$

**22.** $(2\sqrt{3} + 2)(3\sqrt{3} - 1)$

**23.** $(a + \sqrt{b})(2a - 4\sqrt{b})$

**24.** $(3a + \sqrt{ab})(2a + 3\sqrt{ab})$

**25.** $(2\sqrt{7} + 2\sqrt{3})(-2\sqrt{7} - \sqrt{3})$

**26.** $(4\sqrt{5} - \sqrt{2})(\sqrt{5} - 4\sqrt{2})$

**27.** $(2\sqrt{a} - 3\sqrt{b})(\sqrt{a} + 2\sqrt{b})$

**28.** $(3\sqrt{a} - 4\sqrt{b})(\sqrt{a} - \sqrt{b})$

**29.** $(2 - \sqrt{3})^2$

**30.** $(4 + 2\sqrt{5})^2$

**31.** $(2\sqrt{3} - \sqrt{5})^2$

**32.** $(3\sqrt{5} - 2\sqrt{7})^2$

**33.** $(a + \sqrt{b})^2$

**34.** $(3a + 2\sqrt{b})^2$

**35.** $(2\sqrt{a} + 3\sqrt{b})^2$

**36.** $(4\sqrt{a} - 2\sqrt{b})^2$

**37.** $\dfrac{\sqrt{2}}{3 + \sqrt{2}}$

**38.** $\dfrac{-1}{2 - \sqrt{3}}$

**39.** $\dfrac{2 + \sqrt{5}}{3 - 2\sqrt{5}}$

**40.** $\dfrac{4 - \sqrt{6}}{2 + 4\sqrt{6}}$

**41.** $\dfrac{4\sqrt{6} + 2\sqrt{3}}{2\sqrt{6} + 5\sqrt{3}}$

**42.** $\dfrac{2\sqrt{3} - \sqrt{2}}{3\sqrt{3} - 2\sqrt{2}}$

**43.** $\sqrt{3 + \sqrt{5}} \cdot \sqrt{3 - \sqrt{5}}$

**44.** $\sqrt{4 - \sqrt{6}} \cdot \sqrt{4 + \sqrt{6}}$

**45.** $\dfrac{\sqrt{3 + \sqrt{5}}}{\sqrt{3 - \sqrt{5}}}$

**46.** $\dfrac{\sqrt{6 - \sqrt{11}}}{\sqrt{6 + \sqrt{11}}}$

**47.** $\dfrac{a}{a + \sqrt{b}}$

**48.** $\dfrac{\sqrt{a} + \sqrt{b}}{\sqrt{a} - \sqrt{b}}$

## 8.6   EQUATIONS WITH RADICALS

Equations with radicals are usually solved by first raising each side of the equation to the same power. For equations with square roots, we would square each side; for equations with cube roots, we would cube each side; and so on.

This process of raising each side to a power has a risk involved; that is, we run the risk of introducing extraneous roots. Recall that an extraneous root is one that is introduced by means of an equation-solving procedure, but it does not check in the original equation. The only way to discard or cull out the extraneous solutions is by checking. For this reason, it is necessary that *all solutions be checked in the original equation.*

The potential for introducing extraneous roots may be shown using the following very simple equation:

$$x = 3$$

Square each side.

$$x^2 = 9$$

Solve the resulting equation.

$$x = \pm 3$$

By checking both solutions in the *original* equation, we see that we have introduced the extraneous root, $-3$, by using the equation-solving procedure of squaring each side.

**EXAMPLE 1**

Solve $\sqrt{x + 5} = 4$.

$$\sqrt{x + 5} = 4$$
$$x + 5 = 16 \qquad \text{(Square each side.)}$$
$$x = 11$$

*Check:*

$$\sqrt{(11) + 5} = 4$$
$$\sqrt{16} = 4$$
$$4 = 4$$

Therefore, 11 is a solution.

**EXAMPLE 2**

Solve $\sqrt[3]{2x - 5} = 5$.

$$\sqrt[3]{2x - 5} = 5$$
$$2x - 5 = 125 \qquad \text{(Cube each side.)}$$
$$2x = 130$$
$$x = 65$$

*Check:*

$$\sqrt[3]{2(65) - 5} = 5$$
$$\sqrt[3]{130 - 5} = 5$$
$$\sqrt[3]{125} = 5$$
$$5 = 5$$

Therefore, 65 is a solution.

**EXAMPLE 3**

Solve $\sqrt{8x + 17} - 2x = 3$.

The radical must be isolated on one side before each side is squared.

$$\sqrt{8x + 17} = 2x + 3$$
$$8x + 17 = 4x^2 + 12x + 9 \qquad \text{(Square each side.)}$$
$$0 = 4x^2 + 4x - 8$$
$$0 = x^2 + x - 2$$
$$0 = (x + 2)(x - 1)$$
$$x + 2 = 0 \qquad \text{or} \qquad x - 1 = 0$$
$$x = -2 \qquad \text{or} \qquad x = 1$$

*Check:*

$$\sqrt{8(-2) + 17} - 2(-2) = 3 \qquad \qquad \sqrt{8(1) + 17} - 2(1) = 3$$
$$\sqrt{-16 + 17} + 4 = 3 \qquad \qquad \qquad \sqrt{25} - 2 = 3$$
$$5 \neq 3 \qquad \qquad \qquad \qquad \qquad 3 = 3$$

Therefore, the apparent root,     Therefore, 1 is a solution.
$-2$, is extraneous and is not
a solution.

**EXAMPLE 4**

Solve $\sqrt{x + 16} - \sqrt{x - 4} = 2$.

This type of radical equation can be solved most easily by first placing one radical on each side of the equation. The remaining steps are as follows:

$$\sqrt{x + 16} - \sqrt{x - 4} = 2$$
$$\sqrt{x + 16} = 2 + \sqrt{x - 4}$$
$$x + 16 = 4 + 4\sqrt{x - 4} + x - 4 \qquad \text{(Square each side.)}$$
$$16 = 4\sqrt{x - 4} \qquad \text{(Simplify.)}$$
$$4 = \sqrt{x - 4} \qquad \text{(Divide each side by 4.)}$$
$$16 = x - 4 \qquad \text{(Square each side again.)}$$
$$20 = x$$

*Check:*

$$\sqrt{(20) + 16} - \sqrt{(20) - 4} = 2$$
$$\sqrt{36} - \sqrt{16} = 2$$
$$6 - 4 = 2$$
$$2 = 2$$

Therefore, 20 is a solution.

## Exercises 8.6

*Solve and check.*

**1.** $\sqrt{x - 4} = 7$

**2.** $4\sqrt{x - 2} = 22$

**3.** $\sqrt{3 - x} = 2x - 3$

**4.** $\sqrt{3x + 4} = x$

**5.** $\sqrt[3]{3 - x} = 5$

**6.** $3\sqrt[3]{x + 1} = 12$

**7.** $\sqrt[4]{2x - 3} = 3$

**8.** $\sqrt[5]{x - 1} = 2$

**9.** $\sqrt{x + 6} = x$

**10.** $\sqrt{x + 1} - 4 = 7 - x$

**11.** $\sqrt{x^2 + 2x + 6} = x + 2$

**12.** $\sqrt{x^2 - 4x + 29} = 2x + 1$

**13.** $\sqrt{a^2 - 5a + 20} - 4a = 0$

**14.** $\sqrt{t^2 + 10t + 12} - 3t = 0$

**15.** $\sqrt{m^2 + 3m} - 3m + 1 = 0$

**16.** $\sqrt{s^2 - 4s + 28} - 4s - 7 = 0$

**17.** $\sqrt{3x^2 + 4x + 2} = \sqrt{x^2 + x + 11}$

**18.** $\sqrt{4x^2 + 2x + 9} = \sqrt{3x^2 - 2x + 5}$

**19.** $\sqrt[3]{3x - 4} = \sqrt[3]{5x + 8}$

**20.** $\sqrt[4]{x^2 + x} = \sqrt{x + 1}$

**21.** $\sqrt{x + 6} - \sqrt{x} = 4$

**22.** $\sqrt{x + 7} + \sqrt{x + 4} = 3$

**23.** $\sqrt{x + 9} - \sqrt{x + 2} = \sqrt{4x - 27}$

**24.** $\sqrt{5x + 1} + \sqrt{3x + 4} = \sqrt{16x + 9}$

**25.** $\sqrt{13 + \sqrt{x}} = \sqrt{x} + 1$

**26.** $\sqrt{4 + \sqrt{x}} = \sqrt{x} - 2$

*Given each formula, solve for the indicated letter.*

**27.** $v = \sqrt{v_0^2 - 2gh}$   for $h$

**28.** $v = \sqrt{\dfrac{2GM}{r}}$   for $r$

**29.** $P = 2\pi\sqrt{\dfrac{l}{g}}$   for $l$

**30.** $v = \sqrt{\dfrac{RT}{mN}}$   for $m$

**31.** $T = 2\pi\sqrt{\dfrac{R_2 C}{R_1 + R_2}}$   for $C$

**32.** $T = 2\pi\sqrt{\dfrac{R_2 C}{R_1 + R_2}}$   for $R_2$

**33.** $v = \sqrt{Gm\left(\dfrac{1}{D} - \dfrac{1}{r}\right)}$   for $m$

**34.** $v = \sqrt{Gm\left(\dfrac{1}{D} - \dfrac{1}{r}\right)}$   for $r$

**35.** Find the lengths of the sides of the right triangle in Fig. 8.1.

**36.** Find the base and the height of the triangle in Fig. 8.2 if its area is 10.

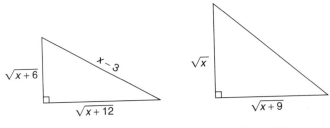

**Figure 8.1**          **Figure 8.2**

## 8.7   EQUATIONS IN QUADRATIC FORM

Equations in the form $ax^{2n} + bx^n + c = 0$ are said to be in **quadratic form.** The following equations are in quadratic form:

$$x^4 - 9x^2 + 20 = 0$$
$$x^{-4} - 20x^{-2} + 64 = 0$$
$$(x + 1)^{2/3} + 3(x + 1)^{1/3} - 4 = 0$$
$$x - 2\sqrt{x} - 3 = 0$$

Such equations may be solved using any of the quadratic methods studied earlier.

**EXAMPLE 1**

Solve $x^4 - 9x^2 + 20 = 0$.

$$x^4 - 9x^2 + 20 = 0$$
$$(x^2 - 5)(x^2 - 4) = 0 \qquad (\text{Factor.})$$

$$x^2 - 5 = 0 \qquad \text{or} \quad x^2 - 4 = 0$$
$$x^2 = 5 \qquad\qquad\qquad x^2 = 4$$
$$x = \pm\sqrt{5} \qquad\qquad\quad x = \pm 2$$

**EXAMPLE 2**

Solve $x^{-4} - 20x^{-2} + 64 = 0$.

$$x^{-4} - 20x^{-2} + 64 = 0$$
$$(x^{-2} - 4)(x^{-2} - 16) = 0 \qquad (\text{Factor.})$$

$$x^{-2} - 4 = 0 \quad \text{or} \quad x^{-2} - 16 = 0$$
$$x^{-2} = 4 \qquad\qquad\quad x^{-2} = 16$$
$$\dfrac{1}{x^2} = 4 \qquad\qquad\quad \dfrac{1}{x^2} = 16 \qquad \left(x^{-2} = \dfrac{1}{x^2}\right)$$

$$x^2 = \frac{1}{4} \qquad\qquad x^2 = \frac{1}{16}$$

$$x = \pm\frac{1}{2} \qquad\qquad x = \pm\frac{1}{4}$$

Sometimes a substitution may be helpful.

### EXAMPLE 3

Solve $(x + 1)^{2/3} + 3(x + 1)^{1/3} - 4 = 0$.

If we let $p = x + 1$, then

$$(x + 1)^{2/3} + 3(x + 1)^{1/3} - 4 = 0$$
$$p^{2/3} + 3p^{1/3} - 4 = 0$$
$$(p^{1/3} + 4)(p^{1/3} - 1) = 0 \qquad (\text{Factor.})$$
$$p^{1/3} + 4 = 0 \quad \text{or} \quad p^{1/3} - 1 = 0$$
$$p^{1/3} = -4 \qquad\qquad p^{1/3} = 1$$
$$p = -64 \qquad\qquad p = 1$$

Recall that $p = x + 1$. Thus,

$$x + 1 = -64 \qquad x + 1 = 1$$
$$x = -65 \qquad\quad x = 0$$

If the equation does not factor, as before we can use the quadratic formula.

### EXAMPLE 4

Solve $3x^4 - 8x^2 + 2 = 0$.

*Note:* This is a quadratic equation in $x^2$.

$$x^2 = \frac{-b \pm \sqrt{b^2 - 4ac}}{2a}$$

$$x^2 = \frac{-(-8) \pm \sqrt{(-8)^2 - 4(3)(2)}}{2(3)}$$

$$x^2 = \frac{8 \pm \sqrt{64 - 24}}{6}$$

$$x^2 = \frac{8 \pm 2\sqrt{10}}{6} = \frac{4 \pm \sqrt{10}}{3}$$

Therefore,

$$x = \pm\sqrt{\frac{4 \pm \sqrt{10}}{3}} \quad \text{or} \quad \pm 1.55, \pm 0.528$$

## Exercises 8.7

*Solve each equation.*

**1.** $x^4 - 11x^2 + 18 = 0$

**2.** $x^4 - 10x^2 + 24 = 0$

**3.** $x^{-4} - 17x^{-2} + 16 = 0$

**4.** $x^{-4} - 13x^{-2} + 36 = 0$

**5.** $(x + 2)^2 + 3(x + 2) + 2 = 0$

**6.** $(x + 1)^{-2} + 8(x + 1)^{-1} + 15 = 0$

**7.** $(x - 1)^4 - 5(x - 1)^2 + 4 = 0$

**8.** $(2x - 5)^4 - (2x - 5)^2 = 0$

**9.** $x - 2\sqrt{x} - 3 = 0$

**10.** $4x - 4\sqrt{x} + 1 = 0$

**11.** $(3x + 2)^{-4} - 1 = 0$

**12.** $(2x - 1)^4 - 9 = 0$

**13.** $x^{2/3} + 2x^{1/3} - 8 = 0$

**14.** $(x - 1)^{2/3} - 5(x - 1)^{1/3} + 6 = 0$

**15.** $x^4 - 3x^2 + 1 = 0$

**16.** $3x^4 - 6x^2 + 2 = 0$

## CHAPTER SUMMARY

**1.** *Laws of exponents:*
  (a) $a^m \cdot a^n = a^{m+n}$
  (b) $\dfrac{a^m}{a^n} = a^{m-n}$  $(a \neq 0)$
  (c) $(a^m)^n = a^{mn}$
  (d) $(ab)^n = a^n b^n$
  (e) $\left(\dfrac{a}{b}\right)^n = \dfrac{a^n}{b^n}$  $(b \neq 0)$
  (f) $a^0 = 1$  $(a \neq 0)$
  (g) $a^{-m} = \dfrac{1}{a^m}$  $(a \neq 0)$

**2.** *Fractional exponents:*
  (a) $a^{1/n} = \sqrt[n]{a}$
  (b) $a^{m/n} = \sqrt[n]{a^m} = (\sqrt[n]{a})^m$
  *Note: m is an integer, n is a positive integer, and a is a real number. If n is even, $a \geq 0$.*

**3.** *Operations with radical expressions:*
  (a) $\sqrt[n]{a} \cdot \sqrt[n]{b} = \sqrt[n]{ab}$
  (b) $\dfrac{\sqrt[n]{a}}{\sqrt[n]{b}} = \sqrt[n]{\dfrac{a}{b}}$  $(b \neq 0)$
  (c) $\sqrt[m]{\sqrt[n]{a}} = \sqrt[mn]{a}$
  (d) $\sqrt[cn]{a^{cm}} = \sqrt[n]{a^m}$

**4.** *Simplest radical form:*
  (a) In a radical with index $n$, the radicand contains no factor with exponent greater than or equal to $n$.
  (b) No radical appears in the denominator of any fraction.
  (c) No fractions are under a radical sign.
  (d) The index of a radical is as small as possible.

**5.** The expressions $a + \sqrt{b}$ and $a - \sqrt{b}$ are **conjugates.**

**6.** An equation in the form $ax^{2n} + bx^n + c = 0$ is in **quadratic form.**

# CHAPTER 8 REVIEW

*Simplify and rewrite using only positive exponents.*

**1.** $3a^{-2}$

**2.** $(2a)^{-4}$

**3.** $a^{-5} \cdot a^{10}$

**4.** $(a^{-2})^{-4}$

**5.** $\dfrac{a^3 b^0 c^{-3}}{a^5 b^{-2} c^{-9}}$

**6.** $a^{-2} + a^{-1}$

*Evaluate.*

**7.** $49^{1/2}$

**8.** $16^{3/2}$

**9.** $8^{-2/3}$

*Perform the indicated operations. Simplify and express with positive exponents.*

**10.** $(x^{2/3} y^{-1/3})^{-3/5}$

**11.** $\dfrac{x^{-2/3}}{x^{-1/4}}$

**12.** $(x^{-3/4})^{-2/3}$

**13.** Evaluate $w = 2t^{-2/3}$ for $t = 27$.

*Simplify.*

**14.** $\sqrt{80}$

**15.** $\sqrt{72a^2 b^3}$

**16.** $\sqrt{80x^5 y^6}$

**17.** $\sqrt{48x^3 y}$

**18.** $\sqrt{\dfrac{5}{54}}$

**19.** $\sqrt{\dfrac{45a^2}{7b^4}}$

**20.** $\sqrt[3]{250}$

**21.** $\sqrt[3]{108a^4 b^2}$

**22.** $\sqrt[3]{256a^5 b^{10}}$

**23.** $\sqrt[3]{\dfrac{9a}{20b^2}}$

**24.** $\dfrac{8}{\sqrt[3]{40}}$

**25.** $\sqrt[4]{9a^2}$

**26.** $\dfrac{6}{\sqrt{12a}}$

**27.** $\dfrac{6}{\sqrt[3]{12a}}$

**28.** $\sqrt{\sqrt{5}}$

**29.** $\sqrt[4]{\sqrt[3]{10}}$

*Perform the indicated operations and simplify when possible.*

**30.** $\sqrt{12} + \sqrt{27} - \sqrt[4]{9}$

**31.** $\sqrt{\dfrac{2}{5}} + \dfrac{1}{\sqrt{10}} - \sqrt{40}$

**32.** $\sqrt[3]{54} - \sqrt[3]{250} + \sqrt[3]{16}$

**33.** $2\sqrt[3]{\dfrac{3}{4}} - \dfrac{3}{\sqrt[3]{36}} + 5\sqrt[3]{\dfrac{2}{9}}$

**34.** $\sqrt{8} \cdot \sqrt{20}$

**35.** $\sqrt{3} \cdot \sqrt[4]{9}$

**36.** $\sqrt{2} \cdot \sqrt[3]{4}$

**37.** $(2 + \sqrt{3})(-4 - \sqrt{3})$

**38.** $(4 - 3\sqrt{5})^2$

**39.** $\dfrac{\sqrt{3}}{2 - 3\sqrt{3}}$

*Solve each equation.*

**40.** $\sqrt{x + 2} = 8$

**41.** $\sqrt[3]{x} + 2 = -1$

**42.** $\sqrt{x - 5} + \sqrt{x} = 5$

**43.** $\sqrt{x + 9} = \sqrt[4]{x^2 + 9x}$

**44.** $4x^4 - 41x^2 + 45 = 0$

**45.** $x^{2/3} - 2x^{1/3} - 8 = 0$

# 9
# Exponentials and Logarithms

## INTRODUCTION

A construction crew has placed flashing caution lights around its site and has timed them to operate 24 hours per day. A battery for each bank of lights diminishes in power by 0.65% for each hour it is operated. The lights will cease to operate when the battery power is reduced to $\frac{1}{10}$ of its original power. How often will the batteries need to be changed?

We use an **exponential function** to calculate that the batteries will last about 354 hours. Thus, they need to be changed every 14.7 days, or for practical purposes, every 2 weeks. Exponential functions are also used to express rates of growth and decay for items such as money, bacteria, and populations. The inverses of such functions, called **logarithms,** are used in formulas for decibels and pH values and in determining flow of heat in an insulated pipe.

In this chapter, we learn the rules of operating with exponential and logarithmic functions.

## Objectives

- Graph exponential functions.
- Graph logarithmic functions.
- Use a calculator to evaluate exponential and logarithmic functions.
- Use the properties of logarithms.
- Evaluate expressions containing logarithms and exponential functions.
- Solve exponential equations.
- Solve logarithmic equations.
- Use logarithmic and semilogarithmic graph paper.

## 9.1 THE EXPONENTIAL FUNCTION

D.3, D.4

We have previously considered equations with a constant exponent in the form

$$y = x^n$$

These are called **power functions**; two examples are $y = x^2$ and $y = x^3$.

---

**EXPONENTIAL FUNCTION**

Equations with a variable exponent in the form

$$y = b^x$$

where $b > 0$ and $b \neq 1$ are called **exponential functions.**

---

Two examples of exponential functions are $y = 2^x$ and $y = \left(\frac{3}{4}\right)^x$.

**EXAMPLE 1**

Graph $y = 2^x$ by plotting points.

| $x$ | $y$ | $y = 2^x$ |
|-----|-----|-----------|
| 0 | 1 | $y = 2^0 = 1$ |
| 1 | 2 | $y = 2^1 = 2$ |
| 2 | 4 | $y = 2^2 = 4$ |
| 3 | 8 | $y = 2^3 = 8$ |
| $-1$ | $\dfrac{1}{2}$ | $y = 2^{-1} = \dfrac{1}{2}$ |
| $-2$ | $\dfrac{1}{4}$ | $y = 2^{-2} = \dfrac{1}{4}$ |
| $-3$ | $\dfrac{1}{8}$ | $y = 2^{-3} = \dfrac{1}{8}$ |

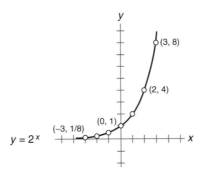

**Figure 9.1**

Now plot the points as in Fig. 9.1.

In general, for $b > 1$, $y = b^x$ is an **increasing** function. That is, as $x$ increases, $y$ increases.

## EXAMPLE 2

Graph $y = \left(\dfrac{1}{2}\right)^x$ by plotting points.

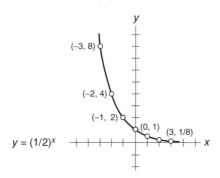

**Figure 9.2**

| $x$ | $y$ | $y = \left(\dfrac{1}{2}\right)^x$ |
|---|---|---|
| 0 | 1 | $y = \left(\dfrac{1}{2}\right)^0 = 1$ |
| 1 | $\dfrac{1}{2}$ | $y = \left(\dfrac{1}{2}\right)^1 = \dfrac{1}{2}$ |
| 2 | $\dfrac{1}{4}$ | $y = \left(\dfrac{1}{2}\right)^2 = \dfrac{1}{4}$ |
| 3 | $\dfrac{1}{8}$ | $y = \left(\dfrac{1}{2}\right)^3 = \dfrac{1}{8}$ |
| $-1$ | 2 | $y = \left(\dfrac{1}{2}\right)^{-1} = 2$ |
| $-2$ | 4 | $y = \left(\dfrac{1}{2}\right)^{-2} = 4$ |
| $-3$ | 8 | $y = \left(\dfrac{1}{2}\right)^{-3} = 8$ |

Plot the points as in Fig. 9.2.

In general, for $0 < b < 1$, $y = b^x$ is a **decreasing** function. That is, as $x$ increases, $y$ decreases.

## EXAMPLE 3

Graph $y = 3^{-x}$ by plotting points.

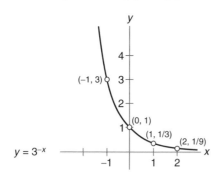

**Figure 9.3**

| $x$ | $y$ | $y = 3^{-x}$ |
|---|---|---|
| 0 | 1 | $y = 3^{-0} = 3^0 = 1$ |
| 1 | $\dfrac{1}{3}$ | $y = 3^{-1} = \dfrac{1}{3}$ |
| 2 | $\dfrac{1}{9}$ | $y = 3^{-2} = \dfrac{1}{3^2} = \dfrac{1}{9}$ |
| 3 | $\dfrac{1}{27}$ | $y = 3^{-3} = \dfrac{1}{3^3} = \dfrac{1}{27}$ |
| $-1$ | 3 | $y = 3^{-(-1)} = 3^1 = 3$ |
| $-2$ | 9 | $y = 3^{-(-2)} = 3^2 = 9$ |
| $-3$ | 27 | $y = 3^{-(-3)} = 3^3 = 27$ |

Plot the points as in Fig. 9.3.

What is the graph of $y = \left(\dfrac{1}{3}\right)^x$?

A calculator may be used to raise a number to a power. Use the $y^x$ button as shown in Examples 4–6.

**EXAMPLE 4**

Find the value of $2.5^{1.4}$, rounded to two significant digits.

| Flow chart | Buttons pushed | Display |
|---|---|---|
| Enter 2.5 | ②．⑤ | 2.5 |
| Push $y^x$ | $y^x$ | 2.5^* |
| Enter 1.4 | ①．④ | 1.4 |
| Push equals | = | 3.60675 |

*Not all calculators display this number as shown.

That is, $2.5^{1.4} = 3.6$ rounded to two significant digits.

**EXAMPLE 5**

Find the value of $2^\pi$, rounded to three significant digits.

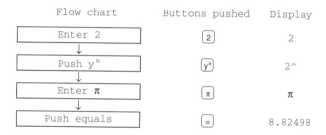

| Flow chart | Buttons pushed | Display |
|---|---|---|
| Enter 2 | ② | 2 |
| Push $y^x$ | $y^x$ | 2^ |
| Enter $\pi$ | $\pi$ | $\pi$ |
| Push equals | = | 8.82498 |

That is, $2^\pi = 8.82$ rounded to three significant digits.

**EXAMPLE 6**

Find the value of $150^{2/3}$, rounded to three significant digits.

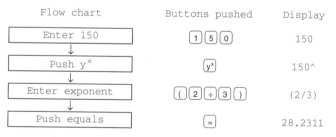

| Flow chart | Buttons pushed | Display |
|---|---|---|
| Enter 150 | ①⑤⓪ | 150 |
| Push $y^x$ | $y^x$ | 150^ |
| Enter exponent | （②÷③） | (2/3) |
| Push equals | = | 28.2311 |

That is, $150^{2/3} = 28.2$ rounded to three significant digits.

In growth situations, we could use the exponential function

$$y = A(1 + r)^n$$

where

$r =$ the rate of growth    (in decimal form)

$n =$ the time interval

$$A = \text{the initial amount}$$
$$y = \text{the new amount}$$

## EXAMPLE 7

According to the records of a utility company, the demand for electricity in its area is growing at a constant annual rate of 12%. During this current year, its customers used 750 billion kilowatt-hours ($7.5 \times 10^{11}$ kWh) of electricity. Assuming no conservation efforts on the part of its customers, how much electrical power will be needed in 8 years?

In this example,

$$r = 12\% = 0.12$$
$$n = 8 \text{ years}$$
$$A = 7.5 \times 10^{11} \text{ kWh}$$

We are to find $y$.

$$y = A(1 + r)^n$$
$$y = (7.5 \times 10^{11} \text{ kWh})(1 + 0.12)^8$$
$$= 1.9 \times 10^{12} \text{ kWh} \qquad \text{(Rounded to two significant digits)}$$

or 1900 billion kWh.

## EXAMPLE 8

Bill and Mary plan to retire this year with a combined annual pension of $50,000. Because of inflation, their purchasing power is constantly decreasing. Assuming a 7% annual rate of inflation, what will their purchasing power be in 10 years?

In this example,

$$r = -7\% = -0.07 \qquad \text{(Negative growth)}$$
$$n = 10 \text{ years}$$
$$A = \$50,000$$

We are to find $y$.

$$y = A(1 + r)^n$$
$$y = \$50,000[1 + (-0.07)]^{10}$$
$$= \$24,000 \qquad \text{(Rounded to two significant digits)}$$

## Exercises 9.1

*Graph each equation.*

**1.** $y = 4^x$

**2.** $y = 3^x$

**3.** $y = 10^x$

**4.** $y = 5^x$

**5.** $y = \left(\dfrac{1}{3}\right)^x$

**6.** $y = \left(\dfrac{1}{4}\right)^x$

**7.** $y = \left(\dfrac{1}{10}\right)^x$

**8.** $y = \left(\dfrac{1}{5}\right)^x$

**9.** $y = \left(\dfrac{3}{4}\right)^x$

**10.** $y = \left(\dfrac{2}{3}\right)^x$

**11.** $y = \left(\dfrac{5}{6}\right)^x$

**12.** $y = \left(\dfrac{2}{5}\right)^x$

**13.** $y = 4^{-x}$

**14.** $y = 5^{-x}$

**15.** $y = \left(\dfrac{4}{3}\right)^{-x}$

**16.** $y = \left(\dfrac{3}{4}\right)^{-x}$

**17.** $y = (1.2)^x$

**18.** $y = (5.5)^{-x}$

**19.** $y = 3^{-x+2}$

**20.** $y = 4^{2x}$

**21.** $y = 2^{x-3}$

**22.** $y = 2^{-2x}$

**23.** $y = 2^{x^2}$

**24.** $y = 2^x - 3$

**25.** $y = 3^x + 2$

*Find the value of each power to three significant digits.*

**26.** $12^{0.3}$

**27.** $5^{0.2}$

**28.** $3^{2.7}$

**29.** $10^{5.5}$

**30.** $4^\pi$

**31.** $5^{2\pi}$

**32.** $6^{2/3}$

**33.** $8^{\pi/3}$

**34.** $15^{\pi/4}$

**35.** $9^{3/4}$

**36.** $\sqrt[3]{12}$

**37.** $\sqrt[4]{6}$

**38.** $\sqrt[5]{9}$

**39.** $\sqrt[6]{140}$

**40.** $\sqrt[3]{46{,}656}$

**41.** According to the records of a utility company, the demand for electricity in its area is constantly growing at an annual rate of 8%. During this current year, its customers used $5\overline{0}0$ billion kWh of electricity. Assuming no conservation efforts on the part of its customers, how much electrical power will be needed in 6 years?

**42.** Doris plans to retire this year with an annual pension of $68,000. Because of inflation, her purchasing power is constantly decreasing. Assuming a 7% annual rate of inflation, find her purchasing power in 10 years.

**43.** The formula for compound interest is $A = P\left(1 + \dfrac{r}{x}\right)^{xn}$ where

$A =$ the amount of money in the account (principal and interest)

$P =$ the original principal (amount invested)

$r =$ the yearly rate of interest (in decimal form)

$x =$ the number of times that interest is compounded per year

$n =$ the number of years that the money is invested

Assume that you have $1000 to invest at 8% interest. Find to the nearest dollar the amount in the account after three years if the interest is compounded **(a)** annually, **(b)** semiannually, **(c)** quarterly, and **(d)** daily.

**44.** Assume that you owe a credit company $1000. The interest rate is $1\frac{1}{2}\%$ per month. Assume that you do not charge any more for the next year.
   **(a)** If you pay nothing for 6 months, what is your balance?
   **(b)** If you pay $100 per month, what do you owe in 6 months?

**45.** When a gas undergoes an adiabatic (constant heat) process, the final and initial absolute temperatures and pressures are related according to the formula

$$\frac{T_2}{T_1} = \left(\frac{P_1}{P_2}\right)^{(1-\gamma)/\gamma}$$

where $\gamma$ is the ratio of specific heats at constant pressure and volume. Find $T_2$ when $T_1 = 575°R$ absolute, $P_1 = 25.0$ psi (pounds per square inch) absolute, $P_2 = 65\overline{0}$ psi absolute, and $\gamma = 1.50$.

**46.** Given the formula for heat flow by convection

$$h = 0.0230 \frac{k}{D} \left(\frac{DV\rho}{\mu}\right)^{0.8} \left(\frac{\mu c}{k}\right)^{n}$$

Find $h$ when $k = 0.0650$, $D = 2.00$, $V = 16\overline{0}$, $\rho = 1.45$, $\mu = 0.108$, $c = 0.720$, and $n = 0.3$.

## 9.2  THE LOGARITHM

D.1, D.4

When the values of $x$ and $y$ are interchanged in an equation, the resulting equation is called the **inverse** of the given equation. The inverse of the exponential equation, $y = b^x$, is the exponential equation, $x = b^y$. We define this inverse equation to be the logarithmic equation. The following middle and right equations show how to express this logarithmic equation in either exponential form or logarithmic form:

| Exponential equation | Logarithmic equation in exponential form | Logarithmic equation in logarithmic form |
|:---:|:---:|:---:|
| $y = b^x$ | $x = b^y$ | $y = \log_b x$ |

That is, $x = b^y$ and $y = \log_b x$ are equivalent equations for $b > 0$ but $b \neq 1$.

The logarithm of a number is the *exponent* indicating the power to which the base must be raised to equal that number. The expression $\log_b x$ is read "the logarithm of $x$ to the base $b$."

---

*Remember:* A logarithm is an exponent.

---

**EXAMPLE 1**

Write each equation in logarithmic form.

|  | *Exponential form* | *Logarithmic form* |
|:---:|:---:|:---:|
| (a) | $2^3 = 8$ | $\log_2 8 = 3$ |
| (b) | $5^2 = 25$ | $\log_5 25 = 2$ |
| (c) | $4^{-2} = \dfrac{1}{16}$ | $\log_4\left(\dfrac{1}{16}\right) = -2$ |
| (d) | $36^{1/2} = 6$ | $\log_{36} 6 = \dfrac{1}{2}$ |
| (e) | $p^q = r$ | $\log_p r = q$ |

**EXAMPLE 2**

Write each equation in exponential form.

|  | *Logarithmic form* | *Exponential form* |
|:---:|:---:|:---:|
| (a) | $\log_7 49 = 2$ | $7^2 = 49$ |
| (b) | $\log_4 64 = 3$ | $4^3 = 64$ |
| (c) | $\log_{10} 0.01 = -2$ | $10^{-2} = 0.01$ |
| (d) | $\log_{27} 3 = \dfrac{1}{3}$ | $27^{1/3} = 3$ |
| (e) | $\log_m p = n$ | $m^n = p$ |

**EXAMPLE 3**

Graph $y = \log_2 x$ by plotting points.

First, change the equation from logarithmic form to exponential form. That is, $y = \log_2 x$ is equivalent to $x = 2^y$. Then choose values for $y$ and compute values for $x$.

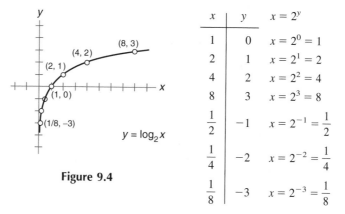

**Figure 9.4**

| $x$ | $y$ | $x = 2^y$ |
|---|---|---|
| 1 | 0 | $x = 2^0 = 1$ |
| 2 | 1 | $x = 2^1 = 2$ |
| 4 | 2 | $x = 2^2 = 4$ |
| 8 | 3 | $x = 2^3 = 8$ |
| $\dfrac{1}{2}$ | $-1$ | $x = 2^{-1} = \dfrac{1}{2}$ |
| $\dfrac{1}{4}$ | $-2$ | $x = 2^{-2} = \dfrac{1}{4}$ |
| $\dfrac{1}{8}$ | $-3$ | $x = 2^{-3} = \dfrac{1}{8}$ |

Plot the points as in Fig. 9.4.

**EXAMPLE 4**

Graph $y = \log_{1/2} x$ by plotting points.

Again, change the equation from logarithmic form to exponential form. That is, $y = \log_{1/2} x$ is equivalent to $x = \left(\dfrac{1}{2}\right)^y$.

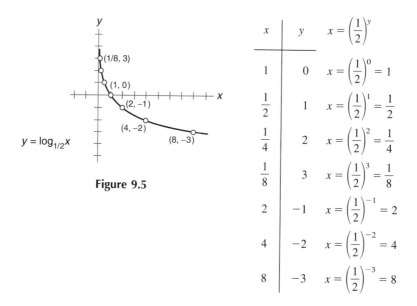

**Figure 9.5**

| $x$ | $y$ | $x = \left(\dfrac{1}{2}\right)^y$ |
|---|---|---|
| 1 | 0 | $x = \left(\dfrac{1}{2}\right)^0 = 1$ |
| $\dfrac{1}{2}$ | 1 | $x = \left(\dfrac{1}{2}\right)^1 = \dfrac{1}{2}$ |
| $\dfrac{1}{4}$ | 2 | $x = \left(\dfrac{1}{2}\right)^2 = \dfrac{1}{4}$ |
| $\dfrac{1}{8}$ | 3 | $x = \left(\dfrac{1}{2}\right)^3 = \dfrac{1}{8}$ |
| 2 | $-1$ | $x = \left(\dfrac{1}{2}\right)^{-1} = 2$ |
| 4 | $-2$ | $x = \left(\dfrac{1}{2}\right)^{-2} = 4$ |
| 8 | $-3$ | $x = \left(\dfrac{1}{2}\right)^{-3} = 8$ |

Plot the points as in Fig. 9.5.

**EXAMPLE 5**

Given $\log_3 81 = x$, find $x$.

The exponential form of $\log_3 81 = x$ is

$$3^x = 81$$

We know that

$$3^4 = 81$$

Therefore,

$$x = 4$$

**EXAMPLE 6**

If $\log_3 x = -2$, find $x$.

$$\log_3 x = -2 \quad \text{or} \quad 3^{-2} = x$$

Therefore,

$$x = \frac{1}{9}$$

**EXAMPLE 7**

If $\log_x 32 = \frac{5}{3}$, find $x$.

$$\log_x 32 = \frac{5}{3} \quad \text{or} \quad x^{5/3} = 32$$

$$x^{1/3} = 2 \qquad \text{(Take the fifth root of each side.)}$$

Therefore,

$$x = 8 \qquad \text{(Cube each side.)}$$

Or begin with

$$x^{5/3} = 32$$

Raise each side to the $\frac{3}{5}$ power.

$$(x^{5/3})^{3/5} = 32^{3/5}$$

$$x = 8 \qquad (\textit{Note: } 32^{3/5} = (2^5)^{3/5} = 2^3 = 8)$$

# Exercises 9.2

*Write each equation in logarithmic form.*

1. $3^2 = 9$

2. $7^2 = 49$

3. $5^3 = 125$

4. $10^3 = 1000$

5. $2^5 = 32$

6. $3^4 = 81$

7. $9^{1/2} = 3$

8. $16^{1/2} = 4$

9. $5^{-2} = \dfrac{1}{25}$

10. $4^0 = 1$

11. $10^{-5} = 0.00001$

12. $d^e = f$

*Write each equation in exponential form.*

**13.** $\log_5 25 = 2$

**14.** $\log_8 64 = 2$

**15.** $\log_2 16 = 4$

**16.** $\log_5 125 = 3$

**17.** $\log_{25} 5 = \dfrac{1}{2}$

**18.** $\log_{81} 9 = \dfrac{1}{2}$

**19.** $\log_8 2 = \dfrac{1}{3}$

**20.** $\log_{27} 3 = \dfrac{1}{3}$

**21.** $\log_2\left(\dfrac{1}{4}\right) = -2$

**22.** $\log_2\left(\dfrac{1}{8}\right) = -3$

**23.** $\log_{10} 0.01 = -2$

**24.** $\log_g h = k$

*Graph each equation.*

**25.** $y = \log_4 x$

**26.** $y = \log_3 x$

**27.** $y = \log_{10} x$

**28.** $y = \log_5 x$

**29.** $y = \log_{1/4} x$

**30.** $y = \log_{1/3} x$

*Solve for x.*

**31.** $\log_4 x = 3$

**32.** $\log_2 x = -1$

**33.** $\log_9 3 = x$

**34.** $\log_6 36 = x$

**35.** $\log_2 8 = x$

**36.** $\log_3 27 = x$

**37.** $\log_{25} 5 = x$

**38.** $\log_{27} 3 = x$

**39.** $\log_x 25 = 2$

**40.** $\log_x\left(\dfrac{1}{27}\right) = -3$

**41.** $\log_{1/2}\left(\dfrac{1}{8}\right) = x$

**42.** $\log_x 3 = \dfrac{1}{2}$

**43.** $\log_{12} x = 2$

**44.** $\log_8\left(\dfrac{1}{64}\right) = x$

**45.** $\log_x 9 = \dfrac{2}{3}$

**46.** $\log_x 64 = \dfrac{3}{2}$

**47.** $\log_x\left(\dfrac{1}{8}\right) = -\dfrac{3}{2}$

**48.** $\log_x\left(\dfrac{1}{27}\right) = -\dfrac{3}{4}$

# 9.3   PROPERTIES OF LOGARITHMS

The most common uses of logarithms today include expressing the exponential and log-arithmic relationships in business and between certain natural phenomena in electronics, biology, and radioactivity; solving exponential and logarithmic equations; and rewriting certain algebraic expressions.

Before studying these concepts, we need to develop the following three basic log-arithmic properties.

**1. Multiplication:** If $M$ and $N$ are positive real numbers,

$$\boxed{\log_a(M \cdot N) = \log_a M + \log_a N \quad \text{where } a > 0 \quad \text{and} \quad a \neq 1}$$

To prove this, let $p = \log_a M$ and $q = \log_a N$. Writing each in exponential form, we have $a^p = M$ and $a^q = N$. Forming the product $MN$, we have $M \cdot N = a^p \cdot a^q = a^{p+q}$.

Now write $M \cdot N = a^{p+q}$ in logarithmic form.

$$\log_a(M \cdot N) = p + q$$
$$= \log_a M + \log_a N$$

That is, the logarithm of a product equals the sum of the logarithms of its factors.

2. **Division:** If $M$ and $N$ are positive real numbers,

$$log_a\left(\frac{M}{N}\right) = log_aM - log_aN \quad \text{where } a > 0 \quad \text{and} \quad a \neq 1$$

Again let $p = log_aM$ and $q = log_aN$. Writing each in exponential form, we have $a^p = M$ and $a^q = N$. Forming the quotient $\dfrac{M}{N}$, we have $\dfrac{M}{N} = \dfrac{a^p}{a^q} = a^{p-q}$. Now write $\dfrac{M}{N} = a^{p-q}$ in logarithmic form.

$$log_a\left(\frac{M}{N}\right) = p - q$$
$$= log_aM - log_aN$$

That is, the logarithm of a quotient equals the difference of the logarithms of its factors.

3. **Powers:** If $M$ is a positive real number and $n$ is any real number,

$$log_aM^n = n \, log_aM \quad \text{where } a > 0 \quad \text{and} \quad a \neq 1$$

Let $p = log_aM$, which in exponential form is

$$a^p = M$$

Taking the $n$th power of each side, we have

$$(a^p)^n = M^n$$

or

$$a^{np} = M^n$$

which, in logarithmic form, is

$$log_aM^n = np$$
$$= n \, log_aM$$

That is, the logarithm of a power of a number equals the product of the exponent times the logarithm of the number.

There are three special cases of the power property that are helpful.

(a) **Roots:** If $M$ is any positive real number and $n$ is any positive integer,

$$log_a\sqrt[n]{M} = \frac{1}{n} \cdot log_aM$$

Note that this is a special case where $\sqrt[n]{M} = M^{1/n}$. That is, the logarithm of the root of a number equals the logarithm of the number divided by the index of the root.

**(b) For $n = 0$:**

$$\log_a M^0 = \log_a 1 \qquad (M^0 = 1)$$
$$\log_a M^0 = 0 \cdot \log_a M \qquad \text{(Property 3)}$$
$$= 0$$

Therefore,

$$\boxed{\log_a 1 = 0}$$

That is, the logarithm of one to any base is zero.

**(c) For $n = -1$:**

$$\log_a M^{-1} = \log_a \frac{1}{M} \qquad \left(M^{-1} = \frac{1}{M}\right)$$
$$\log_a M^{-1} = (-1)\log_a M \qquad \text{(Property 3)}$$

Therefore,

$$\boxed{\log_a \frac{1}{M} = -\log_a M}$$

That is, the logarithm of the reciprocal of a number is the negative of the logarithm of the number.

**EXAMPLE 1**

Write $\log_4 2x^5 y^2$ as a sum of multiples of single logarithms.

$$\log_4 2x^5 y^2 = \log_4 2 + \log_4 x^5 + \log_4 y^2 \qquad \text{Property 1}$$
$$= \log_4 2 + 5\log_4 x + 2\log_4 y \qquad \text{Property 3}$$

**EXAMPLE 2**

Write $\log_3 \dfrac{\sqrt{x(x-2)}}{(x+3)^2}$ as a sum or difference of multiples of the logarithms of $x$, $x - 2$, and $x + 3$.

$$\log_3 \frac{\sqrt{x(x-2)}}{(x+3)^2} = \log_3 \frac{[x(x-2)]^{1/2}}{(x+3)^2}$$

$$= \log_3 [x(x-2)]^{1/2} - \log_3 (x+3)^2 \qquad \text{Property 2}$$

$$= \frac{1}{2}\log_3 [x(x-2)] - 2\log_3 (x+3) \qquad \text{Property 3}$$

$$= \frac{1}{2}[\log_3 x + \log_3 (x-2)] - 2\log_3 (x+3) \qquad \text{Property 1}$$

$$= \frac{1}{2}\log_3 x + \frac{1}{2}\log_3 (x-2) - 2\log_3 (x+3)$$

**EXAMPLE 3**

Write $3 \log_2 x + 4 \log_2 y - 2 \log_2 z$ as a single logarithmic expression.

$$3 \log_2 x + 4 \log_2 y - 2 \log_2 z = \log_2 x^3 + \log_2 y^4 - \log_2 z^2 \quad \text{Property 3}$$
$$= \log_2 (x^3 y^4) - \log_2 z^2 \quad \text{Property 1}$$
$$= \log_2 \frac{x^3 y^4}{z^2} \quad \text{Property 2}$$

**EXAMPLE 4**

Write $3 \log_{10}(x - 1) - \dfrac{1}{3} \log_{10} x - \log_{10}(2x + 3)$ as a single logarithmic expression.

$$3 \log_{10}(x - 1) - \frac{1}{3} \log_{10} x - \log_{10}(2x + 3)$$

$$= \log_{10}(x - 1)^3 - \log_{10} x^{1/3} - \log_{10}(2x + 3) \quad \text{Property 3}$$
$$= \log_{10}(x - 1)^3 - [\log_{10} x^{1/3} + \log_{10}(2x + 3)]$$
$$= \log_{10}(x - 1)^3 - \log_{10}[(x^{1/3})(2x + 3)] \quad \text{Property 1}$$
$$= \log_{10}\frac{(x - 1)^3}{x^{1/3}(2x + 3)} \quad \text{Property 2}$$
$$= \log_{10}\frac{(x - 1)^3}{\sqrt[3]{x}(2x + 3)}$$

There are two other logarithmic properties that are useful in simplifying expressions:

$$\log_a a^x = x \quad \text{and} \quad a^{\log_a x} = x$$

To show the first one, we begin with the identity

$$(a^x) = a^x$$

Then, writing this exponential equation in logarithmic form, we have

$$\boxed{\log_a(a^x) = x}$$

To show the second one, we begin with the identity

$$\log_a x = (\log_a x)$$

Then, writing this logarithmic equation in exponential form, we have

$$\boxed{a^{(\log_a x)} = x}$$

**EXAMPLE 5**

Find the value of $\log_2 16$.

$$\log_2 16 = \log_2 2^4 \qquad \text{(Note that this simplification is possible}$$
$$\text{because 16 is a power of 2.)}$$
$$= 4$$

**EXAMPLE 6**

Find the value of $\log_{10} 0.01$.

$$\log_{10} 0.01 = \log_{10} 10^{-2}$$

$$= -2 \qquad (\textit{Note: } 0.01 = \frac{1}{100} = 10^{-2})$$

**EXAMPLE 7**

Find the value of $5^{\log_5 8}$.

$$5^{\log_5 8} = 8$$

**EXAMPLE 8**

Find the value of $9^{\log_3 25}$.

$$9^{\log_3 25} = (3^2)^{\log_3 25}$$

$$= 3^{2 \log_3 25}$$

$$= 3^{\log_3 25^2} \qquad \text{Property 3}$$

$$= 25^2$$

$$= 625$$

**EXAMPLE 9**

Find the value of $\log_2(8^{\log_2 8})$.

$$\log_2(8^{\log_2 8}) = \log_2(8^{\log_2 2^3})$$

$$= \log_2(8^3) \longleftarrow \qquad (\log_2 2^3 = 3)$$

$$= \log_2(2^3)^3$$

$$= \log_2 2^9$$

$$= 9$$

## Exercises 9.3

*Write each expression as a sum or difference of multiples of single logarithms.*

**1.** $\log_2 5x^3 y$

**2.** $\log_3 \dfrac{8x^2 y^3}{z^4}$

**3.** $\log_{10} \dfrac{2x^2}{y^3 z}$

**4.** $\log_4 \dfrac{y^3}{x\sqrt{z}}$

**5.** $\log_b \dfrac{y^3 \sqrt{x}}{z^2}$

**6.** $\log_b \dfrac{7xy}{\sqrt[3]{z}}$

**7.** $\log_b \sqrt[3]{\dfrac{x^2}{y}}$

**8.** $\log_5 \sqrt[4]{xy^2 z}$

**9.** $\log_2 \dfrac{1}{x} \sqrt{\dfrac{y}{z}}$

**10.** $\log_b \dfrac{1}{z^2} \sqrt[3]{\dfrac{x^2}{y}}$

**11.** $\log_b \dfrac{z^3 \sqrt{x}}{\sqrt[3]{y}}$

**12.** $\log_b \dfrac{\sqrt{y}\sqrt{x}}{z^2}$

**13.** $\log_b \dfrac{x^2(x+1)}{\sqrt{x+2}}$

**14.** $\log_b \dfrac{\sqrt{x}(x+4)}{x^2}$

*Write each as a single logarithmic expression.*

**15.** $\log_b x + 2 \log_b y$

**16.** $2 \log_b z - 3 \log_b x$

**17.** $\log_b x + 2 \log_b y - 3 \log_b z$

**18.** $3 \log_7 x - 4 \log_7 y - 5 \log_7 z$

**19.** $\log_3 x + \dfrac{1}{3} \log_3 y - \dfrac{1}{2} \log_3 z$

**20.** $\dfrac{1}{2} \log_2 x - \dfrac{1}{3} \log_2 y - \log_2 z$

**21.** $2 \log_{10} x - \dfrac{1}{2} \log_{10}(x - 3) - \log_{10}(x + 1)$

**22.** $\log_3(x + 1) + \dfrac{1}{2} \log_3(x + 2) - 3 \log_3(x - 1)$

**23.** $5 \log_b x + \dfrac{1}{3} \log_b(x - 1) - \log_b(x + 2)$

**24.** $\log_b(x + 1) + \dfrac{1}{3} \log_b(x - 7) - 2 \log_b x$

**25.** $\log_{10} x + 2 \log_{10}(x - 1) - \dfrac{1}{3}[\log_{10}(x + 2) + \log_{10}(x - 5)]$

**26.** $\dfrac{1}{2} \log_b(x + 1) - 3[\log_b x + \log_b(x - 1) + \log_b(2x - 1)]$

*Find the value of each expression.*

**27.** $\log_b b^3$

**28.** $\log_2 2^5$

**29.** $\log_3 9$

**30.** $\log_2 16$

**31.** $\log_5 125$

**32.** $\log_4 64$

**33.** $\log_2 \dfrac{1}{4}$

**34.** $\log_3 \dfrac{1}{27}$

**35.** $\log_{10} 0.001$

**36.** $\log_{10} 0.1$

**37.** $\log_3 1$

**38.** $\log_{10} 1$

**39.** $6^{\log_6 5}$

**40.** $3^{\log_3 9}$

**41.** $25^{\log_5 6}$

**42.** $27^{\log_3 2}$

**43.** $4^{\log_2(1/5)}$

**44.** $8^{\log_2(1/3)}$

**45.** $\log_3 9^{\log_3 27}$

**46.** $\log_4 16^{\log_4 64}$

**47.** $\log_2 16^{\log_4 16}$

# 9.4 COMMON LOGARITHMS

In the previous sections we used a variety of bases of logarithms. Actually, only two bases are in general use: base 10 and base $e$, where $e$ is an irrational number approximately equal to 2.71828. Logarithms that use 10 as a base are called **common logarithms.** Logarithms that use $e$ as a base are called **natural logarithms** and are discussed in the next section.

Before the electronic handheld calculator, common logarithms were routinely used for a wide variety of computations, such as multiplication, division, and finding powers and roots of numbers. A variety of technical and scientific measurements using common logarithms remain.

Let's list some powers of 10 and compare each with its equivalent logarithmic form:

| *Exponential form* | *Logarithmic form* |
|---|---|
| $10^2 = 100$ | $\log_{10} 100 = 2$ |
| $10^4 = 10000$ | $\log_{10} 10000 = 4$ |

$$10^{-3} = 0.001 \qquad \log_{10}0.001 = -3$$
$$10^{-1} = 0.1 \qquad \log_{10}0.1 = -1$$
$$10^{0} = 1 \qquad \log_{10}1 = 0$$

For nonintegral powers of 10, approximations can be calculated. For example:

| *Exponential form* | *Logarithmic form* |
|---|---|
| $10^{0.3010} = 2.00$ | $\log_{10}2.00 = 0.3010$ |
| $10^{0.5263} = 3.36$ | $\log_{10}3.36 = 0.5263$ |
| $10^{0.8451} = 7.00$ | $\log_{10}7.00 = 0.8451$ |
| $10^{0.9258} = 8.43$ | $\log_{10}8.43 = 0.9258$ |

A calculator may be used to find the common logarithm of a number as shown in the following examples.

**EXAMPLE 1**

Find $\log_{10}456$, rounded to four significant digits.

| Flow chart | Buttons pushed | Display |
|---|---|---|
| Enter log | log | log |
| ↓ | | |
| Enter 456 | 4 5 6 | 456 |
| ↓ | | |
| Push equals | = | 2.65896 |

That is, log 456 = 2.659 rounded to four significant digits.

*Note:* When working exclusively with common logarithms, we shall follow the common practice of not writing the base 10; for example, $\log_{10}456 = \log 456 = 2.659$.

**EXAMPLE 2**

Find log 0.0596, rounded to four significant digits.

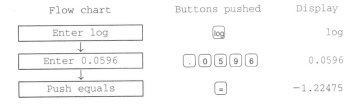

| Flow chart | Buttons pushed | Display |
|---|---|---|
| Enter log | log | log |
| ↓ | | |
| Enter 0.0596 | . 0 5 9 6 | 0.0596 |
| ↓ | | |
| Push equals | = | −1.22475 |

That is, log 0.0596 = −1.225 rounded to four significant digits.

A calculator may also be used to find the number, $N$, when its logarithm is known. Here $N$ is called the **antilogarithm.** Finding the antilogarithm of a number is the reverse process of finding the logarithm of a number.

**EXAMPLE 3**

Given log $N = 1.6845$, find $N$, rounded to three significant digits.

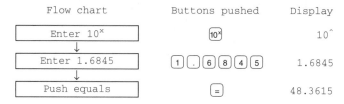

| Flow chart | Buttons pushed | Display |
|---|---|---|
| Enter $10^x$ | $\boxed{10^x}$ | $10^{\wedge}$ |
| Enter 1.6845 | $\boxed{1}\boxed{.}\boxed{6}\boxed{8}\boxed{4}\boxed{5}$ | 1.6845 |
| Push equals | $\boxed{=}$ | 48.3615 |

That is, $N = 48.4$ rounded to three significant digits.

**EXAMPLE 4**

Find the value of $\dfrac{\log 275}{\log 5}$, rounded to three significant digits.

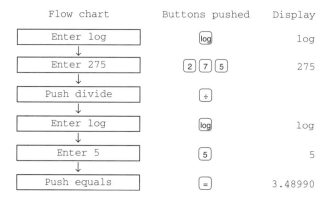

| Flow chart | Buttons pushed | Display |
|---|---|---|
| Enter log | $\boxed{\text{log}}$ | log |
| Enter 275 | $\boxed{2}\boxed{7}\boxed{5}$ | 275 |
| Push divide | $\boxed{\div}$ | |
| Enter log | $\boxed{\text{log}}$ | log |
| Enter 5 | $\boxed{5}$ | 5 |
| Push equals | $\boxed{=}$ | 3.48990 |

The result rounded to three significant digits is 3.49.

**EXAMPLE 5**

In chemistry, the pH (hydrogen potential) of a solution is a measure of its acidity and is defined as

$$pH = -\log(H^+)$$

where $H^+$ is a numerical value for the concentration of hydrogen ions in moles per litre. Water has a pH of 7. Acids have pH numbers less than 7, and alkaline solutions (bases) have pH values greater than 7. For beer, if $H^+ = 6.3 \times 10^{-5}$ moles/litre (M/L), find its pH.

$$\begin{aligned} pH &= -\log(H^+) \\ &= -\log(6.3 \times 10^{-5}) \\ &= 4.2 \qquad \text{(pH values are usually rounded to the nearest tenth.)} \end{aligned}$$

**EXAMPLE 6**

The intensity level of sound is given by the formula

$$\beta = 10 \log \frac{I}{I_0}$$

where $\beta$ is the intensity level in decibels (dB) of a sound of intensity $I$, measured in watts per square centimetre (W/cm$^2$), and $I_0$ is the intensity of the threshold of hearing,

$10^{-16}$ W/cm$^2$. What is the intensity level of a sound with an intensity of $10^{-11}$ W/cm$^2$? (Normal hearing ranges between 0–120 dB.)

$$\beta = 10 \log \frac{I}{I_0}$$

$$\beta = 10 \log \left( \frac{10^{-11} \text{ W/cm}^2}{10^{-16} \text{ W/cm}^2} \right)$$

$$= 10 \log 10^5$$

$$= 10 \cdot 5$$

$$= 50 \text{ dB}$$

**EXAMPLE 7**

The power gain or loss of an amplifier or other electronic device is given by

$$n = 10 \log \left( \frac{P_0}{P_i} \right)$$

where $n$ is the power gain or loss in decibels (dB), $P_0$ is the power output, and $P_i$ is the power input. Find the power gain when $P_0 = 8.0$ W and $P_i = 0.25$ W.

$$n = 10 \log \left( \frac{P_0}{P_i} \right)$$

$$n = 10 \log \left( \frac{8.0 \text{ W}}{0.25 \text{ W}} \right)$$

$$= 10 \log 32 = 15 \text{ dB}$$

# Exercises 9.4

*Find the common logarithm of each number, rounded to four significant digits.*

1. log 68.1
2. log 928
3. log 45
4. log 14,000
5. log 0.142
6. log 0.026
7. log 0.00621
8. log 0.0000497
9. log 805
10. log 0.608
11. log 9.25
12. log 1

*Find N, the antilogarithm, rounded to three significant digits.*

13. $\log N = 1.4048$
14. $\log N = 2.6191$
15. $\log N = 2.8484$
16. $\log N = 4.7400$
17. $\log N = 0.2782$
18. $\log N = 0.5690$
19. $\log N = -1.6050$
20. $\log N = -2.7376$
21. $\log N = -3.6345$
22. $\log N = -4.805$
23. $\log N = -4.8145$
24. $\log N = -6.8163$

*Find the value of each expression, rounded to three significant digits.*

25. $\dfrac{\log 685}{\log 6}$
26. $\dfrac{\log 984}{\log 4}$
27. $\dfrac{\log 1675}{\log 12.5}$

28. $\dfrac{\log 64.5}{\log 207}$
29. $\dfrac{\log 16.5}{\log 1350}$
30. $\dfrac{\log 8}{\log 12.5}$

**31.** $\log\left(\dfrac{596}{45}\right)$    **32.** $\log\left(\dfrac{654}{7}\right)$    **33.** $\log\left(\dfrac{4}{15}\right)$

**34.** $\log\left(\dfrac{16.5}{2.7}\right)$    **35.** $\log 14.6 - \log 3.75$    **36.** $\log 45 + \log 20$

**37.** $\log 145 + \log 25$    **38.** $\log 9.46 - \log 0.48$    **39.** $\dfrac{\log 486 + \log 680}{\log 14}$

**40.** $\dfrac{\log 276 + \log 98.4}{\log 2 - \log 14.5}$

*We defined pH in Example 5. Compute the pH for each value.*

**41.** $H^+ = 10^{-6}$    **42.** $H^+ = 10^{-8}$    **43.** $H^+ = 3.2 \times 10^{-7}$

**44.** $H^+ = 2.0 \times 10^{-6}$    **45.** $H^+ = 5.5 \times 10^{-8}$    **46.** $H^+ = 6.1 \times 10^{-3}$

*The intensity of sound level is defined in Example 6. Compute β for each value.*

**47.** $10^{-6}$ W/cm$^2$    **48.** $10^{-14}$ W/cm$^2$

**49.** $10^{-10}$ W/cm$^2$    **50.** $10^{-9}$ W/cm$^2$

*We defined electronic power gain or loss in Example 7. Find each power gain or loss.*

**51.** $P_0 = 12$ W, $P_i = 0.50$ W    **52.** $P_0 = 150$ W, $P_i = 3.0$ W

**53.** $P_0 = 0.60$ W, $P_i = 0.80$ W    **54.** $P_0 = 1.0$ W, $P_i = 1.2$ W

## 9.5   NATURAL LOGARITHMS

D.1, D.3, D.4

While common logarithms have a base of 10, **natural logarithms** have a base of $e$. The number $e$ is irrational and is approximately equal to 2.71828. Although it may seem strange to have a system of logarithms based on such a number, many applications are based on powers of $e$, especially those involving growth and decay relationships. The form $\log_e x$ is usually written in its special notation $\ln x$.

To raise the natural number, $e$, to a power on a calculator, use the $e^x$ button, as illustrated in Examples 1–2.

**EXAMPLE 1**

Find the value of $e^5$, rounded to three significant digits.

| Flow chart | Buttons pushed | Display |
|---|---|---|
| Enter $e^x$ | $e^x$ | e^ |
| ↓ | | |
| Enter 5 | 5 | e^5 |
| ↓ | | |
| Push equals | = | 148.41316 |

That is, $e^5 = 148$ rounded to three significant digits.

**EXAMPLE 2**

Find the value of $e^{-1.75}$, rounded to three significant digits.

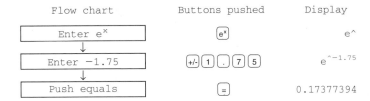

| Flow chart | Buttons pushed | Display |
|---|---|---|
| Enter $e^x$ | $e^x$ | e^ |
| Enter $-1.75$ | +/- 1 . 7 5 | e^−1.75 |
| Push equals | = | 0.17377394 |

That is, $e^{-1.75} = 0.174$ rounded to three significant digits.

A calculator is an easy way to find natural logs and antilogs. The steps are outlined in Examples 3–5.

**EXAMPLE 3**

Find ln 4350, rounded to four significant digits.

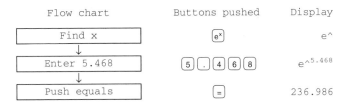

| Flow chart | Buttons pushed | Display |
|---|---|---|
| Enter ln | ln | ln |
| Enter 4350 | 4 3 5 0 | ln 4350 |
| Push equals | = | 8.37793 |

That is, ln 4350 = 8.378 rounded to four significant digits.

**EXAMPLE 4**

Given ln $x = 5.468$, find $x$, rounded to three significant digits.

| Flow chart | Buttons pushed | Display |
|---|---|---|
| Find x | $e^x$ | e^ |
| Enter 5.468 | 5 . 4 6 8 | e^5.468 |
| Push equals | = | 236.986 |

That is, $x = 237$ rounded to three significant digits.

**EXAMPLE 5**

Given ln $x = -3.45$, find $x$, rounded to three significant digits.

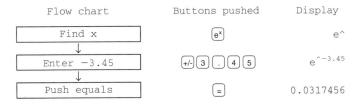

| Flow chart | Buttons pushed | Display |
|---|---|---|
| Find x | $e^x$ | e^ |
| Enter $-3.45$ | +/- 3 . 4 5 | e^−3.45 |
| Push equals | = | 0.0317456 |

That is, $x = 0.0317$ rounded to three significant digits.

If money is invested in an account that pays interest compounded continuously, the amount accumulated is given by the exponential function

$$y = Ae^{rn}$$

where $A$ is the initial invested amount, $r$ is the interest rate, $n$ is the number of years that the initial amount is invested, $e$ is the natural base for logarithms and $y$ is the amount accumulated.

The exponential function $y = Ae^{rn}$ is also an alternate growth function.

**EXAMPLE 6**

Joan invests $1000 at $7\frac{3}{4}$% compounded continuously. How much money will she have in the account after 5 years?

Here

$$A = \$1000$$
$$r = 7\frac{3}{4}\% = 7.75\% = 0.0775$$
$$n = 5$$

We are to find $y$.

$$y = Ae^{rn}$$
$$y = \$1000e^{(0.0775)(5)}$$
$$= \$1473 \qquad \text{(Rounded to the nearest dollar)}$$

**EXAMPLE 7**

A city has a population of 850,000, and a nearby suburb has a population of 65,000. Studies have determined that the city will lose population at the rate of 2%, while the suburb will gain population at the rate of 4%. Find the population of each in 5 years.

| *City* | *Suburb* |
|---|---|
| $A = 850{,}000$ | $A = 65{,}000$ |
| $r = -2\% = -0.02$ | $r = 4\% = 0.04$ |
| $n = 5$ | $n = 5$ |
| Find $y$. | Find $y$. |
| $y = Ae^{rn}$ | $y = Ae^{rn}$ |
| $y = 850{,}000e^{(-0.02)(5)}$ | $y = 65{,}000e^{(0.04)(5)}$ |
| $= 770{,}000$ | $= 79{,}000$ |

(Each has been rounded to two significant digits.)

**EXAMPLE 8**

A bacteria culture contains $10\overline{0}{,}000$ bacteria. It grows at the rate of 15% per hour. How many bacteria will be present in 12 h?

Using $y = Ae^{rn}$, we have

$$A = 10\overline{0}{,}000$$
$$r = 15\% = 0.15$$
$$n = 12 \text{ h}$$

Find $y$.

$$y = Ae^{rn}$$
$$y = 10\bar{0},000e^{(0.15)(12)}$$
$$= 6\bar{0}0,000 \text{ bacteria} \qquad \text{(Rounded to two significant digits)}$$

**EXAMPLE 9**

The current in an electric circuit is given by

$$i = 2.4e^{-4.0t}$$

where $i$ is the current in amperes (A) and $t$ is the time in seconds. Find the current when $t = 0.35$ s.

$$i = 2.4e^{-4.0t}$$
$$= 2.4e^{(-4.0)(0.35)}$$
$$= 0.59 \text{ A} \qquad \text{(Rounded to two significant digits)}$$

## Logarithms with Other Bases

To find the logarithm with another base, use the following formula.

$$\log_b x = \frac{\log_a x}{\log_a b}$$

where $a$ is base 10 or base $e$ and $b$ is another base. To verify this formula, let

$$u = \log_b x$$

Then

$$b^u = x$$
$$\log_a b^u = \log_a x$$
$$u \log_a b = \log_a x$$
$$u = \frac{\log_a x}{\log_a b}$$

Since $u = \log_b x$,

$$\log_b x = \frac{\log_a x}{\log_a b}$$

**EXAMPLE 10**

Find $\log_3 370$ to three significant digits.
Let us use base 10.

$$\log_3 370 = \frac{\log_{10} 370}{\log_{10} 3} = \frac{2.5682}{0.4771} = 5.38$$

What is the relationship between the common logarithm and the natural logarithm of a number? In general,

$$\ln x = \frac{\log x}{\log e} = \frac{\log x}{\log 2.718}$$

$$= \frac{\log x}{0.4343}$$

$$= 2.303 \log x$$

That is,

$$\ln x = (2.303)\log x$$

## Exercises 9.5

*Find the value of each power of e rounded to three significant digits.*

| | | | |
|---|---|---|---|
| **1.** $e^2$ | **2.** $e^3$ | **3.** $e^6$ | **4.** $e^{10}$ |
| **5.** $e^{-2}$ | **6.** $e^{-3}$ | **7.** $e^{-6}$ | **8.** $e^{-10}$ |
| **9.** $e^{3.5}$ | **10.** $e^{2.1}$ | **11.** $e^{0.15}$ | **12.** $e^{0.75}$ |
| **13.** $e^{-2.5}$ | **14.** $e^{-1.4}$ | **15.** $e^{-0.08}$ | **16.** $e^{-0.65}$ |
| **17.** $e^{2/3}$ | **18.** $e^{5/8}$ | **19.** $e^{-1/3}$ | **20.** $e^{-5/6}$ |

*Find the natural logarithm of each number rounded to four significant digits.*

| | | |
|---|---|---|
| **21.** ln 56 | **22.** ln 92 | **23.** ln 406 |
| **24.** ln 1845 | **25.** ln 4.3 | **26.** ln 0.705 |
| **27.** ln 0.00582 | **28.** ln 0.00000114 | **29.** ln 1 |

*Find x rounded to three significant digits.*

| | | |
|---|---|---|
| **30.** $\ln x = 1.605$ | **31.** $\ln x = 0.475$ | **32.** $\ln x = -0.1463$ |
| **33.** $\ln x = -1.445$ | **34.** $\ln x = -3.77$ | **35.** $\ln x = 14.75$ |
| **36.** $\ln x = -25$ | | |

**37.** Rework Exercise 41 in Section 9.1 using the growth function $y = Ae^{rn}$. Compare the results.

**38.** Rework Exercise 42 in Section 9.1 using the growth function $y = Ae^{rn}$. Compare the results.

**39.** Vera Alice invests $3500 at $8\frac{1}{4}\%$ compounded continuously. How much money will she have in the account after 6 years?

**40.** Many savings institutions quote both *interest rates* and *effective interest rates*. The effective interest rate is defined as

$$e^r - 1$$

where $r$ is the rate of interest compounded continuously. Find the effective interest rate in Exercise 39.

**41.** The population of a city is 95,000. Assuming a growth rate of 3.1%, what will its population be in 10 years?

**42.** The public school enrollment of a city is 14,300. Assuming a decline in enrollment at the rate of 4.5%, what will the enrollment be in 5 years?

**43.** The amount of bacterial growth in a given culture is given by

$$N = N_0 e^{0.04t}$$

where $t$ is the time in hours, 0.04 is the growth rate per hour, $N_0$ is the initial amount, and $N$ is the amount after time $t$. If we begin with a culture of $\overline{3}000$ bacteria, how many do we have after 5.0 hours? Give your answer to two significant digits.

**44.** A bacteria culture contains 25,000 bacteria and grows at the rate of 7.5% per hour. How many bacteria will be present in 24 hours?

**45.** A person invests $10,000 at 8% for 20 years compounded continuously. Assuming an annual 12% inflation rate, what will the person's purchasing power be after 20 years?

**46.** The amount of decay of a certain radioactive element is given by

$$y = y_0 e^{-0.4t}$$

where $t$ is the time in seconds, $-0.4$ is the decay rate per second, $y_0$ is the initial amount, and $y$ is the amount remaining. If a given sample has a mass of $15\overline{0}$ g, how much remains after 5.00 min?

**47.** A given radioactive sample of mass 27.0 g decays at the rate of 2.50% per second. How much remains after 1.00 min?

**48.** In certain types of dc circuits, current increases exponentially according to the formula

$$i = \frac{E}{R}(1 - e^{-Rt/L})$$

where

$$i = \text{the instantaneous current in amperes (A)}$$
$$E = \text{the voltage in volts (V)}$$
$$R = \text{the resistance in ohms } (\Omega)$$
$$t = \text{the time}$$
$$L = \text{the inductance in henries (H)}$$

Find $i$ when $E = 12.0$ V, $R = 90.0\ \Omega$, $t = 0.0120$ s, and $L = 8.50$ H.

*An $8\overline{0}$-mg sample of radioactive radium decays at the rate of $\ln A - \ln 8\overline{0} = kt$, where*

$$A = \text{the amount remaining}$$
$$t = \text{the time in years}$$
$$k = \text{a constant } (-4.10 \times 10^{-4})$$

*Find how long it takes the sample to decay to each amount in Exercises 49–53.*

**49.** 79 mg **50.** 75 mg **51.** $6\overline{0}$ mg

**52.** $4\overline{0}$ mg **53.** $2\overline{0}$ mg

**54.** In determination of the flow of heat in an insulated pipe, the heat loss is given by the formula

$$Q = \frac{2\pi k L(\Delta T)}{\ln\left(\dfrac{D_2}{D_1}\right)}$$

where

$Q$ = the heat loss in British thermal units per hours (Btu/h)

$L$ = length of pipe in feet

$\Delta T$ = the difference in temperature between the inner and outer surfaces of insulation in degrees Fahrenheit

$D_2$ = the outer diameter of insulation in inches

$D_1$ = the inner diameter of insulation in inches

$k$ = a constant

Find the heat loss in Btu per hour from a pipe of 12-in. outside diameter. The pipe is $10\overline{0}$ ft long and is covered with 3.0 in. of insulation with a thermal conductivity of $k = 0.045$ Btu/h °F ft. The inner temperature is 790°F, and the outer temperature is 140°F.

*Use the following equation for Exercises 55–58:*

$$V_o = E_{AS} - (E_{AS} - E_o)e^{-t/TC}$$

**55.** Given:

$E_{AS} = 10.0$ V

$E_o = 0$ V

$TC = 20.0$ μs

$t = 50.0$ μs

Find $V_o$.

**56.** Given:

$E_{AS} = 30.0$ V

$E_o = 10.0$ V

$TC = 50.0$ μs

$t = 75.0$ μs

Find $V_o$.

**57.** Given:

$E_{AS} = -40.0$ V

$E_o = 60.0$ V

$TC = 18.0$ μs

$t = 10.0$ μs

Find $V_o$.

**58.** Given:

$E_{AS} = -75.0$ V

$E_o = -32.0$ V

$TC = 175$ μs

$t = 225$ μs

Find $V_o$.

**59.** Soil permeability as derived from a field pumping test is given by the formula

$$k = \frac{q}{\pi(H_2^2 - H_1^2)} \ln\left(\frac{R_2}{R_1}\right)$$

where $k$ is the coefficient of permeability. Find the coefficient of permeability in ft/min for $H_2 = 48.5$ ft, $H_1 = 44.5$ ft, $R_2 = 24$ ft, $R_1 = 11$ ft, and $q = 51$ gal/min.

**60.** Using the formula in Exercise 59, find the coefficient of permeability in cm/min for $H_2 = 15.0$ m, $H_1 = 12.5$ m, $R_2 = 7.5$ m, $R_1 = 3.5$ m, and $q = 185$ L/min.

*Find each logarithm, rounded to three significant digits.*

**61.** $\log_3 84.1$

**62.** $\log_2 297$

**63.** $\log_4 2360$

**64.** $\log_{12} 5.72$

**65.** $\log_5 374$

**66.** $\log_4 4.19$

**67.** $\log_6 9600$

**68.** $\log_7 16.5$

## 9.6 SOLVING EXPONENTIAL EQUATIONS

The solution of the exponential equation $2^x = 8$ may be done by inspection ($x = 3$). However, the solution of $2^x = 6$ is not integral, and trial-and-error attempts at a solution would be complicated, to say the least. Exponential equations are used in such diverse fields as electronics, biology, chemistry, psychology, and economics.

The solution of the general exponential equation $b^x = a$ ($a > 0$, $b > 0$) is based on the fact that if two numbers are equal, their logarithms to the same base (any base) are equal.

$$\begin{array}{ll} \text{If} & x = y \\ \text{then} & log_b x = log_b y \end{array}$$

Let us consider the following example.

**EXAMPLE 1**

Solve $2^x = 60$ to three significant digits.

$$2^x = 60$$
$$\log 2^x = \log 60 \qquad \text{(Take the common log of each side.)}$$
$$x \cdot \log 2 = \log 60 \qquad \text{(Property 3)}$$
$$x = \frac{\log 60}{\log 2} = \frac{1.7782}{0.3010} = 5.91$$

Now, let's solve the same equation by using natural logarithms.

$$2^x = 60$$
$$\ln 2^x = \ln 60 \qquad \text{(Take the natural log of each side.)}$$
$$x \cdot \ln 2 = \ln 60 \qquad \text{(Property 3)}$$
$$x = \frac{\ln 60}{\ln 2} = \frac{4.0943}{0.6931} = 5.91$$

As you can see, the solution to this equation can be done quite easily using either common or natural logarithms.

**EXAMPLE 2**

Solve $4^{x+2} = 36$ by using common logarithms. (Give the result to three significant digits.)

$$4^{x+2} = 36$$
$$\log 4^{x+2} = \log 36 \qquad \text{(Take the common log of each side.)}$$
$$(x + 2)\log 4 = \log 36$$
$$x + 2 = \frac{\log 36}{\log 4}$$
$$x = \frac{\log 36}{\log 4} - 2$$
$$= 0.585$$

**EXAMPLE 3**

Solve $4^{2x} = 12^{x+1}$ by using natural logarithms. (Give the result to three significant digits.)

$$4^{2x} = 12^{x+1}$$
$$\ln 4^{2x} = \ln 12^{x+1} \qquad \text{(Take the natural log of each side.)}$$
$$2x \ln 4 = (x + 1)\ln 12$$
$$2x \ln 4 = x \ln 12 + \ln 12$$
$$2x \ln 4 - x \ln 12 = \ln 12$$
$$x(2 \ln 4 - \ln 12) = \ln 12$$
$$x = \frac{\ln 12}{2 \ln 4 - \ln 12}$$
$$= 8.64$$

Since $\log 10^x = x$ and $\ln e^x = x$, using common logs is easier when solving an exponential equation involving a power of 10, and using natural logs is easier when solving an exponential equation involving a power of $e$.

**EXAMPLE 4**

Solve $10^{2x} = 1450$. (Give the result to three significant digits.)

$$10^{2x} = 1450$$
$$\log 10^{2x} = \log 1450 \qquad \text{(Take the common log of each side.)}$$
$$2x = \log 1450$$
$$x = \frac{\log 1450}{2}$$
$$= 1.58$$

**EXAMPLE 5**

Solve $e^{-3x} = 0.725$. (Give the result to three significant digits.)

$$e^{-3x} = 0.725$$
$$\ln e^{-3x} = \ln 0.725 \qquad \text{(Take the natural log of each side.)}$$
$$-3x = \ln 0.725$$
$$x = \frac{\ln 0.725}{-3}$$
$$= 0.107$$

**EXAMPLE 6**

Solve $36 = 45(1 - e^{x/2})$. (Give the result to three significant digits.)

$$36 = 45(1 - e^{x/2})$$
$$\frac{36}{45} = 1 - e^{x/2}$$
$$e^{x/2} = 1 - \frac{36}{45}$$
$$e^{x/2} = 1 - 0.8 = 0.2$$
$$\ln e^{x/2} = \ln 0.2$$

$$\frac{x}{2} = \ln 0.2$$

$$x = 2 \ln 0.2$$

$$= -3.22$$

## Exercises 9.6

*Solve each exponential equation. (Give each result to three significant digits.)*

**1.** $3^x = 12$

**2.** $5^x = 38.1$

**3.** $2^{-x} = 43.7$

**4.** $4^x = 0.439$

**5.** $3^{2x} = 0.21$

**6.** $5^{-3x} = 100$

**7.** $5^{x+1} = 3^x$

**8.** $2^{2x+1} = 5^x$

*Solve for x. (Give each result to four significant digits.)*

**9.** $4^{-20x} = 50$

**10.** $(5)2^{x+16} = 382$

**11.** $3^{2x+1} = 5^{-x}$

**12.** $8^{-3x} = 3^{2x-1}$

**13.** $e^x = 23$

**14.** $10^x = 1.35$

**15.** $10^{-x} = 0.146$

**16.** $e^{-x} = 14.7$

**17.** $e^{2x} = 40.5$

**18.** $10^{5x} = 2.63$

**19.** $e^{-3x} = 850$

**20.** $e^{-2x} = 0.00448$

**21.** $10^{x/2} = 0.45$

**22.** $e^{x/3} = 47.2$

**23.** $4e^x = 94.7$

**24.** $5.3e^{-4x} = 49.7$

**25.** $e^x = 5^{x-1}$

**26.** $e^{-2x} = 10^{x+1}$

**27.** $e^x = 4^{2x+1}$

**28.** $4e^{3x} = (10)2^{x+3}$

**29.** $5e^{-2x} = (12)3^{x-1}$

**30.** $(40)5^{2x+1} = (250)e^{x+1}$

**31.** $e^{x^2} = 600$

**32.** $10^{x^2} = 4650$

**33.** $3 = 4(1 - e^x)$

**34.** $8 = 20(1 - e^x)$

**35.** $35 = 80(1 - e^{-x})$

**36.** $125 = 175(1 - e^{-x})$

**37.** $8 = 10(1 - e^{x/2})$

**38.** $150 = 425(1 - e^{x/5})$

**39.** $175 = 225(1 - e^{-x/3})$

**40.** $28 = 56(1 - e^{-x/10})$

**41.** $48 = 64(1 - e^{-3x})$

**42.** $120 = 125(1 - e^{3x})$

**43.** $135 = 145(1 - e^{-4x/3})$

**44.** $225 = 375(1 - e^{-5x/6})$

**45.** $50 = 75(1 + e)^x$

**46.** $16 = 18(1 + e)^{4x}$

**47.** $9 = 15(1 + e)^{-x}$

**48.** $1 = 2(1 + e)^{-2x}$

**49.** $7 = 10(1 + e)^{x/2}$

**50.** $8 = 12(1 + e)^{-3x/2}$

*Use the following equation for Exercises 51–56:*

$$V_o = E_{AS} - (E_{AS} - E_o)e^{-t/TC}$$

**51.** Given:

$E_{AS} = 20.0$ V

$E_o = -20.0$ V

$TC = 50.0$ μs

$V_o = 0$

Find $t$.

**52.** Given:

$E_{AS} = 10.0$ V

$E_o = 20.0$ V

$TC = 15.0$ μs

$V_o = 15.0$ V

Find $t$.

**53.** Given:

$E_{AS} = 27.0$ V

$E_o = 30.0$ V

$V_o = 28.5$ V

$t = 145$ μs

Find $TC$.

**54.** Given:

$E_{AS} = 72.0$ V

$E_o = 95.0$ V

$V_o = 90.0$ V

$t = 10.5$ ms

Find $TC$.

**55.** Given:

$$E_o = 15.0 \text{ V}$$
$$V_o = 50.0 \text{ V}$$
$$TC = 10.0 \text{ ms}$$
$$t = 32.0 \text{ ms}$$

Find $E_{AS}$.

**56.** Given:

$$E_{AS} = 50.0 \text{ V}$$
$$V_o = 30.0 \text{ V}$$
$$TC = 90.0 \text{ } \mu\text{s}$$
$$t = 15\overline{0} \text{ } \mu\text{s}$$

Find $E_o$.

## 9.7 SOLVING LOGARITHMIC EQUATIONS

One principle used in solving a logarithmic equation is as follows.

> If $\quad log_b x = log_b y$
> then $\quad\quad x = y$

**EXAMPLE 1**

Solve $\log(x + 7) = \log(3x - 5)$.

$$\log(x + 7) = \log(3x - 5)$$
$$x + 7 = 3x - 5$$
$$12 = 2x$$
$$6 = x$$

**EXAMPLE 2**

Solve $\ln 2 + \ln(x - 4) = \ln(x + 6)$.

$$\ln 2 + \ln(x - 4) = \ln(x + 6)$$
$$\ln 2(x - 4) = \ln(x + 6)$$
$$2(x - 4) = x + 6$$
$$2x - 8 = x + 6$$
$$x = 14$$

Another technique used in solving a logarithmic equation is changing the logarithmic equation to exponential form.

> If $\quad y = log_b x$
> then $\quad x = b^y$

Then solve the resulting equation.

**EXAMPLE 3**

Solve $\log(x - 3) = 2$.

$$\log(x - 3) = 2$$
$$x - 3 = 10^2 \quad \text{(Write in exponential form.)}$$
$$x - 3 = 100$$
$$x = 103$$

**EXAMPLE 4**

Solve $\log(x + 3) + \log x = 1$.

$$\log(x + 3) + \log x = 1$$
$$\log x(x + 3) = 1 \qquad \text{(Property 1)}$$
$$x(x + 3) = 10^1 \qquad \text{(Write in exponential form.)}$$
$$x^2 + 3x - 10 = 0$$
$$(x + 5)(x - 2) = 0$$
$$x + 5 = 0 \quad \text{or} \quad x - 2 = 0$$
$$x = -5 \qquad\qquad x = 2$$

Since logarithms of negative numbers are not defined, the solution is $x = 2$.

**EXAMPLE 5**

Solve $\ln(x + 1) - \ln x = 3$.

$$\ln(x + 1) - \ln x = 3$$
$$\ln\frac{x + 1}{x} = 3 \qquad \text{(Property 2)}$$
$$\frac{x + 1}{x} = e^3 \qquad \text{(Write in exponential form.)}$$
$$x + 1 = xe^3$$
$$1 = xe^3 - x$$
$$1 = x(e^3 - 1)$$
$$\frac{1}{e^3 - 1} = x$$
$$x = 0.0524$$

## Exercises 9.7

*Solve each equation.*

**1.** $\log(3x - 4) = \log(x + 6)$

**2.** $\log(2x + 1) = \log(4x - 5)$

**3.** $\ln(x - 4) = \ln(2x + 5)$

**4.** $\ln(5x - 100) = \ln x$

**5.** $\log 2 + \log(x - 3) = \log(x + 1)$

**6.** $\log 3 + \log(x + 1) = \log 5 + \log(x - 3)$

**7.** $2 \ln x = \ln 49$

**8.** $3 \ln x = \ln 8$

**9.** $\log x + \log(2x) = \log 72$

**10.** $\log(3x) + \log(2x) = \log 150$

**11.** $\log(x + 4) - \log(x - 2) = \log 2$

**12.** $\log(2x + 1) - \log(x - 4) = \log 3$

**13.** $\ln x + \ln(x - 2) = \ln 3$

**14.** $\ln(2x) + \ln(x + 4) = \ln 24$

**15.** $\ln x + \ln(x - 4) = \ln(x + 6)$

**16.** $2 \ln x = \ln(12 - 5x)$

**17.** $2 \log x = 2$

**18.** $2 \ln x = 1$

**19.** $\log(x + 1) = 1$

**20.** $\log(2x + 4) = 2$

**21.** $2 \log(x - 1) = 1$

**22.** $\ln(x - 1) = 2$

**23.** $\log(x + 1) - 2 \log 3 = 2$

**24.** $\log(2x - 3) - 3 \log 2 = 1$

**25.** $\ln(x - 3) - \ln 2 = 1$

**26.** $\ln(2x - 1) + \ln 2 = 1$

**27.** $\ln(2x + 1) + \ln 3 = 2$

**28.** $\ln(2x + 3) - \ln 4 = 2$

**29.** $\log(x + 3) + 2 \log 4 = 2$

**30.** $\log(2x + 1) + 3 \log 3 = 3$

**31.** $\log(2x + 1) + \log(x - 1) = 2$

**32.** $\log(x + 3) + \log(x + 1) = 1$

**33.** $\ln x + \ln(x + 2) = 1$

**34.** $\ln(2x) + \ln(x - 3) = 2$

**35.** $\ln x + \ln(2x - 3e) - 2 = \ln 2$

**36.** $\ln x + \ln(2x - e) - 2 = \ln 3$

## 9.8 APPLICATIONS: SOLVING EXPONENTIAL AND LOGARITHMIC EQUATIONS

Of the many applications involving exponential and logarithmic equations, we present a variety involving several subject fields.

### EXAMPLE 1

The rate of growth of money compounded quarterly is given by the formula

$$A = P\left(1 + \frac{r}{4}\right)^{4n}$$

where

$A =$ the amount of original principal plus accrued interest

$P =$ the original principal invested

$r =$ the rate of interest (expressed as a decimal)

$n =$ the number of years the principal is invested

How long will it take $5000 invested quarterly at 6.4% to accrue to $8000? Here,

$$A = \$8000$$
$$P = \$5000$$
$$r = 6.4\% = 0.064$$

We are to find $n$.

$$A = P\left(1 + \frac{r}{4}\right)^{4n}$$

$$8000 = 5000\left(1 + \frac{0.064}{4}\right)^{4n}$$

$$\frac{8000}{5000} = (1 + 0.016)^{4n} \qquad \text{(Divide each side by 5000.)}$$

$$1.6 = 1.016^{4n}$$

$$\log 1.6 = \log 1.016^{4n} \qquad \text{(Take the common log of each side.)}$$

$$\log 1.6 = 4n \log 1.016$$

$$n = \frac{\log 1.6}{4 \log 1.016}$$

$$= 7.4 \text{ years} \qquad \text{(Rounded to two significant digits)}$$

## EXAMPLE 2

Mary and Sid deposit $2500 in a savings account that pays $7\frac{1}{4}\%$ compounded continuously. In how many years will the amount in the account double?

Using $y = Ae^{rn}$, we know

$$A = \$2500$$
$$y = \$5000$$
$$r = 7\tfrac{1}{4}\% = 7.25\% = 0.0725$$

We are to find $n$.

$$y = Ae^{rn}$$
$$5000 = 2500e^{(0.0725)n}$$
$$\frac{5000}{2500} = e^{0.0725n} \qquad \text{(Divide each side by 2500.)}$$
$$2 = e^{0.0725n}$$
$$\ln 2 = \ln e^{0.0725n} \qquad \text{(Take the natural log of each side.)}$$
$$\ln 2 = 0.0725n$$
$$n = \frac{\ln 2}{0.0725}$$
$$= 9.6 \text{ years} \qquad \text{(Rounded to two significant digits)}$$

## EXAMPLE 3

The amount of decay of a certain radioactive element is given by

$$y = y_0 e^{-0.4t}$$

where

$$t = \text{the time in seconds (s)}$$
$$y_0 = \text{the initial amount}$$
$$y = \text{the amount remaining}$$

How long will it take $5\overline{0}$ g to decay to 25 g?

Here,

$$y_0 = 5\overline{0} \text{ g}$$
$$y = 25 \text{ g}$$

We are to find $t$.

$$y = y_0 e^{-0.4t}$$
$$25 = 50e^{-0.4t}$$
$$\frac{25}{50} = e^{-0.4t} \qquad \text{(Divide each side by 50.)}$$
$$0.5 = e^{-0.4t}$$
$$\ln 0.5 = \ln e^{-0.4t} \qquad \text{(Take the natural log of each side.)}$$
$$\ln 0.5 = -0.4t$$
$$t = \frac{\ln 0.5}{-0.4}$$
$$= 1.7 \text{ s} \qquad \text{(Rounded to two significant digits)}$$

*Note:* This time is called the **half-life** of the substance.

**EXAMPLE 4**

In Section 9.4 we defined pH as

$$pH = -\log(H^+)$$

where $H^+$ is the numerical value for the concentration of hydrogen ions in moles per litre (M/L). Vinegar has a pH of 2.2. Find its $H^+$.

$$pH = -\log(H^+)$$

$$2.2 = -\log(H^+)$$

$$-2.2 = \log(H^+) \qquad \text{(Multiply each side by } -1.)$$

$$H^+ = 10^{-2.2} \qquad \text{(Write in exponential form.)}$$

$$H^+ = 6.3 \times 10^{-3} \text{ M/L}$$

**EXAMPLE 5**

In Section 9.4 we defined the intensity level of sound as

$$\beta = 10 \log \frac{I}{I_0}$$

where

$$\beta = \text{the intensity level in decibels (dB)}$$

$$I = \text{the intensity of a given sound (W/cm}^2)$$

$$I_0 = \text{the intensity of the threshold of hearing, } 10^{-16} \text{ W/cm}^2$$

What is the intensity (in W/cm$^2$) of street traffic sound that has an intensity level of 75 dB?
Here

$$\beta = 75 \text{ dB}$$

$$I_0 = 10^{-16} \text{ W/cm}^2$$

We are to find $I$.

$$\beta = 10 \log \frac{I}{I_0}$$

$$75 = 10 \log \frac{I}{10^{-16}}$$

$$7.5 = \log \frac{I}{10^{-16}}$$

$$7.5 = \log I - \log 10^{-16}$$

$$7.5 = \log I - (-16)$$

$$-8.5 = \log I$$

$$I = 10^{-8.5} \qquad \text{(Write in exponential form.)}$$

$$= 3.2 \times 10^{-9} \text{ W/cm}^2 \qquad \text{(Rounded to two significant digits)}$$

# Exercises 9.8

**1.** Rework Example 1 to find how long it will take **(a)** $2500 invested quarterly at 7.5% to accrue to $4500 and **(b)** $2500 to triple.

2. Rework Example 2 to find how long it will take (a) $5000 compounded continuously at $8\frac{1}{4}\%$ to accrue to $12,500 and (b) $5000 to double.

3. How long will it take 175 g of the element in Example 3 to decay to 25 g?

4. The amount of bacterial growth in a given culture is given by $N = N_0 e^{0.04t}$, where $t$ is the time in hours, $N_0$ is the initial amount, and $N$ is the amount after time $t$. How long will it take a bacteria culture with 6500 members to grow to $10,\overline{0}00$?

*Approximate the hydrogen-ion concentration in moles per litre ($H^+$) for each of the following (see Example 4).*

5. pH = 3.5        6. pH = 4.6        7. pH = 9.5

8. pH = 8.1        9. pH = 2.7        10. pH = 6.5

*Compute the intensity of sound (in watts per square centimetre) for each of the following (see Example 5).*

11. Whisper: 15 dB        12. Normal conversation: $6\overline{0}$ dB

13. Pain threshold: $12\overline{0}$ dB        14. Thunder: $11\overline{0}$ dB

15. Soft music: $3\overline{0}$ dB        16. Hard rock: $10\overline{0}$ dB

17. A city with a population of 75,000 has a growth rate of 4%.
    (a) How long will it take the population to reach $10\overline{0},000$?
    (b) How long will it take the population of the city to double?

18. Harry retires with an annual pension of $65,000.
    (a) With an 8% inflation rate, how long will it take his purchasing power to fall to $50,000?
    (b) How long will it take Harry's purchasing power to be halved?

19. A current flows in a given circuit according to $i = 2.7e^{-0.2t}$. Find $t$ in seconds when $i = 0.35$ A.

20. The discharge current of a capacitor is $i = 0.01e^{-75t}$. Find $t$ in seconds when $i = 0.0040$ A.

21. For a given circuit the instantaneous current, $i$, at any time, $t$ in ms, is given by the formula

$$i = \frac{E}{R}e^{-t/RC}$$

where $E$ is the voltage in volts (V), $R$ is the resistance in ohms ($\Omega$), and $C$ is the capacitance in farads (F). Find $t$ when $i = 3.91 \times 10^{-5}$ A, $E = 10\overline{0}$ V, $R = 2.00 \times 10^4$ $\Omega$, and $C = 2.00 \times 10^{-8}$ F.

22. When an electric capacitor is discharged through a resistor, the voltage across the capacitor decays according to the formula

$$E = E_0 e^{-t/RC}$$

where $E_0$ is the original voltage, $t$ is the time for $E_0$ to decrease to $E$, $R$ is the resistance in ohms ($\Omega$), and $C$ is the capacitance in farads (F). Find the time, $t$ in ms, for the voltage across a given capacitor to decay to 1.0% of its original value when $R = 30,\overline{0}00$ $\Omega$ and $C = 5.00$ $\mu$F.

23. A capacitance, $C$ in farads (F), is charged through a resistance, $R$, by a source having a constant voltage, $V$. The capacitor voltage, $V_c$, varies according to the formula

$$V_c = V(1 - e^{-t/RC})$$

Find $t$ in $\mu$s when $R = 2.00 \times 10^6$ $\Omega$, $C = 4.00$ $\mu$F, $V_c = 75.0$ V, and $V = 10\overline{0}$ V.

*Under certain conditions, the pressure and volume of a gas are related as*

$$\ln P = C - \gamma \ln V$$

*where C and g are constants.*

**24.** Find $P$ if $C = 2.5$, $\gamma = 1.6$, and $V = 2.5$.

**25.** Find $V$ if $C = 3.2$, $\gamma = 1.4$, and $P = 8.5$.

*A 50.0-mg sample of radioactive radium decays at the rate of*

$$\ln A - \ln 50.0 = kt$$

*where A is the amount remaining, t is the time in years, and k is a constant $(-4.10 \times 10^{-4})$. Find the amount remaining after each interval of time in Exercises 26–29.*

**26.** 1.00 year

**27.** 25.0 years

**28.** $10\overline{0}$ years

**29.** $50\overline{0}$ years

## 9.9 DATA ALONG A STRAIGHT LINE

### Using Rectangular Graph Paper

Sometimes the graph of data results in a set of points that closely approximates a straight line. We can approximate this straight line by drawing the "best straight line" through the given points. The slope of the line can be found graphically by choosing two convenient points and by computing $m = \dfrac{y_2 - y_1}{x_2 - x_1}$. The y-intercept can be read directly where the line crosses the y-axis. The equation of this straight line can be written using the slope-intercept form, $y = mx + b$.

**EXAMPLE 1**

Given the following data, graph the best straight line and find its equation:

| $x$ | 1 | 2 | 3 | 4 | 5 | 6 |
|---|---|---|---|---|---|---|
| $y$ | 2 | 5 | 4 | 7 | 8 | 10 |

Choose two points on the line in Fig. 9.6, such as (1, 2) and (5, 8). The slope of the line is found by

$$m = \frac{y_2 - y_1}{x_2 - x_1} = \frac{8 - 2}{5 - 1} = \frac{6}{4} = 1.5$$

From the graph the y-intercept is approximately 1. Therefore, the equation of the preceding data is

$$y = 1.5x + 1$$

Obviously, the difficulty of this method lies with the inaccuracy involved in drawing the best straight line. Different individuals will probably get slightly different results for the equation of the straight line, but any one of these results is usually accurate enough for most work.

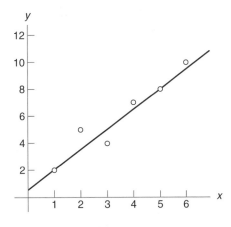

**Figure 9.6**

Graphics calculators have built-in routines that are designed to accurately calculate the slope and the $y$-intercept of this line of best fit (see Fig. 9.7). Statisticians call this process linear regression, which explains why the name of the routine is LinReg(ax + b).

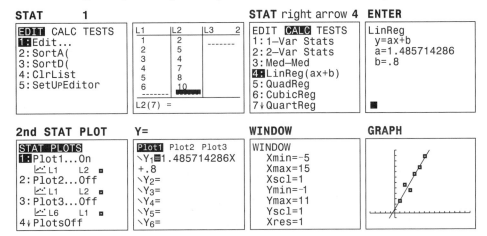

**Figure 9.7**

## Using Logarithmic Graph Paper

Logarithmic graph paper is ruled or scaled logarithmically both vertically and horizontally rather than linearly, as have been all our previous graphs.

If you plot a set of given data on logarithmic graph paper and it yields a straight line, the equation of the given data may be expressed in the form $y = ax^k$, where $a > 0$ and $x > 0$. To see why $y = ax^k$ is a straight line on logarithmic graph paper, take the logarithm of each side.

$$y = ax^k$$
$$\log y = \log ax^k$$
$$\log y = \log a + \log x^k$$
$$\log y = k \log x + \log a$$

which is a linear equation in slope-intercept form where $k$, the exponent, is the slope of the line and log $a$ is the $y$-intercept.

On logarithmic graph paper we plot log $y$ versus log $x$. Therefore, the slope of a straight line on logarithmic graph paper is found by

$$m = \frac{\log y_2 - \log y_1}{\log x_2 - \log x_1}$$

Also, the value of $a$ can be read directly from the graph where the line crosses the log $y$-axis.

**EXAMPLE 2**

Plot the graph of $y = 4x^2$ on logarithmic graph paper.

Generate some ordered pairs that satisfy the equation and then plot them as in Fig. 9.8.

| $x$ | 2 | 3 | 4 | 5 | 7 | 8 | 10 |
|---|---|---|---|---|---|---|---|
| $y$ | 16 | 36 | 64 | 100 | 196 | 256 | 400 |

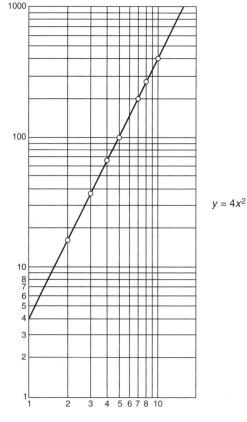

**Figure 9.8**

To verify our graph, note that if we chose two points on the line, such as (4, 64) and (8, 256), we would compute the slope as follows:

$$m = \frac{\log y_2 - \log y_1}{\log x_2 - \log x_1}$$

$$= \frac{\log 256 - \log 64}{\log 8 - \log 4} = 2$$

which is the value of the exponent, $k$.

*Note:* The line crosses the log $y$-axis at 4, which is the value of $a$.

**EXAMPLE 3**

Plot the following set of data on logarithmic graph paper, graph the best straight line, and find its equation.

| $x$ | 2 | 3 | 5 | 7 | 9 | 11 | 13 |
|---|---|---|---|---|---|---|---|
| $y$ | 18 | 51 | 130 | 275 | 408 | 650 | 890 |

Choose two points *on the line* in Fig. 9.9, such as (2, 20) and (5, 130). The value of $k$, the exponent, equals the slope of the line,

$$m = \frac{\log y_2 - \log y_1}{\log x_2 - \log x_1}$$

$$= \frac{\log 130 - \log 20}{\log 5 - \log 2} = 2$$

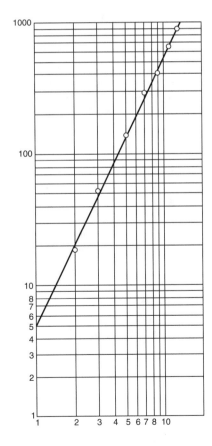

**Figure 9.9**

which is $k$. The line crosses the log $y$-axis at approximately 5, which is $a$. The equation of the preceding data is

$$y = 5x^2$$

## Using Semilogarithmic Graph Paper

Semilogarithmic graph paper has one axis ruled or scaled logarithmically and the other axis ruled linearly.

If you plot a set of given data on semilogarithmic graph paper and it yields a straight line, its equation may be expressed in the form $y = ak^x$, where $a > 0$ and $k > 0$.

Given $y = ak^x$, take the logarithm of each side.

$$\log y = \log ak^x$$
$$= \log a + \log k^x$$
$$\log y = x \log k + \log a$$

which is a linear equation in slope-intercept form, where $\log k$ (the logarithm of the exponent) is the slope and $\log a$ is the $y$-intercept.

On semilogarithmic graph paper we plot $\log y$ versus $x$. Therefore, the slope of a straight line on semilogarithmic graph paper is found by

$$m = \frac{\log y_2 - \log y_1}{x_2 - x_1}$$

**EXAMPLE 4**

Plot the graph of $y = 3^x$ on semilogarithmic graph paper.

Generate some ordered pairs that satisfy the equation and then plot them as in Fig. 9.10.

| $x$ | 1 | 3 | 4 | 6 |
|-----|---|----|----|-----|
| $y$ | 3 | 27 | 81 | 729 |

To verify our graph, note that if we chose two points on the line, such as $(3, 27)$ and $(6, 729)$, we would find the slope as follows:

$$m = \frac{\log y_2 - \log y_1}{x_2 - x_1}$$
$$= \frac{\log 729 - \log 27}{6 - 3}$$
$$= 0.4771 = \log k$$

We find the antilogarithm, $k = 3$, as above.

From the graph of Fig. 9.10, note that at $x = 1$, $y = 3$. Substituting the known values into

$$y = ak^x$$
$$3 = a3^1$$

we find $1 = a$ as above.

**EXAMPLE 5**

Plot the following set of data on semilogarithmic graph paper, graph the best straight line, and find its equation.

| $x$ | 2 | 4 | 5 | 7 | 9 | 10 |
|---|---|---|---|---|---|---|
| $y$ | 3.3 | 10.6 | 19.8 | 64.2 | 205 | 346 |

Choose two points on the line in Fig. 9.11, such as (3, 6) and (8, 110). The value of $m$ is

$$\log k = m = \frac{\log y_2 - \log y_1}{x_2 - x_1}$$

$$= \frac{\log 110 - \log 6}{8 - 3}$$

$$= 0.253 = \log k$$

We find the antilogarithm, $k = 1.8$.

From the graph of Fig. 9.11, note that at $x = 3$, $y = 6$. Substituting the known values into

$$y = ak^x$$

$$6 = a(1.8^3)$$

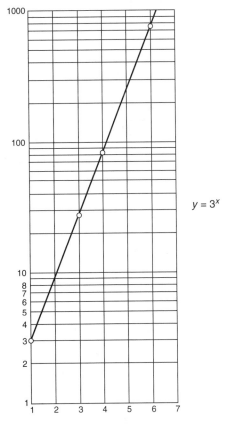

$y = 3^x$

**Figure 9.10**

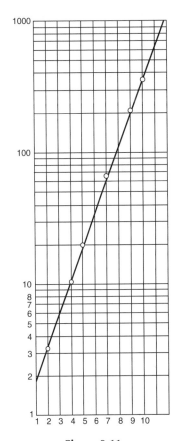

**Figure 9.11**

we find $1.0 = a$. The equation of the preceding data is

$$y = 1.8^x$$

## Exercises 9.9

*Given each set of data on rectangular graph paper, graph the best straight line, and find its equation.*

**1.**

| $x$ | 2 | 4 | 5 | 11 | 15 | 24 |
|---|---|---|---|---|---|---|
| $y$ | 21 | 30 | 31 | 57 | 74 | 104 |

**2.**

| $x$ | 6 | 9 | 15 | 24 | 36 | 54 |
|---|---|---|---|---|---|---|
| $y$ | 10 | 38 | 88 | 170 | 240 | 410 |

**3.**

| $x$ | 20 | 35 | 55 | 75 | 85 | 120 |
|---|---|---|---|---|---|---|
| $y$ | 295 | 605 | 920 | 1270 | 1550 | 2450 |

**4.**

| $x$ | 0.08 | 0.14 | 0.21 | 0.28 | 0.40 | 0.52 | 0.61 | 0.72 |
|---|---|---|---|---|---|---|---|---|
| $y$ | 0.21 | 0.26 | 0.27 | 0.31 | 0.32 | 0.41 | 0.44 | 0.49 |

*Plot the graph of each equation on logarithmic graph paper.*

**5.** $y = 3x^2$

**6.** $y = \frac{3}{2}x^2$

**7.** $y = 2x^3$

**8.** $y = 1.7x^4$

*Plot each set of data on logarithmic graph paper, graph the best straight line, and find its equation.*

**9.**

| $x$ | 2 | 3.7 | 8 | 9.5 | 11 | 14.6 |
|---|---|---|---|---|---|---|
| $y$ | 7.5 | 29 | 125 | 170 | 247 | 438 |

**10.**

| $x$ | 3 | 4.7 | 6.4 | 9.0 | 12.5 |
|---|---|---|---|---|---|
| $y$ | 32 | 74 | 151 | 268 | 561 |

**11.**

| $x$ | 2.1 | 3.9 | 6.4 | 8.7 |
|---|---|---|---|---|
| $y$ | 53 | 177 | 510 | 915 |

**12.**

| $x$ | 3.0 | 4.3 | 5.4 | 6.2 | 7.1 |
|---|---|---|---|---|---|
| $y$ | 61 | 192 | 349 | 550 | 841 |

*Plot the graph of each equation on semilogarithmic graph paper.*

**13.** $y = 4^x$

**14.** $y = 2^x$

**15.** $y = e^x$

**16.** $y = (\frac{3}{2})^x$

**17.** $y = 4(3^x)$

**18.** $y = 2.5(4^x)$

*Plot each set of data on semilogarithmic graph paper, graph the best straight line, and find its equation.*

**19.**

| $x$ | 2 | 4 | 5 | 6 | 7 |
|---|---|---|---|---|---|
| $y$ | 7.4 | 54 | 152 | 410 | 995 |

**20.**

| $x$ | 2 | 3 | 4 | 5 |
|---|---|---|---|---|
| $y$ | 13.1 | 57.2 | 174 | 610 |

**21.**

| $x$ | 2 | 4 | 8 | 10 | 12 | 15 |
|---|---|---|---|---|---|---|
| $y$ | 0.82 | 0.63 | 0.45 | 0.35 | 0.27 | 0.21 |

**22.**

| $x$ | 2 | 4 | 8 | 10 | 12 | 15 | 20 |
|---|---|---|---|---|---|---|---|
| $y$ | 1.5 | 2.0 | 4.4 | 6.0 | 8.8 | 15.5 | 38.0 |

**23.**

| $x$ | 2 | 3 | 4 | 5 |
|---|---|---|---|---|
| $y$ | 19 | 60 | 160 | 515 |

**24.**

| $x$ | 2 | 3 | 4 | 5 | 6 |
|---|---|---|---|---|---|
| $y$ | 13 | 33 | 81 | 235 | 612 |

# CHAPTER SUMMARY

1. *Exponential function:*

$$y = b^x \quad \text{where } b > 0 \quad \text{and} \quad b \neq 1$$

   (a) For $b > 1$, $y = b^x$ is an increasing function.
   (b) For $b < 1$, $y = b^x$ is a decreasing function.

2. The logarithmic function is the inverse of the exponential function. The following middle and right equations show how to express this logarithmic equation in either exponential form or logarithmic form:

| Exponential equation | Logarithmic equation in exponential form | Logarithmic equation in logarithmic form |
|---|---|---|
| $y = b^x$ | $x = b^y$ | $y = \log_b x$ |

   That is, $x = b^y$ and $y = \log_b x$ are equivalent equations for $b > 0$ but $b \neq 1$.
   *Remember:* A logarithm is an exponent.

3. *Properties of logarithms: M, N,* and *a* are positive real numbers and $a \neq 1$.
   (a) *Multiplication:*

$$\log_a(MN) = \log_a M + \log_a N$$

   (b) *Division:*

$$\log_a\left(\frac{M}{N}\right) = \log_a M - \log_a N$$

   (c) *Powers:* For real numbers $n$,

$$\log_a M^n = n \log_a M$$

   (i) *Roots:*

$$\log_a \sqrt[n]{M} = \frac{1}{n} \cdot \log_a M$$

(ii) For $n = 0$,

$$\log_a 1 = 0$$

(iii) For $n = -1$,

$$\log_a \frac{1}{M} = -\log_a M$$

(d) $\log_a(a^x) = x$
(e) $a^{(\log_a x)} = x$

4. *Common logarithms:* Base 10, written $\log_{10} x$ or $\log x$.

5. *Natural logarithms:* Base $e$, written $\log_e x$ or $\ln x$, where $e$ is an irrational number approximately equal to 2.71828.

6. *Growth functions:* $y = A(1 + r)^n$ or $y = Ae^{rn}$, where $r$ is the rate of growth in decimal form, $n$ is the time interval, $A$ is the initial amount, and $y$ is the new amount. For $n > 0$, the function is increasing. For $n < 0$, the function is decreasing.

7. To find the logarithm with another base, use the formula

$$\log_b x = \frac{\log_a x}{\log_a b}$$

where $a$ is base 10 or base $e$, and $b$ is another base.

8. Finding the solution of the general exponential equation $b^x = a$ is usually begun by taking the common logarithm or the natural logarithm of each side of the equation.

9. The solution of a logarithmic equation is usually based on one of the following:
   (a) If $\log_b x = \log_b y$, then $x = y$.
   (b) If $y = \log_b x$, then $x = b^y$.

10. Review Section 9.9 for using rectangular graph paper, logarithmic graph paper, and semilogarithmic graph paper.

# CHAPTER 9 REVIEW

*Graph each equation.*

**1.** $y = 3^x$

**2.** $y = \log_3 x$

*Write each equation in logarithmic form.*

**3.** $2^4 = 16$

**4.** $10^{-3} = 0.001$

*Write each equation in exponential form.*

**5.** $\log_{10} 7.389 = 0.8686$

**6.** $\log_4\left(\frac{1}{16}\right) = -2$

*Solve for x.*

**7.** $\log_9 x = 2$

**8.** $\log_x 8 = 3$

**9.** $\log_2 32 = x$

*Write each expression as a sum or difference of multiples of single logarithms.*

**10.** $\log_4 6x^2 y$

**11.** $\log_3 \dfrac{5x\sqrt{y}}{z^3}$

**12.** $\log \dfrac{x^2(x+1)^3}{\sqrt{x-4}}$

**13.** $\ln \dfrac{[x(x-1)]^3}{\sqrt{x+1}}$

*Write each expression as a single logarithmic expression.*

**14.** $\log_2 x + 3 \log_2 y - 2 \log_2 z$

**15.** $\dfrac{1}{2} \log(x+1) - 3 \log(x-2)$

**16.** $4 \ln x - 5 \ln(x+1) - \ln(x+2)$

**17.** $\dfrac{1}{2}[\ln x + \ln(x+2)] - 2 \ln(x-5)$

*Simplify.*

**18.** $\log 1000$

**19.** $\log 10^{x^2}$

**20.** $\ln e^2$

**21.** $\ln e^x$

*Find the common logarithm of each number. (Round to four significant digits.)*

**22.** $\log 664.8$

**23.** $\log 0.04046$

**24.** $\log 14{,}420$

*Find N, the antilogarithm. (Round to three significant digits.)*

**25.** $\log N = 3.0737$

**26.** $\log N = -2.4289$

**27.** $\log N = -1.7522$

*Find the natural logarithm of each number, rounded to four significant digits.*

**28.** $\ln 72$

**29.** $\ln 421$

**30.** $\ln 0.00185$

*Find x rounded to three significant digits.*

**31.** $\ln x = 1.315$

**32.** $\ln x = 3.45$

**33.** $\ln x = -0.24$

**34.** Evaluate $\log_4 20$.

*Solve for x. (Round to three significant digits.)*

**35.** $6^{-2x} = 48.1$

**36.** $3^{4x-1} = 14^x$

**37.** $26.5 = 3.81e^{4x}$

**38.** $48 = 72(1 - e^{-x/2})$

*Solve each logarithmic equation.*

**39.** $\log(x+4) = 2$

**40.** $\log(2x+3) - 3 \log 2 = 2 \log 2$

**41.** $\log(x+1) + \log(x-2) = 1$

**42.** $\ln x = \ln(3x-2)$

**43.** $\ln(x+1) - \ln x = \ln 3$

**44.** $2 \ln x = 3$

**45.** Suppose the population of a certain city is given by

$$y = 125{,}000e^{-0.03t}$$

where $t$ is the time in years. Find its population in 5.0 years.

**46.** If the annual inflation rate is 8%, how long would it take the average price level to double?

**47.** The energy of an expanding gas at a constant temperature is given by

$$E = P_0 V_0 \ln\left(\frac{V_1}{V_0}\right)$$

where $E$ is the energy, $P_0$ is a constant, $V_0$ is the initial volume, and $V_1$ is the new volume. Find $V_1$ if $E = 15{,}100$, $P_0 = 85.0$, and $V_0 = 265$.

*For each set of data, graph the best straight line on rectangular graph paper and find its equation.*

**48.**

| $x$ | 6 | 11 | 17 | 24 | 28 | 39 | 45 |
|---|---|---|---|---|---|---|---|
| $y$ | 53 | 52 | 44 | 41 | 35 | 32 | 20 |

**49.**

| $x$ | 20 | 55 | 71 | 102 | 110 | 139 | 150 |
|---|---|---|---|---|---|---|---|
| $y$ | 110 | 210 | 300 | 360 | 420 | 510 | 525 |

*Plot the graph of each equation on logarithmic graph paper.*

**50.** $y = 4x^2$  **51.** $y = 2.5x^3$

*Plot each set of data on logarithmic graph paper, graph the best straight line, and find its equation.*

**52.**

| $x$ | 1.5 | 2 | 2.5 | 3 | 4 | 5 | 6 |
|---|---|---|---|---|---|---|---|
| $y$ | 9.8 | 13 | 21 | 35 | 50 | 118 | 112 |

**53.**

| $x$ | 1.5 | 1.8 | 2 | 2.5 | 3 | 4 | 5 | 7 |
|---|---|---|---|---|---|---|---|---|
| $y$ | 4.9 | 6.2 | 11.9 | 12.2 | 21.5 | 31 | 49.5 | 111 |

*Plot the graph of each equation on semilogarithmic graph paper.*

**54.** $y = 5^x$  **55.** $y = 3(4^x)$

*Plot each set of data on semilogarithmic graph paper, graph the best straight line, and find its equation.*

**56.**

| $x$ | 1 | 1.5 | 2 | 3 | 3.6 | 4 |
|---|---|---|---|---|---|---|
| $y$ | 3.7 | 9 | 14.8 | 64 | 130 | 230 |

**57.**

| $x$ | 1 | 1.6 | 2 | 3 | 4 |
|---|---|---|---|---|---|
| $y$ | 4.8 | 8 | 9.5 | 28 | 38.2 |

# Application
## Airbags

According to the National Highway Traffic Safety Administration, between 1987 and 1995 airbags have saved the lives of approximately 1500 people. The concept is straightforward: A cushion of air in front of a passenger should decrease or eliminate crash injuries. In the beginning, airbags used compressed gas that was stored in the vehicle. But storing a gas canister and getting it to release gas quickly at a variety of temperatures was quite expensive. It wasn't until solid-propellant inflators became available in the 1970s that widespread, inexpensive use of airbags was envisioned.

The first use of automobile airbags was to protect from frontal collisions. An impact of about 19 km/h (12 mph) triggers deployment. The engine and other components under the hood receive the initial impact, using about 30 to 40 milliseconds before the impact reaches the occupants. Although that doesn't seem like much time, it's enough. The crash closes a mechanical switch which causes an electrical contact to be made. This ignites a chemical reaction that causes the release of nitrogen gas. As soon as the bag is full, the gas flows out through holes in the back so the bag is quickly out of the way.

Because side impacts result in 40% of all serious injuries, engineers are currently developing airbags that will protect occupants from side collisions. Since only the door is between an oncoming vehicle and the occupant, the collision sensors need to react in about 2.5 to 5 milliseconds, approximately $\frac{1}{10}$ the time needed in a frontal collision. In addition, sensors need to recognize false alarms, such as a rough road. For these reasons most sensors use a mathematical formula that takes into account many pieces of information, including acceleration and velocity. The airbags can be stored in the door, the seat cushion, or the seat back.

Wonderful as airbags seem to be, there are some drawbacks. Having an airbag explode toward an occupant at approximately 300 km/h (200 mph) sometimes causes rather than prevents harm. Cuts, broken bones, and broken noses are some of the documented injuries. In addition, at least 23 infants and small children have been killed by airbags. Employing on and off switches and slowing the acceleration of the airbag are two improvements that have already been implemented by some automakers. Also, engineers are developing detection systems that will differentiate between a rear-facing child seat, a forward-facing child seat, a book bag, or other objects. Passive infrared systems, laser optical systems, radar systems, and ultrasonic systems are being studied. They are programmed to not deploy if they sense a situation in which the potential danger of an airbag outweighs its benefits.

### Bibliography

"Air-Bag Sensor Algorithm." *Automotive Engineering* 103 (Sept. 1995): 49.

Brennan, Richard P. *Dictionary of Scientific Literacy.* New York: John Wiley & Sons, 1992, pp. 4–5.

"Child-Seat and Occupant-Presence Detection." *Automotive Engineering* 102 (May 1994): 47.

*Consumers' Research Magazine* 74 (Jan. 1991): 11.

*The Economist* 332 (Aug. 6, 1994): 66.

Jost, Kevin. "Side Impact and Sensing." *Automotive Engineering* 103 (May 1995): 62.

Resh, Robert E. "Working Knowledge—Air Bags." *Scientific American* (June 1996): 116.

Shuldiner, Herb. "Airbag Explosion." *Popular Mechanics* 173 (April 1996): 49.

Smith, Frances. "The Airbag Controversy." *Consumer Comments* 20 (Fall 1996): 3.

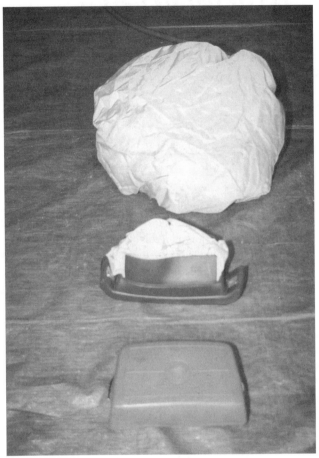

Automobile airbags open and closed (not in car).

# 10

# Trigonometric Functions

## INTRODUCTION

A technician is trying to determine whether the acceleration of an airplane propeller is related to the density of engine lubricant needed. An instrument reading determines that in 10.0 seconds, the angular velocity of a propeller increased from $18\overline{0}0$ rpm to $22\overline{0}0$ rpm. She uses her knowledge of the relationship between angular speed and revolutions per minute and a formula to determine the angular acceleration. (This is problem 15 in Exercises 10.4.)

## Objectives

- Given an angle, find a coterminal angle.
- Find trigonometric functions of an angle in standard position, given a point on its terminal side.
- Find an angle given a trigonometric function of the angle and its quadrant.
- Convert the measure of angles between radians and degrees.
- Work with angular speed.

## 10.1  THE TRIGONOMETRIC FUNCTIONS

In Chapter 3, we defined six trigonometric ratios in terms of the relationships between an acute angle of a right triangle and the lengths of two of its sides. Before extending these six definitions to all angles, we need to discuss the following.

Using a coordinate plane, an angle is in **standard position** when its vertex is located at the origin and its initial side is lying on the positive $x$-axis. An angle resulting from a counterclockwise rotation, as indicated by the direction of the arrow, is a **positive angle.** But if the rotation is clockwise, the angle is **negative** (see Fig. 10.1).

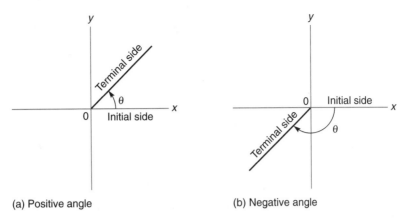

(a) Positive angle               (b) Negative angle

**Figure 10.1**

### EXAMPLE 1

Draw angles of (a) 120° and (b) −240° in standard position.
    The angles are shown in Fig. 10.2.

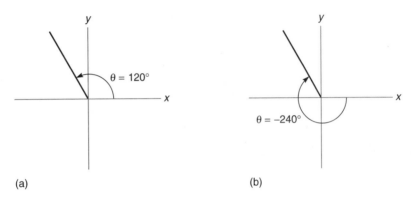

(a)                      (b)

**Figure 10.2**

    Note that the two angles in Example 1 have the same initial side and the same terminal side. However, one was formed by a counterclockwise rotation (120°), and the other by a clockwise rotation (−240°). Angles that share the same initial side and terminal side are **coterminal.** For example, 320° and −40° are coterminal angles, as are 60° and 420° as in Fig. 10.3.

    An angle in standard position is determined by the position of its terminal side in the *xy*-plane. In fact, from geometry we know the position of the terminal side once we know a point on the terminal side other than the origin. Thus, knowing the coordinates $(a, b)$ of the point $P$ determines the angle $\theta$ (see Fig. 10.4).

### EXAMPLE 2

Draw the graph of an angle in standard position whose terminal side passes through the point $(-2, 5)$.
    The graph is shown in Fig. 10.5.

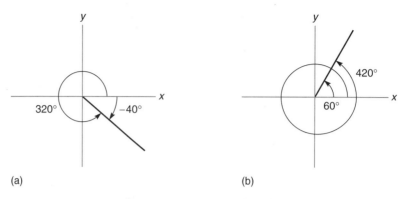

(a)                        (b)

**Figure 10.3**   Coterminal angles

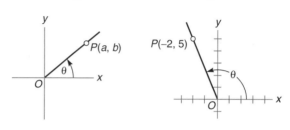

**Figure 10.4**                  **Figure 10.5**

Now we define the trigonometric functions as functions of angles. Consider an angle $\theta$ in standard position and a point $P$ with coordinates $(x, y)$ lying on the terminal side of the angle (see Fig. 10.6). Points $O$, $P$, and $Q$ form the vertices of a right triangle. Angle $Q$ is a right angle. In a right triangle, the side $r$ opposite the right angle $Q$ is called the **hypotenuse** of the right triangle. We can find $r$ if we know the coordinates of $P$.

$$r = \sqrt{x^2 + y^2}$$

*Note:* $r > 0$.

This formula is a direct application of the **Pythagorean theorem,** which states that the square of the hypotenuse of a right triangle is equal to the sum of the squares of the lengths of the other two sides.

There are six trigonometric functions associated with angle $\theta$ in standard position. They are expressed in terms of the coordinates of point $P$, where point $P$ is on the terminal side of angle $\theta$, as follows.

---

**TRIGONOMETRIC FUNCTIONS**

$$\text{sine } \theta = \frac{y}{r} = \frac{\text{ordinate of } P}{r}$$

$$\text{cosine } \theta = \frac{x}{r} = \frac{\text{abscissa of } P}{r}$$

$$\text{tangent } \theta = \frac{y}{x} = \frac{\text{ordinate of } P}{\text{abscissa of } P} \qquad (x \neq 0)$$

---

$$\cot\text{angent } \theta = \frac{x}{y} = \frac{\text{abscissa of } P}{\text{ordinate of } P} \qquad (y \neq 0)$$

$$\sec\text{ant } \theta = \frac{r}{x} = \frac{r}{\text{abscissa of } P} \qquad (x \neq 0)$$

$$\cos\text{ecant } \theta = \frac{r}{y} = \frac{r}{\text{ordinate of } P} \qquad (y \neq 0)$$

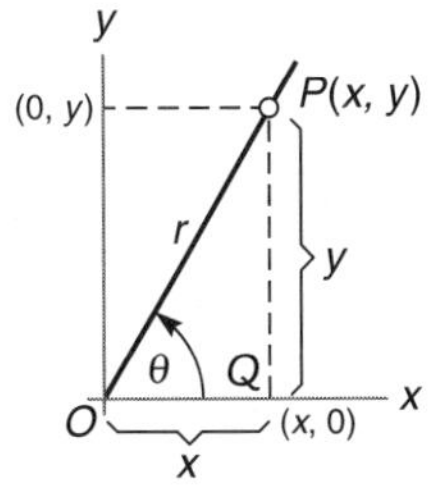

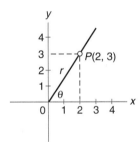

**Figure 10.6**

**EXAMPLE 3**

Find the values of the trigonometric functions for the angle $\theta$ in standard position where the point (2, 3) lies on the terminal side.

First, find $r$ in Fig. 10.7.

$$r = \sqrt{2^2 + 3^2} = \sqrt{4 + 9} = \sqrt{13}$$

**Figure 10.7**

Then we have, by the definitions,

$$\sin\theta = \frac{y}{r} = \frac{3}{\sqrt{13}} \qquad \csc\theta = \frac{r}{y} = \frac{\sqrt{13}}{3}$$

$$\cos\theta = \frac{x}{r} = \frac{2}{\sqrt{13}} \qquad \sec\theta = \frac{r}{x} = \frac{\sqrt{13}}{2}$$

$$\tan\theta = \frac{y}{x} = \frac{3}{2} \qquad \cot\theta = \frac{x}{y} = \frac{2}{3}$$

We have defined functions of angles in this chapter. This means, for example, that $\tan\theta$ is determined by $\theta$ and does not depend upon the choice of the point $P$ lying on the terminal side. Let $(a, b)$ and $(c, d)$ be two different points on the terminal side of an angle $\theta$ as in Fig. 10.8. Triangles $OPQ$ and $ORS$ are similar. From geometry we know that the corresponding sides of similar triangles are proportional. This means that $\dfrac{a}{b} = \dfrac{c}{d}$.

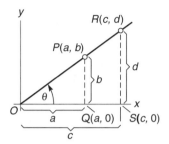

**Figure 10.8**

Thus, no matter which point we use, $P$ or $R$, the numerical value of the ratio used to define $\tan \theta$ is the same. Similar arguments can be given to show that the other five trigonometric functions of $\theta$ do not depend upon the choice of the point used on the terminal side to compute the defining ratios.

Let's demonstrate this fact for two particular points: $P$ with coordinates $(3, 4)$ and $R$ with coordinates $(6, 8)$, which both lie on the terminal side of the same angle $\theta$ in standard position (see Fig. 10.9).

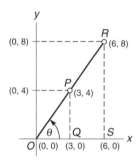

**Figure 10.9**

We shall first use $P$ to find the trigonometric functions of $\theta$. Find the hypotenuse $r$ of the right triangle $OPQ$ as follows:

$$r = \sqrt{3^2 + 4^2} = \sqrt{9 + 16} = \sqrt{25} = 5$$

Then

$$\sin \theta = \frac{y}{r} = \frac{4}{5} \qquad \csc \theta = \frac{r}{y} = \frac{5}{4}$$

$$\cos \theta = \frac{x}{r} = \frac{3}{5} \qquad \sec \theta = \frac{r}{x} = \frac{5}{3}$$

$$\tan \theta = \frac{y}{x} = \frac{4}{3} \qquad \cot \theta = \frac{x}{y} = \frac{3}{4}$$

Similarly, using $R$, we find the hypotenuse $r$ of the right triangle $ORS$.

$$r = \sqrt{6^2 + 8^2} = \sqrt{36 + 64} = \sqrt{100} = 10$$

Again,

$$\sin \theta = \frac{y}{r} = \frac{8}{10} = \frac{4}{5} \qquad \csc \theta = \frac{r}{y} = \frac{10}{8} = \frac{5}{4}$$

$$\cos \theta = \frac{x}{r} = \frac{6}{10} = \frac{3}{5} \qquad \sec \theta = \frac{r}{x} = \frac{10}{6} = \frac{5}{3}$$

$$\tan \theta = \frac{y}{x} = \frac{8}{6} = \frac{4}{3} \qquad \cot \theta = \frac{x}{y} = \frac{6}{8} = \frac{3}{4}$$

Since the ratios computed using the point $R$ reduce to the values of the ratios computed using the point $P$, we see that the choice between $P$ and $R$ does not affect the ultimate value of the trigonometric functions of $\theta$.

We know from algebra that $\dfrac{y}{r}$ and $\dfrac{r}{y}$ are reciprocals of each other; that is,

$$\sin \theta = \frac{y}{r} = \frac{1}{\dfrac{r}{y}} = \frac{1}{\csc \theta}$$

For this reason, $\sin \theta$ and $\csc \theta$ are called **reciprocal trigonometric functions.** In much the same way, we can complete the following table using the defining ratios.

| RECIPROCAL TRIGONOMETRIC FUNCTIONS | |
| --- | --- |
| $\sin \theta = \dfrac{1}{\csc \theta}$ | $\csc \theta = \dfrac{1}{\sin \theta}$ |
| $\cos \theta = \dfrac{1}{\sec \theta}$ | $\sec \theta = \dfrac{1}{\cos \theta}$ |
| $\tan \theta = \dfrac{1}{\cot \theta}$ | $\cot \theta = \dfrac{1}{\tan \theta}$ |

If a given angle $\theta$ in standard position is in the second quadrant as in Fig. 10.10, then $x < 0$ and $y > 0$. Of course, distance $r > 0$. The sign of each of the trigonometric functions is given as follows:

$$\sin \theta = \frac{y}{r} > 0 \qquad \csc \theta = \frac{r}{y} > 0$$

$$\cos \theta = \frac{x}{r} < 0 \qquad \sec \theta = \frac{r}{x} < 0$$

$$\tan \theta = \frac{y}{x} < 0 \qquad \cot \theta = \frac{x}{y} < 0$$

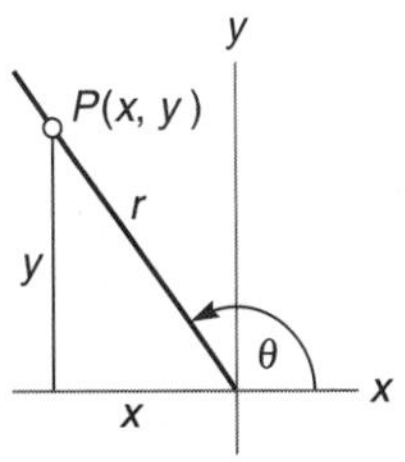

**Figure 10.10**

Note that $r$, a distance, is always a positive quantity.

**TABLE 10.1**  **Signs of the trigonometric functions of nonquadrantal angles**

| Quadrant | $\sin\theta = \dfrac{y}{r}$ | $\cos\theta = \dfrac{x}{r}$ | $\tan\theta = \dfrac{y}{x}$ | $\cot\theta = \dfrac{x}{y}$ | $\sec\theta = \dfrac{r}{x}$ | $\csc\theta = \dfrac{r}{y}$ |
|---|---|---|---|---|---|---|
| **I** $x > 0$ $y > 0$ | + | + | + | + | + | + |
| **II** $x < 0$ $y > 0$ | + | − | − | − | − | + |
| **III** $x < 0$ $y < 0$ | − | − | + | + | − | − |
| **IV** $x > 0$ $y < 0$ | − | + | − | − | + | − |

In summary, if angle $\theta$ is in standard position and its terminal side lies in a given quadrant, the sign of each trigonometric function according to its definition may be tabulated as shown in Table 10.1.

**EXAMPLE 4**

Find the values of $\sin\theta$, $\cos\theta$, and $\tan\theta$ if $\theta$ is in standard position and its terminal side passes through the point $(-3, 4)$ (see Fig. 10.11).

$$r = \sqrt{x^2 + y^2} = \sqrt{9 + 16} = 5$$

$$\sin\theta = \frac{y}{r} = \frac{4}{5}$$

$$\cos\theta = \frac{x}{r} = \frac{-3}{5} = -\frac{3}{5}$$

$$\tan\theta = \frac{y}{x} = \frac{4}{-3} = -\frac{4}{3}$$

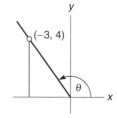

**Figure 10.11**

A **quadrantal angle** is one which, when in standard position, has its terminal side coinciding with one of the axes. Again using the definitions and Fig. 10.12, we can generate the values of the trigonometric functions of $\theta$ when $\theta$ is a quadrantal angle. (Remember, $r > 0$.) The values are given in Table 10.2. For example,

$$\sin 180° = \frac{y}{r} = \frac{0}{r} = 0$$

$$\cos 180° = \frac{x}{r} = \frac{-r}{r} = -1$$

$$\tan 90° = \frac{y}{x} = \frac{r}{0} \qquad \text{(Undefined)}$$

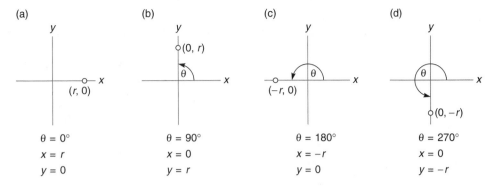

Figure 10.12

TABLE 10.2   Values of quadrantal angles

| $\theta$ | 0° | 90° | 180° | 270° | 360° |
|---|---|---|---|---|---|
| sin $\theta$ | 0 | 1 | 0 | −1 | 0 |
| cos $\theta$ | 1 | 0 | −1 | 0 | 1 |
| tan $\theta$ | 0 | undefined | 0 | undefined | 0 |
| cot $\theta$ | undefined | 0 | undefined | 0 | undefined |
| sec $\theta$ | 1 | undefined | −1 | undefined | 1 |
| csc $\theta$ | undefined | 1 | undefined | −1 | undefined |

## Exercises 10.1

*Draw a graph for each angle in standard position.*

**1.** 30°  **2.** −60°  **3.** 225°  **4.** 390°

**5.** 540°  **6.** −420°

*Find the smallest positive and the largest negative coterminal angle in standard position for each angle.*

**7.** 60°  **8.** 175°  **9.** −86°  **10.** −270°

**11.** 225°  **12.** 300°  **13.** 412°  **14.** −500°

*Draw a graph for the angle in standard position whose terminal side passes through each point.*

**15.** (2, 3)  **16.** (4, −2)  **17.** (−1, 3)  **18.** (−3, −1)

**19.** (0, −3)  **20.** (2, 0)

*Find the values of sin $\theta$, cos $\theta$, tan $\theta$, cot $\theta$, sec $\theta$, and csc $\theta$ if $\theta$ is in standard position and its terminal side passes through the given point.*

**21.** (3, −4)  **22.** (−4, −3)  **23.** (1, 1)  **24.** (−3, 3)

**25.** (−1, −$\sqrt{3}$)  **26.** (5, 12)  **27.** (−4, 5)  **28.** (6, −2)

**29.** (0, −3)  **30.** (2, 0)

**31.** Given that points $P(8, 6)$ and $R(12, 9)$ both lie on the terminal side of an angle $\theta$, show that the trigonometric functions of $\theta$ determined by point $P$ are the same as those determined by point $R$.

**32.** Given that points $P(3, 6)$ and $R(9, 18)$ both lie on the terminal side of an angle $\theta$, show that the trigonometric functions of $\theta$ determined by point $P$ are the same as those determined by point $R$.

*Using the defining ratios, show each equality.*

**33.** $\sec \theta = \dfrac{1}{\cos \theta}$

**34.** $\tan \theta = \dfrac{1}{\cot \theta}$

**35.** $\sin \theta = \dfrac{1}{\csc \theta}$

**36.** $\csc \theta = \dfrac{1}{\sin \theta}$

**37.** $\tan \theta = \dfrac{\sin \theta}{\cos \theta}$

**38.** $\cot \theta = \dfrac{\cos \theta}{\sin \theta}$

## 10.2  TRIGONOMETRIC FUNCTIONS OF ANY ANGLE

D.1, D.5

The **reference angle,** $\alpha$, of any nonquadrantal angle, $\theta$, in standard position is the *acute* angle between the terminal side of $\theta$ and the $x$-axis. Angle $\alpha$ is always considered to be a positive angle less than 90°. *Note:* $0° < \alpha < 90°$.

**EXAMPLE 1**

Find the reference angle $\alpha$ for each given angle $\theta$ in Fig. 10.13.

(a)

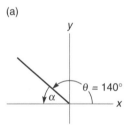

$\alpha = 180° - 140° = 40°$

(b)

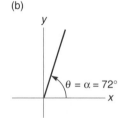

(c)

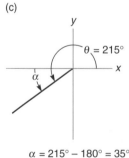

$\alpha = 215° - 180° = 35°$

(d)

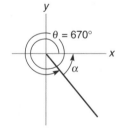

$\alpha = 720° - 670° = 50°$

**Figure 10.13**

Note that if angle $\theta$ is in standard position and

1. $0° < \theta < 90°$, then $\alpha = \theta$.
2. $90° < \theta < 180°$, then $\alpha = 180° - \theta$.
3. $180° < \theta < 270°$, then $\alpha = \theta - 180°$.
4. $270° < \theta < 360°$, then $\alpha = 360° - \theta$.

Consider the four angles shown in Figs. 10.14, 10.15, 10.16, and 10.17 in Table 10.3. The angles are in standard position, where the terminal sides pass through the points $(a, b)$, $(-a, b)$, $(-a, -b)$, and $(a, -b)$, respectively, with $a > 0$ and $b > 0$. Find the values of each of the trigonometric functions for each of the angles.

Note the following:

1. All four triangles in the four cases shown in Table 10.3 are congruent; that is, all corresponding sides and angles are equal. Thus, the reference angle, $\alpha$, is the same in all four quadrants.

2. The absolute value of each corresponding trigonometric function is the same in all four cases.

3. The sign of each trigonometric ratio is determined by the quadrant in which the terminal side of angle $\theta$ lies, as in Table 10.3.

Calculators are used to evaluate the trigonometric function of *any* angle in the same way that we evaluated the trigonometric ratios in Section 3.2. Make certain that your calculator is in the degree mode.

**EXAMPLE 2**

Find tan 350.6°, rounded to four significant digits.

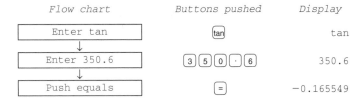

That is, tan 350.6° = −0.1655 rounded to four significant digits.

**EXAMPLE 3**

Find sin(−114°), rounded to four significant digits.

That is, sin(−114°) = −0.9135 rounded to four significant digits.

## TABLE 10.3

| First-quadrant angle | Second-quadrant angle |
|---|---|
| $\sin \theta = \dfrac{b}{r}$ | $\sin \theta = \dfrac{b}{r} = \sin \alpha$ |
| $\cos \theta = \dfrac{a}{r}$ | $\cos \theta = \dfrac{-a}{r} = -\dfrac{a}{r} = -\cos \alpha$ |
| $\tan \theta = \dfrac{b}{a}$ | $\tan \theta = \dfrac{b}{-a} = -\dfrac{b}{a} = -\tan \alpha$ |
| $\cot \theta = \dfrac{a}{b}$ | $\cot \theta = \dfrac{-a}{b} = -\dfrac{a}{b} = -\cot \alpha$ |
| $\sec \theta = \dfrac{r}{a}$ | $\sec \theta = \dfrac{r}{-a} = -\dfrac{r}{a} = -\sec \alpha$ |
| $\csc \theta = \dfrac{r}{b}$ | $\csc \theta = \dfrac{r}{b} = \csc \alpha$ |

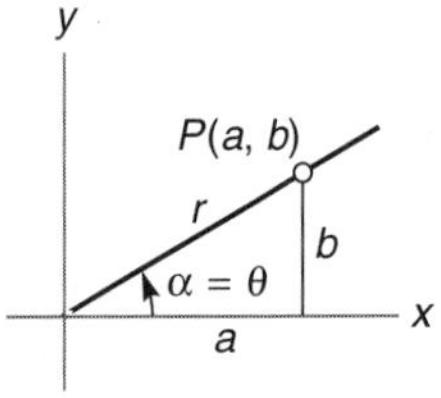

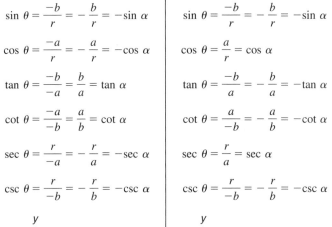

Figure 10.14

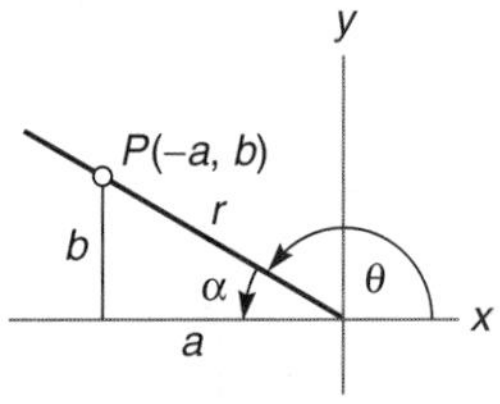

Figure 10.15

| Third-quadrant angle | Fourth-quadrant angle |
|---|---|
| $\sin \theta = \dfrac{-b}{r} = -\dfrac{b}{r} = -\sin \alpha$ | $\sin \theta = \dfrac{-b}{r} = -\dfrac{b}{r} = -\sin \alpha$ |
| $\cos \theta = \dfrac{-a}{r} = -\dfrac{a}{r} = -\cos \alpha$ | $\cos \theta = \dfrac{a}{r} = \cos \alpha$ |
| $\tan \theta = \dfrac{-b}{-a} = \dfrac{b}{a} = \tan \alpha$ | $\tan \theta = \dfrac{-b}{a} = -\dfrac{b}{a} = -\tan \alpha$ |
| $\cot \theta = \dfrac{-a}{-b} = \dfrac{a}{b} = \cot \alpha$ | $\cot \theta = \dfrac{a}{-b} = -\dfrac{a}{b} = -\cot \alpha$ |
| $\sec \theta = \dfrac{r}{-a} = -\dfrac{r}{a} = -\sec \alpha$ | $\sec \theta = \dfrac{r}{a} = \sec \alpha$ |
| $\csc \theta = \dfrac{r}{-b} = -\dfrac{r}{b} = -\csc \alpha$ | $\csc \theta = \dfrac{r}{-b} = -\dfrac{r}{b} = -\csc \alpha$ |

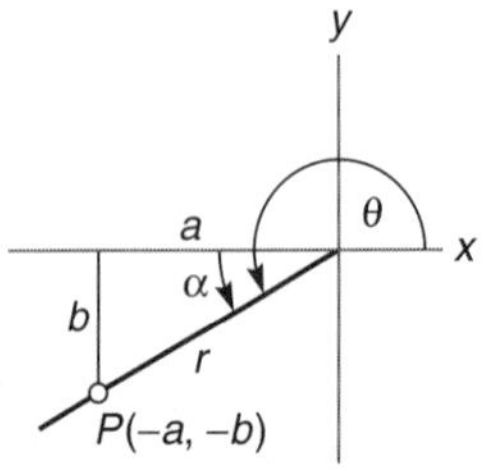

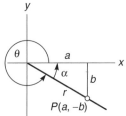

Figure 10.16

Figure 10.17

**EXAMPLE 4**

Find sec 250°, rounded to four significant digits.

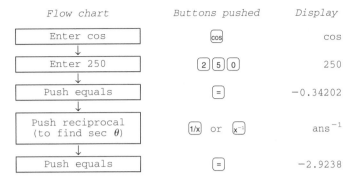

| Flow chart | Buttons pushed | Display |
|---|---|---|
| Enter cos | [cos] | cos |
| ↓ | | |
| Enter 250 | [2][5][0] | 250 |
| ↓ | | |
| Push equals | [=] | −0.34202 |
| ↓ | | |
| Push reciprocal (to find sec θ) | [1/x] or [x⁻¹] | ans$^{-1}$ |
| ↓ | | |
| Push equals | [=] | −2.9238 |

That is, sec 250° = −2.924 rounded to four significant digits.

*Note:* Since calculators do not have secant buttons, we must use the cosine button and the reciprocal relationship $\sec \theta = \dfrac{1}{\cos \theta}$.

To use a calculator to find angles when the value of the trigonometric function is given:

1. Find the reference angle. Enter the inverse trigonometric function followed by the absolute value of the given trigonometric ratio. Then press equals. The angle displayed is the reference angle, $\alpha$.

2. From the sign of the given value, determine the quadrants in which the angles lie.

3. Knowing the reference angle and the quadrants, find the angles.

**EXAMPLE 5**

Given $\sin \theta = 0.4772$, find $\theta$ for $0° \leq \theta < 360°$, rounded to the nearest tenth of a degree.

| Flow chart | Buttons pushed | Display |
|---|---|---|
| Enter sin$^{-1}$ | [sin⁻¹] | sin$^{-1}$ |
| ↓ | | |
| Enter 0.4772 | [.][4][7][7][2] | 0.4772 |
| ↓ | | |
| Push equals | [=] | 28.5027 |

Thus, the reference angle $\alpha = 28.5°$ rounded to the nearest tenth of a degree.
The sine function is positive in Quadrants I and II.
The first-quadrant angle is 28.5° (see Fig. 10.18).
The second-quadrant angle is 180° − 28.5° = 151.5°.

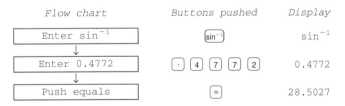

**Figure 10.18**

**EXAMPLE 6**

Given tan $\theta = -0.3172$, find $\theta$ for $0° \le \theta < 360°$, rounded to the nearest tenth of a degree.

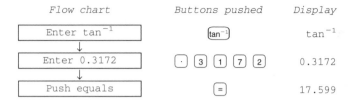

Thus, the reference angle $\alpha = 17.6°$ rounded to the nearest tenth of a degree.
The tangent function is negative in Quadrants II and IV.
The second-quadrant angle is $180° - 17.6° = 162.4°$ (see Fig. 10.19).
The fourth-quadrant angle is $360° - 17.6° = 342.4°$.

**EXAMPLE 7**

Given cos $\theta = -0.2405$, find $\theta$ for $0° \le \theta < 360°$, rounded to the nearest tenth of a degree.

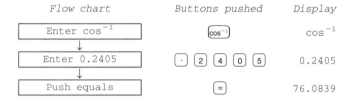

Thus, the reference angle $\alpha = 76.1°$ rounded to the nearest tenth of a degree.
The cosine function is negative in Quadrants II and III.
The second-quadrant angle is $180° - 76.1° = 103.9°$ (see Fig. 10.20).
The third-quadrant angle is $180° + 76.1° = 256.1°$.

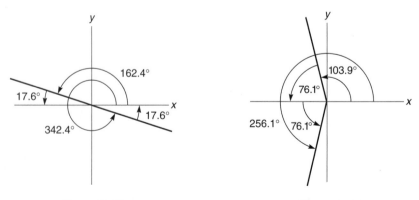

**Figure 10.19**　　　　　　　　　　　　**Figure 10.20**

**EXAMPLE 8**

Given cot $\theta = -1.650$, find $\theta$ for $0° \leq \theta < 360°$, rounded to the nearest tenth of a degree.

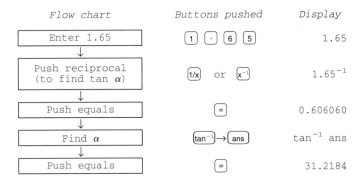

Thus, the reference angle $\alpha = 31.2°$ rounded to the nearest tenth of a degree.

*Note:* Since calculators do not have inverse cotangent buttons, we must use the inverse tangent button and the reciprocal relationship $\cot \theta = \dfrac{1}{\tan \theta}$.

The cotangent function is negative in Quadrants II and IV.
The second-quadrant angle is $180° - 31.2° = 148.8°$ (see Fig. 10.21).
The fourth-quadrant angle is $360° - 31.2° = 328.8°$.

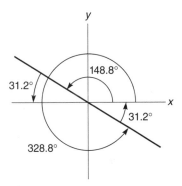

**Figure 10.21**

## Exercises 10.2

*For each angle, find the reference angle.*

1. 120°
2. 312°
3. 253°
4. 19°
5. 293.4°
6. 192.5°
7. −116.7°
8. −274.8°
9. 462°4′
10. 597°13′
11. 1920°
12. 2134°

*Find the value of each trigonometric function, rounded to four significant digits.*

13. sin 125.7°
14. cos 217.4°
15. tan 349.7°
16. tan 98.3°
17. cos 265.7°
18. sin 293.9°
19. cos(−143.5°)
20. sin(−275.6°)
21. sec 192.0°
22. csc 318.3°
23. cot(−36.5°)
24. sec(−105.0°)

*Find θ for 0° ≤ θ < 360°, rounded to the nearest tenth of a degree.*

**25.** $\sin \theta = 0.3684$  **26.** $\cos \theta = 0.1849$  **27.** $\tan \theta = 0.7250$

**28.** $\tan \theta = -1.8605$  **29.** $\cos \theta = -0.1050$  **30.** $\sin \theta = -0.8760$

**31.** $\sin \theta = -0.9111$  **32.** $\sin \theta = 0.5009$  **33.** $\sec \theta = -1.7632$

**34.** $\csc \theta = -2.4105$  **35.** $\cot \theta = 3.6994$  **36.** $\sec \theta = 2.8200$

**37.** $\csc \theta = -1.3250$  **38.** $\cot \theta = -0.1365$  **39.** $\sec \theta = 2.3766$

**40.** $\csc \theta = -1.9130$  **41.** $\cos \theta = 0.7140$  **42.** $\tan \theta = 2.3670$

## 10.3 RADIAN MEASURE

As we have already seen, many trigonometric problems may be solved in terms of the degree measure of angles. However, in many applications, as well as in our development of mathematics, another angular measurement is needed, namely, the radian. A **radian** is the measure of an angle with its vertex at the center of a circle whose intercepted arc is equal in length to the radius of the circle (see Fig. 10.22). That is, the angle $\theta$ is defined as the ratio of the length of arc $PQ$ to the length of the radius, $r$. As a result, the unit *radian* has no physical dimensions since it is the ratio of two lengths.

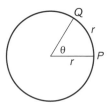

$\theta = 1$ radian        **Figure 10.22**

The circumference of any circle is given by the formula $C = 2\pi r$. The ratio of the circumference of a circle to its radius is $\dfrac{2\pi r}{r} = 2\pi$ (about 6.28). That is, $2\pi$ radians is the measure of the central angle of any complete circle (one complete revolution).

For comparison purposes,

$$2\pi \text{ radians} = 360°$$
$$\pi \text{ rad} = 180°$$

When we divide each side of the above relation by $\pi$, we find

$$1 \text{ rad} = \frac{180°}{\pi} = 57.2958° \quad \text{(Approximately)}$$

Or, when we divide each side of the same relation by 180, we find

$$1° = \frac{\pi}{180} \text{ rad} = 0.01745 \text{ rad} \quad \text{(Approximately)}$$

Recall from Chapter 1 that to convert from one unit of measure to another, multiply the first unit by a conversion factor where the numerator given in one unit equals the denominator given in another unit. That is, multiplying by this fraction (conversion factor) equal to one does not change the quantity but changes only the units. From the relation $\pi$ rad $= 180°$, we can form the conversion factors

$$\frac{\pi \text{ rad}}{180°} \quad \text{and} \quad \frac{180°}{\pi \text{ rad}}$$

### EXAMPLE 1

Change each angle measure from degrees to radians.

(a) $30° = 30° \times \dfrac{\pi \text{ rad}}{180°} = \dfrac{\pi}{6} \text{ rad} = 0.524 \text{ rad}$

(b) $45° = 45° \times \dfrac{\pi \text{ rad}}{180°} = \dfrac{\pi}{4} \text{ rad} = 0.785 \text{ rad}$

(c) $120° = 120° \times \dfrac{\pi \text{ rad}}{180°} = \dfrac{2\pi}{3} \text{ rad} = 2.09 \text{ rad}$

(d) $36° = 36° \times \dfrac{\pi \text{ rad}}{180°} = \dfrac{\pi}{5} \text{ rad} = 0.628 \text{ rad}$

For some of our work, we find it convenient to express radian measure in terms of $\pi$. At other times the decimal expression is more beneficial. *Note: When no unit of angle measure is given, it is understood that the angle is expressed in radians. Thus, $\theta = \dfrac{\pi}{6}$ is understood to be $\theta = \dfrac{\pi}{6}$ rad.*

### EXAMPLE 2

Change each angle measure from radians to degrees.

(a) $\dfrac{\pi}{3} \text{ rad} = \dfrac{\pi}{3} \text{ rad} \times \dfrac{180°}{\pi \text{ rad}} = 60°$

(b) $\dfrac{5\pi}{4} \text{ rad} = \dfrac{5\pi}{4} \text{ rad} \times \dfrac{180°}{\pi \text{ rad}} = 225°$

(c) $6\pi \text{ rad} = 6\pi \text{ rad} \times \dfrac{180°}{\pi \text{ rad}} = 1080°$

(d) $1.2 \text{ rad} = 1.2 \text{ rad} \times \dfrac{180°}{\pi \text{ rad}} = 68.8°$

### EXAMPLE 3

Given each angle in Fig. 10.23 in radians, find the reference angle, $\alpha$.

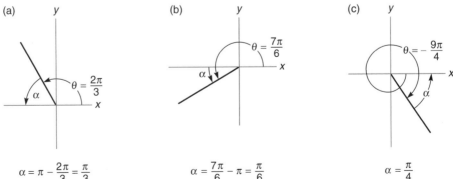

(a)

$\theta = \dfrac{2\pi}{3}$

$\alpha = \pi - \dfrac{2\pi}{3} = \dfrac{\pi}{3}$

(b)

$\theta = \dfrac{7\pi}{6}$

$\alpha = \dfrac{7\pi}{6} - \pi = \dfrac{\pi}{6}$

(c)

$\theta = -\dfrac{9\pi}{4}$

$\alpha = \dfrac{\pi}{4}$

**Figure 10.23**

**EXAMPLE 4**

Find $\cos \dfrac{4\pi}{3}$.

Draw $\theta = \dfrac{4\pi}{3}$ as in Fig. 10.24.

$$\cos \frac{4\pi}{3} = -\cos \frac{\pi}{3} \qquad \text{(The cosine function is negative in Quadrant III.)}$$

$$= -\frac{1}{2}$$

**EXAMPLE 5**

Find $\tan \dfrac{3\pi}{4}$.

Draw $\theta = \dfrac{3\pi}{4}$ as in Fig. 10.25.

$$\tan \frac{3\pi}{4} = -\tan \frac{\pi}{4} \qquad \text{(The tangent function is negative in Quadrant II.)}$$

$$= -1$$

**EXAMPLE 6**

If $\tan \theta = -\sqrt{3}$, find $\theta$ for $0 \le \theta < 2\pi$.

$$\left| -\sqrt{3} \right| = \sqrt{3} = \tan \frac{\pi}{3}$$

The reference angle is $\alpha = \dfrac{\pi}{3}$.

The tangent function is negative in Quadrants II and IV.

The second-quadrant angle is $\pi - \dfrac{\pi}{3} = \dfrac{2\pi}{3}$ (see Fig. 10.26).

The fourth-quadrant angle is $2\pi - \dfrac{\pi}{3} = \dfrac{5\pi}{3}$.

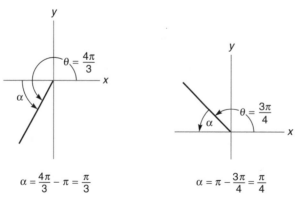

$$\alpha = \frac{4\pi}{3} - \pi = \frac{\pi}{3}$$

**Figure 10.24**

$$\alpha = \pi - \frac{3\pi}{4} = \frac{\pi}{4}$$

**Figure 10.25**

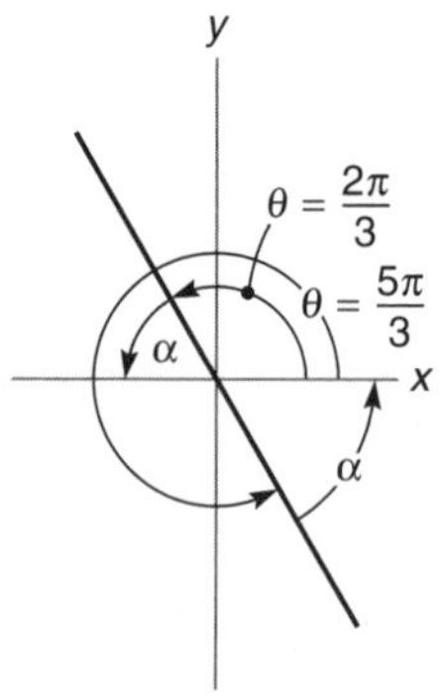

**Figure 10.26**

**TABLE 10.4**

| | $0$ <br> $0°$ | $\dfrac{\pi}{6}$ <br> $30°$ | $\dfrac{\pi}{4}$ <br> $45°$ | $\dfrac{\pi}{3}$ <br> $60°$ | $\dfrac{\pi}{2}$ <br> $90°$ | $\pi$ <br> $180°$ |
|---|---|---|---|---|---|---|
| $\sin \theta$ | 0 | $\dfrac{1}{2}$ | $\dfrac{\sqrt{2}}{2}$ | $\dfrac{\sqrt{3}}{2}$ | 1 | 0 |
| $\cos \theta$ | 1 | $\dfrac{\sqrt{3}}{2}$ | $\dfrac{\sqrt{2}}{2}$ | $\dfrac{1}{2}$ | 0 | $-1$ |
| $\tan \theta$ | 0 | $\dfrac{\sqrt{3}}{3}$ | 1 | $\sqrt{3}$ | undefined | 0 |
| $\cot \theta$ | undefined | $\sqrt{3}$ | 1 | $\dfrac{\sqrt{3}}{3}$ | 0 | undefined |
| $\sec \theta$ | 1 | $\dfrac{2\sqrt{3}}{3}$ | $\sqrt{2}$ | 2 | undefined | $-1$ |
| $\csc \theta$ | undefined | 2 | $\sqrt{2}$ | $\dfrac{2\sqrt{3}}{3}$ | 1 | undefined |

The trigonometric functions of the most used angles are tabulated in both degree and radian measures in Table 10.4.

If your calculator has a Grad button, it contains the **grad** angle system. In this system, the circle is divided into 400 equal parts, and the right angle has 100 parts (see Fig. 10.27). Each of these parts is called one **grad,** or one **new degree,** or one **grade,** written $1^g$. Each grad is divided into 100 **new minutes,** written $1^c$. Each new minute is divided into 100 **new seconds,** written $1^{cc}$. This system is also called the **centesimal** system of measuring angles.

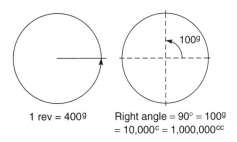

1 rev = 400⁹        Right angle = 90° = 100⁹
                    = 10,000ᶜ = 1,000,000ᶜᶜ

**Figure 10.27**   Grad angle system

A calculator may also be used to evaluate trigonometric functions in radians. Make certain that your calculator is in the radian mode.

**EXAMPLE 7**

Find sin 0.4, rounded to four significant digits.

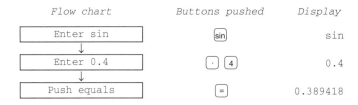

That is, sin 0.4 = 0.3894 rounded to four significant digits.

**EXAMPLE 8**

Find cos 1.684, rounded to four significant digits.

Thus, cos 1.684 = −0.1130 rounded to four significant digits.

**EXAMPLE 9**

Find $\tan\left(\dfrac{\pi}{12}\right)$, rounded to four significant digits.

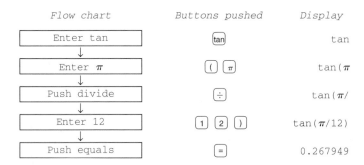

Thus, tan($\pi$/12) = 0.2679 rounded to four significant digits.

To evaluate the cotangent, secant, and cosecant functions, follow the same procedure as in Section 10.2. You may find it helpful to remember the quadrantal angles in radians as shown in Fig. 10.28.

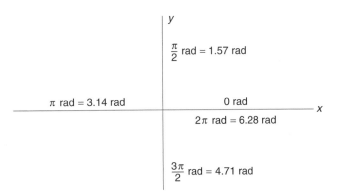

$\dfrac{\pi}{2}$ rad = 1.57 rad

$\pi$ rad = 3.14 rad

0 rad

$2\pi$ rad = 6.28 rad

$\dfrac{3\pi}{2}$ rad = 4.71 rad

**Figure 10.28**

**EXAMPLE 10**

Given $\sin \theta = 0.7565$, find $\theta$ in radians for $0 \le \theta < 2\pi$, rounded to four significant digits.

| *Flow chart* | *Buttons pushed* | *Display* |
|---|---|---|
| Enter $\sin^{-1}$ | [sin⁻¹] | $\sin^{-1}$ |
| Enter 0.7565 | [.] [7] [5] [6] [5] | 0.7565 |
| Find $\alpha$ | [=] | 0.857945 |

The reference angle is $\alpha = 0.8579$.
The sine function is positive in Quadrants I and II.
The first-quadrant angle is 0.8579 (see Fig. 10.29).
The second-quadrant angle is $\pi - 0.8579 = 2.284$ rounded to four significant digits.

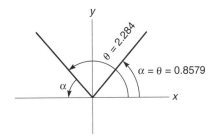

$\theta = 2.284$

$\alpha = \theta = 0.8579$

$\alpha$

**Figure 10.29**

**EXAMPLE 11**

Given $\tan \theta = -1.632$, find $\theta$ in radians for $0 \le \theta < 2\pi$, rounded to four significant digits.

| *Flow chart* | *Buttons pushed* | *Display* |
|---|---|---|
| Enter $\tan^{-1}$ | [tan⁻¹] | $\tan^{-1}$ |
| Enter 1.632 | [1] [.] [6] [3] [2] | 1.632 |
| Find $\alpha$ | [=] | 1.02106 |

The reference angle is $\alpha = 1.021$.
The tangent function is negative in Quadrants II and IV.

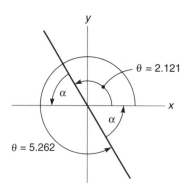

**Figure 10.30**

The second-quadrant angle is $\pi - 1.021 = 2.121$ (see Fig. 10.30).
The fourth-quadrant angle is $2\pi - 1.021 = 5.262$.

## Exercises 10.3

*Change each angle measure from degrees to radians. Express in terms of $\pi$.*

**1.** 135°  **2.** 210°  **3.** 90°  **4.** 270°

**5.** −75°  **6.** −230°  **7.** 1260°  **8.** 2490°

*Change each angle measure from radians to degrees.*

**9.** $\dfrac{7\pi}{4}$  **10.** $\dfrac{7\pi}{6}$  **11.** $\dfrac{19\pi}{4}$  **12.** $\dfrac{27\pi}{8}$

**13.** $9\pi$  **14.** $12\pi$  **15.** 3.7  **16.** 1.82

*Given each angle in radians, find the reference angle in radians.*

**17.** $\dfrac{7\pi}{4}$  **18.** $\dfrac{5\pi}{6}$  **19.** $\dfrac{9\pi}{4}$  **20.** $\dfrac{4\pi}{3}$

**21.** $\dfrac{11\pi}{12}$  **22.** $\dfrac{17\pi}{12}$  **23.** $-\dfrac{8\pi}{5}$  **24.** $-\dfrac{15\pi}{7}$

*Using Table 10.4, find the value of each trigonometric function.*

**25.** $\sin \dfrac{3\pi}{4}$  **26.** $\cos \dfrac{7\pi}{6}$  **27.** $\tan \dfrac{4\pi}{3}$

**28.** $\sec \dfrac{9\pi}{4}$  **29.** $\csc\left(-\dfrac{3\pi}{4}\right)$  **30.** $\cot\left(-\dfrac{5\pi}{4}\right)$

*Using Table 10.4, find $\theta$ for $0 \le \theta < 2\pi$.*

**31.** $\cos \theta = -\dfrac{1}{2}$  **32.** $\sin \theta = \dfrac{\sqrt{3}}{2}$  **33.** $\sec \theta = 2$

**34.** $\tan \theta = -1$  **35.** $\cot \theta = 1$  **36.** $\csc \theta = -\dfrac{2\sqrt{3}}{3}$

*Find the value of each trigonometric function (expressed in radians), rounded to four significant digits.*

**37.** $\sin 0.8$

**38.** $\cos 0.4$

**39.** $\tan 1.2$

**40.** $\tan 0.15$

**41.** $\cos 1.0$

**42.** $\sin 0.75$

**43.** $\sin(-1.65)$

**44.** $\cos(-2.11)$

**45.** $\tan 18.7$

**46.** $\cot 23.6$

**47.** $\sec(-5.6)$

**48.** $\csc(-21.9)$

**49.** $\sin\left(\dfrac{3\pi}{4}\right)$

**50.** $\cos\left(\dfrac{5\pi}{8}\right)$

**51.** $\tan\left(\dfrac{3\pi}{5}\right)$

**52.** $\sin 8\pi$

**53.** $\cos\left(-\dfrac{\pi}{24}\right)$

**54.** $\sin\left(-\dfrac{5\pi}{6}\right)$

*Find $\theta$ for $0 \le \theta < 2\pi$, rounded to four significant digits.*

**55.** $\sin \theta = 0.9845$

**56.** $\cos \theta = 0.3554$

**57.** $\tan \theta = 1.685$

**58.** $\sin \theta = -0.6825$

**59.** $\cos \theta = -0.7540$

**60.** $\tan \theta = -0.1652$

**61.** $\cos \theta = 0.6924$

**62.** $\sin \theta = 0.1876$

**63.** $\tan \theta = -4.672$

**64.** $\sec \theta = -4.006$

**65.** $\csc \theta = -2.140$

**66.** $\cot \theta = -0.1066$

**67.** $\cos \theta = -\dfrac{\sqrt{3}}{2}$

**68.** $\tan \theta = -\dfrac{1}{\sqrt{3}}$

**69.** $\sin \theta = \dfrac{1}{2}$

**70.** $\cot \theta = -1$

**71.** $\sec \theta = -\dfrac{2}{\sqrt{3}}$

**72.** $\csc \theta = -2$

# 10.4 USE OF RADIAN MEASURE

Radian measure has many applications in mathematics and the various technologies. We shall illustrate some of these applications by examples and exercises.

From mathematics, the length of an intercepted arc of a circle, $s$, equals the product of the radius, $r$, and the measure of the central angle, $\theta$, in radians (see Fig. 10.31). That is,

$$s = r\theta$$

Note that when $\theta = 2\pi$, $s = 2\pi r$, where $s$ is the circumference.

**Figure 10.31**

**EXAMPLE 1**

Find the length of the arc of a circle of radius 4.00 ft and with a central angle of $24\overline{0}°$.

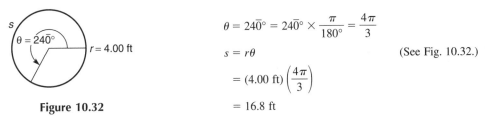

$$\theta = 24\overline{0}° = 24\overline{0}° \times \dfrac{\pi}{180°} = \dfrac{4\pi}{3}$$

$$s = r\theta \qquad \text{(See Fig. 10.32.)}$$

$$= (4.00 \text{ ft})\left(\dfrac{4\pi}{3}\right)$$

$$= 16.8 \text{ ft}$$

**Figure 10.32**

Also from mathematics, the area of a sector of a circle (see Fig. 10.33) is given by the formula

$$A = \frac{1}{2} r^2 \theta$$

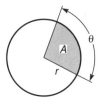

**Figure 10.33**

where $r$ is the radius and $\theta$ is the measure of the central angle in radians.

**EXAMPLE 2**

Find the area of the sector of the circle from Example 1.

$$A = \frac{1}{2} r^2 \theta$$

$$= \frac{1}{2} (4.00 \text{ ft})^2 \left(\frac{4\pi}{3}\right)$$

$$= 33.5 \text{ ft}^2$$

Other illustrations of the use of radian measure from the technologies are given in the following examples.

**EXAMPLE 3**

An automobile engine is rated at $20\overline{0}$ hp at $400\overline{0}$ revolutions per minute (rpm). Find its torque given by the formula

$$G = \frac{P}{\omega}$$

where $G$ is the torque, $P$ is the power, and $\omega$ is the angular speed in radians per unit time, usually radians per s.

The horsepower measure needs to be converted to foot-pounds per second, as follows:

$$P = 20\overline{0} \text{ hp} = 20\overline{0} \text{ hp} \times \frac{55\overline{0} \frac{\text{ft-lb}}{\text{s}}}{1 \text{ hp}}$$

$$= 11\overline{0},000 \frac{\text{ft-lb}}{\text{s}}$$

The angular speed needs to be converted to radians per second, as follows:

$$\omega = 400\overline{0} \frac{\text{rev}}{\text{min}} \times \frac{2\pi \text{ rad}}{\text{rev}} \times \frac{1 \text{ min}}{60 \text{ s}} = 419 \frac{\text{rad}}{\text{s}}$$

Therefore,

$$G = \frac{P}{\omega} = \frac{1\overline{1}0{,}000 \ \frac{\text{ft-lb}}{\cancel{s}}}{419 \ \frac{\text{rad}}{\cancel{s}}} = 263 \ \text{ft-lb}$$

*Note:* Recall that the *radian* unit has no physical dimensions.

**EXAMPLE 4**

The flywheel of a gasoline engine rotates at an angular speed of $30\overline{0}0$ rpm. Find its angular displacement in 5.00 s.

The angular displacement of a revolving body is given by the formula

$$\theta = \omega t$$

where $\theta$ is the angular displacement in radians, $\omega$ is angular speed in radians per second, and $t$ is the time in seconds.

$$\omega = 30\overline{0}0 \ \frac{\cancel{\text{rev}}}{\cancel{\text{min}}} \times \frac{2\pi \ \text{rad}}{\cancel{\text{rev}}} \times \frac{1 \ \cancel{\text{min}}}{60 \ \text{s}} = 314 \ \frac{\text{rad}}{\text{s}}$$

$$\theta = \omega t$$

$$= \left(314 \ \frac{\text{rad}}{\cancel{s}}\right)(5.00 \ \cancel{s})$$

$$= 1570 \ \text{rad}$$

or

$$1570 \ \text{rad} \times \frac{1 \ \text{rev}}{2\pi \ \text{rad}} = 25\overline{0} \ \text{revolutions}$$

**EXAMPLE 5**

An airplane propeller is rotating at $18\overline{0}0$ rpm. If the blades of the propeller are 2.00 m long, find the linear speed of each point:
(a) 1.00 m from the axis of rotation.
(b) On the end of a blade.

The linear speed of a revolving body is given by the formula

$$v = \omega r$$

where $v$ is the linear speed, $\omega$ is the angular speed, and $r$ is the distance from the axis of rotation.

$$\omega = 18\overline{0}0 \ \frac{\cancel{\text{rev}}}{\cancel{\text{min}}} \times \frac{2\pi \ \text{rad}}{\cancel{\text{rev}}} \times \frac{1 \ \cancel{\text{min}}}{60 \ \text{s}} = 188 \ \frac{\text{rad}}{\text{s}}$$

(a) $v = \omega r = \left(188 \ \dfrac{\text{rad}}{\text{s}}\right)(1.00 \ \text{m}) = 188 \ \text{m/s}$

(b) $v = \omega r = \left(188 \ \dfrac{\text{rad}}{\text{s}}\right)(2.00 \ \text{m}) = 376 \ \text{m/s}$

To illustrate the difference between linear and angular velocity, consider a common ice-skating routine performed by many traveling shows. A line of skaters moves around in a circle as shown in Fig. 10.34. Who skates the fastest? As you may recall, the person

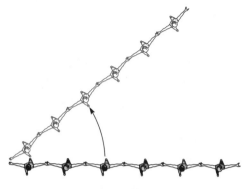

**Figure 10.34**

on the end usually has difficulty in keeping up and in maintaining a straight line. Thus, each successive person out from the center skates faster (has a greater linear velocity) than the person nearer the center. Note that the angular velocity of the line of skaters and of each skater remains the same as they each make the same number of rotations per unit of time.

## Exercises 10.4

1. From a circle of radius 12.0 in., find the length of intercepted arc when the central angle is $\dfrac{2\pi}{3}$ radians.

2. From a circle of radius 8.00 in., find the length of intercepted arc when the central angle is 48°.

3. Find the area of the sector of the circle from Exercise 1.

4. Find the area of the sector of the circle from Exercise 2.

5. Find the central angle, in degrees, of a circle of radius 6.00 in. when the intercepted arc length is 5.00 in.

6. Find the radius of a circle in which a central angle of $\dfrac{3\pi}{4}$ radians intercepts an arc of 24.0 in.

7. Given the two concentric circles in Fig. 10.35 where $\theta = \dfrac{\pi}{6}$ rad, $r_1 = 3.00$ m, and $r_2 = 5.00$ m, find the shaded area of the figure.

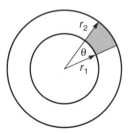

**Figure 10.35**

8. If a sector of a given circle has an area of 36.0 in² and its radius is 6.00 in., find the measure of the central angle.

9. A pendulum of length 5.00 m swings through an angle of 6°. Find the length of arc through which the pendulum swings in one complete swing.

10. An airplane travels in a circular path at $40\overline{0}$ mi/h for 3.00 min. What is the radius of the circle when the central angle is $1\overline{0}°$?

11. An automobile engine is rated at 275 hp at $420\overline{0}$ rpm. Find the torque developed.

12. A gasoline engine develops a torque of $30\overline{0}$ ft-lb at $360\overline{0}$ rpm. What is its horsepower rating?

13. The flywheel of a steam engine is rotating at $42\overline{0}$ rpm.
    (a) Express this angular speed in radians per second.
    (b) Find the angular displacement of the wheel in 10.0 s.
    (c) Find the linear speed of a point on the rim of the wheel which has a radius of 1.75 ft.

14. An airplane propeller whose blades are 6.00 ft long is rotating at $220\overline{0}$ rpm.
    (a) Express its angular speed in radians per second.
    (b) Find the angular displacement in 3.00 s.
    (c) Find the linear speed of a point on the end of the blade.

15. The angular velocity of an airplane propeller is increased from $180\overline{0}$ rpm to $220\overline{0}$ rpm in 10.0 s. Find its angular acceleration. The angular acceleration of a rotating body is given by the formula

$$\alpha = \frac{\Delta \omega}{\Delta t}$$

where $\alpha$ is the angular acceleration, $\Delta \omega$ is the change in angular velocity, and $\Delta t$ is the change in time.

16. Find the angular acceleration of the airplane propeller in Exercise 15 when its angular velocity is increased as given:
    (a) From $180\overline{0}$ rpm to $220\overline{0}$ rpm in 6.00 s.
    (b) From $180\overline{0}$ rpm to $260\overline{0}$ rpm in 10.0 s.

17. The earth rotates on its axis at 1 rev per 24 h. Assume that the average radius of the earth is 3960 mi. Find the linear speed in mi/h of a point on the equator.

18. The earth is revolving about the sun in 1 rev per 365 days. Assume an average circular orbit of radius $1.50 \times 10^8$ km. Find the linear speed in kilometres per second of the earth about the sun.

19. In the pulley system in Fig. 10.36, find the following.
    (a) The linear velocity of a point on the rim of pulley $A$ in m/s.
    (b) The linear velocity of a point on the rim of pulley $B$ in m/s.
    (c) The angular velocity of pulley $B$ in rad/s.

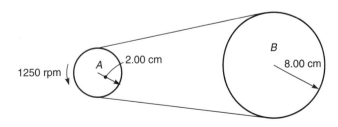

**Figure 10.36**

**20.** An automobile is traveling at 55.0 mi/h. Its tires have a radius of 13.0 in. Find the following.

    **(a)** The tires' angular velocity in rad/s.

    **(b)** The number of revolutions each tire completes in 10.0 s.

    **(c)** The linear distance traveled in feet by a point on the tread in 10.0 s.

## CHAPTER SUMMARY

1. An angle is in **standard position** when its vertex is located at the origin and its initial side is lying on the positive $x$-axis. A positive angle is measured counterclockwise; a negative angle is measured clockwise.

2. Angles that share the same initial side and the same terminal side are called **coterminal.**

3. When angle $\theta$ is in standard position and $P(x, y)$ is on the terminal side, the six trigonometric functions are defined as follows:

$$\sin \theta = \frac{y}{r} \qquad \cos \theta = \frac{x}{r} \qquad \tan \theta = \frac{y}{x}$$

$$\csc \theta = \frac{r}{y} \qquad \sec \theta = \frac{r}{x} \qquad \cot \theta = \frac{x}{y}$$

4. The reciprocal trigonometric functions are as follows:

$$\csc \theta = \frac{1}{\sin \theta} \qquad \sec \theta = \frac{1}{\cos \theta} \qquad \cot \theta = \frac{1}{\tan \theta}$$

5. A **quadrantal angle** is an angle in standard position with its terminal side on one of the coordinate axes.

6. The **reference angle** $\alpha$ of any nonquadrantal angle $\theta$ in standard position is the positive acute angle between the terminal side of $\theta$ and the $x$-axis.

7. To use a calculator to find angles when the trigonometric function is given, review the flow charts and examples in Section 10.2 for angles in degrees and in Section 10.3 for angles in radians.

8. A **radian** is the measure of an angle with its vertex at the center of a circle whose intercepted arc is equal to the length of the radius of the circle.

9. To form conversion factors to change from degrees to radians or from radians to degrees, use the relation $\pi$ rad $= 180°$.

10. The length of an intercepted arc of a circle, $s$, equals the product of the radius, $r$, and the measure of the central angle, $\theta$, in radians:

$$s = r\theta$$

11. The area of a sector of a circle is given by the formula

$$A = \frac{1}{2} r^2 \theta$$

12. The angular velocity of a revolving body is given by

$$\omega = \frac{\theta}{t}$$

where $\theta$ is the angular displacement in radians, $\omega$ is the angular speed in rad/s, and $t$ is the time.

13. The linear speed of a revolving body is given by

$$v = \omega r$$

where $v$ is the linear speed, $\omega$ is the angular speed, and $r$ is the distance from the axis of rotation.

14. Another useful conversion is 1 revolution $= 2\pi$ radians.

## CHAPTER 10 REVIEW

*Find the values of sin $\theta$, cos $\theta$, tan $\theta$, cot $\theta$, sec $\theta$, and csc $\theta$ if $\theta$ is in standard position and its terminal side passes through the given point.*

**1.** $(4, 3)$      **2.** $(-\sqrt{3}, -1)$      **3.** $(-4, 0)$

*For each angle, find the reference angle.*

**4.** $135°$      **5.** $208°20'$      **6.** $-125°$      **7.** $1250°$

*Find the value of each trigonometric function rounded to four significant digits.*

**8.** $\sin 244.3°$      **9.** $\tan 337.5°$      **10.** $\sec 98.7°$

**11.** $\cos(-297.4°)$      **12.** $\cot 402.1°$      **13.** $\csc(-168.0°)$

*Find $\theta$ for $0° \leq \theta < 360°$ rounded to the nearest tenth of a degree.*

**14.** $\sin \theta = 0.3448$      **15.** $\cos \theta = -0.5495$      **16.** $\tan \theta = -1.050$

**17.** $\sec \theta = 1.956$      **18.** $\cot \theta = -1.855$      **19.** $\csc \theta = 1.353$

*Change each angular measurement from degrees to radians. Express in terms of $\pi$.*

**20.** $72°$      **21.** $315°$

*Change each angular measurement from radians to degrees.*

**22.** $\dfrac{5\pi}{6}$      **23.** $\dfrac{3\pi}{4}$

*Given each angle in radians, find the reference angle in radians.*

**24.** $\dfrac{5\pi}{3}$      **25.** $\dfrac{3\pi}{5}$

*Using Table 10.4, find the value for each trigonometric function.*

**26.** $\cos \dfrac{5\pi}{6}$      **27.** $\tan \dfrac{2\pi}{3}$      **28.** $\sec \dfrac{7\pi}{4}$

**29.** $\sin \dfrac{7\pi}{6}$      **30.** $\cot\left(-\dfrac{\pi}{4}\right)$      **31.** $\cos\left(-\dfrac{11\pi}{3}\right)$

*Using Table 10.4, find each θ for 0 ≤ θ < 2π.*

**32.** $\cos \theta = -\dfrac{\sqrt{2}}{2}$

**33.** $\tan \theta = -1$

**34.** $\csc \theta = \dfrac{2\sqrt{3}}{3}$

**35.** $\sin \theta = -\dfrac{\sqrt{3}}{2}$

**36.** $\sec \theta = -2$

**37.** $\tan \theta = \dfrac{\sqrt{3}}{3}$

*Using a calculator, find the value of each trigonometric function (expressed in radians) rounded to four significant digits.*

**38.** $\sin 1.5$

**39.** $\cos 0.25$

**40.** $\tan \dfrac{\pi}{6}$

**41.** $\sin\left(-\dfrac{2\pi}{3}\right)$

*Using a calculator, find each θ for 0 ≤ θ < 2π rounded to four significant digits.*

**42.** $\cos \theta = 0.1981$

**43.** $\sin \theta = -0.6472$

**44.** $\tan \theta = 1.6182$

**45.** $\sec \theta = -2.8061$

**46.** Given a sector with a central angle of 56.0° of a circle of radius 9.00 in. Find the length of arc and area of the sector.

**47.** An automobile engine is rated at 325 hp at $40\overline{0}0$ rpm. Find the torque developed.

**48.** A flywheel is rotating at $63\overline{0}$ rpm.
   (a) Express this angular speed in rad/s.
   (b) Find the angular displacement in 5.00 s.
   (c) Find the linear speed of a point on the wheel that is 1.50 ft from the axis of rotation.

# 11
# Oblique Triangles and Vectors

## INTRODUCTION

A lack of rain for six weeks, a wind speed of 19 mph, and an abundance of campers had increased the likelihood of a forest fire in the Smoky Mountains. Rangers in observation towers and helicopter crews were on the alert. The ranger at tower *A* was the first to spot a fire in a direction 51.5° east of north. Minutes later the ranger in tower *B*, 5.0 miles due east of *A*, sighted the fire at 17.2° west of north. The ranger in station *A* radioed the helicopter crews and determined that there was a helicopter 4.5 miles north of station *A*. Because of cloud cover, the pilot could not see the fire. Did the ranger have enough information to give the pilot the location of the fire?

Solving the preceding problem requires the use of the law of sines and the law of cosines. Part of the information needed is requested in Problem 7 in Exercises 11.4. The ranger radioed the pilot to fly in a direction of 72.1° east of south for 4.2 miles to find the fire.

### Objectives

- Find the sides and angles of a triangle given three parts, one of which is a side.
- Use the law of sines.
- Use the law of cosines.
- Add vectors, using trigonometric methods.
- Express vectors in terms of components.
- Use vectors to solve problems involving displacement, velocity, and force.

## 11.1 LAW OF SINES

An **oblique,** or **general,** triangle is a triangle that contains no right angles. We shall use the standard notation of labeling the vertices of a triangle by the capital letters *A*, *B*, and

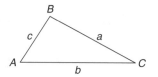

Figure 11.1

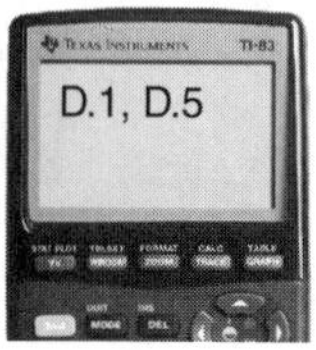

D.1, D.5

$C$ and using the small letters $a$, $b$, and $c$ as the labels for the sides opposite angles $A$, $B$, and $C$, respectively (see Fig. 11.1).

**Solving a triangle** means finding all those sides and angles that are not given or known. To solve a triangle, we need three parts (including at least one side). Solving any oblique triangle falls into one of four cases where the following parts of a triangle are known:

**1.** Two sides and an angle opposite one of them (SSA).

**2.** Two angles and a side opposite one of them (AAS).

**3.** Two sides and the included angle (SAS).

**4.** Three sides (SSS).

One law that we use to solve triangles is called the **law of sines.** In words, for any triangle the ratio of any side to the sine of the opposite angle is a constant. The formula for the law of sines is as follows.

**LAW OF SINES**

$$\frac{a}{\sin A} = \frac{b}{\sin B} = \frac{c}{\sin C}$$

To derive the law of sines, let's take any general triangle $ABC$. From $C$ draw $CD$ perpendicular to $AB$. Note that every oblique triangle is either acute (all three angles are between 0° and 90°) or obtuse (one angle is between 90° and 180°). Both cases are shown in Fig. 11.2. Line $CD$ is called the altitude, $h$, and forms two right triangles in each case.

In right triangle $ADC$ in Figs. 11.2(a) and (b),

$$\sin A = \frac{h}{b} \quad \text{or} \quad h = b \sin A$$

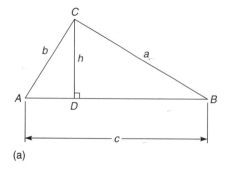

(a)          (b)

Figure 11.2

In right triangle *BCD* in Fig. 11.2(a),

$$\sin B = \frac{h}{a} \quad \text{or} \quad h = a \sin B$$

In Fig. 11.2(b), note that angle *DBC* is the reference angle for angle *ABC* or angle *B*. Then in right triangle *BDC*,

$$\sin B = \sin DBC = \frac{h}{a} \quad \text{or} \quad h = a \sin B$$

In either case we have

$$h = a \sin B = b \sin A$$

Dividing each side by (sin *B*)(sin *A*), we have

$$\frac{a}{\sin A} = \frac{b}{\sin B}$$

By drawing a perpendicular from *A* and *B* to the opposite side, we can show in a similar manner that $\frac{c}{\sin C} = \frac{b}{\sin B}$ and $\frac{c}{\sin C} = \frac{a}{\sin A}$. By combining these equations, we have the law of sines:

$$\frac{a}{\sin A} = \frac{b}{\sin B} = \frac{c}{\sin C}$$

---

In order to use the law of sines, we must know either of the following:

**1.** Two sides and an angle opposite one of them (SSA).

**2.** Two angles and a side opposite one of them (AAS). *Note:* Knowing two angles and any side is sufficient because knowing two angles, we can easily find the third.

You must select the proportion that contains three parts that are known and the unknown part.

---

As in Chapter 3, when calculations with measurements involve a trigonometric function, we shall use the following rule of thumb.

| *Angles expressed to the nearest:* | *Lengths of sides of a triangle will contain:* |
|---|---|
| 1° | Two significant digits |
| 0.1°  or  1′ | Three significant digits |
| 0.01°  or  1″ | Four significant digits |

**EXAMPLE 1**

If $A = 65.0°$, $a = 20.0$ m, and $b = 15.0$ m, solve the triangle.

First, draw a triangle as in Fig. 11.3 and find angle $B$ by using the law of sines.

$$\frac{a}{\sin A} = \frac{b}{\sin B}$$

$$\frac{20.0 \text{ m}}{\sin 65.0°} = \frac{15.0 \text{ m}}{\sin B}$$

$$\sin B = \frac{(15.0 \text{ m})(\sin 65.0°)}{20.0 \text{ m}} = 0.6797$$

$$B = 42.8°$$

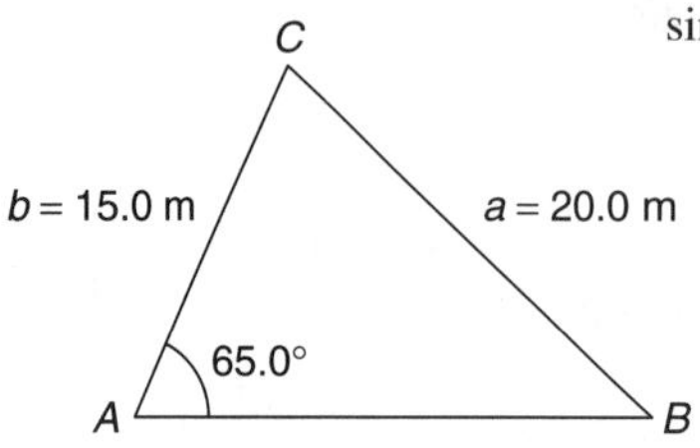

**Figure 11.3**

This angle may be found using a calculator as follows:

| Flow chart | Buttons pushed | Display |
|---|---|---|
| Enter 15 | 15 | 15 |
| Push times | x | |
| Enter sin 65° | sin 6 5 | sin 65 |
| Push divide | ÷ | |
| Enter 20 | 2 0 | 20 |
| Push equals | = | 0.679731 |
| Find angle B | sin⁻¹→ans | sin⁻¹ (ans) |
| Push equals | = | 42.8226 |

So $B = 42.8°$ rounded to the nearest tenth of a degree.

To find $C$, use the fact that the sum of the angles of any triangle is $180°$. Therefore,

$$C = 180° - 65.0° - 42.8°$$
$$= 72.2°$$

Finally, find $c$ using the law of sines.

$$\frac{a}{\sin A} = \frac{c}{\sin C}$$

$$\frac{20.0 \text{ m}}{\sin 65.0°} = \frac{c}{\sin 72.2°}$$

$$c = \frac{(20.0 \text{ m})(\sin 72.2°)}{\sin 65.0°} = 21.0 \text{ m} \qquad \text{(Rounded to three significant digits)}$$

This side may be found using a calculator as follows:

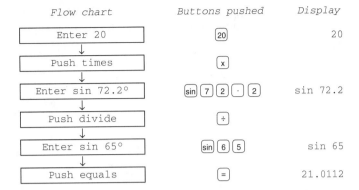

| Flow chart | Buttons pushed | Display |
|---|---|---|
| Enter 20 | [20] | 20 |
| Push times | [×] | |
| Enter sin 72.2° | [sin][7][2][.][2] | sin 72.2 |
| Push divide | [÷] | |
| Enter sin 65° | [sin][6][5] | sin 65 |
| Push equals | [=] | 21.0112 |

That is, side $c = 21.0$ m rounded to three significant digits.
The solution is $B = 42.8°$, $C = 72.2°$, and $c = 21.0$ m.

**EXAMPLE 2**

If $C = 25°$, $c = 59$ ft, and $B = 108°$, solve the triangle.
First, draw a triangle as in Fig. 11.4 and find $b$.

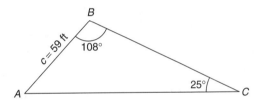

**Figure 11.4**

$$\frac{c}{\sin C} = \frac{b}{\sin B}$$

$$\frac{59 \text{ ft}}{\sin 25°} = \frac{b}{\sin 108°}$$

$$b = \frac{(59 \text{ ft})(\sin 108°)}{\sin 25°} = 130 \text{ ft} \qquad \text{(Rounded to two significant digits)}$$

$$A = 180° - 25° - 108° = 47°$$

Find $a$.

$$\frac{a}{\sin A} = \frac{c}{\sin C}$$

$$\frac{a}{\sin 47°} = \frac{59 \text{ ft}}{\sin 25°}$$

$$a = \frac{(59 \text{ ft})(\sin 47°)}{\sin 25°} = 1\overline{0}0 \text{ ft}$$

The solution is $A = 47°$, $a = 1\overline{0}0$ ft, and $b = 130$ ft.

**EXAMPLE 3**

If $B = 36°21'45''$, $a = 3745$ m, and $b = 4551$ m, solve the triangle.

Draw a triangle as in Fig. 11.5 and find angle $A$.

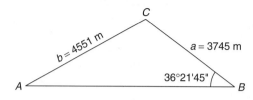

**Figure 11.5**

$$\frac{b}{\sin B} = \frac{a}{\sin A}$$

$$\frac{4551 \text{ m}}{\sin 36°21'45''} = \frac{3745 \text{ m}}{\sin A}$$

$$\sin A = \frac{(\sin 36°21'45'')(3745 \text{ m})}{4551 \text{ m}} = 0.4879$$

$$A = 29°12'7'' \qquad \text{(Rounded to the nearest second)}$$

This angle may be found using a calculator with DMS-DD buttons as follows:

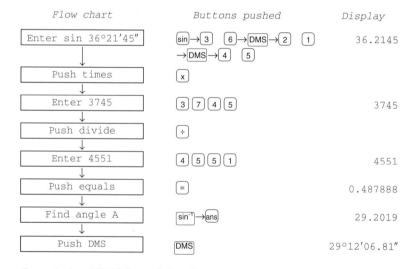

So angle $A = 29°12'7''$ rounded to the nearest second. Then

$$C = 180° - A - B$$

$$= 180° - 29°12'7'' - 36°21'45'' = 114°26'8''$$

Find side $c$.

$$\frac{c}{\sin C} = \frac{b}{\sin B}$$

$$\frac{c}{\sin 114°26'8''} = \frac{4551 \text{ m}}{\sin 36°21'45''}$$

$$c = \frac{(\sin 114°26'8'')(4551 \text{ m})}{\sin 36°21'45''}$$

$$= 6988 \text{ m} \qquad \text{(Rounded to four significant digits)}$$

The solution is $A = 29°12'7''$, $C = 114°26'8''$, and $c = 6988$ m.

## Exercises 11.1

*Solve each triangle using the labels as shown in Fig. 11.6.\**

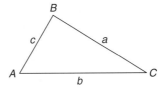

**Figure 11.6**

*Express the lengths of sides to three significant digits and the angles to the nearest tenth of a degree.*

**1.** $A = 69.0°$, $a = 25.0$ m, $b = 16.5$ m      **2.** $C = 57.5°$, $c = 166$ mi, $b = 151$ mi

**3.** $B = 61.4°$, $b = 124$ cm, $c = 112$ cm      **4.** $A = 19.5°$, $a = 487$ km, $c = 365$ km

**5.** $B = 75.3°$, $A = 57.1°$, $b = 257$ ft      **6.** $C = 59.6°$, $B = 43.9°$, $b = 4760$ m

**7.** $A = 115.0°$, $a = 5870$ m, $b = 4850$ m      **8.** $A = 16.4°$, $a = 205$ ft, $b = 187$ ft

*Express the lengths of sides to four significant digits and the angles to the nearest hundredth of a degree.*

**9.** $C = 72.58°$, $b = 28.63$ cm, $c = 42.19$ cm

**10.** $A = 58.95°$, $a = 3874$ m, $c = 2644$ m

**11.** $B = 28.76°$, $C = 19.30°$, $c = 39,750$ mi

**12.** $A = 35.09°$, $B = 48.64°$, $a = 8.362$ km

*Express the lengths of sides to two significant digits and the angles to the nearest degree.*

**13.** $A = 25°$, $a = 5\overline{0}$ cm, $b = 4\overline{0}$ cm      **14.** $B = 42°$, $b = 5.3$ km, $c = 4.6$ km

**15.** $C = 8°$, $c = 16$ m, $a = 12$ m      **16.** $A = 105°$, $a = 460$ mi, $c = 380$ mi

*Express the lengths of sides to four significant digits and the angles to the nearest second.*

**17.** $B = 51°17''$, $b = 1948$ ft, $c = 1525$ ft

**18.** $A = 49°31'50''$, $a = 37,560$ ft, $b = 24,350$ ft

**19.** $A = 31°14'35''$, $B = 85°45'15''$, $c = 4.575$ mi

**20.** $B = 75°30'6''$, $C = 70°12'18''$, $c = 93.45$ m

*\*Because of differences in rounding, your answers may differ slightly from the answers in the text if you choose to solve for the parts of a triangle in an order different from that chosen by the authors.*

# 11.2 THE AMBIGUOUS CASE

D.1, D.5

The solution of a triangle when two sides and an angle opposite one of the sides (SSA) are given requires special care. There may be one, two, or no triangles formed from the given data. By construction and discussion, let's study the possibilities.

### EXAMPLE 1

Construct a triangle given that $A = 35°$, $b = 10$, and $a = 7$.

As you can see from Fig. 11.7, two triangles that satisfy the given information can be drawn: triangles $ACB$ and $ACB'$. Note that in one triangle angle $B$ is acute, and in the other triangle angle $B$ is obtuse.

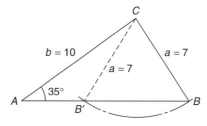

**Figure 11.7**

### EXAMPLE 2

Construct a triangle given that $A = 45°$, $b = 10$, and $a = 5$.

As you can see from Fig. 11.8, no triangle can be drawn that satisfies the given information. Side $a$ is simply not long enough to reach the side opposite angle $C$.

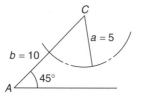

**Figure 11.8**

### EXAMPLE 3

Construct a triangle given that $A = 60°$, $b = 6$, and $a = 10$.

As you can see from Fig. 11.9, only one triangle that satisfies the given information can be drawn. Side $a$ is too long for two solutions.

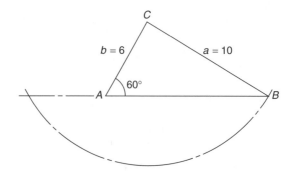

**Figure 11.9**

In summary, let's list the possible cases when two sides and an angle opposite one of the sides are given. Assume that *acute* angle $A$ and adjacent side $b$ are given. As a result of $h = b \sin A$, $h$ is also determined. Depending on the length of the opposite side, $a$, we have the four cases shown in Fig. 11.10.

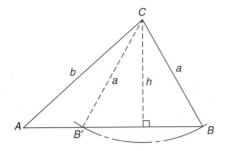

(a) When $h < a < b$, there are two possible triangles. In words, when the side opposite the given *acute* angle is less than the known adjacent side but greater than the altitude, there are two possible triangles.

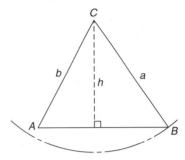

(b) When $h < b < a$, there is only one possible triangle. In words, when the side opposite the given *acute* angle is greater than the known adjacent side, there is only one possible triangle.

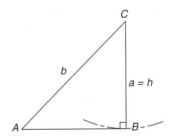

(c) When $a = h$, there is one possible (right) triangle. In words, when the side opposite the given *acute* angle equals the length of the altitude, there is only one possible (right) triangle.

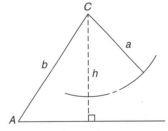

(d) When $a < h$, there is no possible triangle. In words, when the side opposite the given *acute* angle is less than the length of the altitude, there is no possible triangle.

**Figure 11.10**

If angle $A$ is *obtuse,* we have two possible cases (Fig. 11.11).

*Note:* If the given parts are not angle $A$, side opposite $a$, and side adjacent $b$ as in our preceding discussions, then you must substitute the given angle and sides accordingly. This is why it is so important to understand the general word description corresponding to each case.

**EXAMPLE 4**

If $A = 26°$, $a = 25$ cm, and $b = 41$ cm, solve the triangle.
First, find $h$.

$$h = b \sin A = (41 \text{ cm})(\sin 26°) = 18 \text{ cm}$$

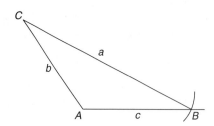 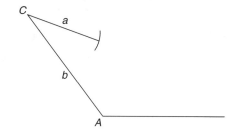

(a) When $a > b$, there is one possible triangle. In words, when the side opposite the given *obtuse* angle is greater than the known adjacent side, there is only one possible triangle.

(b) When $a \leq b$, there is no possible triangle. In words, when the side opposite the given *obtuse* angle is less than or equal to the known adjacent side, there is no possible triangle.

**Figure 11.11**

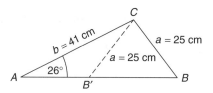

**Figure 11.12**

Since $h < a < b$, there are two solutions. First, let's find $B$ in triangle $ACB$ in Fig. 11.12.

$$\frac{a}{\sin A} = \frac{b}{\sin B}$$

$$\frac{25 \text{ cm}}{\sin 26°} = \frac{41 \text{ cm}}{\sin B}$$

$$\sin B = \frac{(41 \text{ cm})(\sin 26°)}{25 \text{ cm}} = 0.7189$$

$$B = 46°$$

$$C = 180° - 26° - 46° = 108°$$

Find $c$.

$$\frac{c}{\sin C} = \frac{a}{\sin A}$$

$$\frac{c}{\sin 108°} = \frac{25 \text{ cm}}{\sin 26°}$$

$$c = \frac{(25 \text{ cm})(\sin 108°)}{\sin 26°} = 54 \text{ cm}$$

Therefore, the first solution is $B = 46°$, $C = 108°$, and $c = 54$ cm.

The second solution occurs when $B$ is obtuse, as in triangle $ACB'$. That is, find the obtuse angle whose sine is 0.7189.

$$B' = 180° - 46° = 134°$$

Then $C = 180° - 26° - 134° = 2\overline{0}°$.

For $c$,

$$\frac{c}{\sin C} = \frac{a}{\sin A}$$

$$\frac{c}{\sin 20°} = \frac{25 \text{ cm}}{\sin 26°}$$

$$c = \frac{(25 \text{ cm})(\sin 2\overline{0}°)}{\sin 26°} = 2\overline{0} \text{ cm}$$

The second solution is $B' = 134°$, $C = 2\overline{0}°$, and $c = 2\overline{0}$ cm.

*Note:* The triangles in Section 11.1 were carefully chosen so that they had only one solution.

**EXAMPLE 5**

If $A = 62.0°$, $a = 415$ m, and $b = 855$ m, solve the triangle.
First, find $h$.

$$h = b \sin A$$

$$h = (855 \text{ m})(\sin 62.0°)$$

$$= 755 \text{ m}$$

Since $a < h$, there is no possible solution. What would happen if you applied the law of sines anyway?

$$\frac{a}{\sin A} = \frac{b}{\sin B}$$

$$\frac{415 \text{ m}}{\sin 62.0°} = \frac{855 \text{ m}}{\sin B}$$

$$\sin B = \frac{(855 \text{ m})(\sin 62.0°)}{415 \text{ m}} = 1.819 \qquad \text{(Tilt!)}$$

*Note:* $\sin B = 1.819$ is impossible because $-1 \leq \sin B \leq 1$.

In summary,

1. **Given two angles and one side (AAS):** There is only one possible triangle.
2. **Given two sides and an angle opposite one of them (SSA):** There are three possibilities. If the side opposite the given angle is
   (a) greater than the known adjacent side, there is only one possible triangle.
   (b) less than the known adjacent side but greater than the altitude, there are two possible triangles.
   (c) less than the altitude, there is no possible triangle.

Since solving a general triangle requires several operations, errors are often introduced. The following points may be helpful in avoiding some of these errors:

1. Always choose a given value over a calculated value when doing calculations.
2. Always check your results to see that the largest angle is opposite the largest side, and the smallest angle is opposite the smallest side.

## Exercises 11.2

For each general triangle as in Fig. 11.13,
(**a**) determine the number of solutions, and
(**b**) solve the triangle, if possible.

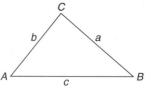

**Figure 11.13**

Express the lengths of sides to three significant digits and the angles to the nearest tenth of a degree.

1. $A = 37.0°$, $a = 21.5$ cm, $b = 16.4$ cm
2. $B = 55.0°$, $b = 182$ m, $c = 203$ m
3. $C = 26.5°$, $c = 42.7$ km, $a = 47.2$ km
4. $B = 40.4°$, $b = 81.4$ m, $c = 144$ m
5. $A = 71.5°$, $a = 3.45$ m, $c = 3.50$ m
6. $C = 17.2°$, $c = 2.20$ m, $b = 2.00$ m
7. $B = 105.0°$, $b = 16.5$ mi, $a = 12.0$ mi
8. $A = 98.8°$, $a = 707$ ft, $b = 585$ ft

Express the lengths of sides to two significant digits and the angles to the nearest degree.

9. $C = 18°$, $c = 24$ mi, $a = 45$ mi
10. $B = 36°$, $b = 75$ cm, $a = 95$ cm
11. $C = \overline{60}°$, $c = 150$ m, $b = 180$ m
12. $A = 30°$, $a = 4800$ ft, $c = 3600$ ft
13. $B = 8°$, $b = 450$ m, $c = 850$ m
14. $B = 45°$, $c = 2.5$ m, $b = 3.2$ m

Express the lengths of sides to four significant digits and the angles to the nearest hundredth of a degree.

15. $B = 41.50°$, $b = 14.25$ km, $a = 18.50$ km
16. $A = 15.75°$, $a = 642.5$ m, $c = 592.7$ m
17. $C = 63.85°$, $c = 29.50$ cm, $b = 38.75$ cm
18. $B = 50.00°$, $b = 41,250$ km, $c = 45,650$ km
19. $C = 8.75°$, $c = 89.30$ m, $a = 61.93$ m
20. $A = 31.50°$, $a = 375.0$ mm, $b = 405.5$ mm

Express the lengths of sides to four significant digits and the angles to the nearest second.

21. $B = 29°16'37''$, $b = 215.6$ m, $c = 304.5$ m
22. $A = 61°12'30''$, $a = 3457$ ft, $c = 2535$ ft
23. $C = 25°45''$, $a = 524.5$ ft, $c = 485.6$ ft
24. $A = 21°45'$, $a = 1785$ m, $b = 2025$ m

## 11.3  LAW OF COSINES

D.1, D.5

When the law of sines cannot be used, we use the **law of cosines.** In words, the square of any side of a triangle is equal to the sum of the squares of the other two sides minus twice the product of these two sides and the cosine of their included angle (see Fig. 11.14). By formula, the law is stated as follows.

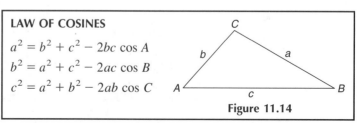

**LAW OF COSINES**

$$a^2 = b^2 + c^2 - 2bc \cos A$$
$$b^2 = a^2 + c^2 - 2ac \cos B$$
$$c^2 = a^2 + b^2 - 2ab \cos C$$

**Figure 11.14**

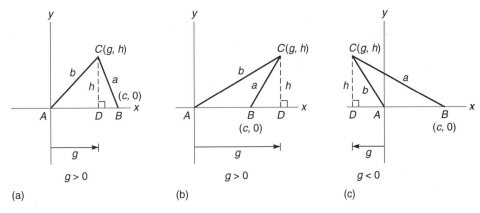

Figure 11.15

To derive the law of cosines, let triangle $ABC$ be any general triangle. Figure 11.15 shows the three possible triangles or cases: point $C$ between $A$ and $B$, point $C$ to the right of $B$, and point $C$ to the left of $A$.

The length of side $a$ in any of the three triangles in Fig. 11.15 is found using the formula for the distance between two points:

$$a = \sqrt{(g - c)^2 + (h - 0)^2}$$
$$a^2 = (g - c)^2 + (h - 0)^2 \qquad \text{(Square each side.)}$$
$$a^2 = g^2 - 2gc + c^2 + h^2 \qquad\qquad\qquad \textbf{(11.1)}$$

From Fig. 11.15, note that $b^2 = g^2 + h^2$, and make this substitution in Equation (11.1).

$$a^2 = b^2 + c^2 - 2gc \qquad\qquad\qquad \textbf{(11.2)}$$

Note also in each triangle that $\cos A = \dfrac{g}{b}$ or $g = b \cos A$. Now, make this substitution in Equation (11.2).

$$a^2 = b^2 + c^2 - 2bc \cos A$$

*Note:* If angle $A$ is acute, $\cos A > 0$; if angle $A$ is obtuse, $\cos A < 0$.

The other two forms of the law of cosines may be derived in a similar manner by relabeling the vertices.

---

There are two cases when the law of sines does not apply and we use the law of cosines to solve triangles:

**1.** Two sides and the included angle are known (SAS).

**2.** All three sides are known (SSS).

---

Do you see that when the law of cosines is used, there is no possibility of an ambiguous case? If not, draw a few triangles for each of these two cases (SAS and SSS) to convince yourself intuitively.

If $A = 90°$, then $\cos A = 0$, and the law of cosines

$$a^2 = b^2 + c^2 - 2bc \cos A$$

reduces to

$$a^2 = b^2 + c^2$$

which is the Pythagorean theorem. The Pythagorean theorem is thus a special case of the law of cosines.

**EXAMPLE 1**

If $a = 112$ m, $b = 135$ m, and $C = 104.3°$, solve the triangle.

First, draw a triangle as in Fig. 11.16 and find $c$ by using the law of cosines.

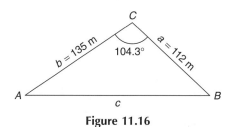

**Figure 11.16**

$$c^2 = a^2 + b^2 - 2ab \cos C$$
$$c^2 = (112 \text{ m})^2 + (135 \text{ m})^2 - 2(112 \text{ m})(135 \text{ m})(\cos 104.3°)$$
$$c = 196 \text{ m}$$

This side may be found using a calculator as follows:

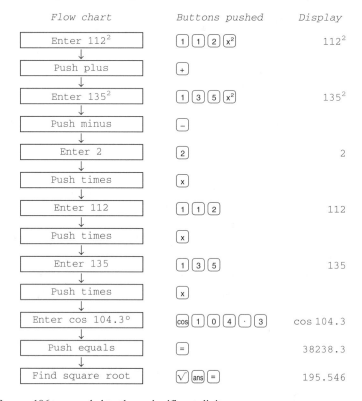

So $c = 196$ m rounded to three significant digits.

To find $A$, use the law of sines since it requires less computation.

$$\frac{a}{\sin A} = \frac{c}{\sin C}$$

$$\frac{112 \text{ m}}{\sin A} = \frac{196 \text{ m}}{\sin 104.3°}$$

$$\sin A = \frac{(112 \text{ m})(\sin 104.3°)}{196 \text{ m}} = 0.5537$$

$$A = 33.6°$$

$$B = 180° - 104.3° - 33.6° = 42.1°$$

The solution is $A = 33.6°$, $B = 42.1°$, and $c = 196$ m.

**EXAMPLE 2**

If $a = 375.0$ ft, $b = 282.0$ ft, and $c = 114.0$ ft, solve the triangle.

First, draw a triangle as in Fig. 11.17 and find $A$ by using the law of cosines.

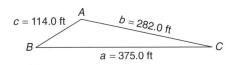

**Figure 11.17**

$$a^2 = b^2 + c^2 - 2bc \cos A$$

$$(375.0 \text{ ft})^2 = (282.0 \text{ ft})^2 + (114.0 \text{ ft})^2 - 2(282.0 \text{ ft})(114.0 \text{ ft}) \cos A$$

$$\cos A = \frac{(375.0 \text{ ft})^2 - (282.0 \text{ ft})^2 - (114.0 \text{ ft})^2}{-2(282.0 \text{ ft})(114.0 \text{ ft})}$$

$$\cos A = -0.7482$$

$$A = 138.43° \qquad \text{(Rounded to the nearest hundredth of a degree)}$$

Next, to find $B$, let's use the law of sines.

$$\frac{a}{\sin A} = \frac{b}{\sin B}$$

$$\frac{375.0 \text{ ft}}{\sin 138.43°} = \frac{282.0 \text{ ft}}{\sin B}$$

$$\sin B = \frac{(282.0 \text{ ft})(\sin 138.43°)}{375.0 \text{ ft}} = 0.4990$$

$$B = 29.93°$$

$$C = 180° - 138.43° - 29.93° = 11.64°$$

The solution is $A = 138.43°$, $B = 29.93°$, and $C = 11.64°$.

# Exercises 11.3

*Solve each triangle using the labels shown in Fig. 11.18.*

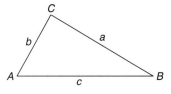

**Figure 11.18**

*Express the lengths of sides to three significant digits and the angles to the nearest tenth of a degree.*

**1.** $A = 60.0°$, $b = 19.5$ m, $c = 25.0$ m  **2.** $B = 19.5°$, $a = 21.5$ ft, $c = 12.5$ ft
**3.** $C = 109.0°$, $a = 14\overline{0}$ km, $b = 215$ km  **4.** $A = 94.7°$, $c = 875$ yd, $b = 185$ yd
**5.** $a = 19.2$ m, $b = 21.3$ m, $c = 27.2$ m  **6.** $a = 125$ km, $b = 195$ km, $c = 145$ km
**7.** $a = 4.25$ ft, $b = 7.75$ ft, $c = 5.50$ ft  **8.** $a = 3590$ m, $b = 7950$ m, $c = 4650$ m

*Express the lengths of sides to two significant digits and the angles to the nearest degree.*

**9.** $A = 45°$, $b = 51$ m, $c = 39$ m  **10.** $B = 6\overline{0}°$, $a = 160$ cm, $c = 230$ cm
**11.** $a = 7\overline{0}00$ m, $b = 5600$ m, $c = 4800$ m  **12.** $a = 5.8$ cm, $b = 5.8$ cm, $c = 9.6$ cm
**13.** $C = 135°$, $a = 36$ ft, $b = 48$ ft  **14.** $A = 5°$, $b = 19$ m, $c = 25$ m

*Express the lengths of sides to four significant digits and the angles to the nearest hundredth of a degree.*

**15.** $B = 19.25°$, $a = 4815$ m, $c = 1925$ m  **16.** $C = 75.00°$, $a = 37{,}550$ mi, $b = 45{,}250$ mi
**17.** $C = 108.75°$, $a = 405.0$ mm, $b = 325.0$ mm  **18.** $A = 111.05°$, $b = 1976$ ft, $c = 325\overline{0}$ ft
**19.** $a = 207.5$ km, $b = 105.6$ km, $c = 141.5$ km  **20.** $a = 19.45$ m, $b = 36.50$ m, $c = 25.60$ m

*Express the lengths of sides to four significant digits and the angles to the nearest second.*

**21.** $A = 72°18'0''$, $b = 1074$ m, $c = 1375$ m  **22.** $C = 101°25'30''$, $a = 685.0$ ft, $b = 515.0$ ft
**23.** $a = 1.250$ mi, $b = 1.975$ mi, $c = 1.250$ mi  **24.** $a = 375.1$ m, $b = 286.0$ m, $c = 305.0$ m

## 11.4 APPLICATIONS OF OBLIQUE TRIANGLES

In Section 3.4, we saw several applications of right triangles. Similarly, there are many applications of oblique, or general, triangles in the world around us. The following examples and exercises illustrate a few of these applications.

### EXAMPLE 1

The ground of a subdivision lot slopes upward from the street at an angle of 8°. The builder wants to build a house on a level lot that is $6\overline{0}$ ft deep. To control erosion, the back of the lot must be cut to have a slope of 25°. How far from the street, measured along the present slope, will the excavation extend?

First, draw a diagram as in Fig. 11.19 and find $AC$. In $\triangle ABC$, $A = 8°$ and

$$\angle ABC = 180° - 25° = 155°$$

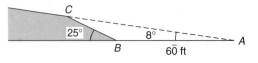

**Figure 11.19**

Thus, $C = 180° - 8° - 155° = 17°$.

Using the law of sines, we have

$$\frac{AC}{\sin(\angle ABC)} = \frac{AB}{\sin C}$$

$$\frac{AC}{\sin 155°} = \frac{60 \text{ ft}}{\sin 17°}$$

$$AC = \frac{(\sin 155°)(60 \text{ ft})}{\sin 17°} = 87 \text{ ft}$$

**EXAMPLE 2**

The Sanchez family wants a northern skylight built into the roof line of their house, as shown in Fig. 11.20. The windows for the skylight are to be 4.00 m long and make an angle of 40.0° with the horizontal.

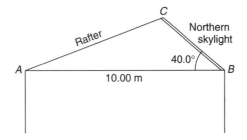

**Figure 11.20**

(a) How long must the rafters on the south side be?

(b) What angle does the roof make with the horizontal on the south?

(a) We need to find $AC$. In $\triangle ABC$, $B = 40.0°$, $AB = 10.00$ m, and $BC = 4.00$ m. Using the law of cosines, we have

$$(AC)^2 = (AB)^2 + (BC)^2 - 2(AB)(BC)\cos B$$

$$(AC)^2 = (10.00 \text{ m})^2 + (4.00 \text{ m})^2 - 2(10.00 \text{ m})(4.00 \text{ m})\cos 40.0°$$

$$AC = 7.40 \text{ m}$$

(b) We need to find $A$. Using the law of sines, we have

$$\frac{BC}{\sin A} = \frac{AC}{\sin B}$$

$$\frac{4.00 \text{ m}}{\sin A} = \frac{7.40 \text{ m}}{\sin 40.0°}$$

$$\sin A = \frac{(\sin 40.0°)(4.00 \text{ m})}{7.40 \text{ m}} = 0.3475$$

$$A = 20.3°$$

## Exercises 11.4

1. Find the lengths of rafters $AC$ and $BC$ for the roof in Fig. 11.21.

2. The sides of a triangular metal plate measure 17.5 in., 12.3 in., and 21.3 in. Find the measure of the largest angle.

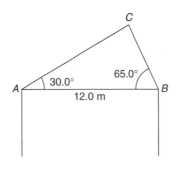

**Figure 11.21**

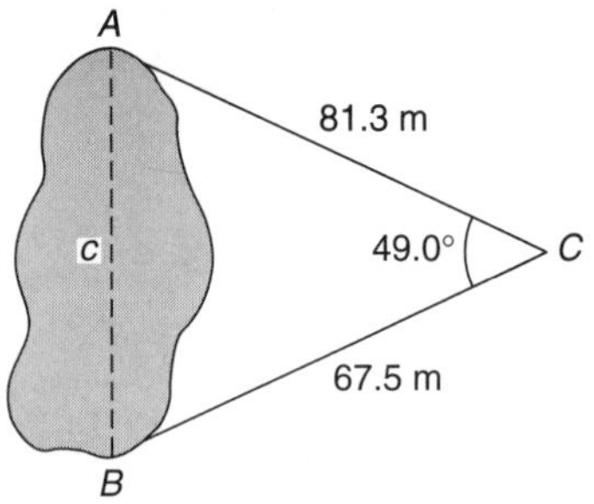

**Figure 11.22**

3. Find the distance $c$ across the pond in Fig. 11.22.

4. A surveyor on the side of a hill, whose surface makes an angle of 25.0° with the horizontal, measures the angle of depression to the top of a tree at the bottom of the hill as 10.0°. Find the height of the tree if its base is 90.0 ft from the surveyor.

5. To find the distance between two points $A$ and $B$ lying on opposite banks of a river, a length $AC$ of 300 m is measured. Angle $BAC$ measures 58° and $\angle ACB$ measures 49°. Find the distance between $A$ and $B$.

6. A weather balloon is sighted between points $A$ and $B$, which are 3.6 mi apart on level ground. The angle of elevation of the balloon from $A$ is 28°, and its angle of elevation from $B$ is 49°. Find the height (in ft) of the balloon above the ground.

7. A forest ranger in an observation tower at $A$ sights a fire in a direction 51.5° east of north. Another ranger in a tower at $B$ 5.00 mi due east of $A$ sights the fire at 17.2° west of north. How far is the fire from each observation tower?

8. The angle at one corner of a triangular plot of ground measures 65.5°. If the sides that meet at this corner measure 225 m and 320 m, what is the length of the third side?

9. Two automobiles depart at the same time from the intersection of two straight highways, which intersect at 75°. If the automobiles' speeds are 60 mph and 45 mph, how far apart will they be after 1.5 h?

10. A vertical cable television tower as in Fig. 11.23 is standing on the side of a hill, which makes an angle of 25.0° with the horizontal. Guy wires are attached 160 ft up the tower. What lengths of guy wires are needed to reach points 50.0 ft uphill and 80.0 ft downhill from the base of the tower?

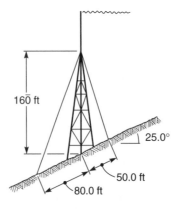

**Figure 11.23**

11. From point $A$ the angle of elevation to the top of a cliff is $37°$ as in Fig. 11.24. On level ground the angle of elevation from point $B$ to the top of the cliff is $24°$. Point $A$ is 270 ft closer to the cliff than point $B$. Find the height of the cliff.

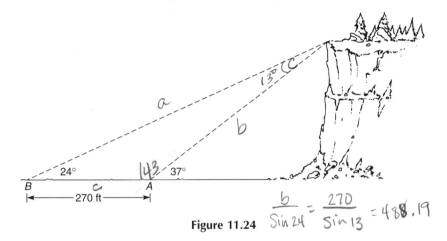

**Figure 11.24** $\dfrac{b}{\sin 24} = \dfrac{270}{\sin 13} = 488.19$

12. A lighthouse $15\overline{0}$ ft high is on the edge of a vertical cliff overlooking the ocean. The angle of elevation from a ship to the bottom of the lighthouse measures $15°$. The angle of elevation from the ship to the top of the lighthouse measures $24°$.
   (a) Find the distance from the ship to the cliff.
   (b) Find the height of the cliff.

13. A tower 75 m high is on a vertical cliff on the bank of a river. From the top of the tower, the angle of depression to a piece of driftwood on the opposite bank of the river is $28°$. From the bottom of the tower, the angle of depression to the same piece of driftwood is $19°$.
   (a) Find the width of the river.
   (b) Find the height of the cliff.

14. Find the lengths $L_1$ and $L_2$ of the rafters in Fig. 11.25.

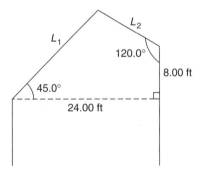

**Figure 11.25**

15. An owner of a triangular lot wishes to fence it in along the lot lines. Lot markers at $A$ and $B$ have been located, but the lot marker at $C$ cannot be located. The owner's attorney gives the following information by phone: $AB = 245$ ft, $BC = 185$ ft, and $A = 35.0°$. What is the length of $AC$?

16. The average distance from the sun to Earth is $1.5 \times 10^8$ km, and from the sun to Venus it is $1.1 \times 10^8$ km. Find the distance between Earth and Venus when the angle between Earth and

the sun and Earth and Venus is 25°. (Assume that Earth and Venus have circular orbits around the sun.)

**17.** A farmer wants to extend the roof of a barn in Fig. 11.26 to add a storage area for machinery. If 20.0 ft of clearance is needed at the lowest point, how wide can the addition be?

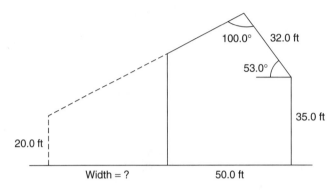

**Figure 11.26**

**18.** In the framework in Fig. 11.27, we know that $AE = CD$, $AB = BC$, $BD = BE$, and $AC \parallel ED$. Find the measure of each.

(a) $\angle AEB$      (b) $\angle A$      (c) $BE$      (d) $DE$

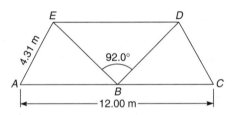

**Figure 11.27**

**19.** In the roof truss in Fig. 11.28, we know that $AB = CD$, $AG = DE$, $GF = FE$, $BG = CE$, and $BF = CF$. Find the measure of each.

(a) $BG$      (b) $\angle ABG$      (c) $BF$      (d) $\angle GFB$      (e) $\angle FBC$

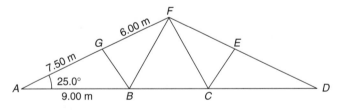

**Figure 11.28**

**20.** Find the distance between the peaks of two hills across the gorge in Fig. 11.29. Points $A$ and $B$ are trees on the peaks of the two hills. Point $C$ is where you stand. Measure a length of $10\overline{0}$ m from point $C$ to point $D$. Then $\angle BCD$ measures 115.0°, $\angle CDA$ measures 120.5°, $\angle BCA$ measures 86.5°, and $\angle ADB$ measures 98.1°.

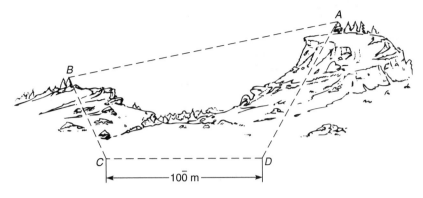

**Figure 11.29**

**21.** A deck is to be built on the house in Fig. 11.30. How long will the stairway, *l*, be?

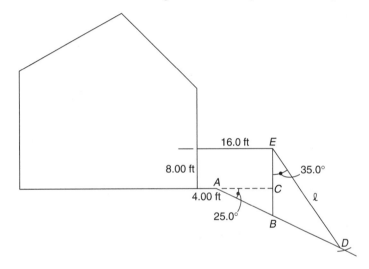

**Figure 11.30**

**22.** A dental office building in the shape of a regular pentagon is planned as in Fig. 11.31. The central records department is located at point *A*. The various offices receive the dental records by a conveyor system.
 (a) Find the length of the conveyor from *A* to *D*.
 (b) Find the length of the conveyor from *A* to *N*, the midpoint of side *BC*.
 (c) Find the length of the conveyor from *A* to *M*, the midpoint of side *CD*.
 *Note:* Each interior angle of a regular pentagon is 108°.

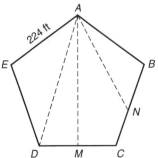

**Figure 11.31**

*Find the value of x in each diagram. Find angles to the nearest tenth of a degree and sides to three significant digits.*

**23.**

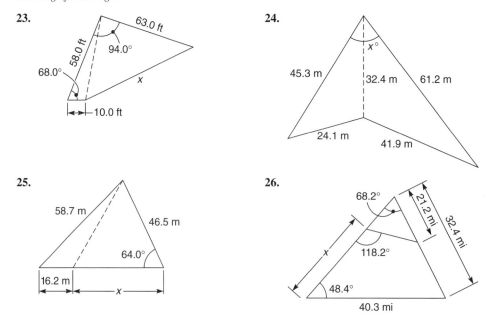

**24.**

**25.**

**26.**

## 11.5 ADDITION OF VECTORS: GRAPHICAL METHODS

A quantity such as length, volume, time, or temperature is completely described when its magnitude (size) is given. These quantities are called **scalars.** Other physical quantities in science and technology require both magnitude and direction in order to be completely described. Examples of such quantities include force, velocity, torque, and certain quantities from electricity. More specifically, to completely describe wind velocity, it requires not only a speed, such as 25 mph, but also the direction, 20° east of north. When a lawnmower is pushed with a given force of 20 lb, the direction of the force (angle of the handle with the ground) determines the ease of mowing. These are called **vector** quantities.

Graphically, vectors are usually represented by directed line segments. The length of a segment indicates the magnitude of the quantity. An arrowhead is used to indicate the direction (see Fig. 11.32). If $A$ and $B$ are the end points of a line segment, the symbol **AB** denotes the **vector from $A$ to $B$;** point $A$ is called the **initial point,** and point $B$ is called the **terminal point.** The vector **BA** has the same length as **AB** but has the opposite direction. Vectors may also be denoted by a single lowercase letter, such as **v, u,** or **w**. The magnitude or length of a vector **v** is denoted as $|\mathbf{v}|$.

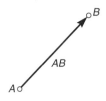

**Figure 11.32**

When writing vectors on paper or a chalkboard, we use a small arrow above the vector quantity, such as $\vec{v}$, $\vec{R}$, or $\overrightarrow{AB}$ as in Fig. 11.33.

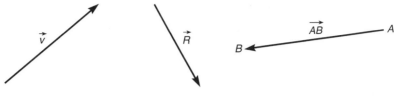

**Figure 11.33**

Two vectors are equal when they have the same magnitude and the same direction. Two vectors are negatives of each other when they have the same magnitude but opposite directions (see Fig. 11.34).

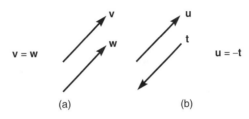

(a)                    (b)

**Figure 11.34**

Note that a given vector **v** may be placed in any position as long as its magnitude and direction are not changed. Such vectors are called *free vectors* (see Fig. 11.35).

A vector is in **standard position** when its initial point is at the origin of the *xy*-coordinate system. A vector in standard position is expressed in terms of its length and angle $\theta$, where $\theta$ is measured counterclockwise from the positive *x*-axis to the vector. The vector in Fig. 11.36 is expressed in standard position.

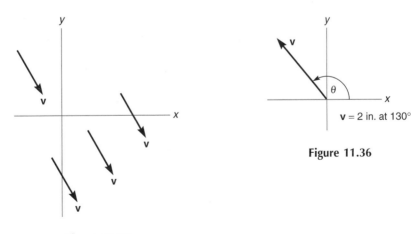

**v** = 2 in. at 130°

**Figure 11.36**

**Figure 11.35**

The sum of two or more vectors is called the **resultant.** This sum may be obtained graphically using one of two methods: the parallelogram method or the vector triangle method, defined as follows.

---

**PARALLELOGRAM METHOD**

To add two vectors, **v** and **w,** construct a parallelogram using **v** as one pair of parallel sides and **w** for the other pair. The diagonal of the parallelogram, as shown in Fig. 11.37, is the resultant or sum of the two vectors.

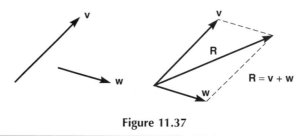

**Figure 11.37**

---

**VECTOR TRIANGLE METHOD**

To add two vectors, **v** and **w,** construct the second vector **w** with its initial point on the terminal point of the first vector **v.** The resultant vector is the vector joining the initial point of the first vector to the terminal point of the second vector (see Fig. 11.38).

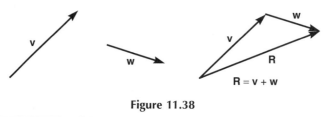

**Figure 11.38**

---

The triangle method is particularly useful when several vectors are to be added (see Fig. 11.39).

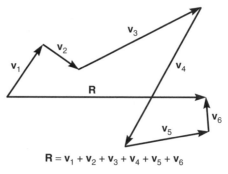

$$R = v_1 + v_2 + v_3 + v_4 + v_5 + v_6$$

**Figure 11.39**

A vector may be subtracted by adding its negative. That is, $\mathbf{v} - \mathbf{w} = \mathbf{v} + (-\mathbf{w})$. Construct $\mathbf{v}$ as usual, construct the negative of $\mathbf{w}$, and find the resultant as in Fig. 11.40.

Suppose a friend offers to fly you from Parkland to Kampsville and asks you how to get there. You reply, "It is 250 km." Have you given the friend enough information to find Kampsville? Obviously not! You must also tell him or her in what direction to go. If you reply, "Go 250 km due west," the friend can find Kampsville. This change in position is represented by the vector **PK** in Fig. 11.41.

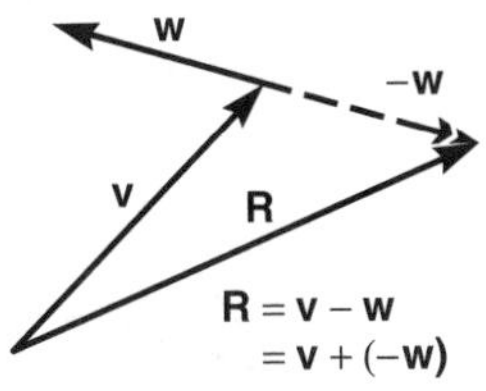

Figure 11.40

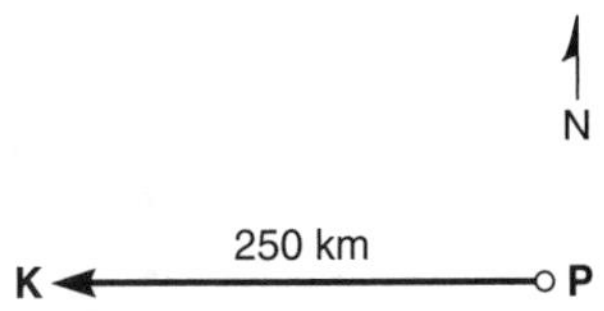

Figure 11.41

Perhaps the simplest vector is **displacement,** which is the net change in position. Displacement is a vector because it requires both a magnitude and a direction for its complete description.

Suppose your friend needs to stop at Hillsfield on the way to Kampsville, as shown in the flight plan in Fig. 11.42. Which is the displacement vector now? Since displacement is the net change in position, the displacement is the shortest distance between the beginning point and the ending point, or vector **PK**. The displacement vector is the same no matter which route is taken. This second situation may be expressed by vectors as follows:

$$\mathbf{PH} + \mathbf{HK} = \mathbf{PK}$$

This does not mean the distances are the same. In fact, we know that

$$|\mathbf{PH}| + |\mathbf{HK}| > |\mathbf{PK}|$$

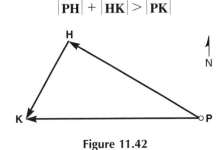

Figure 11.42

### EXAMPLE 1

Find the vector sum of the following two vectors, each given in standard position, by using each graphic method:

$$\mathbf{v} = 2.4 \text{ km at } 15°$$

$$\mathbf{w} = 3.3 \text{ km at } 75°$$

Choose a suitable scale as in Fig. 11.43. Using ruler and protractor for measuring, we find that

$$\mathbf{v} + \mathbf{w} = \mathbf{R} = 5.0 \text{ km at } \overline{5}0°$$

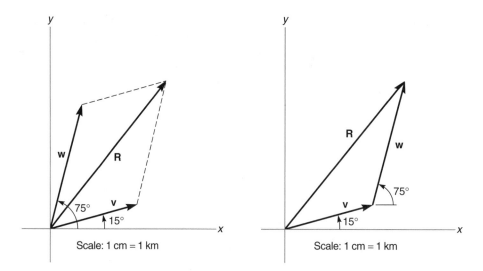

**Figure 11.43**

**EXAMPLE 2**

Given the following three vectors, each in standard position, find the vector sum graphically:

$$\mathbf{v}_1 = 38 \text{ mi at } 12\overline{0}°$$
$$\mathbf{v}_2 = 48 \text{ mi at } 35°$$
$$\mathbf{v}_3 = 82 \text{ mi at } 195°$$

Choose a suitable scale as in Fig. 11.44. Using ruler and protractor for measuring, we find that

$$\mathbf{v}_1 + \mathbf{v}_2 + \mathbf{v}_3 = \mathbf{R} = 71 \text{ mi at } 146°$$

**EXAMPLE 3**

An airplane is flying at 250 mi/h (air speed) on a compass heading of $4\overline{0}°$ east of north. A wind of $5\overline{0}$ mi/h is blowing from the north. What is the true course (the heading with respect to the ground) of the airplane, and what is its ground speed?

Let's use the scale 1 cm = $5\overline{0}$ mph as in Fig. 11.45. Let

$$\left.\begin{array}{l} \mathbf{v} = 250 \text{ mi/h at } 5\overline{0}° \\ \mathbf{w} = 5\overline{0} \text{ mi/h at } 27\overline{0}° \end{array}\right\} \text{ (Standard position)}$$

By measuring the length of $\mathbf{R}$ with a ruler, we find

$$(4.2 \text{ cm}) \times \frac{5\overline{0} \text{ mi/h}}{\text{cm}} = 210 \text{ mi/h}$$

Now use a protractor to measure the angle that $\mathbf{R}$ makes with the north line, 51°. Therefore,

$$\mathbf{v} + \mathbf{w} = \mathbf{R} = 210 \text{ mi/h at } 51° \text{ east of north}$$
$$= 210 \text{ mi/h at } 39° \text{ (in standard position)}$$

**Velocity** may be defined as the time rate of change of displacement. Thus, velocity is a vector, and both its magnitude (speed) and its direction are required for its complete description.

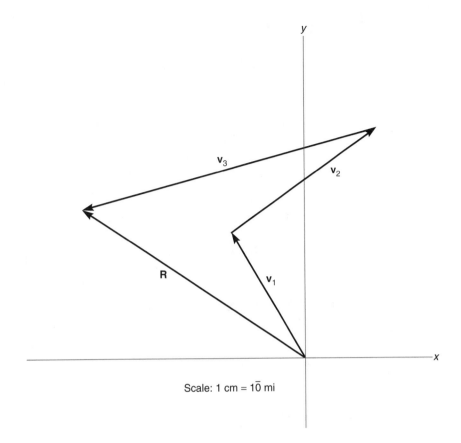

Scale: 1 cm = 1$\bar{0}$ mi

**Figure 11.44**

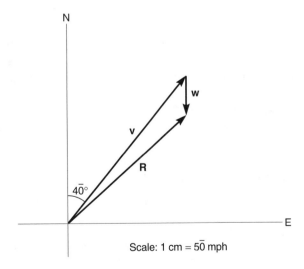

Scale: 1 cm = 5$\bar{0}$ mph

**Figure 11.45**

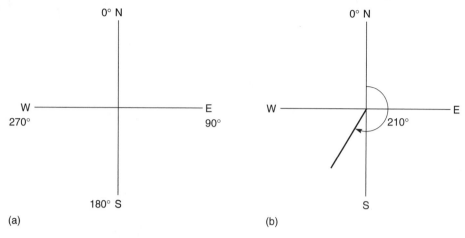

**Figure 11.46**

Bearing is another system of measuring angles as in Fig. 11.46. Angles are measured *clockwise* from north. An angle with bearing 210° is shown in Fig. 11.46(b). To avoid confusion, we shall consistently refer to angles in standard position.

## Exercises 11.5

*(Angles are given in standard position unless stated otherwise.)*

*Find the sum of each set of vectors by using the parallelogram method. Choose a suitable scale.*

**1.** $\mathbf{v}$ = 4.5 km at 67°
$\mathbf{w}$ = 6.5 km at 105°

**2.** $\mathbf{v}$ = 150 mi at 175°
$\mathbf{w}$ = 270 mi at 215°

**3.** $\mathbf{v}$ = 75 mi/h at 345°
$\mathbf{w}$ = 25 mi/h at 27$\overline{0}$°

**4.** $\mathbf{v}$ = 85 km/h at 25°
$\mathbf{w}$ = 25 km/h at 165°

*Find the sum of each set of vectors by using the vector triangle method. Choose a suitable scale.*

**5.** $\mathbf{v}$ = 75 mi at 30$\overline{0}$°
$\mathbf{w}$ = 25 mi at 195°

**6.** $\mathbf{v}$ = 250 km at 25°
$\mathbf{w}$ = 450 km at 345°

**7.** $\mathbf{v}$ = 18 mi at 45°
$\mathbf{w}$ = 32 mi at 11$\overline{0}$°
$\mathbf{t}$ = 28 mi at 20$\overline{0}$°

**8.** $\mathbf{v}$ = 95 km at 35$\overline{0}$°
$\mathbf{w}$ = 65 km at 30$\overline{0}$°
$\mathbf{t}$ = 25 km at 255°
$\mathbf{u}$ = 75 km at 18$\overline{0}$°

**9.** Using the vectors in Exercise 1, find $\mathbf{v} - \mathbf{w}$ graphically.

**10.** Using the vectors in Exercise 2, find $\mathbf{w} - \mathbf{v}$ graphically.

**11.** A ship sails 55 mi due north and then sails 3$\overline{0}$ mi at an angle of 45° east of north. How far is the ship from its starting point? What is the angle of the resultant path with respect to the starting point?

**12.** A plane is flying at 3$\overline{0}$0 mi/h (air speed) on a compass heading of 4$\overline{0}$° west of south. A wind of 6$\overline{0}$ mi/h is blowing from the east. What are its true course and its ground speed?

**13.** Do Exercise 12 when the wind is blowing from the west.

# 11.6 ADDITION OF VECTORS: TRIGONOMETRIC METHODS

D.1, D.5

Vector sums may be found more accurately using trigonometry.

**EXAMPLE 1**

Using trigonometry, find the sum of the given vectors.

$$\mathbf{v} = 19.5 \text{ km due west}$$
$$\mathbf{w} = 45.0 \text{ km due north}$$

The vector triangle method gives the sketch in Fig. 11.47.

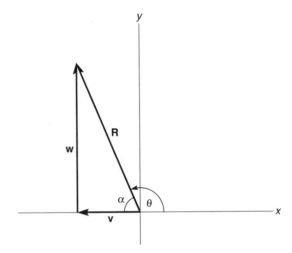

**Figure 11.47**

Using the Pythagorean theorem to find $|\mathbf{R}|$, we have

$$|\mathbf{R}| = \sqrt{|\mathbf{v}|^2 + |\mathbf{w}|^2}$$
$$|\mathbf{R}| = \sqrt{(19.5 \text{ km})^2 + (45.0 \text{ km})^2}$$
$$= 49.0 \text{ km}$$

Next, find $\alpha$.

$$\tan \alpha = \frac{\text{side opposite } \alpha}{\text{side adjacent to } \alpha} = \frac{|\mathbf{w}|}{|\mathbf{v}|}$$

$$\tan \alpha = \frac{45.0 \text{ km}}{19.5 \text{ km}} = 2.308$$

$$\alpha = 66.6°$$
$$\theta = 180° - \alpha = 180° - 66.6° = 113.4°$$

Therefore,

$$\mathbf{R} = 49.0 \text{ km at } 113.4°$$

Vector sums may also be found by using the law of sines and the law of cosines.

**EXAMPLE 2**

Using trigonometry, find the sum of the given vectors.

$$\mathbf{v} = 2.40 \text{ km at } 15.0°$$
$$\mathbf{w} = 3.30 \text{ km at } 75.0°$$

First, draw the vectors as in Fig. 11.48. Use the law of cosines to find the length of **R**.

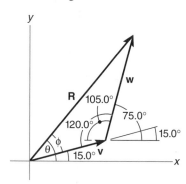

**Figure 11.48**

$$|\mathbf{R}|^2 = |\mathbf{v}|^2 + |\mathbf{w}|^2 - 2|\mathbf{v}||\mathbf{w}|\cos 120.0°$$
$$|\mathbf{R}|^2 = (2.40 \text{ km})^2 + (3.30 \text{ km})^2 - 2(2.40 \text{ km})(3.30 \text{ km})\cos 120.0°$$
$$|\mathbf{R}| = 4.96 \text{ km}$$

Using the law of sines to find $\phi$, we have

$$\frac{|\mathbf{R}|}{\sin 120.0°} = \frac{|\mathbf{w}|}{\sin \phi}$$

$$\frac{4.96 \text{ km}}{\sin 120.0°} = \frac{3.30 \text{ km}}{\sin \phi}$$

$$\sin \phi = \frac{(\sin 120.0°)(3.30 \text{ km})}{4.96 \text{ km}} = 0.5762$$

$$\phi = 35.2°$$
$$\theta = \phi + 15.0° = 35.2° + 15.0° = 50.2°$$

Therefore,

$$\mathbf{R} = 4.96 \text{ km at } 50.2°$$

## Exercises 11.6

*(Angles are given in standard position unless stated otherwise.)*

*Using trigonometry, find the sum of each set of vectors.*

**1.** $\mathbf{v} = 65.3$ km/h at $270.0°$
 $\mathbf{w} = 40.5$ km/h at $180.0°$

**2.** $\mathbf{v} = 6150$ m due south
 $\mathbf{w} = 1780$ m due east

**3.** $\mathbf{v} = 4.50$ km at $67.0°$
 $\mathbf{w} = 6.50$ km at $105.0°$

**4.** $\mathbf{v} = 15\overline{0}$ mi at $175.0°$
 $\mathbf{w} = 27\overline{0}$ mi at $215.0°$

**5.** $\mathbf{v} = 87.1$ mi/h at $130.5°$
$\mathbf{w} = 46.7$ mi/h at $207.0°$

**6.** $\mathbf{v} = 60.0$ km/h at $286.0°$
$\mathbf{w} = 60.0$ km/h at $254.0°$

**7.** $\mathbf{v} = 605$ m at $60.0°$
$\mathbf{w} = 415$ m at $120.0°$
$\mathbf{t} = 295$ m at $90.0°$

**8.** $\mathbf{v} = 15.3$ mi at $135.0°$
$\mathbf{w} = 24.5$ mi at $75.0°$
$\mathbf{t} = 19.7$ mi at $180.0°$

**9.** Using the vectors in Exercise 3, find $\mathbf{v} - \mathbf{w}$.

**10.** Using the vectors in Exercise 8, find $\mathbf{w} - \mathbf{v}$.

**11.** An airplane takes off and flies 175 km on a course of $15.0°$ west of north and then changes course and flies 105 km due north to where it lands. Find the displacement from the starting point to the landing point.

**12.** A ship travels 75.0 mi on a course $25.0°$ north of east; then it travels 45.0 mi on a course $15.0°$ east of north to a point where it lands. Find the displacement from the starting point to the landing point.

**13.** An automobile is driven 25.0 km due east, then 10.0 km due north, then 15.0 km due east, then 10.0 km due south. Find the displacement from the starting point to the ending point.

**14.** En route between two cities, a car travels $10\overline{0}$ km due west, then 125 km southwest, then $15\overline{0}$ km due south. Find the displacement from the starting point to the ending point.

**15.** A plane is traveling south at 175 mi/h in still air. What is its velocity with a wind of 65.0 mi/h blowing from the east?

**16.** A plane is traveling due north at 175 mi/h in still air. What is its velocity with a wind of 65.0 mi/h blowing from $45.0°$ east of south?

## 11.7 VECTOR COMPONENTS

We shall find it desirable to express a given vector as the sum of two vectors, especially two vectors along the coordinate axes. In Fig. 11.49,

$$\mathbf{v} = \mathbf{v}_1 + \mathbf{v}_2$$
$$\mathbf{v} = \mathbf{u}_1 + \mathbf{u}_2$$
$$\mathbf{v} = \mathbf{v}_x + \mathbf{v}_y$$

The vectors $\mathbf{v}_1$, $\mathbf{v}_2$, $\mathbf{u}_1$, $\mathbf{u}_2$, $\mathbf{v}_x$, and $\mathbf{v}_y$ are called **components** of vector $\mathbf{v}$. That is, if two or more vectors are added and their sum is the resultant vector $\mathbf{v}$, then each of these vectors is called a component of $\mathbf{v}$. We call $\mathbf{v}_x$ the **horizontal component** and $\mathbf{v}_y$ the **vertical component.**

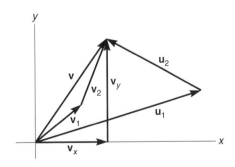

**Figure 11.49**

In general, the horizontal and vertical components may be found using the definitions for sine and cosine.

$$\mathbf{v}_x = |\mathbf{v}|\cos\theta$$
$$\mathbf{v}_y = |\mathbf{v}|\sin\theta$$

where $\theta$ is in standard position (see Fig. 11.50).

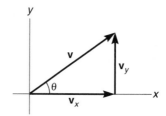

**Figure 11.50**

### EXAMPLE 1

Find the horizontal and vertical components of the following vector in standard position: 50.0 mi/h at 60.0°.

First, draw the vector as in Fig. 11.51. Thus,

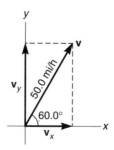

**Figure 11.51**

$$\mathbf{v}_x = |\mathbf{v}|\cos\theta$$
$$= (50.0 \text{ mi/h})(\cos 60.0°)$$
$$= 25.0 \text{ mi/h} \qquad \text{(At } 0° \text{ is understood.)}$$
$$\mathbf{v}_y = |\mathbf{v}|\sin\theta$$
$$= (50.0 \text{ mi/h})(\sin 60.0°)$$
$$= 43.3 \text{ mi/h} \qquad \text{(At } 90° \text{ is understood.)}$$

### EXAMPLE 2

Find the horizontal and vertical components of the following vector in standard position: $1\overline{0}$ m at 153°.

First, draw the vector as in Fig. 11.52. Then

$$\mathbf{v}_x = |\mathbf{v}|\cos\theta$$
$$= (1\overline{0} \text{ m})(\cos 153°)$$
$$= -8.9 \text{ m}$$

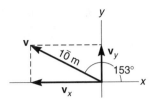

**Figure 11.52**

$$\mathbf{v}_y = |\mathbf{v}|\sin\theta$$
$$= (1\overline{0}\text{ m})(\sin 153°)$$
$$= 4.5\text{ m}$$

**EXAMPLE 3**

The person in Fig. 11.53 exerts a $5\overline{0}$-lb force on the handle of a lawnmower that is at an angle of $4\overline{0}°$ with the ground. What is the net horizontal component of the force that pushes the mower ahead? What is the net vertical component of the force that pushes the mower into the ground?

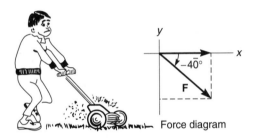

Force diagram

**Figure 11.53**

$$\mathbf{F}_x = |\mathbf{F}|\cos\theta$$
$$= (5\overline{0}\text{ lb})[\cos(-40°)]$$
$$= 38\text{ lb}$$
$$\mathbf{F}_y = |\mathbf{F}|\sin\theta$$
$$= (5\overline{0}\text{ lb})[\sin(-40°)]$$
$$= -32\text{ lb}$$

That is, a $5\overline{0}$-lb force at $-4\overline{0}°$ and a 38-lb force at $0°$ produce the same force in keeping the lawnmower moving forward.

We must also be able to find a vector $\mathbf{v}$ when its horizontal and vertical components, $\mathbf{v}_x$ and $\mathbf{v}_y$, are given. In general, to express $\mathbf{v}$ in standard position, use these steps:

**1.** The length of $\mathbf{v}$ may be found by using the Pythagorean theorem.

$$\boxed{|\mathbf{v}| = \sqrt{|\mathbf{v}_x|^2 + |\mathbf{v}_y|^2}}$$

**2.** To find angle $\theta$, we first need to find $\alpha$, the reference angle, by using the definition of tangent.

$$\tan \alpha = \frac{|\mathbf{v}_y|}{|\mathbf{v}_x|}$$

**EXAMPLE 4**

If $\mathbf{v}_x = 36$ km/h and $\mathbf{v}_y = 52$ km/h, find $\mathbf{v}$.

First, draw the components and $\mathbf{v}$ as in Fig. 11.54. Then

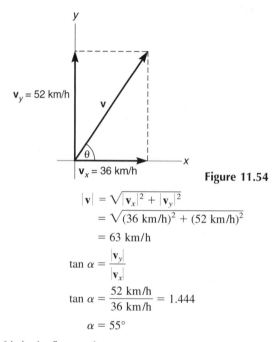

**Figure 11.54**

$$|\mathbf{v}| = \sqrt{|\mathbf{v}_x|^2 + |\mathbf{v}_y|^2}$$
$$= \sqrt{(36 \text{ km/h})^2 + (52 \text{ km/h})^2}$$
$$= 63 \text{ km/h}$$

$$\tan \alpha = \frac{|\mathbf{v}_y|}{|\mathbf{v}_x|}$$

$$\tan \alpha = \frac{52 \text{ km/h}}{36 \text{ km/h}} = 1.444$$

$$\alpha = 55°$$

Since $\theta$ is in the first quadrant,

$$\alpha = \theta = 55°$$

Therefore, $\mathbf{v} = 63$ km/h at $55°$

**EXAMPLE 5**

If $\mathbf{v}_x = -26.4$ N and $\mathbf{v}_y = -15.3$ N, find $\mathbf{v}$.

First, draw the components and $\mathbf{v}$ as in Fig. 11.55.

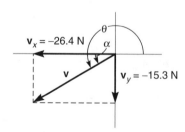

**Figure 11.55**

$$\mathbf{v} = \sqrt{|\mathbf{v}_x|^2 + |\mathbf{v}_y|^2}$$
$$= \sqrt{|-26.4 \text{ N}|^2 + |-15.3 \text{ N}|^2}$$
$$= 30.5 \text{ N}$$

$$\tan \alpha = \frac{|\mathbf{v}_y|}{|\mathbf{v}_x|}$$

$$\tan \alpha = \frac{15.3 \text{ N}}{26.4 \text{ N}} = 0.5795$$

$$\alpha = 30.1°$$

Since $\theta$ is in the third quadrant,

$$\theta = 180° + \alpha = 180° + 30.1° = 210.1°$$

Therefore,

$$\mathbf{v} = 30.5 \text{ N at } 210.1°$$

The **impedance** of a series circuit containing a **resistance** and an **inductance** can be represented by the vector diagram in Fig. 11.56, where $\phi$ is the phase angle that equals the amount by which the *current lags behind the voltage*. The resistance is always drawn as a vector pointing in the positive x-direction, and the inductive reactance is always drawn as a vector pointing in the positive y-direction.

305.5

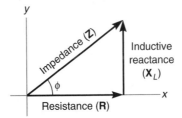

**Figure 11.56**

#### EXAMPLE 6

If the resistance is $6\overline{0}$ ohms ($\Omega$) and the inductive reactance is 36 $\Omega$, find the impedance.
Using Fig. 11.56, we have

$$|\mathbf{Z}| = \sqrt{(6\overline{0} \text{ }\Omega)^2 + (36 \text{ }\Omega)^2} = 7\overline{0} \text{ }\Omega$$

$$\tan \phi = \frac{36 \text{ }\Omega}{6\overline{0} \text{ }\Omega} = 0.6000$$

$$\phi = 31°$$

Therefore,

$$\mathbf{Z} = 7\overline{0} \text{ }\Omega \text{ at } 31°$$

The **impedance** of a series circuit containing a **resistance** and a **capacitance** can be represented by the vector diagram in Fig. 11.57, where $\phi$ is the phase angle that equals the amount by which the *voltage lags behind the current*. The resistance is always drawn as a vector pointing in the positive x-direction, and the capacitive reactance is drawn as a vector pointing in the negative y-direction.

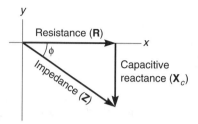

**Figure 11.57**

**EXAMPLE 7**

If the impedance is 120 $\Omega$ and $\phi = 4\bar{0}°$, find the resistance and the capacitive reactance.
Using Fig. 11.57, we have

$$|\mathbf{R}| = |\mathbf{Z}|\cos \phi = (120\ \Omega)(\cos 4\bar{0}°) = 92\ \Omega$$
$$|\mathbf{X}_c| = |\mathbf{Z}|\sin \phi = (120\ \Omega)(\sin 4\bar{0}°) = 77\ \Omega$$

Addition of vectors using graphic methods has a limited degree of accuracy. Also, addition of vectors using the law of sines and the law of cosines, while accurate, is sometimes cumbersome. The following method of adding vectors is accurate and rather efficient.

---

**COMPONENT METHOD**

To find the resultant vector **R** of two or more vectors using the component method:

1. Find the horizontal component, $\mathbf{R}_x$, of vector **R** by finding the algebraic sum of the horizontal components of each of the vectors being added.

2. Find the vertical component, $\mathbf{R}_y$, of vector **R** by finding the algebraic sum of the vertical components of each of the vectors being added.

3. Find the length of **R**:

$$|\mathbf{R}| = \sqrt{|\mathbf{R}_x|^2 + |\mathbf{R}_y|^2}$$

4. To find angle $\theta$, first find $\alpha$, the reference angle.

$$\tan \alpha = \frac{|\mathbf{R}_y|}{|\mathbf{R}_x|}$$

---

**EXAMPLE 8**

Using the component method, find the sum of the given vectors.

$$\mathbf{v} = 25.0 \text{ km at } 121.0°$$
$$\mathbf{w} = 66.0 \text{ km at } 245.0°$$

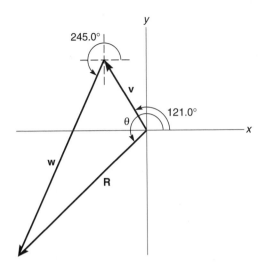

**Figure 11.58**

First, draw the components and **R** as in Fig. 11.58.

$$\mathbf{v}_x = |\mathbf{v}|\cos\theta$$
$$= (25.0 \text{ km})(\cos 121.0°)$$
$$= \qquad -12.9 \text{ km}$$

$$\mathbf{w}_x = |\mathbf{w}|\cos\theta$$
$$= (66.0 \text{ km})(\cos 245.0°)$$
$$= \qquad -27.9 \text{ km}$$

$\mathbf{R}_x$: Sum of $x$-components $= -40.8$ km

$$\mathbf{v}_y = |\mathbf{v}|\sin\theta$$
$$= (25.0 \text{ km})(\sin 121.0°)$$
$$= \qquad 21.4 \text{ km}$$

$$\mathbf{w}_y = |\mathbf{w}|\sin\theta$$
$$= (66.0 \text{ km})(\sin 245.0°)$$
$$= \qquad -59.8 \text{ km}$$

$\mathbf{R}_y$: Sum of $y$-components $= -38.4$ km

Then

$$\mathbf{R} = \sqrt{|\mathbf{R}_x|^2 + |\mathbf{R}_y|^2}$$
$$= \sqrt{|-40.8 \text{ km}|^2 + |-38.4 \text{ km}|^2}$$
$$= 56.0 \text{ km}$$

$$\tan\alpha = \frac{|\mathbf{R}_y|}{|\mathbf{R}_x|}$$

$$\tan\alpha = \frac{38.4 \text{ km}}{40.8 \text{ km}} = 0.9412$$

$$\alpha = 43.3°$$

Since $\mathbf{R}_x < 0$ and $\mathbf{R}_y < 0$, **R** is in the third quadrant. So

$$\theta = 180° + \alpha = 180° + 43.3° = 223.3°$$

Therefore,

$$\mathbf{R} = 56.0 \text{ km at } 223.3°$$

## Exercises 11.7

*(Angles are given in standard position unless stated otherwise.)*

*Find the horizontal and vertical components of each vector given in standard position.*

1. $\mathbf{v} = 18.2$ km at $85.0°$
2. $\mathbf{v} = 27.9$ mi at $138.0°$
3. $\mathbf{v} = 135$ mi/h at $270.0°$
4. $\mathbf{v} = 448$ m at $319.0°$
5. $\mathbf{v} = 2680$ ft at $152.5°$
6. $\mathbf{v} = 3620$ ft at $187.3°$

*For each pair of horizontal and vertical components, find the vector.*

7. $\mathbf{v}_x = 8.70$ m, $\mathbf{v}_y = 6.40$ m
8. $\mathbf{v}_x = -2.10$ ft, $\mathbf{v}_y = 3.20$ ft
9. $\mathbf{v}_x = 4.70$ m/s, $\mathbf{v}_y = -6.60$ m/s
10. $\mathbf{v}_x = -925$ m, $\mathbf{v}_y = 125$ m
11. $\mathbf{v}_x = -14.7$ km, $\mathbf{v}_y = 0$
12. $\mathbf{v}_x = 427$ mi/h, $\mathbf{v}_y = 381$ mi/h

13. A series circuit containing a resistance and an inductance has an impedance of $9\overline{0}$ Ω, and the phase angle is $2\overline{0}°$. Find the resistance and the inductive reactance.

14. A series circuit contains a resistance of 85 Ω and has an inductive reactance of 45 Ω. Find the impedance and the phase angle.

15. A series circuit contains a resistance of 240 Ω and a capacitive reactance of 140 Ω. Find the impedance and the phase angle.

16. A series circuit containing a resistance and a capacitance has an impedance of 110 Ω, and $\phi = 49°$. Find the resistance and the capacitive reactance.

*Using the component method, find the sum of each set of vectors. Find the magnitudes to three significant digits and the angles to the nearest tenth of a degree.*

17. $\mathbf{v} = 324$ ft at $0°$
    $\mathbf{w} = 576$ ft at $90.0°$
18. $\mathbf{v} = 91.2$ km/h at $180.0°$
    $\mathbf{w} = 84.7$ km/h at $270.0°$

19. $\mathbf{v} = 28.9$ mi/h at $52.0°$
    $\mathbf{w} = 16.2$ mi/h at $310.0°$
20. $\mathbf{v} = 59.7$ km at $125.0°$
    $\mathbf{w} = 86.4$ km at $298.0°$

21. $\mathbf{v} = 655$ km at $108.0°$
    $\mathbf{w} = 655$ km at $27.0°$
    $\mathbf{u} = 655$ km at $270.0°$
22. $\mathbf{v} = 29.7$ mi at $237.0°$
    $\mathbf{w} = 16.4$ mi at $180.0°$
    $\mathbf{u} = 18.5$ mi at $15.0°$

23. $\mathbf{v} = 5020$ m at $0°$
    $\mathbf{w} = 3130$ m at $148.0°$
    $\mathbf{u} = 6250$ m at $65.0°$
    $\mathbf{t} = 4620$ m at $335.0°$
24. $\mathbf{v} = 5760$ ft at $90.0°$
    $\mathbf{w} = 3940$ ft at $205.0°$
    $\mathbf{u} = 6140$ ft at $150.0°$
    $\mathbf{t} = 1230$ ft at $330.0°$

25. Using the vectors in Exercise 19, find $\mathbf{v} - \mathbf{w}$.

26. Using the vectors in Exercise 20, find $\mathbf{w} - \mathbf{v}$.

27. An airplane flies 165 mi on a course of $25.0°$ west of south; then it changes course and flies 125 mi on a course of $15.0°$ north of west to where it lands. Find the displacement from the starting point to the landing point.

28. A ship travels 75.0 mi on a course of $35.0°$ west of north; then it travels 45.0 mi on a course of $60.0°$ south of west to where it docks. Find the displacement from the starting point to the docking point.

## 11.8 VECTOR APPLICATIONS

D.1, D.5

Many applications involve vector quantities. Displacement, velocity, and force have been chosen here to illustrate some basic applications.

### Displacement

Recall that **displacement** is a change of position; that is, the difference between the initial position of a body and any later position.

#### EXAMPLE 1

A ship travels 175 mi at 52.0°, then 295 mi at 141.0°, and then 225 mi due west. Find the displacement; that is, find the net distance between the starting point and the end point and the angle.

Using the component method, let

$$\mathbf{u} = 175 \text{ mi at } 52.0°$$
$$\mathbf{v} = 295 \text{ mi at } 141.0°$$
$$\mathbf{w} = 225 \text{ mi at } 180.0°$$

Draw the components and **R** as in Fig. 11.59. Then we have the following:

| | |
|---|---|
| $\mathbf{u}_x = \|\mathbf{u}\|\cos\theta$ | $\mathbf{u}_y = \|\mathbf{u}\|\sin\theta$ |
| $\quad = (175 \text{ mi})(\cos 52.0°)$ | $\quad = (175 \text{ mi})(\sin 52.0°)$ |
| $\quad = \qquad\qquad 108 \text{ mi}$ | $\quad = \qquad\qquad 138 \text{ mi}$ |
| $\mathbf{v}_x = \|\mathbf{v}\|\cos\theta$ | $\mathbf{v}_y = \|\mathbf{v}\|\sin\theta$ |
| $\quad = (295 \text{ mi})(\cos 141.0°)$ | $\quad = (295 \text{ mi})(\sin 141.0°)$ |
| $\quad = \qquad\qquad -229 \text{ mi}$ | $\quad = \qquad\qquad 186 \text{ mi}$ |
| $\mathbf{w}_x = \|\mathbf{w}\|\cos\theta$ | $\mathbf{w}_y = \|\mathbf{w}\|\sin\theta$ |
| $\quad = (225 \text{ mi})(\cos 180.0°)$ | $\quad = (225 \text{ mi})(\sin 180.0°)$ |
| $\quad = \qquad\qquad \underline{-225 \text{ mi}}$ | $\quad = \qquad\qquad \underline{0 \text{ mi}}$ |
| $\mathbf{R}_x$: Sum of $x$-components $= -346 \text{ mi}$ | $\mathbf{R}_y$: Sum of $y$-components $= 324 \text{ mi}$ |

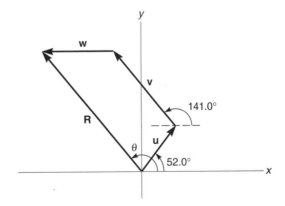

**Figure 11.59**

Then

$$\mathbf{R} = \sqrt{|\mathbf{R}_x|^2 + |\mathbf{R}_y|^2}$$
$$= \sqrt{|-346 \text{ mi}|^2 + |324 \text{ mi}|^2}$$
$$= 474 \text{ mi}$$

$$\tan \alpha = \frac{|\mathbf{R}_y|}{|\mathbf{R}_x|}$$

$$\tan \alpha = \frac{324 \text{ mi}}{346 \text{ mi}} = 0.9364$$

$$\alpha = 43.1°$$

Since $\mathbf{R}_x < 0$ and $\mathbf{R}_y > 0$, $\mathbf{R}$ is in the second quadrant. So

$$\theta = 180° - \alpha = 180° - 43.1° = 136.9°$$

Therefore, the displacement is

$$\mathbf{R} = 474 \text{ mi at } 136.9°$$

## Velocity

**Velocity** is a vector quantity that is described in terms of both speed and direction. For example, wind typically has an effect on both the speed and the direction of a plane in flight. The *heading* of an airplane is the direction the plane is pointed with respect to the compass. The *course* of the airplane is the direction the plane is actually flying with respect to the ground. The *air speed* is the speed of the plane with respect to the air; the *ground speed* is the speed of the plane with respect to the ground.

**EXAMPLE 2**

An airplane is flying on a heading of 25.0° west of north at an air speed of 95.0 mi/h. The wind is from the east at 30.0 mi/h. Find the plane's course and ground speed.

First, draw a vector diagram as in Fig. 11.60. You often have a choice of vector methods to use to solve a particular problem. Here, let's use trigonometry.

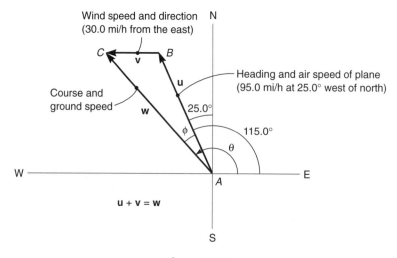

**Figure 11.60**

Since **BC** is parallel to the $x$-axis, angle $B = 115.0°$. Then use the law of cosines in triangle $ABC$.

$$|\mathbf{w}|^2 = |\mathbf{u}|^2 + |\mathbf{v}|^2 - 2|\mathbf{u}||\mathbf{v}|\cos B$$
$$|\mathbf{w}| = \sqrt{(95.0 \text{ mi/h})^2 + (30.0 \text{ mi/h})^2 - 2(95.0 \text{ mi/h})(30.0 \text{ mi/h})(\cos 115.0°)}$$
$$= 111 \text{ mi/h}$$

Next, find angle $\phi$ using the law of sines.

$$\frac{|\mathbf{w}|}{\sin B} = \frac{|\mathbf{v}|}{\sin \phi}$$
$$\frac{111 \text{ mi/h}}{\sin 115.0°} = \frac{30.0 \text{ mi/h}}{\sin \phi}$$
$$\sin \phi = 0.2449$$
$$\phi = 14.2°$$

Then

$$\theta = 115.0° + 14.2° = 129.2°$$

The ground speed is 111 mi/h, and the course is 129.2° in standard position or 39.2° west of north.

## Force

In physics a **force** is defined as a push or a pull that tends to cause or prevent motion. When a force is applied, it must be applied in some direction. Thus, force is a vector quantity.

**EXAMPLE 3**

Two forces act on a point at an angle of 70.0°. One force is 1850 newtons (N). The resultant force is 2250 N. Find the second force and the angle it makes with the resultant.

First, draw a force diagram as in Fig. 11.61.

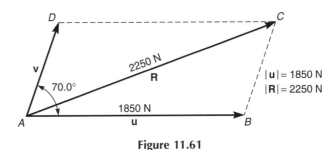

**Figure 11.61**

The consecutive angles of a parallelogram are supplementary; angle $B = 110.0°$. Let's first find angle $ACB$ using the law of sines.

$$\frac{\overline{AC}}{\sin B} = \frac{|\mathbf{u}|}{\sin ACB}$$
$$\frac{2250 \text{ N}}{\sin 110.0°} = \frac{1850 \text{ N}}{\sin ACB}$$

$$\sin ACB = \frac{(\sin 110.0°)(1850 \text{ N})}{2250 \text{ N}} = 0.7726$$

$$\text{angle } ACB = 50.6°$$

Then angle $CAB = 180° - 110.0° - 50.6° = 19.4°$.

Now find $|\mathbf{v}|$ using the law of sines and noting that $\overline{BC} = |\mathbf{v}|$ in triangle $ABC$.

$$\frac{\overline{BC}}{\sin CAB} = \frac{\overline{AC}}{\sin B}$$

$$\frac{|\mathbf{v}|}{\sin 19.4°} = \frac{2250 \text{ N}}{\sin 110.0°} \qquad (\textit{Note: } \overline{BC} = |\mathbf{v}|.)$$

$$|\mathbf{v}| = \frac{(\sin 19.4°)(2250 \text{ N})}{\sin 110.0°} = 795 \text{ N}$$

Thus, $\mathbf{v} = 795$ N at $50.6°$ from $\mathbf{R}$.

When two or more forces act at a point, the force applied at the same point that produces equilibrium is called the *equilibrant force*. Thus, the equilibrant force is *equal* in magnitude but *opposite* in direction to the resultant force.

### EXAMPLE 4

A 1250-lb weight hangs from two ropes of equal lengths suspended from a support as shown in Fig. 11.62. Find the tension in each rope.

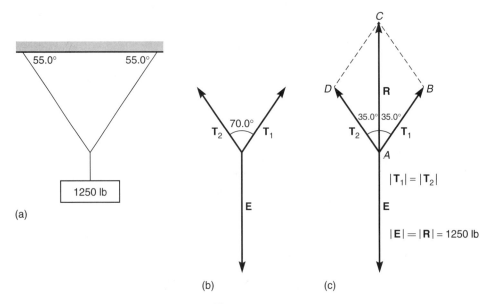

**Figure 11.62**

First, draw the force diagram as in Fig. 11.62(b). Note that the equilibrant force is 1250 lb; the resultant force is shown in Fig. 11.62(c). Find $|\mathbf{T}_1|$ using the law of sines in triangle *ABC* as follows:

$$\frac{|\mathbf{T}_1|}{\sin ACB} = \frac{|\mathbf{R}|}{\sin B}$$

$$\frac{|\mathbf{T}_1|}{\sin 35.0°} = \frac{1250 \text{ lb}}{\sin 110.0°} \qquad \text{(\textit{Note:} Angle } B = 110.0°.)$$

$$|\mathbf{T}_1| = \frac{(\sin 35.0°)(1250 \text{ lb})}{\sin 110.0°} = 763 \text{ lb} = |\mathbf{T}_2|$$

The tension in each rope is 763 lb.

## Exercises 11.8

1. An airplane takes off and flies 125 mi on a course of 61.5° south of west; then it changes course and flies 185 mi due south, where it lands. Find the displacement from the starting point to the landing point.

2. A ship travels 18.5 km on a course of 31.2° south of east, then 12.7 km due south, then 21.5 km on a course of 61.3° west of south, where it lands. Find the displacement from the starting point to the landing point.

3. En route between two cities, a person travels by automobile 115 km due south, then 195 km at 45.0° west of south by plane, and then 45.0 km due west by boat. Find the displacement from the starting point to the ending point.

4. Natasha jogs 2.00 km due west, then 1.00 km due north, then 3.00 km due east, and then 2.00 km due south. Find the displacement from her starting point to her ending point.

*In Exercises 5–8, given displacements with magnitudes of 10.0 mi and 15.0 mi:*

5. Find the magnitude of the maximum resultant displacement.

6. Find the magnitude of the minimum resultant displacement.

7. Find the angle between the two original displacements if the magnitude of the resultant displacement is 12.0 mi.

8. Find the angle between the two original displacements if the magnitude of the resultant displacements is 7.50 mi.

*In Exercises 9–10, given three displacements of magnitudes 10.0 km, 15.0 km, and 20.0 km:*

9. Find the magnitude of the maximum resultant displacement.

10. Find the magnitude of the minimum resultant displacement.

11. A boat travels 12 mi/h in still water. The flow of the current is 3 mi/h.
    (a) What is the boat's speed going downstream?
    (b) What is its speed going upstream?

12. A plane's air speed is 125 mi/h. The wind speed is 21 mi/h.
    (a) What is the plane's ground speed when it is flying into the wind?
    (b) What is the plane's ground speed when it is flying with the wind?

13. A plane is flying due west at 175 mi/h. Suddenly a wind of 25.0 mi/h from the south develops. Find the plane's new course and ground speed.

**14.** A plane is flying north at 145 mi/h in still air. Find the velocity of the plane when the wind is blowing 55.0 mi/h from the east.

**15.** A plane is flying on a heading of 35.0° west of north at an air speed of 315 km/h. The wind is blowing from the west at 50.0 km/h. Find the plane's course and ground speed.

**16.** A plane is flying on a heading of 60.0° east of south at an air speed of 165 mi/h. The wind is blowing at 45.0 mi/h from 25.0° south of west. Find the plane's course and ground speed.

**17.** A pilot is flying on a course of 25.5° north of west at 215 mi/h. The wind is blowing from the east at 45.6 mi/h. Find the plane's heading and air speed.

**18.** During a model boating competition, the following circumstances are presented to the contestants: width of stream, 125 ft; speed of current, 4.00 ft/s; speed of boat in still water, 12.0 ft/s. At what angle must the boat be steered to reach a point on the opposite side of the stream **(a)** that is directly across the stream, **(b)** that is 45.0 ft upstream, and **(c)** that is 45.0 ft downstream?

**19.** DeCarlos applies a 25.0-lb force to the handle of a lawnmower, which makes a 35.0° angle with the ground. Find the horizontal and vertical components of this force.

**20.** Cindy pulls a sled with a rope with a force of 50.0 lb. The rope makes an angle of 30.0° with the ground. Find the horizontal and vertical components of this force.

*In Exercises 21–24, find the vector sum of each set of forces.*

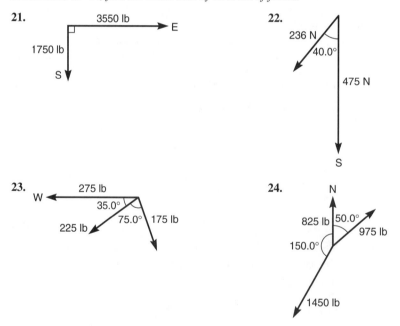

**21.**

3550 lb E

1750 lb

S

**22.**

236 N

40.0°

475 N

S

**23.**

W

275 lb

35.0°

75.0° 175 lb

225 lb

**24.**

N

825 lb 50.0°

975 lb

150.0°

1450 lb

**25.** Two forces act on a point at an angle of 125.0°. One force is 225 lb. The resultant force is 195 lb. Find the second force and the angle it makes with the resultant.

**26.** Two forces act on a point at an angle of 65.0°. One force is 175 lb. The resultant force is 215 lb. Find the second force and the angle it makes with the resultant.

*In Exercises 27–30, find each equilibrant force in standard position.*

**27.**

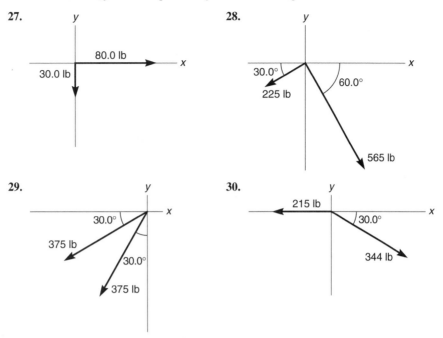

**28.**

**29.**

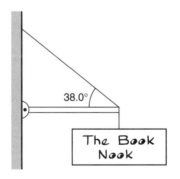

**30.**

**31.** A sign weighing 615 lb is supported as shown in Fig. 11.63. Find the tension in the cable.

The Book Nook

**Figure 11.63**

**32.** A sign weighing 275 lb is suspended by two cables attached between two buildings at equal heights above the ground. Each cable makes an angle of 65.0° with its building. Find the tension in each cable.

**33.** A 475-lb sign hangs from two cables as shown in Fig. 11.64. Find the tension in each cable.

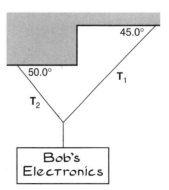

**Figure 11.64**

Bob's Electronics

**34.** A 1250-lb weight hangs from two cables of equal length as shown in Fig. 11.65. If each cable can withstand a maximum tension of 825 lb, at what angle $A$ are the cables unsafe?

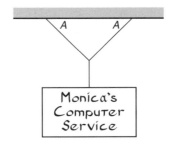

**Figure 11.65**

## CHAPTER SUMMARY

**1.** *For any general triangle as in Fig. 11.66:*
   (a) We use the law of sines

$$\frac{a}{\sin A} = \frac{b}{\sin B} = \frac{c}{\sin C}$$

   for cases involving two sides and an angle opposite one of them (SSA) and for cases involving two angles and a side opposite one of them (AAS).
   (b) We use the law of cosines

$$a^2 = b^2 + c^2 - 2bc \cos A$$
$$b^2 = a^2 + c^2 - 2ac \cos B$$
$$c^2 = a^2 + b^2 - 2ab \cos C$$

   for cases involving two sides and the included angle (SAS) and for cases involving three sides (SSS).

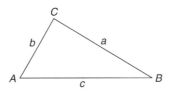

**Figure 11.66**

**2.** The SSA case requires special consideration. If the side opposite the given angle is:
   (a) greater than the known adjacent side, there is only one possible triangle.
   (b) less than the known adjacent side but greater than the altitude, there are two possible triangles.
   (c) less than the altitude, there is no possible triangle.

**3.** *As a final check:*
   (a) always choose a given value over a calculated value for doing calculations; and
   (b) always check your results to see that the largest angle is opposite the largest side and that the smallest angle is opposite the smallest side.

4. Vector problems may be solved:
   (a) graphically using
        (i) the parallelogram method or
        (ii) the vector triangle method;
   (b) algebraically using the law of sines and/or the law of cosines; or
   (c) by the component method.
5. Given $|\mathbf{v}|$ and angle $\theta$, the horizontal and vertical components are found as follows (see Fig. 11.67):

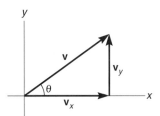

$$\mathbf{v}_x = |\mathbf{v}| \cos \theta$$
$$\mathbf{v}_y = |\mathbf{v}| \sin \theta$$

**Figure 11.67**

6. Given the horizontal and vertical components $\mathbf{v}_x$ and $\mathbf{v}_y$, the magnitude of $\mathbf{v}$ may be found as follows:

$$|\mathbf{v}| = \sqrt{|\mathbf{v}_x|^2 + |\mathbf{v}_y|^2}$$

The reference angle $\alpha$ may be found as follows:

$$\tan \alpha = \frac{|\mathbf{v}_y|}{|\mathbf{v}_x|}$$

Then angle $\theta$ in standard position is determined from angle $\alpha$ and the quadrant in which $\mathbf{v}$ lies.

7. *Component method of adding vectors:* To find the resultant vector $\mathbf{R}$ of two or more vectors using the component method:
   (a) Find the horizontal component, $\mathbf{R}_x$, of vector $\mathbf{R}$ by finding the algebraic sum of the horizontal components of each of the vectors being added.
   (b) Find the vertical component, $\mathbf{R}_y$, of vector $\mathbf{R}$ by finding the algebraic sum of the vertical components of each of the vectors being added.
   (c) Find the length of $\mathbf{R}$: $|\mathbf{R}| = \sqrt{|\mathbf{R}_x|^2 + |\mathbf{R}_y|^2}$.
   (d) To find angle $\theta$, first find $\alpha$, the reference angle.

$$\tan \alpha = \frac{|\mathbf{R}_y|}{|\mathbf{R}_x|}$$

## CHAPTER 11 REVIEW

*Solve each triangle using the labels as in Fig. 11.68.*

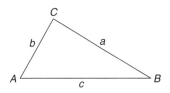

**Figure 11.68**

*Express the lengths of sides to three significant digits and the angles to the nearest tenth of a degree.*

**1.** $B = 71.4°$, $b = 409$ ft, $c = 327$ ft      **2.** $A = 25.1°$, $C = 37.7°$, $a = 15.7$ m

**3.** $A = 15.5°$, $b = 236$ cm, $c = 209$ cm      **4.** $a = 25.6$ m, $b = 42.2$ m, $c = 35.2$ m

*Express the lengths of sides to two significant digits and the angles to the nearest degree.*

**5.** $B = 44°$, $b = 150$ mi, $c = 240$ mi      **6.** $A = 29°$, $a = 41$ cm, $b = 49$ cm

**7.** $C = 36°$, $a = 2100$ ft, $b = 3600$ ft      **8.** $C = 58°$, $a = 450$ m, $c = 410$ m

*Express the lengths of sides to four significant digits and the angles to the nearest hundredth of a degree.*

**9.** $B = 105.15°$, $a = 231.1$ m, $c = 190.7$ m

**10.** $A = 74.75°$, $a = 22.19$ cm, $c = 15.28$ cm

**11.** $B = 18.25°$, $a = 1675$ ft, $b = 1525$ ft

**12.** $C = 40.16°$, $b = 25{,}870$ ft, $c = 10{,}250$ ft

*Express the lengths of sides to four significant digits and the angles to the nearest second.*

**13.** $A = 48°15'35''$, $B = 68°7'18''$, $a = 2755$ ft

**14.** $C = 29°25'16''$, $a = 13{,}560$ ft, $b = 24{,}140$ ft

**15.** Find the total length of fence needed to enclose the triangular field shown in Fig. 11.69.

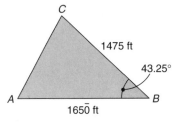

**Figure 11.69**

**16.** In surveying a tract of timber, it is necessary to find the distance to an inaccessible, but visible, point $C$. A distance between two points $A$ and $B$ of 300.0 ft is measured. Then $\angle ABC$ is measured as $85.00°$ and $\angle BAC$ is measured as $79.55°$.
   **(a)** Find $AC$.
   **(b)** Find the perpendicular distance from $C$ to line $AB$.

**17.** A toolmaker needs to lay out three holes in a plate, as shown in Fig. 11.70. Find the distance between holes $A$ and $C$.

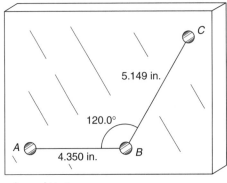

**Figure 11.70**

18. The centers of five holes are equally spaced around a 5.000-in.-diameter circle, as shown in Fig. 11.71.
   (a) Find the distance between the centers of holes A and B.
   (b) Find the distance between the centers of holes A and C.

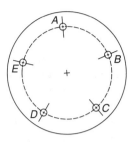

**Figure 11.71**

19. In the roof truss in Fig. 11.72., we know that $AB = DE$, $BC = CD$, and $AE = 20.0$ m. Find each length.

   (a) *AF*          (b) *BF*          (c) *CF*          (d) *BC*

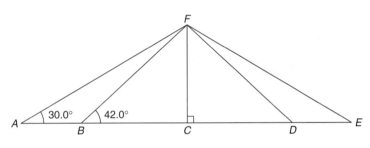

**Figure 11.72**

*(Angles are given in standard position unless stated otherwise.)*

20. Find the sum of the given vectors using a graphical method. Choose a suitable scale.

$$\mathbf{v} = 75 \text{ km/h at } 87°$$

$$\mathbf{w} = 25 \text{ km/h at } 142°$$

*Using trigonometry, find the sum of each set of vectors.*

21. $\mathbf{v} = 126$ mi at $35.0°$
   $\mathbf{w} = 306$ mi at $180.0°$

22. $\mathbf{v} = 89.4$ mi/h at $142.0°$
   $\mathbf{w} = 44.7$ mi/h at $322.0°$

*Find the horizontal and vertical components of each vector.*

23. 258 km at $135.0°$

24. 42.2 mi/h at $303.0°$

25. A man pushes with 160 N of force on the handle of a pushcart. If the angle between the handle and the ground is 45°, find the horizontal and vertical components of this force.

26. A woman pulls a loaded cart with a pulling force of 110 lb when the handle makes an angle of 23° with the ground. Find the horizontal and vertical components of this force.

27. If $\mathbf{v}_x = 18.5$ N and $\mathbf{v}_y = -31.0$ N, find $\mathbf{v}$.

*Using the component method, find the sum of each set of vectors. Find magnitudes to three significant digits and angles to the nearest tenth of a degree.*

**28.** $\mathbf{v}$ = 87.1 mi/h at 120.0°
  $\mathbf{w}$ = 25.6 mi/h at 247.0°

**29.** $\mathbf{v}$ = 2560 N at 237.1°
  $\mathbf{w}$ = 3890 N at 346.7°

**30.** $\mathbf{u}$ = 325 N at 90.0°
  $\mathbf{v}$ = 325 N at 162.0°
  $\mathbf{w}$ = 325 N at 270.0°

**31.** $\mathbf{v}$ = 19.7 km at 144.5°
  $\mathbf{w}$ = 28.5 km at 180.0°
  $\mathbf{u}$ = 10.3 km at 225.5°
  $\mathbf{t}$ = 31.7 km at 90.0°

**32.** An airplane is to maintain a velocity of 550 km/h on a true course of $\overline{60}°$ east of north. A wind of 90 km/h is blowing from the east. What should the air speed and compass reading be to offset the wind?

**33.** A ship travels 16.5 mi at 13.5° east of north, then 24.7 mi at 34.5° west of north, then 30.5 mi due north, where it lands. Find the displacement from the starting point to the landing point.

**34.** A weight of 850 lb is suspended by two ropes attached to opposite walls at equal heights above the floor. The first rope makes an angle of 25° with the first wall, and the second rope makes an angle of 45° with the second wall. Find the tension in each rope.

# 12
# Graphing the Trigonometric Functions

## INTRODUCTION

Harmonic motion can be described using the equation $y = r \sin(\omega t + \theta)$. Voltage or current can be described using the equation $y = a \cos(\omega t + \theta)$. In this chapter we will graph these trigonometric functions and identify their amplitude, period, and phase shift.

### Objectives

- Sketch the graphs of the trigonometric functions.
- Identify the amplitude, period, and phase shift for functions of the form

$$y = a \cos(\omega t + \theta) \quad \text{and} \quad y = a \sin(\omega t + \theta)$$

- Graph composite curves.

## 12.1  GRAPHING THE SINE AND COSINE FUNCTIONS

We shall first consider the graphs of the trigonometric equations $y = \sin x$ and $y = \cos x$. We find it convenient to express $x$ in radian measure. (Be careful not to confuse the variables $x$ and $y$ in the equation $y = \sin x$ with the coordinates of a point on the terminal side of an angle. That is, in the equation of each trigonometric function, $x$ is an angle in radian measure, and $y$ is the trigonometric value for that angle.)

To graph $y = \sin x$, find a large number of values of $x$ and $y$ that satisfy the equation, and plot them in the $xy$-plane. It is convenient to scale the $x$-axis in multiples of $\pi$ radians. A table of ordered pairs is as follows:

| $x$ | 0 | $\frac{\pi}{6}$ | $\frac{\pi}{4}$ | $\frac{\pi}{3}$ | $\frac{\pi}{2}$ | $\frac{2\pi}{3}$ | $\frac{3\pi}{4}$ | $\frac{5\pi}{6}$ | $\pi$ | $\frac{7\pi}{6}$ | $\frac{5\pi}{4}$ | $\frac{4\pi}{3}$ | $\frac{3\pi}{2}$ | $\frac{5\pi}{3}$ | $\frac{7\pi}{4}$ | $\frac{11\pi}{6}$ | $2\pi$ |
|---|---|---|---|---|---|---|---|---|---|---|---|---|---|---|---|---|---|
| $y$ | 0 | 0.5 | 0.71 | 0.87 | 1 | 0.87 | 0.71 | 0.5 | 0 | $-0.5$ | $-0.71$ | $-0.87$ | $-1$ | $-0.87$ | $-0.71$ | $-0.5$ | 0 |

The graph is shown in Fig. 12.1. Note that the graph is a smooth, continuous curve. The domain (set of $x$-replacements or input) is the set of all real numbers (angles measured in radians). The range (set of output) is $-1 \leq y \leq 1$.

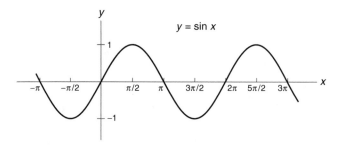

**Figure 12.1**

The graph of $y = \cos x$ is found in the same manner and is shown in Fig. 12.2.

| $x$ | 0 | $\dfrac{\pi}{6}$ | $\dfrac{\pi}{4}$ | $\dfrac{\pi}{3}$ | $\dfrac{\pi}{2}$ | $\dfrac{2\pi}{3}$ | $\dfrac{3\pi}{4}$ | $\dfrac{5\pi}{6}$ | $\pi$ | $\dfrac{7\pi}{6}$ | $\dfrac{5\pi}{4}$ | $\dfrac{4\pi}{3}$ | $\dfrac{3\pi}{2}$ | $\dfrac{5\pi}{3}$ | $\dfrac{7\pi}{4}$ | $\dfrac{11\pi}{6}$ | $2\pi$ |
|---|---|---|---|---|---|---|---|---|---|---|---|---|---|---|---|---|---|
| $y$ | 1 | 0.87 | 0.71 | 0.5 | 0 | $-0.5$ | $-0.71$ | $-0.87$ | $-1$ | $-0.87$ | $-0.71$ | $-0.5$ | 0 | 0.5 | 0.71 | 0.87 | 1 |

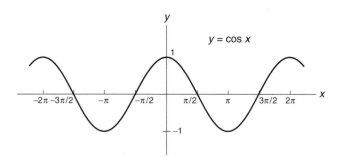

**Figure 12.2**

The domain and the range for $y = \cos x$ are the same as for $y = \sin x$. Note how the two graphs differ. Could we shift the cosine curve so that it coincides with the sine curve? Show $\sin\left(x + \dfrac{\pi}{2}\right) = \cos x$ for all values of $x$ by choosing various values of $x$ and checking them in each side of the equation.

It is important to note the points at which these two curves cross the axes. We shall "sketch" similar curves based on the fact that the basic shape of the sine and cosine curves always remains the same. This will save time and work in that it will be unnecessary to plot a large number of points each time we need to sketch a curve.

The basic sine curve is an equation of the form $y = a \sin x$, where $a$ is a real number. We have already graphed this equation for $a = 1$. For $y = 2 \sin x$, each ordered pair satisfying this relationship has a $y$-value that is two times the corresponding $y$ value for $y = \sin x$. Likewise, for $y = \frac{1}{2} \sin x$, each ordered pair satisfying this relationship has a $y$-value that is one-half the corresponding $y$-value for $y = \sin x$.

**EXAMPLE 1**

Graph $y = \sin x$, $y = 2 \sin x$, $y = \frac{1}{2} \sin x$, and $y = 4 \sin x$ on the same set of coordinate axes. The graph is drawn in Fig. 12.3.

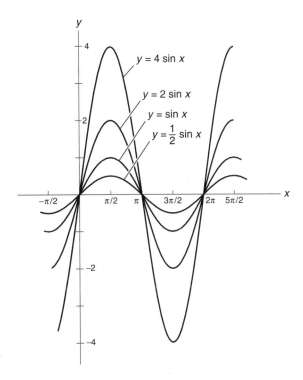

**Figure 12.3**

The coefficient, $a$, of $a \sin x$ or $a \cos x$ in the equations $y = a \sin x$ and $y = a \cos x$ determines the amplitude of the curve. The **amplitude** is the maximum $y$-value of the curve. The amplitudes for the curves in Example 1 are 1, 2, $\frac{1}{2}$, and 4, respectively. The effect of a negative value of $a$ in $y = a \sin x$ or $y = a \cos x$ is to "flip," or invert, the curve about the $x$-axis. That is, for $y = -\cos x$, each ordered pair satisfying this relationship has a $y$-value that is opposite the corresponding $y$-value for $y = \cos x$.

**EXAMPLE 2**

Graph $y = \cos x$, $y = -\cos x$, and $y = -3 \cos x$ on the same set of coordinate axes. The graph is shown in Fig. 12.4. The amplitudes for the curves in this example are 1, 1, and 3, respectively.

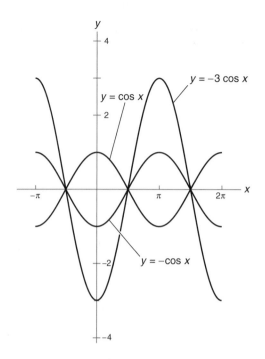

**Figure 12.4**

EXAMPLE 3

Graph $y = \sin x$ and $y = \sin 2x$ on the same set of coordinate axes.

To find ordered pairs that satisfy the equation $y = \sin 2x$, first choose a value for $x$, second multiply this value by 2, and third find the sine of this result. Some of the values are tabulated as follows:

| $x$ | 0 | $\dfrac{\pi}{6}$ | $\dfrac{\pi}{4}$ | $\dfrac{\pi}{3}$ | $\dfrac{\pi}{2}$ | $\dfrac{2\pi}{3}$ | $\dfrac{3\pi}{4}$ | $\dfrac{5\pi}{6}$ | $\pi$ | $\dfrac{7\pi}{6}$ | $\dfrac{5\pi}{4}$ | $\dfrac{4\pi}{3}$ | $\dfrac{3\pi}{2}$ | $\dfrac{5\pi}{3}$ | $\dfrac{7\pi}{4}$ | $\dfrac{11\pi}{6}$ | $2\pi$ |
|---|---|---|---|---|---|---|---|---|---|---|---|---|---|---|---|---|---|
| $\sin x$ | 0 | 0.5 | 0.71 | 0.87 | 1 | 0.87 | 0.71 | 0.5 | 0 | −0.5 | −0.71 | −0.87 | −1 | −0.87 | −0.71 | −0.5 | 0 |
| $2x$ | 0 | $\dfrac{\pi}{3}$ | $\dfrac{\pi}{2}$ | $\dfrac{2\pi}{3}$ | $\pi$ | $\dfrac{4\pi}{3}$ | $\dfrac{3\pi}{2}$ | $\dfrac{5\pi}{3}$ | $2\pi$ | $\dfrac{7\pi}{3}$ | $\dfrac{5\pi}{2}$ | $\dfrac{8\pi}{3}$ | $3\pi$ | $\dfrac{10\pi}{3}$ | $\dfrac{7\pi}{2}$ | $\dfrac{11\pi}{3}$ | $4\pi$ |
| $\sin 2x$ | 0 | 0.87 | 1 | 0.87 | 0 | −0.87 | −1 | −0.87 | 0 | 0.87 | 1 | 0.87 | 0 | −0.87 | −1 | −0.87 | 0 |

Plot these points as in Fig. 12.5.

Note that the two curves in Fig. 12.5 have the same basic shape, but $y = \sin 2x$ goes through a complete cycle twice as $y = \sin x$ completes one cycle. The length of each of these cycles is called a *period*. The **period** is the $x$-distance between any point on the curve and the next corresponding point in the next cycle where the graph starts repeating itself (see Fig. 12.6). The period for $y = \sin x$ is $2\pi$; for $y = \cos x$, $2\pi$; and for $y = \sin 2x$, $\pi$.

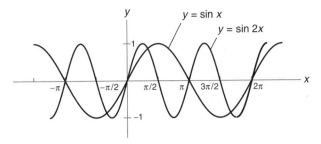

Figure 12.5

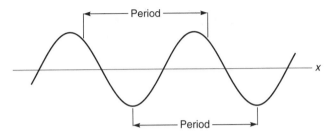

Figure 12.6

In general, the period for $y = a \sin bx$ or $y = a \cos bx$ may be found by the following formula.

$$P = \frac{2\pi}{b}$$

**EXAMPLE 4**

Graph $y = 2 \cos 3x$.

The amplitude is 2 and the period is

$$P = \frac{2\pi}{b} = \frac{2\pi}{3}$$

Sketch a cosine graph with amplitude 2 that completes one complete cycle each $\frac{2\pi}{3}$ radians as in Fig. 12.7.

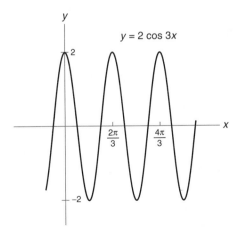

Figure 12.7

**EXAMPLE 5**

Graph $y = -3 \sin \frac{1}{2}x$.

   The amplitude is 3; the period is

$$P = \frac{2\pi}{b} = \frac{2\pi}{\frac{1}{2}} = 4\pi$$

The effect of the negative sign is to flip, or invert, the curve $y = 3 \sin \frac{1}{2}x$ about the $x$-axis. Sketch a sine graph with amplitude 3 that completes one complete cycle each $4\pi$ radians as shown by the dashed graph in Fig. 12.8. Then flip the dashed curve about the $x$-axis to obtain the solid-line graph of the given function.

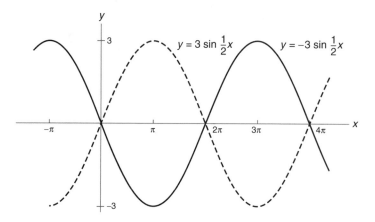

**Figure 12.8**

**EXAMPLE 6**

Graph $y = 25 \cos 120\pi x$.

   The amplitude is 25; the period is

$$P = \frac{2\pi}{b} = \frac{2\pi}{120\pi} = \frac{1}{60}$$

Sketch a cosine graph with amplitude 25 that completes one complete cycle each $\dfrac{1}{60}$ radian as shown in Fig. 12.9.

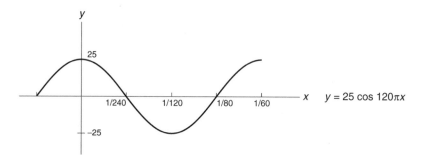

**Figure 12.9**

## Exercises 12.1

*Find the amplitude and the period of each function, and sketch its graph through at least two cycles.*

**1.** $y = 2 \cos x$

**2.** $y = 5 \sin x$

**3.** $y = -3 \sin x$

**4.** $y = -4 \cos x$

**5.** $y = \sin 3x$

**6.** $y = \sin 4x$

**7.** $y = \cos 2x$

**8.** $y = \cos 3x$

**9.** $y = 2 \sin 4x$

**10.** $y = 3 \cos 6x$

**11.** $y = \dfrac{5}{2} \cos \dfrac{1}{2}x$

**12.** $y = \dfrac{3}{2} \sin \dfrac{3}{2}x$

**13.** $y = -\dfrac{1}{2} \sin \dfrac{2}{3}x$

**14.** $y = -\dfrac{5}{2} \cos \dfrac{3}{4}x$

**15.** $y = 2 \sin 3\pi x$

**16.** $y = 6 \cos \dfrac{4\pi x}{3}$

**17.** $y = -3 \cos \pi x$

**18.** $y = -\sin \dfrac{\pi x}{2}$

**19.** $y = 6.5 \sin 120\pi x$

**20.** $y = 12 \cos 160\pi x$

**21.** $y = 40 \cos 60x$

**22.** $y = 60 \sin 40x$

**23.** $y = -60 \sin 80\pi x$

**24.** $y = -240 \cos 120x$

## 12.2 PHASE SHIFT

If the graph of a sine curve does not pass through the origin $(0, 0)$, or if the graph of a cosine curve does not pass through the point $(0, a)$ where $a$ is the amplitude, the curve is **out of phase**. If a curve is out of phase, the **phase shift** is the directed distance between two successive corresponding $x$-intercepts of the curve $y = a \sin bx$ or $y = a \cos bx$ and the out-of-phase curve (see Fig. 12.10).

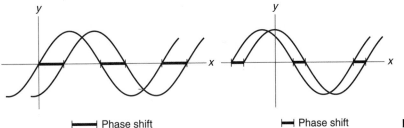

Phase shift        Phase shift        **Figure 12.10**

### EXAMPLE 1

Graph $y = \sin x$ and $y = \sin\left(x - \dfrac{\pi}{4}\right)$ on the same set of coordinate axes.

For $y = \sin\left(x - \dfrac{\pi}{4}\right)$, complete the following table:

| $x$ | $-\dfrac{\pi}{4}$ | $0$ | $\dfrac{\pi}{4}$ | $\dfrac{\pi}{2}$ | $\dfrac{3\pi}{4}$ | $\pi$ | $\dfrac{5\pi}{4}$ | $\dfrac{3\pi}{2}$ | $\dfrac{7\pi}{4}$ | $2\pi$ | $\dfrac{9\pi}{4}$ | $\dfrac{5\pi}{2}$ | $\dfrac{11\pi}{4}$ |
|---|---|---|---|---|---|---|---|---|---|---|---|---|---|
| $y$ | $-1$ | $-0.71$ | $0$ | $0.71$ | $1$ | $0.71$ | $0$ | $-0.71$ | $-1$ | $-0.71$ | $0$ | $0.71$ | $1$ |

Plot the points as in Fig. 12.11.

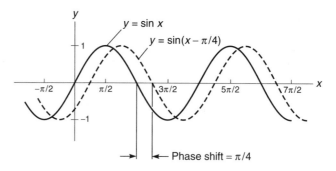

**Figure 12.11**

A consideration of phase shifts is helpful when graphing equations in the form $y = a \sin(bx + c)$ or $y = a \cos(bx + c)$. The effect of $c$ in these equations is to shift the curve $y = a \sin bx$ or $y = a \cos bx$ as follows:

**1.** To the *left* $\dfrac{c}{b}$ units if $\dfrac{c}{b} > 0$.

**2.** To the *right* $\dfrac{c}{b}$ units if $\dfrac{c}{b} < 0$.

**EXAMPLE 2**

Graph $y = 3 \cos\left(2x + \dfrac{\pi}{4}\right)$.

The amplitude is 3. The period is $\dfrac{2\pi}{b} = \dfrac{2\pi}{2} = \pi$. The phase shift is $\dfrac{c}{b} = \dfrac{\pi/4}{2} = \dfrac{\pi}{8}$, or $\dfrac{\pi}{8}$ to the left.

Sketch a cosine graph with amplitude 3 and period $\pi$, as shown by the dashed graph in Fig. 12.12. Then shift the dashed curve $\dfrac{\pi}{8}$ units to the left, which gives the graph of the given function.

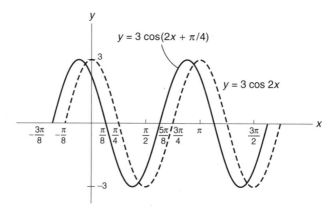

**Figure 12.12**

**EXAMPLE 3**

Graph $y = 4 \sin\left(3x + \dfrac{\pi}{2}\right)$.

The amplitude is 4. The period is $\dfrac{2\pi}{b} = \dfrac{2\pi}{3}$. The phase shift is $\dfrac{c}{b} = \dfrac{\pi/2}{3} = \dfrac{\pi}{6}$, or $\dfrac{\pi}{6}$ to the left.

Sketch a sine graph with amplitude 4 and period $\dfrac{2\pi}{3}$ as shown by the dashed graph in Fig. 12.13. Then shift the dashed curve $\dfrac{\pi}{6}$ units to the left, which gives the graph of the given function.

**EXAMPLE 4**

Graph $y = -3 \sin(\tfrac{1}{2}x - \pi)$.

The amplitude is 3. The period is $\dfrac{2\pi}{b} = \dfrac{2\pi}{\frac{1}{2}} = 4\pi$. The phase shift is $\dfrac{c}{b} = \dfrac{-\pi}{\frac{1}{2}} = -2\pi$, or $2\pi$ to the right.

First, graph $y = -3 \sin \tfrac{1}{2}x$, the dashed curve in Fig. 12.14. Then shift the dashed curve $2\pi$ units to the right, which gives the graph of the given function.

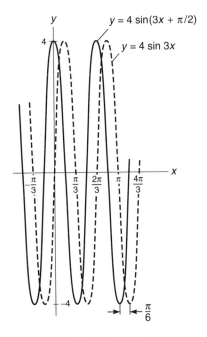

**Figure 12.13**

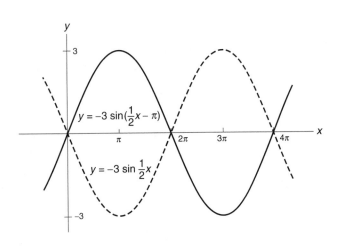

**Figure 12.14**

## AMPLITUDE, PERIOD, AND PHASE SHIFT

In summary, the values of $a$, $b$, and $c$ in the following equations determine amplitude, period, and phase shift as follows:

$$y = a \sin(bx + c)$$
$$y = a \cos(bx + c)$$

Amplitude $= |a|$    Phase shift $= \dfrac{c}{b}$    $\left(\text{To the } \textit{left} \text{ if } \dfrac{c}{b} > 0\right)$

Period $= \dfrac{2\pi}{b}$    $\left(\text{To the } \textit{right} \text{ if } \dfrac{c}{b} < 0\right)$

## Exercises 12.2

*Find the amplitude, period, and phase shift of each function, and sketch its graph.*

**1.** $y = \sin\left(x + \dfrac{\pi}{3}\right)$

**2.** $y = \cos\left(x + \dfrac{\pi}{4}\right)$

**3.** $y = 2 \cos\left(x - \dfrac{\pi}{6}\right)$

**4.** $y = 3 \sin\left(x - \dfrac{\pi}{3}\right)$

**5.** $y = \sin(3x - \pi)$

**6.** $y = \cos\left(2x + \dfrac{\pi}{3}\right)$

**7.** $y = -\cos(4x + \pi)$

**8.** $y = -\sin\left(4x - \dfrac{2\pi}{3}\right)$

**9.** $y = 3 \sin\left(\dfrac{1}{2}x - \dfrac{\pi}{4}\right)$

**10.** $y = 4 \sin\left(\dfrac{1}{2}x - \dfrac{4\pi}{3}\right)$

**11.** $y = 2 \sin\left(\dfrac{4}{3}x + \dfrac{\pi}{3}\right)$

**12.** $y = 5 \cos\left(\dfrac{2}{3}x + \pi\right)$

**13.** $y = 3 \sin(\pi x + \pi)$

**14.** $y = 3 \cos\left(\dfrac{\pi x}{4} + \dfrac{\pi}{8}\right)$

**15.** $y = 2 \cos\left(\dfrac{\pi x}{2} - \dfrac{\pi}{2}\right)$

**16.** $y = 4 \sin\left(\dfrac{\pi x}{6} - \dfrac{\pi}{3}\right)$

**17.** $y = 40 \cos\left(60x - \dfrac{\pi}{3}\right)$

**18.** $y = 80 \sin\left(40x - \dfrac{\pi}{4}\right)$

**19.** $y = 120 \cos\left(40\pi x - \dfrac{\pi}{2}\right)$

**20.** $y = 40 \sin\left(20\pi x + \dfrac{\pi}{3}\right)$

**21.** $y = 7.5 \sin(220x + \pi)$

**22.** $y = 4.5 \cos\left(180x - \dfrac{\pi}{3}\right)$

**23.** $y = 20 \sin(120\pi x + 4\pi)$

**24.** $y = 30 \cos(160\pi x + 8\pi)$

## 12.3 GRAPHING THE OTHER TRIGONOMETRIC FUNCTIONS

The remaining trigonometric functions have interesting graphs, but they do not have as many technical applications. We shall graph these remaining functions and discuss them briefly.

To graph $y = \tan x$, we shall find a large number of values of $x$ and $y$ that satisfy the equation and plot them in the $xy$-plane as in Fig. 12.15(a).

| $x$ | $-\dfrac{\pi}{2}$ | $-\dfrac{\pi}{3}$ | $-\dfrac{\pi}{4}$ | $-\dfrac{\pi}{6}$ | $0$ | $\dfrac{\pi}{6}$ | $\dfrac{\pi}{4}$ | $\dfrac{\pi}{3}$ | $\dfrac{\pi}{2}$ | $\dfrac{2\pi}{3}$ | $\dfrac{3\pi}{4}$ | $\dfrac{5\pi}{6}$ | $\pi$ | $\dfrac{7\pi}{6}$ | $\dfrac{5\pi}{4}$ | $\dfrac{4\pi}{3}$ | $\dfrac{3\pi}{2}$ |
|---|---|---|---|---|---|---|---|---|---|---|---|---|---|---|---|---|---|
| $y$ | — | $-1.7$ | $-1$ | $-0.58$ | $0$ | $0.58$ | $1$ | $1.7$ | — | $-1.7$ | $-1$ | $-0.58$ | $0$ | $0.58$ | $1$ | $1.7$ | — |

Note that $y = \tan x$ is a cyclic curve but not continuous. At the points

$$x = \cdots, -\frac{\pi}{2}, \frac{\pi}{2}, \frac{3\pi}{2}, \frac{5\pi}{2}, \cdots$$

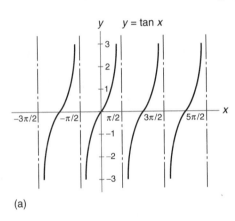

(a)

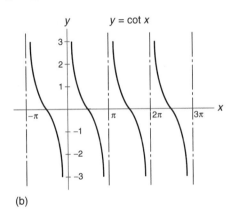

(b)

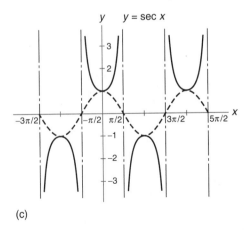

(c)

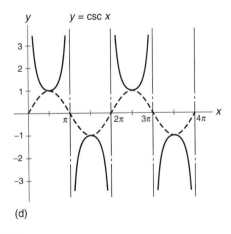

(d)

**Figure 12.15**

there are vertical lines called **asymptotes,** which form guidelines to the curve. That is, as $x$ gets larger and closer to $\dfrac{\pi}{2}$, $y$ becomes larger and larger and the graph approaches the asymptote. The graph never crosses the asymptote because at $x = \dfrac{\pi}{2}$, the tangent function is undefined.

Likewise, the three other trigonometric functions may be graphed. These graphs are shown in Figs. 12.15(b)–(d).

Since the secant and cosecant functions are reciprocals of the cosine and sine, respectively, each may be sketched by plotting reciprocals of the cosine and sine functions, which are designated by the dotted graphs in Figs. 12.15(c)–(d). Note that they, too, are cyclic but not continuous. They also have vertical asymptotes, as noted by the vertical dashed lines.

The period for both $y = \tan x$ and $y = \cot x$ is $\pi$. The period for both $y = \sec x$ and $y = \csc x$ is $2\pi$. The amplitude for each of the four functions is undefined, since each has no maximum $y$-value.

Given equations in the form $y = a \tan bx$ and $y = a \cot bx$, the period may be found by the following formula.

$$P = \frac{\pi}{b}$$

If $a > 1$, each branch intersects the $x$-axis at a greater angle than for $y = \tan x$. If $0 < a < 1$, each branch intersects the $x$-axis at a smaller angle than for $y = \tan x$. If $a$ is negative, each branch is flipped or inverted about the $x$-axis.

**EXAMPLE 1**

Graph $y = 3 \tan 2x$, $y = \dfrac{1}{2} \tan 2x$, and $y = -\tan 2x$ on the same set of coordinate axes.

The period for each is $\dfrac{\pi}{b} = \dfrac{\pi}{2}$ (see Fig. 12.16).

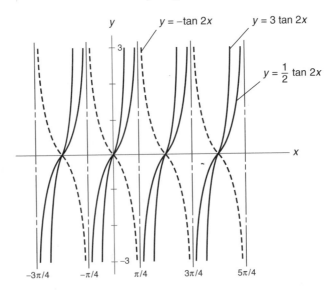

**Figure 12.16**

Given equations in the form $y = a \sec bx$ and $y = a \csc bx$, the period may be found by the following formula.

$$P = \frac{2\pi}{b}$$

The quantity $|a|$ is the vertical distance between the $x$-axis and the low point of each of the branches above the $x$-axis. If $a$ is negative, each branch is flipped or inverted about the $x$-axis.

**EXAMPLE 2**

Graph $y = 2 \sec 4x$.

The period is $\frac{2\pi}{4} = \frac{\pi}{2}$ and $a = 2$. As a graphical aid, you may first graph $y = 2 \cos 4x$ as in Fig. 12.17.

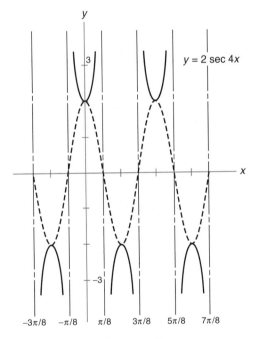

**Figure 12.17**

# Exercises 12.3

*Find the period of each function and sketch its graph.*

**1.** $y = \tan 3x$

**2.** $y = \tan 4x$

**3.** $y = -2 \tan \frac{1}{2}x$

**4.** $y = -\tan \frac{3}{2}x$

**5.** $y = \cot 6x$

**6.** $y = 2 \cot \frac{1}{3}x$

**7.** $y = 3 \sec x$

**8.** $y = 4 \csc x$

**9.** $y = -2 \csc x$

**10.** $y = -5 \sec x$    **11.** $y = 2 \sec 6x$    **12.** $y = 4 \sec 2x$

**13.** $y = -3 \sec \dfrac{1}{2}x$    **14.** $y = -5 \sec \dfrac{3}{2}x$    **15.** $y = \csc 2x$

**16.** $y = 2 \csc 3x$

*Sketch the graph of each function.*

**17.** $y = \tan\left(x + \dfrac{\pi}{2}\right)$    **18.** $y = \cot\left(2x - \dfrac{\pi}{2}\right)$

**19.** $y = 3 \sec\left(2x - \dfrac{\pi}{2}\right)$    **20.** $y = 2 \csc\left(x + \dfrac{\pi}{3}\right)$

## 12.4   GRAPHING COMPOSITE CURVES

In applications it is common to find functions that are composites of sums or differences of expressions, some of which are trigonometric, such as $y = \cos x + \sin 2x$ or $y = x - \sin x$. To graph such functions, a graphical technique called **addition of ordinates** is useful.

---

### ADDITION OF ORDINATES

**1.** Graph each of the functions that make up the composite on the same set of coordinate axes.

**2.** It may be helpful to draw several vertical lines perpendicular to the $x$-axis.

**3.** If the composite is
   (a) a sum, find the algebraic sum of the $y$-values where each vertical line intersects each of the graphs; or
   (b) a difference, find the algebraic difference of the $y$-values where each vertical line intersects each of the graphs, or graph the negative (or "flipped") curve and find the algebraic sum as in (a).

**4.** Plot each resultant $y$-value from Step 3 on each vertical line and connect the points with a smooth curve.

---

Or, instead of Step 3 above, add or subtract the $y$-values on the graph itself by means of a compass or ruler. Remember, the $y$-values of points above the $x$-axis are positive, and the $y$-values of points below the $x$-axis are negative.

#### EXAMPLE 1

Graph $y = \cos x + \sin 2x$ using the method of addition of ordinates.
   The graph is shown in Fig. 12.18. *Note:* One period of $y = \sin 2x$ is not sufficient since $\cos x$ has a longer period.

#### EXAMPLE 2

Graph $y = x - \sin x$ using the method of addition of ordinates.
   The graph is shown in Fig. 12.19.

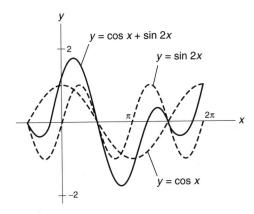

Figure 12.18

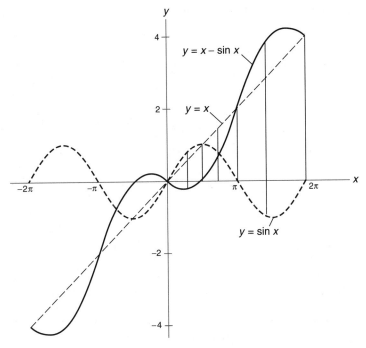

Figure 12.19

# Exercises 12.4

*Using the method of addition of ordinates, graph each equation for one complete period of the composite curve.*

1. $y = \sin x + \cos x$

2. $y = 2 \cos x + \sin x$

3. $y = 2 \cos x + 2 \sin x$

4. $y = \sin x + 3 \cos x$

5. $y = 2 \sin x + \sin \dfrac{x}{2}$

6. $y = 2 \cos \dfrac{x}{2} + \sin x$

7. $y = 3 \sin x + 2 \cos x$

8. $y = 2 \sin 2x + 2 \cos x$

9. $y = 2 \sin x - \cos x$

10. $y = \cos x - 2 \sin x$

**11.** $y = 2 \sin 2x - \cos 2x$

**12.** $y = \cos 2x - \cos x$

**13.** $y = x + \sin x$

**14.** $y = x - 2 \cos x$

**15.** $y = \sin 4x + \cos\left(2x + \dfrac{\pi}{3}\right)$

**16.** $y = \sin 2x + \cos\left(2x - \dfrac{\pi}{4}\right)$

## 12.5   SIMPLE HARMONIC MOTION

One of the most important and most widely used applications of the sine function and its graph is that of simple harmonic motion. For example, consider a weight suspended on a spring. Now, pull down on the weight and then let go; the weight bobs up and down. Next, graph the vertical displacement of the weight over equal units of time as in Fig. 12.20, where successive vertical positions are displaced horizontally over equal time intervals. Our result is the familiar sine wave.

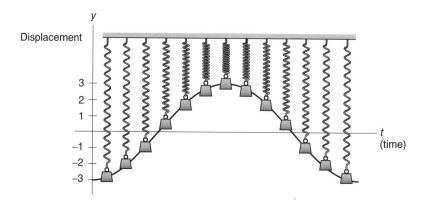

**Figure 12.20**

We find it quite helpful to analyze simple harmonic motion in terms of circular motion. Consider a crank 1 unit long rotating counterclockwise at a constant rate of 1 revolution per second. Set up the rotating crank and the $xy$-axes, as follows. The motion of the projection of the crank handle on the $y$-axis (think of the motion of its shadow projected on the $y$-axis as a function of time when a light is placed far out on the negative $x$-axis) is also an example of **simple harmonic motion** (see Fig. 12.21).

**Figure 12.21**

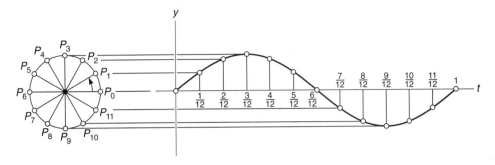

**Figure 12.22**

Let's assume that the crank starts from 0°. At the end of $\frac{1}{12}$ s, it will have rotated 30°, or $\frac{\pi}{6}$ rad, to position $P_1$ as in Fig. 12.22; at the end of $\frac{2}{12}$ s, it will have rotated 60°, or $\frac{\pi}{3}$ rad, to position $P_2$; and so forth; until after 1 s, or 1 rev, it will have rotated back to its original position $P_0$.

The projection of the crank handle on the y-axis (that is, its y-value per unit time) can be plotted as a curve. We do so by letting the x-axis be the time axis and dividing it into as many intervals as there are angles to be plotted—in this case 12, every $\frac{\pi}{6}$ rad from 0 to $2\pi$ rad. Through the points of division on the time axis (x-axis), construct vertical lines. Through the position points $P_0$, $P_1$, $P_2$, . . . , construct horizontal lines. Draw a small circle around the intersection of the corresponding vertical and horizontal lines; that is, $P_1$ and $\frac{1}{12}$, $P_2$ and $\frac{2}{12}$, and so on. Draw a smooth curve through these points of intersection which, as you can see, results in a sine curve. If we were to continue rotating the crank, successive rotations would generate successive periods of the sine curve. The equation is $y = \sin x$. Since the crank rotates through $2\pi$ rad in 1 s, $x = 2\pi$ in 1 s, $x = 4\pi$ in 2 s, and $x = 2\pi t$ in t s. Therefore, the y-value at any time t is given by $y = \sin 2\pi t$.

More generally, we let r be the length of a vector **B** rotating at a uniform angular velocity of $\omega$ rad per unit time. If we further assume a horizontal starting position at $t = 0$, the projections of vector **B** onto the y-axis, which are the vertical components of **B,** will be in simple harmonic motion. Their motion may be represented by the equation

$$y = r \sin \omega t$$

This rotating position vector is sometimes called a **phasor.**

This equation is of fundamental importance in that it describes the motion of any object or quantity that is in simple harmonic motion. Familiar examples of motion, that is simple harmonic motion (or very nearly so), are (1) the motion of a satellite in a circular orbit; (2) the motion of a pendulum bob, if the displacement is small; (3) the motion of the prongs of a vibrating tuning fork (sound waves); (4) the motion of a rotating wire through a magnetic field—alternating current; and (5) the motion of a mass vibrating up and down on a spring.

If the rotating vector is not in position $P_0$ at $t = 0$, the simple harmonic motion equation becomes

$$y = r \sin(\omega t + \theta)$$

where $\theta$ is the initial position of the vector at $t = 0$. The graph results in a phase shift of $\dfrac{\theta}{\omega}$.

**EXAMPLE 1**

A crank 15 in. long starts from a horizontal position (0 rad) and rotates in a counterclock-wise direction at a uniform angular velocity of 3 revolutions per second ($6\pi$ rad/s), as illustrated in Fig. 12.23.

(a) Plot the curve (as shown previously in Fig. 12.22) that shows the projection of the crank handle on the y-axis per unit time.

(b) Find the equation of this curve.

(c) Through how many radians does the crank turn in 0.15 s?

(d) What is the distance of the crank handle from the horizontal axis (t-axis) at $\frac{1}{9}$ s? That is, what is the length of the projection of the crank on the y-axis at $\frac{1}{9}$ s?

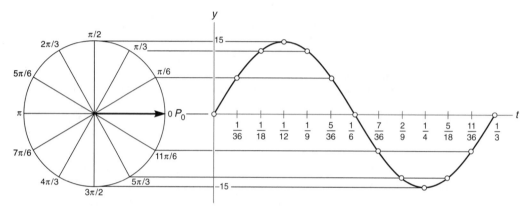

$P_0$ = Position of
crank at $t = 0$

**Figure 12.23**

(a) The amplitude, $r$, is 15. The period is $\dfrac{2\pi}{6\pi} = \dfrac{1}{3}$. There is no phase shift, and $\omega = 6\pi$ rad/s.

(b) $y = r \sin \omega t$
$y = 15 \sin 6\pi t$

(c) $\theta = \omega t$
$\theta = (6\pi \text{ rad/s})(0.15 \text{ s}) = 2.8$ rad

(d) Length of projection $= (15 \text{ in.})\sin \dfrac{2\pi}{3} = 13$ in.

**EXAMPLE 2**

A wire rotating through a magnetic field generates an alternating current. The instantaneous current, $i$, is given by the equation

$$i = I_{max} \sin(\omega t + \theta)$$

where $I_{max}$ is the maximum instantaneous current, $\omega$ is the angular velocity of the rotating wire, $t$ is the time, and $\theta$ is the phase angle. Given $I_{max} = 12$ A, $\omega = 120\pi$ rad/s (60 cycles/s), and $\theta = \dfrac{\pi}{3}$ (see Fig. 12.24).

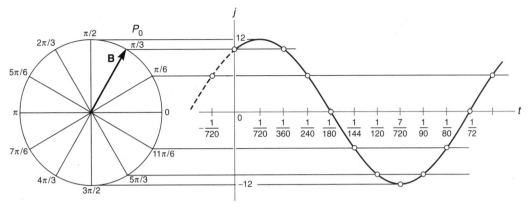

$P_0$ = Position of vector **B**
at $t = 0$

**Figure 12.24**

(a) Graph the resulting equation $i = 12 \sin\left(120\pi t + \dfrac{\pi}{3}\right)$.

(b) From the graph find the time at which the current is $+10$ A, $-6$ A, and $0$ A.

(c) From the graph find the current at $t = \frac{1}{240}$ s and $\frac{1}{120}$ s.

(a) The amplitude is 12. The period is $\dfrac{2\pi}{120\pi} = \dfrac{1}{60}$. The phase shift is $\dfrac{\pi/3}{120\pi} = \dfrac{1}{360}$ to the left.

(b) At $i = +10$ A, $t = 0, \dfrac{1}{360}, \dfrac{1}{60}, \dfrac{7}{360}, \cdots$

At $i = -6$ A, $t = \dfrac{1}{144}, \dfrac{1}{80}, \dfrac{17}{720}, \cdots$

At $i = 0$, $t = \dfrac{1}{180}, \dfrac{1}{72}, \dfrac{1}{45}, \cdots$

(c) At $t = \dfrac{1}{240}$ s, $i = +6$ A.

At $t = \dfrac{1}{120}$ s, $i = -10$ A.

## Exercises 12.5

1. A crank 8 in. long starts from a horizontal position ($0°$) and rotates in a counterclockwise direction at a uniform angular velocity of 5 revolutions per second ($10\pi$ rad/s). Plot the curve that shows the simple harmonic motion of the crank handle, and find the equation of this curve.

2. Complete Exercise 1 for when the crank starts from a vertical position ($90°$).

3. If the rotating wire from Example 2 produces a maximum current of 15 A at 60 cycles/s, graph the resulting equation when $\theta = \dfrac{\pi}{4}$.

4. Complete Exercise 3 for $\theta = 0$.

5. A rotating wire through a magnetic field generates an alternating current whose instantaneous voltage, $e$, is given by the equation

$$e = E_{max} \cos(\omega t + \theta)$$

where $E_{max}$ is the maximum instantaneous voltage, $\omega$ is the angular velocity of the rotating wire, $t$ is the time, and $\theta$ is the phase angle. Graph the resulting equation when $E_{max} = 110$ V, $\omega = 60$ cycles/s, and $\theta = 0°$.

6. Complete Exercise 5 for $\theta = -\dfrac{\pi}{3}$.

7. A mass hanging on a spring is pulled down 10 cm from its equilibrium position and then released $\left( \theta = -\dfrac{\pi}{2} \right)$. The resulting oscillation of the mass is simple harmonic motion. The period of one complete cycle is observed to be 0.80 s. Find and graph the resulting simple harmonic motion equation.

8. A weight is attached to a spring and vibrates at a frequency of 4 vibrations per second and an observed amplitude of 3 in. Find and graph the resulting equation beginning with $t = 0$ at $\theta = 0$.

9. A thin reed fixed at one end vibrates at 200 vibrations per second with an amplitude of 0.25 cm. Find and graph the resulting equation beginning with $t = 0$ at $\theta = 0$.

10. The equation of motion for a particle in simple harmonic motion is given by

$$y = 5.0 \cos 8.0t$$

where $y$ is in metres and $t$ is in seconds. Graph the equation. Find the distance from the horizontal at $t = \frac{\pi}{4}$.

## CHAPTER SUMMARY

1. The values of $a$, $b$, and $c$ determine amplitude, period, and phase shift for the sine and cosine functions as follows:

$$y = a \sin(bx + c)$$
$$y = a \cos(bx + c)$$

$$\text{Amplitude} = |a| \qquad \text{Phase shift} = \frac{c}{b} \qquad \left( \text{To the } \textit{left} \text{ if } \frac{c}{b} > 0 \right)$$

$$\text{Period} = \frac{2\pi}{b} \qquad \qquad \left( \text{To the } \textit{right} \text{ if } \frac{c}{b} < 0 \right)$$

If $a$ is negative, each branch is flipped or inverted about the $x$-axis.

2. For $y = a \tan bx$ and $y = a \cot bx$, the period may be found by

$$P = \frac{\pi}{b}$$

If $a > 1$, each branch intersects the $x$-axis at a greater angle than for $y = \tan x$. If $0 < a < 1$, each branch intersects the $x$-axis at a smaller angle than for $y = \tan x$. If $a$ is negative, each branch is flipped or inverted about the $x$-axis.

**3.** For $y = a \sec bx$ and $y = a \csc bx$, the period may be found by

$$P = \frac{2\pi}{b}$$

The quantity $|a|$ is the vertical distance between the $x$-axis and the low point of each of the branches above the $x$-axis. If $a$ is negative, each branch is flipped or inverted about the $x$-axis.

**4.** To add or subtract two functions graphically, use the method of addition of ordinates as follows:

(a) Graph each of the functions that make up the composite on the same set of co-ordinate axes.

(b) It may be helpful to draw several vertical lines perpendicular to the $x$-axis.

(c) If the composite is

    (i) a sum, find the algebraic sum of the $y$-values where each vertical line inter-sects each of the graphs; or

    (ii) a difference, find the algebraic difference of the $y$-values where each vertical line intersects each of the graphs, or graph the negative (or "flipped") curve and find the algebraic sum as in (i).

(d) Plot each resultant $y$-value from Step (c) on each vertical line and connect with a smooth curve.

**5.** Simple harmonic motion is described by the equations

$$y = r \sin(\omega t + \theta) \quad \text{or} \quad y = r \cos(\omega t + \theta)$$

# CHAPTER 12 REVIEW

*Find the amplitude and period of each function and sketch its graph.*

**1.** $y = 4 \cos 6x$        **2.** $y = -2 \sin \dfrac{1}{3}x$        **3.** $y = 3 \cos 2\pi x$

*Find the amplitude, period, and phase shift of each function and sketch its graph.*

**4.** $y = 3 \sin\left(x - \dfrac{\pi}{4}\right)$      **5.** $y = \cos\left(2x + \dfrac{2\pi}{3}\right)$      **6.** $y = 4 \sin\left(\pi x + \dfrac{\pi}{2}\right)$

*Find the period of each function and sketch its graph.*

**7.** $y = \tan 5x$        **8.** $y = -\cot 3x$        **9.** $y = 2 \sec 4x$

**10.** Graph $y = 3 \sin x + 2 \cos 2x$ using the method of addition of ordinates for one complete period of the composite curve.

**11.** A pendulum has a period of 4 s and an observed amplitude of 3 ft. The motion of the projection of the pendulum bob on the floor is in simple harmonic motion. Find and graph the resulting equation. (Assume that the phase angle is 0.)

# Application
## Improving Color Television

The technology of gas-plasma television allows a TV set to be lightweight and thin—so much so that the TV could be hung on a wall. Currently, most color TV sets use a cathode-ray tube, which requires an electron-gun assembly to be located behind the inside screen. The bigger the screen, the further back the gun assembly must be. Gas-plasma or thin TVs use charged gas rather than electrons to light up the color pixels, thereby eliminating the distance requirement.

Both systems use red, green, and blue phosphors placed in some pattern on the inside of the screen. In the cathode-ray tube a variety of electrodes in the gun assembly and conductive coatings around the funnel tube accelerate the electrons and keep them focused. A magnetic deflection yoke guides the electrons horizontally and vertically.

A perforated sheet of metal called a *shadow mask* is the last thing to guide the electrons before they hit the screen. The screen is curved so that the electrons travel about the same distance no matter where they strike it. It helps to think of the path of the electrons as radii of a sphere. Flattening the faceplate would mean that the distance between the gun and points on the faceplate would differ, and the beam would not be focused at all points. However, we really don't want the front of a TV screen to be spherical; imagine how the picture would look to the viewer. We would like the faceplate to be as flat as possible without distorting the picture.

Engineers and technologists are continuing to find ways to improve the picture. New focusing technology converts the focus on the edges and corners to permit a flatter faceplate. For example, the Sony Trinitron™ uses a cylindrical faceplate which is flat from top to bottom and curved in the horizontal direction. This construction prevents vertical lines from bowing at the edges. Other industry improvements involve using shadow masks of invar (a steel alloy) instead of iron. The invar expands less at high temperatures and thus does not bulge. A bulge causes colors to bleed into one another. Rare-earth elements have been added to the phosphors to increase their brightness. The faceplate has been made darker which gives more detail and depth. Also, the contrast is increased by reducing the amount of light the faceplate reflects. Finally, antiglare coating is added on the outside of the faceplate.

For a television to be thin, a totally different approach is needed. Two methods for applying the gas-plasma technology are being developed by competing companies. One combines flat-panel display technology with plasma-addressed liquid-crystal technology (PALC) to activate each pixel. The liquid crystals are opened and closed when the gas plasma operates as an electronic on/off switch. The other uses two glass panels to enclose the gas plasma. Then a voltage is applied between electrodes which creates ultraviolet ra-

458

diation that in turn produces visible light by activating the phosphors. Both methods result in a system that weighs approximately 12 pounds, is about 4 inches thick, and can have a viewing screen from 20 to 60 inches.

Flat TVs are available to the public, but the cost can be as high as $15,000 depending on the size. When mass production becomes more efficient, the cost is expected to be around $2500.

### Bibliography

Barker, Dennis P. "High-Tech Tubes." *Popular Mechanics* (April 1997): 60–63.
Normile, Dennis. "Flat TV Arrives." *Popular Science* (May 1997): 69–73.

# 13

# Trigonometric Formulas and Identities

## INTRODUCTION

Thus far, we have solved algebraic, exponential, and logarithmic equations. Trigonometric equations are more complicated than any we have done. They may involve more than one trigonometric function or contain functions with different (but related) arguments. In this chapter we develop many trigonometric identities that will be used to simplify expressions and convert equations into a form that we can readily solve.

### Objectives

- Develop basic trigonometric identities.
- Simplify trigonometric expressions.
- Develop a strategy for proving trigonometric identities.
- Use trigonometric identities to solve equations.
- Work with inverse trigonometric functions.
- Identify the domain of an inverse trigonometric function.

## 13.1 BASIC TRIGONOMETRIC IDENTITIES

An equation that is true for all values of the variables for which both sides are defined is called an **identity.** The following are examples of algebraic identities:

**1.** $2(x - 3) = 2x - 6$     (True for all $x$)

**2.** $4(x + y) = 4x + 4y$     (True for all $x$ and all $y$)

**3.** $\dfrac{4x + y}{x} = 4 + \dfrac{y}{x}$     (True for all $y$ and all $x$ except 0)

The following are examples of trigonometric identities. (Some have been shown previously; the rest will be shown later.)

**4.** $\sin \theta = \dfrac{1}{\csc \theta}$        (True for all $\theta$ except where $\csc \theta = 0$)

**5.** $\sin^2 \theta + \cos^2 \theta = 1$        (True for all $\theta$)

**6.** $\sin(\theta + \phi) = \sin \theta \cos \phi + \cos \theta \sin \phi$        (True for all $\theta$ and all $\phi$)

Identities are used to prove other identities or to simplify a given trigonometric expression. They are also used to change a given trigonometric expression into a different but equivalent expression which is more useful for a particular problem, either technical or mathematical.

A rather long list of trigonometric identities exists. We shall first prove (show the validity of) the basic identities. Before you use an identity, you must be sure that it comes from a list of valid identities or be able to show that it is valid. At times, just a change in the form of an identity will greatly simplify the solution of a given problem.

A given identity may be proven by any of a number of ways, as long as all the steps are valid. We shall illustrate five basic ways to prove trigonometric identities. When a number of valid ways may be used to prove a given identity, each is *correct;* the one most preferred is the one that is shortest and most efficient. Generally, a helpful suggestion is to try to simplify the more complicated side to the less complicated side of the identity.

## Method 1: Use of Trigonometric Definitions

The trigonometric definitions from Chapter 10 are shown in the following box with Fig. 13.1; they may be used to prove identities.

$$\sin \theta = \frac{y}{r}$$

$$\cos \theta = \frac{x}{r}$$

$$\tan \theta = \frac{y}{x} \quad (x \neq 0)$$

$$\cot \theta = \frac{x}{y} \quad (y \neq 0)$$

$$\sec \theta = \frac{r}{x} \quad (x \neq 0)$$

$$\csc \theta = \frac{r}{y} \quad (y \neq 0)$$

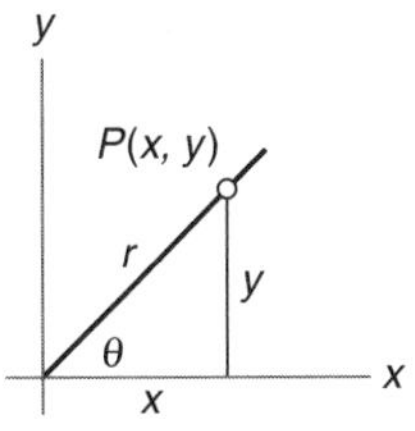

**Figure 13.1**

**EXAMPLE 1**

Prove $\cos \theta = \dfrac{1}{\sec \theta}$.

$$\frac{1}{\sec \theta} = \frac{1}{\dfrac{r}{x}} = \frac{x}{r} = \cos \theta$$

Each of the reciprocal identities may be proven in a similar way and is listed in the following box. All basic identities are listed and numbered for easy reference.

$$\sin \theta = \frac{1}{\csc \theta} \tag{1}$$

$$\cos \theta = \frac{1}{\sec \theta} \tag{2}$$

$$\tan \theta = \frac{1}{\cot \theta} \tag{3}$$

$$\cot \theta = \frac{1}{\tan \theta} \tag{4}$$

$$\sec \theta = \frac{1}{\cos \theta} \tag{5}$$

$$\csc \theta = \frac{1}{\sin \theta} \tag{6}$$

**EXAMPLE 2**

Prove $\tan \theta = \dfrac{\sin \theta}{\cos \theta}$.

$$\frac{\sin \theta}{\cos \theta} = \frac{\dfrac{y}{r}}{\dfrac{x}{r}} = \frac{y}{x} = \tan \theta$$

$$\tan \theta = \frac{\sin \theta}{\cos \theta} \tag{7}$$

$$\cot \theta = \frac{\cos \theta}{\sin \theta} \tag{8}$$

The next three identities are called the **Pythagorean identities** because they use the Pythagorean relationship, $x^2 + y^2 = r^2$.

**EXAMPLE 3**

Prove $\sin^2 \theta + \cos^2 \theta = 1$.

Given $x^2 + y^2 = r^2$, divide each side by $r^2$.

$$\left(\frac{x}{r}\right)^2 + \left(\frac{y}{r}\right)^2 = 1$$

From the trigonometric definitions, $\dfrac{x}{r} = \cos \theta$ and $\dfrac{y}{r} = \sin \theta$. Therefore,

$$\cos^2 \theta + \sin^2 \theta = 1$$

*Note:* $\cos^2 \theta$ may also be written $(\cos \theta)^2$.

**EXAMPLE 4**

Prove $1 + \tan^2 \theta = \sec^2 \theta$.

Given $x^2 + y^2 = r^2$, divide each side by $x^2$.

$$1 + \left(\frac{y}{x}\right)^2 = \left(\frac{r}{x}\right)^2$$

Then

$$1 + \tan^2 \theta = \sec^2 \theta$$

The three Pythagorean identities are as follows.

| | |
|---|---|
| $\sin^2 \theta + \cos^2 \theta = 1$ | **(9)** |
| $1 + \tan^2 \theta = \sec^2 \theta$ | **(10)** |
| $\cot^2 \theta + 1 = \csc^2 \theta$ | **(11)** |

Method 1 is typically used only to develop or prove the first 11 identities. Other methods are then needed to prove the more complex identities because method 1 commonly results in cumbersome fractional expressions. We suggest that you use method 1 only in Exercises 1–4 at the end of this section.

## Method 2: Substitution of Known Identities

We may also show that an identity is valid by changing one of the sides of the given expression or any of its parts by substitution of a known identity or any of its forms one or more times.

**EXAMPLE 5**

Prove $\tan x = \sin x \sec x$.

Let's start with the right-hand side and show that it is equal to the left-hand side.

$$\sin x \sec x = \sin x \left(\frac{1}{\cos x}\right) \qquad \text{(Identity 5)}$$

$$= \frac{\sin x}{\cos x}$$

$$= \tan x \qquad \text{(Identity 7)}$$

Therefore, $\tan x = \sin x \sec x$.

**EXAMPLE 6**

Prove $(1 - \sin^2 \theta)\tan^2 \theta = \sin^2 \theta$.

Let's start with the left-hand side and show that it is equal to the right-hand side.

$$(1 - \sin^2 \theta)\tan^2 \theta = \cos^2 \theta \tan^2 \theta \qquad \text{(Form of Identity 9)}$$

$$= \cos^2 \theta \left(\frac{\sin^2 \theta}{\cos^2 \theta}\right) \qquad \text{(Identity 7)}$$

$$= \sin^2 \theta$$

Therefore, $(1 - \sin^2 \theta)\tan^2 \theta = \sin^2 \theta$.

**EXAMPLE 7**

Prove $\cos^2 x(1 + \tan^2 x) = 1$.

Let's start with the left-hand side and show that it is equal to the right-hand side.

$$\cos^2 x(1 + \tan^2 x) = \cos^2 x \sec^2 x \qquad \text{(Identity 10)}$$

$$= \cos^2 x \left( \frac{1}{\cos^2 x} \right) \qquad \text{(Identity 5)}$$

$$= 1$$

Therefore, $\cos^2 x(1 + \tan^2 x) = 1$.

**EXAMPLE 8**

Prove $\sin \theta + \cot \theta \cos \theta = \csc \theta$.

Again, let's start with the left-hand side and show that it is equal to the right-hand side.

$$\sin \theta + \cot \theta \cos \theta = \sin \theta + \frac{\cos \theta}{\sin \theta} \cdot \cos \theta \qquad \text{(Identity 8)}$$

$$= \sin \theta + \frac{\cos^2 \theta}{\sin \theta}$$

$$= \frac{\sin^2 \theta + \cos^2 \theta}{\sin \theta}$$

$$= \frac{1}{\sin \theta} \qquad \text{(Identity 9)}$$

$$= \csc \theta \qquad \text{(Identity 6)}$$

Therefore, $\sin \theta + \cot \theta \cos \theta = \csc \theta$.

## Method 3: Factoring One Side of the Identity

**EXAMPLE 9**

Prove $\sin^2 \theta + \sin^2 \theta \tan^2 \theta = \tan^2 \theta$.

$$\sin^2 \theta + \sin^2 \theta \tan^2 \theta = \sin^2 \theta (1 + \tan^2 \theta) \qquad \text{(Factor.)}$$

$$= \sin^2 \theta \sec^2 \theta \qquad \text{(Identity 10)}$$

$$= \sin^2 \theta \left( \frac{1}{\cos^2 \theta} \right) \qquad \text{(Identity 5)}$$

$$= \tan^2 \theta \qquad \text{(Identity 7)}$$

Therefore, $\sin^2 \theta + \sin^2 \theta \tan^2 \theta = \tan^2 \theta$.

**EXAMPLE 10**

Prove $\sin^4 \theta - \cos^4 \theta = 2 \sin^2 \theta - 1$.

$$\sin^4 \theta - \cos^4 \theta = (\sin^2 \theta + \cos^2 \theta)(\sin^2 \theta - \cos^2 \theta) \qquad \text{(Factor.)}$$

$$= (1)(\sin^2 \theta - \cos^2 \theta) \qquad \text{(Identity 9)}$$

$$= \sin^2 \theta - (1 - \sin^2 \theta) \qquad \text{(Identity 9)}$$

$$= 2 \sin^2 \theta - 1$$

Therefore, $\sin^4 \theta - \cos^4 \theta = 2 \sin^2 \theta - 1$.

# Method 4: Multiplication of the Numerator and Denominator of a Fractional Expression

Multiplication of the numerator and denominator by some trigonometric quantity may change the form of one side of the identity so that you may then use one of the other methods to complete the proof.

**EXAMPLE 11**

Prove $\dfrac{\cos x}{1 + \sin x} = \dfrac{1 - \sin x}{\cos x}$.

Multiply the numerator and the denominator of the right-hand side by $1 + \sin x$.

$$\frac{1 - \sin x}{\cos x} \cdot \frac{1 + \sin x}{1 + \sin x} = \frac{1 - \sin^2 x}{\cos x(1 + \sin x)}$$

$$= \frac{\cos^2 x}{\cos x(1 + \sin x)} \qquad \text{(Identity 9)}$$

$$= \frac{\cos x}{1 + \sin x}$$

Therefore, $\dfrac{\cos x}{1 + \sin x} = \dfrac{1 - \sin x}{\cos x}$.

# Method 5: Expressing All Functions on One Side in Terms of Sines and Cosines and Using Any of Methods 1–4

**EXAMPLE 12**

Prove $\tan x + \cot x = \sec x \csc x$.

$$\tan x + \cot x = \frac{\sin x}{\cos x} + \frac{\cos x}{\sin x} \qquad \text{(Identities 7 and 8)}$$

$$= \frac{\sin^2 x + \cos^2 x}{\cos x \sin x}$$

$$= \frac{1}{\cos x \sin x} \qquad \text{(Identity 9)}$$

$$= \sec x \csc x \qquad \text{(Identities 5 and 6)}$$

Therefore, $\tan x + \cot x = \sec x \csc x$.

*Note:* In any method, introduce radicals only as a last resort!

Proving identities and using identities to simplify a trigonometric expression are quite different from solving an equation. You must work from only one side of the identity to show that it is equivalent to the other. When you use identities to simplify a trigonometric expression, you begin with the expression and use identities to change from one equivalent expression to another until the expression is simplified or in the form that is most useful to you. That is, you may *not* use equation-solving principles such as adding the same quantity to each side or multiplying each side by the same quantity.

## Exercises 13.1

*Use the trigonometric definitions opposite Fig. 13.1 to prove each identity.*

**1.** $\dfrac{1}{\csc \theta} = \sin \theta$

**2.** $\dfrac{1}{\tan \theta} = \cot \theta$

**3.** $\dfrac{\cos \theta}{\sin \theta} = \cot \theta$

**4.** $\cot^2 \theta + 1 = \csc^2 \theta$

*Prove each identity without using the trigonometric definitions.*

**5.** $\cos \theta \sec \theta = 1$

**6.** $\tan \theta \cot \theta = 1$

**7.** $\cos x \tan x = \sin x$

**8.** $\csc x \tan x = \sec x$

**9.** $\dfrac{\csc \theta}{\sec \theta} = \cot \theta$

**10.** $\dfrac{\tan \theta}{\cot \theta} = \tan^2 \theta$

**11.** $\dfrac{\tan \theta}{\sin \theta} = \sec \theta$

**12.** $\dfrac{\cot \theta}{\cos \theta} = \csc \theta$

**13.** $\sec \theta \cot \theta = \csc \theta$

**14.** $\sin \theta \cot \theta = \cos \theta$

**15.** $(1 - \cos^2 x)\csc^2 x = 1$

**16.** $(\cot^2 x + 1)\tan^2 x = \sec^2 x$

**17.** $(1 - \sin^2 \theta)\cos^2 \theta = \cos^4 \theta$

**18.** $(1 + \tan^2 x)\sin^2 x = \tan^2 x$

**19.** $\dfrac{\cos^2 x - 1}{\sin x} = -\sin x$

**20.** $\dfrac{\sec^2 x - 1}{\sin^2 x} = \sec^2 x$

**21.** $\cos \theta (\csc \theta - \sec \theta) = \cot \theta - 1$

**22.** $\cot \theta - \cos \theta = \cot \theta(1 - \sin \theta)$

**23.** $\tan^2 \theta - \tan^2 \theta \sin^2 \theta = \sin^2 \theta$

**24.** $\tan \theta \sin \theta + \tan \theta \cot \theta \cos \theta = \sec \theta$

**25.** $\dfrac{\sec x - \cos x}{\sin x} = \tan x$

**26.** $\dfrac{\csc x - \sin x}{\cot x} = \cos x$

**27.** $\dfrac{\sec^2 \theta - 1}{\sec^2 \theta} = \sin^2 \theta$

**28.** $\sin^2 \theta - \cos^2 \theta = 1 - 2 \cos^2 \theta$

**29.** $\dfrac{\csc^2 x}{1 + \tan^2 x} = \cot^2 x$

**30.** $\dfrac{1 + \tan^2 x}{\tan^2 x} = \csc^2 x$

**31.** $\dfrac{(1 + \sin \theta)(1 - \sin \theta)}{\sin^2 \theta} = \cot^2 \theta$

**32.** $\dfrac{\sin x \cos x}{(1 + \cos x)(1 - \cos x)} = \cot x$

**33.** $\dfrac{\sin^2 x}{1 + \cos x} = 1 - \cos x$

**34.** $\dfrac{\cos^2 x}{1 + \sin x} = 1 - \sin x$

**35.** $\cos^4 \theta - \sin^4 \theta = 2 \cos^2 \theta - 1$

**36.** $\sec^4 x - \tan^4 x = 2 \tan^2 x + 1$

**37.** $\dfrac{1 + \sec x}{\csc x} = \sin x + \tan x$

**38.** $\dfrac{\cos x \tan x + \sin x}{\tan x} = 2 \cos x$

**39.** $(\sec x - \tan x)(\csc x + 1) = \cot x$

**40.** $\sec x + \tan x + \sin x = \dfrac{1 + \sin x + \sin x \cos x}{\cos x}$

**41.** $\dfrac{1 - \sin^2 x}{1 - \cos^2 x} = \cot^2 x$

**42.** $\dfrac{1 - \tan x}{1 + \tan x} = \dfrac{\cot x - 1}{\cot x + 1}$

**43.** $\dfrac{\sin x}{1 + \cos x} = \dfrac{1 - \cos x}{\sin x}$

**44.** $\dfrac{\sin \theta}{1 + \cos \theta} = \dfrac{1 - \cos \theta}{\cos \theta \tan \theta}$

**45.** $\dfrac{\cos^2 x}{1 - \sin x} = 1 + \sin x$

**46.** $\dfrac{\sin \theta + \cos \theta}{\cos \theta - \sin \theta} = \dfrac{\cot \theta + 1}{\cot \theta - 1}$

**47.** $\dfrac{1}{\sec x - 1} - \dfrac{1}{\sec x + 1} = 2 \cot^2 x$

**48.** $\dfrac{1}{\sin x + 1} - \dfrac{1}{\sin x - 1} = 2 \sec^2 x$

**49.** $\cos^4 x - \sin^4 x = 1 - 2 \sin^2 x$

**50.** $\sec^4 x - 1 = \tan^2 x(\sec^2 x + 1)$

**51.** $\dfrac{\tan^2 x - 1}{1 - \cot^2 x} = \tan^2 x$

**52.** $\sec \theta - \tan \theta = \dfrac{1}{\sec \theta + \tan \theta}$

**53.** $\dfrac{\tan x + \tan y}{\cot x + \cot y} = \tan x \tan y$

**54.** $\dfrac{\tan \theta}{\sec \theta - \cos \theta} = \csc \theta$

## 13.2 FORMULAS FOR THE SUM AND THE DIFFERENCE OF TWO ANGLES

The sine, cosine, and tangent of the sum and the difference of two angles have practical applications. These formulas are also used to develop the double- and half-angle formulas in the next section.

First, we want to show

$$\sin(\theta + \phi) = \sin \theta \cos \phi + \cos \theta \sin \phi$$

In Fig. 13.2, angles $\theta$ and $\theta + \phi$ are constructed so that they are in standard position. The initial side of $\phi$ coincides with the terminal side of $\theta$; its terminal side coincides with the terminal side of $\theta + \phi$. Now choose any point $P$ on the terminal side of $\theta + \phi$. Then drop perpendicular lines to the terminal side of $\theta$ at $Q$ and to the initial side of $\theta$ at $S$. Next, draw perpendiculars $QR$ and $QT$. Because $RQ$ and $OT$ are parallel, $\theta = \angle 1$. (They are alternate interior angles.)

$$\angle 1 + \angle 2 = 90° \qquad (PQ \perp OQ)$$

$$\underline{\angle 3 + \angle 2 = 90°} \qquad \text{(The sum of the two acute angles of a right triangle equals 90°.)}$$

$$\angle 1 - \angle 3 = 0 \qquad \text{(Subtract.)}$$

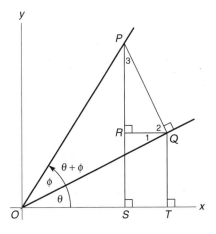

**Figure 13.2**

or

$$\angle 1 = \angle 3$$

We have already shown that

$$\angle 1 = \theta$$

Therefore, $\theta = \angle 3$. That is, $\angle SPQ$ and $\theta$ are equal. In right triangle $OSP$ of Fig. 13.2,

$$\sin(\theta + \phi) = \frac{SP}{OP} = \frac{SR + RP}{OP}$$

$$= \frac{SR}{OP} + \frac{RP}{OP}$$

$$= \frac{TQ}{OP} + \frac{RP}{OP} \quad \text{(In rectangle } SRQT, SR = TQ.)$$

Next, multiply numerator and denominator of the first term by $OQ$ and rearrange the terms; then multiply numerator and denominator of the second term by $PQ$ and rearrange the terms.

$$\frac{TQ}{OP} + \frac{RP}{OP} = \frac{TQ}{OP} \cdot \frac{OQ}{OQ} + \frac{RP}{OP} \cdot \frac{PQ}{PQ}$$

$$= \frac{TQ}{OQ} \cdot \frac{OQ}{OP} + \frac{RP}{PQ} \cdot \frac{PQ}{OP}$$

From Fig. 13.2 and each of the following right triangles, note that

$$\text{Right } \triangle OTQ: \qquad \sin \theta = \frac{TQ}{OQ}$$

$$\text{Right } \triangle OPQ: \qquad \cos \phi = \frac{OQ}{OP}$$

$$\text{Right } \triangle PRQ: \qquad \cos \angle 3 = \frac{RP}{PQ}$$

$$\text{Right } \triangle OPQ: \qquad \sin \phi = \frac{PQ}{OP}$$

Then

$$\sin(\theta + \phi) = \frac{TQ}{OQ} \cdot \frac{OQ}{OP} + \frac{RP}{PQ} \cdot \frac{PQ}{OP}$$

$$= \sin \theta \cos \phi + \cos \angle 3 \sin \phi$$

$$= \sin \theta \cos \phi + \cos \theta \sin \phi \qquad (\angle 3 = \theta)$$

Therefore, we have the following.

$$\boxed{\sin(\theta + \phi) = \sin \theta \cos \phi + \cos \theta \sin \phi} \tag{12}$$

Using Fig. 13.2 and a similar procedure, we can show the following.

$$\boxed{\cos(\theta + \phi) = \cos \theta \cos \phi - \sin \theta \sin \phi} \tag{13}$$

To illustrate these formulas let's use the following examples.

**EXAMPLE 1**

Simplify $\cos(\theta + 90°)$ using Formula (13).

$$\cos(\theta + 90°) = \cos\theta \cos 90° - \sin\theta \sin 90°$$
$$= \cos\theta \quad (0) \quad - \sin\theta \quad (1)$$
$$= -\sin\theta$$

**EXAMPLE 2**

Simplify $\sin(\theta + 90°)$ using Formula (12).

$$\sin(\theta + 90°) = \sin\theta \cos 90° + \cos\theta \sin 90°$$
$$= \sin\theta \quad (0) \quad + \cos\theta \quad (1)$$
$$= \cos\theta$$

Before proceeding, we need to show the following identities.

$$\sin(-\theta) = -\sin\theta \tag{14}$$
$$\cos(-\theta) = \cos\theta \tag{15}$$

First, construct any angle $\theta$ and its negative, $-\theta$, as in Fig. 13.3. Label any point $P_1$ on the terminal side of $\theta$ with coordinates $(x_1, y_1)$. Let $r$ be the distance from $P_1$ to the origin. On the terminal side of $-\theta$, label a point at a distance of $r$ from the origin as $P_2$. Note that its coordinates are $(x_1, -y_1)$.

Then

$$\sin\theta = \frac{y_1}{r} \quad \text{and} \quad \sin(-\theta) = \frac{-y_1}{r} \qquad \text{Thus, } \sin(-\theta) = -\sin\theta$$

$$\cos\theta = \frac{x_1}{r} \quad \text{and} \quad \cos(-\theta) = \frac{x_1}{r} \qquad \text{Thus, } \cos(-\theta) = \cos\theta$$

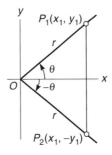

**Figure 13.3**

Similarly, the following may also be shown.

$$\tan(-\theta) = -\tan\theta \tag{16}$$
$$\cot(-\theta) = -\cot\theta \tag{17}$$
$$\sec(-\theta) = \sec\theta \tag{18}$$
$$\csc(-\theta) = -\csc\theta \tag{19}$$

To find the formula for $\sin(\theta - \phi)$, use Formula (12), and write $\theta - \phi$ as $\theta + (-\phi)$.

$$\sin[\theta + (-\phi)] = \sin\theta\cos(-\phi) + \cos\theta\sin(-\phi)$$
$$= \sin\theta\cos\phi - \cos\theta\sin\phi$$

Therefore:

$$\sin(\theta - \phi) = \sin\theta\cos\phi - \cos\theta\sin\phi \tag{20}$$

Likewise, to find the formula for $\cos(\theta - \phi)$, use Formula (13), and again write $\theta - \phi$ as $\theta + (-\phi)$.

$$\cos[\theta + (-\phi)] = \cos\theta\cos(-\phi) - \sin\theta\sin(-\phi)$$
$$= \cos\theta\cos\phi + \sin\theta\sin\phi$$

Therefore:

$$\cos(\theta - \phi) = \cos\theta\cos\phi + \sin\theta\sin\phi \tag{21}$$

To find a formula for $\tan(\theta + \phi)$, observe the following:

$$\tan(\theta + \phi) = \frac{\sin(\theta + \phi)}{\cos(\theta + \phi)} \qquad \text{(Identity 7)}$$

$$= \frac{\sin\theta\cos\phi + \cos\theta\sin\phi}{\cos\theta\cos\phi - \sin\theta\sin\phi} \qquad \text{(Identities 12 and 13)}$$

$$= \frac{\dfrac{\sin\theta\cos\phi}{\cos\theta\cos\phi} + \dfrac{\cos\theta\sin\phi}{\cos\theta\cos\phi}}{\dfrac{\cos\theta\cos\phi}{\cos\theta\cos\phi} - \dfrac{\sin\theta\sin\phi}{\cos\theta\cos\phi}} \qquad \begin{array}{l}\text{(Divide numerator and denominator}\\ \text{by } \cos\theta\cos\phi.)\end{array}$$

$$= \frac{\tan\theta + \tan\phi}{1 - \tan\theta\tan\phi} \qquad \text{(Identity 7)}$$

Therefore:

$$\tan(\theta + \phi) = \frac{\tan\theta + \tan\phi}{1 - \tan\theta\tan\phi} \tag{22}$$

Similarly, we can show the following.

$$\tan(\theta - \phi) = \frac{\tan\theta - \tan\phi}{1 + \tan\theta\tan\phi} \tag{23}$$

The cotangent, secant, and cosecant of the sum and the difference of two angles formulas are not as frequently used but may be derived similarly.

### EXAMPLE 3

Prove $\sin(x - 180°) = -\sin x$

$$\sin(x - 180°) = \sin x \cos 180° - \cos x \sin 180° \qquad \text{(Identity 20)}$$
$$= (\sin x)(-1) - (\cos x)(0)$$
$$= -\sin x$$

### EXAMPLE 4

Simplify $\cos 2\theta \cos 3\theta - \sin 2\theta \sin 3\theta$.
By Formula (13),

$$\cos 2\theta \cos 3\theta - \sin 2\theta \sin 3\theta = \cos (2\theta + 3\theta) = \cos 5\theta$$

## Exercises 13.2

*Prove each identity.*

**1.** $\sin(x + \pi) = -\sin x$

**2.** $\cos(x + 180°) = -\cos x$

**3.** $\sin(x + 2\pi) = \sin x$

**4.** $\cos(x + 2\pi) = \cos x$

**5.** $\tan(x + \pi) = \tan x$

**6.** $\tan(\pi - x) = -\tan x$

**7.** $\sin(90° - \theta) = \cos \theta$

**8.** $\cos\left(\dfrac{\pi}{2} - \theta\right) = \sin \theta$

**9.** $\cos\left(\dfrac{\pi}{2} + \theta\right) = -\sin \theta$

**10.** $\sin(90° + \theta) = \cos \theta$

**11.** $\tan(180° - \theta) = -\tan \theta$

**12.** $\cos(2\pi - \theta) = \cos \theta$

**13.** $\cos\left(\dfrac{\pi}{4} + \theta\right) = \dfrac{\cos \theta - \sin \theta}{\sqrt{2}}$

**14.** $\sin\left(\dfrac{\pi}{3} + \theta\right) = \dfrac{\sqrt{3} \cos \theta + \sin \theta}{2}$

**15.** $\tan(90° - x) = \cot x$

**16.** $\sec\left(\dfrac{\pi}{2} - x\right) = \csc x$

**17.** $\tan(x + 45°) = \dfrac{1 + \tan x}{1 - \tan x}$

**18.** $\tan(\theta + 270°) = -\cot \theta$

**19.** $\cos(x + y)\cos(x - y) = \cos^2 x - \sin^2 y$

**20.** $\sin(x + y)\sin(x - y) = \sin^2 x - \sin^2 y$

*Simplify each expression.*

**21.** $\cos M \cos N - \sin M \sin N$

**22.** $\sin 4C \cos C - \cos 4C \sin C$

**23.** $\sin \theta \cos 3\theta + \cos \theta \sin 3\theta$

**24.** $\cos 2\theta \cos \theta - \sin 2\theta \sin \theta$

**25.** $\cos 4\theta \cos 3\theta + \sin 4\theta \sin 3\theta$

**26.** $\sin 2\theta \cos 3\theta - \cos 2\theta \sin 3\theta$

**27.** $\dfrac{\tan 3\theta + \tan 2\theta}{1 - \tan 3\theta \tan 2\theta}$

**28.** $\dfrac{\tan \theta - \tan 2\theta}{1 + \tan \theta \tan 2\theta}$

**29.** $\sin(\theta + \phi) + \sin(\theta - \phi)$

**30.** $\cos(\theta + \phi) + \cos(\theta - \phi)$

**31.** $\sin(A + B)\cos B + \cos(A + B)\sin B$

**32.** $\cos(A + B)\cos B + \sin(A + B)\sin B$

## 13.3 DOUBLE- AND HALF-ANGLE FORMULAS

To find a formula for $\sin 2\theta$, use Formula (12) and let $\phi = \theta$.

$$\sin 2\theta = \sin(\theta + \theta) = \sin \theta \cos \theta + \cos \theta \sin \theta$$
$$= 2 \sin \theta \cos \theta$$

Therefore, we have the following formula.

$$\boxed{\sin 2\theta = 2 \sin \theta \cos \theta} \tag{24}$$

Likewise,

$$\cos 2\theta = \cos(\theta + \theta) = \cos \theta \cos \theta - \sin \theta \sin \theta \qquad \text{(Identity 13)}$$
$$= \cos^2 \theta - \sin^2 \theta$$

This formula has two other forms:

$$\cos^2 \theta - \sin^2 \theta = \cos^2 \theta - (1 - \cos^2 \theta) \qquad \text{(Identity 9)}$$
$$= 2 \cos^2 \theta - 1$$

and

$$\cos^2 \theta - \sin^2 \theta = (1 - \sin^2 \theta) - \sin^2 \theta \qquad \text{(Identity 9)}$$
$$= 1 - 2 \sin^2 \theta$$

Therefore:

$$\boxed{\begin{aligned} \cos 2\theta &= \cos^2 \theta - \sin^2 \theta \\ &= 2 \cos^2 \theta - 1 \\ &= 1 - 2 \sin^2 \theta \end{aligned}} \qquad \begin{aligned} &\text{(25a)} \\ &\text{(25b)} \\ &\text{(25c)} \end{aligned}$$

Similarly,

$$\tan 2\theta = \tan(\theta + \theta) = \frac{\tan \theta + \tan \theta}{1 - \tan \theta \tan \theta} \qquad \text{(Identity 22)}$$
$$= \frac{2 \tan \theta}{1 - \tan^2 \theta}$$

Therefore:

$$\boxed{\tan 2\theta = \frac{2 \tan \theta}{1 - \tan^2 \theta}} \tag{26}$$

**EXAMPLE 1**

Find $\sin 120°$ using Formula (24).

$$\sin 120° = \sin 2(60°) = 2 \sin 60° \cos 60°$$
$$= 2 \left(\frac{\sqrt{3}}{2}\right)\left(\frac{1}{2}\right)$$
$$= 0.8660$$

**EXAMPLE 2**

Find cos 120° using Formula (25a).

$$\cos 120° = \cos 2(60°) = \cos^2 60° - \sin^2 60°$$
$$= \left(\frac{1}{2}\right)^2 - \left(\frac{\sqrt{3}}{2}\right)^2$$
$$= \frac{1}{4} - \frac{3}{4} = -\frac{1}{2}$$

**EXAMPLE 3**

Simplify $2 \sin 3x \cos 3x$.

Using Formula (24), we obtain

$$2 \sin 3x \cos 3x = \sin 2(3x) = \sin 6x$$

**EXAMPLE 4**

Simplify $1 - 2 \cos^2 5x$.

Using Formula (25b), we have

$$1 - 2 \cos^2 5x = -\cos 2(5x) = -\cos 10x$$

**EXAMPLE 5**

Find $\sin 2\theta$ when $\cos \theta = \frac{4}{5}$ ($\theta$ in the fourth quadrant).

If $\cos \theta = \frac{4}{5}$ in the fourth quadrant (see Fig. 13.4), then $\sin \theta = -\frac{3}{5}$.

$$\sin 2\theta = 2 \sin \theta \cos \theta \qquad \text{(Identity 24)}$$
$$= 2\left(-\frac{3}{5}\right)\left(\frac{4}{5}\right)$$
$$= -\frac{24}{25}$$

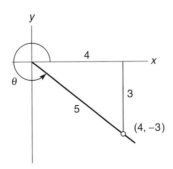

**Figure 13.4**

**EXAMPLE 6**

Find $\cos 2\theta$ when $\sin \theta = \dfrac{1}{2}$ .

Using Formula (25c), we have

$$\cos 2\theta = 1 - 2\sin^2 \theta$$

$$= 1 - 2\left(\frac{1}{2}\right)^2 = 1 - \frac{1}{2} = \frac{1}{2}$$

In an identity the choice of specific variables does not matter. For example, the identity

$$\sin^2 \theta + \cos^2 \theta = 1$$

may also be written in the form

$$\sin^2 m + \cos^2 m = 1$$

To find a formula for $\sin \dfrac{\theta}{2}$ , let's begin with Formula (25c) in the form

$$\cos 2m = 1 - 2\sin^2 m$$

Since this identity is true for all values of $m$, let $m = \dfrac{\theta}{2}$ and then solve for $\sin \dfrac{\theta}{2}$ .

$$\cos 2\left(\frac{\theta}{2}\right) = 1 - 2\sin^2 \frac{\theta}{2}$$

$$\cos \theta = 1 - 2\sin^2 \frac{\theta}{2}$$

$$2\sin^2 \frac{\theta}{2} = 1 - \cos \theta$$

$$\sin^2 \frac{\theta}{2} = \frac{1 - \cos \theta}{2}$$

$$\boxed{\sin \frac{\theta}{2} = \pm \sqrt{\frac{1 - \cos \theta}{2}}} \qquad (27)$$

To find a formula for $\cos \dfrac{\theta}{2}$ , let's begin with Formula (25b) in the form

$$\cos 2m = 2\cos^2 m - 1$$

Let $m = \dfrac{\theta}{2}$ and then solve for $\cos \dfrac{\theta}{2}$ .

$$\cos 2\left(\frac{\theta}{2}\right) = 2\cos^2 \frac{\theta}{2} - 1$$

$$\cos \theta = 2\cos^2 \frac{\theta}{2} - 1$$

$$\frac{1 + \cos \theta}{2} = \cos^2 \frac{\theta}{2}$$

$$\boxed{\cos \frac{\theta}{2} = \pm \sqrt{\frac{1 + \cos \theta}{2}}} \qquad \text{(28)}$$

*Note:* In both Formulas (27) and (28), the sign used depends on the quadrant in which $\frac{\theta}{2}$ lies.

Now find $\tan \frac{\theta}{2}$.

$$\tan \frac{\theta}{2} = \frac{\sin \frac{\theta}{2}}{\cos \frac{\theta}{2}} \qquad \text{(Identity 7)}$$

Multiply numerator and denominator by $2 \sin \frac{\theta}{2}$.

$$\tan \frac{\theta}{2} = \frac{2 \sin^2 \frac{\theta}{2}}{2 \sin \frac{\theta}{2} \cos \frac{\theta}{2}}$$

Note that

$$\cos 2\left(\frac{\theta}{2}\right) = 1 - 2 \sin^2 \frac{\theta}{2}$$

That is,

$$2 \sin^2 \frac{\theta}{2} = 1 - \cos \theta$$

Also note that

$$\sin 2\left(\frac{\theta}{2}\right) = 2 \sin \frac{\theta}{2} \cos \frac{\theta}{2}$$

That is,

$$\sin \theta = 2 \sin \frac{\theta}{2} \cos \frac{\theta}{2}$$

Thus:

$$\boxed{\tan \frac{\theta}{2} = \frac{1 - \cos \theta}{\sin \theta}} \qquad \text{(29)}$$

**EXAMPLE 7**

Find $\cos\dfrac{\theta}{2}$ when $\sin\theta = -\dfrac{3}{5}$ ($\theta$ in the third quadrant).

If $\sin\theta = -\dfrac{3}{5}$ in the third quadrant as in Fig. 13.5, then $\cos\theta = -\dfrac{4}{5}$.

$$\cos\frac{\theta}{2} = -\sqrt{\frac{1+\cos\theta}{2}}$$

$$= -\sqrt{\frac{1+\left(-\dfrac{4}{5}\right)}{2}}$$

$$= -\sqrt{\frac{1}{10}} \quad\text{or}\quad -\frac{\sqrt{10}}{10}$$

The sign is negative because if $180° < \theta < 270°$, then $90° < \dfrac{\theta}{2} < 135°$ where $\cos\dfrac{\theta}{2}$ is negative.

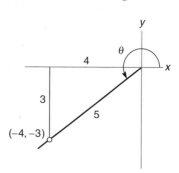

**Figure 13.5**

**EXAMPLE 8**

Prove $\dfrac{1-\cos 2x}{\sin 2x} = \tan x.$

$$\frac{1-\cos 2x}{\sin 2x} = \frac{1-(1-2\sin^2 x)}{2\sin x\cos x} \qquad \text{(Identities 25c and 24)}$$

$$= \frac{2\sin^2 x}{2\sin x\cos x}$$

$$= \frac{\sin x}{\cos x}$$

$$= \tan x \qquad\qquad\qquad \text{(Identity 7)}$$

Therefore, $\dfrac{1-\cos 2x}{\sin 2x} = \tan x.$

The following trigonometric formulas are given without proof for reference.

$$\sin \theta + \sin \phi = 2 \sin \left(\frac{\theta + \phi}{2}\right) \cos \left(\frac{\theta - \phi}{2}\right) \tag{30}$$

$$\sin \theta - \sin \phi = 2 \cos \left(\frac{\theta + \phi}{2}\right) \sin \left(\frac{\theta - \phi}{2}\right) \tag{31}$$

$$\cos \theta + \cos \phi = 2 \cos \left(\frac{\theta + \phi}{2}\right) \cos \left(\frac{\theta - \phi}{2}\right) \tag{32}$$

$$\cos \theta - \cos \phi = -2 \sin \left(\frac{\theta + \phi}{2}\right) \sin \left(\frac{\theta - \phi}{2}\right) \tag{33}$$

$$\sin(\theta + \phi) + \sin(\theta - \phi) = 2 \sin \theta \cos \phi \tag{34}$$
$$\sin(\theta + \phi) - \sin(\theta - \phi) = 2 \cos \theta \sin \phi \tag{35}$$
$$\cos(\theta + \phi) + \cos(\theta - \phi) = 2 \cos \theta \cos \phi \tag{36}$$
$$\cos(\theta + \phi) - \cos(\theta - \phi) = -2 \sin \theta \sin \phi \tag{37}$$

## Exercises 13.3

*Simplify each expression.*

**1.** $2 \sin \dfrac{x}{4} \cos \dfrac{x}{4}$

**2.** $20 \sin^2 x \cos^2 x$

**3.** $1 - 2 \sin^2 3x$

**4.** $\sqrt{\dfrac{1 - \cos 6\theta}{2}}$

**5.** $\sqrt{\dfrac{1 + \cos \dfrac{\theta}{4}}{2}}$

**6.** $\cos 2x + 2 \sin^2 x$

**7.** $\cos^2 \dfrac{x}{6} - \sin^2 \dfrac{x}{6}$

**8.** $10 \cos^2 44° - 10 \sin^2 44°$

**9.** $20 \sin 4\theta \cos 4\theta$

**10.** $2 \sin 3x \cos 3x$

**11.** $-\sqrt{\dfrac{1 + \cos 250°}{2}}$

**12.** $\sqrt{\dfrac{1 - \cos 16\theta}{2}}$

**13.** $4 - 8 \sin^2 \theta$

**14.** $1 - 2 \sin^2 7t$

**15.** $100 \sin 30t \cos 30t$

**16.** $15 \sin \dfrac{x}{6} \cos \dfrac{x}{6}$

**17.** Find $\sin 2\theta$ when $\cos \theta = \dfrac{3}{5}$ ($\theta$ in the first quadrant).

**18.** Find $\cos 2\theta$ when $\sin \theta = \dfrac{3}{5}$ ($\theta$ in the second quadrant).

**19.** Find $\tan 2\theta$ when $\cos \theta = -\dfrac{12}{13}$ ($\theta$ in the third quadrant).

**20.** Find $\sin \dfrac{\theta}{2}$ when $\sin \theta = -\dfrac{5}{13}$ ($\theta$ in the fourth quadrant).

**21.** Find $\cos \dfrac{\theta}{2}$ when $\sin \theta = -\dfrac{2}{3}$ ($\theta$ in the third quadrant).

**22.** Find $\tan \dfrac{\theta}{2}$ when $\tan \theta = -\dfrac{3}{4}$ ($\theta$ in the fourth quadrant).

*Prove each identity.*

**23.** $(\sin x + \cos x)^2 = 1 + \sin 2x$

**24.** $\dfrac{2 \tan x}{\sin 2x} = 1 + \tan^2 x$

**25.** $\cos^4 x - \sin^4 x = \cos 2x$

**26.** $\sin 2x = \dfrac{2 \tan x}{1 + \tan^2 x}$

**27.** $\dfrac{1 - \tan^2 x}{1 + \tan^2 x} = \cos 2x$

**28.** $2 \tan x \csc 2x = \sec^2 x$

**29.** $\cot 2x = \dfrac{\cot^2 x - 1}{2 \cot x}$

**30.** $\sec 2x = \dfrac{\sec^2 x}{2 - \sec^2 x}$

**31.** $\tan x + \cot 2x = \csc 2x$

**32.** $\tan(x + 45°) + \tan(x - 45°) = 2 \tan 2x$

**33.** $\sin^2 \dfrac{x}{2} = \dfrac{\sec x - 1}{2 \sec x}$

**34.** $2 \cos^2 \dfrac{\theta}{2} = \dfrac{1 + \sec \theta}{\sec \theta}$

**35.** $\sec^2 \dfrac{x}{2} = \dfrac{2}{1 + \cos x}$

**36.** $\csc^2 \dfrac{x}{2} = \dfrac{2 \sec x}{\sec x - 1}$

**37.** $\tan \dfrac{x}{2} = \dfrac{\sin x}{1 + \cos x}$

**38.** $\tan \left( \dfrac{\theta}{2} + \dfrac{\pi}{4} \right) = \dfrac{1 + \sin \theta}{\cos \theta}$

**39.** $2 \cos \dfrac{x}{2} = (1 + \cos x)\sec \dfrac{x}{2}$

**40.** $\dfrac{\sin 3x}{\sin x} - \dfrac{\cos 3x}{\cos x} = 2$

**41.** $\tan \dfrac{x}{2} + \cot \dfrac{x}{2} = 2 \csc x$

**42.** $\dfrac{\sin^3 x - \cos^3 x}{\sin x - \cos x} = 1 + \dfrac{1}{2} \sin 2x$

**43.** $\left( \sin \dfrac{\theta}{2} - \cos \dfrac{\theta}{2} \right)^2 = 1 - \sin \theta$

**44.** $8 \sin^2 \dfrac{x}{2} \cos^2 \dfrac{x}{2} = 1 - \cos 2x$

**45.** $\sin 3x = 3 \sin x - 4 \sin^3 x$

**46.** $\cos 3x = 4 \cos^3 x - 3 \cos x$

## 13.4 TRIGONOMETRIC EQUATIONS

D.5, D.6

A trigonometric equation that is not an identity may have a number of solutions. The solutions may be given in terms of degrees or radians. There are no general procedures to follow when solving a trigonometric equation. You may try algebraic methods or use identities to write the equation in terms of a single trigonometric function.

**EXAMPLE 1**

Solve $2 \sin \theta + 1 = 0$ for $0° \le \theta < 360°$.

$$2 \sin \theta + 1 = 0$$

$$\sin \theta = -\dfrac{1}{2}$$

$$\theta = 210°, 330° \quad \text{(Sine is negative in third and fourth quadrants.)}$$

*Note:* If we did not have the restriction on $\theta$, $0° \le \theta < 360°$, there would be an infinite number of solutions; that is, angles that are coterminal with $210°$ and $330°$.

**EXAMPLE 2**

Solve $\sin 2\theta - \sin \theta = 0$ for $0 \le \theta < 2\pi$.

Replace $\sin 2\theta$ by $2 \sin \theta \cos \theta$ [Formula (24)] and factor.

$$2 \sin \theta \cos \theta - \sin \theta = 0$$

$$\sin \theta (2 \cos \theta - 1) = 0$$

$$\sin \theta = 0 \quad \text{or} \quad \cos \theta = \frac{1}{2}$$

$$\theta = 0, \pi \quad \text{or} \quad \theta = \frac{\pi}{3}, \frac{5\pi}{3}$$

**EXAMPLE 3**

Solve $2 \cos^2 x + 5 \sin x = 4$ for $0° \le x < 360°$.

Replace $\cos^2 x$ by $1 - \sin^2 x$ [from Formula (9)].

$$2(1 - \sin^2 x) + 5 \sin x = 4$$

$$0 = 2 \sin^2 x - 5 \sin x + 2$$

$$0 = (2 \sin x - 1)(\sin x - 2) \quad \text{(Factor.)}$$

$$\sin x = \frac{1}{2} \quad \text{or} \quad \sin x = 2$$

$$x = 30°, 150° \quad \text{No solution because } -1 \le \sin x \le 1$$

**EXAMPLE 4**

Solve $2 \cos^2 x + 6 \cos x + 3 = 0$ for $0° \le x < 360°$.

Since the left-hand side does not factor, we must use the quadratic formula with $a = 2$, $b = 6$, and $c = 3$.

$$\cos x = \frac{-b \pm \sqrt{b^2 - 4ac}}{2a}$$

$$\cos x = \frac{-6 \pm \sqrt{6^2 - 4(2)(3)}}{2(2)}$$

$$= \frac{-6 \pm \sqrt{12}}{4}$$

$$= \frac{-6 \pm 2\sqrt{3}}{4}$$

$$= \frac{-3 \pm \sqrt{3}}{2}$$

Changing each solution to a decimal value, we have

$$\cos x = \frac{-3 + \sqrt{3}}{2} = -0.6340 \quad \text{or} \quad \cos x = \frac{-3 - \sqrt{3}}{2} = -2.366$$

$$x = 129.3°, 230.7° \quad \text{No solution because } -1 \le \cos x \le 1$$

**EXAMPLE 5**

Solve $\tan 2\theta + \cot 2\theta + 2 = 0$ for $0 \leq \theta < 2\pi$.

Replace $\cot 2\theta$ by $\dfrac{1}{\tan 2\theta}$ [from Formula (4)].

$$\tan 2\theta + \frac{1}{\tan 2\theta} + 2 = 0$$

$$\tan^2 2\theta + 1 + 2 \tan 2\theta = 0 \qquad \text{(Multiply each side by } \tan 2\theta.)$$

$$(\tan 2\theta + 1)^2 = 0$$

$$\tan 2\theta + 1 = 0$$

$$\tan 2\theta = -1$$

Hence,

$$2\theta = \frac{3\pi}{4}, \frac{7\pi}{4}, \frac{11\pi}{4}, \frac{15\pi}{4}$$

$$\theta = \frac{3\pi}{8}, \frac{7\pi}{8}, \frac{11\pi}{8}, \frac{15\pi}{8}$$

*Note:* The restriction $0 \leq \theta < 2\pi$ is on $\theta$ not $2\theta$.

**EXAMPLE 6**

Solve $\cos x = \sin \dfrac{x}{2}$ for $0 \leq x < 2\pi$.

Replace $\sin \dfrac{x}{2}$ by $\pm \sqrt{\dfrac{1 - \cos x}{2}}$ [from Formula (27)].

$$\cos x = \pm \sqrt{\frac{1 - \cos x}{2}}$$

$$\cos^2 x = \frac{1 - \cos x}{2} \qquad \text{(Square each side.)}$$

$$2 \cos^2 x + \cos x - 1 = 0$$

$$(2 \cos x - 1)(\cos x + 1) = 0 \qquad \text{(Factor.)}$$

$$\cos x = \frac{1}{2} \qquad \text{or} \quad \cos x = -1$$

$$x = \frac{\pi}{3}, \frac{5\pi}{3} \quad \text{or} \qquad x = \pi$$

Since we squared each side, we *must* check for possible extraneous roots.

(a) $\cos \dfrac{\pi}{3} = \sin \dfrac{\dfrac{\pi}{3}}{2}$ 　　 (b) $\cos \dfrac{5\pi}{3} = \sin \dfrac{\dfrac{5\pi}{3}}{2}$ 　　 (c) $\cos \pi = \sin \dfrac{\pi}{2}$

　　 $\cos \dfrac{\pi}{3} = \sin \dfrac{\pi}{6}$ 　　　　 $\cos \dfrac{5\pi}{3} = \sin \dfrac{5\pi}{6}$ 　　　　 $-1 \neq 1$

　　 $\dfrac{1}{2} = \dfrac{1}{2}$ 　　　　　　 $\dfrac{1}{2} = \dfrac{1}{2}$ 　　　　　 $\pi$ is not a solution.

$\dfrac{\pi}{3}$ is a solution. 　　 $\dfrac{5\pi}{3}$ is a solution.

Therefore, the solutions are $\dfrac{\pi}{3}$ and $\dfrac{5\pi}{3}$.

## Exercises 13.4

*Solve each trigonometric equation for $0° \leq x < 360°$.*

1. $\cos x - 1 = 0$

2. $2 \sin x + \sqrt{3} = 0$

3. $\tan x - 1 = 0$

4. $2 \sin^2 x - 1 = 0$

5. $4 \cos^2 x - 3 = 0$

6. $\tan^2 x - 3 = 0$

7. $\sin 2x = 1$

8. $2 \cos 3x + 1 = 0$

9. $3 \tan^2 3x - 1 = 0$

10. $4 \sin^2 2x = 3$

11. $\sin 2x + \cos x = 0$

12. $\cos 2x - \sin x = 0$

13. $\sin^2 x + \sin x = 0$

14. $\cos x \tan x + \cos x = 0$

15. $2 \sin^2 x = 1 - 2 \sin x$

16. $\cos^2 x - 3 \cos x + 1 = 0$

*Solve each trigonometric equation for $0 \leq x < 2\pi$.*

17. $2 \cos^2 x + \sin x = 1$

18. $\tan x + 2 \cos x = \sec x$

19. $\cos^2 x - \cos x \sec x = 0$

20. $\sin^2 x + \sin x \cos x = 0$

21. $2 \sin^2 x + 2 \cos 2x = 1$

22. $6 \sin^2 x + \cos 2x = 4$

23. $2 \sin^2 2x - \sin 2x - 1 = 0$

24. $\cos 2x - 2 \sin^2 2x + 1 = 0$

25. $4 \tan^2 x = 3 \sec^2 x$

26. $2 \sin x - \tan x = 0$

27. $4 \sin 2x \cos 2x = 1$

28. $\cos^2 3x - \sin^2 3x = 1$

29. $\cos x = \cos \dfrac{x}{2}$

30. $\sin x = \cos \dfrac{x}{2}$

31. $\cos \dfrac{x}{2} = 1 + \cos x$

32. $\sin \dfrac{x}{2} = 1 - \cos x$

33. $1 + \cos^2 \dfrac{x}{2} = 2 \cos x$

34. $1 + \sin^2 \dfrac{x}{2} = \cos x$

## 13.5  INVERSE TRIGONOMETRIC RELATIONS

D.6

In Section 4.1 we defined a relation as a set of ordered pairs of the form $(x, y)$. In Section 9.2 we defined the inverse of a given equation to be the resulting equation when the variables $x$ and $y$ are interchanged. Recall that

$$\text{the inverse of } \quad y = b^x \quad \text{is} \quad x = b^y$$

Similar relationships are as follows:

| *The inverse of:* | *Is:* |
|---|---|
| $y = x^2$ | $x = y^2$ |
| $y = 3x^2 + 4x + 7$ | $x = 3y^2 + 4y + 7$ |
| $y = \dfrac{x + 4}{3x - 1}$ | $x = \dfrac{y + 4}{3y - 1}$ |
| $y = \log_b x$ | $x = \log_b y$ |

Likewise, each basic trigonometric equation has an inverse:

| The inverse of: | Is: |
|---|---|
| $y = \sin x$ | $x = \sin y$ |
| $y = \cos x$ | $x = \cos y$ |
| $y = \tan x$ | $x = \tan y$ |
| $y = \cot x$ | $x = \cot y$ |
| $y = \sec x$ | $x = \sec y$ |
| $y = \csc x$ | $x = \csc y$ |

Also, recall from the study of logarithms that we found it necessary to write the logarithmic equation solved for $y$ as follows.

| | |
|---|---|
| Exponential equation: | $y = b^x$ |
| Inverse equation (logarithmic equation in exponential form): | $x = b^y$ |
| Inverse equation (logarithmic equation in logarithmic form): | $y = \log_b x$ |

There are two common forms of the inverse trigonometric equations solved for $y$.

|  |  | Solved for y: | |
|---|---|---|---|
| *The inverse of:* | *Is:* | *Is:* | *Is:* |
| $y = \sin x$ | $x = \sin y$ | $y = \arcsin x$* | $y = \sin^{-1} x$† |
| $y = \cos x$ | $x = \cos y$ | $y = \arccos x$ | $y = \cos^{-1} x$ |
| $y = \tan x$ | $x = \tan y$ | $y = \arctan x$ | $y = \tan^{-1} x$ |
| $y = \cot x$ | $x = \cot y$ | $y = \text{arccot}\, x$ | $y = \cot^{-1} x$ |
| $y = \sec x$ | $x = \sec y$ | $y = \text{arcsec}\, x$ | $y = \sec^{-1} x$ |
| $y = \csc x$ | $x = \csc y$ | $y = \text{arccsc}\, x$ | $y = \csc^{-1} x$ |

**EXAMPLE 1**

What is the meaning of each equation?
(a) $y = \arctan x$
(b) $y = \arccos 3x$
(c) $y = 4 \,\text{arccsc}\, 5x$

(a) $y$ is the angle whose tangent is $x$.
(b) $y$ is the angle whose cosine is $3x$.
(c) $y$ is four times the angle whose cosecant is $5x$.

*This is read, "$y$ equals the arcsine of $x$" and means that $y$ is the angle whose sine is $x$.
†This notation will not be used here because of the confusion caused by the fact that $-1$ is not an exponent.

Remember, $x = \sin y$ and $y = \arcsin x$ express the same relationship. The first form expresses the relationship in terms of the function (sine) of the angle; the second form expresses the relationship in terms of the angle itself.

**EXAMPLE 2**

Given the equation $y = \arcsin x$, find $y$ when $x = \dfrac{1}{2}$. (Give the answer in degrees.)

Substituting $x = \dfrac{1}{2}$, we have

$$y = \arcsin \frac{1}{2}$$

which means that $y$ is the angle whose sine is $\dfrac{1}{2}$. We know that $\sin 30° = \dfrac{1}{2}$, so $y = 30°$.
But we also know that

$$\sin 150° = \frac{1}{2} \quad \text{so} \quad y = 150°$$

$$\sin 390° = \frac{1}{2} \quad \text{so} \quad y = 390° \qquad (390° = 30° + 360°)$$

$$\sin 510° = \frac{1}{2} \quad \text{so} \quad y = 510° \qquad (510° = 150° + 360°)$$

$$\begin{array}{cc} \cdot & \cdot \\ \cdot & \cdot \\ \cdot & \cdot \end{array}$$

$$\sin(-210°) = \frac{1}{2} \quad \text{so} \quad y = -210°$$

$$\sin(-330°) = \frac{1}{2} \quad \text{so} \quad y = -330°$$

$$\begin{array}{cc} \cdot & \cdot \\ \cdot & \cdot \\ \cdot & \cdot \end{array}$$

Thus, there are infinitely many angles whose sine is $\dfrac{1}{2}$. We saw this when we graphed $y = \sin x$. That is,

$$y = 30° + n \cdot 360°$$
$$y = 150° + n \cdot 360°$$

for every integer $n$.

**EXAMPLE 3**

Given the equation $y = \arctan x$, find $y$ when $x = -1$ for $0 \leq y < 2\pi$.
Substituting $x = -1$, we have

$$y = \arctan(-1)$$

which means that $y$ is the angle whose tangent is $-1$. We know that

$$\tan \frac{3\pi}{4} = -1 \quad \text{so} \quad y = \frac{3\pi}{4}$$

and

$$\tan\frac{7\pi}{4} = -1 \quad \text{so} \quad y = \frac{7\pi}{4}$$

for $0 \le y < 2\pi$.

**EXAMPLE 4**

Find arccos $\dfrac{\sqrt{3}}{2}$ for the smallest positive angle in degrees.

Let $y = \arccos \dfrac{\sqrt{3}}{2}$, which means that $y$ is the (smallest positive) angle whose cosine is $\dfrac{\sqrt{3}}{2}$. We know that

$$\cos 30° = \frac{\sqrt{3}}{2}$$

Thus, arccos $\dfrac{\sqrt{3}}{2} = 30°$ (for the smallest positive angle).

**EXAMPLE 5**

Find arcsec$(-2)$ for the smallest positive angle in radians.

Let $y = \text{arcsec}(-2)$, which means that $y$ is the (smallest positive) angle whose secant is $-2$. We know that

$$\sec 120° = -2$$

Therefore, arcsec$(-2) = \dfrac{2\pi}{3}$ ( for the smallest positive angle).

**EXAMPLE 6**

Find arccot $\dfrac{1}{\sqrt{3}}$ for all angles in radians.

Let $y = \text{arccot} \dfrac{1}{\sqrt{3}}$, which means that $y$ is the angle whose cotangent is $\dfrac{1}{\sqrt{3}}$. We know that

$$\cot\frac{\pi}{3} = \frac{1}{\sqrt{3}} \quad \text{and} \quad \cot\frac{4\pi}{3} = \frac{1}{\sqrt{3}}$$

for $0 \le y < 2\pi$. Thus,

$$\text{arccot}\frac{1}{\sqrt{3}} = \begin{cases} \dfrac{\pi}{3} + 2n\pi \\[2mm] \dfrac{4\pi}{3} + 2n\pi \end{cases}$$

for every integer $n$.

**EXAMPLE 7**

Solve the equation $y = \cos 2x$ for $x$.

The equation $y = \cos 2x$ is equivalent to

$$\arccos y = 2x$$

So

$$x = \frac{1}{2} \arccos y$$

**EXAMPLE 8**

Solve the equation $2y = \arctan 3x$ for $x$.

The equation $2y = \arctan 3x$ is equivalent to

$$\tan 2y = 3x$$

So

$$x = \frac{1}{3} \tan 2y$$

**EXAMPLE 9**

Solve the equation $y = \frac{1}{3} \text{arcsec } 2x$ for $x$.

First, multiply both sides by 3.

$$3y = \text{arcsec } 2x$$

The equation $3y = \text{arcsec } 2x$ is equivalent to

$$\sec 3y = 2x$$

Thus,

$$x = \frac{1}{2} \sec 3y$$

## Exercises 13.5

*Write the meaning of each equation.*

**1.** $y = \arcsin x$      **2.** $y = \text{arcsec } x$      **3.** $y = \text{arccot } 4x$

**4.** $y = \arccos 2x$      **5.** $y = 3 \text{ arccsc } \frac{1}{2} x$      **6.** $y = \frac{1}{2} \arctan 3x$

*Solve each equation for $0 \le y < 2\pi$.*

**7.** $y = \arcsin\left(\frac{\sqrt{3}}{2}\right)$      **8.** $y = \arccos\left(\frac{1}{2}\right)$      **9.** $y = \arctan 1$

**10.** $y = \text{arccot}(-1)$      **11.** $y = \text{arcsec}\left(-\frac{2}{\sqrt{3}}\right)$      **12.** $y = \text{arccsc } \sqrt{2}$

**13.** $y = \arccos\left(-\frac{1}{2}\right)$      **14.** $y = \arcsin 0$

*Solve each equation for 0° ≤ y < 360°.*

**15.** $y = \arccos\left(-\dfrac{1}{\sqrt{2}}\right)$

**16.** $y = \arctan(-\sqrt{3})$

**17.** $y = \arcsin\left(-\dfrac{1}{\sqrt{2}}\right)$

**18.** $y = \text{arccot }\sqrt{3}$

**19.** $y = \text{arcsec }1$

**20.** $y = \arcsin(-1)$

**21.** $y = \arctan(1.963)$

**22.** $y = \arccos(-0.9063)$

*Find the smallest positive angle in degrees for each expression.*

**23.** $\arcsin\left(\dfrac{1}{2}\right)$

**24.** $\arccos 0$

**25.** $\text{arccot}\left(-\dfrac{1}{\sqrt{3}}\right)$

**26.** $\arctan \sqrt{3}$

**27.** $\text{arcsec }\sqrt{2}$

**28.** $\text{arccsc}(-\sqrt{2})$

**29.** $\arccos\left(-\dfrac{\sqrt{3}}{2}\right)$

**30.** $\arcsin\left(\dfrac{1}{\sqrt{2}}\right)$

*Find the smallest positive angle in radians for each expression.*

**31.** $\arctan(-\sqrt{3})$

**32.** $\text{arccsc }2$

**33.** $\arccos\left(-\dfrac{\sqrt{3}}{2}\right)$

**34.** $\text{arcsec}\left(\dfrac{2}{\sqrt{3}}\right)$

**35.** $\arcsin(-1)$

**36.** $\arctan 0$

**37.** $\text{arcsec}(-1)$

**38.** $\arcsin\left(\dfrac{1}{\sqrt{2}}\right)$

*Find all angles in degrees for each expression.*

**39.** $\arcsin\left(\dfrac{1}{\sqrt{2}}\right)$

**40.** $\arctan\left(\dfrac{1}{\sqrt{3}}\right)$

**41.** $\text{arccot }\sqrt{3}$

**42.** $\arcsin\left(-\dfrac{\sqrt{3}}{2}\right)$

**43.** $\text{arcsec}\left(\dfrac{2}{\sqrt{3}}\right)$

**44.** $\arctan(-\sqrt{3})$

**45.** $\arcsin(0.7431)$

**46.** $\arccos(-0.9397)$

*Find all angles in radians for each expression.*

**47.** $\arccos\left(\dfrac{1}{2}\right)$

**48.** $\arctan 1$

**49.** $\arccos 1$

**50.** $\arcsin 1$

**51.** $\text{arcsec }2$

**52.** $\arccos\left(-\dfrac{1}{\sqrt{2}}\right)$

**53.** $\arcsin(-0.8572)$

**54.** $\text{arccot}(-0.8195)$

*Solve each equation for x.*

**55.** $y = \sin 3x$

**56.** $y = \tan 4x$

**57.** $y = 4\cos x$

**58.** $y = 3\sec x$

**59.** $y = 5\tan\dfrac{x}{2}$

**60.** $y = \dfrac{1}{2}\cos 3x$

**61.** $y = \dfrac{3}{2}\cot\dfrac{x}{4}$

**62.** $y = \dfrac{5}{2}\sin\dfrac{2x}{3}$

**63.** $y = 3\sin(x - 1)$

**64.** $y = 4\tan(2x + 1)$

**65.** $y = \dfrac{1}{2}\cos(3x + 1)$

**66.** $y = \dfrac{1}{3}\sec(1 - 4x)$

## 13.6 INVERSE TRIGONOMETRIC FUNCTIONS

To graph the inverse trigonometric relation

$$y = \arcsin x$$

first solve for $x$; that is,

$$x = \sin y$$

Then plot ordered pairs of solutions of this equation as in Fig. 13.6. In a similar manner, we can graph all six inverse trigonometric relations. They are shown in Fig. 13.7.

As you can see from the graphs, each $x$-value in each domain corresponds to (infinitely) many values of $y$. Thus, none of the inverse trigonometric relations is a function.

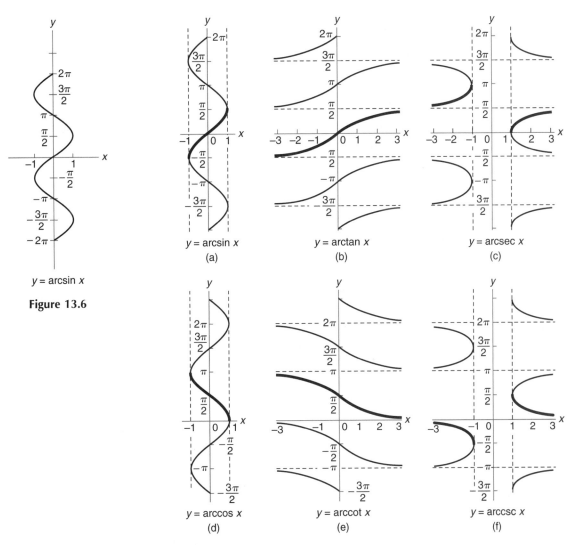

$y = \arcsin x$

**Figure 13.6**

$y = \arcsin x$
(a)

$y = \arctan x$
(b)

$y = \text{arcsec } x$
(c)

$y = \arccos x$
(d)

$y = \text{arccot } x$
(e)

$y = \text{arccsc } x$
(f)

**Figure 13.7**

However, if we restrict the *y*-values of each inverse trigonometric relation, we can define an inverse that is also a function. While this could be done in any of several ways, it is customary to restrict the *y*-values as follows.

---

**INVERSE TRIGONOMETRIC FUNCTIONS**

$y = \text{Arcsin } x \qquad -\dfrac{\pi}{2} \leq y \leq \dfrac{\pi}{2}$

$y = \text{Arccos } x \qquad 0 \leq y \leq \pi$

$y = \text{Arctan } x \qquad -\dfrac{\pi}{2} < y < \dfrac{\pi}{2}$

$y = \text{Arccot } x \qquad 0 < y < \pi$

$y = \text{Arcsec } x \qquad 0 \leq y \leq \pi \quad y \neq \dfrac{\pi}{2}$

$y = \text{Arccsc } x \qquad -\dfrac{\pi}{2} \leq y \leq \dfrac{\pi}{2} \quad y \neq 0$

---

*Note:* The inverse trigonometric *functions* are *capitalized* to distinguish them from the inverse trigonometric relations.

Look once again at the graphs of the inverse trigonometric relations in Fig. 13.7. The extra thick lines indicate the portions of the graphs that correspond to the inverse trigonometric *functions*. These are also shown in Fig. 13.8.

*Note:* The three inverse trigonometric functions on calculators are programmed to these same restricted ranges. When using a calculator to find the value of Arccot $x$, Arcsec $x$, or Arccsc $x$, use the following:

**1.** $\text{Arccot } x = \begin{cases} \text{Arctan } \dfrac{1}{x} & \text{if } x > 0 \\[2ex] \pi + \text{Arctan } \dfrac{1}{x} & \text{if } x < 0 \end{cases}$

**2.** $\text{Arcsec } x = \text{Arccos } \dfrac{1}{x} \quad$ where $x \geq 1 \quad$ or $\quad x \leq -1$

**3.** $\text{Arccsc } x = \text{Arcsin } \dfrac{1}{x} \quad$ where $x \geq 1 \quad$ or $\quad x \leq -1$

**EXAMPLE 1**

Find $\text{Arcsin}\left(\dfrac{1}{2}\right)$.

$$\text{Arcsin}\left(\dfrac{1}{2}\right) = \dfrac{\pi}{6}$$

This is the only value in the defined range of $-\dfrac{\pi}{2} \leq y \leq \dfrac{\pi}{2}$.

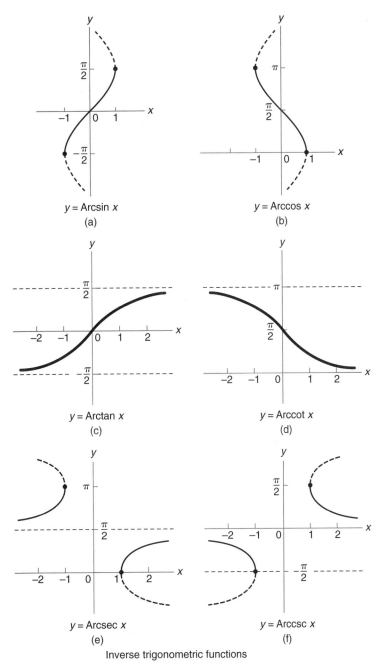

$y = \text{Arcsin } x$
(a)

$y = \text{Arccos } x$
(b)

$y = \text{Arctan } x$
(c)

$y = \text{Arccot } x$
(d)

$y = \text{Arcsec } x$
(e)

$y = \text{Arccsc } x$
(f)

Inverse trigonometric functions

**Figure 13.8**

**EXAMPLE 2**

Find Arctan$(-1)$.

$$\text{Arctan}(-1) = -\frac{\pi}{4}$$

This is the only value in the defined range of $-\dfrac{\pi}{2} < y < \dfrac{\pi}{2}$.

**EXAMPLE 3**

Find $\text{Arccos}\left(-\dfrac{1}{2}\right)$.

$$\text{Arccos}\left(-\frac{1}{2}\right) = \frac{2\pi}{3}$$

This is the only value in the defined range of $0 \le y \le \pi$.

**EXAMPLE 4**

Find $\tan[\text{Arccos}(-1)]$.

$$\tan[\text{Arccos}(-1)] = \tan \pi = 0$$

**EXAMPLE 5**

Find $\cos(\text{Arcsec } 2)$.

$$\cos(\text{Arcsec } 2) = \cos \frac{\pi}{3} = \frac{1}{2}$$

**EXAMPLE 6**

Find $\sin\left[\text{Arctan}\left(-\dfrac{1}{\sqrt{3}}\right)\right]$.

$$\sin\left[\text{Arctan}\left(-\frac{1}{\sqrt{3}}\right)\right] = \sin\left(-\frac{\pi}{6}\right) = -\frac{1}{2}$$

**EXAMPLE 7**

Find an algebraic expression for $\sin(\text{Arccos } x)$.

Let $\theta = \text{Arccos } x$. Then

$$\cos \theta = x = \frac{x}{1}$$

Draw a right triangle with $\theta$ as an acute angle, $x$ as the adjacent side, and 1 as the hypotenuse as in Fig. 13.9(a).

Using the Pythagorean theorem, we have

$$c^2 = a^2 + b^2$$
$$1^2 = x^2 + (\text{side opposite } \theta)^2$$

and

$$\text{side opposite } \theta = \sqrt{1 - x^2} \qquad [\text{Fig. 13.9(b)}]$$

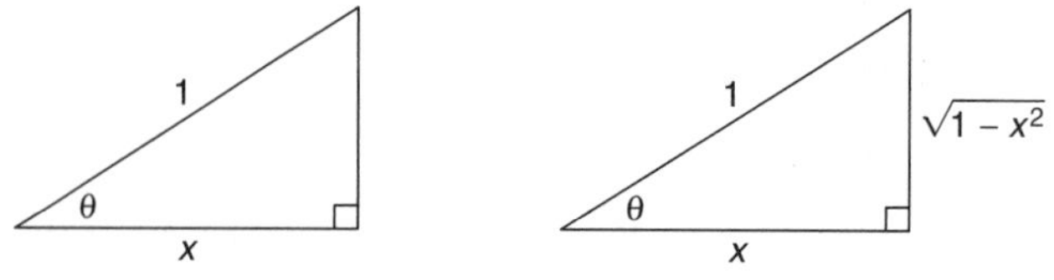

(a)                    (b)

**Figure 13.9**

Now we see that

$$\sin(\text{Arccos } x) = \sin \theta$$
$$= \frac{\text{side opposite } \theta}{\text{hypotenuse}}$$
$$= \frac{\sqrt{1 - x^2}}{1}$$
$$= \sqrt{1 - x^2}$$

**EXAMPLE 8**

Find an algebraic expression for $\sec(\text{Arctan } x)$.

Let $\theta = \text{Arctan } x$. Then

$$\tan \theta = x = \frac{x}{1}$$

Draw a right triangle with $\theta$ as an acute angle, $x$ as the opposite side, and 1 as the adjacent side as in Fig. 13.10. Using the Pythagorean theorem, we find that the hypotenuse is $\sqrt{x^2 + 1}$.

$$\sec(\text{Arctan } x) = \sec \theta$$
$$= \frac{\text{hypotenuse}}{\text{side adjacent to } \theta}$$
$$= \frac{\sqrt{x^2 + 1}}{1}$$
$$= \sqrt{x^2 + 1}$$

**EXAMPLE 9**

Find an algebraic expression for $\cos(2 \text{ Arcsin } x)$.

Let $\theta = \text{Arcsin } x$. Then

$$\sin \theta = \frac{x}{1}$$

Draw a right triangle with $\theta$ as an acute angle, $x$ as the opposite side, and 1 as the hypotenuse as in Fig. 13.11. Using the Pythagorean theorem, we find that the side adjacent to $\theta$ is $\sqrt{1 - x^2}$.

$$\cos(2 \text{ Arcsin } x) = \cos 2\theta$$
$$= 1 - 2 \sin^2 \theta \qquad \text{(Formula 25c)}$$
$$= 1 - 2x^2$$

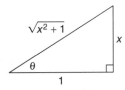

**Figure 13.10**

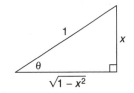

**Figure 13.11**

## Exercises 13.6

*Find the value of each expression in radians.*

**1.** $\text{Arcsin}\left(\dfrac{\sqrt{3}}{2}\right)$      **2.** $\text{Arccos}\left(\dfrac{1}{2}\right)$      **3.** $\text{Arctan}\left(-\dfrac{1}{\sqrt{3}}\right)$

**4.** $\text{Arcsin}\left(-\dfrac{1}{2}\right)$      **5.** $\text{Arccos}\left(-\dfrac{\sqrt{3}}{2}\right)$      **6.** $\text{Arctan}\left(\dfrac{1}{\sqrt{3}}\right)$

**7.** $\text{Arcsec}(-2)$      **8.** $\text{Arccot}(-1)$      **9.** $\text{Arccsc}\sqrt{2}$

**10.** $\text{Arcsin}(-1)$      **11.** $\text{Arctan}\sqrt{3}$      **12.** $\text{Arccos } 0$

**13.** $\text{Arccos}\left(\dfrac{1}{\sqrt{2}}\right)$      **14.** $\text{Arcsin } 1$      **15.** $\text{Arcsin}\left(-\dfrac{\sqrt{3}}{2}\right)$

**16.** $\text{Arctan}(-\sqrt{3})$      **17.** $\text{Arccot } 0.35$      **18.** $\text{Arcsec } 1.357$

**19.** $\text{Arccot}(-1.5)$      **20.** $\text{Arccsc } 2.5$      **21.** $\text{Arcsec}(-3.2)$

**22.** $\text{Arcsin } 0.75$      **23.** $\text{Arccsc}(-1.15)$      **24.** $\text{Arccos}(-0.55)$

*Find the value of each expression.*

**25.** $\cos(\text{Arctan}\sqrt{3})$      **26.** $\tan\left[\text{Arcsin}\left(\dfrac{1}{\sqrt{2}}\right)\right]$      **27.** $\sin\left[\text{Arccos}\left(-\dfrac{1}{\sqrt{2}}\right)\right]$

**28.** $\sin[\text{Arctan}(-1)]$      **29.** $\tan[\text{Arccos}(-1)]$      **30.** $\sec\left[\text{Arccos}\left(-\dfrac{1}{2}\right)\right]$

**31.** $\sin\left[\text{Arcsin}\left(\dfrac{\sqrt{3}}{2}\right)\right]$      **32.** $\tan[\text{Arctan}(-\sqrt{3})]$      **33.** $\cos\left[\text{Arcsin}\left(\dfrac{3}{5}\right)\right]$

**34.** $\tan\left[\text{Arcsin}\left(\dfrac{12}{13}\right)\right]$      **35.** $\tan[\text{Arcsin}(-0.1560)]$      **36.** $\sin[\text{Arccot}(1.635)]$

*Find an algebraic expression for each.*

**37.** $\cos(\text{Arcsin } x)$      **38.** $\tan(\text{Arccos } x)$      **39.** $\sin(\text{Arcsec } x)$

**40.** $\cot(\text{Arcsec } x)$      **41.** $\sec(\text{Arccos } x)$      **42.** $\sin(\text{Arctan } x)$

**43.** $\tan(\text{Arctan } x)$      **44.** $\sin(\text{Arcsin } x)$      **45.** $\cos(\text{Arcsin } 2x)$

**46.** $\tan(\text{Arccos } 3x)$      **47.** $\sin(2\,\text{Arcsin } x)$      **48.** $\cos(2\,\text{Arctan } x)$

*Graph the following.*

**49.** $y = \text{Arcsin } 2x$      **50.** $y = 3\,\text{Arccos } x$      **51.** $y = 2\,\text{Arctan } 3x$

## CHAPTER SUMMARY

**1.** The **basic identities** developed in this chapter are as follows:

$$\sin\theta = \frac{1}{\csc\theta} \qquad\qquad \textbf{(1)}$$

$$\cos\theta = \frac{1}{\sec\theta} \qquad\qquad \textbf{(2)}$$

$$\tan \theta = \frac{1}{\cot \theta} \qquad \qquad \textbf{(3)}$$

$$\cot \theta = \frac{1}{\tan \theta} \qquad \qquad \textbf{(4)}$$

$$\sec \theta = \frac{1}{\cos \theta} \qquad \qquad \textbf{(5)}$$

$$\csc \theta = \frac{1}{\sin \theta} \qquad \qquad \textbf{(6)}$$

$$\tan \theta = \frac{\sin \theta}{\cos \theta} \qquad \qquad \textbf{(7)}$$

$$\cot \theta = \frac{\cos \theta}{\sin \theta} \qquad \qquad \textbf{(8)}$$

$$\sin^2 \theta + \cos^2 \theta = 1 \qquad \qquad \textbf{(9)}$$

$$1 + \tan^2 \theta = \sec^2 \theta \qquad \qquad \textbf{(10)}$$

$$\cot^2 \theta + 1 = \csc^2 \theta \qquad \qquad \textbf{(11)}$$

$$\sin(\theta + \phi) = \sin \theta \cos \phi + \cos \theta \sin \phi \qquad \textbf{(12)}$$

$$\cos(\theta + \phi) = \cos \theta \cos \phi - \sin \theta \sin \phi \qquad \textbf{(13)}$$

$$\sin(-\theta) = -\sin \theta \qquad \qquad \textbf{(14)}$$

$$\cos(-\theta) = \cos \theta \qquad \qquad \textbf{(15)}$$

$$\tan(-\theta) = -\tan \theta \qquad \qquad \textbf{(16)}$$

$$\cot(-\theta) = -\cot \theta \qquad \qquad \textbf{(17)}$$

$$\sec(-\theta) = \sec \theta \qquad \qquad \textbf{(18)}$$

$$\csc(-\theta) = -\csc \theta \qquad \qquad \textbf{(19)}$$

$$\sin(\theta - \phi) = \sin \theta \cos \phi - \cos \theta \sin \phi \qquad \textbf{(20)}$$

$$\cos(\theta - \phi) = \cos \theta \cos \phi + \sin \theta \sin \phi \qquad \textbf{(21)}$$

$$\tan(\theta + \phi) = \frac{\tan \theta + \tan \phi}{1 - \tan \theta \tan \phi} \qquad \qquad \textbf{(22)}$$

$$\tan(\theta - \phi) = \frac{\tan \theta - \tan \phi}{1 + \tan \theta \tan \phi} \qquad \qquad \textbf{(23)}$$

$$\sin 2\theta = 2 \sin \theta \cos \theta \qquad \qquad \textbf{(24)}$$

$$\cos 2\theta = \cos^2 \theta - \sin^2 \theta \qquad \qquad \textbf{(25a)}$$
$$= 2 \cos^2 \theta - 1 \qquad \qquad \textbf{(25b)}$$
$$= 1 - 2 \sin^2 \theta \qquad \qquad \textbf{(25c)}$$

$$\tan 2\theta = \frac{2 \tan \theta}{1 - \tan^2 \theta} \qquad \qquad \textbf{(26)}$$

$$\sin \frac{\theta}{2} = \pm \sqrt{\frac{1 - \cos \theta}{2}} \qquad \qquad \textbf{(27)}$$

$$\cos \frac{\theta}{2} = \pm \sqrt{\frac{1 + \cos \theta}{2}} \qquad \qquad \textbf{(28)}$$

$$\tan \frac{\theta}{2} = \frac{1 - \cos \theta}{\sin \theta} \qquad \qquad \textbf{(29)}$$

**2.** The general methods for proving a trigonometric identity are as follows:
   (a) Substitute one expression for another using a known identity.
   (b) Factor one side of the identity.
   (c) Multiply numerator and denominator of a fractional expression by some trigonometric expression.
   (d) Express all functions on one side in terms of sines and cosines, and use one of the preceding methods.

**3.** The **inverse trigonometric functions** are defined as follows:

$$y = \text{Arcsin } x \qquad -\frac{\pi}{2} \le y \le \frac{\pi}{2}$$

$$y = \text{Arccos } x \qquad 0 \le y \le \pi$$

$$y = \text{Arctan } x \qquad -\frac{\pi}{2} < y < \frac{\pi}{2}$$

$$y = \text{Arccot } x \qquad 0 < y < \pi$$

$$y = \text{Arcsec } x \qquad 0 \le y \le \pi \quad y \ne \frac{\pi}{2}$$

$$y = \text{Arccsc } x \qquad -\frac{\pi}{2} \le y \le \frac{\pi}{2} \quad y \ne 0$$

See Fig. 13.8 for the graphs of the inverse trigonometric functions.

## CHAPTER 13 REVIEW

*Prove each identity.*

**1.** $\sec x \cot x = \csc x$

**2.** $\sec^2 \theta + \tan^2 \theta + 1 = \dfrac{2}{\cos^2 \theta}$

**3.** $\dfrac{\cos \theta}{\cos \theta + \sin \theta} = \dfrac{\cot \theta}{1 + \cot \theta}$

**4.** $\cos\left(\theta - \dfrac{3\pi}{2}\right) = -\sin \theta$

**5.** $\left(\sin \dfrac{1}{2}x + \cos \dfrac{1}{2}x\right)^2 = 1 + \sin x$

**6.** $2 \cos^2 \dfrac{\theta}{2} = \dfrac{1 + \sec \theta}{\sec \theta}$

**7.** $\dfrac{2 \cot \theta}{1 + \cot^2 \theta} = \sin 2\theta$

**8.** $\csc x - \cot x = \tan \dfrac{1}{2}x$

**9.** $\tan 2x = \dfrac{2 \cos x}{\csc x - 2 \sin x}$

**10.** $\tan^2 \dfrac{x}{2} + 1 = 2 \tan \dfrac{x}{2} \csc x$

*Simplify each expression.*

**11.** $\sin \theta \cos \theta$

**12.** $\cos^2 3\theta - \sin^2 3\theta$

**13.** $\dfrac{1 + \cos 4\theta}{2}$

**14.** $1 - 2 \sin^2 \dfrac{\theta}{3}$

**15.** $\cos 2x \cos 3x - \sin 2x \sin 3x$

**16.** $\sin 2x \cos x - \cos 2x \sin x$

**17.** Find $\sin 2\theta$ when $\cos \theta = -\dfrac{5}{13}$ ($\theta$ in the second quadrant).

**18.** Find $\cos \dfrac{\theta}{2}$ when $\tan \theta = \dfrac{4}{3}$ ($\theta$ in the third quadrant).

*Solve each trigonometric equation for $0 \leq x < 2\pi$.*

**19.** $2 \cos^2 x = \cos x$

**20.** $4 \sin^2 x - 1 = 0$

**21.** $2 \cos^2 (3x) + \sin (3x) - 1 = 0$

**22.** $\tan^2 x = \sin^2 x$

**23.** $\sin \dfrac{x}{2} + \cos \dfrac{x}{2} = 0$   (*Hint:* Square each side.)

**24.** $\sin 2x = \cos^2 x - \sin^2 x$

*Solve each equation for $0 \leq y < 2\pi$.*

**25.** $y = \arcsin \left( \dfrac{1}{2} \right)$

**26.** $y = \arctan \left( -\dfrac{1}{\sqrt{3}} \right)$

*Solve each equation for $0° \leq y < 360°$.*

**27.** $y = \arccos \left( \dfrac{1}{\sqrt{2}} \right)$

**28.** $y = \text{arcsec}(-2)$

**29.** Find the smallest positive angle in degrees for $\arcsin \left( -\dfrac{1}{2} \right)$.

**30.** Find the smallest positive angle in radians for $\arctan(-\sqrt{3})$.

**31.** Find all angles in degrees for $\text{arccot}(-\sqrt{3})$.

**32.** Find all angles in radians for $\arcsin(-1)$.

**33.** Solve for $x$:   $y = \dfrac{1}{2} \sin \dfrac{3x}{4}$ .

*Find the value of each expression in radians.*

**34.** $\text{Arcsin} \left( \dfrac{1}{\sqrt{2}} \right)$

**35.** $\text{Arctan} \left( -\dfrac{1}{\sqrt{3}} \right)$

**36.** $\text{Arcsec}(-1)$

**37.** $\text{Arccos} \left( -\dfrac{1}{2} \right)$

*Find the value of each expression.*

**38.** $\sin \left[ \text{Arccos} \left( -\dfrac{1}{2} \right) \right]$

**39.** $\tan(\text{Arctan } \sqrt{3})$

**40.** Find an algebraic expression for $\sin(\text{Arccot } x)$.

**41.** Graph $y = 1.5 \text{ Arccos } 2x$

# 14
# Complex
# Numbers

## INTRODUCTION

Mathematicians apply the laws of logic to a few assumptions and definitions to produce theorems and truths about mathematical systems. Often these systems have no basis in our physical world when they are first studied, but as our knowledge of the physical world increases, scientists often find real phenomena that match the mathematics already discovered! Such is the case with complex numbers which are used extensively in the study of electricity and electronics.

### Objectives

- Know when two complex numbers are equal.
- Write complex numbers in trigonometric form.
- Write complex numbers in exponential form.
- Add, subtract, multiply, and divide complex numbers in all forms.

## 14.1   COMPLEX NUMBERS IN RECTANGULAR FORM

Up to this point we have considered only problems with real-number solutions. When we studied roots, we restricted the radicand of even roots so that they were non-negative. In order to study equations whose solutions are not real, such as $x^2 + 1 = 0$, or to study even roots that have negative radicands, we must extend our number system to include the imaginary and complex numbers.

The square root of a negative number is an imaginary number. (Historically, this term *imaginary* was, indeed, a poor choice of terms. It was meant to distinguish such numbers from the "real" numbers—also a poor choice.) The imaginary unit is defined as $\sqrt{-1}$ and in many mathematics texts is denoted by the symbol $i$. However, in technical work $i$

is used to denote current. To avoid confusion, many technical books use $j$ to denote $\sqrt{-1}$, which is what we shall do.

> **IMAGINARY UNIT**
>
> $$j = \sqrt{-1}$$

Thus, we may write $\sqrt{-16}$ as $4j$ [since $\sqrt{-16} = \sqrt{(-1)(16)} = \sqrt{-1}\sqrt{16} = 4j$].

We now define an **imaginary number** to be any number in the form $bj$ where $b$ is a real number. We also define a **complex number** to be any number in the form $a + bj$ where $a$ and $b$ are real numbers. Note that when $a = 0$, we have an imaginary number; when $b = 0$, we have a real number.

> **RECTANGULAR FORM OF A COMPLEX NUMBER**
>
> $$a + bj$$
>
> is called the *rectangular form* of a complex number, where $a$ is the real part and $bj$ is the imaginary part.

Two complex numbers $a + bj$ and $c + dj$ are **equal** only when $a = c$ and $b = d$.

### EXAMPLE 1

Express each number in terms of $j$ and simplify.

(a) $\sqrt{-36} = \sqrt{(-1)(36)}$
$= \sqrt{(-1)}\sqrt{36}$
$= 6j$

(b) $\sqrt{-45} = \sqrt{(-1)(9)(5)}$
$= \sqrt{-1}\sqrt{9}\sqrt{5}$
$= 3\sqrt{5}j$

Next, we need to consider powers of $j$, or $\sqrt{-1}$. Using the properties of exponents and the definition of $j$, some powers of $j$ are

$$j = j$$
$$j^2 = (\sqrt{-1})^2 = -1$$
$$j^3 = j^2 \cdot j = (-1)j = -j$$
$$j^4 = j^2 \cdot j^2 = (-1)(-1) = 1$$
$$j^5 = j^4 \cdot j = (1)j = j$$
$$j^6 = j^4 \cdot j^2 = (1)(-1) = -1$$
$$j^7 = j^4 \cdot j^3 = (1)(-j) = -j$$
$$j^8 = (j^4)^2 = 1^2 = 1$$

That is, the integral powers of $j$ are cyclic in the order of $j, -1, -j, 1, j, -1, -j, 1, \ldots$. It is helpful to note that any power of $j$ evenly divisible by four is equal to one.

### EXAMPLE 2

Simplify.

(a) $j^{14} = j^{12} \cdot j^2$
$= (1)(-1)$
$= -1$

(b) $j^{37} = j^{36} \cdot j$
$= (1)j$
$= j$

(c) $j^{323} = j^{320} \cdot j^3$
$= (1)(-j)$
$= -j$

The imaginary numbers and the complex numbers, unlike the real numbers, are not ordered. That is, given two unequal complex numbers, one is not larger or smaller than the other.

Complex numbers may be added by finding the sum of the real parts and the sum of the imaginary parts.

$$(a + bj) + (c + dj) = (a + c) + (b + d)j$$

**EXAMPLE 3**

Add $(2 + 3j) + (4 - 2j)$.

$$(2 + 3j) + (4 - 2j) = (2 + 4) + [3 + (-2)]j$$
$$= 6 + j$$

Complex numbers may be subtracted by finding the difference of the real parts and the difference of the imaginary parts.

$$(a + bj) - (c + dj) = (a - c) + (b - d)j$$

**EXAMPLE 4**

Subtract $(-3 + 4j) - (7 - 2j)$.

$$(-3 + 4j) - (7 - 2j) = (-3 - 7) + [4 - (-2)]j$$
$$= -10 + 6j$$

The product of two complex numbers may be found as if they were two ordinary binomials. Simplifying, we have

$$(a + bj)(c + dj) = ac + (ad + bc)j + bdj^2 \qquad (j^2 = -1)$$
$$= (ac - bd) + (ad + bc)j$$

**EXAMPLE 5**

Multiply $(5 - 3j)(-2 + 7j)$.

$$(5 - 3j)(-2 + 7j) = -10 + 41j - 21j^2 \qquad (j^2 = -1)$$
$$= -10 + 41j + 21$$
$$= 11 + 41j$$

Before we consider the division of complex numbers, we need to define and discuss the conjugate of a complex number. The **conjugate** of the complex number $a + bj$ is $a - bj$, and the conjugate of $a - bj$ is $a + bj$. That is, the conjugate is formed by changing the sign of the imaginary part.

The product of two conjugates is always a real number.

$$(a + bj)(a - bj) = a^2 - b^2j^2 = a^2 + b^2$$

*Complex numbers may be divided by multiplying numerator and denominator by the conjugate of the denominator.*

$$\frac{a + bj}{c + dj} = \frac{a + bj}{c + dj} \cdot \frac{c - dj}{c - dj}$$

$$= \frac{(ac + bd) + (bc - ad)j}{c^2 + d^2} \quad \text{or} \quad \frac{ac + bd}{c^2 + d^2} + \left(\frac{bc - ad}{c^2 + d^2}\right)j$$

**EXAMPLE 6**

Divide $\dfrac{4 + j}{2 - 3j}$.

$$\frac{4 + j}{2 - 3j} = \frac{4 + j}{2 - 3j} \cdot \frac{2 + 3j}{2 + 3j} = \frac{8 + 14j + 3j^2}{4 - 9j^2}$$

$$= \frac{5 + 14j}{13} \quad \text{or} \quad \frac{5}{13} + \frac{14}{13}j$$

*Note:* This is the same technique used in rationalizing binomial denominators in Section 8.5.

When we solved the quadratic equation $ax^2 + bx + c = 0$, its solutions were $x = \dfrac{-b \pm \sqrt{b^2 - 4ac}}{2a}$. You may recall the restriction that the discriminant, $b^2 - 4ac$, could not be negative. We can now remove this restriction. If $b^2 - 4ac < 0$, the solutions are complex.

**EXAMPLE 7**

Solve $x^2 + 1 = 0$.

$$x^2 + 1 = 0$$
$$x^2 = -1$$
$$x = \pm\sqrt{-1} = \pm j$$

**EXAMPLE 8**

Solve $3x^2 + 4x + 2 = 0$.

Using the quadratic formula, we have

$$x = \frac{-b \pm \sqrt{b^2 - 4ac}}{2a}$$

$$x = \frac{-4 \pm \sqrt{4^2 - 4(3)(2)}}{2(3)}$$

$$= \frac{-4 \pm \sqrt{-8}}{6}$$

$$= \frac{-4 \pm 2j\sqrt{2}}{6}$$

$$= \frac{-2 \pm j\sqrt{2}}{3} \quad \text{or} \quad -\frac{2}{3} \pm \frac{\sqrt{2}}{3}j$$

The complex number $a + bj$ may also be written as an ordered pair of real numbers $(a, b)$. This allows us to associate points in the coordinate plane with the complex numbers. To do so, we modify the coordinate axes by letting the horizontal axis be the real axis and the vertical axis be the imaginary axis. This plane is called the **complex plane.**

To find the point that corresponds to a given complex number $a + bj$, plot the ordered pair $(a, b)$. For example, the complex numbers $4 + 2j$ and $-3 - j$ are plotted in Fig. 14.1.

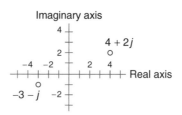

**Figure 14.1**  Complex plane

The sum of two complex numbers may also be found graphically, which is shown by example.

**EXAMPLE 9**

Add $5 + 4j$ and $3 - j$ graphically.

First plot $5 + 4j$ and $3 - j$ in the complex plane, as in Fig. 14.2. Each may be drawn as a vector from the origin to the points $(5, 4)$ and $(3, -1)$. The graphical sum is the diagonal of the parallelogram which is the resultant of the two vectors. The end point of the resultant is $(8, 3)$, which corresponds to the complex number $8 + 3j$, which is also the algebraic sum.

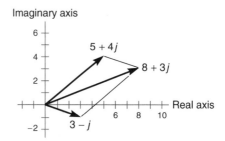

**Figure 14.2**

To subtract one complex number from another graphically, add the negative of the number being subtracted (the second number or subtrahend).

**EXAMPLE 10**

Find the difference $(-2 - 4j) - (-3 + 5j)$ graphically.

First, plot $-2 - 4j$ and $-3 + 5j$ in the complex plane as in Fig. 14.3. Then plot the negative of $-3 + 5j$, which is $3 - 5j$. The end point of the resultant is $(1, -9)$, which corresponds to the complex number $1 - 9j$, which is also the algebraic difference.

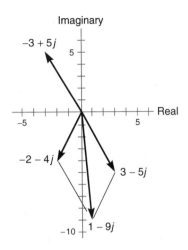

Imaginary

$-3 + 5j$

$-2 - 4j$

$3 - 5j$

$1 - 9j$

Real

**Figure 14.3**

## Exercises 14.1

*Express each number in terms of j and simplify.*

**1.** $\sqrt{-49}$  **2.** $\sqrt{-100}$  **3.** $\sqrt{-64}$  **4.** $\sqrt{-121}$

**5.** $\sqrt{-12}$  **6.** $\sqrt{-72}$  **7.** $\sqrt{-54}$  **8.** $\sqrt{-500}$

*Simplify.*

**9.** $j^{19}$  **10.** $j^{34}$  **11.** $j^{22}$  **12.** $j^{39}$

**13.** $j^{81}$  **14.** $j^{97}$  **15.** $j^{246}$  **16.** $j^{308}$

*Perform the indicated operations and simplify.*

**17.** $(3 + 4j) + (9 + 2j)$  **18.** $(-2 + 5j) + (6 - 7j)$  **19.** $(4 - 9j) - (2 - j)$

**20.** $(-6 + 3j) - (9 - 2j)$  **21.** $(4 + 2j) + (-4 - 3j)$  **22.** $(6 - j) - (1 - j)$

**23.** $(2 + j)(8 - 3j)$  **24.** $(4 - 6j)(9 - 2j)$  **25.** $(-4 + 5j)(3 + 2j)$

**26.** $(-3 - j)(8 + j)$  **27.** $(2 + 5j)(2 - 5j)$  **28.** $(-6 - 2j)(-6 + 2j)$

**29.** $(-3 + 4j)^2$  **30.** $(5 - 3j)^2$  **31.** $\dfrac{3 + 7j}{4 - j}$

**32.** $\dfrac{-3 + j}{2 + j}$  **33.** $\dfrac{6 - 3j}{4 + 8j}$  **34.** $\dfrac{1 - 4j}{2 - 3j}$

**35.** $\dfrac{-9 + 8j}{6 - 2j}$  **36.** $\dfrac{4 + 3j}{4 - 3j}$

*Solve each equation.*

**37.** $x^2 + 4 = 0$  **38.** $x^2 + 12 = 0$  **39.** $3x^2 + 4x + 9 = 0$

**40.** $2x^2 - x + 6 = 0$  **41.** $5x^2 - 2x + 5 = 0$  **42.** $9x^2 + 9x + 9 = 0$

**43.** $1 - 2x + 3x^2 = 0$  **44.** $10 + 4x + 2x^2 = 0$  **45.** $x^3 + 1 = 0$

**46.** $x^3 - 1 = 0$  **47.** $x^4 - 1 = 0$  **48.** $x^3 + 9x = 0$

**49.** $x^4 + 80x^2 = 0$  **50.** $x^4 + 25x^2 = 0$  **51.** $2x^4 + 54x = 0$

**52.** $4x^4 - 32x = 0$  **53.** $x^5 = x^2$  **54.** $2x^5 + 128x^2 = 0$

*Plot each complex number in the complex plane.*

**55.** $4 + 2j$  **56.** $-1 - 5j$  **57.** $-2 + 3j$

**58.** $3 - 3j$  **59.** $-4j$ or $0 - 4j$  **60.** $5$ or $5 + 0j$

*Add or subtract the complex numbers graphically.*

**61.** $(3 + 2j) + (-2 + j)$  **62.** $(-4 - 3j) + (-3 + 2j)$  **63.** $(5 - 3j) - (1 + 5j)$

**64.** $(-6 - 3j) - (5 + 3j)$  **65.** $(-3 - 4j) - (-6 + 2j)$  **66.** $(8 + j) + (0 - 4j)$

**67.** $(2 + 4j) + (-2 + 3j)$  **68.** $(6 - 4j) - (2 - 4j)$

## 14.2  TRIGONOMETRIC AND EXPONENTIAL FORMS OF COMPLEX NUMBERS

The graphical representation of a complex number $a + bj$ leads us into a very useful trigonometric relationship. From Fig. 14.4, we can see that

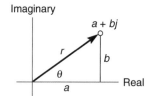

**Figure 14.4**

$$\cos \theta = \frac{a}{r} \qquad \text{(From the definition of cosine)}$$

$$a = r \cos \theta \tag{14.1}$$

and

$$\sin \theta = \frac{b}{r} \qquad \text{(From the definition of sine)}$$

$$b = r \sin \theta \tag{14.2}$$

Substituting these values into $a + bj$, we have the following.

$$a + bj = r \cos \theta + (r \sin \theta)j$$
$$= r(\cos \theta + j \sin \theta) \tag{14.3}$$

---

**TRIGONOMETRIC FORM OF A COMPLEX NUMBER**

$$r(\cos \theta + j \sin \theta)$$

is called the *trigonometric form* of a complex number. Here, $r$ is the *absolute value* or the *modulus* and corresponds to the length of the vector when the complex number is expressed as a vector. Angle $\theta$ is called the *argument* of the complex number and is given in standard position.

---

Sometimes this form is called the *polar form* of a complex number. Other notations for trigonometric form include $r$ cis $\theta$ and $r\angle\theta$. (Recall that $a + bj$ was called the rectangular form of a complex number.)

If we know $r$ and $\theta$, we can find $a$ and $b$ using Equations (14.1) and (14.2). If we know $a$ and $b$, we can find $r$ by using the Pythagorean theorem.

$$r = \sqrt{a^2 + b^2} \qquad\qquad (14.4)$$

Also, we can find $\theta$ using the definition of tangent.

$$\tan \theta = \frac{b}{a} \qquad\qquad (14.5)$$

Using Equations (14.1)–(14.5), we can change from rectangular to trigonometric form, and vice versa.

**EXAMPLE 1**

Write $2 - 2j$ in trigonometric form.

First, graph $2 - 2j$ as in Fig. 14.5. Then

$$r = \sqrt{a^2 + b^2} = \sqrt{2^2 + (-2)^2} = \sqrt{8} = 2\sqrt{2}$$

$$\tan \theta = \frac{b}{a} = \frac{-2}{2} = -1$$

$$\theta = 315° \qquad\qquad \text{(Normally we choose } \theta \text{ so that } 0 \le \theta < 360°.)$$

Therefore,

$$2 - 2j = 2\sqrt{2}(\cos 315° + j \sin 315°)$$

**EXAMPLE 2**

Write $4(\cos 120° + j \sin 120°)$ in rectangular form.

First, graph $4(\cos 120° + j \sin 120°)$ as in Fig. 14.6.

$$a = r \cos \theta = 4 \cos 120° = 4\left(-\frac{1}{2}\right) = -2$$

$$b = r \sin \theta = 4 \sin 120° = 4\left(\frac{\sqrt{3}}{2}\right) = 2\sqrt{3}$$

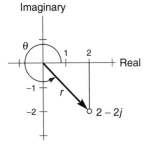

**Figure 14.5**

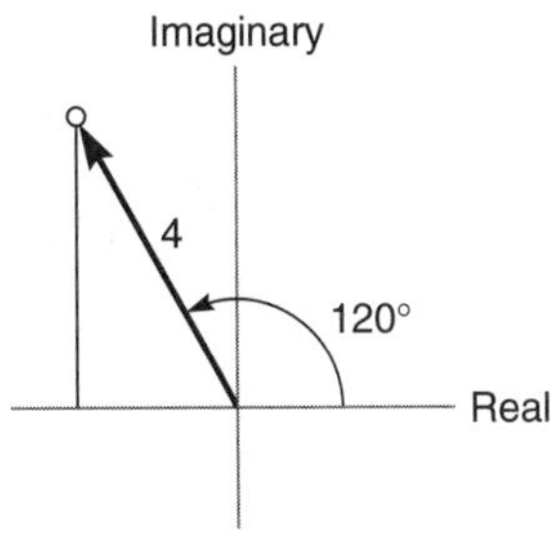

**Figure 14.6**

Therefore,

$$4(\cos 120° + j \sin 120°) = -2 + 2\sqrt{3}\, j$$

Another very useful form of a complex number was discovered by the Swiss mathematician Euler. In equation form,

$$e^{j\theta} = \cos\theta + j\sin\theta*$$  (14.6)

where $e$ is an irrational number whose approximation is 2.71828. If we multiply each side of Equation (14.6) by $r$, we have the following.

$$re^{j\theta} = r(\cos\theta + j\sin\theta)$$

**EXPONENTIAL FORM OF A COMPLEX NUMBER**

$$re^{j\theta}$$

is called the *exponential form* of a complex number, where $r$ and $\theta$ are defined the same as for trigonometric form. When $\theta$ is expressed in radians, $j\theta$ is the actual exponent of the complex number and follows all the laws of exponents. Here, as in most applications, $\theta$ is expressed in radians when using exponential form.

**EXAMPLE 3**

Write $-4 - 3j$ in exponential form.

$$r = \sqrt{a^2 + b^2} = \sqrt{(-4)^2 + (-3)^2} = \sqrt{25} = 5$$

$$\tan\theta = \frac{b}{a} = \frac{-3}{-4} = 0.7500 \qquad \text{(Note: } \theta \text{ in third quadrant)}$$

$$\theta = 217° \qquad\qquad \text{(To the nearest degree)}$$

$$\theta = 217° \times \frac{\pi\,\text{rad}}{180°} = 3.79\text{ rad}$$

Therefore, the exponential form is

$$-4 - 3j = 5e^{3.79j}$$

**EXAMPLE 4**

Write $3e^{2.86j}$ in trigonometric and rectangular forms.

Note that $r = 3$ and $\theta = 2.86\text{ rad} \times \dfrac{180°}{\pi\,\text{rad}} = 164°$ (to the nearest degree). Thus, the trigonometric form is $3(\cos 164° + j \sin 164°)$.

The rectangular form is:

$$3(\cos 164° + j \sin 164°) = -2.88 + 0.827j$$

*This relationship is demonstrated in Section 10.6 of *Technical Calculus,* 3d ed., by D. Ewen et al. (Columbus, OH: Prentice Hall, 1998).

## Exercises 14.2

Write each complex number in trigonometric form. Give angles to the nearest degree.

**1.** $2 + 2j$         **2.** $4 - 4j$         **3.** $-1 - \sqrt{3}\,j$

**4.** $-4\sqrt{3} + 4j$     **5.** $4j$         **6.** $-3$

**7.** $-6 - 6j$        **8.** $5 + 12j$       **9.** $-2 + 3j$

**10.** $5 - 3j$

Write each complex number in rectangular form. Use simplest radical form, where possible; otherwise use decimal values rounded to three significant digits.

**11.** $4(\cos 60° + j \sin 60°)$        **12.** $6(\cos 150° + j \sin 150°)$

**13.** $2(\cos 330° + j \sin 330°)$      **14.** $3(\cos 240° + j \sin 240°)$

**15.** $3\sqrt{2}(\cos 135° + j \sin 135°)$    **16.** $\sqrt{3}(\cos 225° + j \sin 225°)$

**17.** $3 \angle 270°$                 **18.** $4 \angle 65°$

**19.** $\sqrt{53} \angle 344°$           **20.** $2\sqrt{11} \angle 243.5°$

Write each complex number in exponential form.

**21.** $\sqrt{3} - j$        **22.** $-3 + 3j$       **23.** $-\sqrt{2} - \sqrt{2}\,j$

**24.** $2 + 2\sqrt{3}\,j$     **25.** $4 + 6j$        **26.** $5 - 7j$

Write each complex number in trigonometric and rectangular forms.

**27.** $3e^{1.35j}$       **28.** $5e^{3.02j}$       **29.** $4e^{5.76j}$

**30.** $e^{6.91j}$        **31.** $2e^{j}$         **32.** $6e^{\pi j}$

## 14.3 MULTIPLICATION AND DIVISION OF COMPLEX NUMBERS IN EXPONENTIAL AND TRIGONOMETRIC FORMS

The product of two complex numbers given in exponential form is given as follows.

$$r_1 e^{j\theta_1} \cdot r_2 e^{j\theta_2} = r_1 r_2 e^{j\theta_1 + j\theta_2} = r_1 r_2 e^{j(\theta_1 + \theta_2)}$$

The product of two complex numbers given in trigonometric form is as follows.

$$r_1(\cos \theta_1 + j \sin \theta_1) \cdot r_2(\cos \theta_2 + j \sin \theta_2) = r_1 r_2[\cos(\theta_1 + \theta_2) + j \sin(\theta_1 + \theta_2)]$$
$$\text{or} \quad (r_1 \angle \theta_1)(r_2 \angle \theta_2) = r_1 r_2 \angle \theta_1 + \theta_2$$

This formula may be proven as follows:

$$[r_1(\cos \theta_1 + j \sin \theta_1)][r_2(\cos \theta_2 + j \sin \theta_2)]$$
$$= r_1 r_2[\cos \theta_1 \cos \theta_2 + j \sin \theta_1 \cos \theta_2 + j \cos \theta_1 \sin \theta_2 + j^2 \sin \theta_1 \sin \theta_2]$$
$$= r_1 r_2[\cos \theta_1 \cos \theta_2 - \sin \theta_1 \sin \theta_2 + j(\sin \theta_1 \cos \theta_2 + \cos \theta_1 \sin \theta_2)]$$
$$= r_1 r_2[\cos(\theta_1 + \theta_2) + j \sin(\theta_1 + \theta_2)] \quad \text{[Formulas (13) and (12) in Section 13.2]}$$

**EXAMPLE 1**

Find the product $(3e^{2.5j})(4e^{1.7j})$ and write the result in exponential form.

$$(3e^{2.5j})(4e^{1.7j}) = (3)(4)e^{2.5j+1.7j}$$
$$= 12e^{4.2j}$$

**EXAMPLE 2**

Find the product $(5e^{\pi j})(7e^{\pi j/2})$ and write the result in rectangular form.

$$(5e^{\pi j})(7e^{\pi j/2}) = (5)(7)e^{\pi j+\pi j/2}$$
$$= 35e^{3\pi j/2}$$

Note that $r = 35$ and $\theta = \dfrac{3\pi}{2} = 270°$. Thus, the trigonometric form is

$$35(\cos 270° + j \sin 270°) = 35[0 + j(-1)]$$
$$= -35j$$

**EXAMPLE 3**

Find the product

$$[3(\cos 235° + j \sin 235°)][6(\cos 175° + j \sin 175°)]$$

and write the result in trigonometric form.

$[3(\cos 235° + j \sin 235°)][6(\cos 175° + j \sin 175°)]$
$$= (3)(6)[\cos(235° + 175°) + j \sin(235° + 175°)]$$
$$= 18(\cos 410° + j \sin 410°)$$
$$= 18(\cos 50° + j \sin 50°)$$

Note that $18(\cos 410° + j \sin 410°)$ and $18(\cos 50° + j \sin 50°)$ both have the same coordinates when plotted in the complex plane.

In general,

$$r(\cos \theta + j \sin \theta) = r[\cos (\theta \pm 360°) + j \sin(\theta \pm 360°)]$$

since they have the same coordinates when plotted in the complex plane. Unless stated otherwise, we shall write $0° \leq \theta < 360°$.

**EXAMPLE 4**

Find the product

$$[5(\cos 60° + j \sin 60°)][4(\cos 120° + j \sin 120°)]$$

and write the result in rectangular form.

$[5(\cos 60° + j \sin 60°)][4(\cos 120° + j \sin 120°)]$
$$= (5)(4)[\cos(60° + 120°) + j \sin(60° + 120°)]$$
$$= 20(\cos 180° + j \sin 180°)$$
$$= 20[-1 + j(0)]$$
$$= -20$$

The quotient of two complex numbers in exponential form is as follows.

$$\frac{r_1 e^{j\theta_1}}{r_2 e^{j\theta_2}} = \frac{r_1}{r_2} e^{j(\theta_1 - \theta_2)}$$

The quotient of two complex numbers given in trigonometric form is as follows.

$$\frac{r_1(\cos\theta_1 + j\sin\theta_1)}{r_2(\cos\theta_2 + j\sin\theta_2)} = \frac{r_1}{r_2}[\cos(\theta_1 - \theta_2) + j\sin(\theta_1 - \theta_2)]$$

$$\text{or} \quad \frac{r_1 \angle \theta_1}{r_2 \angle \theta_2} = \frac{r_1}{r_2} \angle \theta_1 - \theta_2$$

This formula may be proven as follows:

$$\frac{r_1(\cos\theta_1 + j\sin\theta_1)}{r_2(\cos\theta_2 + j\sin\theta_2)} = \frac{r_1(\cos\theta_1 + j\sin\theta_1)}{r_2(\cos\theta_2 + j\sin\theta_2)} \cdot \frac{(\cos\theta_2 - j\sin\theta_2)}{(\cos\theta_2 - j\sin\theta_2)}$$

$$= \frac{r_1}{r_2} \cdot \frac{\cos\theta_1 \cos\theta_2 + j\sin\theta_1 \cos\theta_2 - j\cos\theta_1 \sin\theta_2 - j^2 \sin\theta_1 \sin\theta_2}{\cos^2\theta_2 - j^2 \sin^2\theta_2}$$

$$= \frac{r_1}{r_2} \cdot \frac{\cos\theta_1 \cos\theta_2 + \sin\theta_1 \sin\theta_2 + j(\sin\theta_1 \cos\theta_2 - \cos\theta_1 \sin\theta_2)}{\cos^2\theta_2 + \sin^2\theta_2}$$

$$= \frac{r_1}{r_2}[\cos(\theta_1 - \theta_2) + j\sin(\theta_1 - \theta_2)] \quad \text{[Formulas (21) and (20) in Section 13.2]}$$

**EXAMPLE 5**

Find the quotient $\dfrac{18e^{3.10j}}{6e^{5.50j}}$ and write the result in exponential form.

$$\frac{18e^{3.10j}}{6e^{5.50j}} = \frac{18}{6} e^{3.10j - 5.50j}$$

$$= 3e^{-2.40j}$$

$$= 3e^{3.88j} \quad \text{(\textit{Note:} } -2.40 + 2\pi = 3.88\text{)}$$

Note that $3e^{-2.40j}$ and $3e^{3.88j}$ both have the same coordinates when plotted in the complex plane.

In general,

$$re^{j\theta} = re^{(\theta \pm 2\pi)j}$$

since they have the same coordinates when plotted in the complex plane.

Unless stated otherwise, we shall write $0 \le \theta < 2\pi$.

**EXAMPLE 6**

Find the quotient $\dfrac{18e^{3.94j}}{24e^{-1.72j}}$ and write the result in rectangular form.

$$\frac{18e^{3.94j}}{24e^{-1.72j}} = \frac{18}{24} e^{3.94j - (-1.72j)}$$

$$= \frac{3}{4} e^{5.66j}$$

Note that $r = \dfrac{3}{4}$ and $\theta = 5.66$ rad $\times \dfrac{180°}{\pi \text{ rad}} = 324°$ (rounded to the nearest degree). Thus, the trigonometric form is

$$\frac{3}{4}(\cos 324° + j \sin 324°) = 0.607 - 0.441j$$

**EXAMPLE 7**

Find the quotient $\dfrac{28(\cos 59° + j \sin 59°)}{16(\cos 135° + j \sin 135°)}$ and write the result in trigonometric form.

$$\frac{28(\cos 59° + j \sin 59°)}{16(\cos 135° + j \sin 135°)} = \frac{28}{16}[\cos(59° - 135°) + j \sin(59° - 135°)]$$

$$= 1.75[\cos(-76°) + j \sin(-76°)]$$

$$= 1.75(\cos 284° + j \sin 284°)$$

$$(\textit{Note: } -76° + 360° = 284°)$$

**EXAMPLE 8**

Find the quotient $\dfrac{18(\cos 300° + j \sin 300°)}{9(\cos 60° + j \sin 60°)}$ and write the result in rectangular form.

$$\frac{18(\cos 300° + j \sin 300°)}{9(\cos 60° + j \sin 60°)} = \frac{18}{9}[\cos(300° - 60°) + j \sin(300° - 60°)]$$

$$= 2(\cos 240° + j \sin 240°)$$

$$= 2\left[-\frac{1}{2} + j\left(-\frac{\sqrt{3}}{2}\right)\right]$$

$$= -1 - j\sqrt{3}$$

## Exercises 14.3

*Find each product and write the result in exponential form.*

**1.** $(4e^{j})(7e^{3j})$      **2.** $(6e^{2j})(8e^{3j})$      **3.** $(9e^{5j})(3e^{-3j})$

**4.** $(2e^{4j})(12e^{2j})$      **5.** $(6e^{5.6j})(4e^{3.7j})$      **6.** $(3e^{-9.4j})(7e^{6.7j})$

*Find each product and write the result in rectangular form.*

**7.** $(3e^{\pi j})(8e^{\pi j/3})$      **8.** $(3e^{\pi j/2})(5e^{2\pi j/3})$      **9.** $(4e^{3.7j})(20e^{6.1j})$

**10.** $(5e^{1.2j})(9e^{3.6j})$      **11.** $(6e^{-1.4j})(9e^{-2.5j})$      **12.** $(3e^{-j})(1e^{4.3j})$

*Find each product and write the result in trigonometric form.*

**13.** $[3(\cos 75° + j \sin 75°)][4(\cos 38° + j \sin 38°)]$

**14.** $[5(\cos 145° + j \sin 145°)][2(\cos 153° + j \sin 153°)]$

**15.** $[3(\cos 150° + j \sin 150°)][3(\cos 150° + j \sin 150°)]$

**16.** $[6(\cos 240° + j \sin 240°)][3(\cos 300° + j \sin 300°)]$

**17.** $(1 \angle 180°)(7 \angle 315°)$

**18.** $(8 \angle 168°)(9 \angle -215°)$

*Find each product and write the result in rectangular form.*

**19.** $[5(\cos 50° + j \sin 50°)][5(\cos 10° + j \sin 10°)]$

**20.** $[4(\cos 105° + j \sin 105°)][1(\cos 120° + j \sin 120°)]$

**21.** $[2(\cos 120° + j \sin 120°)][6(\cos 60° + j \sin 60°)]$

**22.** $[3(\cos 145° + j \sin 145°)][9(\cos 125° + j \sin 125°)]$

**23.** $(7 \angle 162°)(8 \angle 213°)$

**24.** $(3 \angle 305°)(7 \angle 215°)$

*Find each quotient and write the result in exponential form.*

**25.** $\dfrac{3e^{6j}}{9e^{2j}}$
  **26.** $\dfrac{6e^{3j}}{4e^{j}}$
  **27.** $\dfrac{20e^{-4j}}{5e^{3j}}$

**28.** $\dfrac{24e^{-2j}}{2e^{-4j}}$
  **29.** $\dfrac{8e^{1.6j}}{24e^{3.8j}}$
  **30.** $\dfrac{12e^{-2.1j}}{3e^{4.8j}}$

*Find each quotient and write the result in rectangular form.*

**31.** $\dfrac{10e^{\pi j/6}}{5e^{\pi j}}$
  **32.** $\dfrac{6e^{\pi j}}{9e^{3\pi j}}$
  **33.** $\dfrac{14e^{4.6j}}{2e^{1.3j}}$

**34.** $\dfrac{36e^{5.2j}}{4e^{-1.8j}}$
  **35.** $\dfrac{35e^{6.7j}}{7e^{-5.2j}}$
  **36.** $\dfrac{15e^{-3.6j}}{12e^{-5.8j}}$

*Find each quotient and write the result in trigonometric form.*

**37.** $\dfrac{25(\cos 120° + j \sin 120°)}{5(\cos 50° + j \sin 50°)}$
  **38.** $\dfrac{18(\cos 170° + j \sin 170°)}{2(\cos 70° + j \sin 70°)}$

**39.** $\dfrac{42(\cos 275° + j \sin 275°)}{7(\cos 156° + j \sin 156°)}$
  **40.** $\dfrac{49(\cos 318° + j \sin 318°)}{14(\cos 251° + j \sin 251°)}$

**41.** $\dfrac{40 \angle 86°}{5 \angle 215°}$
  **42.** $\dfrac{54 \angle 140°}{9 \angle 350°}$

*Find each quotient and write the result in rectangular form.*

**43.** $\dfrac{72(\cos 240° + j \sin 240°)}{8(\cos 120° + j \sin 120°)}$
  **44.** $\dfrac{6(\cos 295° + j \sin 295°)}{24(\cos 160° + j \sin 160°)}$

**45.** $\dfrac{8(\cos 185° + j \sin 185°)}{40(\cos 35° + j \sin 35°)}$
  **46.** $\dfrac{60(\cos 240° + j \sin 240°)}{12(\cos 330° + j \sin 330°)}$

**47.** $\dfrac{96 \angle 85°}{16 \angle 145°}$
  **48.** $\dfrac{80 \angle 30°}{25 \angle 300°}$

**49.** Find $[2(\cos 60° + j \sin 60°)]^3$ and write the result in rectangular form.

**50.** Find $[3(\cos 150° + j \sin 150°)]^3$ and write the result in rectangular form.

**51.** Find $[3(\cos 157.5° + j \sin 157.5°)]^4$ and write the result in rectangular form.

**52.** Find $[2(\cos 315° + j \sin 315°)]^4$ and write the result in rectangular form.

## 14.4 POWERS AND ROOTS

D.1, D.2

The *n*th power of a complex number in exponential form is given by the following equation.

$$(re^{j\theta})^n = r^n e^{jn\theta}$$

The trigonometric form of $(re^{j\theta})^n$ is

$$[r(\cos\,\theta + j\,\sin\,\theta)]^n$$

The trigonometric form of $r^n e^{jn\theta}$ is

$$r^n(\cos\,n\theta + j\,\sin\,n\theta)$$

Therefore:

---

**DEMOIVRE'S THEOREM**

$$[r(\cos\,\theta + j\,\sin\,\theta)]^n = r^n(\cos\,n\theta + j\,\sin\,n\theta)$$

$$\text{or} \quad (r\,\angle\,\theta)^n = r^n\,\angle\,n\theta$$

---

which is the *n*th power of a complex number in trigonometric form. This theorem is valid for all real values of *n*.

### EXAMPLE 1

Find $(3e^{2j})^4$ and write the result in exponential form.

$$\begin{aligned}
(3e^{2j})^4 &= 3^4(e^{2j})^4 \\
&= 81e^{8j} \\
&= 81e^{1.72j} \qquad (Note:\ 8 - 2\pi = 1.72)
\end{aligned}$$

### EXAMPLE 2

Find $[2(\cos\,125° + j\,\sin\,125°)]^5$ and write the result in trigonometric form.

$$\begin{aligned}
[2(\cos\,125° + j\,\sin\,125°)]^5 \\
= 2^5(\cos\,5 \cdot 125° + j\,\sin\,5 \cdot 125°) \\
= 32(\cos\,625° + j\,\sin\,625°) \\
= 32(\cos\,265° + j\,\sin\,265°) \qquad (Note:\ 625° - 360° = 265°)
\end{aligned}$$

### EXAMPLE 3

Find $(-2 + 2j)^6$ and write the result in rectangular form.

First,

$$r = \sqrt{a^2 + b^2} = \sqrt{(-2)^2 + 2^2} = \sqrt{8} = 2\sqrt{2}$$

$$\tan\,\theta = \frac{b}{a} = \frac{2}{-2} = -1$$

Since $\theta$ is in the second quadrant, $\theta = 135°$. Thus,

$$-2 + 2j = 2\sqrt{2}(\cos 135° + j \sin 135°)$$

and

$$
\begin{aligned}
(-2 + 2j)^6 &= [2\sqrt{2}(\cos 135° + j \sin 135°)]^6 \\
&= (2\sqrt{2})^6(\cos 6 \cdot 135° + j \sin 6 \cdot 135°) \\
&= 512(\cos 810° + j \sin 810°) \\
&= 512(\cos 90° + j \sin 90°) \qquad \text{(\textit{Note:} } 810° - 2 \cdot 360° = 90°) \\
&= 512(0 + j \cdot 1) \\
&= 512j
\end{aligned}
$$

DeMoivre's theorem is also valid for negative integers; that is,

$$(a + bj)^{-n} = [r(\cos \theta + j \sin \theta)]^{-n} = r^{-n}[\cos(-n\theta) + j \sin(-n\theta)]$$

where $n$ is a positive integer.

**EXAMPLE 4**

Find $(-1 - j\sqrt{3})^{-4}$ and write the result in rectangular form.
First,

$$r = \sqrt{a^2 + b^2} = \sqrt{(-1)^2 + (-\sqrt{3})^2} = \sqrt{4} = 2$$

$$\tan \theta = \frac{b}{a} = \frac{-\sqrt{3}}{-1} = \sqrt{3}$$

Since $\theta$ is in the third quadrant, $\theta = 240°$. Thus,

$$-1 - j\sqrt{3} = 2(\cos 240° + j \sin 240°)$$

and

$$
\begin{aligned}
(-1 - j\sqrt{3})^{-4} &= [2(\cos 240° + j \sin 240°)]^{-4} \\
&= 2^{-4}[\cos(-4 \cdot 240°) + j \sin(-4 \cdot 240°)] \\
&= \frac{1}{16}[\cos(-960°) + j \sin(-960°)] \\
&= \frac{1}{16}(\cos 120° + j \sin 120°) \qquad \text{(\textit{Note:} } -960° + 3 \cdot 360° = 120°) \\
&= \frac{1}{16}\left(-\frac{1}{2} + j\frac{\sqrt{3}}{2}\right) \\
&= -\frac{1}{32} + j\frac{\sqrt{3}}{32}
\end{aligned}
$$

To find an $n$th root of the complex number $z$ means to find a number $a + bj$ such that

$$(a + bj)^n = [r(\cos \theta + j \sin \theta)]^n = z$$

For example, consider the complex number

$$r^{1/n}\left(\cos \frac{\theta}{n} + j \sin \frac{\theta}{n}\right)$$

where $r > 0$ and $n$ is a positive integer. By DeMoivre's theorem, we have

$$\left[ r^{1/n} \left( \cos \frac{\theta}{n} + j \sin \frac{\theta}{n} \right) \right]^n = (r^{1/n})^n \left[ \cos \left( n \cdot \frac{\theta}{n} \right) + j \sin \left( n \cdot \frac{\theta}{n} \right) \right]$$

$$= r(\cos \theta + j \sin \theta)$$

Thus, $r^{1/n} \left( \cos \dfrac{\theta}{n} + j \sin \dfrac{\theta}{n} \right)$ is an $n$th root of $z$.

In general, there are $n$ distinct $n$th roots of a given complex number. The following formula may be used to find the $n$ distinct roots of a complex number written in trigonometric form.

---

**DEMOIVRE'S THEOREM FOR ROOTS**

$$[r(\cos \theta + j \sin \theta)]^{1/n} = r^{1/n} \left[ \cos \left( \frac{\theta + k \cdot 360°}{n} \right) + j \sin \left( \frac{\theta + k \cdot 360°}{n} \right) \right]$$

where $k = 0, 1, 2, \ldots, n - 1$

---

**EXAMPLE 5**

Find the four fourth roots of 1. That is, solve the equation $x^4 = 1$.

In trigonometric form, $1 = 1(\cos 0° + j \sin 0°)$

For the first root ($k = 0$),

$$1^{1/4} = 1^{1/4} \left[ \cos \frac{1}{4}(0°) + j \sin \frac{1}{4}(0°) \right]$$

$$= \cos 0° + j \sin 0° = 1$$

For the second root ($k = 1$), $\theta = 0° + (1)360° = 360°$,

$$1^{1/4} = 1^{1/4} \left[ \cos \frac{1}{4}(360°) + j \sin \frac{1}{4}(360°) \right]$$

$$= \cos 90° + j \sin 90°$$

$$= 0 + j = j$$

For the third root ($k = 2$), $\theta = 0° + 2(360°) = 720°$,

$$1^{1/4} = 1^{1/4} \left[ \cos \frac{1}{4}(720°) + j \sin \frac{1}{4}(720°) \right]$$

$$= \cos 180° + j \sin 180°$$

$$= -1 + 0j = -1$$

For the fourth root ($k = 3$), $\theta = 0° + 3(360°) = 1080°$,

$$1^{1/4} = 1^{1/4} \left[ \cos \frac{1}{4}(1080°) + j \sin \frac{1}{4}(1080°) \right]$$

$$= \cos 270° + j \sin 270°$$

$$= 0 - j = -j$$

As you can see in Fig. 14.7, when the four roots of Example 5 are shown graphically in the complex plane, they differ by $\dfrac{360°}{4} = 90°$. In general, the roots differ by $\dfrac{360°}{n}$, where $n$ is the number of roots.

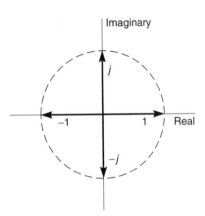

**Figure 14.7**

**EXAMPLE 6**

Find all complex solutions of $x^5 = -1 + j\sqrt{3}$.

In trigonometric form,

$$-1 + j\sqrt{3} = 2(\cos 120° + j \sin 120°)$$

For $k = 0$,

$$2^{1/5}\left[\cos\left(\frac{120° + 0 \cdot 360°}{5}\right) + j \sin\left(\frac{120° + 0 \cdot 360°}{5}\right)\right] = 2^{1/5}(\cos 24° + j \sin 24°)$$

By letting $k = 1, 2, 3,$ and 4, we obtain the other roots as follows:

$$k = 1: \quad 2^{1/5}(\cos 96° + j \sin 96°)$$
$$k = 2: \quad 2^{1/5}(\cos 168° + j \sin 168°)$$
$$k = 3: \quad 2^{1/5}(\cos 240° + j \sin 240°)$$
$$k = 4: \quad 2^{1/5}(\cos 312° + j \sin 312°)$$

*Note:* These last four roots could also have been found by adding $\dfrac{360°}{n} = \dfrac{360°}{5} = 72°$ to the root for $k = 0$ and then to each successive root through $k = 4$. These five roots are shown graphically in Fig. 14.8.

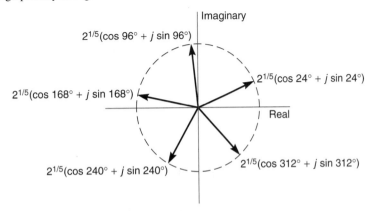

**Figure 14.8**

Approximations of the five roots in rectangular form are

$$2^{1/5}(\cos 24° + j \sin 24°) = 1.05 + 0.467j$$
$$2^{1/5}(\cos 96° + j \sin 96°) = -0.120 + 1.14j$$
$$2^{1/5}(\cos 168° + j \sin 168°) = -1.12 + 0.239j$$
$$2^{1/5}(\cos 240° + j \sin 240°) = -0.574 - 0.995j$$
$$2^{1/5}(\cos 312° + j \sin 312°) = 0.769 - 0.854j$$

## Exercises 14.4

*Find each power and write the result in exponential form.*

**1.** $(3e^{1.4j})^5$     **2.** $(2e^{2.75j})^3$     **3.** $(5e^{4.6j})^2$     **4.** $(2e^{1.7j})^6$

*Find each power and write the result in trigonometric form.*

**5.** $[3(\cos 20° + j \sin 20°)]^4$     **6.** $[1(\cos 25° + j \sin 25°)]^7$

**7.** $(2\angle 150°)^5$     **8.** $(5\angle 275°)^3$

**9.** $[2(\cos 240° + j \sin 240°)]^{-3}$     **10.** $[1(\cos 220° + j \sin 220°)]^{-4}$

*Find each power and write the result in rectangular form.*

**11.** $(1 - j)^8$     **12.** $(\sqrt{3} + j)^4$     **13.** $(-2\sqrt{3} - 2j)^3$

**14.** $(-2 + 2j)^6$     **15.** $(1 + j\sqrt{3})^5$     **16.** $(-4 - 3j)^5$

**17.** $(-3 + 3j)^{-4}$     **18.** $(-1 - j\sqrt{3})^{-3}$

*In Exercises 19–22, write each result in rectangular form.*

**19.** Find the three cube roots of 1.     **20.** Find the four fourth roots of $j$.

**21.** Find the five fifth roots of $j$.     **22.** Find the six sixth roots of $-1$.

*Find all complex solutions of each equation. Write each solution in both trigonometric and rectangular forms.*

**23.** $x^3 = 27(\cos 405° + j \sin 405°)$     **24.** $x^4 = 81(\cos 180° + j \sin 180°)$

**25.** $x^2 = j$     **26.** $x^3 = -1$

**27.** $x^5 = 1$     **28.** $x^3 = -j$

**29.** $x^5 = -1$     **30.** $x^6 = j$

**31.** $x^4 = -16$     **32.** $x^3 = 8$

**33.** $x^3 = -2 + 11j$     **34.** $x^2 = 1 + j$

## CHAPTER SUMMARY

1. A **complex number** is any number in the form $a + bj$, where $a$ and $b$ are real numbers and $j = \sqrt{-1}$. If $a = 0$, $a + bj$ is an imaginary number; if $b = 0$, $a + bj$ is a real number.

2. The **rectangular form** of a complex number is $a + bj$, where $a$ and $b$ are real numbers; $a$ is called the *real part* and $bj$ is called the *imaginary part*.

3. Two complex numbers $a + bj$ and $c + dj$ are equal only when $a = c$ and $b = d$.

4. *Operations with complex numbers in rectangular form:*
   (a) *Addition:* $(a + bj) + (c + dj) = (a + c) + (b + d)j$
   (b) *Subtraction:* $(a + bj) - (c + dj) = (a - c) + (b - d)j$
   (c) *Multiplication:* $(a + bj)(c + dj) = (ac - bd) + (ad + bc)j$; or multiply the two complex numbers as if they were binomials, and simplify.
   (d) *Division:* To divide two complex numbers, multiply both numerator and denominator by the conjugate of the denominator and simplify.

   *Note:* The complex numbers $a + bj$ and $a - bj$ are **conjugates** of each other.

5. The **complex plane** is shown in Fig. 14.9. Complex numbers may be added or subtracted graphically as vectors.

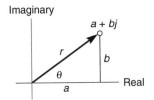

**Figure 14.9**

6. The trigonometric form of

$$a + bj = r(\cos \theta + j \sin \theta) \quad \text{or} \quad r\angle\theta$$

7. The exponential form of the complex number

$$r(\cos \theta + j \sin \theta) = re^{j\theta} \quad \text{where } \theta \text{ is in radians}$$

8. *The product of two complex numbers:*
   (a) *In exponential form:*

$$r_1 e^{j\theta_1} \cdot r_2 e^{j\theta_2} = r_1 r_2 e^{j(\theta_1 + \theta_2)}$$

   (b) *In trigonometric form:*

$$r_1(\cos \theta_1 + j \sin \theta_1) \cdot r_2(\cos \theta_2 + j \sin \theta_2) = r_1 r_2[\cos (\theta_1 + \theta_2) + j \sin(\theta_1 + \theta_2)]$$
$$\text{or} \quad (r_1 \angle \theta_1)(r_2 \angle \theta_2) = r_1 r_2 \angle \theta_1 + \theta_2$$

9. *The quotient of two complex numbers:*
   (a) *In exponential form:*

$$\frac{r_1 e^{j\theta_1}}{r_2 e^{j\theta_2}} = \frac{r_1}{r_2} e^{j(\theta_1 - \theta_2)}$$

   (b) *In trigonometric form:*

$$\frac{r_1(\cos \theta_1 + j \sin \theta_1)}{r_2(\cos \theta_2 + j \sin \theta_2)} = \frac{r_1}{r_2}[\cos(\theta_1 - \theta_2) + j \sin(\theta_1 - \theta_2)]$$

$$\text{or} \quad \frac{r_1 \angle \theta_1}{r_2 \angle \theta_2} = \frac{r_1}{r_2} \angle \theta_1 - \theta_2$$

**10.** *The nth power of a complex number:*
   (a) *In exponential form:*

$$(re^{j\theta})^n = r^n e^{jn\theta}$$

   (b) *In trigonometric form:*

$$[r(\cos\theta + j\sin\theta)]^n = r^n(\cos n\theta + j\sin n\theta) \qquad \text{(DeMoivre's theorem)}$$
$$\text{or} \quad (r\angle\theta)^n = r^n\,\angle n\theta$$

**11.** *DeMoivre's theorem for roots:*

$$[r(\cos\theta + j\sin\theta)]^{1/n} = r^{1/n}\left[\cos\left(\frac{\theta + k\cdot 360°}{n}\right) + j\sin\left(\frac{\theta + k\cdot 360°}{n}\right)\right]$$
$$\text{where } k = 0, 1, 2, \ldots, n-1$$

## CHAPTER 14 REVIEW

*Express in terms of j and simplify.*

**1.** $\sqrt{-81}$

**2.** $\sqrt{-18}$

*Simplify.*

**3.** $j^{18}$   **4.** $j^{23}$   **5.** $j^{48}$   **6.** $j^{145}$

*Perform the indicated operations in rectangular form and simplify.*

**7.** $(9 + 3j) + (-4 + 7j)$   **8.** $(-1 + j) - (-4 + 5j)$

**9.** $(5 + 2j)(6 - 7j)$   **10.** $(3 - 2j)^2$

**11.** $\dfrac{1 - 2j}{4 + j}$   **12.** $\dfrac{5 + j}{3 - 7j}$

*Solve.*

**13.** $x^2 + 36 = 0$   **14.** $2x^2 + 3x + 2 = 0$

*Write in trigonometric and exponential forms.*

**15.** $-1 + j$   **16.** $1 - \sqrt{3}j$

*Write in rectangular and exponential forms.*

**17.** $6(\cos 315° + j\sin 315°)$   **18.** $4(\cos 210° + j\sin 210°)$

*Write in trigonometric and rectangular forms.*

**19.** $2e^{0.489j}$   **20.** $3e^{8.75j}$

*Do as indicated and write the result in the given form.*

**21.** $(5e^{2j})(3e^{3j})$   **22.** $[2(\cos 150° + j\sin 150°)][4(\cos 300° + j\sin 300°)]$

**23.** $\dfrac{12e^{3j}}{3e^{-2j}}$   **24.** $\dfrac{24\angle 150°}{8\angle 275°}$

**25.** $(4e^{2j})^3$     **26.** $[2(\cos 60° + j \sin 60°)]^7$

*Find each power and write the result in rectangular form.*

**27.** $(-2 + 2j)^4$     **28.** $(1 + j\sqrt{3})^6$     **29.** $(1 + j)^{-4}$

*Find all complex solutions of each equation. Write each solution in both trigonometric form and rectangular form.*

**30.** $x^3 = j$     **31.** $x^4 = -1$     **32.** $x^4 = 16$     **33.** $x^5 = 4 - 4j$

# 15
# Analytic Geometry

## INTRODUCTION

Parabolas, circles, ellipses, and hyperbolas are conic sections; that is, curves formed by the intersection of a plane with a right circular cone. These curves occur often in nature and play an important role in applied mathematics. For example, the existence of the focus of a parabola is what makes flashlights, microphones, and satellite dishes work. Planets orbit the sun in elliptical paths, and wheels and gears help to keep us mobile. Analytic geometry, which we use to study these curves, evolved from the work of René Descartes, a French mathematician and philosopher in the seventeenth century.

### Objectives

- Graph circles, ellipses, parabolas, and hyperbolas.
- Find the center and radius of a circle.
- Find the vertex, directrix, and focus of a parabola.
- Find the vertices, foci, and lengths of the major and minor axes of an ellipse.
- Find the vertices, foci, and lengths of the transverse and conjugate axes of a hyperbola.
- Use translation of axes in sketching graphs and identifying key features.
- Solve systems of quadratic equations.
- Graph using polar coordinates.
- Convert between polar and rectangular coordinates.

## 15.1 THE CIRCLE

Recall that **analytic geometry** is the study of the relationships between algebra and geometry. We began developing these relationships in Chapter 4 and now extend the development to general second-degree equations.

Equations in two variables of second degree in the form

$$Ax^2 + Bxy + Cy^2 + Dx + Ey + F = 0$$

are called **conics.** We now begin a systematic study of conics with the circle.

The **circle** consists of the set of points located the same distance from a given point, called the **center.** The distance at which all points are located from the center is called the **radius.** A circle may thus be graphed in the plane given its center and radius.

**EXAMPLE 1**

Graph the circle whose center is $(1, -2)$ and whose radius $r = 3$.

Plot all points in the plane located 3 units away from the point $(1, -2)$ as in Fig. 15.1. (You may wish to use a compass.)

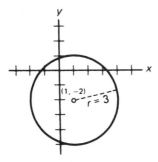

**Figure 15.1**

From the definition of a circle we can determine the equation of a circle. Let $(h, k)$ be the coordinates of the center, and let $r$ represent the radius. If any point $(x, y)$ is located on the circle, it must be a distance $r$ from the center $(h, k)$ as in Fig. 15.2.

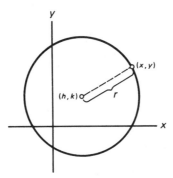

**Figure 15.2**

Using the distance formula,

$$\sqrt{(x_2 - x_1)^2 + (y_2 - y_1)^2} = d$$

we find that $\sqrt{(x - h)^2 + (y - k)^2} = r$. Squaring each side, we have the following.

**STANDARD FORM OF A CIRCLE**

$$(x - h)^2 + (y - k)^2 = r^2$$

where $r$ is the radius and $(h, k)$ is the center

Any point $(x, y)$ satisfying this equation must lie on the circle.

**EXAMPLE 2**

Find the equation of the circle with radius 3 and center $(1, -2)$. (See Example 1.)

Using the standard form of the equation of a circle, we have

$$(x - h)^2 + (y - k)^2 = r^2$$
$$(x - 1)^2 + [y - (-2)]^2 = (3)^2$$
$$(x - 1)^2 + (y + 2)^2 = 9$$

**EXAMPLE 3**

Find the equation of the circle with center at $(3, -2)$ and passing through $(-1, 1)$.

To write the equation, we need to know the radius $r$ of the circle. While $r$ has not been stated, we do know that every point on the circle is a distance $r$ from the center, $(3, -2)$. In particular, the point $(-1, 1)$ is a distance $r$ from $(3, -2)$ (see Fig. 15.3).

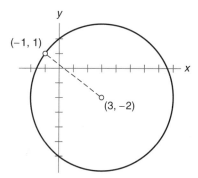

**Figure 15.3**

Using the distance formula,

$$d = \sqrt{(x_2 - x_1)^2 + (y_2 - y_1)^2}$$
$$d = r = \sqrt{(x - h)^2 + (y - k)^2}$$
$$r = \sqrt{(-1 - 3)^2 + [1 - (-2)]^2}$$
$$= \sqrt{(-4)^2 + 3^2} = \sqrt{16 + 9} = \sqrt{25} = 5$$

Now write the equation of the circle.

$$(x - h)^2 + (y - k)^2 = r^2$$
$$(x - 3)^2 + [y - (-2)]^2 = 5^2$$
$$(x - 3)^2 + (y + 2)^2 = 25$$

If we perform the multiplications in the equation

$$(x - h)^2 + (y - k)^2 = r^2$$

we have

$$x^2 - 2xh + h^2 + y^2 - 2yk + k^2 = r^2$$

Rearranging terms, we have

$$x^2 + y^2 - 2hx - 2ky + h^2 + k^2 - r^2 = 0$$

If we let $D = -2h$, $E = -2k$, and $F = h^2 + k^2 - r^2$, we obtain the following equation.

> **GENERAL FORM OF A CIRCLE**
>
> $$x^2 + y^2 + Dx + Ey + F = 0$$

Any equation in this form represents a circle.

### EXAMPLE 4

Write the equation $(x - 3)^2 + (y + 2)^2 = 25$ obtained in Example 3 in general form.

$$(x - 3)^2 + (y + 2)^2 = 25$$
$$x^2 - 6x + 9 + y^2 + 4y + 4 = 25$$
$$x^2 + y^2 - 6x + 4y - 12 = 0$$

### EXAMPLE 5

Find the center and the radius of the circle given by the equation

$$x^2 + y^2 - 4x + 2y - 11 = 0$$

Looking back at how we arrived at the general equation of a circle, we see that if we rearrange the terms of the equation

$$(x^2 - 4x \quad) + (y^2 + 2y \quad) = 11$$

then $(x^2 - 4x \quad)$ represents the first two terms of

$$(x - h)^2 = x^2 - 2hx + h^2$$

and $(y^2 + 2y \quad)$ represents the first two terms of

$$(y - k)^2 = y^2 - 2ky + k^2$$

This means that

$$-4 = -2h \quad \text{or} \quad h = 2$$

and

$$2 = -2k \quad \text{or} \quad k = -1$$

To complete the squares $(x - h)^2$ and $(y - k)^2$, we must add $h^2 = 2^2 = 4$ and $k^2 = (-1)^2 = 1$ to each side of the equation.

$$(x^2 - 4x \quad) + (y^2 + 2y \quad) = 11$$
$$(x^2 - 4x + 4) + (y^2 + 2y + 1) = 11 + 4 + 1$$
$$(x - 2)^2 + (y + 1)^2 = 16$$
$$(x - 2)^2 + [y - (-1)]^2 = 4^2$$

From this we see that we have the standard form of the equation of a circle with radius 4 and center at the point $(2, -1)$.

This process is called **completing the square** of the x- and y-terms. In general, if the coefficients of $x^2$ and $y^2$ are both equal to 1, then these values can be found as follows: Add $h^2$ and $k^2$ to each side of the equation, where

$$h^2 = (\tfrac{1}{2} \text{ the coefficient of } x)^2 = (\tfrac{1}{2}D)^2$$
$$k^2 = (\tfrac{1}{2} \text{ the coefficient of } y)^2 = (\tfrac{1}{2}E)^2$$

**EXAMPLE 6**

Find the center and the radius of the circle given by the equation

$$x^2 + y^2 + 6x - 4y - 12 = 0$$

Also sketch the graph of the circle.

$$h^2 = (\tfrac{1}{2} \cdot 6)^2 = 3^2 = 9$$
$$k^2 = [\tfrac{1}{2}(-4)]^2 = (-2)^2 = 4$$

Rewrite the equation and add 9 and 4 to each side.

$$(x^2 + 6x \quad) + (y^2 - 4y \quad) = 12$$
$$(x^2 + 6x + 9) + (y^2 - 4y + 4) = 12 + 9 + 4$$
$$(x + 3)^2 + (y - 2)^2 = 25 = 5^2$$

The center is at $(-3, 2)$ and the radius is 5. Plot all points that are at a distance of 5 from the point $(-3, 2)$ as in Fig. 15.4.

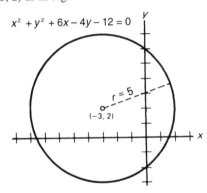

Figure 15.4

If the center of a circle is at the origin, then $h = 0$ and $k = 0$, and its standard equation becomes the following.

$$x^2 + y^2 = r^2$$

where $r$ is the radius and the center is at the origin

**EXAMPLE 7**

Find the equation of the circle with radius 3 and center at the origin. Also, graph the circle.

$$x^2 + y^2 = r^2$$
$$x^2 + y^2 = (3)^2$$
$$x^2 + y^2 = 9$$

The graph is shown in Fig. 15.5.

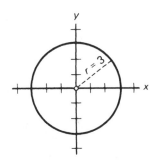

**Figure 15.5**

## Exercises 15.1

*Graph the circle with the given center and radius.*

**1.** Center at $(2, -1)$, $r = 3$.

**2.** Center at $(3, 3)$, $r = 2$.

**3.** Center at $(0, 2)$, $r = 4$.

**4.** Center at $(-4, -5)$, $r = 3$.

*Find the equation of the circle (in standard form) with the given properties.*

**5.** Center at $(1, -1)$, radius 4.

**6.** Center at $(-2, 3)$, radius $\sqrt{5}$.

**7.** Center at $(-2, -4)$, passing through $(1, -9)$.

**8.** Center at $(5, 2)$, passing through $(-2, -6)$.

**9.** Center at $(0, 0)$, radius 6.

**10.** Center at $(0, 0)$, passing through $(3, -4)$.

*Find the center and radius of the given circle.*

**11.** $x^2 + y^2 = 16$

**12.** $x^2 + y^2 - 4x - 5 = 0$

**13.** $x^2 + y^2 + 6x - 8y - 39 = 0$

**14.** $x^2 + y^2 - 6x + 14y + 42 = 0$

**15.** $x^2 + y^2 - 8x + 12y - 8 = 0$

**16.** $x^2 + y^2 + 10x + 2y - 14 = 0$

**17.** $x^2 + y^2 - 12x - 2y - 12 = 0$

**18.** $x^2 + y^2 + 4x - 9y + 4 = 0$

**19.** $x^2 + y^2 + 7x + 3y - 9 = 0$

**20.** $x^2 + y^2 - 5x - 8y = 0$

**21.** Find the equation of the circle or circles whose center is on the $y$-axis and contains the points $(1, 4)$ and $(-3, 2)$. Give its center and radius.

**22.** Find the equation of the circle with center in the first quadrant on the line $y = 2x$, tangent to the $x$-axis, and radius 6. Give its center.

**23.** Find the equation of the circle containing the points $(3, 1)$, $(0, 0)$, and $(8, 4)$. Give its center and radius.

**24.** Find the equation of the circle containing the points $(1, -4)$, $(-3, 4)$, and $(4, 5)$. Give its center and radius.

## 15.2   THE PARABOLA

While the parabola may not be as familiar a geometric curve as the circle, examples of the parabola are found in many technical applications. A **parabola** consists of all points that are the same distance from a given fixed point and a given fixed line. The fixed point

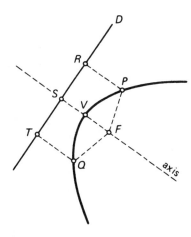

**Figure 15.6**

is called the **focus.** The fixed line is called the **directrix.** This relationship is shown in Fig. 15.6 for the points $P$, $Q$, and $V$, which lie on a parabola with focus $F$ and directrix $D$.

*Note:*

$$RP = PF$$
$$SV = VF$$
$$TQ = QF$$

The point $V$ midway between the directrix and the focus is called the **vertex.** The vertex and the focus lie on a line perpendicular to the directrix, which is called the **axis of symmetry.**

There are two standard forms for the equation of a parabola. The form depends on the position of the parabola in the plane. We first discuss the parabola with focus on the x-axis at $(p, 0)$ and directrix the line $x = -p$ as in Fig. 15.7. Let $P(x, y)$ represent any point on this parabola. Vertex $V$ is then at the origin, and the axis of symmetry is the x-axis.

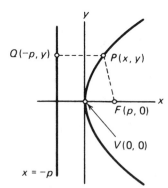

**Figure 15.7**

By the way we have described the parabola, the distance between $P$ and $F$ must equal the distance between $P$ and $Q$. Using the distance formula, we have

$$PF = PQ$$
$$\sqrt{(x - p)^2 + (y - 0)^2} = \sqrt{[x - (-p)]^2 + (y - y)^2}$$

Squaring each side yields

$$(x - p)^2 + y^2 = (x + p)^2$$
$$x^2 - 2px + p^2 + y^2 = x^2 + 2px + p^2$$

---

**STANDARD FORM OF PARABOLA**

$$y^2 = 4px$$

with focus at $(p, 0)$ and with the line $x = -p$ as the directrix

---

Note that in Fig. 15.7, $p > 0$.

**EXAMPLE 1**

Find the equation of the parabola with focus at $(3, 0)$ and directrix $x = -3$.
In this case $p = 3$, so we have

$$y^2 = 4(3)x$$
$$y^2 = 12x$$

**EXAMPLE 2**

Find the focus and the equation of the directrix of the parabola $y^2 = 24x$.

$$y^2 = 24x$$
$$y^2 = 4(6)x$$
$$y^2 = 4px$$

Since $p$ must be 6, the focus is $(6, 0)$ and the directrix is the line $x = -6$.

**EXAMPLE 3**

Find the equation of the parabola with focus at $(-2, 0)$ and with directrix $x = 2$. Sketch the graph.
Here $p = -2$. The equation becomes

$$y^2 = 4(-2)x$$
$$y^2 = -8x$$

The graph is shown in Fig. 15.8.

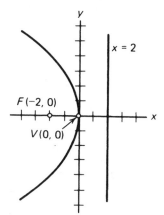

**Figure 15.8**

Note that when $p < 0$, the parabola opens to the left and that the coefficient of $x$ in the equation is negative.

1. If $p > 0$, the coefficient of $x$ in the equation $y^2 = 4px$ is *positive* and the parabola opens to the *right* [see Fig. 15.9(a)].

2. If $p < 0$, the coefficient of $x$ in the equation $y^2 = 4px$ is *negative* and the parabola opens to the *left* [see Fig. 15.9(b)].

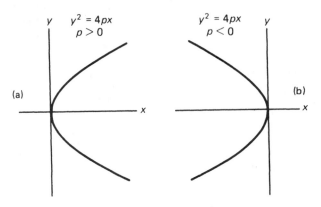

Figure 15.9

We obtain the other standard form of the parabola when the focus lies on the $y$-axis and the directrix is parallel to the $x$-axis. Let $(0, p)$ be the focus $F$ and $y = -p$ be the directrix. The vertex is still at the origin, but the axis of symmetry is now the $y$-axis (see Fig. 15.10).

---

**STANDARD FORM OF PARABOLA**

$$x^2 = 4py$$

with focus at $(0, p)$ and with the line $y = -p$ as the directrix (see Fig. 15.10).

---

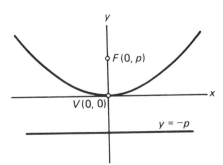

Figure 15.10

**EXAMPLE 4**

Find the equation of the parabola with focus at $(0, 3)$ and with directrix $y = -3$. Sketch the graph.

Since the focus lies on the $y$-axis and the directrix is parallel to the $x$-axis, we use the equation $x^2 = 4py$ with $p = 3$ (see Fig. 15.11).

$$x^2 = 4(3)y$$
$$x^2 = 12y$$

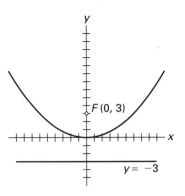

**Figure 15.11**

### EXAMPLE 5

Find the equation of the parabola with focus at $(0, -1)$ and with directrix $y = 1$. Sketch the graph.

Again, the focus lies on the $y$-axis with directrix parallel to the $x$-axis, so we use the equation $x^2 = 4py$ with $p = -1$ (see Fig. 15.12).

$$x^2 = 4(-1)y$$
$$x^2 = -4y$$

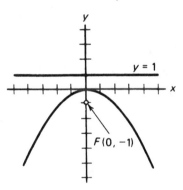

**Figure 15.12**

Observe the following:

1. If $p > 0$, the coefficient of $y$ in the equation $x^2 = 4py$ is *positive* and the parabola opens *upward* [see Fig. 15.13(a)].

2. If $p < 0$, the coefficient of $y$ in the equation $x^2 = 4py$ is *negative* and the parabola opens *downward* [see Fig. 15.13(b)].

We are now able to describe the graph of a parabola with an equation in standard form by inspection. We can also find the focus and the directrix.

### EXAMPLE 6

Describe the graph of the equation $y^2 = 20x$.

This is an equation of a parabola in the form $y^2 = 4px$. Since $p = 5$, this parabola has its focus at $(5, 0)$, and its directrix is the line $x = -5$. The parabola opens to the right (since $p > 0$).

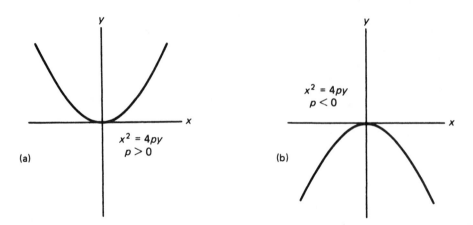

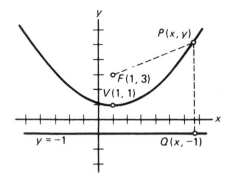

**Figure 15.13**

**EXAMPLE 7**

Describe the graph of the equation $x^2 = -2y$.

This is an equation of a parabola in the form $x^2 = 4py$, where $p = -\frac{1}{2}$ [as $4(-\frac{1}{2}) = -2$]. The focus is at $(0, -\frac{1}{2})$, and the directrix is the line $y = \frac{1}{2}$. The parabola opens downward (since $p < 0$).

Of course, not all parabolas are given in standard position.

**EXAMPLE 8**

Find the equation of the parabola with focus at $(1, 3)$ and with the line $y = -1$ as directrix.

We must use the definition of the parabola (see Fig. 15.14).

$$PF = PQ$$
$$\sqrt{(x - 1)^2 + (y - 3)^2} = \sqrt{(x - x)^2 + [y - (-1)]^2}$$

**Figure 15.14**

This simplifies to

$$x^2 - 2x - 8y + 9 = 0$$

In fact, any equation of the form

$$Ax^2 + Dx + Ey + F = 0$$

or

$$Cy^2 + Dx + Ey + F = 0$$

represents a parabola.

In graphing a parabola in the form $y = f(x) = ax^2 + bx + c$, $a \neq 0$, it is most helpful to graph the x-intercepts, if any, and the vertex. To find the x-intercepts, let $y = 0$ and solve $ax^2 + bx + c = 0$ for $x$. The solutions for this equation are given by the quadratic formula:

$$x = \frac{-b \pm \sqrt{b^2 - 4ac}}{2a}$$

Recall that the solutions are real numbers only if the discriminant $b^2 - 4ac$ is non-negative and that the solutions are imaginary if $b^2 - 4ac < 0$. Thus, the graph of the parabola $y = f(x) = ax^2 + bx + c$, $a \neq 0$, has

1. two different x-intercepts if $b^2 - 4ac > 0$,

2. only one x-intercept if $b^2 - 4ac = 0$ (and the graph is tangent to the x-axis), and

3. no x-intercepts if $b^2 - 4ac < 0$.

The **axis** of the parabola in the form $y = f(x) = ax^2 + bx + c$ is a vertical line halfway between the x-intercepts. The equation of the axis is the vertical line passing through the midpoint of the line segment joining the two x-intercepts (see Fig. 15.15). This midpoint is

$$\frac{x_1 + x_2}{2} = \frac{\dfrac{-b - \sqrt{b^2 - 4ac}}{2a} + \dfrac{-b + \sqrt{b^2 - 4ac}}{2a}}{2}$$

$$= \frac{\dfrac{-2b}{2a}}{2} = -\frac{b}{2a}$$

Thus, the equation of the axis is $x = -\dfrac{b}{2a}$.

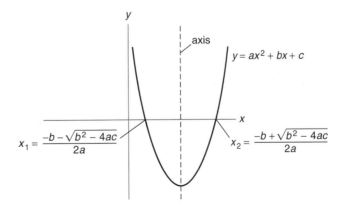

**Figure 15.15**

Since the parabola contains the vertex, its $x$-coordinate is $-\dfrac{b}{2a}$. Its $y$-coordinate is then $f\left(-\dfrac{b}{2a}\right)$. To find the $y$-coordinate, evaluate the following:

$$f(x) = ax^2 + bx + c$$

$$f\left(-\frac{b}{2a}\right) = a\left(-\frac{b}{2a}\right)^2 + b\left(-\frac{b}{2a}\right) + c$$

$$= \frac{b^2}{4a} - \frac{b^2}{2a} + c$$

$$= \frac{b^2}{4a} - \frac{2b^2}{4a} + \frac{4ac}{4a}$$

$$= \frac{-b^2 + 4ac}{4a}$$

---

**AXIS AND VERTEX OF A PARABOLA**

Given the parabola $y = f(x) = ax^2 + bx + c$, its axis is the vertical line $x = -\dfrac{b}{2a}$ and its vertex is the point

$$\left(-\frac{b}{2a}, f\left(-\frac{b}{2a}\right)\right) = \left(-\frac{b}{2a}, \frac{-b^2 + 4ac}{4a}\right)$$

The vertex is a maximum if $a < 0$ and a minimum if $a > 0$.

---

**EXAMPLE 9**

Graph $y = f(x) = 2x^2 - 8x + 11$. Find its vertex and the equation of the axis.

First, note that $b^2 - 4ac = (-8)^2 - 4(2)(11) = -24 < 0$, which means that the graph has no $x$-intercepts. The equation of the axis is

$$x = -\frac{b}{2a} = -\frac{-8}{2(2)} = 2$$

The vertex is the point $(2, f(2))$ or $(2, 2 \cdot 2^2 - 8 \cdot 2 + 11) = (2, 3)$. This $y$-coordinate may also be found using the formula

$$\frac{-b^2 + 4ac}{4a} = \frac{-(-8)^2 + 4 \cdot 2 \cdot 11}{4 \cdot 2} = \frac{24}{8} = 3$$

Since $a > 0$, the vertex $(2, 3)$ is a minimum point, and the graph opens upward. You may also find some additional ordered pairs to graph this equation, depending on whether you need only a rough sketch or a fairly accurate graph. The graph is shown in Fig. 15.16.

Since the vertex of a parabola in the form $y = f(x) = ax^2 + bx + c$ is the highest point or the lowest point on the graph, we can use this fact to find a maximum or a minimum value of a quadratic function.

**EXAMPLE 10**

An object is thrown upward with an initial velocity of 48 ft/s. Its height after $t$ seconds is given by $h = f(t) = 48t - 16t^2$. Find its maximum height and the time it takes the object to hit the ground.

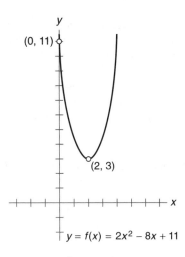

$(0, 11)$

$(2, 3)$

$y = f(x) = 2x^2 - 8x + 11$

**Figure 15.16**

First, find the vertex.

$$-\frac{b}{2a} = -\frac{48}{2(-16)} = \frac{3}{2}$$

Then the vertex is

$$\left(\frac{3}{2}, f\left(\frac{3}{2}\right)\right) = \left(\frac{3}{2}, 48\left(\frac{3}{2}\right) - 16\left(\frac{3}{2}\right)^2\right) = \left(\frac{3}{2}, 36\right)$$

Since $a = -16 < 0$, the vertex is a maximum value, and the maximum height is 36 ft.

The first coordinate of the vertex gives the amount of time it takes the object to reach its maximum height. The time it takes such a projectile to reach its maximum height is the same as the time it takes to drop back to the ground. Thus, the object hits the ground $2 \cdot \frac{3}{2}$ or 3 s after it is thrown.

## Exercises 15.2

*Find the focus and the directrix of each parabola. Sketch each graph.*

**1.** $x^2 = 4y$        **2.** $x^2 = -8y$        **3.** $y^2 = -16x$

**4.** $x^2 = -6y$       **5.** $y^2 = x$        **6.** $y^2 = -4x$

**7.** $x^2 = 16y$       **8.** $y^2 = -12x$       **9.** $y^2 = 8x$

**10.** $x^2 = -y$

*Find the equation of the parabola with given focus and directrix.*

**11.** $(2, 0)$, $x = -2$      **12.** $(0, -3)$, $y = 3$      **13.** $(-8, 0)$, $x = 8$

**14.** $(5, 0)$, $x = -5$      **15.** $(0, 6)$, $y = -6$      **16.** $(0, -1)$, $y = 1$

**17.** Find the equation of the parabola with focus at $(-4, 0)$ and vertex at $(0, 0)$.

**18.** Find the equation of the parabola with vertex at $(0, 0)$ and directrix $y = -2$.

**19.** Find the equation of the parabola with focus $(-1, 3)$ and directrix $x = 3$.

**20.** Find the equation of the parabola with focus $(2, -5)$ and directrix $y = -1$.

21. The surface of a roadway over a bridge follows a parabolic curve with vertex at the middle of the bridge. The span of the bridge is 400 m. The roadway is 16 m higher in the middle than at the end supports. How far above the end supports is a point 50 m from the middle? 150 m from the middle?

22. The shape of a wire hanging between two poles closely approximates a parabola. Find the equation of a wire which is suspended between two poles 40 m apart and whose lowest point is 10 m below the level of the insulators. (Choose the lowest point as the origin of your coordinate system.)

23. A suspension bridge is supported by two cables that hang between two supports. The curve of these cables is approximately parabolic. Find the equation of this curve if the focus lies 8 m above the lowest point of the cable. (Set up the $xy$-coordinate system so that the vertex is at the origin.)

24. A culvert is shaped like a parabola, 120 cm across the top and 80 cm deep. How wide is the culvert 50 cm from the top?

*Graph each parabola. Find its vertex and the equation of the axis.*

25. $y = 2x^2 + 7x - 15$

26. $y = -x^2 - 6x - 8$

27. $f(x) = -2x^2 + 4x + 16$

28. $f(x) = 3x^2 + 6x + 10$

29. Starting at (0, 0), a projectile travels along the path $y = f(x) = -\dfrac{1}{256}x^2 + 4x$, where $x$ is in metres. Find **(a)** the maximum height and **(b)** the range of the projectile.

30. The height of a bullet fired vertically upward is given by $h = f(t) = 1200t - 16t^2$ (the initial velocity is 1200 ft/s). Find **(a)** its maximum height and **(b)** the time it takes to hit the ground.

31. Enclose a rectangular area with 240 m of fencing. Find the largest possible area that can be enclosed.

32. A 36-in.-wide sheet of metal is bent into a rectangular trough with a cross section as shown in Fig. 15.17. What dimensions will maximize the flow of water? That is, what dimensions will maximize the cross-sectional area?

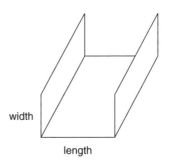

width

length

**Figure 15.17**

## 15.3 THE ELLIPSE

An **ellipse** consists of the set of points in a plane, the *sum* of whose distances from two fixed points is a positive constant. These two fixed points are called **foci**. As in Fig. 15.18, let the foci lie on the $x$-axis at $(-c, 0)$ and $(c, 0)$. Then any point $P(x, y)$ lies on the el-

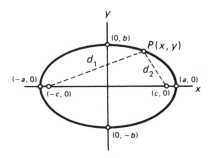

**Figure 15.18**

lipse if its distance $d_1$ from $P$ to the point $(-c, 0)$ plus its distance $d_2$ from $P$ to the point $(c, 0)$ is equal to a given constant $k$. Let the constant be written as $k = 2a$; then

$$d_1 + d_2 = 2a$$

Again using the formula for computing the distance between two points, we have

$$\sqrt{[x - (-c)]^2 + (y - 0)^2} + \sqrt{(x - c)^2 + (y - 0)^2} = 2a \qquad \textbf{(15.1)}$$

Rewrite Equation (15.1) as follows:

$$\sqrt{(x + c)^2 + y^2} = 2a - \sqrt{(x - c)^2 + y^2}$$
$$(x + c)^2 + y^2 = 4a^2 - 4a\sqrt{(x - c)^2 + y^2} + (x - c)^2 + y^2 \qquad \text{(Square each side.)}$$
$$x^2 + 2cx + c^2 + y^2 = 4a^2 - 4a\sqrt{(x - c)^2 + y^2} + x^2 - 2cx + c^2 + y^2$$
$$4cx - 4a^2 = -4a\sqrt{(x - c)^2 + y^2}$$
$$a^2 - cx = a\sqrt{(x - c)^2 + y^2} \qquad \text{(Divide each side by } -4.)$$
$$(a^2 - cx)^2 = a^2[(x - c)^2 + y^2] \qquad \text{(Square each side.)}$$
$$a^4 - 2a^2cx + c^2x^2 = a^2(x^2 - 2cx + c^2 + y^2)$$
$$a^4 - 2a^2cx + c^2x^2 = a^2x^2 - 2a^2cx + a^2c^2 + a^2y^2$$
$$a^4 - a^2c^2 = a^2x^2 - c^2x^2 + a^2y^2$$
$$a^2(a^2 - c^2) = (a^2 - c^2)x^2 + a^2y^2 \qquad \text{(Factor.)}$$
$$1 = \frac{x^2}{a^2} + \frac{y^2}{a^2 - c^2} \qquad \text{[Divide each side by } a^2(a^2 - c^2).] \qquad \textbf{(15.2)}$$

If we now let $y = 0$ in this equation, we find that $x^2 = a^2$. The points $(-a, 0)$ and $(a, 0)$ which lie on the graph are called **vertices** of the ellipse. Observe that $a > c$.

If we let $b^2 = a^2 - c^2$, Equation (15.2) then becomes

$$\frac{x^2}{a^2} + \frac{y^2}{b^2} = 1 \qquad \textbf{(15.3)}$$

The line segment connecting the vertices $(a, 0)$ and $(-a, 0)$ is called the **major axis.** The point midway between the vertices is called the **center** of the ellipse. In this case the major axis lies on the $x$-axis, and the center is at the origin. If we let $x = 0$ in Equation (15.3), we find $y^2 = b^2$. The line connecting $(0, b)$ and $(0, -b)$ is perpendicular to the major axis and passes through the center (see Fig. 15.18). This line is called the **minor axis** of the ellipse. In this case the minor axis lies on the $y$-axis. Note that $2a$ is the length of the major axis and $2b$ is the length of the minor axis.

**STANDARD FORM OF ELLIPSE**

$$\frac{x^2}{a^2} + \frac{y^2}{b^2} = 1$$

with center at the origin and with the major axis lying on the *x*-axis

*Note: a > b.*

One easy way to approximate the curve of an ellipse is to fix a string at two points ( foci) on a piece of paper as in Fig. 15.19. Then using a pencil to keep the string taut, trace out the curve as illustrated. Note that $d_1 + d_2$ is always constant—namely, the length of the string. Detach the string and compare the length of the string with the length of the major axis; note that the lengths are the same, $2a$.

The relationship $b^2 = a^2 - c^2$ or $a^2 = b^2 + c^2$ can also be seen from this string demonstration as in Fig. 15.20. Put a pencil inside the taut string and on an end of the minor axis; this bisects the length of string and sets up a right triangle with $a$ as its hypotenuse and $b$ and $c$ as the legs.

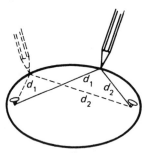

**Figure 15.19**

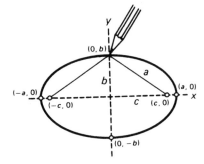

**Figure 15.20**

**EXAMPLE 1**

Find the vertices, the foci, and the lengths of the major and minor axes of the following ellipse:

$$\frac{x^2}{25} + \frac{y^2}{9} = 1$$

Sketch the graph.

Since $a^2 = 25$, the vertices are at $(5, 0)$ and $(-5, 0)$. The length of the major axis is $2a = 2(5) = 10$. Since $b^2 = 9$, the length of the minor axis is $2b = 2(3) = 6$. We need the value of $c$ to determine the foci. Since $b^2 = a^2 - c^2$, we can write

$$c^2 = a^2 - b^2 = 25 - 9 = 16$$

$$c = 4$$

The foci are thus $(4, 0)$ and $(-4, 0)$. The graph of the ellipse is shown in Fig. 15.21.

You will want to remember the equation relating $a$, $b$, and $c$ for the ellipse.

$$\boxed{c^2 = a^2 - b^2}$$

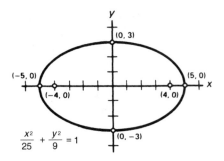

Figure 15.21

When the major axis lies on the $y$-axis with center at the origin as in Fig. 15.22, the **standard form** of the equation of the ellipse becomes the following.

---

**STANDARD FORM OF ELLIPSE**

$$\frac{y^2}{a^2} + \frac{x^2}{b^2} = 1$$

with center at the origin and with the major axis lying on the $y$-axis
*Note: $a > b$.*

---

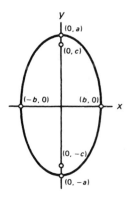

Figure 15.22

This result may be shown similarly as the derivation of the first standard form. Notice that the larger denominator now lies below $y^2$ instead of below $x^2$ as in the first case. The vertices are now $(0, a)$ and $(0, -a)$.

**EXAMPLE 2**

Given the ellipse $25x^2 + 9y^2 = 225$, find the foci, vertices, and lengths of the major and minor axes. Sketch the graph.

First divide each side of the equation by 225 to put the equation in standard form.

$$\frac{x^2}{9} + \frac{y^2}{25} = 1$$

Since the larger denominator belongs to the $y^2$-term, this ellipse has its major axis on the $y$-axis, and $a^2$ must then be 25. So $a = 5$ and $b = 3$. The vertices are $(0, 5)$ and $(0, -5)$. The length of the major axis is $2a = 10$, and the length of the minor axis is $2b = 6$.

$$c^2 = a^2 - b^2 = 25 - 9 = 16$$
$$c = 4$$

Thus, the foci are $(0, 4)$ and $(0, -4)$ (see Fig. 15.23).

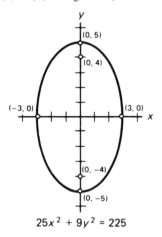

$$25x^2 + 9y^2 = 225$$

**Figure 15.23**

In general, $a$ is always greater than $b$ for an ellipse. The following are also true:

1. If the larger denominator belongs to the $x^2$-term, then this denominator is $a^2$; the major axis lies on the $x$-axis; and the vertices are $(a, 0)$ and $(-a, 0)$.

2. If the larger denominator belongs to the $y^2$-term, then this denominator is $a^2$; the major axis lies on the $y$-axis; and the vertices are $(0, a)$ and $(0, -a)$.

**EXAMPLE 3**

Find the equation of the ellipse with vertices at $(6, 0)$ and $(-6, 0)$ and foci at $(4, 0)$ and $(-4, 0)$.

Since $a = 6$ and $c = 4$, we have $a^2 = 36$ and $c^2 = 16$. Thus,

$$b^2 = a^2 - c^2 = 36 - 16 = 20$$

Since the major axis lies on the $x$-axis, the equation in standard form is

$$\frac{x^2}{a^2} + \frac{y^2}{b^2} = 1$$
$$\frac{x^2}{36} + \frac{y^2}{20} = 1$$

Ellipses with centers not located at the origin will be presented in Section 15.5.

If we were to determine the equation of the ellipse where the sum of the distances of all points from the foci $(-2, 3)$ and $(6, 3)$ is always 10, we would have

$$9x^2 + 25y^2 - 36x - 150y + 36 = 0$$

In general, an equation of the form

$$Ax^2 + Cy^2 + Dx + Ey + F = 0$$

represents an ellipse with axes parallel to the coordinate axes, where $A$ and $C$ are both positive (or both negative) and, unlike the circle, $A \neq C$.

## Exercises 15.3

*Find the vertices, foci, and lengths of the major and minor axes of each ellipse. Sketch each graph.*

1. $\dfrac{x^2}{25} + \dfrac{y^2}{16} = 1$

2. $\dfrac{x^2}{36} + \dfrac{y^2}{64} = 1$

3. $9x^2 + 16y^2 = 144$

4. $25x^2 + 16y^2 = 400$

5. $36x^2 + y^2 = 36$

6. $4x^2 + 3y^2 = 12$

7. $16x^2 + 9y^2 = 144$

8. $x^2 + 4y^2 = 16$

*Find the equation of each ellipse satisfying the given conditions.*

9. Vertices at $(4, 0)$ and $(-4, 0)$; foci at $(2, 0)$ and $(-2, 0)$.

10. Vertices at $(0, 7)$ and $(0, -7)$; foci at $(0, 5)$ and $(0, -5)$.

11. Vertices at $(0, 9)$ and $(0, -9)$; foci at $(0, 6)$ and $(0, -6)$.

12. Vertices at $(12, 0)$ and $(-12, 0)$; foci at $(10, 0)$ and $(-10, 0)$.

13. Vertices at $(6, 0)$ and $(-6, 0)$; length of minor axis is 10.

14. Vertices at $(0, 10)$ and $(0, -10)$; length of minor axis is 18.

15. Foci at $(0, 5)$ and $(0, -5)$; length of major axis is 16.

16. Foci at $(3, 0)$ and $(-3, 0)$; length of major axis is 8.

17. A weather satellite with an orbit about the earth reaches a minimum altitude of 1000 mi and a maximum altitude of 1600 mi. The path of its orbit is approximately an ellipse with the center of the earth at one focus. Find the equation of this curve. Assume that the radius of the earth is 4000 mi and the $x$-axis is the major axis.

18. An arch is in the shape of the upper half of an ellipse with a horizontal major axis supporting a foot bridge 40 m long over a stream in a park. The center of the arch is 8 m above the bridge supports. Find an equation of the ellipse. (Choose the point midway between the bridge supports as the origin.)

## 15.4 THE HYPERBOLA

A **hyperbola** consists of the set of points, the *difference* of whose distances from two fixed points is a positive constant. The two fixed points are called the **foci.**

Assume now as in Fig. 15.24 that the foci lie on the $x$-axis at $(-c, 0)$ and $(c, 0)$. Then a point $P(x, y)$ lies on the hyperbola if the difference between its distances to the foci is equal to a given constant $k$. That is, $d_1 - d_2 = k$ or $d_2 - d_1 = k$. Again, this constant $k$ equals $2a$; that is,

$$d_2 - d_1 = 2a$$

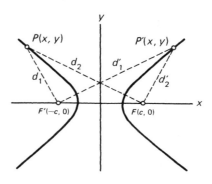

**Figure 15.24**

To obtain the equation of the hyperbola, use the distance formula.

$$d_2 - d_1 = 2a$$

$$\sqrt{(x-c)^2 + (y-0)^2} - \sqrt{[x-(-c)]^2 + (y-0)^2} = 2a \qquad \textbf{(15.4)}$$

Rewrite Equation (15.4) as follows:

$$\sqrt{(x-c)^2 + y^2} = 2a + \sqrt{(x+c)^2 + y^2}$$

$$(x-c)^2 + y^2 = 4a^2 + 4a\sqrt{(x+c)^2 + y^2} + (x+c)^2 + y^2 \qquad \text{(Square each side.)}$$

$$x^2 - 2cx + c^2 + y^2 = 4a^2 + 4a\sqrt{(x+c)^2 + y^2} + x^2 + 2cx + c^2 + y^2$$

$$-4a^2 - 4cx = 4a\sqrt{(x+c)^2 + y^2}$$

$$-a^2 - cx = a\sqrt{(x+c)^2 + y^2} \qquad \text{(Divide each side by 4.)}$$

$$a^4 + 2a^2cx + c^2x^2 = a^2[(x+c)^2 + y^2] \qquad \text{(Square each side.)}$$

$$a^4 + 2a^2cx + c^2x^2 = a^2(x^2 + 2cx + c^2 + y^2)$$

$$a^4 + 2a^2cx + c^2x^2 = a^2x^2 + 2a^2cx + a^2c^2 + a^2y^2$$

$$a^4 - a^2c^2 = a^2x^2 - c^2x^2 + a^2y^2$$

$$a^2(a^2 - c^2) = (a^2 - c^2)x^2 + a^2y^2 \qquad \text{(Factor.)}$$

$$1 = \frac{x^2}{a^2} + \frac{y^2}{a^2 - c^2} \qquad \text{[Divide each side by } a^2(a^2 - c^2).\text{]}$$

$$1 = \frac{x^2}{a^2} - \frac{y^2}{c^2 - a^2}$$

In triangle $F'PF$,

$$\overline{PF'} < \overline{PF} + \overline{FF'} \qquad \text{(The sum of any two sides of a triangle is greater than the third side.)}$$

$$\overline{PF'} - \overline{PF} < \overline{FF'}$$

$$2a < 2c \qquad (\overline{PF'} - \overline{PF} = 2a \text{ by the definition of a hyperbola and } \overline{FF'} = 2c.)$$

$$a < c$$

$$a^2 < c^2 \qquad \text{(Since } a > 0 \text{ and } c > 0)$$

$$0 < c^2 - a^2$$

Since $c^2 - a^2$ is positive, we may replace it by the positive number $b^2$ as follows:

$$1 = \frac{x^2}{a^2} - \frac{y^2}{b^2}$$

where $b^2 = c^2 - a^2$.

The equation of the hyperbola with foci on the $x$-axis at $(c, 0)$ and $(-c, 0)$ is as follows.

$$\boxed{\frac{x^2}{a^2} - \frac{y^2}{b^2} = 1}$$

The points $(a, 0)$ and $(-a, 0)$ are called the **vertices.** The line segment connecting the vertices is called the **transverse axis.** The vertices and the transverse axis in this case lie on the $x$-axis. The length of the transverse axis is $2a$. The line segment connecting the points $(0, b)$ and $(0, -b)$ is called the **conjugate axis** and in this case lies on the $y$-axis. The length of the conjugate axis is $2b$. The **center** lies at the intersection of the conjugate and the transverse axes.

---

**STANDARD FORM OF HYPERBOLA**

$$\frac{x^2}{a^2} - \frac{y^2}{b^2} = 1$$

with center at the origin and with the transverse axis lying on the $x$-axis

---

If we draw the central rectangle as in Fig. 15.25 and draw lines passing through opposite vertices of the rectangle, we obtain lines called the **asymptotes** of the hyperbola. In this case the equations of these lines are as follows.

$$\boxed{\begin{array}{c} y = \dfrac{b}{a}x \\[2mm] y = -\dfrac{b}{a}x \end{array}}$$

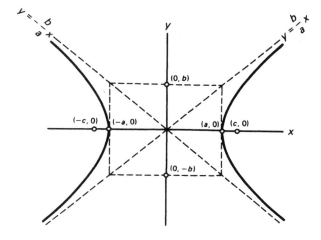

Figure 15.25

Asymptotes serve as guidelines to the branches of the hyperbola. That is, as the distance from the center of the hyperbola increases, the points on the branches get closer and closer to the asymptotes but never cross or touch.

To sketch the graph of the hyperbola:

1. Locate the vertices $(a, 0)$ and $(-a, 0)$.
2. Locate the points $(0, b)$ and $(0, -b)$.
3. Sketch the central rectangle as in Fig. 15.25. [The coordinates of the vertices are $(a, b)$, $(a, -b)$, $(-a, b)$, and $(-a, -b)$.]
4. Sketch the two asymptotes: the lines passing through the pairs of opposite vertices of the rectangle.
5. Sketch the branches of the hyperbola.

**EXAMPLE 1**

Find the vertices, foci, and lengths of the transverse and the conjugate axes of the following hyperbola.

$$\frac{x^2}{9} - \frac{y^2}{16} = 1$$

Sketch the graph. Find the equations of the asymptotes.

Since 9 is the denominator of the $x^2$-term, $a^2 = 9$ and $a = 3$. The vertices are therefore $(3, 0)$ and $(-3, 0)$, and the length of the transverse axis is $2a = 2(3) = 6$. Since 16 is the denominator of the $y^2$-term, $b^2 = 16$ and $b = 4$. So the length of the conjugate axis is $2b = 2(4) = 8$.

To find the foci we need to know $c^2$. Since $b^2 = c^2 - a^2$, we have

$$c^2 = a^2 + b^2 = (3)^2 + (4)^2 = 25$$
$$c = 5$$

The foci are $(5, 0)$ and $(-5, 0)$. The asymptotes are $y = \frac{4}{3}x$ and $y = -\frac{4}{3}x$ (see Fig. 15.26).

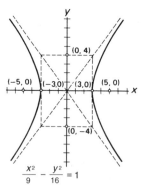

**Figure 15.26**

You will want to remember the equation relating $a$, $b$, and $c$ for the hyperbola.

$$c^2 = a^2 + b^2$$

**EXAMPLE 2**

Write the equation of the hyperbola with foci at $(5, 0)$ and $(-5, 0)$ and whose transverse axis is 8 units in length.

Here we have $c = 5$. Since $2a = 8$, $a = 4$,

$$c^2 = a^2 + b^2$$
$$25 = 16 + b^2$$
$$b^2 = 9$$

The equation is then

$$\frac{x^2}{16} - \frac{y^2}{9} = 1$$

---

**STANDARD FORM OF HYPERBOLA**

$$\frac{y^2}{a^2} - \frac{x^2}{b^2} = 1$$

with center at the origin and with the transverse axis lying on the $y$-axis

---

We obtain a graph as shown in Fig. 15.27.

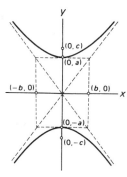

**Figure 15.27**

Note that the difference between this equation and the first equation is that $a^2$ is now the denominator of the $y^2$-term, which is the positive term. This means that the vertices (and the transverse axis) now lie on the $y$-axis.

The equations of the asymptotes are as follows.

$$y = \frac{a}{b}x$$

$$y = -\frac{a}{b}x$$

In general, the positive term indicates on which axis the vertices, foci, and transverse axis lie.

**1.** If the $x^2$-term is positive, then the denominator of $x^2$ is $a^2$ and the denominator of $y^2$ is $b^2$. The transverse axis lies along the $x$-axis and the vertices are $(a, 0)$ and $(-a, 0)$.

**2.** If the $y^2$-term is positive, then the denominator of $y^2$ is $a^2$ and the denominator of $x^2$ is $b^2$. The transverse axis lies along the $y$-axis and the vertices are $(0, a)$ and $(0, -a)$.

### EXAMPLE 3

Sketch the graph of the following hyperbola:

$$\frac{y^2}{36} - \frac{x^2}{49} = 1$$

Since the $y^2$-term is positive, the vertices lie on the $y$-axis and $a^2 = 36$. Then $b^2 = 49$, $a = 6$, and $b = 7$. The graph is sketched in Fig. 15.28.

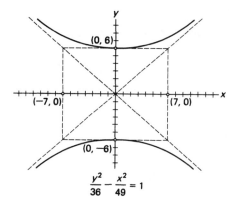

$$\frac{y^2}{36} - \frac{x^2}{49} = 1$$

**Figure 15.28**

### EXAMPLE 4

Write the equation of the hyperbola with foci at $(0, 8)$ and $(0, -8)$ and vertices at $(0, 6)$ and $(0, -6)$.

In this case, $a = 6$ and $c = 8$, so $b^2 = c^2 - a^2 = 64 - 36 = 28$. Since the vertices and foci lie on the $y$-axis, the $y^2$-term is positive with denominator $a^2$. The equation is

$$\frac{y^2}{36} - \frac{x^2}{28} = 1$$

As with the ellipse, not all hyperbolas are located with their centers at the origin. We have seen the standard forms of the equation of the hyperbola with center at the origin and whose transverse and conjugate axes lie on the $x$-axis and $y$-axis. In general, however, the equation of a hyperbola is of the form

$$Ax^2 + Bxy + Cy^2 + Dx + Ey + F = 0$$

where either **(1)** $B = 0$ and $A$ and $C$ differ in sign, or **(2)** $A = 0$, $C = 0$, and $B \neq 0$.

A simple example of this last case is the equation $xy = k$. The foci and vertices lie on the line $y = x$ if $k > 0$ or on the line $y = -x$ if $k < 0$.

### EXAMPLE 5

Sketch the graph of the hyperbola $xy = -6$.

Since there are no easy clues for sketching this equation (unlike hyperbolas in standard position), we must set up a table of values for $x$ and $y$. We then plot the corresponding points in the plane as in Fig. 15.29.

| $x$ | $y$ |
|---|---|
| 6 | $-1$ |
| 3 | $-2$ |
| 2 | $-3$ |
| 1 | $-6$ |
| $-1$ | 6 |
| $-2$ | 3 |
| $-3$ | 2 |
| $-6$ | 1 |

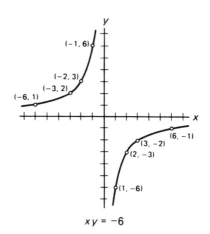

**Figure 15.29**

## Exercises 15.4

*Find the vertices, foci, and lengths of the transverse and conjugate axes of each hyperbola. Also find the equations of the asymptotes and sketch each graph.*

1. $\dfrac{x^2}{25} - \dfrac{y^2}{144} = 1$    2. $\dfrac{x^2}{144} - \dfrac{y^2}{25} = 1$    3. $\dfrac{y^2}{9} - \dfrac{x^2}{16} = 1$

4. $\dfrac{y^2}{16} - \dfrac{x^2}{9} = 1$    5. $5x^2 - 2y^2 = 10$    6. $3y^2 - 2x^2 = 6$

7. $4y^2 - x^2 = 4$    8. $4x^2 - y^2 = 4$

*Find the equation of the hyperbola satisfying each of the given conditions.*

9. Vertices at $(4, 0)$ and $(-4, 0)$; foci at $(6, 0)$ and $(-6, 0)$.
10. Vertices at $(0, 5)$ and $(0, -5)$; foci at $(0, 7)$ and $(0, -7)$.
11. Vertices at $(0, 6)$ and $(0, -6)$; foci at $(0, 8)$ and $(0, -8)$.
12. Vertices at $(2, 0)$ and $(-2, 0)$; foci at $(5, 0)$ and $(-5, 0)$.
13. Vertices at $(3, 0)$ and $(-3, 0)$; length of conjugate axis is 10.
14. Vertices at $(0, 6)$ and $(0, -6)$; length of conjugate axis is 8.
15. Foci at $(6, 0)$ and $(-6, 0)$; length of transverse axis is 10.
16. Foci at $(0, 8)$ and $(0, -8)$; length of transverse axis is 12.
17. Sketch the graph of the hyperbola given by $xy = 8$.
18. Sketch the graph of the hyperbola given by $xy = -4$.

## 15.5 TRANSLATION OF AXES

D.3, D.4

We have seen the difficulty in determining the equations of the parabola, ellipse, and hyperbola when these are not in standard position in the plane. It is still possible to find the equations of these curves fairly easily if the axes of these curves lie on lines parallel to the coordinate axes. This is accomplished by the method of translation of axes. We shall demonstrate this method with four examples.

## EXAMPLE 1

Find the equation of the ellipse with foci at $(-2, 3)$, and $(6, 3)$ and vertices at $(-3, 3)$ and $(7, 3)$.

The center of the ellipse is at $(2, 3)$, which is midway between the foci or the vertices. The distance between the foci $(-2, 3)$ and $(6, 3)$ is 8. So $c = 4$. The distance between $(-3, 3)$ and $(7, 3)$ is 10. So $a = 5$.

$$c^2 = a^2 - b^2$$
$$16 = 25 - b^2$$
$$b^2 = 9$$
$$b = 3$$

Sketch the graph as in Fig. 15.30. Next, plot the same ellipse in another coordinate system with center at the origin as in Fig. 15.31. Label the coordinate axes of this new system by $x'$ and $y'$. We know that in this $x'y'$-coordinate system, the equation for this ellipse is

$$\frac{(x')^2}{a^2} + \frac{(y')^2}{b^2} = 1$$

And, since $a = 5$ and $b = 3$, we have

$$\frac{(x')^2}{25} + \frac{(y')^2}{9} = 1$$

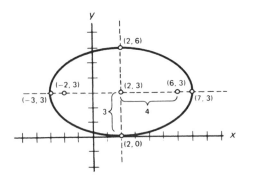

**Figure 15.30**

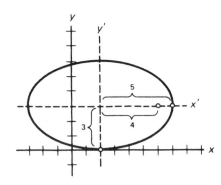

**Figure 15.31**

Each point on the ellipse can now be seen as having coordinates $(x, y)$ in the $xy$-plane and coordinates $(x', y')$ in the $x'y'$-plane. If we compare coordinates in the two coordinate systems, we see, for example, that the right-hand vertex has coordinates $(7, 3)$ in the $xy$-plane, but the same point has coordinates $(5, 0)$ in the $x'y'$-plane. Likewise, the point at the upper end of the minor axis has coordinates $(2, 6)$ in the $xy$-plane, but the same point has coordinates $(0, 3)$ in the $x'y'$-plane.

In general, the $x$- and $x'$-coordinates are related as follows:

$$x = x' + 2$$

That is, the original $x$-coordinates are 2 larger than the new $x'$-coordinates. Note that this is the distance that the new origin was moved along the $x$-axis: the $x$-coordinate of the center of the ellipse (see Fig. 15.31).

Similarly, the $y$- and $y'$-coordinates are related as follows:

$$y = y' + 3$$

Note that 3 is the distance that the new origin was moved along the $y$-axis: the $y$-coordinate of the center of the ellipse (see Fig. 15.31). We now rearrange terms and have

$$x' = x - 2$$
$$y' = y - 3$$

Now replace $x'$ by $x - 2$ and $y'$ by $y - 3$ in the equation

$$\frac{(x')^2}{25} + \frac{(y')^2}{9} = 1$$

$$\frac{(x - 2)^2}{25} + \frac{(y - 3)^2}{9} = 1$$

This is the equation of the ellipse with center at $(2, 3)$ in the $xy$-plane.

---

To write an equation for a parabola, an ellipse, or a hyperbola whose axes are parallel to the $x$-axis and the $y$-axis:

1. For a parabola, identify $(h, k)$ as the vertex; for an ellipse or a hyperbola identify $(h, k)$ as the center.

2. Translate $xy$-coordinates to a new $x'y'$-coordinate system by using the translation equations

$$x' = x - h$$
$$y' = y - k$$

where $(h, k)$ has been identified as in Step 1.

3. Write the equation of the conic, which is now in standard position in the $x'y'$-coordinate system.

4. Translate the equation derived in Step 3 back into the original coordinate system by making the following substitutions for $x$ and $y$ into the derived equation:

$$x' = x - h$$
$$y' = y - k$$

The resulting equation is an equation for the conic in the $xy$-coordinate system.

---

**EXAMPLE 2**

Find the equation of the parabola with focus $(-1, 2)$ and directrix $x = -7$.

**Step 1:** The vertex of this parabola is halfway along the line $y = 2$ between the focus $(-1, 2)$, and the directrix $x = -7$ (see Fig. 15.32). Thus, the vertex has coordinates $(-4, 2)$. This becomes the origin of the new coordinate system. So $h = -4$ and $k = 2$.

**Step 2:**

$$x' = x - h = x - (-4) = x + 4$$
$$y' = y - k = y - 2$$

The new $x'y'$-coordinates of the focus become

$$x' = x - h = -1 - (-4) = 3$$
$$y' = y - k = 2 - 2 = 0$$

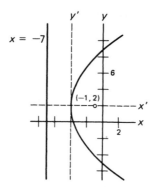

**Figure 15.32**

and the equation of the directrix $x = -7$ becomes

$$x' = x - h = -7 - (-4)$$
$$x' = -3$$

**Step 3:** Since the parabola is now in standard position in the new coordinate system with focus $(3, 0)$, we have $p = 3$. The equation in this system becomes

$$(y')^2 = 4px'$$
$$(y')^2 = 4(3)x'$$
$$(y')^2 = 12x'$$

**Step 4:** Replace $x'$ with $x + 4$ and $y'$ with $y - 2$.

$$(y')^2 = 12x'$$
$$(y - 2)^2 = 12(x + 4)$$
$$y^2 - 4y - 12x - 44 = 0$$

We sometimes know the equation of a curve and need to identify the curve and sketch its graph, as in the following example.

**EXAMPLE 3**

Describe and sketch the graph of the equation

$$\frac{(y - 4)^2}{9} - \frac{(x + 2)^2}{16} = 1$$

If we let

$$x' = x - h = x + 2 = x - (-2)$$
$$y' = y - k = y - 4$$

we have

$$\frac{(y')^2}{9} - \frac{(x')^2}{16} = 1$$

This is the equation of a hyperbola with center at $(-2, 4)$. Since $a^2 = 9$ and $b^2 = 16$, we have

$$c^2 = a^2 + b^2 = 9 + 16 = 25$$

so

$$a = 3 \qquad b = 4 \quad \text{and} \quad c = 5$$

In terms of the $x'y'$-coordinates, the foci are $(0, 5)$ and $(0, -5)$; the vertices are at $(0, 3)$ and $(0, -3)$; the length of the transverse axis is 6; and the length of the conjugate axis is 8.

To translate the $x'y'$-coordinates to $xy$-coordinates, we use the equations

$$x = x' + h \qquad y = y' + k$$

In this case,

$$x = x' + (-2) \qquad y = y' + 4$$

So in the $xy$-plane the foci are at $(-2, 9)$ and $(-2, -1)$; the vertices are at $(-2, 7)$ and $(-2, 1)$; the length of the transverse axis is 6; and the length of the conjugate axis is 8 (see Fig. 15.33).

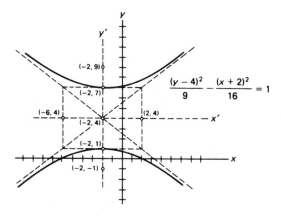

$$\frac{(y - 4)^2}{9} - \frac{(x + 2)^2}{16} = 1$$

**Figure 15.33**

**EXAMPLE 4**

Name the equation $16x^2 + 9y^2 + 64x + 54y + 1 = 0$. Locate the vertex if it is a parabola or the center if it is an ellipse or a hyperbola.

First complete the square for $x$ and $y$ (see Section 15.1).

$$(16x^2 + 64x \quad) + (9y^2 + 54y \quad) = -1$$

$$16(x^2 + 4x \quad) + 9(y^2 + 6y \quad) = -1 \qquad \text{( Factor out the coefficients of } x^2$$
$$\text{and } y^2 \text{ before completing the}$$
$$\text{square. The coefficients of}$$
$$x^2 \text{ and } y^2 \text{ must be \textbf{one}.)}$$

$$16(x^2 + 4x + 4) + 9(y^2 + 6y + 9) = -1 + 16(4) + 9(9)$$

$$16(x + 2)^2 + 9(y + 3)^2 = 144$$

$$\frac{(x + 2)^2}{9} + \frac{(y + 3)^2}{16} = 1 \qquad \text{(Divide each side by 144.)}$$

This is an equation of an ellipse. Noting that

$$x' = x - h = x + 2 = x - (-2)$$
$$y' = y - k = y + 3 = y - (-3)$$

we see that the center is at $(-2, -3)$.

## Exercises 15.5

*Find the equation of each curve determined by the given information.*

1. Ellipse with center at $(1, -1)$; vertices at $(5, -1)$ and $(-3, -1)$; and foci at $(3, -1)$ and $(-1, -1)$.

2. Parabola with vertex at $(-1, 3)$; focus at $(-1, 4)$; and directrix $y = 2$.

3. Hyperbola with center at $(1, 1)$; vertices at $(1, 7)$ and $(1, -5)$; and foci at $(1, 9)$ and $(1, -7)$.

4. Ellipse with center at $(-2, -3)$; vertices at $(4, -3)$ and $(-8, -3)$; and length of the minor axis is 10.

5. Parabola with vertex at $(3, -1)$; focus at $(5, -1)$; and directrix $x = 1$.

6. Hyperbola with center at $(-2, -2)$; vertices at $(1, -2)$ and $(-5, -2)$; and length of the conjugate axis is 10.

*Name and graph each equation.*

7. $(x - 2)^2 = 4(y + 3)$

8. $\dfrac{(x + 1)^2}{36} + \dfrac{(y - 2)^2}{64} = 1$

9. $\dfrac{y^2}{9} - \dfrac{(x + 2)^2}{16} = 1$

10. $y^2 = 8(x + 1)$

11. $9(x - 2)^2 + 16y^2 = 144$

12. $\dfrac{(x + 1)^2}{9} - \dfrac{(y + 3)^2}{16} = 1$

13. $\dfrac{(x - 3)^2}{36} + \dfrac{(y - 1)^2}{16} = 1$

14. $\dfrac{(x - 3)^2}{36} - \dfrac{(y - 1)^2}{16} = 1$

15. $(y + 3)^2 = 8(x - 1)$

16. $(x - 5)^2 = 12(y + 2)$

17. $\dfrac{(y + 1)^2}{9} - \dfrac{(x + 1)^2}{9} = 1$

18. $\dfrac{(y + 4)^2}{4} + \dfrac{(x - 2)^2}{9} = 1$

*Name and sketch the graph of each equation. Locate the vertex if it is a parabola, or locate the center if it is an ellipse or a hyperbola.*

**19.** $x^2 - 4x + 2y + 6 = 0$

**20.** $9x^2 + 4y^2 - 18x + 24y + 9 = 0$

**21.** $x^2 + 4y^2 + 4x - 8y - 8 = 0$

**22.** $-2x^2 + 3y^2 + 8x - 14 = 0$

**23.** $4x^2 - y^2 - 8x + 2y + 3 = 0$

**24.** $y^2 + 6y - x + 12 = 0$

**25.** $25y^2 - 4x^2 - 24x - 150y + 89 = 0$

**26.** $25x^2 + 9y^2 - 100x - 54y - 44 = 0$

**27.** $x^2 + 16x - 12y + 40 = 0$

**28.** $9x^2 - 4y^2 + 54x + 40y - 55 = 0$

**29.** $4x^2 + y^2 + 48x + 4y + 84 = 0$

**30.** $y^2 - 10x - 6y + 39 = 0$

# 15.6 THE GENERAL SECOND-DEGREE EQUATION

The circle, parabola, ellipse, and hyperbola are all special cases of the second-degree equation

$$Ax^2 + Bxy + Cy^2 + Dx + Ey + F = 0$$

When $B = 0$ and at least one of the coefficients $A$ or $C$ is not zero, the following summarizes the conditions for each curve:

1. If $A = C$, we have a *circle*.
   In special cases, the graph of the equation may be a point, or there may be no graph. (The equation may have only one or no solution.)

2. If $A = 0$ and $C \neq 0$, or if $C = 0$ and $A \neq 0$, then we have a *parabola*.

3. If $A \neq C$, and if $A$ and $C$ are either both positive or both negative, then we have an *ellipse*.
   In special cases, the graph of the equation may be a point, or there may be no graph. (The equation may have only one or no solution.)

4. If $A$ and $C$ differ in sign, then we have a *hyperbola*.
   In some special cases, the graph may be a pair of intersecting lines.

If $D \neq 0$ or $E \neq 0$ or both are not equal to zero, the curve does not have its center (or vertex in the case of the parabola) at the origin (see Section 15.5). If $B \neq 0$, then the axis of the curve does not lie along the $x$-axis or the $y$-axis. The hyperbola $xy = k$ is the only such example we have studied (see Section 15.4).

**EXAMPLE**

Identify the curve

$$x^2 + 3y^2 - 2x + 4y - 7 = 0$$

Since $A \neq C$, $A$ and $C$ are both positive, and $B = 0$, the curve is an ellipse. (The center is not the origin since $D \neq 0$ and $E \neq 0$.)

The curves represented by the second-degree equation

$$Ax^2 + Bxy + Cy^2 + Dx + Ey + F = 0$$

are called **conic sections** since they can be obtained by cutting the cones with a plane, as in Fig. 15.34.

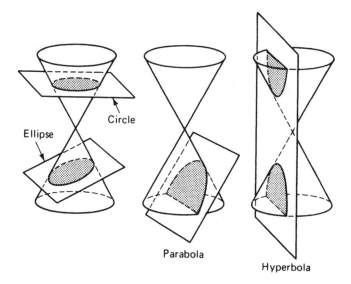

Figure 15.34

## Exercises 15.6

*Determine whether each equation represents a circle, a parabola, an ellipse, or a hyperbola.*

**1.** $x^2 + 3y^2 + 4x - 5y - 40 = 0$

**2.** $x^2 + y^2 + 4x - 6y - 12 = 0$

**3.** $4y^2 - 8y + 3x - 2 = 0$

**4.** $9x^2 + 4y^2 + 36x - 8y + 4 = 0$

**5.** $4x^2 - 5y^2 - 16x + 10y + 20 = 0$

**6.** $x^2 + y^2 + 3x - 2y - 14 = 0$

**7.** $3x^2 + 3y^2 + x - y - 6 = 0$

**8.** $x^2 + 4x - 3y - 52 = 0$

**9.** $x^2 + y^2 + 2x - 3y - 21 = 0$

**10.** $x^2 - y^2 - 6x + 3y - 100 = 0$

**11.** $9x^2 + 4y^2 - 18x + 8y + 4 = 0$

**12.** $3x^2 - 2y^2 + 6x - 8y - 17 = 0$

**13.** $3x^2 - 3y^2 - 2x - 4y - 13 = 0$

**14.** $4x^2 + 4y^2 - 16x - 4y - 5 = 0$

**15.** $x^2 - 6x - 6y + 3 = 0$

**16.** $4x^2 - 4x - 4y - 5 = 0$

## 15.7 SYSTEMS OF QUADRATIC EQUATIONS

We now look at how we can solve systems of equations involving conics. The basic idea is still to eliminate a variable.

### Substitution Method

**EXAMPLE 1**

Solve the following system of equations using the substitution method:

$$y^2 = x$$
$$y = x - 2$$

Since $y^2 = x$, we can substitute $y^2$ for $x$ in the second equation.

$$y = x - 2$$
$$y = (y^2) - 2$$
$$y^2 - y - 2 = 0$$
$$(y - 2)(y + 1) = 0$$

So

$$y = 2 \quad \text{or} \quad y = -1$$

Substituting these values for $y$ in the second equation $y = x - 2$, we have $x = 1$ when $y = -1$ and $x = 4$ when $y = 2$. The solutions of the system are then $(1, -1)$ and $(4, 2)$.

*Check:* Substitute the solutions in each original equation.

|  | $y^2 = x$ | $y = x - 2$ |
|---|---|---|
| $(1, -1)$ | $(-1)^2 = (1)$ | $(-1) = (1) - 2$ |
|  | $1 = 1$ | $-1 = -1$ |
| $(4, 2)$ | $(2)^2 = (4)$ | $(2) = (4) - 2$ |
|  | $4 = 4$ | $2 = 2$ |

A graphical solution is given in Fig. 15.35.

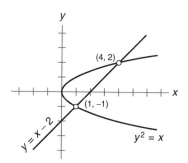

**Figure 15.35**

**EXAMPLE 2**

Solve the following system of equations using the substitution method:

$$4x^2 - 3y^2 = 4$$
$$x^2 - 4x + y^2 = 0$$

If we solve the second equation for $y^2$, we have $y^2 = 4x - x^2$. We can now substitute $4x - x^2$ for $y^2$ in the first equation.

$$4x^2 - 3(4x - x^2) = 4$$
$$4x^2 - 12x + 3x^2 = 4$$
$$7x^2 - 12x - 4 = 0$$
$$(x - 2)(7x + 2) = 0$$
$$x = 2 \quad \text{or} \quad x = -\tfrac{2}{7}$$

To find $y$, we substitute each of these values for $x$ in one of the original equations. We use the second equation.

For $x = 2$:

$$(2)^2 - 4(2) + y^2 = 0$$
$$4 - 8 + y^2 = 0$$
$$y^2 = 4$$

Thus, $y = 2$ or $-2$ when $x = 2$.

For $x = -\frac{2}{7}$:

$$\left(-\frac{2}{7}\right)^2 - 4\left(-\frac{2}{7}\right) + y^2 = 0$$

$$\frac{4}{49} + \frac{8}{7} + y^2 = 0$$

$$y^2 = -\frac{60}{49}$$

Since $y^2$ can never be negative, we conclude that there are no real solutions when $x = -\frac{2}{7}$. (This is what we call an *extraneous root*.) The solutions of the system are $(2, 2)$ and $(2, -2)$. The solutions should be checked in each original equation. A graphical solution is shown in Fig. 15.36.

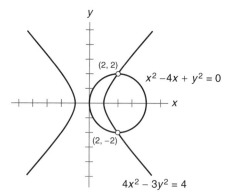

**Figure 15.36**

## Addition-Subtraction Method

### EXAMPLE 3

Solve the following system of equations using the addition-subtraction method:

$$3x^2 + y^2 = 14$$
$$x^2 - y^2 = 2$$

By adding the two equations, we can eliminate the $y^2$-term.

$$\begin{array}{r} 3x^2 + y^2 = 14 \\ \underline{x^2 - y^2 = 2} \\ 4x^2 \quad\quad = 16 \end{array}$$

Solve for $x$.

$$x^2 = 4$$
$$x = 2 \quad \text{or} \quad x = -2$$

Using the second equation, $x^2 - y^2 = 2$, we find the corresponding values for $y$.

For $x = 2$:

$$(2)^2 - y^2 = 2$$
$$4 - y^2 = 2$$
$$y^2 = 2$$
$$y = \sqrt{2} \quad \text{or} \quad y = -\sqrt{2}$$

For $x = -2$:

$$(-2)^2 - y^2 = 2$$
$$4 - y^2 = 2$$
$$y^2 = 2$$
$$y = \sqrt{2} \quad \text{or} \quad y = -\sqrt{2}$$

The solutions are $(2, \sqrt{2})$, $(2, -\sqrt{2})$, $(-2, \sqrt{2})$, and $(-2, -\sqrt{2})$. These should be checked in each original equation.

When you are solving a system of two equations where one represents a conic and the other a line, the substitution method is usually preferred. Example 4 shows how the addition-subtraction method may be helpful in some cases.

**EXAMPLE 4**

Solve the following system of equations:

$$y + 6x = 2$$
$$y^2 = 6x$$

We use the addition-subtraction method.

$$
\begin{array}{ll}
y + 6x = 2 & \\
\underline{y^2 \quad\quad - 6x = 0} & \\
y^2 + y \quad\quad = 2 & \text{(Add.)} \\
y^2 + y - 2 = 0 & \\
(y + 2)(y - 1) = 0 & \\
y = -2 \quad \text{or} \quad y = 1 &
\end{array}
$$

Using the first equation, we find that when $y = -2$, $x = \frac{2}{3}$ and that when $y = 1$, $x = \frac{1}{6}$. The solutions are $(\frac{2}{3}, -2)$ and $(\frac{1}{6}, 1)$. These solutions should be checked in each original equation.

## Exercises 15.7

*Solve each system of equations.*

**1.** $x^2 = 3y$
$\quad y = 2x - 3$

**2.** $x^2 - 2y^2 = 1$
$\quad 3x^2 + 2y^2 = 3$

**3.** $x^2 + 4x + y^2 - 8 = 0$
$\quad x^2 \quad\quad + y^2 \quad\quad = 4$

**4.** $x^2 + 2y^2 = 12$
$\quad\quad y = -x$

**5.** $y^2 - x^2 = 12$
    $x^2 = 4y$

**6.** $x^2 + y^2 = 9$
    $y = 4$

**7.** $x^2 + y^2 = 4$
    $x^2 - y^2 = 4$

**8.** $x^2 + y^2 - 6y = 0$
    $y = x$

**9.** $x^2 = 6y$

    $y = 6$

**10.** $\dfrac{y^2}{16} + \dfrac{x^2}{9} = 1$

    $4x + 3y = 12$

**11.** $y^2 = 4x + 12$
    $y^2 = -4x - 4$

**12.** $x^2 - y^2 = 2$
    $y^2 = x$

**13.** $x^2 + y^2 = 36$
    $y = x^2$

**14.** $y = x^2 - 3x - 10$
    $2x + y + 4 = 0$

**15.** $x^2 - y^2 = 9$
    $x^2 + 9y^2 = 169$

**16.** $x^2 + 4y^2 = 36$
    $x^2 + y^2 = 16$

**17.** $x^2 + y^2 = 17$
    $xy = 4$

**18.** $3x^2 + 4y^2 = 48$
    $xy = 6$

## 15.8 POLAR COORDINATES

Each point in the number plane has been associated with an ordered pair of real numbers $(x, y)$, which are called *rectangular* or *Cartesian coordinates*. Point $P(x, y)$ is shown in Fig. 15.37. Point $P$ can also be located by specifying an angle $\theta$ from the positive $x$-axis and a directed distance $r$ from the origin, and it can be described by the ordered pair $(r, \theta)$ called *polar coordinates*. The polar coordinate system has a fixed point in the number plane called the *pole* or *origin*. From the pole draw a horizontal ray directed to the right, which is called the *polar axis* (see Fig. 15.38).

Angle $\theta$ is a directed angle: $\theta > 0$ is measured counterclockwise and $\theta < 0$ is measured clockwise. Angle $\theta$ is commonly expressed in either degrees or radians. Distance $r$ is a directed distance: $r > 0$ is measured in the direction of the ray (terminal side of $\theta$), and $r < 0$ is measured in the direction opposite the direction of the ray.

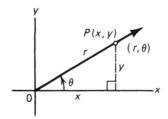

**Figure 15.37**

**Figure 15.38**

### EXAMPLE 1

Graph each point whose polar coordinates are given:
(a) $(2, 120°)$        (b) $(4, 4\pi/3)$        (c) $(4, -2\pi/3)$
(d) $(-5, 135°)$       (e) $(-2, -60°)$      (f) $(-3, 570°)$
    The graphs are shown in Fig. 15.39.

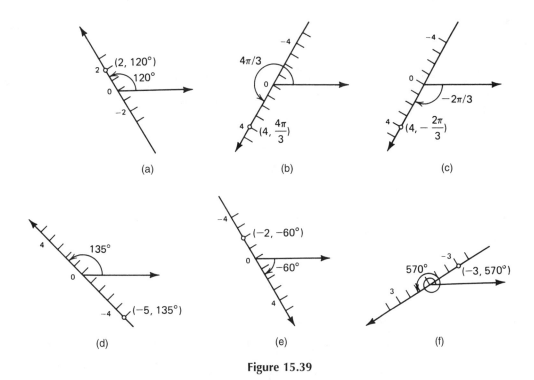

**Figure 15.39**

From the results of Example 1, you can see that there is a major difference between the rectangular coordinate system and the polar coordinate system. In the rectangular system there is a one-to-one correspondence between points in the plane and ordered pairs of real numbers. That is, each point is named by exactly one ordered pair, and each ordered pair corresponds to exactly one point. This one-to-one correspondence is not a property of the polar coordinate system. In Example 1, parts (a) and (e) describe the same point in the plane, and parts (b) and (c) describe the same point. In fact, each point may be named by infinitely many polar coordinates. In general, the point $P(r, \theta)$ may be represented by

$$(r, \theta + k \cdot 360°) \quad \text{or} \quad (r, \theta + k \cdot 2\pi)$$

where $k$ is any integer. $P(r, \theta)$ may also be represented by

$$(-r, \theta + k \cdot 180°) \quad \text{or} \quad (-r, \theta + k\pi)$$

where $k$ is any odd integer.

### EXAMPLE 2

Name an ordered pair of polar coordinates that corresponds to the pole or origin.

Any set of coordinates in the form $(0, \theta)$, where $\theta$ is any angle, corresponds to the pole. For example, $(0, 64°)$, $(0, 2\pi/3)$, and $(0, -\pi/6)$ name the pole.

Polar graph paper is available for working with polar coordinates. Figure 15.40 shows graph paper in both degrees and radians.

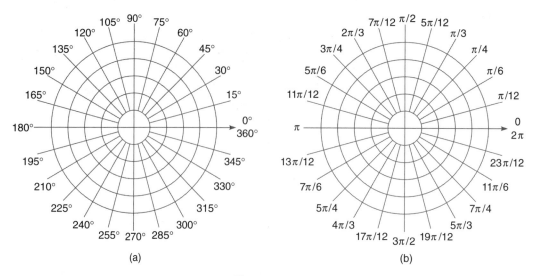

(a)                                     (b)

**Figure 15.40**

**EXAMPLE 3**

Plot each point whose polar coordinates are given. Use polar graph paper in degrees.

$A(6, 60°)$   $B(4, 270°)$   $C(3, -210°)$   $D(-6, 45°)$   $E(-2, -150°)$   $F(8, 480°)$

The graph is shown in Fig. 15.41.

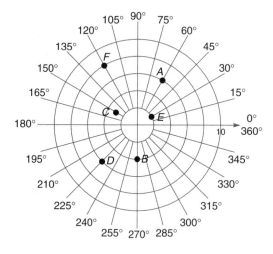

**Figure 15.41**

**EXAMPLE 4**

Plot each point whose polar coordinates are given. Use polar graph paper in radians.

$$A\left(3, \frac{3\pi}{4}\right) \quad B\left(5, \frac{11\pi}{6}\right) \quad C\left(6, -\frac{\pi}{4}\right)$$

$$D\left(-2, \frac{\pi}{3}\right) \quad E(-4, -\pi) \quad F\left(7, \frac{13\pi}{2}\right)$$

The graph is shown in Fig. 15.42.

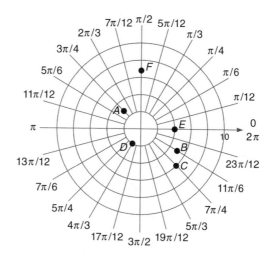

**Figure 15.42**

**EXAMPLE 5**

Given the point $P(4, 150°)$, name three other sets of polar coordinates for $P$ such that $-360° \leq \theta \leq 360°$.

$$\text{For } r > 0 \text{ and } \theta < 0: \quad (4, -210°)$$
$$\text{For } r < 0 \text{ and } \theta > 0: \quad (-4, 330°)$$
$$\text{For } r < 0 \text{ and } \theta < 0: \quad (-4, -30°)$$

**EXAMPLE 6**

Graph $r = 10 \cos \theta$ by plotting points. Assign $\theta$ values of $0°$, $30°$, $45°$, $60°$, and so on, until you have a smooth curve.

Make a table for the ordered pairs as follows. [*Note:* Although $r$ and $\theta$ are given in the same order as the ordered pair $(r, \theta)$, $\theta$ is actually the independent variable.]

| $r$ | $\theta$ | $r = 10 \cos \theta$ |
|-----|----------|----------------------|
| 10 | 0° | $r = 10 \cos 0° = 10$ |
| 8.7 | 30° | $r = 10 \cos 30° = 8.7$ |
| 7.1 | 45° | $r = 10 \cos 45° = 7.1$ |
| 5 | 60° | $r = 10 \cos 60° = 5$ |
| 0 | 90° | $r = 10 \cos 90° = 0$ |
| −5 | 120° | $r = 10 \cos 120° = -5$ |
| −7.1 | 135° | $r = 10 \cos 135° = -7.1$ |
| −8.7 | 150° | $r = 10 \cos 150° = -8.7$ |
| −10 | 180° | $r = 10 \cos 180° = -10$ |

Then plot the points as shown in Fig. 15.43.

*Note:* You should plot values of $\theta$ from $0°$ to $360°$, since the period of the cosine function is $360°$. In this case, choosing values of $\theta$ between $180°$ and $360°$ will give ordered pairs that duplicate those in Fig. 15.43.

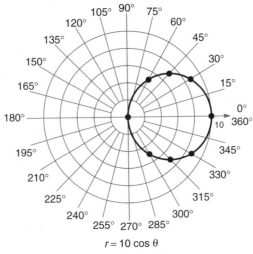

$$r = 10 \cos \theta$$

**Figure 15.43**

Let point $P(x, y)$ be any point in the rectangular plane. Let the polar plane coincide with the rectangular plane so that $P(x, y)$ and $P(r, \theta)$ represent the same point, as shown in Fig. 15.44. Note the following relationships.

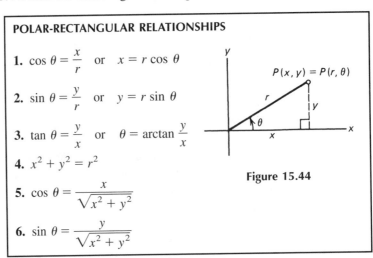

**POLAR-RECTANGULAR RELATIONSHIPS**

**1.** $\cos \theta = \dfrac{x}{r}$  or  $x = r \cos \theta$

**2.** $\sin \theta = \dfrac{y}{r}$  or  $y = r \sin \theta$

**3.** $\tan \theta = \dfrac{y}{x}$  or  $\theta = \arctan \dfrac{y}{x}$

**4.** $x^2 + y^2 = r^2$

**5.** $\cos \theta = \dfrac{x}{\sqrt{x^2 + y^2}}$

**6.** $\sin \theta = \dfrac{y}{\sqrt{x^2 + y^2}}$

**Figure 15.44**

Suppose that we wish to change coordinates from one system to the other. The following example illustrates the procedure.

**EXAMPLE 7**

Change $A(4, 60°)$ and $B(8, 7\pi/6)$ to rectangular coordinates.
    For point $A$:

$$x = r \cos \theta \qquad y = r \sin \theta$$
$$x = 4 \cos 60° \qquad y = 4 \sin 60°$$
$$= 4\left(\frac{1}{2}\right) \qquad = 4\left(\frac{\sqrt{3}}{2}\right)$$
$$= 2 \qquad = 2\sqrt{3}$$

Thus, $A(4, 60°) = (2, 2\sqrt{3})$.

For point $B$:

$$x = r \cos \theta \qquad\qquad y = r \sin \theta$$

$$x = 8 \cos \frac{7\pi}{6} \qquad\qquad y = 8 \sin \frac{7\pi}{6}$$

$$= 8\left(-\frac{\sqrt{3}}{2}\right) \qquad\qquad = 8\left(-\frac{1}{2}\right)$$

$$= -4\sqrt{3} \qquad\qquad = -4$$

Thus, $B(8, 7\pi/6) = (-4\sqrt{3}, -4)$.

### EXAMPLE 8

Find polar coordinates for each point: $C(2\sqrt{3}, 2)$ in degrees $0° \le \theta < 360°$ and $D(6, -6)$ in radians $0 \le \theta < 2\pi$.

*Note:* The signs of $x$ and $y$ determine the quadrant for $\theta$. That is, the signs of $x$ and $y$ determine in which quadrant the point lies and hence the quadrant in which $\theta$ must lie.

For point $C$:

$$r^2 = x^2 + y^2 \qquad\qquad \theta = \arctan \frac{y}{x}$$

$$r^2 = (2\sqrt{3})^2 + 2^2 = 16 \qquad \theta = \arctan \frac{2}{2\sqrt{3}}$$

$$r = 4 \qquad\qquad \theta = 30°$$

Thus, $C(2\sqrt{3}, 2) = (4, 30°)$.

For point $D$:

$$r^2 = x^2 + y^2 \qquad\qquad \theta = \arctan \frac{y}{x}$$

$$r^2 = 6^2 + (-6)^2 = 72 \qquad \theta = \arctan\left(\frac{-6}{6}\right) = \arctan(-1)$$

$$r = 6\sqrt{2} \qquad\qquad \theta = \frac{7\pi}{4} \qquad \left(\textit{Note: } \alpha = \frac{\pi}{4}.\right)$$

Thus, $D(6, -6) = (6\sqrt{2}, 7\pi/4)$.

Some curves are most simply expressed and easiest to work with in rectangular coordinates; others are most simply expressed and easiest to work with in polar coordinates. As a result, you must be able to change a polar equation to a rectangular equation and to change a rectangular equation to a polar equation.

### EXAMPLE 9

Change $x^2 + y^2 - 4x = 0$ to polar form.

Substituting $x^2 + y^2 = r^2$ and $x = r \cos \theta$, we have

$$r^2 - 4r \cos \theta = 0$$

$$r(r - 4 \cos \theta) = 0 \qquad \text{(Factor.)}$$

So

$$r = 0 \quad \text{or} \quad r - 4 \cos \theta = 0$$

But $r = 0$ (the pole) is a point that is included in the graph of the equation $r - 4 \cos \theta = 0$. Note that $(0, \pi/2)$ is an ordered pair that satisfies the second equation and names the pole. Thus, the simplest polar equation is

$$r = 4 \cos \theta$$

**EXAMPLE 10**

Change $r = 4 \sin \theta$ to rectangular form.
　　　Multiply both sides of the equation by $r$:

$$r^2 = 4r \sin \theta$$

Substituting $r^2 = x^2 + y^2$ and $r \sin \theta = y$, we have

$$x^2 + y^2 = 4y$$

Note that by multiplying both sides of the given equation by $r$, we added the root $r = 0$. But the point represented by that root is already included in the original equation. So no new points are added to those represented by the original equation.

**EXAMPLE 11**

Change $r \cos^2 \theta = 6 \sin \theta$ to rectangular form.
　　　First multiply both sides by $r$:

$$r^2 \cos^2 \theta = 6r \sin \theta$$
$$(r \cos \theta)^2 = 6r \sin \theta$$

Substituting $r \cos \theta = x$ and $r \sin \theta = y$, we have

$$x^2 = 6y$$

**EXAMPLE 12**

Change $r = \dfrac{2}{1 - \cos \theta}$ to rectangular form.

$$r = \frac{2}{1 - \cos \theta} \tag{15.5}$$

First multiply both sides by $1 - \cos \theta$:

$$r(1 - \cos \theta) = 2$$
$$r - r \cos \theta = 2$$
$$r = 2 + r \cos \theta \tag{15.6}$$

Substituting $r = \pm\sqrt{x^2 + y^2}$ and $r \cos \theta = x$, we have

$$\pm\sqrt{x^2 + y^2} = 2 + x$$

Squaring both sides, we have

$$x^2 + y^2 = 4 + 4x + x^2$$
$$y^2 = 4x + 4$$

　　　Note that squaring both sides was a risky operation because we introduced the possible extraneous solutions

$$r = -(2 + r \cos \theta) \tag{15.7}$$

However, in this case, both Equations (15.6) and (15.7) have the same graph. To show this, solve Equation (15.7) for $r$:

$$r = \frac{-2}{1 + \cos \theta} \qquad (15.8)$$

Recall that the ordered pairs $(r, \theta)$ and $(-r, \theta + \pi)$ represent the same point. Let us replace $(r, \theta)$ by $(-r, \theta + \pi)$ in Equation (15.8).

$$-r = \frac{-2}{1 + \cos (\theta + \pi)}$$

$$r = \frac{2}{1 - \cos \theta} \qquad \text{[Recall that } \cos (\theta + \pi) = -\cos \theta.\text{]}$$

Equations (15.6) and (15.7) and thus Equations (15.5) and (15.8) have the same graph, and no extraneous solutions were introduced when we squared both sides. So our result $y^2 = 4x + 4$ is correct.

## Exercises 15.8

*Plot each point whose polar coordinates are given.*

1. $A(3, 150°)$, $B(7, -45°)$, $C(2, -120°)$, $D(-4, 225°)$

2. $A(5, -90°)$, $B(2, -210°)$, $C(6, -270°)$, $D(-5, 30°)$

3. $A\left(4, \frac{\pi}{3}\right)$, $B\left(5, -\frac{\pi}{4}\right)$, $C\left(3, -\frac{7\pi}{6}\right)$, $D\left(-6, \frac{11\pi}{6}\right)$

4. $A\left(4, \frac{5\pi}{3}\right)$, $B\left(5, -\frac{3\pi}{2}\right)$, $C\left(3, -\frac{19\pi}{12}\right)$, $D\left(-6, -\frac{2\pi}{3}\right)$

*For each point name three other sets of polar coordinates such that* $-360° \le \theta \le 360°$.

5. $(3, 60°)$          6. $(2, 240°)$          7. $(-5, 315°)$

8. $(-6, 90°)$          9. $(4, -135°)$          10. $(-1, -180°)$

*For each point name three other sets of polar coordinates such that* $-2\pi \le \theta \le 2\pi$.

11. $\left(3, \frac{\pi}{6}\right)$          12. $\left(-7, \frac{\pi}{2}\right)$          13. $\left(-9, \frac{2\pi}{3}\right)$

14. $\left(-2, -\frac{5\pi}{6}\right)$          15. $\left(-4, -\frac{7\pi}{4}\right)$          16. $\left(5, -\frac{5\pi}{3}\right)$

*Graph each equation by plotting points. Assign $\theta$ values of 0°, 30°, 45°, 60°, and so on, until you have a smooth curve.*

17. $r = 10 \sin \theta$          18. $r = -10 \sin \theta$          19. $r = 4 + 4 \cos \theta$

20. $r = 4 + 4 \sin \theta$          21. $r \cos \theta = 4$          22. $r \sin \theta = -4$

*Graph each equation by plotting points. Assign $\theta$ values of 0, $\pi/6$, $\pi/4$, $\pi/3$, and so on, until you have a smooth curve.*

23. $r = -10 \cos \theta$          24. $r = 6 \sin \theta$          25. $r = 4 - 4 \sin \theta$

26. $r = 4 - 4 \cos \theta$          27. $r = \theta, 0 \le \theta \le 4\pi$          28. $r = 2\theta, 0 \le \theta \le 2\pi$

*Change each set of polar coordinates to rectangular coordinates.*

**29.** $(3, 30°)$

**30.** $(2, 180°)$

**31.** $\left(2, \dfrac{\pi}{3}\right)$

**32.** $\left(7, \dfrac{5\pi}{6}\right)$

**33.** $(-4, 150°)$

**34.** $(1, 420°)$

**35.** $\left(-6, \dfrac{3\pi}{2}\right)$

**36.** $(3, -\pi)$

**37.** $(-5, -240°)$

**38.** $(2, -120°)$

**39.** $\left(2, -\dfrac{7\pi}{4}\right)$

**40.** $\left(-1, -\dfrac{5\pi}{3}\right)$

*Change each set of rectangular coordinates to polar coordinates in degrees $0° \leq \theta < 360°$.*

**41.** $(5, 5)$

**42.** $(-\sqrt{3}, 1)$

**43.** $(0, 4)$

**44.** $(-3, 0)$

**45.** $(-2, -2\sqrt{3})$

**46.** $(-1, 1)$

*Change each set of rectangular coordinates to polar coordinates in radians $0 \leq \theta < 2\pi$.*

**47.** $(-4, 4)$

**48.** $(-1, -\sqrt{3})$

**49.** $(-\sqrt{6}, \sqrt{2})$

**50.** $(5\sqrt{2}, -5\sqrt{2})$

**51.** $(0, -4)$

**52.** $(0, 0)$

*Change each equation to polar form.*

**53.** $x = 3$

**54.** $y = 5$

**55.** $x^2 + y^2 = 36$

**56.** $y^2 = 5x$

**57.** $x^2 + y^2 + 2x + 5y = 0$

**58.** $2x + 3y = 6$

**59.** $4x - 3y = 12$

**60.** $ax + by = c$

**61.** $9x^2 + 4y^2 = 36$

**62.** $4x^2 - 9y^2 = 36$

**63.** $x^3 = 4y^2$

**64.** $x^4 - 2x^2y^2 + y^4 = 0$

*Change each equation to rectangular form.*

**65.** $r \sin \theta = -3$

**66.** $r \cos \theta = 7$

**67.** $r = 5$

**68.** $r = 3 \sec \theta$

**69.** $\theta = \dfrac{\pi}{4}$

**70.** $\theta = -\dfrac{2\pi}{3}$

**71.** $r = 5 \cos \theta$

**72.** $r = 6 \sin \theta$

**73.** $r = 6 \cos\left(\theta + \dfrac{\pi}{3}\right)$

**74.** $r = 4 \sin\left(\theta - \dfrac{\pi}{4}\right)$

**75.** $r \sin^2 \theta = 3 \cos \theta$

**76.** $r^2 = \tan^2 \theta$

**77.** $r^2 \sin 2\theta = 2$

**78.** $r^2 \cos 2\theta = 6$

**79.** $r^2 = \sin 2\theta$

**80.** $r^2 = \cos 2\theta$

**81.** $r = \tan \theta$

**82.** $r = 4 \tan \theta \sec \theta$

**83.** $r = \dfrac{3}{1 + \sin \theta}$

**84.** $r = \dfrac{-4}{1 + \cos \theta}$

**85.** $r = 4 \sin 3\theta$

**86.** $r = 4 \cos 2\theta$

**87.** $r = 2 + 4 \sin \theta$

**88.** $r = 1 - \cos \theta$

**89.** Find the distance between the points whose polar coordinates are $(3, 60°)$ and $(2, 330°)$.

**90.** Find the distance between the points whose polar coordinates are $(5, \pi/2)$ and $(1, 7\pi/6)$.

**91.** Find a formula for the distance between two points whose polar coordinates are $P_1(r_1, \theta_1)$ and $P_2(r_2, \theta_2)$.

# 15.9 GRAPHS IN POLAR COORDINATES

As you undoubtedly know, a graph of any equation may be made by finding and plotting "enough" ordered pairs that satisfy the equation and connecting them with a curve. As you also undoubtedly know, this process is often tedious and time-consuming at best. We need a method for sketching the graph of a polar equation that minimizes the number of ordered pairs that must be found and plotted. One such method involves symmetry. We shall present tests for three kinds of symmetry.

---

**SYMMETRY WITH RESPECT TO THE:**

1. *Horizontal axis:* Replace $\theta$ by $-\theta$ in the original equation. If the resulting equation is equivalent to the original equation, then the graph of the original equation is symmetric with respect to the *horizontal* axis.

2. *Vertical axis:* Replace $\theta$ by $\pi - \theta$ in the original equation. If the resulting equation is equivalent to the original equation, then the graph of the original equation is symmetric with respect to the *vertical* axis.

3. *Pole:*
   (a) Replace $r$ by $-r$ in the original equation. If the resulting equation is equivalent to the original equation, then the graph of the original equation is symmetric with respect to the *pole*.
   (b) Replace $\theta$ by $\pi + \theta$ in the original equation. If the resulting equation is equivalent to the original equation, then the graph of the original equation is symmetric with respect to the *pole*.

---

You should note that these tests for symmetry are sufficient conditions for symmetry; that is, they are sufficient to assure symmetry. You should also note that these are not necessary conditions for symmetry; that is, symmetry may exist even though the test fails.

If either Test 3(a) or Test 3(b) is satisfied, then the graph is symmetric with respect to the pole. It is also true that if any two of the three kinds of symmetry hold, then the remaining third symmetry automatically holds. Can you explain why?

In order to help you quickly test for symmetry, the following identities are listed for your convenience.

---

**POLAR COORDINATE IDENTITIES FOR TESTING SYMMETRY**

$$\sin(-\theta) = -\sin\theta$$
$$\cos(-\theta) = \cos\theta$$
$$\tan(-\theta) = -\tan\theta$$
$$\sin(\pi - \theta) = \sin\theta$$
$$\cos(\pi - \theta) = -\cos\theta$$
$$\tan(\pi - \theta) = -\tan\theta$$
$$\sin(\pi + \theta) = -\sin\theta$$
$$\cos(\pi + \theta) = -\cos\theta$$
$$\tan(\pi + \theta) = \tan\theta$$

---

**EXAMPLE 1**

Graph $r = 4 + 2 \cos \theta$.

Replacing $\theta$ by $-\theta$, we see that the graph is symmetric with respect to the horizontal axis. The other tests fail. Thus, we need to make a table as follows. (Note that because of symmetry with respect to the horizontal axis, we need to generate ordered pairs only for $0° \leq \theta \leq 180°$.)

| $r$ | $\theta$ | $r = 4 + 2 \cos \theta$ |
|---|---|---|
| 6 | 0° | $r = 4 + 2 \cos 0° = 6$ |
| 5.7 | 30° | $r = 4 + 2 \cos 30° = 5.7$ |
| 5 | 60° | $r = 4 + 2 \cos 60° = 5$ |
| 4 | 90° | $r = 4 + 2 \cos 90° = 4$ |
| 3 | 120° | $r = 4 + 2 \cos 120° = 3$ |
| 2.3 | 150° | $r = 4 + 2 \cos 150° = 2.3$ |
| 2 | 180° | $r = 4 + 2 \cos 180° = 2$ |

Plot the points as shown in Fig. 15.45(a). Because of the symmetry with respect to the horizontal axis, plot the corresponding mirror-image points below the horizontal axis [see Fig. 15.45(b)].

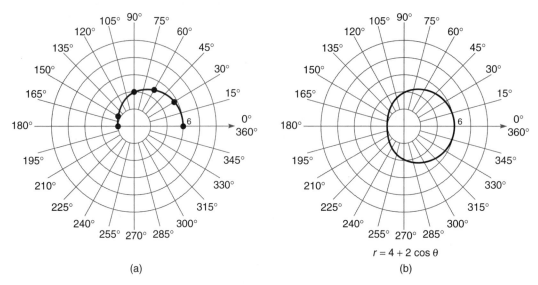

(a)

(b)

$r = 4 + 2 \cos \theta$

**Figure 15.45**

**EXAMPLE 2**

Graph $r = 4 + 4 \sin \theta$.

Replacing $\theta$ by $\pi - \theta$, we see that the graph is symmetric with respect to the vertical axis. The other tests fail. Thus, make a table as follows. (Note that because of symmetry with respect to the vertical axis, we need to generate ordered pairs only for $-\pi/2 \leq \theta \leq \pi/2$.)

| $r$ | $\theta$ | $r = 4 + 4 \sin \theta$ |
|---|---|---|
| 4 | 0 | $r = 4 + 4 \sin 0 = 4$ |
| 6 | $\pi/6$ | $r = 4 + 4 \sin \pi/6 = 6$ |
| 7.5 | $\pi/3$ | $r = 4 + 4 \sin \pi/3 = 7.5$ |
| 8 | $\pi/2$ | $r = 4 + 4 \sin \pi/2 = 8$ |
| 2 | $-\pi/6$ | $r = 4 + 4 \sin(-\pi/6) = 2$ |
| 0.54 | $-\pi/3$ | $r = 4 + 4 \sin(-\pi/3) = 0.54$ |
| 0 | $-\pi/2$ | $r = 4 + 4 \sin(-\pi/2) = 0$ |

Plot the points as shown in Fig. 15.46(a). Because of the symmetry with respect to the vertical axis, plot the corresponding mirror-image points to the left of the vertical axis [see Fig. 15.46(b)].

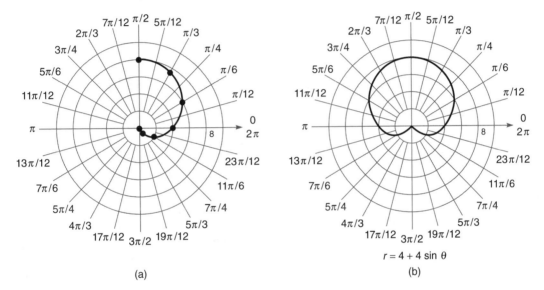

(a)　　　　　　　　　　　　　　$r = 4 + 4 \sin \theta$

(b)

**Figure 15.46**

### EXAMPLE 3

Graph $r^2 = 16 \sin \theta$.

Replacing $r$ by $-r$, we see that the graph is symmetric with respect to the pole. Also, replacing $\theta$ by $\pi - \theta$, we see that the graph is also symmetric with respect to the vertical axis. Since two of the three kinds of symmetry hold, the graph is also symmetric with respect to the horizontal axis. (*Note:* Replacing $\theta$ by $-\theta$ gives the resulting equation $r^2 = -16 \sin \theta$, which is different from the original equation. However, its solutions when graphed give the same curve.)

| $r$ | $\theta$ | $r^2 = 16 \sin \theta$ |
|---|---|---|
| 0 | 0° | $r^2 = 16 \sin 0° = 0; r = 0$ |
| 2.8 | 30° | $r^2 = 16 \sin 30° = 8; r = 2.8$ |
| 3.7 | 60° | $r^2 = 16 \sin 60° = 13.9; r = 3.7$ |
| 4 | 90° | $r^2 = 16 \sin 90° = 16; r = 4$ |

Plot the points as shown in Fig. 15.47(a). Because of the symmetry with respect to the horizontal and vertical axes, plot the corresponding mirror-image points below the horizontal axis. Then plot the mirror image points of all resulting points to the left of the vertical axis [see Fig. 15.47(b)].

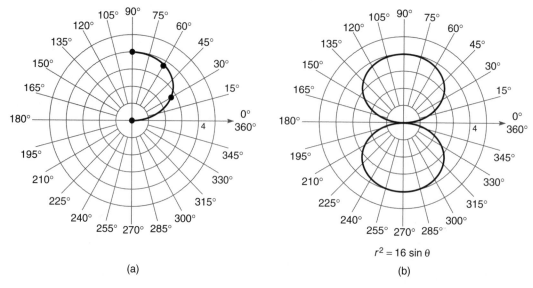

$r^2 = 16 \sin \theta$

(a)                                         (b)

**Figure 15.47**

### EXAMPLE 4

Graph $r^2 = 25 \cos 2\theta$.

By replacing $\theta$ by $-\theta$ and $r$ by $-r$, we have symmetry with respect to the horizontal axis and the pole, respectively. Thus, we also have symmetry with respect to the vertical axis. Working in the first quadrant, we have the following table:

| $r$ | $\theta$ | $r^2 = 25 \cos 2\theta$ |
|---|---|---|
| 5 | 0 | $r^2 = 25 \cos 2(0) = 25; r = 5$ |
| 4.7 | $\pi/12$ | $r^2 = 25 \cos 2(\pi/12) = 21.7; r = 4.7$ |
| 3.5 | $\pi/6$ | $r^2 = 25 \cos 2(\pi/6) = 12.5; r = 3.5$ |
| 0 | $\pi/4$ | $r^2 = 25 \cos 2(\pi/4) = 0; r = 0$ |

*Note:* For the interval $\pi/4 < \theta \le \pi/2$, $r^2 < 0$ and $r$ is undefined.

Plot the points as shown in Fig. 15.48(a). Because of the symmetry with respect to the horizontal and vertical axes, plot the corresponding mirror-image points below the horizontal axis and to the left of the vertical axis [see Fig. 15.48(b)].

There are various general polar equations whose graphs may be classified as shown in Fig. 15.49. What do you think the graphs of the various forms of $r = a + b \sin \theta$ are like?

Equations in the form

$$r = a \sin n\theta \quad \text{or} \quad r = a \cos n\theta$$

where $n$ is a positive integer, are called *petal* or *rose curves*. The number of petals is equal to $n$ if $n$ is an *odd* integer, and the number is equal to $2n$ if $n$ is an *even* integer. The rea-

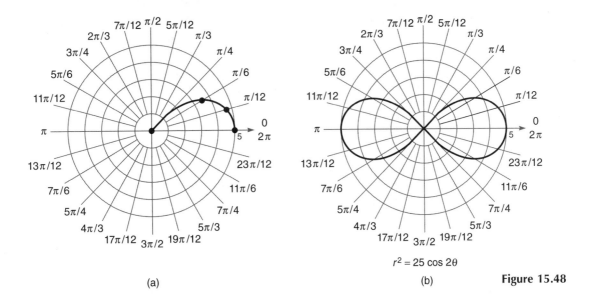

(a)

$r^2 = 25 \cos 2\theta$

(b)

**Figure 15.48**

Limacons ($r = a + b \cos \theta$)

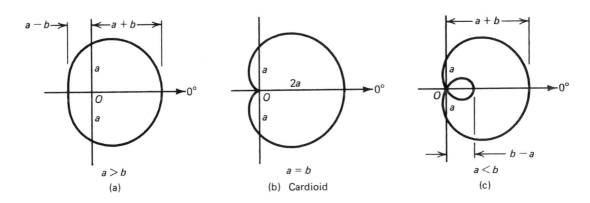

$a > b$

(a)

$a = b$

(b) Cardioid

$a < b$

(c)

Lemniscates

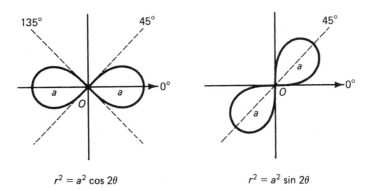

$r^2 = a^2 \cos 2\theta$

$r^2 = a^2 \sin 2\theta$

**Figure 15.49**

son is that the graph "retraces" itself as $\theta$ goes from 0° to 360° when $n$ is odd, so there are only half as many distinct petals. (For $n = 1$ there is one circular petal. See Example 6, Section 15.8.) The value of $a$ corresponds to the length of each petal.

The tests for symmetry may be used to graph petal curves. However, we shall illustrate a somewhat different, as well as easier and quicker, method for graphing petal curves.

### EXAMPLE 5

Graph $r = 6 \cos 2\theta$.

First, note that $n = 2$, which is even. Therefore, we have four petals. The petals are always uniform; each petal occupies 360°/4, or 90°, of the polar coordinate system. Next, find the tip of a petal; this occurs when $r$ is maximum or when

$$\cos 2\theta = 1$$
$$2\theta = 0°$$
$$\theta = 0°$$

That is, $r = 6$ when $\theta = 0°$.

Finally, sketch four petals, each having a maximum length of six and occupying 90° (see Fig. 15.50). For more accuracy, you may graph the ordered pairs corresponding to a "half petal" (0° $\leq \theta \leq$ 45°, in this case).

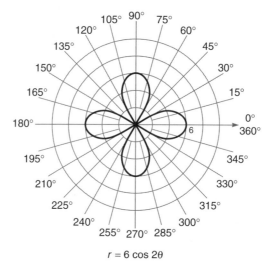

$$r = 6 \cos 2\theta$$

**Figure 15.50**

Polar coordinates are especially useful for the study and graphing of *spirals*. The *spiral of Archimedes* has an equation in the form

$$r = a\theta$$

Its graph is shown in Fig. 15.51; the dashed portion of the graph corresponds to $\theta < 0$.

The *logarithmic spiral* has an equation of the form

$$\log_b r = \log_b a + k\theta \quad \text{or} \quad r = a \cdot b^{k\theta}$$

Its graph is shown in Fig. 15.52.

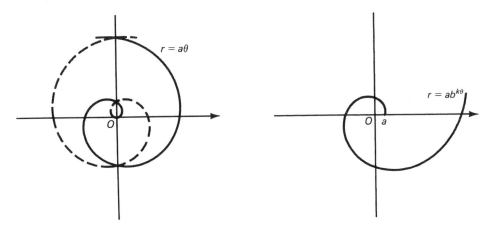

**Figure 15.51**  **Figure 15.52**

## Exercises 15.9

*Graph each equation.*

**1.** $r = 6$    **2.** $r = 3$    **3.** $r = -2$

**4.** $r = -4$    **5.** $\theta = 30°$    **6.** $\theta = -120°$

**7.** $\theta = -\dfrac{\pi}{3}$    **8.** $\theta = \dfrac{7\pi}{6}$    **9.** $r = 5 \sin \theta$

**10.** $r = 8 \cos \theta$    **11.** $r = 6 \cos\left(\theta + \dfrac{\pi}{3}\right)$    **12.** $r = 4 \sin\left(\theta - \dfrac{\pi}{4}\right)$

**13.** $r = 4 + 2 \sin \theta$    **14.** $r = 8 + 2 \cos \theta$    **15.** $r = 4 - 2 \cos \theta$

**16.** $r = 4 - 2 \sin \theta$    **17.** $r = 3 + 3 \cos \theta$    **18.** $r = 5 + 5 \sin \theta$

**19.** $r = 2 + 4 \sin \theta$    **20.** $r = 2 + 8 \cos \theta$    **21.** $r = 2 - 4 \cos \theta$

**22.** $r = 2 - 4 \sin \theta$    **23.** $r = 3 - 3 \cos \theta$    **24.** $r = 5 - 5 \sin \theta$

**25.** $r \cos \theta = 6$    **26.** $r \sin \theta = -4$    **27.** $r^2 = 25 \cos \theta$

**28.** $r^2 = -9 \sin \theta$    **29.** $r^2 = 9 \sin 2\theta$    **30.** $r^2 = 16 \cos 2\theta$

**31.** $r^2 = -36 \cos 2\theta$    **32.** $r = -36 \sin 2\theta$    **33.** $r = 5 \sin 3\theta$

**34.** $r = 4 \cos 5\theta$    **35.** $r = 3 \cos 2\theta$    **36.** $r = 6 \sin 4\theta$

**37.** $r = 9 \sin^2 \theta$    **38.** $r = 16 \cos^2 \theta$    **39.** $r = 4 \cos \dfrac{\theta}{2}$

**40.** $r = 5 \sin^2 \dfrac{\theta}{2}$    **41.** $r = \tan \theta$    **42.** $r = 2 \csc \theta$

**43.** $r = 3\theta,\ \theta > 0$    **44.** $r = \dfrac{3}{\theta},\ \theta > 0$    **45.** $r = 2^{3\theta}$

**46.** $r = 2 \cdot 3^{2\theta}$    **47.** $r = \dfrac{4}{\sin \theta + \cos \theta}$    **48.** $r = \dfrac{-2}{\sin \theta + \cos \theta}$

**49.** $r(1 + \cos \theta) = 4$    **50.** $r(1 + 2 \sin \theta) = -4$

# CHAPTER SUMMARY

1. Equations in the form $Ax^2 + Bxy + Cy^2 + Dx + Ey + F = 0$ are called **conics**.

2. *Circle:*
   (a) *Standard form:* $(x - h)^2 + (y - k)^2 = r^2$, where $r$ is the radius and $(h, k)$ is the center.
   (b) *General form:* $x^2 + y^2 + Dx + Ey + F = 0$.
   (c) *Center at the origin:* $x^2 + y^2 = r^2$, where $r$ is the radius.

3. *Parabola with vertex at the origin:*
   (a) $y^2 = 4px$ with focus at $(p, 0)$ and $x = -p$ as the directrix.
      (i) When $p > 0$, the parabola opens to the right.
      (ii) When $p < 0$, the parabola opens to the left.
   (b) $x^2 = 4py$ with focus at $(0, p)$ and $y = -p$ as the directrix.
      (i) When $p > 0$, the parabola opens upward.
      (ii) When $p < 0$, the parabola opens downward.

4. *Ellipse with center at the origin:*
   (a) $\dfrac{x^2}{a^2} + \dfrac{y^2}{b^2} = 1$ with the major axis on the $x$-axis and $a > b$.
   (b) $\dfrac{y^2}{a^2} + \dfrac{x^2}{b^2} = 1$ with major axis on the $y$-axis and $a > b$.

5. *Hyperbola with center at the origin:*
   (a) $\dfrac{x^2}{a^2} - \dfrac{y^2}{b^2} = 1$ with the transverse axis on the $x$-axis.
   (b) $\dfrac{y^2}{a^2} - \dfrac{x^2}{b^2} = 1$ with the transverse axis on the $y$-axis.

6. *Translation equations:*

$$x' = x - h \quad \text{and} \quad y' = y - k$$

7. *General forms of conics with axes parallel to the coordinate axes:*
   (a) $(y - k)^2 = 4p(x - h)$ is a parabola with vertex at $(h, k)$ and axis parallel to the $x$-axis.
   (b) $(x - h)^2 = 4p(y - k)$ is a parabola with vertex at $(h, k)$ and axis parallel to the $y$-axis.
   (c) $\dfrac{(x - h)^2}{a^2} + \dfrac{(y - k)^2}{b^2} = 1$, $a > b$, is an ellipse with center at $(h, k)$ and major axis parallel to the $x$-axis.
   (d) $\dfrac{(y - k)^2}{a^2} + \dfrac{(x - h)^2}{b^2} = 1$, $a > b$, is an ellipse with center at $(h, k)$ and major axis parallel to the $y$-axis.
   (e) $\dfrac{(x - h)^2}{a^2} - \dfrac{(y - k)^2}{b^2} = 1$ is a hyperbola with center at $(h, k)$ and transverse axis parallel to the $x$-axis.
   (f) $\dfrac{(y - k)^2}{a^2} - \dfrac{(x - h)^2}{b^2} = 1$ is a hyperbola with center at $(h, k)$ and transverse axis parallel to the $y$-axis.

**8.** *The general second-degree equation:* The circle, parabola, ellipse, and hyperbola are all special cases of the second-degree equation

$$Ax^2 + Bxy + Cy^2 + Dx + Ey + F = 0$$

When $B = 0$ and at least one of the coefficients $A$ or $C$ is not zero, the following summarizes the conditions for each curve:

(a) If $A = C$, we have a *circle.*

In special cases, the graph of the equation may be a point, or there may be no graph. (The equation may have only one or no solution.)

(b) If $A = 0$ and $C \neq 0$, or if $C = 0$ and $A \neq 0$, then we have a *parabola.*

(c) If $A \neq C$, and if $A$ and $C$ are either both positive or both negative, then we have an *ellipse.*

In special cases, the graph of the equation may be a point, or there may be no graph. (The equation may have only one or no solution.)

(d) If $A$ and $C$ differ in sign, then we have a *hyperbola.*

In some special cases the graph may be a pair of intersecting lines.

If $D \neq 0$ or $E \neq 0$ or both are not zero, the curve does not have its center (or vertex in the case of the parabola) at the origin. If $B \neq 0$, then the axis of the curve does not lie along the *x*-axis or the *y*-axis.

**9.** Each point $P(x, y)$ in the rectangular coordinate system may be described by the ordered pair $P(r, \theta)$ in the polar coordinate system.

**10.** In the rectangular coordinate system, there is a one-to-one correspondence between points in the plane and ordered pairs of real numbers. This one-to-one correspondence is not a property of the polar coordinate system. In general, the point $P(r, \theta)$ may be represented by

$$(r, \theta + k \cdot 360°) \quad \text{or} \quad (r, \theta + k \cdot 2\pi)$$

where $k$ is any integer. $P(r, \theta)$ may also be represented by

$$(-r, \theta + k \cdot 180°) \quad \text{or} \quad (-r, \theta + k\pi)$$

where $k$ is any odd integer.

**11.** The relationships between the rectangular and polar coordinate systems are as follows:

(a) $x = r \cos \theta$

(b) $y = r \sin \theta$

(c) $\tan \theta = \dfrac{y}{x}$ or $\theta = \arctan \dfrac{y}{x}$

(d) $x^2 + y^2 = r^2$

(e) $\cos \theta = \dfrac{x}{\sqrt{x^2 + y^2}}$

(f) $\sin \theta = \dfrac{y}{\sqrt{x^2 + y^2}}$

**12.** *Symmetry tests for graphing polar equations:*

(a) *Horizontal axis:* Replace $\theta$ by $-\theta$ in the original equation. If the resulting equation is equivalent to the original equation, then the graph of the original equation is symmetric with respect to the *horizontal* axis.

(b) *Vertical axis:* Replace $\theta$ by $\pi - \theta$ in the original equation. If the resulting equation is equivalent to the original equation, then the graph of the original equation is symmetric with respect to the *vertical* axis.

(c) *Pole:*

  (i) Replace $r$ by $-r$ in the original equation. If the resulting equation is equivalent to the original equation, then the graph of the original equation is symmetric with respect to the *pole.*

  (ii) Replace $\theta$ by $\pi + \theta$ in the original equation. If the resulting equation is equivalent to the original equation, then the graph of the original equation is symmetric with respect to the *pole.*

## CHAPTER 15 REVIEW

1. Write the equation of the circle with center at $(5, -7)$ and with radius 6.

2. Find the center and radius of the circle $x^2 + y^2 - 8x + 6y - 24 = 0$.

3. Find the focus and directrix of the parabola $x^2 = 6y$ and sketch its graph.

4. Write the equation of the parabola with focus at $(-4, 0)$ and directrix $x = 4$.

5. Write the equation of the parabola with focus at $(4, 3)$ and directrix $x = 0$.

6. Find the vertices and foci of the ellipse $4x^2 + 49y^2 = 196$ and sketch its graph.

7. Write the equation of the ellipse with vertices $(0, 4)$ and $(0, -4)$ and with foci at $(0, 2\sqrt{3})$ and $(0, -2\sqrt{3})$.

8. Find the vertices and foci of the hyperbola $4x^2 - 9y^2 = 144$ and sketch its graph.

9. Write the equation of the hyperbola with vertices at $(0, 5)$ and $(0, -5)$ and with foci at $(0, \sqrt{41})$ and $(0, -\sqrt{41})$.

10. Write the equation of the ellipse with center at $(3, -4)$, vertices at $(3, 1)$ and $(3, -9)$, and foci at $(3, 0)$ and $(3, -8)$.

11. Write the equation of the hyperbola with center at $(-7, 4)$ and with vertices at $(2, 4)$ and $(-16, 4)$; the length of the conjugate axis is 6.

12. Name and sketch the graph of $16x^2 - 4y^2 - 64x - 24y + 12 = 0$.

13. Solve

$$y^2 + 4y + x = 0$$
$$x = 2y$$

14. Solve.

$$3x^2 - 4y^2 = 36$$
$$5x^2 - 8y^2 = 56$$

*Plot each point whose polar coordinates are given.*

15. $A(6, 60°)$, $B(3, -210°)$, $C(-2, -270°)$, $D(-4, 750°)$

16. $A\left(5, \dfrac{\pi}{6}\right)$, $B\left(2, -\dfrac{5\pi}{4}\right)$, $C\left(-3, -\dfrac{\pi}{2}\right)$, $D\left(-5, \dfrac{19\pi}{2}\right)$

17. For point $A(5, 135°)$, name three other sets of polar coordinates for $-360° \le \theta \le 360°$.

18. For point $B(-2, 7\pi/6)$, name three other sets of polar coordinates for $-2\pi \le \theta \le 2\pi$.

*Change each set of polar coordinates to rectangular coordinates.*

19. $(3, 210°)$   20. $(2, -120°)$   21. $\left(-5, \dfrac{11\pi}{6}\right)$   22. $\left(-6, -\dfrac{\pi}{2}\right)$

*Change each set of rectangular coordinates to polar coordinates in degrees for $0° \leq \theta < 360°$.*

**23.** $(-3, 3)$     **24.** $(0, -6)$     **25.** $(-1, \sqrt{3})$

*Change each set of rectangular coordinates to polar coordinates in radians for $0 \leq \theta < 2\pi$.*

**26.** $(-5, 0)$     **27.** $(-6\sqrt{3}, 6)$     **28.** $(1, -1)$

*Change each equation to polar form.*

**29.** $x^2 + y^2 = 49$     **30.** $y^2 = 9x$     **31.** $5x + 2y = 8$

**32.** $x^2 - 4y^2 = 12$     **33.** $y^3 = 6x^2$     **34.** $y(x^2 + y^2) = x^2$

*Change each equation to rectangular form.*

**35.** $r \cos \theta = 12$     **36.** $r = 9$     **37.** $\theta = \dfrac{2\pi}{3}$

**38.** $r = 8 \cos \theta$     **39.** $r \sin^2 \theta = 5 \cos \theta$     **40.** $r^2 \sin 2\theta = 8$

**41.** $r^2 = 4 \cos 2\theta$     **42.** $r = \csc \theta$     **43.** $r = 1 + \sin \theta$

**44.** $r = \dfrac{2}{1 - \sin \theta}$

*Graph each equation.*

**45.** $r = 7$     **46.** $\theta = -\dfrac{\pi}{4}$     **47.** $r = 5 \cos \theta$

**48.** $r = 6 + 3 \sin \theta$     **49.** $r = 6 - 3 \sin \theta$     **50.** $r = 4 + 4 \cos \theta$

**51.** $r = 3 - 6 \cos \theta$     **52.** $r \sin \theta = 5$     **53.** $r^2 = 36 \cos \theta$

**54.** $r = 6 \sin 5\theta$     **55.** $r^2 = 25 \sin 2\theta$     **56.** $r(1 - \sin \theta) = 6$

# APPENDIX **A**

# Review of Geometry

A basic knowledge of fundamentals of geometry is necessary to solve technical problems as well as to understand mathematical discussions and developments. We shall briefly review the most basic, most often used formulas and theorems.

## A.1 ANGLES AND LINES

**1.** A **right angle** is an angle whose measure is 90°.

∠*AOB* is a right angle. (*Note*: The vertex letter is always written in the middle.)

**2.** An **acute angle** is an angle whose measure is less than 90°.

∠*A* is an acute angle.

**3.** An **obtuse angle** is an angle whose measure is more than 90° but less than 180°.

∠*B* is an obtuse angle.

4. Two **vertical angles** are the opposite angles formed by two intersecting lines. Angles $m$ and $n$ are vertical angles, as are angles $p$ and $q$ as shown below.

5. Vertical angles are equal.

6. Two angles are **supplementary** when their sum is 180°.

7. Two angles are **complementary** when their sum is 90°.

8. Two lines are **perpendicular** when they form a right angle.

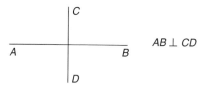

$$AB \perp CD$$

9. The *shortest* distance between a point and a line is the **perpendicular** distance between them.

10. Two lines are **parallel** if they lie in the same plane and do not intersect even when extended.

11. When two parallel lines are intersected by a third line called a **transversal,**
    (a) the alternate interior angles are *equal;*

$$\angle a = \angle g \quad \text{and} \quad \angle d = \angle f$$

   (b) the corresponding angles are *equal;*

$$\angle a = \angle e \qquad \angle b = \angle f \qquad \angle c = \angle g \qquad \angle d = \angle h$$

   (c) the interior angles on the same side of the transversal are *supplementary.*

$$\angle a + \angle f = 180° \quad \text{and} \quad \angle d + \angle g = 180°$$

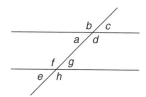

## A.2 TRIANGLES

1. A **polygon** is a closed figure whose sides are all line segments.

2. A **triangle** is a polygon with three sides.

3. A **scalene triangle** is a triangle in which *no two* sides are equal.

4. An **isosceles triangle** is a triangle in which *two* sides are equal.

**5.** An **equilateral triangle** is a triangle in which *all three* sides are equal.

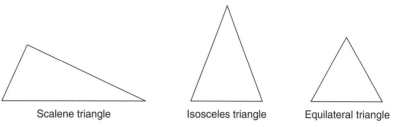

Scalene triangle      Isosceles triangle      Equilateral triangle

**6.** An **acute triangle** is a triangle in which *all three* angles are acute.

**7.** An **obtuse triangle** is a triangle in which there is *one* obtuse angle.

**8.** A **right triangle** is a triangle in which there is *one* right angle.

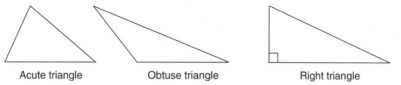

Acute triangle            Obtuse triangle                Right triangle

**9.** An **oblique triangle** is a triangle that does not contain a right angle.

**10.** In a right triangle, the side opposite the right angle is called the **hypotenuse,** *c,* and the other two sides are called the **legs,** *a* and *b,* as shown.

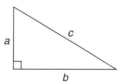

**11. Pythagorean theorem:** The square of the hypotenuse of a right triangle is equal to the sum of the squares of the two legs ($c^2 = a^2 + b^2$).

**12.** A **median** of a triangle is a line segment joining any vertex to the *midpoint* of the opposite side.

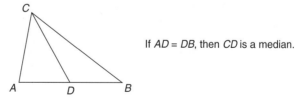

If $AD = DB$, then $CD$ is a median.

**13.** An **altitude** of a triangle is a *perpendicular* line segment from any vertex to the opposite side.

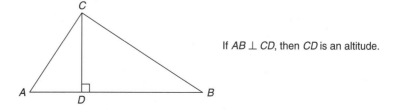

If $AB \perp CD$, then $CD$ is an altitude.

**14.** An **angle bisector** of a triangle is a line segment that bisects any angle and intersects the opposite side.

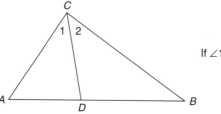

If ∠1 = ∠2, then *CD* is an angle bisector.

**15.** The **sum** of the interior angles of *any* triangle is 180°.

**16.** The **area** of a triangle equals one-half the base times the height ($A = \frac{1}{2}bh$).

**17.** In a 30°–60°–90° triangle,

    **(a)** the side opposite the 30° angle equals one-half the hypotenuse;

    **(b)** the side opposite the 60° angle equals $\dfrac{\sqrt{3}}{2}$ times the length of the hypotenuse.

**18.** Triangles are **similar** (~) if their corresponding angles are equal or if their corresponding sides are proportional.

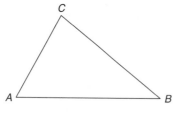

If ∠A = ∠D, ∠B = ∠E, and ∠C = ∠F, then △ABC ~ △DEF.

**19.** Triangles are **congruent** (≅) if their corresponding angles and sides are equal.

## A.3 QUADRILATERALS

**1.** A **quadrilateral** is a polygon with four sides.

**2.** A **parallelogram** is a quadrilateral having two pairs of parallel sides.

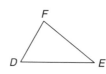

**3.** The **area** of a parallelogram equals the length of the base times the height ($A = bh$).

**4.** The opposite sides and the opposite angles of a parallelogram are equal.

**5.** A **rectangle** is a parallelogram with right angles.

**6.** A **square** is a rectangle with equal sides.

**7.** A **rhombus** is a parallelogram with equal sides.

**8.** A **trapezoid** is a quadrilateral with only one pair of parallel sides.

**9.** The **area** of a trapezoid is given by the formula $A = \frac{1}{2}h(a + b)$.

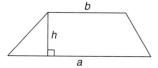

**10.** The diagonal of a parallelogram divides it into two congruent triangles.

**11.** The diagonals of a parallelogram bisect each other.

## A.4 CIRCLES

**1.** A **circle** is the set of all points on a curve equidistant from a given point called the *center.*

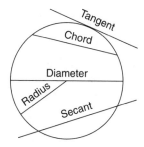

**2.** A **radius** is a line segment joining the center and any point on the circle.

**3.** A **chord** is a line segment joining any two points on the circle.

**4.** A **diameter** is a chord passing through the center.

**5.** A **tangent** is a line intersecting a circle at only one point.

**6.** A **secant** is a line intersecting a circle in two points.

**7.** The **area** of a circle is $A = \pi r^2$, where $r$ is the radius.

**8.** The **circumference** of a circle is $C = 2\pi r$, where $r$ is the radius, or $C = \pi d$, where $d$ is the diameter.

**9.** A line tangent to a circle is perpendicular to the radius at the point of tangency.

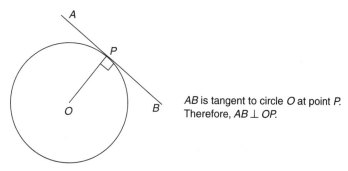

AB is tangent to circle O at point P.
Therefore, AB ⊥ OP.

**10.** A **semicircle** is one-half of a circle.

**11.** An angle inscribed in a semicircle is a right angle.

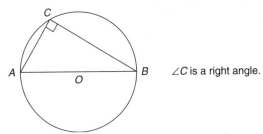

$\angle C$ is a right angle.

## A.5 AREAS AND VOLUMES OF SOLIDS

**1.** The **lateral surface area** of a solid is the sum of the areas of the sides excluding the area of the bases.

**2.** The **total surface area** of a solid is the sum of the lateral surface area plus the area of the bases.

**3.** The **volume** of a solid is the number of cubic units of measure contained in the solid.

Instead of verbally describing each of the following solids, we name and show each and give the formula for its volume and lateral surface area. We shall use $B$, $r$, and $h$ as the area of the base, the length of the radius, and the height, respectively.

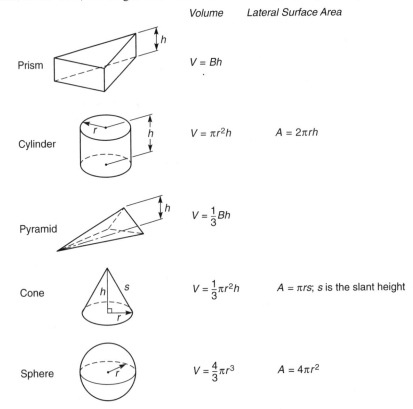

|  | Volume | Lateral Surface Area |
|---|---|---|
| Prism | $V = Bh$ | |
| Cylinder | $V = \pi r^2 h$ | $A = 2\pi rh$ |
| Pyramid | $V = \frac{1}{3}Bh$ | |
| Cone | $V = \frac{1}{3}\pi r^2 h$ | $A = \pi rs$; $s$ is the slant height |
| Sphere | $V = \frac{4}{3}\pi r^3$ | $A = 4\pi r^2$ |

APPENDIX **B**

# Calculator: Degrees, Minutes, Seconds

A calculator makes it very easy to change an angle from decimal degrees to degrees, minutes, and seconds, or vice versa. To change an angle expressed in decimal degrees to degrees, minutes, and seconds, look for a key labeled DMS or DD→DMS. On most current calculator models, pressing this key will perform the conversion, producing the mathematical symbols for degrees and minutes to separate the numbers on the display (some even show the symbol for seconds). There are still some scientific calculators that require the user to interpret the first two digits behind a decimal point as the number of minutes and the next two decimal digits as the number of seconds. Thus, the display for the conversion of 23.46 degrees could look like any of the following:

23°27′36.00

23°27′36″

23.2736

which all symbolize 23 degrees, 27 minutes, and 36 seconds. To change from degrees, minutes, and seconds to decimal degrees, look for a key labeled →DEG, DD, or DMS→DD. Consult your manual for further details.

To change from decimal degrees to degrees, minutes, and seconds if your calculator does not have a DMS or DD→DMS key, perform the following steps:

1. Multiply **only** the decimal portion less than one by **60.** The whole-number part of this result represents the number of minutes.

2. Multiply **only** the decimal portion less than one from the result in Step 1 by **60.** This result represents the number of seconds.

**EXAMPLE 1**

Change 18.3751° to degrees, minutes, and seconds without using a DMS key.

| Flow chart | Buttons pushed | Display |
|---|---|---|
| Enter 0.3751 | `.` `3` `7` `5` `1` | 0.3751 |
| ↓ | | |
| Push times | `x` | 0.3751 |
| ↓ | | |
| Enter 60 | `6` `0` | 60 |
| ↓ | | |
| Push equals | `=` | 22.506* |
| ↓ | | |
| Push minus | `-` | 22.506 |
| ↓ | | |
| Enter 22 | `2` `2` | 22 |
| ↓ | | |
| Push equals | `=` | 0.506 |
| ↓ | | |
| Push times | `x` | 0.506 |
| ↓ | | |
| Enter 60 | `6` `0` | 60 |
| ↓ | | |
| Push equals | `=` | 30.36† |

The result is 18.3751° = 18°22′30″, rounded to the nearest second.

To change from degrees, minutes, and seconds to decimal degrees if your calculator does not have a DD key, use the formula

$$\frac{\dfrac{s}{60} + m}{60}$$

where $s$ is the number of seconds and $m$ is the number of minutes. This result changes minutes and seconds to decimal degrees.

**EXAMPLE 2**

Change 68°7′33″ to decimal degrees without using a DD key, and round the result to four decimal places.

| Flow chart | Buttons pushed | Display |
|---|---|---|
| Enter 33 | `3` `3` | 33 |
| ↓ | | |
| Push divide | `÷` | 33 |
| ↓ | | |
| Enter 60 | `6` `0` | 60 |
| ↓ | | |
| Push plus | `+` | 0.55 |
| ↓ | | |

*The whole-number part represents 22′.
†The whole-number part represents 30″.

| | | |
|---|---|---|
| Enter 7 | 7 | 7 |
| ↓ | | |
| Push equals | = | 7.55 |
| ↓ | | |
| Push divide | ÷ | 7.55 |
| ↓ | | |
| Enter 60 | 6 0 | 60 |
| ↓ | | |
| Push equals | = | 0.12583333 |

The result is 68°7′33″ = 68.1258°, rounded to four decimal places.

# An Alternate Method for Evaluating Third-Order Determinants

The following method is often used to evaluate third-order determinants. **Beware: This method works only for third-order determinants!**

---

**VALUE OF THIRD-ORDER DETERMINANT**

One way to evaluate a third-order determinant is as follows:

**1.** Repeat the first two columns at the right of the determinant.

**2.** Draw diagonal arrows as follows:

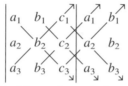

**3.** Find the product of each set of three elements through which an arrow is drawn.
   **(a)** Find the algebraic sum of the products indicated by arrows pointing down.
   **(b)** Find the algebraic sum of the products indicated by arrows pointing up.
   **(c)** Subtract the result of Step 3(b) from the result of Step 3(a).

---

**EXAMPLE 1**

Evaluate.

$$\begin{vmatrix} 3 & -1 & 2 \\ -5 & 6 & 4 \\ 2 & -2 & -3 \end{vmatrix}$$

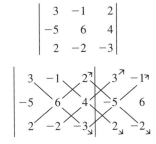

**Step 3(a):**

$$(3)(6)(-3) + (-1)(4)(2) + (2)(-5)(-2) = -54 - 8 + 20 = -42$$

**Step 3(b):**

$$(2)(6)(2) + (-2)(4)(3) + (-3)(-5)(-1) = 24 - 24 - 15 = \underline{-15}$$
$$-27 \quad \text{(Subtract.)}$$

The value of the determinant is $-27$.

**EXAMPLE 2**

Evaluate.

$$\begin{vmatrix} 4 & 5 & -3 \\ 0 & -1 & 12 \\ 2 & 3 & -5 \end{vmatrix}$$

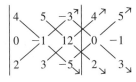

**Step 3(a):**

$$(4)(-1)(-5) + (5)(12)(2) + (-3)(0)(3) = 20 + 120 + 0 = 140$$

**Step 3(b):**

$$(2)(-1)(-3) + (3)(12)(4) + (-5)(0)(5) = 6 + 144 + 0 = \underline{150}$$
$$-10 \quad \text{(Subtract.)}$$

The value of the determinant is $-10$.

**EXAMPLE 3**

Solve using determinants.

$$x - 3y + 2z = -8$$
$$2x + 4y - z = 9$$
$$3x - y + 5z = 5$$

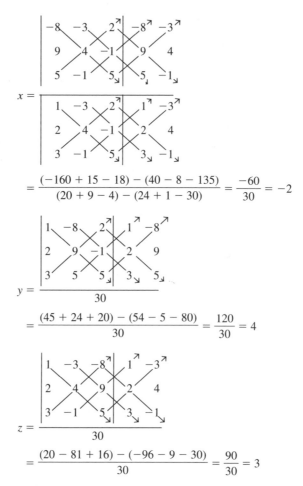

$$x = \frac{(-160 + 15 - 18) - (40 - 8 - 135)}{(20 + 9 - 4) - (24 + 1 - 30)} = \frac{-60}{30} = -2$$

$$y = \frac{(45 + 24 + 20) - (54 - 5 - 80)}{30} = \frac{120}{30} = 4$$

$$z = \frac{(20 - 81 + 16) - (-96 - 9 - 30)}{30} = \frac{90}{30} = 3$$

The solution is $(-2, 4, 3)$.

*Check*: Substitute the solution in all three original equations.

## Exercises C.1

*Evaluate each determinant.*

1. $\begin{vmatrix} 3 & 2 & -1 \\ 5 & 7 & -4 \\ -2 & -3 & 3 \end{vmatrix}$

2. $\begin{vmatrix} 4 & 3 & -6 \\ 1 & 4 & 9 \\ -2 & 0 & 4 \end{vmatrix}$

3. $\begin{vmatrix} 1 & -4 & 8 \\ -6 & -3 & 1 \\ 2 & 1 & 5 \end{vmatrix}$

4. $\begin{vmatrix} -3 & -5 & 2 \\ 0 & 1 & 1 \\ 2 & -7 & 4 \end{vmatrix}$

5. $\begin{vmatrix} 0 & 1 & -4 \\ 2 & -3 & 9 \\ 7 & 6 & -2 \end{vmatrix}$

6. $\begin{vmatrix} 2 & -2 & 5 \\ 9 & -4 & 6 \\ 3 & -1 & 1 \end{vmatrix}$

**7.** $\begin{vmatrix} 1 & 2 & 2 \\ 3 & -4 & -7 \\ 3 & 6 & 6 \end{vmatrix}$

**8.** $\begin{vmatrix} -2 & 3 & -9 \\ 6 & -4 & 7 \\ 2 & 1 & -5 \end{vmatrix}$

**9.** $\begin{vmatrix} 8 & 5 & -3 \\ 7 & 3 & 1 \\ -2 & 4 & -2 \end{vmatrix}$

**10.** $\begin{vmatrix} 5 & -7 & 9 \\ 6 & 1 & 0 \\ 1 & 2 & -3 \end{vmatrix}$

**11.** $\begin{vmatrix} 2 & -4 & 9 \\ 6 & 2 & 3 \\ -5 & 7 & -2 \end{vmatrix}$

**12.** $\begin{vmatrix} -1 & 6 & -7 \\ -2 & 3 & 1 \\ 2 & 9 & -10 \end{vmatrix}$

# APPENDIX $\mathbf{D}$

# Using a Graphing Calculator

## INTRODUCTION

This appendix is included to provide faculty with the flexibility of integrating graphing calculators in their classes. Each section explains and illustrates important features of the Texas Instruments TI-83 calculator. Calculator symbols shown throughout the text refer to specific sections of this appendix, but the material is organized so that an interested student could also study it as a separate chapter.

## D.1 INTRODUCTION TO THE TI-83 KEYBOARD

This section provides a guided tour of the keyboard of the TI-83 graphics calculator. In this and the following sections, please have your calculator in front of you (or see diagram on page 609) and be sure to try out the features as they are discussed.

First notice that the keys forming the bottom six rows of the keyboard perform the standard functions of a scientific calculator. The thin blue keys which form the very top row allow functions to be defined and their graphs to be drawn (see Section D.3 for details). The second, third, and fourth rows of keys provide access to menus full of advanced features and perform special tasks such as **INS**ert, **DEL**ete, **CLEAR**, and **QUIT** (to leave a menu, an editor, or a graph, and return to the home screen). Also found in these rows are the **2nd** and **ALPHA** shift keys, which give additional, color-coded meanings to almost every key on the calculator.

The **ON** key is in the lower left-hand corner. Note that pressing the (golden yellow) **2nd** key followed by the **ON** key will turn the calculator **OFF**. If the calculator is left unattended (or no buttons are pressed for a couple of minutes), the calculator will shut itself off. No work is lost when the unit is turned off. Just turn the calculator back **ON**, and the display will be exactly as you left it. Due to different lighting conditions and battery strengths, the screen contrast needs adjustment from time to time. Press the **2nd** key, then *press and hold* the up (or down) arrow key to darken (or lighten) the screen contrast.

The **ENTER** key in the lower right-hand corner is like the = key on many scientific calculators; it signals the calculator to perform the calculation that you've been typing. Its (shifted) **2nd** meaning, **ENTRY**, gives you access to previously entered formulas, starting with the most recent one. If you continue to press **2nd ENTRY**, you can access previous entries up to an overall memory limit of 128 characters. Depending on the length of your formulas, this means that about 10 to 15 of your most recent entries can be retrieved from the calculator's memory to be reused or modified.

Just above **ENTER** is a column of four other blue keys which perform the standard operations of arithmetic. Please note, though, that the multiplication key indicated by an ×, prints an asterisk on the screen, and the division key prints a slash on the screen. Just above these four is the ^ key which indicates that you're raising something to the power that follows, for example, 2^5 would mean $2^5$. Moving to the left across that row, you will see the keys for the trigonometric functions: **SIN**, **COS**, and **TAN** (note that their standard setting is radians, but you can specify degrees by using the degree symbol which is option 1 in the **ANGLE** menu; or the calculator can be set to always think in degrees by specifying that option in the **MODE** menu). Always press the trig key before typing the angle, as in cos ($\pi$) or sin (30°). Notice that the left-hand parenthesis is automatically included when you press any of the trig keys. To the left of these three is a key labeled $x^{-1}$, which acts as a reciprocal key for ordinary arithmetic. It will also invert a matrix, as in $[A]^{-1}$, which explains why the key isn't labeled $1/x$, as it would be on most scientific calculators. Beneath that key is $x^2$ (the squaring key), whose shifted **2nd** meaning is square root. Below in that column are keys for logs, whose shifted **2nd** versions give exponential functions. Like the trig keys, the square root, **LOG**, **LN**, and exponential keys also precede their arguments. For example, log(2) will find the common logarithm of 2.

Between **LN** and **ON** is the **STO>** key which is used to store a number (possibly the result of a calculation) into any of the 27 memory locations whose names are A, B, C, . . ., Z, and $\theta$. First indicate the number or calculation, then press **STO>** (which just prints an arrow on the screen) followed by the (green) **ALPHA** key, then the (green) letter name you want the stored result to have, and finally press **ENTER**. The computation will be performed, and the result will be stored in the desired memory location as well as being displayed on the screen. If you have just performed a calculation and now wish that you had stored it, don't worry. Just press **STO>** on the next line followed by **ALPHA** and the letter name you want to give this quantity, then press **ENTER**.

Some examples:

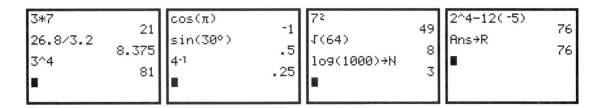

If you watched the last **STO>** example closely, you may have noticed that the calculator prints **Ans** (which stands for the "previous answer") on the screen whenever you don't indicate the first operand on a given line. For example, if you begin a formula with a plus sign, the calculator assumes that you want to add something to the previous result, so it displays "Ans +" instead of just "+." At times, you'll want to refer to the previous result somewhere other than at the beginning of your formula. In that case, press **2nd ANS**

(the shifted version of the key next to **ENTER**) wherever you want the previous answer to appear in the computation.

The shifted **2nd** meaning of the **STO>** key is **RCL** (recall), as in **RCL Z**, which would display the *contents* of memory location Z at the current cursor position in your formula. It is usually easier to write the letter Z itself (press **ALPHA** followed by **Z**) in formulas instead of the current value that's stored there, so this recall feature isn't the best choice in most computations. However, the **RCL** feature is very useful in creating instant copies of functions (Rcl $Y_1$) and programs (Rcl prgmSIMPSON) so that newly modified versions don't have to destroy the old ones.

The key that changes the sign of a number is labeled **(-)** and is located just to the left of the **ENTER** key. Don't confuse this gray key with the dark blue subtraction key! Note also that the TI-83 consistently views the lack of an indicated operation between two quantities as an intended multiplication.

The parentheses keys are just above the 8 and 9 keys. These are used for all levels of parentheses. Do not be confused by symbols such as { } and [ ], which are the shifted **2nd** versions of these and other nearby keys. Braces { } are used to indicate lists on the TI-83, and brackets [ ] are used only for matrices. Once again, these special symbols *cannot* be used to indicate higher levels of parentheses; just nest ordinary parentheses to show several levels of quantification. Also note that the comma key is used only with matrices, lists, multiple-argument functions, and certain commands in the TI-83 programming language. Never use commas to separate digits within a number. The number three thousand should always be typed 3000 (not 3,000). The shifted **2nd** meaning of the comma key is **EE** (enter exponent), which is used to enter data in scientific notation; for example, **1.3** followed by **2nd EE (-)8** would be the keystrokes needed to enter $1.3 \times 10^{-8}$ in a formula. It would be displayed on the screen as 1.3E-8.

The shifted **2nd** versions of the numbers 1 through 9 provide keyboard access to lists and sequences. The shifted **ALPHA** version of the zero key prints a blank space on the display. The shifted **2nd** version of the zero key is **CATALOG**, which provides alphabetical access to every feature of the calculator. Just press the first letter of the desired feature (without pressing **ALPHA**), then scroll from there using the down arrow key. Press **ENTER** when the desired feature is marked by the small arrow. The shifted **2nd** version of the decimal point is "*i*," the imaginary unit (which is often called "*j*" in electronics applications). This symbol can be used in computations involving imaginary and complex numbers even when **MODE Real** has been selected.

The shifted **2nd** version of the plus sign is **MEM** (the memory management menu), which gives you a chance to erase programs, lists, and anything else stored in memory. Use this menu sparingly (remember, the TI-83 has a 32K memory, so you don't usually need to be in a hurry to dispose of things which might prove useful later). If you get into **MEM** by accident, just press **2nd QUIT** to get back to the home screen. **2nd QUIT** always takes you back to the home screen from any menu, editor, or graph, but it will not terminate a running TI-83 program. To interrupt a running program, just press the **ON** button, then choose "quit" in the menu you'll see.

If you're looking for keys that will compute cube roots, absolute values, complex conjugates, permutations, combinations, or factorials, press the **MATH** key, and you'll see four submenus (selectable by using the right or left arrow key) which give you these options and many more. Especially interesting is **>FRAC** (convert to fraction) which will convert a decimal to its simplified fractional form, provided that the denominator would be less than 10,000 (otherwise, it just writes the decimal form of the number). Other examples are also included below to give you a better idea of just how many options are available in the **MATH** menu.

```
MATH NUM CPX PRB
1▶Frac
2:▶Dec
3:3
4:3√(
5:×√
6:fMin(
7↓fMax(
```

```
2.7/3.6▶Frac
              3/4
2/7-3/8▶Frac
            -5/56
2/71+3/541▶Frac
       .0337143006
■
```

```
MATH NUM CPX PRB
1:abs(
2:round(
3:iPart(
4:fPart(
5:int(
6:min(
7↓max(
```

```
abs(-7)
              7
round(π,4)
         3.1416
max(π,22/7)
      3.142857143
■
```

```
MATH NUM CPX PRB
1:conj(
2:real(
3:imag(
4:angle(
5:abs(
6:▶Rect
7:▶Polar
```

```
conj(5+6i)
           5-6i
real(18-7i)
            18
abs(3+4i)
            5
■
```

```
MATH NUM CPX PRB
1:rand
2:nPr
3:nCr
4:!
5:randInt(
6:randNorm(
7:randBin(
```

```
7*6*5*4*3*2*1
            5040
7!
            5040
10 nCr 3
            120
■
```

# D.2 COMPUTATIONAL EXAMPLES

1. Compute the following:

   **a.** $7 \times 6$

   **b.** $3 \times 7 + 6(3 - 5)$

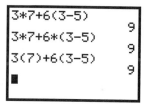

2. Compute $8\{3 + 5[2 - 7(8 - 9)]\}$

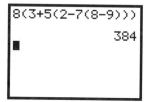

The calculator uses only ordinary parentheses.

3. Express the following as a decimal and as a simplified fraction:

   **a.** $\dfrac{105}{100}$

   **b.** $\dfrac{3}{8} + \dfrac{21}{10} - \dfrac{17}{25}$

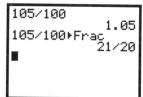

Note that a fraction is an indicated division operation and that the division key always prints a diagonal fraction bar line on the screen. The "convert to fraction" feature is the first item in the **MATH** menu and is accessed by pressing **MATH** then **1** (or **MATH**, then **ENTER**) at the end of a formula. Note also that simplified improper fractions are the intended result. Mixed numbers are not supported. A decimal result would mean that the answer cannot be written as a simplified fraction with a denominator less than 10,000.

**4.** Compute the following, expressing the answer as a simplified fraction:

**a.** $\dfrac{2^5}{6^2}$

**b.** $\dfrac{5 - (-7)}{-2 - 12}$

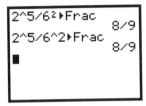

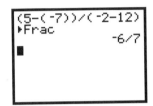

Squares can be computed by pressing the $x^2$ key; similarly, a third power can be indicated by pressing **MATH**, then **3**. Most other exponents require the use of the ^ key (found between **CLEAR** and the division key). In part (b), notice the calculator's need for additional parentheses which enclose the numerator and denominator of the fraction. Also notice the difference between the calculator's negative sign (the gray key below **3**) and its subtraction symbol (the blue key to the right of **6**).

**5.** Compute the following complex numbers:

**a.** $(3 + 4i)(-2 - 5i)$

**b.** $\dfrac{7 + 29i}{30 + 10i}$

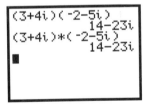

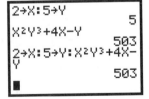

Note that the calculator key for the imaginary unit, "$i$," is the shifted **2nd** version of the decimal point. This imaginary number is often called "$j$" in electronics applications.

**6.** Evaluate the expressions:

**a.** $x^2 + 5x - 8$ when $x = 4$

**b.** $x^2y^3 + 4x - y$ when $x = 2$ and $y = 5$

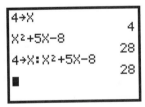

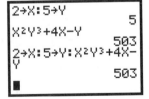

The **STO>** key (just above **ON**) is used to print the arrow symbol on the screen. The letter $x$ can be typed on the screen by pressing the key next to **ALPHA**, labeled

**X,T,θ,n**, or by pressing **ALPHA**, then **X**. Note that several steps can be performed on one line if the steps are separated by a colon (the shifted **ALPHA** version of the decimal point key). In such cases, all steps are performed in sequence, but only the result of the very last step is displayed on the screen.

7. Given that $f(x) = x^4 - 7x + 11$, find $f(3)$, $f(5)$, and $f(-1)$.

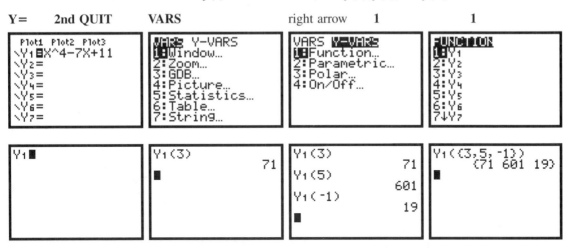

Note that using a stored function requires entering the function in the **Y=** menu, pressing **2nd QUIT** to return to the home screen, then finding the name of that function in the **FUNCTION** submenu under **Y-VARS**. Function evaluation requires the use of parentheses; without parentheses around the argument, multiplication would be assumed. To type the second and third uses of $Y_1$, press **2nd ENTRY**, then left arrow twice, modifying only the argument from the previous formula. The last screen shows how a list of arguments can be used to calculate a list of function values. Your list entries must be enclosed by braces { } and separated by commas (list entries output by the calculator are separated by spaces instead of commas).

## D.3 GRAPHING FEATURES

The thin blue buttons along the top row of the calculator do most of its graphical work. The **Y=** key provides access to the calculator's list of ten functions (assuming that the calculator is set in **MODE Func**). Pressing the **Y=** key will reveal functions $Y_1$ through $Y_7$; the other three functions, $Y_8$, $Y_9$, and $Y_0$, can be seen by pressing the (blue) down arrow key nine times (or just press and hold the down arrow key). These functions are part of the calculator's memory, but the information stored on this page can be easily edited, overwritten, or **CLEAR**ed. Functions are selected for graphing (turned "on") by highlighting their equal sign. This is done automatically when you type in a new function or modify an old one. In other cases, to change the status of a function (from "off" to "on" or vice versa), you will need to use the arrow keys to position the cursor over the equal sign (making it blink); then press **ENTER**. Functions marked with an ordinary equal sign are stored in memory, but will *not* be graphed. To the left of each function name is a symbol indicating how it will be graphed. The normal setting looks like a backslash \ and simply indicates that the graph will be drawn with a thin line. Other settings include a thicker line, shading above the graph, shading below the graph, two animated settings (one marks the path of motion on the screen; the other just shows the motion without marking its path), and finally an option that graphs with a dotted line. To switch from one option to the other, just press the left arrow key until the cursor is over the option marking (at the far left of that function's name), then press **ENTER** repeatedly until the desired option appears. One warning about the **Y=** menu: the names Plot1, Plot2, and Plot3 at the top of the function list refer only to the TI-83's **STAT**istical **PLOT**s. They have nothing to do with ordinary graphing and should *not* be highlighted if you are just trying to graph some functions.

The **WINDOW** key allows you to specify *manually* the extents of the *x* and *y* values that will be visible on the calculator's graphing screen (see **ZOOM** for *automatic* ways of doing this). The Xscl and Yscl options specify the meaning of a mark on the *x*- or *y*-axis. For example Xscl=5 means that each mark shown on the *x*-axis will mean an increment of five units (Xscl=1 is a common setting for algebraic functions; Xscl = $\pi/2$ is commonly used when graphing trigonometric functions). The last option, Xres, allows you to control how many points will actually be calculated when a graph is drawn. Xres = 1 means that an accurate point will be calculated for each pixel on the *x*-axis (somewhat slow, but very accurate). Xres = 2 will calculate only at every other pixel, etc.; Xres = 8 only calculates a point for every eighth pixel (this is the fastest setting, but also the least accurate). In the examples that follow, all graphs are shown with Xres = 1.

Pressing **2nd FORMAT** (the shifted version of the **ZOOM** key) reveals additional graphing options that allow you to change the way coordinates are displayed (polar instead of rectangular), turn coordinates off completely (inhibiting some **TRACE** features), provide a coordinate grid, hide the axes, label the axes, or inhibit printing expressions that describe the graphs. If you find your graphs looking cluttered or notice that axes, coordinates, or algebraic expressions are missing, the "standard" settings are all in the left-hand column. Like other menus where the options aren't numbered (**MODE** is similar), use your arrow keys to make a new option blink, then select it by pressing **ENTER**.

**ZOOM** accesses a menu full of *automatic* ways to set the graphical viewing window. **ZStandard** (option 6) is usually a pretty good place to start, but you should consider option 7, **ZTrig**, if you're graphing trigonometric functions. **ZStandard** shows the origin in the exact center of the screen with *x* and *y* values both ranging from $-10$ to 10. From here you can **Zoom In** or **Zoom Out** (options 2 and 3) or draw a box around a portion of the graph that you would like magnified to fit the entire screen (option 1, **ZBox**). There is also an option to "square up" your graph so that units along the *x*-axis are equal in length to units along the *y*-axis (option 5, **ZSquare**); the *smaller* unit length from the axes of the previous graph will now be used on both axes. This option makes the graph look more like it would on regular graph paper; for example, circles really look like circles. **ZDecimal** and **ZInteger** (options 4 and 8) prepare the screen for **TRACE**s which will utilize *x*-coordinates at exact tenths or integer values, respectively. Option 9, **Zoom-Stat**, makes sure that all of the data in a statistical plot will fit in the viewing window. Option 0, **ZoomFit**, calculates a viewing window using the present *x*-axis but adjusting the *y*-axis so that the function fits neatly within the viewing window. All of these options work by making automatic changes to the **WINDOW** settings. Want to go back to the view you had before? The **MEMORY** submenu (press the right arrow key after pressing **ZOOM**) contains options to go back to your immediately previous view (option 1, **ZPrevious**) or to a window setting you saved a while ago (option 3, **ZoomRcl**). **ZoomSto** (option 2) is the way to save the current window setting for later (note that it can retain only one window setting, so the new information replaces whatever setting you had saved before). Option 4, **SetFactors . . .**, gives you the chance to control how dramatically your calculator will **Zoom In** or **Zoom Out**. These zoom factors are set by Texas Instruments for a magnification ratio of 4 on each axis. Many people prefer smaller factors, such as 2 on each axis. It is possible to set either factor to any number greater than or equal to 1; they don't need to be whole numbers, and they don't necessarily have to be equal.

The **TRACE** key takes you from any screen or menu to the current graph, displaying the $x$- and $y$- coordinates of specific points as you trace along a curve using the left and right arrow keys. Note that in **TRACE**, the up and down arrow keys are used to jump from one curve to another when several curves have been drawn on the same screen. The expression ( formula) for the function you are presently tracing is shown in the upper left-hand corner of the screen (or its subscript number is shown in the upper right-hand corner if you have selected the **ExprOff** option from the **FORMAT** menu). If you press **ENTER** while in **TRACE**, the graph will be redrawn with the currently selected point in the exact center of the screen, even if that point is presently outside the current viewing window. This feature is especially convenient if you need to pan up or down to see higher or lower portions of the graph. To pan left or right, just press and hold the left or right arrow key until new portions of the graph come into view. These useful features change the way the graph is centered on the screen without changing its magnification. Note also that these recentering features work *only* in **TRACE**. To exit **TRACE** without disturbing your view of the graph, just press **GRAPH** (or **CLEAR**). To return to the home screen, abandoning both **TRACE** and the graph, just press **2nd QUIT** (or press **CLEAR** twice). Note that using any **ZOOM** feature also causes an exit from **TRACE** (to resume tracing on the new zoomed version, you must press **TRACE** again).

The **GRAPH** key takes the calculator from any screen or menu to the current graph. Note that the calculator is smart enough that it will redraw the curves only if changes have been made to the function list (**Y=**). As previously mentioned, the **GRAPH** key can be used to turn off **TRACE**. You can also hide an unwanted free cursor by pressing **GRAPH**. When you're finished with viewing a graph, press **CLEAR** or **2nd QUIT** to return to the home screen.

## D.4 EXAMPLES OF GRAPHING

1. To graph $y = x^2 - 5x$, first press **Y=**, then press **CLEAR** to erase the current formula in $Y_1$ (or use the down arrow key to find a blank function); then press the **X,T,θ,n** key (**ALPHA**, then **X** will also work), followed by the $x^2$ button; now press the (blue) minus sign key, then **5**, followed immediately by the **X,T,θ,n** key (a multiplication sign is not needed). Your screen should look very much like the first one shown below. There is no need to press **ENTER** when you have finished typing a function's formula. To set up a good graphing window, press **ZOOM** and then **6** to choose **ZStandard**. This causes the graph to be immediately drawn on axes that range from $-10$ to 10. Notice that you did not have to press the **GRAPH** key; the **ZOOM** menu items and the **TRACE** key also activate the graphing screen. If you have one or more unwanted graphs drawn on top of this one, go back to your function list (**Y=**) and turn off the unwanted functions by placing the cursor over their highlighted equal signs and

pressing **ENTER**. After you have turned off the unwanted functions, just press **GRAPH**, and you will finally see the last screen below.

**2.** To modify this function to be $y = -x^2 + 4$, press **Y=** , then insert the negative sign by pressing **2nd INS** followed by the gray sign change key (-); now press the right arrow key twice to skip over the parts of the formula that are to be preserved. Note that the arrow keys also take you out of insert mode. Now type the plus sign and the 4 (replacing the −5), and finally press **DEL** to delete the extra X at the end of the formula. Press **TRACE** to plot this function. **TRACE** gives the added bonus of a highlighted point, with its coordinates shown at the bottom of the screen. Press the right or left arrow keys to highlight other points on the curve.

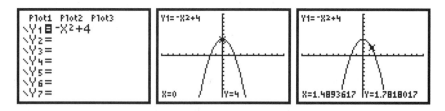

Perhaps X = 1.4893617, Y = 1.7818017 was not a coordinate pair you had expected to investigate. Two special **ZOOM** features (options 4 and 8) can be used to make the **TRACE** option more predictable. Press **ZOOM**, then **4** to select **ZDecimal**; now press **TRACE**.

ZOOM    4                             **TRACE**              right arrows

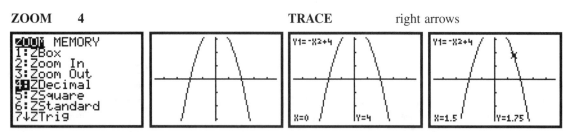

Try pressing the right or left arrow key about 15 times, watching the values at the bottom of the screen. You'll quickly notice that the *x*-values are now all *exact tenths* (**ZOOM** option 8, **ZInteger**, produces **TRACE**able *x*-values which are all integers). Another nice thing about **ZDecimal** is that the graph is "square" in the sense that units on the *x*- and *y*-axes have the same length. The main disadvantage to **ZDecimal** is that the graphing window is "small," displaying only points with *x*-values between −4.7 and 4.7 and *y*-values between −3.1 and 3.1. This disadvantage is apparent on the current graph, which runs off the top of the screen. To demonstrate how this problem can be overcome, **TRACE** the graph to the point X = 1.5, Y = 1.75 and then press **ENTER**. This special feature of **TRACE** causes the graph to be redrawn with the highlighted point in the exact center of the screen (with no change in the magnification of the graph). This is the way to pan up or down from the current viewing window (to pan left or right, see Example 6).

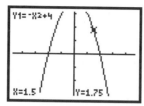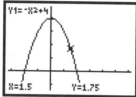

3. Another way to deal with the preceding problem is to **Zoom Out**, but first you'll want to set your ZOOM FACTORS to 2 (the factory setting is 4). Press **ZOOM**, then the right arrow key (**MEMORY**), then press **4** (**SetFactors**). To change the factors to 2, just type **2**, press **ENTER**, and then type another **2**.

**ZOOM**        right arrow    **4**        **2**  **ENTER**  **2**

```
ZOOM MEMORY
1:ZBox
2:Zoom In
3:Zoom Out
4:ZDecimal
5:ZSquare
6:ZStandard
7↓ZTrig
```
```
ZOOM MEMORY
1:ZPrevious
2:ZoomSto
3:ZoomRcl
4:SetFactors...
```
```
ZOOM FACTORS
XFact=2
YFact=2
```

Now to reproduce our problem, press **ZOOM**, then **4** (**ZDecimal**). However, this time correct it by pressing **ZOOM** followed by **3** (**Zoom Out**). At first glance, it looks like nothing has happened, except that X = 0, Y = 0 is displayed at the bottom of the screen. The calculator is waiting for you to use your arrow keys to locate the point in the current window where you would like the exact center of the new graph to be (then press **ENTER**). Of course, if you like the way the graph is already centered, you will still have to press **ENTER** (you'll just skip pressing the arrow keys).

**ZOOM**    **4**                                    **ZOOM**    **3**

```
ZOOM MEMORY
1:ZBox
2:Zoom In
3:Zoom Out
4:ZDecimal
5:ZSquare
6:ZStandard
7↓ZTrig
```

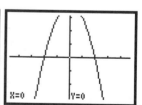

```
ZOOM MEMORY
1:ZBox
2:Zoom In
3:Zoom Out
4:ZDecimal
5:ZSquare
6:ZStandard
7↓ZTrig
```

**ENTER**                         **TRACE**                        right arrows

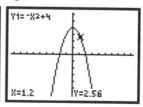

This extra keystroke has proven to be a bit confusing to beginners who think that **Zoom In**, **Zoom Out**, and **ZInteger** should work like the six zoom options (4, 5, 6, 7, 9, and 0) that do their job without pressing **ENTER**. Perhaps more interesting is the fact that you can continue to **Zoom Out** just by pressing **ENTER** again and again (of course, you can also press some arrow keys to re-center between zooms if you wish). Before you experiment with that feature, press **TRACE** and notice, by pressing the left or right arrow key a few times, that the *x*-values are now changing by .2 (instead of .1) and the graph is still "square". Other popular square window settings can be obtained by repeating this example with both zoom factors set to 2.5 or 5. These give "larger" windows where the *x*-values change by .25 or .5, respectively, during a **TRACE**.

4. The only other zoom option that needs extra keystrokes is **ZBox** (option 1). This is a very powerful option that lets you draw a box around a part of the graph that you would like enlarged to fit the entire screen. After selecting this option, use the arrow keys to locate the position of one corner of the box and press **ENTER**. Now use the arrow keys to locate the *opposite* corner, and press **ENTER** again.

**ZOOM    1**     left and up arrows     **ENTER**     right and down arrows

 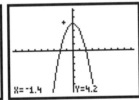 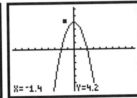 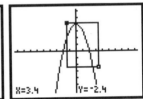

**ENTER**     **GRAPH**

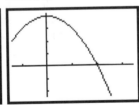

Note that the resulting graph has a free cursor identifying the point in the exact center of the screen. What may not be apparent is that your calculator is ready for you to draw another box if you wish to zoom-in closer. To get rid of this free cursor, just press **GRAPH** (or **CLEAR**).

5. Sometimes, you will know the precise interval on the *x*-axis (the domain) that you want for a graph, but a corresponding interval for the *y*-values (the range) may not be obvious. **ZoomFit** (the tenth **ZOOM** option) is designed for this circumstance. For example, to graph $f(x) = x^3 - 8x + 9$ on the interval $[-3, 2]$, manually set the **WINDOW** so that Xmin $= -3$ and Xmax $= 2$ (the other values shown in the second frame are just leftovers from **ZStandard**). Now press **ZOOM**, then **0** to select **ZoomFit**. There is a noticeable pause while appropriate values of Ymin and Ymax are calculated, then the graph is drawn. To view the values calculated for the range, just press **WINDOW**.

| Y= | WINDOW | GRAPH | ZOOM 0 |
|---|---|---|---|

 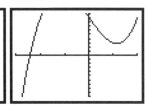 

| | WINDOW | WINDOW | GRAPH |
|---|---|---|---|

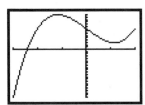 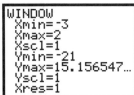  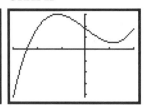

Note that **ZoomFit** changes only Ymin and Ymax. You may also wish to change Yscl.

**6.** Panning to the right:

| Y= | ZOOM 6 | | TRACE |
|---|---|---|---|

  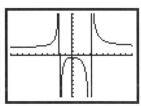 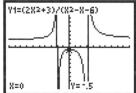

Right arrrow beyond the edge of the screen to pan to the right . . .

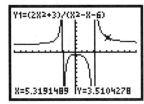 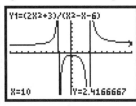 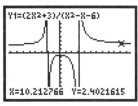 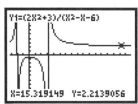

Panning to the left is done similarly. To pan up or down, see the last part of Example 2.

**7.** Creating, storing, and retrieving viewing windows:

| Y= | WINDOW | GRAPH | ZOOM 2 |
|---|---|---|---|

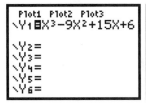  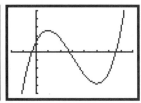

(Let's change it)     **ZOOM**     **3**                              **ENTER**

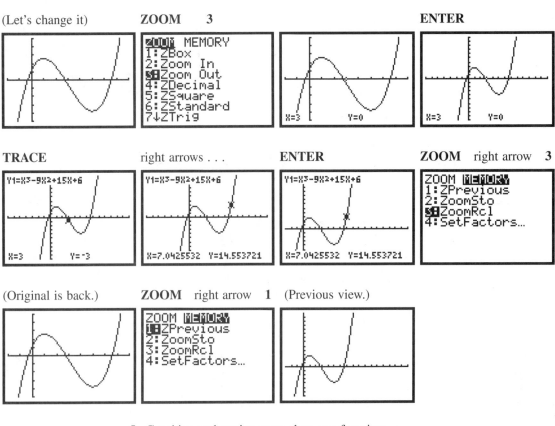

**8.** Graphing and tracing more than one function:

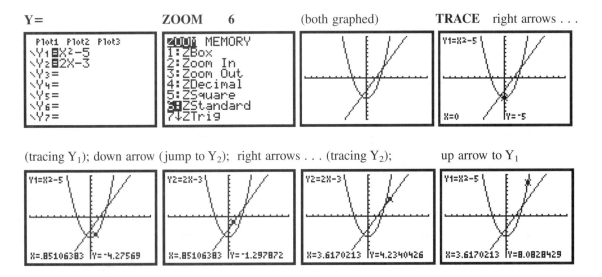

Note that the down arrow increases the subscript of the function being traced
and the up arrow decreases the subscript. The result has nothing to do with
which graph is above or below the other.

**9.** Finding a point of intersection:

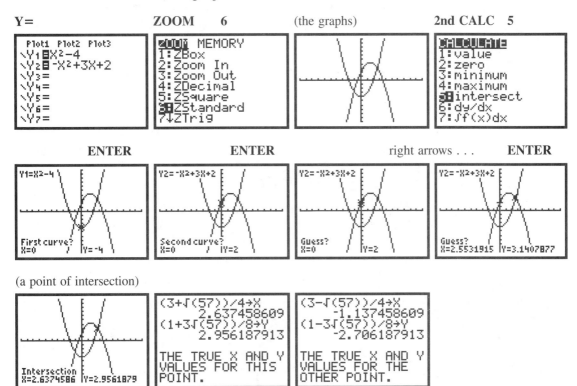

Y=

ZOOM 6

(the graphs)

2nd CALC 5

ENTER

ENTER

right arrows . . .

ENTER

(a point of intersection)

Note that the *Guess* was very important in determining which point of intersection would be calculated. Now find the other intersection point using **intersect**.

**10.** Calculating *y*-values and locating the resulting points on your graphs:
If you have just completed Example 9, please skip to the fourth frame.

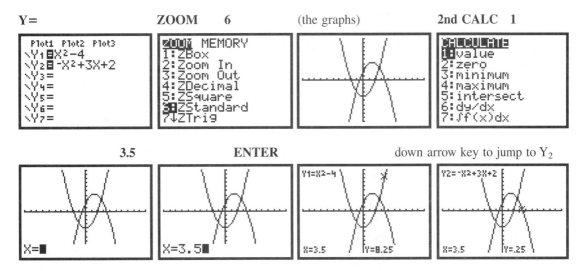

Y=

ZOOM 6

(the graphs)

2nd CALC 1

3.5

ENTER

down arrow key to jump to $Y_2$

Note again that the subscript of a function increases as you jump from one curve to another by pressing the down arrow key. The up arrow key decreases this number, and either key can be used to wrap around and start over.

## D.5 TRIGONOMETRIC FUNCTIONS AND POLAR COORDINATES

1. Setting the **MODE** to **Degree**s (radian mode is the standard factory setting):

**MODE** down arrow twice, right arrow **ENTER** **CLEAR** (or **2nd QUIT**) to exit

2. Evaluating trigonometric functions using degrees, minutes, and seconds:

  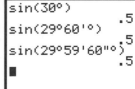

The symbols for degrees and minutes are the first two options in the menu found by pressing **2nd ANGLE** (just to the right of the **MATH** key). The symbol for seconds is the shifted **ALPHA** version of the addition key. Note that in **Radian MODE**, degrees can still be used, but an *additional* degree symbol must follow the angle's measure.

3. To graph $y = \sin x$, first press **Y=**, then press **CLEAR** to erase the current formula in $Y_1$ (or use the arrow keys to find a blank function), then press the **SIN** key followed by the **X,T,θ,n** key (**ALPHA X** will also work) and finally press the right parenthesis key. Your screen should look very much like the first one in the following figure. To set up a good graphing window, press **ZOOM** and then press **7** to choose **ZTrig**. This causes the graph to be immediately drawn on axes that range from roughly $-2\pi$ to $2\pi$ (actually from $-352.5°$ to $352.5°$) in the $x$ direction and from $-4$ to $4$ in the $y$ direction. Each mark along the $x$-axis represents a multiple of $\pi/2$ radians ($90°$). Notice also that we did not need to press the **GRAPH** key; the **ZOOM** menu items and the **TRACE** key also activate the graphing screen. If you have one or more unwanted graphs drawn on top of this one, go back to your function list (**Y=**) and turn off the unwanted functions by placing the cursor over their highlighted equal signs and pressing **ENTER**. After you have turned off the unwanted functions, just press **GRAPH**, and you will finally see the last frame in the following figure.

4. To modify this function to be $y = -3 \sin 2x$, press **Y=**, then insert the $-3$ by pressing **(2nd) INS** followed by the gray sign change key **(-)** then the number **3**. Now press the right arrow key to skip over the part of the formula that's OK. Note that the arrow keys take you out of insert mode. Now type **(2nd) INS**, then the **2**. Press **ZOOM 7**, then **TRACE** to plot this function. **TRACE** gives the added bonus of a highlighted point with its coordinates shown at the bottom of the screen. Press the right or left arrow keys to highlight other points on the curve. The highlighted coordinate pair in the third frame below is $x = 5\pi/12$, $y = -1.5$. **ZTrig** allows **TRACE** to display all points whose $x$-values are multiples of $\pi/24$ (of course, this includes such special values as $0$, $\pi/6$, $\pi/4$, $\pi/3$, $\pi/2$, etc., written in their decimal forms). In degrees, follow the same directions. The only difference is that the traced $x$-values are now multiples of $7.5°$ (which is the equivalent of $\pi/24$ radians). As this example illustrates, **ZTrig** has been carefully designed to produce the same graph for both radians and degrees. Other automatic ways of establishing a viewing window, such as **ZStandard** and **ZDecimal**, ignore the **MODE** setting and are not recommended for graphing trigonometric functions in degrees.

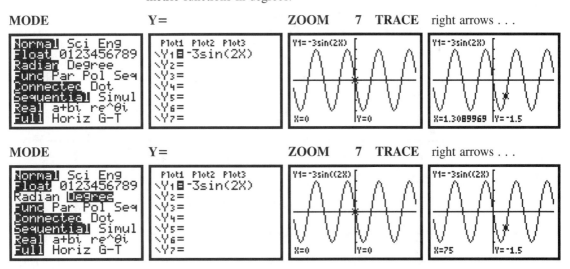

5. Several related trig functions can be drawn on the same screen, either by typing them separately in the function list **Y=** or by using a list of coefficients as shown in the following figure. A list consists of numbers separated by commas which are enclosed by braces { }. The braces are the shifted **2nd** versions of the parentheses keys. The first two frames indicate how to efficiently graph $y = \sin (x)$,

$y = 2 \sin(x)$, and $y = 4 \sin(x)$ on the same screen. The last two frames graph $y = 2 \sin(x)$ and $y = 2 \sin(3x)$.

**Y=**

```
Plot1 Plot2 Plot3
\Y1◻{1,2,4}sin(X
)
\Y2=
\Y3=
\Y4=
\Y5=
\Y6=
```

**GRAPH**

**Y=**

```
Plot1 Plot2 Plot3
\Y1◻2sin({1,3}X)

\Y2=
\Y3=
\Y4=
\Y5=
\Y6=
```

**GRAPH**

6. Multiple lists are allowed but are not highly recommended. For example, to graph the two functions $y = 2 \sin 3x$ and $y = 4 \sin x$, you could do what's shown in the first frame or type them separately as shown in the third frame.

**Y=**

```
Plot1 Plot2 Plot3
\Y1◻{2,4}sin({3,
1}X)
\Y2=
\Y3=
\Y4=
\Y5=
\Y6=
```

**GRAPH**

**Y=**

```
Plot1 Plot2 Plot3
\Y1◻2sin(3X)
\Y2◻4sin(X)
\Y3=
\Y4=
\Y5=
\Y6=
\Y7=
```

**GRAPH**

7. When graphing trig functions that have vertical asymptotes, remember that your calculator just evaluates individual points and arbitrarily assumes that it should connect those points if it is set in **Connected MODE**. The effect is shown in the following graph of $y = \sec x$. Some people like these "vertical asymptotes" being shown on the graph (they are really just nearly vertical lines that are trying to connect two points on the curve). The last frame shows the same graph in **Dot MODE**.

**Y=**    **ZOOM    7**    **MODE**    **GRAPH**

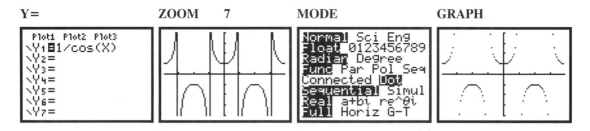

```
Plot1 Plot2 Plot3
\Y1◻1/cos(X)
\Y2=
\Y3=
\Y4=
\Y5=
\Y6=
\Y7=
```

```
Normal Sci Eng
Float 0123456789
Radian Degree
Func Par Pol Seq
Connected Dot
Sequential Simul
Real a+bi re^θi
Full Horiz G-T
```

8. To graph in polar coordinates, the calculator's **MODE** must be changed from **Func** to **Pol**. You may also wish to change your **2nd FORMAT** options from **RectGC** to **PolarGC**, which will show values of $r$ and $\theta$ (instead of $x$ and $y$) when you **TRACE** your polar graphs. Note that the calculator treats these as two completely separate issues (it is possible to graph in one coordinate system and trace in the other). Pressing **Y=** will reveal the calculator's six polar functions $r_1, r_2, \ldots, r_6$. The **X,T,θ,n** key now prints $\theta$ on the screen.

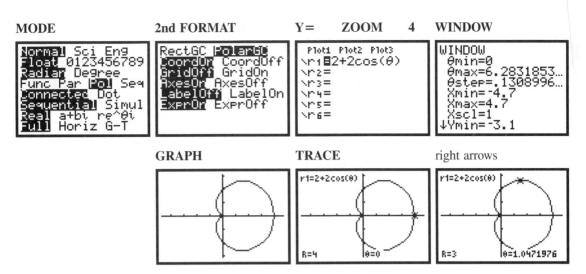

GRAPH        TRACE        right arrows

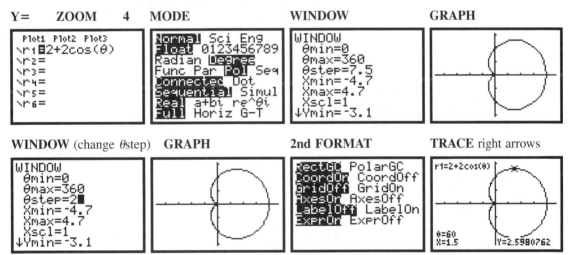

Note in the preceding frames that the standard radian values of $\theta$max, $\theta$min, and $\theta$step are $0$, $2\pi$, and $\pi/24$, respectively. A more accurate, smoother graph can be obtained by using $\theta$step $= \pi/48$ or $\pi/96$. Also note that the *right* arrow key is used to **TRACE** in the standard, counterclockwise direction.

9. Graphing polar equations in **degree MODE**:

Y=    ZOOM    4    MODE        WINDOW        GRAPH

WINDOW (change $\theta$step)  GRAPH        2nd FORMAT        TRACE right arrows

The standard degree values of $\theta$max, $\theta$min, and $\theta$step are $0$, $360$, and $7.5$, respectively. Smoother (but slower) graphs can be obtained by using smaller values of $\theta$step. The last two frames show a polar graph traced in **RectGC FORMAT**.

## D.6 EQUATION-SOLVING AND TABLE FEATURES

1. Solving an equation on the home screen:
   **a.** Rewrite the equation on paper in the form $f(x) = 0$; for example, rewrite

$$x^3 + 15x = 9x^2 - 6$$

as:
$$x^3 - 9x^2 + 15x + 6 = 0$$

**b.** From the home screen, press **2nd CATALOG**, then press the letter **T** (the **4** key), next press the up arrow repeatedly until **solve(** comes into view; then press **ENTER**. You should now see **solve(** on the home screen.

**c.** Finish the statement so that it looks like one of the following:
**solve(X³ − 9X² + 15X + 6 , X , 3)** or **solve(Y₁, X, 3)**, presuming you've entered the function in Y₁ (to type the symbol Y₁ in a formula, press **VARS**, then the right arrow key, then **1**, then **1** again).

| **Y=** | **WINDOW** | **GRAPH** | **2nd CATALOG T** up arrow |
|---|---|---|---|

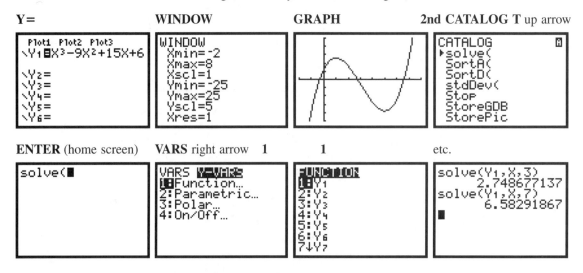

| **ENTER** (home screen) | **VARS** right arrow   **1** | **1** | etc. |
|---|---|---|---|

**d.** Press **ENTER**, and the calculator will try to find a zero of this function near 3 (answer: 2.748677137).

**e.** Press **2nd ENTRY** to bring back your formula, then arrow left and change the 3 to a 7.

**f.** Press **ENTER**, and it will now find the zero near 7 (answer: 6.58291867).

**g.** See if you can use the **solve** feature to find the other zero (answer: −.3315958073).

Notice that the **solve** feature finds solutions of an equation one at a time, with each new solution requiring its own estimate. Graphing the function and noticing where it crosses the *x*-axis is usually the easiest way to discover good estimates. If you prefer the **TABLE** feature (see Example 3), look for sign changes in the list of *y*-values; the corresponding *x*-values should be good estimates. Random guesses, although not recommended, can be effective when the equation has very few solutions.

**2.** Solving an equation on the graphics screen:

**a.** As in (1a), be sure to rewrite the equation as a function set equal to zero.

**b.** Enter this function in your (**Y=**) list of functions and make sure that it is the only one selected for graphing.

**c.** Press **2nd CALC** (the shifted **TRACE** key), and choose option 2, **zero**.

**d.** The prompt "Left Bound?" is asking you to trace the curve using the arrow keys until you are just to the *left* of the desired zero (then press **ENTER**).

Again, "Left Bound" just refers to an $x$-value that's too small to be the solution; do not consider whether the curve is above the axis or below the axis at that point. Similarly, the prompt "Right Bound?" is asking you to trace the curve until you are just to the *right* of the desired zero (then press **ENTER**). You'll notice in each case that a bracketing arrow is displayed near the top of the screen to graphically document the interval that will be searched for a solution.

**e.** The prompt "Guess?" is asking you to trace the curve to a point as close as possible to where it crosses the axis (then press **ENTER**). This Guess is just an approximate solution like the **solve** feature uses (See example 1).

**f.** The solution (the "Zero") is displayed at the bottom of the screen (using 7 or 8 significant digits rather than the 10 digits you get on the home screen). An added bonus is that the $y$-value is also included (it should be exactly zero or extremely close to zero such as "1E-12," which means $10^{-12}$). Two of the solutions are found in the following figure. Try the third one on your own. The **WINDOW** from the previous example is assumed.

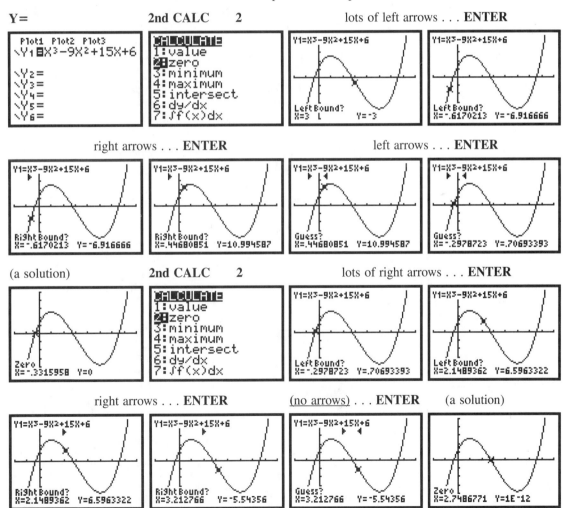

Note that a Right Bound can also be used as the Guess (see the last three frames).

3. Basic **TABLE** features:

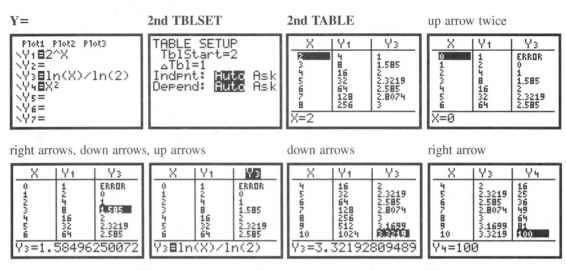

**Y=**  **2nd TBLSET**  **2nd TABLE**  up arrow twice

right arrows, down arrows, up arrows    down arrows    right arrow

Note that functions to be investigated using a **TABLE** need to be entered and turned on in the same sense as those you want to graph. To get to the TABLE SETUP screen, press **2nd TBLSET** (the shifted version of the **WINDOW** key). TblStart is just a beginning *x*-value for the table; you can scroll up or down using the arrow keys. ΔTbl is the incremental change in *x*. You can use ΔTbl = 1 (as in the preceding example) to calculate the function at consecutive integers; in calculus, you could use .001 to investigate what is happening to a function as it approaches a limit; or you could use a larger number like 10 or 100 to study the function's numerical behavior as *x* goes to ∞.

4. Split-screen graphing with a table (**MODE G-T**):

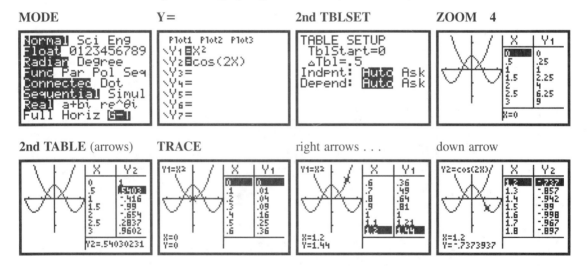

**MODE**  **Y=**  **2nd TBLSET**  **ZOOM  4**

**2nd TABLE** (arrows)  **TRACE**  right arrows . . .  down arrow

In **MODE G-T**, the graph and a corresponding table share the screen, but only one of them is "active" at any given moment. The **ZOOM** commands and the

**GRAPH** key give control to the graphical side of the screen. This just means that the arrow keys refer to the graph rather than the table. Pressing **2nd TABLE** enables the arrow keys to be used to scroll through its values. The **TRACE** key links the table to the graph, with the graph in control (all previous TABLE SETUP specifications are replaced with values related to the **TRACE**). Note in the last frame that jumping to a different function will display a different column of the table (the same value of $x$ is highlighted).

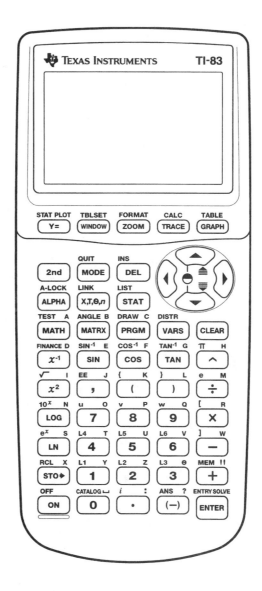

# APPENDIX E
# Tables

**TABLE 1  English weights and measures**

| Units of length | Units of Weight |
|---|---|
| Standard unit = inch (in. or ") | Standard unit = pound (lb) |
| 12 inches = 1 foot (ft or ') | 16 ounces (oz) = 1 pound |
| 3 feet = 1 yard (yd) | 2000 pounds = 1 ton (T) |
| $5\frac{1}{2}$ yards or $16\frac{1}{2}$ feet = 1 rod (rd) | |
| 5280 feet = 1 mile (mi) | |

| Volume measure |
|---|
| *Liquid* |
| 16 ounces (fl oz) = 1 pint (pt) |
| 2 pints = 1 quart (qt) |
| 4 quarts = 1 gallon (gal) |
| *Dry* |
| 2 pints (pt) = 1 quart (qt) |
| 8 quarts = 1 peck (pk) |
| 4 pecks = 1 bushel (bu) |

## TABLE 2 Metric system prefixes

| Multiple or submultiple[a] decimal form | Power of 10 | Prefix[b] | Prefix symbol | Pronun- ciation | Meaning |
|---|---|---|---|---|---|
| 1,000,000,000,000 | $10^{12}$ | tera | T | tĕr′ă | one trillion times |
| 1,000,000,000 | $10^{9}$ | giga | G | jĭg′ă | one billion times |
| 1,000,000 | $10^{6}$ | mega | M | mĕg′ă | one million times |
| 1,000 | $10^{3}$ | kilo | k | kĭl′ō | one thousand times |
| 100 | $10^{2}$ | hecto | h | hĕk′tō | one hundred times |
| 10 | $10^{1}$ | deka | da | dĕk′ă | ten times |
| 0.1 | $10^{-1}$ | deci | d | dĕs′ĭ | one tenth of |
| 0.01 | $10^{-2}$ | centi | c | sĕnt′ĭ | one hundredth of |
| 0.001 | $10^{-3}$ | milli | m | mĭl′ĭ | one thousandth of |
| 0.000001 | $10^{-6}$ | micro | $\mu$ | mī′krō | one millionth of |
| 0.000000001 | $10^{-9}$ | nano | n | năn′ō | one billionth of |
| 0.000000000001 | $10^{-12}$ | pico | p | pē′kō | one trillionth of |

[a] Factor by which the unit is multiplied.

[b] The same prefixes are used with all SI units.

The following prefixes are used with the metric standard unit of length, metre (m).

| | |
|---|---|
| 1 *tera*metre (Tm) = 1,000,000,000,000 m | 1 m = 0.000000000001 Tm |
| 1 *giga*metre (Gm) = 1,000,000,000 m | 1 m = 0.000000001 Gm |
| 1 *mega*metre (Mm) = 1,000,000 m | 1 m = 0.000001 Mm |
| 1 *kilo*metre (km) = 1,000 m | 1 m = 0.001 km |
| 1 *hecto*metre (hm) = 100 m | 1 m = 0.01 hm |
| 1 *deka*metre (dam) = 10 m | 1 m = 0.1 dam |
| 1 *deci*metre (dm) = 0.1 m | 1 m = 10 dm |
| 1 *centi*metre (cm) = 0.01 m | 1 m = 100 cm |
| 1 *milli*metre (mm) = 0.001 m | 1 m = 1,000 mm |
| 1 *micro*metre ($\mu$m) = 0.000001 m | 1 m = 1,000,000 $\mu$m |
| 1 *nano*metre (nm) = 0.000000001 m | 1 m = 1,000,000,000 nm |
| 1 *pico*metre (pm) = 0.000000000001 m | 1 m = 1,000,000,000,000 pm |

## TABLE 3   Conversion tables

### Length

|  | cm | m | km | in. | ft | mi |
|---|---|---|---|---|---|---|
| 1 centimetre | 1 | $10^{-2}$ | $10^{-5}$ | 0.394 | $3.28 \times 10^{-2}$ | $6.21 \times 10^{-6}$ |
| 1 metre | 100 | 1 | $10^{-3}$ | 39.4 | 3.28 | $6.21 \times 10^{-4}$ |
| 1 kilometre | $10^5$ | 1000 | 1 | $3.94 \times 10^4$ | 3280 | 0.621 |
| 1 inch | 2.54 | $2.54 \times 10^{-2}$ | $2.54 \times 10^{-5}$ | 1 | $8.33 \times 10^{-2}$ | $1.58 \times 10^{-5}$ |
| 1 foot | 30.5 | 0.305 | $3.05 \times 10^{-4}$ | 12 | 1 | $1.89 \times 10^{-4}$ |
| 1 mile | $1.61 \times 10^5$ | 1610 | 1.61 | $6.34 \times 10^4$ | 5280 | 1 |

### Area

| Metric | English |
|---|---|
| $1 \ m^2 = 10{,}000 \ cm^2$ | $1 \ ft^2 = 144 \ in^2$ |
| $= 1{,}000{,}000 \ mm^2$ | $1 \ yd^2 = 9 \ ft^2$ |
| $1 \ cm^2 = 100 \ mm^2$ | $1 \ rd^2 = 30.25 \ yd^2$ |
| $= 0.0001 \ m^2$ | $1 \ acre = 160 \ rd^2$ |
| $1 \ km^2 = 1{,}000{,}000 \ m^2$ | $= 4840 \ yd^2$ |
| $1 \ ha = 10{,}000 \ m^2$ | $= 43{,}560 \ ft^2$ |
|  | $1 \ mi^2 = 640 \ acres$ |

|  | $m^2$ | $cm^2$ | $ft^2$ | $in^2$ |
|---|---|---|---|---|
| $1 \ m^2$ | 1 | $10^4$ | 10.8 | 1550 |
| $1 \ cm^2$ | $10^{-4}$ | 1 | $1.08 \times 10^{-3}$ | 0.155 |
| $1 \ ft^2$ | $9.29 \times 10^{-2}$ | 929 | 1 | 144 |
| $1 \ in^2$ | $6.45 \times 10^{-4}$ | 6.45 | $6.94 \times 10^{-3}$ | 1 |

$1 \ mi^2 = 2.79 \times 10^7 \ ft^2 = 640 \ acres$

$1 \ circular \ mil = 5.07 \times 10^{-6} \ cm^2 = 7.85 \times 10^{-7} \ in^2$

$1 \ hectare = 2.47 \ acres$

**TABLE 3,** *Continued*

## Volume

*Metric*       *English*

$1\ m^3\ =\ 10^6\ cm^3$    $1\ ft^3\ =\ 1728\ in^3$
$1\ cm^3\ =\ 10^{-6}\ m^3$    $1\ yd^3\ =\ 27\ ft^3$
$=\ 10^3\ mm^3$

| | $m^3$ | $cm^3$ | L | $ft^3$ | $in^3$ |
|---|---|---|---|---|---|
| $1\ m^3$ | 1 | $10^6$ | 1000 | 35.3 | $6.10 \times 10^4$ |
| $1\ cm^3$ | $10^{-6}$ | 1 | $10^{-3}$ | $3.53 \times 10^{-5}$ | $6.10 \times 10^{-2}$ |
| 1 L | $10^{-3}$ | 1000 | 1 | $3.53 \times 10^{-2}$ | 61.0 |
| $1\ ft^3$ | $2.83 \times 10^{-2}$ | $2.83 \times 10^4$ | 28.3 | 1 | 1728 |
| $1\ in^3$ | $1.64 \times 10^{-5}$ | 16.4 | $1.64 \times 10^{-2}$ | $5.79 \times 10^{-4}$ | 1 |

1 U.S. fluid gallon = 4 U.S. fluid quarts = 8 U.S. pints =
128 U.S. fluid ounces = 231 $in^3$ = 0.134 $ft^3$ = 3.79 litres

1 L = 1000 $cm^3$ = 1.06 qt

## Other useful conversion factors

| | |
|---|---|
| 1 newton (N) = 0.225 lb | 1 atm = 101.32 kPa |
| 1 pound (lb) = 4.45 N | = 14.7 lb/$in^2$ |
| 1 slug = 14.6 kg | 1 Btu = 0.252 kcal |
| 1 joule (J) = 0.738 ft-lb | 1 kcal = 3.97 Btu |
| = $2.39 \times 10^{-4}$ kcal | $F = \frac{9}{5}C + 32°$ |
| 1 calorie (cal) = 4.185 J | $C = \frac{5}{9}(F - 32°)$ |
| 1 kilocalorie (kcal) = 4185 J | 1 kg = 2.20 lb (on the |
| 1 foot-pound (ft-lb) = 1.36 J | earth's surface) |
| 1 watt (W) = 1 J/s = 0.738 ft-lb/s | 1 lb = 454 g |
| 1 kilowatt (kW) = 1000 W | = 16 oz |
| = 1.34 hp | 1 metric ton = 1000 kg |
| 1 hp = 550 ft-lb/s = 746 W | = 2200 lb |

## TABLE 4 Physical quantities and their units

| Quantity | Symbol | Unit | |
|---|---|---|---|
| | | *Metric* | *English* |
| Distance | $s$ | metre (m) | foot (ft) |
| Time | $t$ | second (s) | second (s) |
| Mass | $m$ | kilogram (kg) | slug |
| Force, weight | $F, w$ | newton (N) | pound (lb) |
| Area | $A$ | $m^2$ | $ft^2$ |
| Volume | $V$ | $m^3$ or L | $ft^3$ |
| Velocity | $v$ | m/s | ft/s |
| Acceleration | $a$ | $m/s^2$ | $ft/s^2$ |
| Energy, work | $E, W$ | Nm or joule (J) | ft-lb |
| Power | $P$ | joule/s or watt (W) | ft-lb/s or hp |
| Heat | $Q$ | joule (J) | British thermal unit (Btu) |
| Pressure | $p$ | $N/m^2$ or pascal (Pa) | $lb/in^2$ |
| Electric charge | $q$ | coulomb (C) | coulomb (C) |
| Electric current | $I$ | ampere (A) | ampere (A) |
| Electric potential | $V, E$ | volt (V) | volt (V) |
| Capacitance | $C$ | farad (F) | farad (F) |
| Inductance | $L$ | henry (H) | henry (H) |
| Resistance | $R$ | ohm ($\Omega$) | ohm ($\Omega$) |
| Frequency | $f$ | 1/s or hertz (Hz) | 1/s or Hz |

# Answers to Odd-Numbered Exercises and to Chapter Reviews

**CHAPTER 1**

**Exercises 1.1, Pages 4–6**

**1.** 15 **3.** 11 **5.** 8 **7.** 2 **9.** −9 **11.** −2 **13.** 10 **15.** 3 **17.** $-1\frac{1}{12}$ **19.** $\frac{35}{36}$ **21.** −13
**23.** −7 **25.** −5 **27.** 6 **29.** 9 **31.** −14 **33.** −3 **35.** −15.7 **37.** $-\frac{5}{12}$ **39.** $-1\frac{7}{12}$ **41.** 6
**43.** 11 **45.** −5 **47.** −9 **49.** 30 **51.** −24 **53.** 4 **55.** $-\frac{9}{5}$ **57.** $\frac{3}{8}$ **59.** $-\frac{3}{4}$
**61.** $-\frac{2}{3}$ **63.** −84 **65.** 60 **67.** 7 **69.** $-\frac{27}{8}$ **71.** 20° **73.** 30° **75.** 75 ft

**Exercises 1.2, Pages 9–10**

**1.** −3 **3.** 0 **5.** 0 **7.** Meaningless **9.** 0 **11.** Indeterminate **13.** 2 **15.** $0, -\frac{4}{3}$ **17.** $-1, \frac{1}{2}$
**19.** 0 **21.** $\frac{5}{6}$ **23.** $3, -\frac{1}{2}$ **25.** 23 **27.** −21 **29.** 14 **31.** −29 **33.** 34 **35.** −6 **37.** −4
**39.** 76 **41.** 3 **43.** 21 **45.** $-\frac{12}{5}$ **47.** −2

**Exercises 1.3, Pages 15–16**

**1.** $10^{11}$ **3.** $10^{9}$ **5.** $10^{-8}$ **7.** $10^{3}$ **9.** $10^{4}$ **11.** $10^{-17}$ **13.** $10^{16}$ **15.** $10^{-15}$ **17.** $10^{4}$
**19.** $2.07 \times 10^{3}$ **21.** $9.1 \times 10^{-2}$ **23.** $5.61 \times 10^{0}$ **25.** $8.5 \times 10^{6}$ **27.** $6 \times 10^{-6}$ **29.** $1.006 \times 10^{4}$
**31.** 127 **33.** 0.0000614 **35.** 9,240,000 **37.** 0.00000000696 **39.** 9.66 **41.** 50,300 **43.** $3.32 \times 10^{19}$
**45.** $-6.83 \times 10^{-6}$ **47.** $8.36 \times 10^{-11}$ **49.** $-7.98 \times 10^{19}$ **51.** $-6.85 \times 10^{1}$ **53.** $-4.92 \times 10^{-23}$

**Exercises 1.4, Pages 18–20**

**1.** kilo **3.** hecto **5.** milli **7.** mega **9.** h **11.** m **13.** M **15.** c **17.** 133 mm **19.** 18 kL
**21.** 19 cg **23.** 72 hm **25.** 14 metres **27.** 19 grams **29.** 17 millimetres **31.** 25 dekametres
**33.** 16 megametres **35.** metre **37.** litre and cubic metre **39.** second

**Exercises 1.5, Pages 23–24**

**1.** 1 metre **3.** 1 kilometre **5.** 1 kilometre **7.** cm **9.** m **11.** mm **13.** km **15.** mm **17.** m
**19.** km **21.** cm **23.** km **25.** cm **27.** km **29.** mm **31.** cm **33.** 1000 **35.** 100 **37.** 0.1
**39.** 1000 **41.** 100 **43.** 0.001 **45.** 10 **47.** 0.23 km **49.** 198,000 m **51.** 8.4 m **53.** 4,750 mm
**55.** 4,000,000 $\mu$m

**Exercises 1.6, Pages 24–25**

**1.** 1 gram **3.** 1 kilogram **5.** 1 kilogram **7.** kg **9.** kg **11.** metric ton **13.** g **15.** mg **17.** kg **19.** g **21.** kg **23.** g **25.** g **27.** kg **29.** kg **31.** metric ton **33.** kg **35.** 1000 **37.** 100 **39.** 0.1 **41.** 1000 **43.** 100 **45.** 0.001 **47.** 1,000,000 **49.** 565,000 mg **51.** 0.85 g **53.** 5 g **55.** 80,000,000 mg **57.** 1.5 kg **59.** 0.5 mg

**Exercises 1.7, Pages 30–31**

**1.** 1 litre **3.** 1 cubic centimetre **5.** 1 square kilometre **7.** L **9.** $m^2$ **11.** $m^3$ **13.** ha **15.** mL **17.** $m^3$ **19.** L **21.** mL **23.** L **25.** $m^2$ **27.** L **29.** ha **31.** $m^3$ **33.** $m^2$ **35.** L **37.** 1000 **39.** 0.1 **41.** 0.01 **43.** 10 **45.** 1 **47.** 1,000,000 **49.** 0.001 **51.** 10,000 **53.** 100 **55.** 0.01 **57.** 100 **59.** 6.5 L **61.** 1400 mL **63.** 225,000 $mm^3$ **65.** $2 \times 10^9$ $mm^3$ **67.** 175 mL **69.** 1000 L **71.** 7500 $cm^3$ **73.** 50 $cm^2$ **75.** 50,000 $cm^2$ **77.** 400 $km^2$ **79.** 750 g

**Exercises 1.8, Pages 36–38**

**1.** (c) **3.** (d) **5.** (d) **7.** (b) **9.** (c) **11.** (b) **13.** (d) **15.** (b) **17.** 77 **19.** 60 **21.** $-24.4$ **23.** 1022 **25.** $-223$ **27.** 41°F **29.** $-38.9$°C **31.** 2012 **33.** 1324 **35.** 482 **37.** second; s **39.** watt; W **41.** 1 milliampere **43.** 1 megawatt **45.** 1 A **47.** 2.7 $\mu$A **49.** 9.5 kW **51.** 15 ns **53.** 135 MW **55.** $10^3$ **57.** $10^9$ **59.** $10^6$ **61.** $10^3$ **63.** $10^{12}$ **65.** 6000 mA **67.** 42,000 $\mu$W **69.** 7.8 A **71.** 15,915 s **73.** $3 \times 10^9$ ns **75.** 0.04 MW

**Exercises 1.9, Page 43**

**1.** 24.5 **3.** 75,000 **5.** 650 **7.** 850,000 **9.** 240 **11.** 1.45 **13.** 0.00275 **15.** 18,000 **17.** 1.4 **19.** 0.00045 **21.** 13 **23.** $3.5 \times 10^{-3}$, or 0.035 **25.** 145 kHz **27.** 2.1 MV **29.** 118 MHz **31.** 8.5 ms **33.** 80 $\mu$V **35.** 87.2 ns **37.** 3.25 mW **39.** 48 W **41.** 7.5 k$\Omega$ **43.** $T = 1$ ms **45.** $f = 1$ GHz **47.** $T = 100$ ns **49.** $f = 10$ GHz **51.** $T = 100$ ns **53.** $f = 100$ kHz **55.** $f = 10$ kHz **57.** $T = 100$ $\mu$s **59.** $f = 10$ kHz **61.** $1 \times 10^{14}$ W

**Exercises 1.10, Pages 45–46**

**1.** (a) 7.20 mi (b) 11.6 km (c) 12,700 yd **3.** (a) 6700 m (b) 4.16 mi (c) 7330 yd **5.** (a) 14.0 ft (b) 427 cm (c) 4.27 m (d) 4.67 yd **7.** (a) 27,800 $yd^2$ (b) 23,200 $m^2$ (c) 5.74 acres (d) 2.32 ha **9.** (a) 25,900 $in^3$ (b) 0.425 $m^3$ (c) 425 L **11.** (a) 0.750 L (b) 1.59 pints (c) 0.795 quart **13.** 610 cm/s **15.** 0.036°C/h **17.** $1 \times 10^{-4}$ in./min **19.** 98.4 ft/$s^2$ **21.** 0.362 lb/$in^2$ **23.** 43.3 oz/$in^3$ **25.** 2220 Btu **27.** (a) 2180 J (b) 0.521 kcal **29.** 77.6 mi **31.** 15.9 mm **33.** 284 L **35.** (a) 300 ft (b) 91.5 m **37.** (a) 0.826 cm (b) 8.26 mm **39.** 4.27 fl oz

**Exercises 1.11, Pages 48–49**

**1.** 3 **3.** 3 **5.** 4 **7.** 3 **9.** 2 **11.** 3 **13.** 0.1 cm **15.** 0.01 cm **17.** 1 mm **19.** 0.01 m **21.** 10 $\Omega$ **23.** 0.0001 A **25.** (a) 15.2 m (b) 0.023 m **27.** (a) 14.02 cm (b) 0.642 cm **29.** (a) 0.0270 A (b) 0.00060 A **31.** (a) 305,000 $\Omega$ (b) 305,000 $\Omega$, 38,000 $\Omega$ **33.** (a) 0.08 m (b) 13.2 m **35.** (a) 0.52 km (b) 16.8 km **37.** (a) 0.00009 A (b) 0.41 A **39.** (a) 500,000 $\Omega$ (b) 500,000 $\Omega$

**Exercises 1.12, Pages 51–53**

**1.** 22.1 in. **3.** 84.8 cm **5.** 1.1369 g **7.** 19 V **9.** 25.09 cm **11.** 3.9 cm **13.** 2.4 mm **15.** 5.24 oz **17.** 3.996 in. **19.** 2.35 in. **21.** 0.85 A **23.** 853 $m^2$ **25.** 0.13 $in^2$ **27.** 25,800 $cm^3$ **29.** 2.1 m **31.** 73 cm/$s^2$ **33.** 0.078 N/$m^2$ **35.** 65 kg $\cdot$ m/s **37.** 110 $cm^2$ **39.** 614 $cm^3$ **41.** $1.28 \times 10^8$ kg $m^2$/$s^2$ **43.** 12.48 mm **45.** 1993: 100.7 bu/acre. 1995: 113.5 bu/acre; 12.8 bu/acre

**Chapter 1 Review, Pages 56–58**

**1.** $-5$ **2.** 0 **3.** $-210$ **4.** $-2$ **5.** $-4$ **6.** 9 **7.** 4 **8.** $-1$ **9.** 11 **10.** $-24$ **11.** $\frac{27}{13}$ **12.** $\frac{1}{6}$ **13.** 0 **14.** Indeterminate **15.** Meaningless **16.** $10^8$ **17.** $10^4$ **18.** $10^7$ **19.** $3.42 \times 10^6$ **20.** 0.000561 **21.** $4.24 \times 10^{-6}$ **22.** $3.00 \times 10^{29}$ **23.** centi **24.** kilo **25.** mL **26.** $\mu$g

**27.** 16 kilometres    **28.** 250 milliamperes    **29.** 1.1 hectolitres    **30.** 18 megawatts    **31.** 1 litre
**32.** 1 kilometre    **33.** 1 kilogram    **34.** 1 $m^3$    **35.** 1 $km^2$    **36.** 1 ns    **37.** 0.18    **38.** 0.25    **39.** 5700
**40.** 1500    **41.** 650    **42.** 15,000    **43.** 15,000,000    **44.** 75,000    **45.** 750,000    **46.** 1.8    **47.** 21
**48.** 23    **49.** 100    **50.** 0    **51.** 1200 m    **52.** 1 kg    **53.** 1.5 L    **54.** 170 cm    **55.** 50 kg    **56.** 12 km/L
**57.** 80 km/h    **58.** 8.85    **59.** 77.5    **60.** 48.5 M$\Omega$    **61.** 75 $\mu$A    **62.** $T = 10$ ns    **63.** $f = 100$ Hz
**64.** (a) 12$\overline{0}$0 yd  (b) 11$\overline{0}$0 m    **65.** (a) 53,500 g  (b) 118 lb    **66.** (a) 32,400 $ft^2$  (b) $3.01 \times 10^7$ $cm^2$
**67.** 31.1 mi/h    **68.** 15.6 lb/$ft^3$    **69.** (a) 3 significant digits  (b) 1 m    **70.** (a) 2 significant digits  (b) 0.001 A
**71.** (a) 3 significant digits  (b) 1000 V    **72.** 57.6 L    **73.** 730.9 cm    **74.** 40$\overline{0}$ $m^2$    **75.** 3.11 m    **76.** 2.5 lb/$in^2$

## CHAPTER 2

### Exercises 2.1, Pages 65–67

**1.** Binomial    **3.** Trinomial    **5.** Monomial    **7.** Binomial    **9.** Trinomial    **11.** Binomial    **13.** 2    **15.** 4
**17.** 10    **19.** 5    **21.** 6    **23.** 4    **25.** $3x^2 + 5x + 2$, degree 2    **27.** $9x^8 - 5x^4 + 6x^3 + 5x^2$, degree 8
**29.** $4y^5 + 3y^3 - 3y + 5$, degree 5    **31.** $4x + 2x^3 - 3x^4$, degree 4    **33.** $-7 + c + 5c^3 + 3c^4 - 8c^5$, degree 5
**35.** $2 + 2y + 5y^3 - 8y^4 - 6y^6$, degree 6    **37.** 3    **39.** 5    **41.** 8    **43.** 9    **45.** $9x^2 - 5x + 5$
**47.** $4x^2 + x - 5$    **49.** $-8x^2 - 3x + 4$    **51.** $-10x^2 - x + 10$    **53.** $4x^2 + x - 10$    **55.** $-4x^2 - 7x + 3$
**57.** $-4x^2 + 8x$    **59.** $-6x^2 - 11x + 4$    **61.** $8x^3 + 2x^2 - 5x + 3$    **63.** $-x^3 - 10x^2 + 9x - 2$    **65.** $-2y$
**67.** $x + 6y$    **69.** $-5$    **71.** $-4$    **73.** 324    **75.** $-216$    **77.** 3    **79.** $-30/29$    **81.** $-110,592$    **83.** $\frac{3}{10}$

### Exercises 2.2, Pages 71–72

**1.** $x^{12}$    **3.** $12a^5$    **5.** $m^6$    **7.** $\dfrac{1}{x^4}$    **9.** $3x^4$    **11.** $\dfrac{5}{x^3}$    **13.** $a^6$    **15.** $c^{16}$    **17.** $81a^2$    **19.** $32x^{10}$

**21.** $\frac{9}{16}$    **23.** $\dfrac{16}{a^{12}}$    **25.** 1    **27.** 3    **29.** $9x^2$    **31.** $t^{12}$    **33.** $-a^6$    **35.** $-8a^6b^3$    **37.** $9x^4y^6$

**39.** $-27x^9y^{12}z^3$    **41.** $\dfrac{4x^2}{9y^6}$    **43.** $\dfrac{16x^2}{9y^4}$    **45.** $\dfrac{1}{36y^6}$    **47.** 2    **49.** 8    **51.** 11    **53.** $5^8$    **55.** 5
**57.** $-6$    **59.** 8    **61.** $3\sqrt{5}$    **63.** $5\sqrt{2}$    **65.** $6\sqrt{2}$    **67.** $4\sqrt{3}$    **69.** $4\sqrt{2}$    **71.** 18.1    **73.** 51.0
**75.** 0.0687    **77.** 7.21    **79.** 9.43    **81.** $1.04 \times 10^5$    **83.** 134

### Exercises 2.3, Pages 75–76

**1.** $32x^5$    **3.** $-24a^5b^3$    **5.** $-48a^3bc^5$    **7.** $24a^4b^9c^6$    **9.** $12a^2 - 21ab$    **11.** $6x^3 + 12x^2 - 15x$
**13.** $-15x^4 + 25x^3 - 40x^2$    **15.** $24a^3b^4 - 48a^4b^7$    **17.** $3a^6b^7 - 9a^3b^5 + 3a^2b^6$    **19.** $6x^2 + x - 35$
**21.** $48x^2 + 54x + 15$    **23.** $9x^2 - 16$    **25.** $24x^2 + 2x - 1$    **27.** $6x^2 - 23x + 21$    **29.** $18x^2 + 60xy + 32y^2$
**31.** $40s^2 - 62st - 18t^2$    **33.** $-15x^2 + 2x + 24$    **35.** $6x^4 + 19x^2 - 7$    **37.** $30x^4 - 61x^2 + 30$
**39.** $x^4 + 4x^2 + x + 6$    **41.** $x^2 - y^2 - 3x + 11y - 28$    **43.** $15x^4 - 2x^3 - 41x^2 + 22x + 6$
**45.** $12x^4 + 20x^3 - x^2 - 15x - 6$    **47.** $4x^2 - 20x + 25$    **49.** $9x^2 + 48x + 64$    **51.** $25x^2 - 20x + 4$
**53.** $9x^4 - 42x^3 + 73x^2 - 56x + 16$    **55.** $8x^3 - 12x^2 + 6x - 1$    **57.** $8a^3 + 60a^2b + 150ab^2 + 125b^3$

### Exercises 2.4, Pages 78–79

**1.** $6x$    **3.** $-4ab^3$    **5.** $\dfrac{5x^3}{8y}$    **7.** $-\dfrac{5}{4x^2y^2}$    **9.** $\dfrac{b^2}{12a}$    **11.** $\dfrac{16t^6}{3}$    **13.** $3x^2 - 2x + 1$    **15.** $3x^4 - 4x^3 + 2x$

**17.** $-4x^2 + 5x - 7$    **19.** $-8x^3 + 6x + \dfrac{9}{2x} + \dfrac{3}{x^3}$    **21.** $2a - 3ab + 4$    **23.** $-3mn - 4n^2 + \dfrac{5m^2}{n}$

**25.** $8x^3z^3 - 6x^2yz^2 - 4y^2$    **27.** $2x - 5$    **29.** $x^2 - x + 3$    **31.** $x^2 - 2x + 3 + \dfrac{4}{2x - 1}$    **33.** $2x^2 + 7x - 3$

**35.** $-4x^2 - 3x + 7 - \dfrac{3}{4x - 3}$    **37.** $3x^3 - 2x^2 + 4$    **39.** $x^2 + 3x - 5$    **41.** $x^2 - 4x + 16 - \dfrac{128}{x + 4}$

**43.** $4x^2 - 2x + 1$

## Exercises 2.5, Page 83

**1.** $-2$  **3.** $16$  **5.** $7$  **7.** $-72$  **9.** $5$  **11.** $2$  **13.** $-2$  **15.** $-5$  **17.** $-\frac{17}{11}$  **19.** $14$  **21.** $5$
**23.** $-\frac{8}{15}$  **25.** $2$  **27.** $-2$  **29.** $12$  **31.** $14$  **33.** $-\frac{15}{2}$  **35.** $6$  **37.** $-\frac{21}{2}$  **39.** $\frac{154}{45}$  **41.** $25.6$
**43.** $1.15$  **45.** $6.39$  **47.** $117$  **49.** $1.65 \times 10^{-4}$  **51.** $7.35 \times 10^{-9}$

## Exercises 2.6, Pages 85–86

**1.** $J = \dfrac{W}{Q}$  **3.** $R_2 = R_T - R_1 - R_3$  **5.** $R = \dfrac{E}{I}$  **7.** $Q = CV$  **9.** $Q = \dfrac{W}{V}$  **11.** $R = \dfrac{JQ}{I^2 t}$

**13.** $N = \dfrac{(O.D.) - 2P}{P}$  **15.** $L = \dfrac{RD^2}{k}$  **17.** $P = \dfrac{\pi}{2R}$  **19.** $T = \dfrac{VT'}{V'}$  **21.** $F = \frac{9}{5}C + 32$  **23.** $N_s = \dfrac{N_p I_p}{I_s}$

**25.** $a = \dfrac{1.22\lambda d}{\Delta d}$  **27.** $\lambda = \dfrac{2l}{n}$  **29.** $T = \dfrac{(\Delta L) + \alpha L T_0}{\alpha L}$  **31.** $v = \dfrac{f'v_s + fv_0}{f' - f}$  **33.** $q_1 = \dfrac{4\pi\epsilon_0 r^2 F}{q_2}$

**35.** $R = \dfrac{E}{I} - \dfrac{q}{IC}$ or $R = \dfrac{EC - q}{IC}$  **37.** $R_2 = \dfrac{R_T R_1}{R_1 - R_T}$  **39.** $R_T = \dfrac{R_1 R_2 R_3}{R_1 R_3 + R_1 R_2 + R_2 R_3}$  **41.** $s_0 = \dfrac{fs_i}{s_i - f}$

**43.** $n = \dfrac{R'R'' + fR'' - fR'}{fR'' - fR'}$  **45.** $f = \dfrac{eV + \phi}{h}$  **47.** $Z_2 = \dfrac{Z_3(I_T - I_2)}{I_2}$  **49.** $R_1 = \dfrac{R_A(R_2 + R_3)}{R_3 - R_A}$

## Exercises 2.7, Pages 90–91

**1.** $4.00$ in.  **3.** $36.0$ m  **5.** $3.91$ m  **7.** $28.5$ ft  **9.** $131°$  **11.** $4280$ cal  **13.** $0.199$ $\Omega$/Hz  **15.** $16$ hp
**17.** $960$ cm$^3$  **19.** $5.22$ m  **21.** $94.8$ m$^3$  **23.** $45.3$ cm  **25.** $396$ Hz  **27.** $80.0$ $\Omega$  **29.** $13$ $\Omega$
**31.** $-6\overline{0}$ cm  **33.** $21.9$ m/s  **35.** $0.150$ mA

## Exercises 2.8, Pages 96–97

**1.** $\$54, \$162$  **3.** $12$ in.  **5.** $12$ m, $16$ m  **7.** $12$ acres at \$650/acre, $28$ acres at \$450/acre  **9.** $22$
**11.** **(a)** $4.5$ h  **(b)** $337.5$ mi  **13.** $85$ mi/h, $105$ mi/h  **15.** $7.5$ lb of 70%, $12.5$ lb of 30%  **17.** $4$ qt
**19.** $0.65$ A, $1.40$ A, $2.60$ A  **21.** $40$ m $\times$ $80$ m  **23.** $150$ $\Omega$, $165$ $\Omega$, $225$ $\Omega$

## Exercises 2.9, Pages 100–103

**1.** $\frac{9}{2}$  **3.** $\frac{1}{10}$  **5.** $\frac{72}{1}$  **7.** $\frac{500}{1}$  **9.** $\frac{16}{1}$  **11.** $\frac{30}{1}$  **13.** $\frac{1}{40,000}$  **15.** $\frac{4}{1}$  **17.** $\$43.50$/ft$^2$  **19.** $25$ gal/acre
**21.** $2.25$ or $\frac{9}{4}$  **23.** $\frac{24}{5}$  **25.** $135$ bu/acre  **27.** $\frac{5}{3}$  **29.** $27$  **31.** $168$  **33.** $2$  **35.** $12$  **37.** $\dfrac{mp}{n}$  **39.** $8$
**41.** $782$  **43.** $8670$  **45.** $0.0815$  **47.** $900$ ft by $360$ ft  **49.** $20{,}100$ and $48{,}240$  **51.** $1500$ lb and $1000$ lb
**53.** $\$90, \$336$  **55.** $\$218.88$  **57.** $900$  **59.** $\$88{,}200$  **61.** $\$1106.56$  **63.** $675$ lb  **65.** $62{,}400$ bu
**67.** $120$ V  **69.** $470{,}000$  **71.** $450$ N  **73.** $51$ m, $68$ m  **75.** Hypotenuse: $150$ ft  Sides: $90$ ft, $120$ ft

## Exercises 2.10, Pages 106–108

**1.** Direct: $k = \frac{3}{4}$  **3.** Inverse; $k = \frac{3}{2}$  **5.** Neither  **7.** $y = kz$  **9.** $a = kbc$  **11.** $r = \dfrac{ks}{\sqrt{t}}$  **13.** $f = \dfrac{kgh}{j^2}$
**15.** $k = \frac{1}{3}$; $12$  **17.** $k = 54$; $3$  **19.** $k = 6$; $36$  **21.** $k = 32$; $\frac{64}{3}$  **23.** $k = 6.00$ cm$^3$/K; $18\overline{0}0$ cm$^3$
**25.** $k = 75\overline{0}$ ft$^3$ lb/(in$^2$ °R); $8630$ ft$^3$  **27.** $k = 1.0$ W$\Omega$/V$^2$; $480$ W  **29.** $160$ teeth  **31.** $24$ rpm
**33.** **(a)** $50$ rpm  **(b)** $100$ rpm

## Chapter 2 Review, Pages 112–114

**1.** Trinomial, degree 3  **2.** $16x^2 + 4x - 21$  **3.** $7x^2 - x - 11$  **4.** $x^2 - 5x + 2$  **5.** $5x^2 + 4x$  **6.** $-5a + 8b$
**7.** $-3a - 7b$  **8.** $\frac{18}{5}$  **9.** $y^9$  **10.** $b^9$  **11.** $-15x^6$  **12.** $6m^4$  **13.** $a^{12}$  **14.** $25a^6$  **15.** $\dfrac{y^3}{x^6}$  **16.** $1$
**17.** $-s^9$  **18.** $x^{18}$  **19.** $8a^9 b^6$  **20.** $\dfrac{x^8}{4}$  **21.** $-8x^{18}$  **22.** $7$  **23.** $3$  **24.** $3\sqrt{7}$  **25.** $6\sqrt{3}$
**26.** $56.1$  **27.** $0.143$  **28.** $-24a^5 b^2 c^4$  **29.** $10x^2 - 30xy$  **30.** $-12x^5 + 9x^4 + 3x^3 - 12x^2$
**31.** $-15a^4 b^4 + 9a^5 b - 15a^2 b^3$  **32.** $6x^2 + 13x - 28$  **33.** $30x^2 - 63x + 27$  **34.** $20x^2 + 38x + 12$

**35.** $6x^2 - 5x - 6$    **36.** $64x^2 - 80x + 25$    **37.** $2x^2 + xy - y^2 - 11x + 10y - 21$    **38.** $\dfrac{3y^2}{z}$    **39.** $8a^3$

**40.** $16x - 8 + \dfrac{5}{x}$    **41.** $3x - \dfrac{5}{x} + \dfrac{7}{x^2}$    **42.** $3n^2 + \dfrac{4}{mn^2} - \dfrac{4m^4}{3n}$    **43.** $3x + 4$    **44.** $3x^2 - 3x + 8 - \dfrac{11}{x+1}$

**45.** $x^2 - 2x + 3 + \dfrac{10}{2x-1}$    **46.** $-4$    **47.** $-4$    **48.** $-30$    **49.** $-\frac{2}{3}$    **50.** 23    **51.** $\frac{32}{7}$    **52.** $-\frac{5}{7}$

**53.** $-\frac{11}{42}$    **54.** 8.77    **55.** $D = \dfrac{12S}{\pi}$    **56.** $V = \dfrac{Q}{C}$    **57.** $t = \dfrac{v_0 - v}{g}$    **58.** $T_0 = \dfrac{\beta VT - (\Delta V)}{\beta V}$    **59.** 82 m/s

**60.** 55.0 Ω    **61.** 40.7 Ω    **62.** 18    **63.** 20 m by 25 m    **64.** 12 oz of 15% silver, 18 oz of 20% silver    **65.** $\frac{1}{2}$ h

**66.** $\frac{40}{49}$    **67.** $\frac{2}{5}$    **68.** $\frac{49}{9}$    **69.** 52    **70.** 52    **71.** 24    **72.** $\dfrac{a^2 - 6b}{b}$    **73.** $\dfrac{bc}{a}$    **74.** 56.7

**75.** 100,000 gal    **76.** 711 and 1659    **77.** $y = k\sqrt{z}$    **78.** $y = kvu^2$    **79.** $y = \dfrac{kp}{q}$    **80.** $y = \dfrac{kmn}{p^2}$    **81.** 5184

**82.** 48    **83.** 160 N    **84.** 18.8 lb    **85.** $-270$ N, attractive force

# CHAPTER 3

## Exercises 3.1, Pages 123–125

**1.**

**3.**

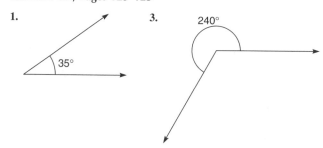

**5.** $\frac{1°}{4}$    **7.** $2°$    **9.** $30'$    **11.** $24'$    **13.** $37.2°$    **15.** $75.78°$    **17.** $69°20'$    **19.** $23°18'$    **21.** $34.4042°$
**23.** $19.3075°$    **25.** $18°12'36''$    **27.** $8°55'30''$    **29.** $a$    **31.** $c$    **33.** $a$    **35.** $B$    **37.** $B$    **39.** 13.0 cm
**41.** 173 mi    **43.** 338 yd    **45.** 39.5 m    **47.** 3.00 cm
**49.** $\sin A = 0.385$, $\cos A = 0.923$, $\tan A = 0.417$, $\cot A = 2.40$, $\sec A = 1.08$, $\csc A = 2.60$
**51.** $\sin A = 0.489$, $\cos A = 0.872$, $\tan A = 0.561$, $\cot A = 1.78$, $\sec A = 1.15$, $\csc A = 2.04$
**53.** $\sin A = 0.867$, $\cos A = 0.500$, $\tan A = 1.73$, $\cot A = 0.577$, $\sec A = 2.00$, $\csc A = 1.15$
**55.** $\sin B = 0.923$, $\cos B = 0.385$, $\tan B = 2.40$, $\cot B = 0.417$, $\sec B = 2.60$, $\csc B = 1.08$
**57.** $\sin B = 0.868$, $\cos B = 0.497$, $\tan B = 1.75$, $\cot B = 0.573$, $\sec B = 2.01$, $\csc B = 1.15$
**59.** $\sin B = 0.500$, $\cos B = 0.866$, $\tan B = 0.578$, $\cot B = 1.73$, $\sec B = 1.16$, $\csc B = 2.00$
**61.**

| | (a) | (b) |
|---|---|---|
| $\sin A$ | 0.385 | 0.385 |
| $\cos A$ | 0.923 | 0.923 |
| $\tan A$ | 0.417 | 0.417 |
| $\cot A$ | 2.40 | 2.40 |
| $\sec A$ | 1.08 | 1.08 |
| $\csc A$ | 2.60 | 2.60 |

## Exercises 3.2, Pages 131–132

**1.** 0.3173    **3.** 0.8816    **5.** 0.2215    **7.** 1.216    **9.** 1.483    **11.** 1.309    **13.** 0.7280    **15.** 3.443
**17.** 1.011    **19.** $25.5°$    **21.** $25.1°$    **23.** $81.6°$    **25.** $72.7°$    **27.** $11.5°$    **29.** $65.0°$    **31.** $59.6°$
**33.** $26.8°$    **35.** $80.4°$    **37.** $61.08°$    **39.** $33.01°$    **41.** $24.54°$    **43.** 0.5934    **45.** 1.314    **47.** 0.9850
**49.** 1.363    **51.** 1.004    **53.** 3.237    **55.** $58°49'34''$    **57.** $80°55'58''$    **59.** $53°22'24''$    **61.** $49°2'55''$
**63.** $24°52'43''$    **65.** $31°5'42''$

## Exercises 3.3, Page 137

**1.** $A = 26.6°$, $B = 63.4°$, $c = 8.94$ ft  **3.** $B = 62.7°$, $a = 9.63$ cm, $b = 18.7$ cm  **5.** $A = 56.0°$, $B = 34.0°$, $a = 11.1$ m
**7.** $A = 58.2°$, $B = 31.8°$, $c = 14.6$ mi  **9.** $A = 53.0°$, $a = 332$ km, $c = 416$ km
**11.** $A = 22.32°$, $B = 67.68°$, $b = 34.62$ cm  **13.** $B = 21.25°$, $b = 2627$ mi, $c = 7248$ mi
**15.** $A = 74.20°$, $a = 43.61$ m, $b = 12.34$ m  **17.** $A = 56.13°$, $B = 33.87°$, $a = 3832$ ft
**19.** $A = 65°$, $B = 25°$, $a = 3200$ mi  **21.** $A = 4\overline{0}°$, $a = 29$ m, $b = 34$ m  **23.** $B = 53°$, $b = 190$ ft, $c = 230$ ft
**25.** $A = 68°$, $b = 1.4$ mi, $c = 3.8$ mi  **27.** $A = 52°18'30''$, $b = 1354$ m, $c = 2215$ m
**29.** $A = 53°25'31''$, $B = 36°34'29''$, $b = 367.7$ ft  **31.** $B = 31°48'35''$, $b = 23.27$ m, $c = 44.15$ m
**33.** $A = 62°54'44''$, $a = 6011$ ft, $b = 3075$ ft

## Exercises 3.4, Pages 140–145

**1.** 52 m  **3.** 4°  **5.** 257 ft  **7.** 92.2 m  **9.** 269 ft  **11. (a)** 139 $\Omega$, 36.4°  **(b)** 270 $\Omega$, 110 $\Omega$
**13.** 2.86 cm, 36.5°  **15.** $x = 2.04$ cm, $y = 5.86$ cm, $z = 5.42$ cm, $\alpha = 68.5°$, $\beta = 63.7°$, $\theta = 149.7°$, $\phi = 78.1°$
**17.** 0.2445 in.  **19.** 12.6°  **21.** $x = 31.21$ m, $\alpha = 125.22°$, $\beta = 144.78°$  **23.** 11.6 ft  **25.** 25.5°  **27.** 43 ft
**29.** 107.2 mm  **31.** 92.1 mm  **33.** 373.1 ft  **35.** 26°33'54''; 63°26'6''

## Chapter 3 Review, Pages 146–147

**1.** 129.5°  **2.** 76.2°  **3.** 35°40'  **4.** 314°18'  **5.** 16.4625°  **6.** 38°24'18''  **7.** 32.2 m  **8.** 31.6 mi
**9.** $\sin A = 0.843$, $\cos A = 0.540$, $\tan A = 1.56$, $\cot A = 0.641$, $\sec A = 1.85$, $\csc A = 1.19$  **10.** 0.9677  **11.** 0.7848
**12.** 0.5658  **13.** 1.044  **14.** 1.797  **15.** 1.018  **16.** 37.4°  **17.** 69.4°  **18.** 51.0°  **19.** 41.2°
**20.** 47.3°  **21.** 54.9°  **22.** 0.6643  **23.** 3.774  **24.** 1.210  **25.** 63°26'55''  **26.** 13°32'21''
**27.** 16°50'12''  **28.** $A = 36.4°$, $B = 53.6°$, $c = 11.8$ m  **29.** $A = 53.50°$, $a = 21.28$ cm, $c = 26.48$ cm
**30.** $B = 7\overline{0}°$, $a = 580$ km, $b = 1600$ km  **31.** $B = 54°45'28''$, $b = 347.8$ m, $c = 425.8$ m  **32.** 4800 ft
**33.** $5\overline{0}00$ ft  **34.** 2.4°  **35.** $x = 22.3$ m, $y = 24.3$ m, $\alpha = 76.3°$, $\beta = 63.1°$  **36.** 10.2 ft
**37. (a)** 86 $\Omega$, $\phi = 61°$  **(b)** $R = 88$ $\Omega$, $Z = 130$ $\Omega$

## CHAPTER 4

## Exercises 4.1, Pages 155–156

| | Function | Domain | Range |
|---|---|---|---|
| **1.** | Yes | {2, 3, 9} | {2, 4, 7} |
| **3.** | No | {1, 2, 7} | {1, 3, 5} |
| **5.** | Yes | {−2, 2, 3, 5} | {2} |
| **7.** | Yes | Real numbers | Real numbers |
| **9.** | Yes | Real numbers | Real numbers where $y \geq 1$ |
| **11.** | No | Real numbers where $x \geq -2$ | Real numbers |
| **13.** | Yes | Real numbers where $x \geq -3$ | Real numbers where $y \geq 0$ |
| **15.** | Yes | Real numbers where $x \geq 4$ | Real numbers where $y \geq 6$ |

**17. (a)** 20  **(b)** $-12$  **(c)** $-28$  **19. (a)** 35  **(b)** 15  **(c)** $-25$  **21. (a)** 95  **(b)** 0  **(c)** 4
**23. (a)** 2  **(b)** $\frac{2}{3}$  **(c)** 0 is not in the domain of $f(t)$.  **25. (a)** $6a + 8$  **(b)** $24a + 8$  **(c)** $6c^2 + 8$
**27. (a)** $4x^2 + 4x - 8$  **(b)** $4x^2 - 36x + 72$  **(c)** $16x^2 - 8x - 8$
**29. (a)** $x^2 - 3x$  **(b)** $-x^2 + 9x - 2$  **(c)** $3x^3 - 19x^2 + 9x - 1$  **(d)** $3x + 3h - 1$  **31.** All real numbers $x \neq 2$
**33.** All real numbers $t \neq 6$ or $t \neq -3$  **35.** All real numbers $x < 5$

**1.**

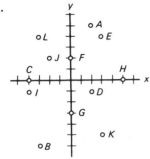

**3.**

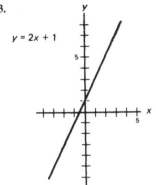

$y = 2x + 1$

**5.**

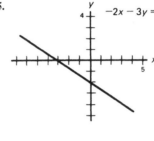

$-2x - 3y = 6$

**7.**

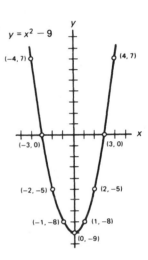

$y = x^2 - 9$

(−4, 7)    (4, 7)

(−3, 0)    (3, 0)

(−2, −5)    (2, −5)

(−1, −8)    (1, −8)

(0, −9)

**9.**

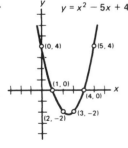

$y = x^2 - 5x + 4$

(0, 4)    (5, 4)

(1, 0)    (4, 0)

(2, −2)    (3, −2)

**11.**

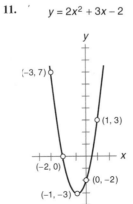

$y = 2x^2 + 3x - 2$

(−3, 7)

(−2, 0)    (1, 3)

(0, −2)

(−1, −3)

**13.**

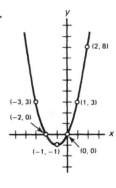

(2, 8)

(−3, 3)    (1, 3)

(−2, 0)

(−1, −1)    (0, 0)

**15.** $y = -2x^2 + 4x$

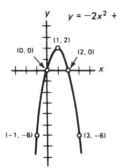

**17.** $y = x^3 - x^2 - 10x + 8$

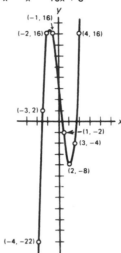

**19.** $y = x^3 + 2x^2 - 7x + 4$

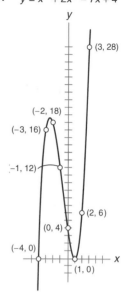

**21.** $y = \sqrt{x + 4}$

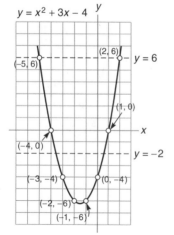

**23.** $y = \sqrt{12 - 6x}$

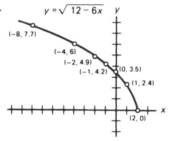

**25.** 3 and $-3$, 2 and $-2$, 3.3 and $-3.3$   **27.** 1 and 4, 4.5 and 0.5, no solution

**29.** $-2$ and $\frac{1}{2}$, 1 and $-2.5$, 1.3 and $-2.8$   **31.** 0 and $-2$, 1 and $-3$, 1.7 and $-3.7$

**33.** 0 and 2, no solution, 2.7 and $-0.7$, 2.3 and $-0.3$   **35.** 3.3, $-3.1$, 0.8; 3.4, $-3.0$, 0.6; 3.2, $-3.2$, 1.0

**37.** $-4$, 1; $-3.8$, 0, 1.8; $-3.6$, $-0.5$, 2.1

**39.** $-4$ and 1, $-5$ and 2, 0.5 and $-3.5$   **41.** 2 and $-2$, no solution, 3.5 and $-3.5$

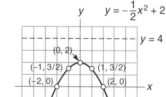

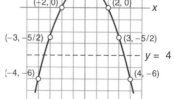

**43.** 2.9, −0.5, 0.7; 2.5, −0.9, 1.3; 2.8, −0.6, 0.8

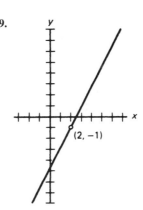

$y = x^3 - 3x^2 + 1$

**45.** 2.6 ms, 4 ms, 5.5 ms    **47.** 0.27 ms, 0.48 ms, 0.94 ms
**49.** 2.7 s, 3.1 s    **51.** A(1.18, −1.62), B(2.35, −3.24), C(3.53, −4.85)

**Exercises 4.3, Pages 172–173**

**1.** 1    **3.** −4    **5.** 0    **7.** $\frac{5}{8}$    **9.**    **11.**

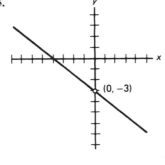

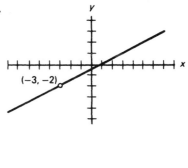

**13.**    **15.**

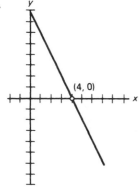

**17.** $3x + y - 2 = 0$    **19.** $x - 2y - 5 = 0$    **21.** $x + y - 5 = 0$    **23.** $x + 2y + 10 = 0$    **25.** $y = -5x - 2$
**27.** $y = 2x + 7$    **29.** $y = 5$    **31.** $x = -2$    **33.** $y = -3$    **35.** $x = -7$    **37.** $m = -\frac{1}{4}, b = 3$
**39.** $m = 2, b = 7$    **41.** $m = 0, b = 6$

**43.**

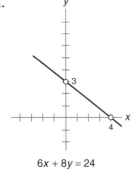

$y = 3x - 2$

**45.**

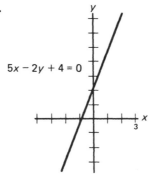

$5x - 2y + 4 = 0$

**47.**

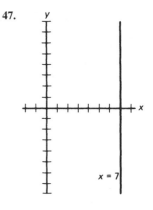

$x = 7$

**49.**

$y = -3$

**51.**

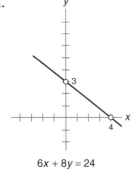

$6x + 8y = 24$

**53.**

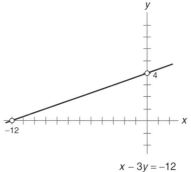

$x - 3y = -12$

**55.** $\frac{1}{350}$

**Exercises 4.4, Pages 176–177**

**1.** Perpendicular    **3.** Neither    **5.** Parallel    **7.** $2x - y + 7 = 0$    **9.** $5x + y + 31 = 0$    **11.** $3x - 4y = 0$
**13.** $3x - 2y = 18$    **15.** $y = 8$    **17.** $x = 7$    **19.** **(a)** Yes, slopes of opposite sides are equal. **(b)** No, slopes of adjacent sides are not negative reciprocals.

**Exercises 4.5, Page 179**

**1.** 15    **3.** 7    **5.** $4\sqrt{2}$    **7.** 7    **9.** $(3\frac{1}{2}, 5)$    **11.** $(1\frac{1}{2}, -1)$    **13.** $(0, -2\frac{1}{2})$
**15.** **(a)** 24 **(b)** yes **(c)** no **(d)** 24    **17.** **(a)** $10 + \sqrt{82} + \sqrt{58}$ or 26.7 **(b)** no **(c)** no    **19.** $4\sqrt{2}$
**21.** $x - 2y = 10$    **23.** $x + 2y = 21$

**Chapter 4 Review, Pages 181–182**

| Function | Domain | Range |
|---|---|---|
| **1.** Yes | $\{2, 3, 4, 5\}$ | $\{3, 4, 5, 6\}$ |
| **2.** No | $\{2, 4, 6\}$ | $\{1, 3, 4, 6\}$ |
| **3.** Yes | Real numbers | Real numbers |
| **4.** Yes | Real numbers | Real numbers where $y \geq -5$ |
| **5.** No | Real numbers where $x \geq 4$ | Real numbers |
| **6.** Yes | Real numbers where $x \leq \frac{1}{2}$ | Real numbers where $y \geq 0$ |

**7.** (a) 24  (b) 14  (c) $-6$   **8.** (a) 10  (b) $-12$  (c) 38   **9.** (a) 5  (b) $\frac{85}{4}$  (c) $-15$ is not in the domain of $h(x)$.
(d) 1 is not in the domain of $h(x)$.   **10.** (a) $a^2 - 6a + 4$  (b) $4x^2 - 12x + 4$  (c) $z^2 - 10z + 20$

**11.**

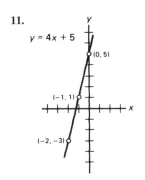

$y = 4x + 5$

(0, 5)
(−1, 1)
(−2, −3)

**12.** $y = x^2 + 4$

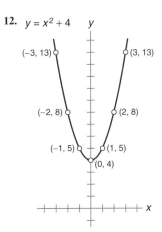

(−3, 13)  (3, 13)
(−2, 8)  (2, 8)
(−1, 5)  (1, 5)
(0, 4)

**13.** $y = x^2 + 2x - 8$

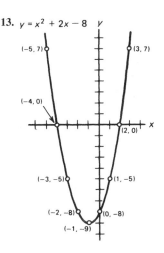

(−5, 7)  (3, 7)
(−4, 0)  (2, 0)
(−3, −5)  (1, −5)
(−2, −8)  (0, −8)
(−1, −9)

**14.** $y = 2x^2 + x - 6$

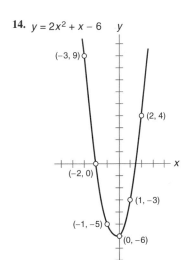

(−3, 9)
(2, 4)
(−2, 0)
(1, −3)
(−1, −5)  (0, −6)

**15.** $y = -x^2 - x + 4$

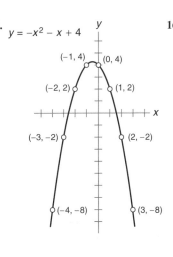

(−1, 4)  (0, 4)
(−2, 2)  (1, 2)
(−3, −2)  (2, −2)
(−4, −8)  (3, −8)

**16.**

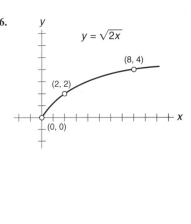

$y = \sqrt{2x}$

(8, 4)
(2, 2)
(0, 0)

ANSWERS TO ODD-NUMBERED EXERCISES AND TO CHAPTER REVIEWS    **625**

**17.**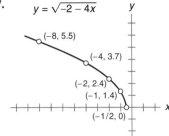

$y = \sqrt{-2 - 4x}$

(−8, 5.5)
(−4, 3.7)
(−2, 2.4)
(−1, 1.4)
(−1/2, 0)

**18.**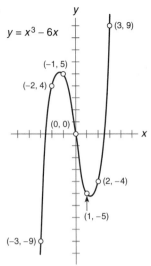

$y = x^3 - 6x$

(3, 9)
(−1, 5)
(−2, 4)
(0, 0)
(2, −4)
(1, −5)
(−3, −9)

**19.** 1 and $-1$, 1.7 and $-1.7$, no solution     **20.** $-4$ and 2, 1.6 and $-3.6$, $-4.5$ and 2.5

**21.** $-2$ and 1, $-2.6$ and 1.6, $-3$ and 2     **22.** 0, 2.4, $-2.4$; 2.6, $-2.3$, $-0.3$; 2.1, $-2.7$, 0.5     **23.** 1, 1.7, 2

**24.** 2.1, 2.4     **25.** $-\frac{2}{9}$     **26.** $\sqrt{85}$     **27.** $(-\frac{3}{2}, -3)$     **28.** $11x + 2y - 58 = 0$     **29.** $2x - 3y + 9 = 0$

**30.** $x + 3y + 9 = 0$     **31.** $x = -3$     **32.** $m = \frac{3}{2}; b = -3$     **33.**

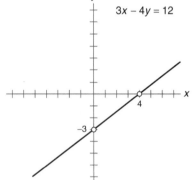

$3x - 4y = 12$

4
−3

**34.** Perpendicular     **35.** Parallel     **36.** Neither     **37.** Perpendicular     **38.** Parallel     **39.** $2x - y - 8 = 0$

**40.** $5x - 3y + 20 = 0$

# CHAPTER 5

## Exercises 5.1, Page 186

**1.** $-40x^5 + 72x^4 - 80x^3$     **3.** $24x^4y^6 - 42x^6y^7 + 54x^7y^5$     **5.** $4x^2 - 49$     **7.** $49x^4 + 28x^2y + 4y^2$

**9.** $4a^4 - 12a^2b + 9b^2$     **11.** $x^2 + 16x + 48$     **13.** $-6x^2 - 31x - 40$     **15.** $4a^2 + b^2 + 9c^2 - 4ab + 12ac - 6bc$

**17.** $8a^3 + 12a^2b + 6ab^2 + b^3$     **19.** $125 - 150x^2 + 60x^4 - 8x^6$     **21.** $42a^2 + 17ab - 15b^2$

**23.** $16x^2 + 56xy + 49y^2$     **25.** $12a^2 + 32ab - 99b^2$     **27.** $27a^9 + 270a^6b^2 + 900a^3b^4 + 1000b^6$

**29.** $-24a^5b^7 + 36a^7b^{10} - 80a^7b^3$     **31.** $81a^6 - 216a^3b^2 + 144b^4$     **33.** $x^2 + 9y^2 + 25z^2 - 6xy - 10xz + 30yz$

**35.** $54a^2 - 114ab + 40b^2$     **37.** $30x^6 - 4x^3 - 16$     **39.** $15x^2y^2 + 21xy^2z - 72y^2z^2$     **41.** $25a^2b^2c^2 - 36$

**43.** $3x^4 + \frac{3}{2}x^2y - \frac{1}{3}y^2$     **45.** $x^2 - \frac{1}{3}xy + \frac{4}{9}y^2$     **47.** $x^2 + 2x^4 + x^6$

## Exercises 5.2, Pages 191–192

**1.** $3(2x + 3y)$ **3.** $5(2x + 5y - 9z)$ **5.** $6x(2x - 5y + z)$ **7.** $2xy^2(4x^2 - 3y + 6z)$ **9.** $(x + 4)(x - 4)$
**11.** $(3x + 5y)(3x - 5y)$ **13.** $2(e + 6)(e - 6)$ **15.** $4(2d + 5)(2d - 5)$ **17.** $4(R + r)(R - r)$ **19.** $(x + 2)(x + 4)$
**21.** $(b + 3)(b + 8)$ **23.** $(x - 3)(x - 6)$ **25.** $(a - 2)(a - 16)$ **27.** $(x + 5)(x - 7)$ **29.** $(a + 4)(a - 1)$
**31.** $(2x + 3)(x - 2)$ **33.** $(3x - 2)(5x - 7)$ **35.** $(5y + 6)(9y + 1)$ **37.** $(7a - 1)(5a + 1)$ **39.** $(3c - 4)^2$
**41.** $5(b + 2)(5b + 2)$ **43.** $(7t - 5s)(5t + 3s)$ **45.** $5(2k - 5t)(3k - 2t)$ **47.** $3(6x + 7)(3x - 2)$ **49.** $(2a + 5)^2$
**51.** $(3x - 8y^2)^2$ **53.** $(5x + y^3)^2$ **55.** $(x^2 + 9)(x + 3)(x - 3)$ **57.** $(a^2 - 8)(a^2 + 5)$
**59.** $(t + 3)(t - 3)(t + 2)(t - 2)$

## Exercises 5.3, Page 195

**1.** $(m + n)(a - b)$ **3.** $(x + y)(m + n)$ **5.** $(3x + y)(1 - 2x)$ **7.** $(3x - 2)(2x^2 + 1)$
**9.** $(x + y + 2z)(x + y - 2z)$ **11.** $(10 + 7x - 7y)(10 - 7x + 7y)$ **13.** $(x - 3 + 2y)(x - 3 - 2y)$
**15.** $(2x + 2y - 1)(2x - 2y + 1)$ **17.** $(x + y + 4)(x + y + 9)$
**19.** $(4(a + b) + 1)(6(a + b) - 5)$, or $(4a + 4b + 1)(6a + 6b - 5)$ **21.** $(a + b)(a^2 - ab + b^2)$
**23.** $(x - 4)(x^2 + 4x + 16)$ **25.** $(ab + c)(a^2b^2 - abc + c^2)$ **27.** $(a - 3b)(a^2 + 3ab + 9b^2)$
**29.** $(3a + 4b)(9a^2 - 12ab + 16b^2)$ **31.** $(3a^2 - 2b^3)(9a^4 + 6a^2b^3 + 4b^6)$
**33.** $(x + y)(x^2 - xy + y^2)(x - y)(x^2 + xy + y^2)$

## Exercises 5.4, Pages 197–198

**1.** $\frac{1}{6}$ **3.** $\frac{5}{9y}$ **5.** $\frac{4}{(x + 4)^2}$ **7.** $2(x - 3)^2$ **9.** $\frac{5}{6}$ **11.** $\frac{6}{m - 1}$ **13.** $x$ **15.** $\frac{x + 4}{x - 7}$ **17.** $\frac{2x + 3}{2x + 9}$

**19.** $-2$ **21.** $-x - 1$ **23.** $\frac{a - 2}{a + 2}$ **25.** $-m$ **27.** $\frac{-t^2}{t + 1}$ **29.** $\frac{x}{x - 2}$ **31.** $y - 2$ **33.** $\frac{x - 2}{y^2 + 3y + 9}$

**35.** $\frac{1}{x - 4}$ **37.** $\frac{-3 - 2x}{2x - 3}$ **39.** $\frac{a^2 - ab + b^2}{a + b}$

## Exercises 5.5, Pages 200–202

**1.** $\frac{3}{4}$ **3.** $\frac{5}{2}$ **5.** $2$ **7.** $\frac{a^5}{3}$ **9.** $\frac{9}{10b^2y}$ **11.** $\frac{8y^3}{5a}$ **13.** $\frac{a^2b}{3}$ **15.** $\frac{1}{4y}$ **17.** $\frac{5}{27}$ **19.** $\frac{y(y + 2)}{y - 2}$

**21.** $-(m + 1)^2$ **23.** $\frac{-x(x + 8)}{x + 2}$ **25.** $\frac{x + y}{x - 2}$ **27.** $\frac{-x - 3}{x^3(x - 3)}$ **29.** $x + 3$ **31.** $\frac{x + 3}{x + 1}$ **33.** $1$

**35.** $\frac{a - 4b}{5a + b}$ **37.** $\frac{y}{x + y}$ **39.** $\frac{(x - 4)(2x + 7)}{(x + 3)(3x + 1)}$ **41.** $(x - 2)(x + 2)$ **43.** $\frac{y - x}{3y}$ **45.** $\frac{(x + 2)(5x + 9)}{x}$

**47.** $\frac{(3x + 4)^2}{2x(x - 5)}$

## Exercises 5.6, Pages 204–206

**1.** $24x$ **3.** $4k^2$ **5.** $30a^2b^2$ **7.** $5x(x - 1)$ **9.** $6(a + 2)$ **11.** $(x + 5)(x - 5)^2$ **13.** $(x + 2)(x + 4)(x - 3)$
**15.** $6x^2(x + 2)(x - 4)$ **17.** $12c(c + 2)(c - 4)^2$ **19.** $\frac{4}{7}$ **21.** $\frac{7}{a}$ **23.** $\frac{15}{x + 1}$ **25.** $\frac{7}{20}$ **27.** $\frac{2x + 5}{30}$

**29.** $\frac{-8}{3y}$ **31.** $\frac{4p + 6}{3p^2}$ **33.** $\frac{-18x^2 + 15x + 4}{12x^3}$ **35.** $\frac{bc + ac + ab}{abc}$ **37.** $\frac{6x^2 - x + 4}{2x}$ **39.** $\frac{3c - 1}{c(c - 1)}$

**41.** $\frac{3t - 18}{(t + 2)(t - 2)}$ **43.** $\frac{3a + 2}{a(a + 1)}$ **45.** $\frac{-d^2 - 13d}{(d + 4)(d - 5)}$ **47.** $\frac{6}{x - 4}$ **49.** $\frac{-2}{12(a + 3)}$ **51.** $\frac{-2a^2 + 10a - 2}{(a + 1)(a - 1)}$

**53.** $\frac{11}{x - 3}$ **55.** $\frac{3r + 4}{(r - 2)^2(r + 3)}$ **57.** $\frac{1}{x - 3}$ **59.** $\frac{5t + 9}{4(t + 1)}$ **61.** $\frac{5x^2 + 6x - 9}{(x - 4)(x + 1)(x + 3)}$ **63.** $0$

**65.** $\frac{7x + 21}{(x + 4)(x - 4)(x - 3)}$ **67.** $\frac{5x^2 + 4x - 5}{2x(4x + 1)}$ **69.** $\frac{x - 2}{x^2 + 2x + 4}$ **71.** $\frac{x^2 + 1}{x^2 - x + 1}$

## Exercises 5.7, Pages 207–208

**1.** $\dfrac{x+2}{3x}$   **3.** $\dfrac{3(t-1)}{2(t+1)}$   **5.** 1   **7.** $\frac{39}{28}$   **9.** $\dfrac{a}{2}$   **11.** $\dfrac{a-5}{a}$   **13.** $\dfrac{15}{xy}$   **15.** $-y$   **17.** $\dfrac{a+1}{a-1}$

**19.** $\dfrac{n+m}{n-m}$   **21.** $\dfrac{x-7}{x+6}$   **23.** $\dfrac{x-y}{(x+y)^2}$   **25.** 1   **27.** $\dfrac{(2a+b)^2}{(a-b)^2}$   **29.** $\dfrac{x}{x+1}$   **31.** $\dfrac{x-1}{2x-1}$

**33.** $\dfrac{x^2}{x^2-x+1}$   **35.** $\dfrac{-6}{x^2(x+5)}$   **37.** $\frac{3}{5}$

## Exercises 5.8, Pages 211–212

**1.** 12   **3.** 10   **5.** 15   **7.** $-1$   **9.** $-7$   **11.** $-1$   **13.** $-19$   **15.** $-\frac{3}{4}$   **17.** $r=\dfrac{mv^2}{F}$

**19.** $R=\dfrac{V-I_LR_L}{I_L}$   **21.** $R_1=\dfrac{RR_2}{R_2-R}$   **23.** $y=\dfrac{x}{x-1}$   **25.** $x=-a-b$   **27.** $Q=\dfrac{VR_1R_2}{R_2-R_1}$   **29.** $\frac{1}{2}$

**31.** 12   **33.** 5   **35.** $-\frac{1}{3}$   **37.** 2   **39.** No solution   **41.** $-22$   **43.** $-10$

## Chapter 5 Review, Pages 214–215

**1.** $-6a^5b^4+72a^3b^6+54a^2b^9$   **2.** $15x^2+53x-52$   **3.** $16x^2-49$   **4.** $25a^4-90a^2b^2+81b^4$
**5.** $64a^3-48a^2b^2+12ab^4-b^6$   **6.** $30x^2+23xyz^2-14y^2z^4$   **7.** $9a^2b(2a-b^3+ab)$   **8.** $(4x+3y)(4x-3y)$
**9.** $(x+4)(5x+8)$   **10.** $(7x-2)(2x+9)$   **11.** $(2a-5)^2$   **12.** $(m-n)(m^2+mn+n^2)$   **13.** $3(x+1)(x-7)$
**14.** $(x^2+9)(x+3)(x-3)$   **15.** $(3x+10)(5x-9)$   **16.** $5(x+4y)(2x-3y)$   **17.** $(x+3+5y)(x+3-5y)$

**18.** $(a+2b)(x-4y)$   **19.** $\dfrac{15y^6}{8x^4}$   **20.** 2   **21.** $\dfrac{x-8}{2x+3}$   **22.** $\dfrac{-(2x+1)}{x+4}$   **23.** $\dfrac{x}{2x+5}$   **24.** $\dfrac{-(2x-1)}{2x+1}$

**25.** $3y$   **26.** $\dfrac{-1}{6(x+2)^2}$   **27.** $\dfrac{16a+27-30a^2}{12a^2}$   **28.** $\dfrac{5x}{2x+1}$   **29.** $\dfrac{3x}{(x-5)(2x-1)}$

**30.** $\dfrac{5x^2-12x+1}{(x+1)(2x+1)(3x-4)}$   **31.** $\dfrac{1}{x^2-x+1}$   **32.** $\dfrac{14y+4}{3y+1}$   **33.** $-8$   **34.** $\dfrac{a}{a+b}$
**35.** No solution   **36.** $-8$

## CHAPTER 6

## Exercises 6.1, Pages 224–226

**1.** $(4,-1)$   **3.** $(-3,5)$   **5.** System dependent, lines coincide   **7.** $(2,7)$   **9.** $(1,-6)$   **11.** $(-1,6)$
**13.** $(-5,0)$   **15.** $(3,1)$   **17.** System inconsistent, lines parallel   **19.** $(-4,6)$   **21.** $(\frac{2}{3},-\frac{7}{3})$   **23.** $(2,8)$
**25.** $(-1,\frac{1}{3})$   **27.** $(-2,-14)$   **29.** $(-6,3)$   **31.** $(-1,-4)$   **33.** $(\frac{1}{6},\frac{2}{3})$   **35.** $(5,-\frac{1}{4})$   **37.** $(5.2,0.6)$
**39.** $(180,240)$   **41.** 40 μF, 15 μF   **43.** Cement: 2.3 m$^3$   Gravel: 9.2 m$^3$
**45.** 1600 L of 3% solution, 400 L of 8% solution   **47.** 0.5 km/h, 4.5 km/h   **49.** Length: 30 m   Width: 18 m
**51.** 120 mA; 720 mA   **53.** $I_2=2$ mA; $I_2=6$ mA   **55.** $\frac{7}{2}$

## Exercises 6.2, Page 228

**1.** $(1,b-a)$   **3.** $(ab,a^2)$   **5.** $\left(\dfrac{b}{a^2+b^2},\dfrac{a}{a^2+b^2}\right)$   **7.** $\left(\dfrac{c}{a+b},\dfrac{c}{a+b}\right)$   **9.** $\left(\dfrac{a-b}{a+b},\dfrac{a-3b}{a+b}\right)$   **11.** $(3,4)$

**13.** $(\frac{1}{2},-\frac{1}{3})$   **15.** $(\frac{2}{3},\frac{3}{2})$   **17.** $(\frac{3}{4},\frac{1}{2})$   **19.** $\left(\dfrac{2}{a+b},\dfrac{2}{a-b}\right)$

## Exercises 6.3, Pages 233–234

**1.** $(3,2,-1)$   **3.** $(2,5,0)$   **5.** $(4,6,-6)$   **7.** $(13,1,2)$   **9.** $(2,-3,1)$   **11.** $(\frac{1}{3},\frac{3}{4},-2)$   **13.** $(2,4,-1,5)$
**15.** $R_1=1200\ \Omega;\ R_2=150\ \Omega,\ R_3=600\ \Omega$   **17.** 29 cm, 24 cm, 12 cm   **19.** 64 mm; 16 mm; 5 mm
**21.** $f(x)=2x^2-5x+7$

**Exercises 6.4, Pages 238–239**

**1.** 14  **3.** $-2$  **5.** 39  **7.** $-42$  **9.** $-12$  **11.** 16  **13.** 1  **15.** 95  **17.** $mn + n^3$  **19.** 29
**21.** 78  **23.** 16  **25.** 0  **27.** 0  **29.** $-138$  **31.** $-1139$  **33.** 24  **35.** $-504$

**Exercises 6.5, Pages 242–243**

**1.** 0  **3.** 0  **5.** $-126$  **7.** $-105$  **9.** $-30$  **11.** 31  **13.** 61  **15.** $-11$  **17.** 450  **19.** 410
**21.** 970  **23.** 744

**Exercises 6.6, Pages 248–249**

**1.** $(3, -2)$  **3.** $(-2, 9)$  **5.** $(0, -4)$  **7.** Dependent  **9.** Inconsistent  **11.** $(-\frac{2}{3}, \frac{5}{6})$  **13.** $(\frac{1}{2}, 2)$
**15.** $\left(\dfrac{2a - 4b}{a^2 - b^2}, \dfrac{4a - 2b}{a^2 - b^2}\right)$  **17.** $\left(\dfrac{a^2 + b^2}{2a + 5b}, \dfrac{2b - 5a}{2a + 5b}\right)$  **19.** $\left(\dfrac{c}{a + b^2}, \dfrac{bc}{a + b^2}\right)$  **21.** 75 V, 135 V  **23.** $(2, 1, -4)$
**25.** $(3, 2, -4)$  **27.** $(-1, 4, -5)$  **29.** $(1, 2, -1)$  **31.** $(2, -1, 5)$  **33.** $(4, 0, -3)$  **35.** 6 cm, 7 cm, 8 cm
**37.** $(1, 3, -2, 5)$  **39.** $(2, -2, 3, 4, -1)$

**Exercises 6.7, Page 256**

**1.** $\dfrac{5}{x + 2} + \dfrac{3}{x - 7}$  **3.** $\dfrac{3}{2x + 3} - \dfrac{2}{x - 4}$  **5.** $\dfrac{7}{x} + \dfrac{2}{3x - 4} + \dfrac{5}{2x + 1}$  **7.** $\dfrac{1}{x + 1} + \dfrac{1}{(x + 3)^2}$
**9.** $\dfrac{3}{4x - 1} + \dfrac{1}{(4x - 1)^2} - \dfrac{7}{(4x - 1)^3}$  **11.** $\dfrac{3}{x} + \dfrac{8}{x - 1} - \dfrac{4}{(x - 1)^2}$  **13.** $\dfrac{x - 1}{x^2 + 1} - \dfrac{x}{x^2 - 3}$
**15.** $\dfrac{4x + 1}{x^2 + x + 1} - \dfrac{1}{x^2 - 5}$  **17.** $\dfrac{5x - 2}{x^2 + 5x + 3} + \dfrac{1}{x + 3} - \dfrac{2}{x - 3}$  **19.** $\dfrac{5}{x} + \dfrac{3x - 1}{x^2 + 1} - \dfrac{5}{(x^2 + 1)^2}$
**21.** $\dfrac{1}{x} - \dfrac{2}{x^2} - \dfrac{4x}{(x^2 + 2)^2}$  **23.** $\dfrac{2}{x + 3} + \dfrac{4}{x - 3} - \dfrac{6x}{x^2 + 9}$  **25.** $x + \dfrac{\frac{1}{2}}{x + 1} + \dfrac{\frac{1}{2}}{x - 1}$  **27.** $x - 1 + \dfrac{3}{x - 2} + \dfrac{1}{x + 2}$
**29.** $3x - 2 + \dfrac{5}{x} - \dfrac{8x}{x^2 + 1}$

**Chapter 6 Review, Pages 260–262**

**1.** $(1, 2)$  **2.** $(0, 2)$  **3.** $(-1, 8)$  **4.** System inconsistent, lines parallel  **5.** $(3, 1)$  **6.** $(\frac{25}{8}, \frac{1}{4})$  **7.** $(2, -2)$
**8.** $(4, \frac{7}{2})$  **9.** $(4, -1)$  **10.** $(-2, -6)$  **11.** $(2, 3)$  **12.** $(-\frac{5}{2}, \frac{1}{8})$  **13.** $\left(\dfrac{a + 3b}{2a + 4b}, \dfrac{-4}{a + 2b}\right)$  **14.** $(1, \frac{1}{2})$
**15.** $(\frac{5}{2}, -\frac{1}{2}, 0)$  **16.** $(2, -2, -4)$  **17.** 22  **18.** $-25$  **19.** 276  **20.** 45  **21.** $(10, 2)$
**22.** System dependent, lines coincide  **23.** $(2, 0)$  **24.** $(\frac{8}{9}, \frac{11}{9})$  **25.** $(-7, 5, -15)$  **26.** $(0, 3, 6)$  **27.** $(-\frac{1}{4}, \frac{1}{2})$
**28.** $(1, 5)$  **29.** $(\frac{7}{2}, -1)$  **30.** $(-1, -24)$  **31.** $(1, 5, 1)$  **32.** $(2, 30, 11)$
**33.** 24 kg of 10% lead, 6 kg of 20% lead  **34.** \$1400 at $5\frac{1}{2}\%$, \$2200 at $6\frac{1}{2}\%$
**35.** Wind velocity: 10 km/h; plane speed: 90 km/h  **36.** Jack, 13 pieces; John, 11 pieces; Bob, 8 pieces  **37.** 18
**38.** 385  **39.** $\dfrac{4}{x - 3} + \dfrac{2}{x + 5}$  **40.** $\dfrac{1}{x} + \dfrac{2}{x + 1} - \dfrac{4}{x - 1}$  **41.** $\dfrac{4}{x} + \dfrac{1}{x^2} - \dfrac{1}{x + 1}$  **42.** $\dfrac{4}{x + 1} + \dfrac{6x}{(x + 1)^2}$
**43.** $\dfrac{3x + 2}{x^2 + 1} + \dfrac{4}{x - 1}$  **44.** $\dfrac{5}{x^2 + 4} + \dfrac{2x + 1}{(x^2 + 4)^2}$

**CHAPTER 7**

**Exercises 7.1, Page 269**

**1.** $7, -4$  **3.** $2, 4$  **5.** $-5, 2$  **7.** $3, -\frac{3}{2}$  **9.** $-\frac{5}{7}, -\frac{1}{2}$  **11.** $\frac{8}{3}, \frac{7}{6}$  **13.** $0, -\frac{1}{8}$  **15.** $0, 3$  **17.** $6, -6$
**19.** $\frac{5}{4}, -\frac{5}{4}$  **21.** $\pm\dfrac{2\sqrt{15}}{5}$  **23.** $\pm\dfrac{\sqrt{21}}{2}$  **25.** $\pm 2\sqrt{2}$  **27.** $\pm\sqrt{6}$  **29.** $-\frac{6}{5}, \frac{2}{3}$  **31.** $-2, -\frac{1}{2}$  **33.** $4, -3$
**35.** $5, -1$  **37.** $3, -\frac{8}{5}$

**Exercises 7.2, Page 272**

**1.** $2, -8$  **3.** $5, 7$  **5.** $\frac{1}{2}, -1$  **7.** $-\frac{3}{5}, \frac{6}{5}$  **9.** $-2 \pm \sqrt{11}$  **11.** $\frac{-1 \pm \sqrt{19}}{2}$  **13.** $\frac{-9 \pm \sqrt{21}}{6}$

**15.** $-\frac{3}{2}, 3$  **17.** $\frac{1}{4}, -\frac{13}{4}$  **19.** $\frac{-5 \pm \sqrt{17}}{4}$

**Exercises 7.3, Pages 276–277**

**1.** $8, -4$  **3.** $3, -\frac{3}{2}$  **5.** $\frac{4}{3}, -\frac{7}{4}$  **7.** $-2, -3$  **9.** $\frac{9}{4}, \frac{3}{2}$  **11.** $0, -\frac{5}{2}$  **13.** $\frac{-5 \pm \sqrt{53}}{2}$  **15.** $2 \pm 3\sqrt{2}$

**17.** $\frac{2 \pm 3\sqrt{3}}{2}$  **19.** $\frac{7 \pm 2\sqrt{6}}{3}$  **21.** $2 \pm \sqrt{3}$  **23.** $\frac{1}{3}, 2$  **25.** $2, -12$  **27.** $-\frac{3}{2}$  **29.** $\frac{5 \pm \sqrt{17}}{4}$

**31.** No solution  **33.** $0, 1; 3, -2;$ no solution  **35.** $0, -\frac{1}{3}; 1, -\frac{4}{3}; \frac{-1 \pm \sqrt{13}}{6}$  **37.** $\frac{1}{m}$  **39.** $-\frac{2}{n}$

**41.** $4a, 4a - 5$  **43.** $\sqrt{\dfrac{A}{\pi}}$

**Exercises 7.4, Pages 280–282**

**1.** $5, 8$  **3.** $3.50 \text{ cm} \times 10.0 \text{ cm}$  **5.** $11.0 \text{ m}, 12.0 \text{ m}$

**7.** $2.00 \text{ s}, 5.00 \text{ s}; 3.00 \text{ s}, 4.00 \text{ s}; \dfrac{7 \pm \sqrt{17}}{2}$ s, or 5.56 s and 1.44 s  **9.** $1.00\ \mu\text{s}; \dfrac{2 \pm \sqrt{2}}{2}\ \mu\text{s}$, or 1.71 $\mu$s and 0.293 $\mu$s

**11.** $0.100 \text{ ms}$  **13.** $15.0 \text{ m} \times 25.0 \text{ m}$  **15.** $18.0 \text{ in.}$  **17.** Two possible solutions: $45\bar{0} \text{ m} \times 160\bar{0} \text{ m}; 80\bar{0} \text{ m} \times 90\bar{0} \text{ m}$
**19.** $6.97 \text{ m}$  **21.** $6.00 \text{ in.} \times 18.0 \text{ in.}$  **23.** $10\bar{0} \text{ m} \times 12\bar{0} \text{ m}$  **25.** $60.0 \ \Omega, 18\bar{0} \ \Omega$  **27.** $50.0 \text{ m}$

**Chapter 7 Review, Page 283**

**1.** $3, -7$  **2.** $\frac{8}{3}, \frac{5}{2}$  **3.** $-\frac{1}{4}, -\frac{4}{5}$  **4.** $\frac{1}{6}, -\frac{1}{6}$  **5.** $-2, -\frac{1}{2}$  **6.** $0, 6$  **7.** $-\frac{3}{4}, \frac{3}{2}$  **8.** $3, -3$  **9.** $-5$

**10.** $\frac{4}{3}$  **11.** $-4, -\frac{5}{2}$  **12.** $-1, \frac{7}{3}$  **13.** $4 \pm \sqrt{7}$  **14.** $\frac{3 \pm \sqrt{14}}{5}$  **15.** $2, \frac{10}{3}$  **16.** $\frac{3 \pm \sqrt{21}}{4}$

**17.** $\frac{-3 \pm \sqrt{17}}{-4}$  **18.** $\frac{-4 \pm \sqrt{46}}{6}$  **19.** $\frac{m \pm \sqrt{m^2 - 4mn^2}}{2mn}$  **20.** $\frac{m^2}{n^2}$  **21.** $t = \frac{-v \pm \sqrt{v^2 + 2as}}{a}$

**22.** $3, 12$  **23.** $60\bar{0} \text{ ft} \times 90\bar{0} \text{ ft}$  **24.** $10.0 \text{ cm}, 90\bar{0} \text{ cm}^3$  **25.** (a) $1.00 \text{ s}, 3.00 \text{ s}$  (b) no  (c) $4.00 \text{ s}$  (d) $64.0 \text{ ft}$
**26.** $275 \pm 5\sqrt{385} \ \Omega$, or $373 \ \Omega$ and $177 \ \Omega$

**CHAPTER 8**

**Exercises 8.1, Pages 287–288**

**1.** $x^5$  **3.** $5^3$  **5.** $7^6$  **7.** $m^3$  **9.** $\frac{1}{d^7}$  **11.** $\frac{1}{2^6}$  **13.** $\frac{1}{y^8}$  **15.** $s^{10}$  **17.** $t^3$  **19.** $\frac{16}{49}$  **21.** $3$

**23.** $\frac{6}{5}$  **25.** $2^3$ or $8$  **27.** $\frac{9}{4}$  **29.** $a^{10}$  **31.** $a^2$  **33.** $\frac{1}{a^{12}}$  **35.** $9a^4$  **37.** $\frac{a^4}{4b^2}$  **39.** $-\frac{48}{k^2}$  **41.** $\frac{15a}{b^2}$

**43.** $\frac{1}{w^5}$  **45.** $x^6$  **47.** $\frac{y^5}{x^5}$  **49.** $t^8$  **51.** $\frac{1}{b^2}$  **53.** $\frac{c^4}{ab^2}$  **55.** $a^{18}$  **57.** $\frac{108}{a^{14}}$  **59.** $\frac{8a^{12}}{c^6}$  **61.** $\frac{49a^{10}}{b^8}$

**63.** $\frac{a+b}{ab}$  **65.** $\frac{b+ab}{a^2}$  **67.** $\frac{2}{9}$  **69.** $\frac{6}{5}$  **71.** $\frac{1}{2}$

**Exercises 8.2, Pages 290–292**

**1.** $6$  **3.** $4$  **5.** $16$  **7.** $32$  **9.** $\frac{1}{4}$  **11.** $\frac{1}{3}$  **13.** $9$  **15.** $\frac{1}{27}$  **17.** $\frac{4}{5}$  **19.** $\frac{8}{27}$  **21.** $\frac{64}{125}$  **23.** $-\frac{1}{2}$

**25.** $9$  **27.** $3$  **29.** $2$  **31.** $x^2$  **33.** $x^{1/4}$  **35.** $x^{1/2}$  **37.** $\frac{1}{x^{1/3}}$  **39.** $\frac{1}{x^{1/3}}$  **41.** $x^{1/4}$  **43.** $\frac{a^3}{b^6 c^{12}}$

**45.** $\frac{a^{16}c^8}{b^{12}}$  **47.** $x + 4x^2$  **49.** $8t + 2$  **51.** $6c^{1/3} + \frac{8}{c}$  **53.** $x - y$  **55.** $x + 2x^{1/2}y^{1/2} + y$  **57.** $56$

**59.** $32$  **61.** $256$  **63.** $640$  **65.** $\frac{2}{\pi}$

**630**  ANSWERS TO ODD-NUMBERED EXERCISES AND TO CHAPTER REVIEWS

## Exercises 8.3, Pages 297–298

**1.** 7 **3.** $5\sqrt{3}$ **5.** $6\sqrt{5}$ **7.** $2a\sqrt{2}$ **9.** $6b\sqrt{2}$ **11.** $4a^2b\sqrt{5a}$ **13.** $4ab^2c^4\sqrt{2c}$ **15.** $\dfrac{\sqrt{3}}{2}$

**17.** $\dfrac{\sqrt{10}}{4}$ **19.** $\dfrac{\sqrt{15}}{5}$ **21.** $\dfrac{5\sqrt{6}}{12}$ **23.** $\dfrac{2\sqrt{15a}}{15b}$ **25.** $\dfrac{\sqrt{2b}}{2b}$ **27.** $\dfrac{a}{2b}$ **29.** 5 **31.** $2a\sqrt[3]{2a}$

**33.** $2a^2\sqrt[3]{5a^2}$ **35.** $3x\sqrt[3]{2x^2}$ **37.** $2x^2yz\sqrt[3]{7xy^2}$ **39.** $\dfrac{\sqrt[3]{3}}{2}$ **41.** $\dfrac{\sqrt[3]{10}}{2}$ **43.** $\dfrac{\sqrt[3]{90}}{6}$ **45.** $\dfrac{\sqrt[4]{4}}{2}$

**47.** $\dfrac{\sqrt[3]{10ab}}{5b}$ **49.** $\dfrac{2\sqrt[3]{147a^2}}{21ab}$ **51.** $\dfrac{\sqrt[3]{2a^2b}}{2b}$ **53.** $2x\sqrt[4]{5x^2}$ **55.** $ab\sqrt[4]{25a}$ **57.** $\dfrac{2ab^2\sqrt[5]{a^3b^2c^4}}{c^2}$ **59.** $\sqrt{a}$

**61.** $b\sqrt{3ab}$ **63.** $\sqrt[4]{3}$ **65.** 2 **67.** $2x\sqrt{2x}$ **69.** $2\sqrt{a^2-b^2}$ **71.** $2(a-b)$

## Exercises 8.4, Pages 299–300

**1.** $-\sqrt{2}$ **3.** $\sqrt{2}$ **5.** $32\sqrt{3}$ **7.** $2\sqrt{5}$ **9.** $-7\sqrt{3}$ **11.** $\sqrt{2}$ **13.** $-22\sqrt{3}$ **15.** 0 **17.** $4\sqrt{2x}$

**19.** $25x\sqrt{2}$ **21.** $-3\sqrt[3]{6}$ **23.** $2\sqrt{2}$ **25.** $3\sqrt[3]{5}$ **27.** $6\sqrt[3]{3}-\sqrt{3}$ **29.** $-\sqrt[3]{2}$ **31.** $2\sqrt{2}$ **33.** $\dfrac{13\sqrt{3}}{6}$

**35.** $-\dfrac{3\sqrt{6}}{2}$ **37.** $\dfrac{2\sqrt{3}}{3}$ **39.** $\dfrac{3\sqrt{2}}{2}$ **41.** $\dfrac{4\sqrt{10}}{5}$ **43.** $2\sqrt[3]{3}$ **45.** $\dfrac{\sqrt[3]{6}}{6}-\dfrac{\sqrt[3]{36}}{6}$ **47.** $-\dfrac{17\sqrt{2}}{4}$

**49.** $(5x+6ax-2a^2)\sqrt{2ax}$ **51.** $(3x^2+2x+5)\sqrt[3]{2x}$ **53.** $\left(\dfrac{a+6b-2}{6ab}\right)\sqrt{6ab}$ **55.** $\left(\dfrac{a+b-1}{ab}\right)\sqrt[3]{ab}$

## Exercises 8.5, Pages 302–303

**1.** $3\sqrt{2}$ **3.** $2\sqrt[3]{9}$ **5.** $\sqrt{2}$ **7.** $\sqrt[6]{500}$ **9.** $2\sqrt[6]{2}$ **11.** $9\sqrt[6]{3}$ **13.** $\sqrt[6]{3}$ **15.** $\dfrac{\sqrt[6]{72}}{3}$ **17.** $-3-\sqrt{5}$

**19.** 22 **21.** $42+17\sqrt{3}$ **23.** $2a^2-2a\sqrt{b}-4b$ **25.** $-34-6\sqrt{21}$ **27.** $2a+\sqrt{ab}-6b$ **29.** $7-4\sqrt{3}$

**31.** $17-4\sqrt{15}$ **33.** $a^2+2a\sqrt{b}+b$ **35.** $4a+12\sqrt{ab}+9b$ **37.** $\dfrac{3\sqrt{2}-2}{7}$ **39.** $\dfrac{16+7\sqrt{5}}{-11}$

**41.** $\dfrac{6-16\sqrt{2}}{-17}$ **43.** 2 **45.** $\dfrac{3+\sqrt{5}}{2}$ **47.** $\dfrac{a^2-a\sqrt{b}}{a^2-b}$

## Exercises 8.6, Pages 305–306

**1.** 53 **3.** 2 **5.** $-122$ **7.** 42 **9.** 9 **11.** 1 **13.** $1,-\dfrac{4}{3}$ **15.** $1,\dfrac{1}{8}$ **17.** $\dfrac{3}{2},-3$ **19.** $-6$

**21.** No solution **23.** 7 **25.** 9 **27.** $h=\dfrac{v_0^2-v^2}{2g}$ **29.** $l=\dfrac{P^2g}{4\pi^2}$ **31.** $C=\dfrac{(R_1+R_2)T^2}{4\pi^2R_2}$

**33.** $m=\dfrac{v^2Dr}{G(r-D)}$ **35.** $6,\sqrt{15},\sqrt{21}$

## Exercises 8.7, Pages 307–308

**1.** $3,-3,\sqrt{2},-\sqrt{2}$ **3.** $1,-1,\dfrac{1}{4},-\dfrac{1}{4}$ **5.** $-3,-4$ **7.** $0,2,3,-1$ **9.** 9 **11.** $-\dfrac{1}{3},-1$ **13.** $8,-64$

**15.** $\pm\sqrt{\dfrac{3\pm\sqrt{5}}{2}}$ or $\pm1.62;\pm0.618$

## Chapter 8 Review, Page 309

**1.** $\dfrac{3}{a^2}$ **2.** $\dfrac{1}{16a^4}$ **3.** $a^5$ **4.** $a^8$ **5.** $\dfrac{b^2c^6}{a^2}$ **6.** $\dfrac{a+1}{a^2}$ **7.** 7 **8.** 64 **9.** $\dfrac{1}{4}$ **10.** $\dfrac{y^{1/5}}{x^{2/5}}$ **11.** $\dfrac{1}{x^{5/12}}$

**12.** $x^{1/2}$ **13.** $\dfrac{2}{9}$ **14.** $4\sqrt{5}$ **15.** $6ab\sqrt{2b}$ **16.** $4x^2y^3\sqrt{5x}$ **17.** $4x\sqrt{3xy}$ **18.** $\dfrac{\sqrt{30}}{18}$ **19.** $\dfrac{3a\sqrt{35}}{7b^2}$

**20.** $5\sqrt[3]{2}$ **21.** $3a\sqrt[3]{4ab^2}$ **22.** $4ab^3\sqrt[3]{4a^2b}$ **23.** $\dfrac{\sqrt[3]{450ab}}{10b}$ **24.** $\dfrac{4\sqrt[3]{25}}{5}$ **25.** $\sqrt{3a}$ **26.** $\dfrac{\sqrt{3a}}{a}$

**27.** $\dfrac{\sqrt[3]{18a^2}}{a}$ **28.** $\sqrt[4]{5}$ **29.** $\sqrt[12]{10}$ **30.** $4\sqrt{3}$ **31.** $-\dfrac{17\sqrt{10}}{10}$ **32.** 0 **33.** $\dfrac{13}{6}\sqrt[3]{6}$ **34.** $4\sqrt{10}$ **35.** 3

**36.** $2\sqrt[6]{2}$    **37.** $-11 - 6\sqrt{3}$    **38.** $61 - 24\sqrt{5}$    **39.** $\dfrac{9 + 2\sqrt{3}}{-23}$    **40.** 62    **41.** $-27$    **42.** 9    **43.** $-9$

**44.** $3, -3, \dfrac{\sqrt{5}}{2} - \dfrac{\sqrt{5}}{2}$    **45.** $-8, 64$

# CHAPTER 9

**Exercises 9.1, Pages 314–315**

**1.**

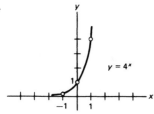

**3.**

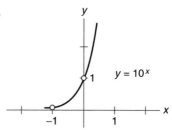

**5.**

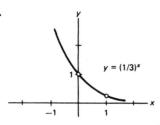

**7.**

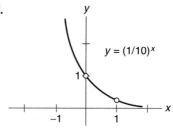

**9.**

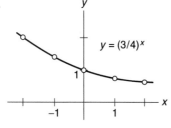

**11.**

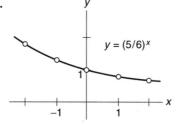

**13.**

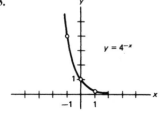

**15.**

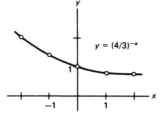

**17.**

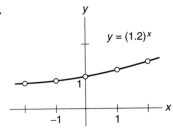

$y = (1.2)^x$

**19.**

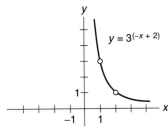

$y = 3^{(-x + 2)}$

**21.**

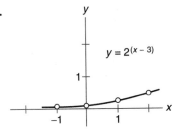

$y = 2^{(x - 3)}$

**23.**

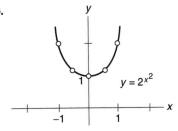

$y = 2^{x^2}$

**25.**

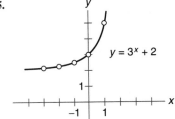

$y = 3^x + 2$

**27.** 1.38    **29.** 316,000    **31.** 24,600    **33.** 8.82    **35.** 5.20    **37.** 1.57    **39.** 2.28    **41.** $7.93 \times 10^{11}$ kWh
**43. (a)** \$1260    **(b)** \$1265    **(c)** \$1268    **(d)** \$1271    **45.** $170\overline{0}°R$

### Exercises 9.2, Pages 318–319

**1.** $\log_3 9 = 2$    **3.** $\log_5 125 = 3$    **5.** $\log_2 32 = 5$    **7.** $\log_9 3 = \frac{1}{2}$    **9.** $\log_5(\frac{1}{25}) = -2$    **11.** $\log_{10} 0.00001 = -5$
**13.** $5^2 = 25$    **15.** $2^4 = 16$    **17.** $25^{1/2} = 5$    **19.** $8^{1/3} = 2$    **21.** $2^{-2} = \frac{1}{4}$    **23.** $10^{-2} = 0.01$
**25.**

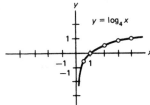

$y = \log_4 x$

**27.**

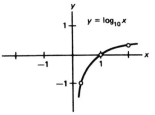

$y = \log_{10} x$

**29.**

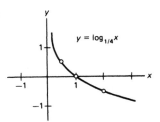

$y = \log_{1/4} x$

**31.** 64    **33.** $\frac{1}{2}$    **35.** 3    **37.** $\frac{1}{2}$    **39.** 5    **41.** 3    **43.** 144    **45.** 27    **47.** 4

## Exercises 9.3, Pages 323–324

**1.** $\log_2 5 + 3\log_2 x + \log_2 y$   **3.** $\log_{10} 2 + 2\log_{10} x - 3\log_{10} y - \log_{10} z$   **5.** $3\log_b y + \frac{1}{2}\log_b x - 2\log_b z$
**7.** $\frac{2}{3}\log_b x - \frac{1}{3}\log_b y$   **9.** $\frac{1}{2}\log_2 y - \log_2 x - \frac{1}{2}\log_2 z$   **11.** $3\log_b z + \frac{1}{2}\log_b x - \frac{1}{3}\log_b y$
**13.** $2\log_b x + \log_b(x+1) - \frac{1}{2}\log_b(x+2)$   **15.** $\log_b xy^2$   **17.** $\log_b \dfrac{xy^2}{z^3}$   **19.** $\log_3 \dfrac{x\sqrt[3]{y}}{\sqrt{z}}$
**21.** $\log_{10} \dfrac{x^2}{(x+1)\sqrt{x-3}}$   **23.** $\log_b \dfrac{x^5\sqrt[3]{x-1}}{x+2}$   **25.** $\log_{10} \dfrac{x(x-1)^2}{\sqrt[3]{(x+2)(x-5)}}$   **27.** 3   **29.** 2   **31.** 3
**33.** $-2$   **35.** $-3$   **37.** 0   **39.** 5   **41.** 36   **43.** $\frac{1}{25}$   **45.** 6   **47.** 8

## Exercises 9.4, Pages 327–328

**1.** 1.833   **3.** 1.653   **5.** $-0.8477$   **7.** $-2.207$   **9.** 2.906   **11.** 0.9661   **13.** 25.4   **15.** 705   **17.** 1.90
**19.** 0.0248   **21.** $2.32 \times 10^{-4}$   **23.** 0.000652   **25.** 3.64   **27.** 2.94   **29.** 0.389   **31.** 1.12   **33.** $-0.574$
**35.** 0.590   **37.** 3.56   **39.** 4.82   **41.** 6.0   **43.** 6.5   **45.** 7.3   **47.** 100 dB   **49.** 60 dB   **51.** 14 dB
**53.** $-1.2$ dB

## Exercises 9.5, Pages 332–334

**1.** 7.39   **3.** 403   **5.** 0.135   **7.** 0.00248   **9.** 33.1   **11.** 1.16   **13.** 0.0821   **15.** 0.923   **17.** 1.95
**19.** 0.717   **21.** 4.025   **23.** 6.006   **25.** 1.459   **27.** $-5.146$   **29.** 0   **31.** 1.61   **33.** 0.236
**35.** 2,550,000   **37.** $8.08 \times 10^{11}$ kWh   **39.** \$5742   **41.** 130,000   **43.** 3700   **45.** \$4500   **47.** 6.02 g
**49.** 31 years   **51.** $7\overline{0}0$ years   **53.** 3400 years   **55.** 9.18 V   **57.** $-97.4$ V   **59.** 0.25 ft/min   **61.** 4.03
**63.** 5.60   **65.** 3.68   **67.** 5.12

## Exercises 9.6, Pages 337–338

**1.** 2.26   **3.** $-5.45$   **5.** $-0.710$   **7.** $-3.15$   **9.** $-0.1411$   **11.** $-0.2886$   **13.** 3.135   **15.** 0.8356
**17.** 1.851   **19.** $-2.248$   **21.** $-0.6936$   **23.** 3.164   **25.** 2.641   **27.** $-0.7821$   **29.** 0.07201
**31.** $\pm 2.529$   **33.** $-1.386$   **35.** 0.5754   **37.** $-3.219$   **39.** 4.512   **41.** 0.4621   **43.** 2.006
**45.** $-0.3087$   **47.** 0.3890   **49.** $-0.5432$   **51.** 34.7 $\mu$s   **53.** 209 $\mu$s   **55.** 51.5 V

## Exercises 9.7, Pages 339–340

**1.** 5   **3.** No solution   **5.** 7   **7.** 7   **9.** 6   **11.** 8   **13.** 3   **15.** 6   **17.** 10   **19.** 9   **21.** 4.162
**23.** 899   **25.** 8.437   **27.** 0.7315   **29.** 3.25   **31.** 7.361   **33.** 0.9283   **35.** $2e$, or 5.437

## Exercises 9.8, Pages 342–344

**1.** (a) 7.9 years   (b) 14.8 years   **3.** 4.9 s   **5.** $3.2 \times 10^{-4}$ M/L   **7.** $3.2 \times 10^{-10}$ M/L   **9.** $2.0 \times 10^{-3}$ M/L
**11.** $3.2 \times 10^{-15}$ W/cm$^2$   **13.** $1.0 \times 10^{-4}$ W/cm$^2$   **15.** $1.0 \times 10^{-13}$ W/cm$^2$   **17.** (a) 7.2 years   (b) 17 years
**19.** $1\overline{0}$ s   **21.** $1.94 \times 10^{-3}$ ms   **23.** 11.1 $\mu$s   **25.** 2.1   **27.** 49.5 mg   **29.** 40.7 mg

## Exercises 9.9, Pages 350–351

**1.** $y = 3.5x + 15$   **3.** $y = 18.2x - 60$

**5.**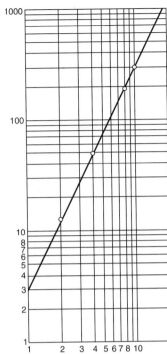

**7.**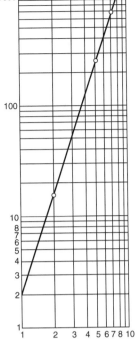

**9.** $y = 2x^2$    **11.** $y = 12x^2$

**13.**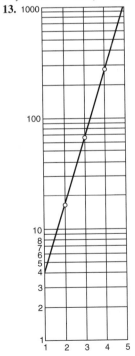

**15.**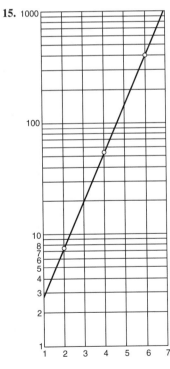

**17.**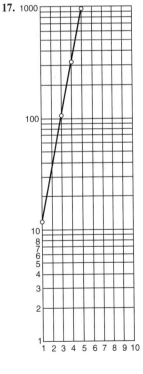

**19.** $y = 2.7^x$    **21.** $y = 0.9^x$    **23.** $y = 2.1(3^x)$

## Chapter 9 Review, Pages 352–354

**1.**

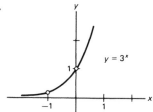

**2.**

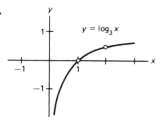

**3.** $\log_2 16 = 4$     **4.** $\log_{10} 0.001 = -3$     **5.** $10^{0.8686} = 7.389$     **6.** $4^{-2} = \frac{1}{16}$     **7.** 81     **8.** 2     **9.** 5

**10.** $\log_4 6 + 2\log_4 x + \log_4 y$     **11.** $\log_3 5 + \log_3 x + \frac{1}{2}\log_3 y - 3\log_3 z$     **12.** $2\log x + 3\log(x+1) - \frac{1}{2}\log(x-4)$

**13.** $3\ln x + 3\ln(x-1) - \frac{1}{2}\ln(x+1)$     **14.** $\log_2 \dfrac{xy^3}{z^2}$     **15.** $\log \dfrac{\sqrt{x+1}}{(x-2)^3}$     **16.** $\ln \dfrac{x^4}{(x+1)^5(x+2)}$

**17.** $\ln \dfrac{\sqrt{x(x+2)}}{(x-5)^2}$     **18.** 3     **19.** $x^2$     **20.** 2     **21.** $x$     **22.** 2.823     **23.** $-1.393$     **24.** 4.159     **25.** 1180

**26.** 0.00372     **27.** 0.565     **28.** 4.277     **29.** 6.043     **30.** $-6.293$     **31.** 3.72     **32.** 31.5     **33.** 0.787

**34.** 2.16     **35.** $-1.08$     **36.** 0.626     **37.** 0.485     **38.** 2.20     **39.** 96     **40.** 14.5     **41.** 4     **42.** 1

**43.** 0.5     **44.** 4.48     **45.** 110,000     **46.** 8.7 years     **47.** 518     **48.** $y = -0.83x + 61$     **49.** $y = 3.1x + 60$

**50.**

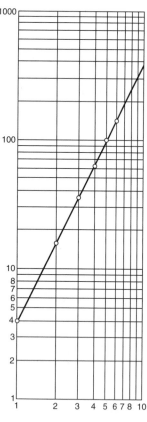

**51.**

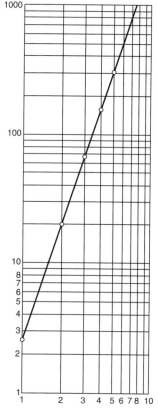

**52.** $y = 3.6x^2$     **53.** $y = 2.1x^2$

**54.**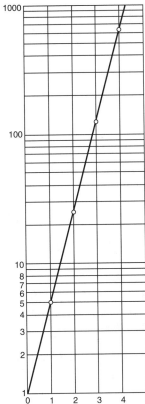

**55.**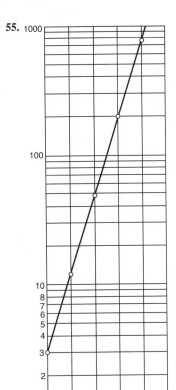

**56.** $y = 3.9^x$  **57.** $y = 2.5(2^x)$

**CHAPTER 10**

**Exercises 10.1, Pages 365–366**

**1.**
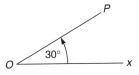

**3.**
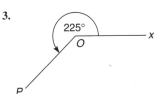

**5.**
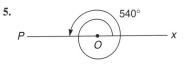

**7.** $420°, -300°$  **9.** $274°, -446°$  **11.** $585°, -135°$  **13.** $52°, -308°$

**15.**

**17.**
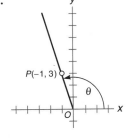

**19.**
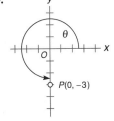

| | sin θ | cos θ | tan θ | cot θ | sec θ | csc θ |
|---|---|---|---|---|---|---|
| **21.** | $-\frac{4}{5}$ | $\frac{3}{5}$ | $-\frac{4}{3}$ | $-\frac{3}{4}$ | $\frac{5}{3}$ | $-\frac{5}{4}$ |
| **23.** | $\frac{1}{\sqrt{2}}$ | $\frac{1}{\sqrt{2}}$ | $1$ | $1$ | $\sqrt{2}$ | $\sqrt{2}$ |
| **25.** | $-\frac{\sqrt{3}}{2}$ | $-\frac{1}{2}$ | $\sqrt{3}$ | $\frac{1}{\sqrt{3}}$ | $-2$ | $-\frac{2}{\sqrt{3}}$ |
| **27.** | $\frac{5}{\sqrt{41}}$ | $-\frac{4}{\sqrt{41}}$ | $-\frac{5}{4}$ | $-\frac{4}{5}$ | $-\frac{\sqrt{41}}{4}$ | $\frac{\sqrt{41}}{5}$ |
| **29.** | $-1$ | $0$ | Undefined | $0$ | Undefined | $-1$ |
| **31.** | $\frac{3}{5}$ | $\frac{4}{5}$ | $\frac{3}{4}$ | $\frac{4}{3}$ | $\frac{5}{4}$ | $\frac{5}{3}$ |

## Exercises 10.2, Pages 371–372

**1.** 60°   **3.** 73°   **5.** 66.6°   **7.** 63.3°   **9.** 77°56′   **11.** 60°   **13.** 0.8121   **15.** −0.1817
**17.** −0.07498   **19.** −0.8039   **21.** −1.022   **23.** −1.351   **25.** 21.6°, 158.4°   **27.** 35.9°, 215.9°
**29.** 96.0°, 264.0°   **31.** 245.7°, 294.3°   **33.** 124.6°, 235.4°   **35.** 15.1°, 195.1°   **37.** 229.0°, 311.0°
**39.** 65.1°, 294.9°   **41.** 44.4°, 315.6°

## Exercises 10.3, Pages 378–379

**1.** $\frac{3\pi}{4}$   **3.** $\frac{\pi}{2}$   **5.** $-\frac{5\pi}{12}$   **7.** $7\pi$   **9.** 315°   **11.** 855°   **13.** 1620°   **15.** 212°   **17.** $\frac{\pi}{4}$   **19.** $\frac{\pi}{4}$
**21.** $\frac{\pi}{12}$   **23.** $\frac{2\pi}{5}$   **25.** $\frac{\sqrt{2}}{2}$   **27.** $\sqrt{3}$   **29.** $-\sqrt{2}$   **31.** $\frac{2\pi}{3}, \frac{4\pi}{3}$   **33.** $\frac{\pi}{3}, \frac{5\pi}{3}$   **35.** $\frac{\pi}{4}, \frac{5\pi}{4}$
**37.** 0.7174   **39.** 2.572   **41.** 0.5403   **43.** −0.9969   **45.** −0.1507   **47.** 1.289   **49.** 0.7071
**51.** −3.078   **53.** 0.9914   **55.** 1.395, 1.747   **57.** 1.035, 4.177   **59.** 2.425, 3.858   **61.** 0.8060, 5.477
**63.** 1.782, 4.923   **65.** 3.628, 5.797   **67.** 2.618, 3.665   **69.** 0.5236, 2.618   **71.** 2.618, 3.665

## Exercises 10.4, Pages 382–383

**1.** 25.1 in.   **3.** 151 in$^2$   **5.** 47.7°   **7.** 4.19 m$^2$   **9.** 0.524 m   **11.** 344 ft-lb   **13. (a)** 44.0 rad/s
**(b)** 44$\overline{0}$ rad or 70.0 rev   **(c)** 77.0 ft/s   **15.** 4.2 rad/s$^2$   **17.** 1040 mi/h   **19. (a)** 2.62 m/s   **(b)** 2.62 m/s
**(c)** 32.8 rad/s

## Chapter 10 Review, Pages 385–386

**1.** $\frac{3}{5}, \frac{4}{5}, \frac{3}{4}, \frac{4}{3}, \frac{5}{4}, \frac{5}{3}$   **2.** $-\frac{1}{2}, -\frac{\sqrt{3}}{2}, \frac{1}{\sqrt{3}}, \sqrt{3}, -\frac{2}{\sqrt{3}}, -2$   **3.** 0, −1, 0, undefined, −1, undefined   **4.** 45°
**5.** 28°20′   **6.** 55°   **7.** 10°   **8.** −0.9011   **9.** −0.4142   **10.** −6.611   **11.** 0.4602   **12.** 1.107
**13.** −4.810   **14.** 20.2°, 159.8°   **15.** 123.3°, 236.7°   **16.** 133.6°, 313.6°   **17.** 59.3°, 300.7°
**18.** 151.7°, 331.7°   **19.** 47.7°, 132.3°   **20.** $\frac{2\pi}{5}$   **21.** $\frac{7\pi}{4}$   **22.** 150°   **23.** 135°   **24.** $\frac{\pi}{3}$   **25.** $\frac{2\pi}{5}$
**26.** $-\frac{\sqrt{3}}{2}$   **27.** $-\sqrt{3}$   **28.** $\sqrt{2}$   **29.** $-\frac{1}{2}$   **30.** −1   **31.** $\frac{1}{2}$   **32.** $\frac{3\pi}{4}, \frac{5\pi}{4}$   **33.** $\frac{3\pi}{4}, \frac{7\pi}{4}$
**34.** $\frac{\pi}{3}, \frac{2\pi}{3}$   **35.** $\frac{4\pi}{3}, \frac{5\pi}{3}$   **36.** $\frac{2\pi}{3}, \frac{4\pi}{3}$   **37.** $\frac{\pi}{6}, \frac{7\pi}{6}$   **38.** 0.9975   **39.** 0.9689   **40.** 0.5774
**41.** −0.8660   **42.** 1.371, 4.912   **43.** 3.845, 5.579   **44.** 1.017, 4.159   **45.** 1.935, 4.348
**46.** 8.80 in., 39.6 in$^2$   **47.** 427 ft-lb   **48. (a)** 66.0 rad/s   **(b)** 33$\overline{0}$ rad   **(c)** 99.0 ft/s

## CHAPTER 11

### Exercises 11.1, Page 393

**1.** $B = 38.0°$, $C = 73.0°$, $c = 25.6$ m   **3.** $C = 52.5°$, $A = 66.1°$, $a = 129$ cm   **5.** $C = 47.6°$, $a = 223$ ft, $c = 196$ ft
**7.** $B = 48.5°$, $C = 16.5°$, $c = 1840$ m   **9.** $A = 67.07°$, $B = 40.35°$, $a = 40.72$ cm
**11.** $A = 131.94°$, $a = 89{,}460$ mi, $b = 57{,}870$ mi   **13.** $B = 2\overline{0}°$, $C = 135°$, $c = 84$ cm
**15.** $A = 6°$, $B = 166°$, $b = 28$ m   **17.** $B = 37°28′35″$, $C = 91° 31′8″$, $b = 2506$ ft
**19.** $C = 63°10″$, $a = 2.663$ mi, $b = 5.120$ mi

## Exercises 11.2, Page 398

**1.** $B = 27.3°$, $C = 115.7°$, $c = 32.2$ cm   **3.** $A = 29.6°$, $B = 123.9°$, $b = 79.4$ km; or $A = 150.4°$, $B = 3.1°$, $b = 5.18$ km
**5.** $B = 34.3°$, $C = 74.2°$, $b = 2.05$ m; or $B = 2.7°$, $C = 105.8°$, $b = 0.171$ m   **7.** $A = 44.6°$, $C = 30.4°$, $c = 8.64$ mi
**9.** $A = 35°$, $B = 127°$, $b = 62$ mi; or $A = 145°$, $B = 17°$, $b = 23$ mi   **11.** No triangle
**13.** $A = 157°$, $C = 15°$, $a = 1300$ m; or $A = 7°$, $C = 165°$, $a = 390$ m
**15.** $A = 59.34°$, $C = 79.16°$, $c = 21.12$ km; or $A = 120.66°$, $C = 17.84°$, $c = 6.588$ km   **17.** No triangle
**19.** $A = 6.06°$, $B = 165.19°$, $b = 150.1$ m
**21.** $A = 107°2'21''$, $C = 43°41'2''$, $a = 421.5$ m; or $A = 14°24'25''$, $C = 136°18'58''$, $a = 109.7$ m
**23.** $A = 27°10'25''$, $B = 127°48'50''$, $b = 907.3$ ft; or $A = 152°49'35''$, $B = 2°9'40''$, $b = 43.31$ ft

## Exercises 11.3, Page 402

**1.** $B = 47.8°$, $C = 72.2°$, $a = 22.8$ m   **3.** $A = 27.0°$, $B = 44.0°$, $c = 292$ km   **5.** $A = 44.6°$, $B = 51.2°$, $C = 84.2°$
**7.** $A = 32.1°$, $B = 104.6°$, $C = 43.3°$   **9.** $B = 85°$, $C = 50°$, $a = 36$ m   **11.** $A = 84°$, $B = 53°$, $C = 43°$
**13.** $A = 19°$, $B = 26°$, $c = 78$ ft   **15.** $A = 148.80°$, $C = 11.95°$, $b = 3064$ m
**17.** $A = 40.11°$, $B = 31.14°$, $c = 595.2$ mm   **19.** $A = 113.43°$, $B = 27.84°$, $C = 38.73°$
**21.** $a = 1465$ m, $B = 44°17'56''$, $C = 63°23'52''$   **23.** $A = C = 37°48'52''$, $B = 104°22'16''$

## Exercises 11.4, Pages 403–408

**1.** $AC = 10.9$ m, $BC = 6.02$ m   **3.** $63.0$ m   **5.** $240$ m   **7.** $5.13$ mi from tower $A$, $3.34$ mi from tower $B$
**9.** $98$ mi   **11.** $290$ ft   **13.** (a) $4\overline{0}0$ m   (b) $140$ m   **15.** $321$ ft or $80.2$ ft   **17.** $48.9$ ft
**19.** (a) $3.86$ m   (b) $55.2°$   (c) $6.56$ m   (d) $35.4°$   (e) $60.4°$   **21.** $24.7$ ft   **23.** $79.4$ ft   **25.** $45.4$ m

## Exercises 11.5, Page 414

**1.** $1\overline{0}$ km at $90°$   **3.** $85$ mi/h at $328°$   **5.** $73$ mi at $281°$   **7.** $41$ mi at $126°$   **9.** $4.0$ km at $328°$
**11.** $79$ mi at $74°$ ($16°$ east of north)   **13.** $270$ mi/h at $24\overline{0}°$ ($3\overline{0}°$ west of south)

## Exercises 11.6, Pages 416–417

**1.** $76.8$ km/h at $238.2°$   **3.** $10.4$ km at $89.6°$   **5.** $108$ mi/h at $155.4°$   **7.** $1180$ m at $85.4°$   **9.** $4.05$ km at $328.1°$
**11.** $278$ km at $99.4°$ ($9.4°$ west of north)   **13.** $40.0$ km due east   **15.** $187$ mi/h at $249.6°$ ($20.4°$ west of south)

## Exercises 11.7, Page 424

**1.** $1.59$ km, $18.1$ km   **3.** $0$, $-135$ mi/h   **5.** $-2380$ ft, $1240$ ft   **7.** $10.8$ m at $36.3°$   **9.** $8.10$ m/s at $305.5°$
**11.** $14.7$ km at $180.0°$   **13.** $|\mathbf{R}| = 85$ Ω, $|\mathbf{X}_L| = 31$ Ω   **15.** $280$ Ω, $3\overline{0}°$   **17.** $661$ ft at $60.6°$
**19.** $30.1$ mi/h at $20.2°$   **21.** $464$ km at $34.8°$   **23.** $10{,}600$ m at $30.3°$   **25.** $36.0$ mi/h at $78.2°$
**27.** $224$ mi at $211.7°$ ($31.7°$ south of west)

## Exercises 11.8, Pages 429–431

**1.** $301$ mi at $258.6°$ ($11.4°$ west of south)   **3.** $312$ km at $234.1°$ ($54.1°$ south of west)   **5.** $25.0$ mi   **7.** $127.1°$
**9.** $45.0$ km   **11.** (a) $12$ mi/h   (b) $9$ mi/h   **13.** $177$ mi/h at $171.9°$ ($8.1°$ north of west)
**15.** $289$ km/h at $116.9°$ ($26.9°$ west of north)   **17.** $175$ mi/h at $148.0°$ ($32.0°$ north of west)   **19.** $-14.3$ lb, $20.5$ lb
**21.** $3960$ lb at $333.8°$ ($26.4°$ south of east)   **23.** $496$ lb at $216.4°$ ($36.4°$ south of west)   **25.** $193$ lb at $70.9°$ from $\mathbf{R}$
**27.** $85.4$ lb at $159.4°$   **29.** $724$ lb at $45.0°$   **31.** $999$ lb   **33.** $\mathbf{T}_1 = 35\overline{0}0$ lb, $\mathbf{T}_2 = 3850$ lb

## Chapter 11 Review, Pages 434–436

**1.** $A = 59.3°$, $C = 49.3°$, $a = 371$ ft   **2.** $B = 117.2°$, $c = 22.6$ m, $b = 32.9$ m
**3.** $B = 106.3°$, $C = 58.2°$, $a = 65.7$ cm   **4.** $A = 37.3°$, $B = 86.4°$, $C = 56.3°$   **5.** No triangle
**6.** $B = 35°$, $C = 116°$, $c = 76$ cm; or $B = 145°$, $C = 6°$, $c = 8.8$ cm
**7.** $A = 32°$, $B = 112°$, $c = 2300$ ft   **8.** $A = 69°$, $B = 53°$, $b = 390$ m; or $A = 111°$, $B = 11°$, $b = 92$ m
**9.** $A = 41.62°$, $C = 33.23°$, $b = 335.9$ m   **10.** $B = 63.62°$, $C = 41.63°$, $b = 20.60$ cm
**11.** $A = 20.12°$, $C = 141.63°$, $c = 3023$ ft; or $A = 159.88°$, $C = 1.87°$, $c = 158.9$ ft   **12.** No triangle

**13.** $C = 63°37'7''$, $b = 3426$ ft, $c = 3308$ ft     **14.** $A = 28°23'18''$, $B = 122°11'26''$, $c = 14{,}010$ ft     **15.** 4288 ft
**16. (a)** 1122 ft   **(b)** 1103 ft     **17.** 8.236 in.     **18. (a)** 2.939 in.   **(b)** 4.755 in.
**19. (a)** 11.5 m   **(b)** 8.62 m   **(c)** 5.77 m   **(d)** 6.41 m     **20.** 92 km/h at $10\overline{0}°$     **21.** 215 mi at 160.4°
**22.** 44.7 mi/h at 142.0°     **23.** $-182$ km, 182 km     **24.** 23.0 mi/h, $-35.4$ mi/h     **25.** 110 N, $-110$ N
**26.** $-1\overline{0}0$ lb, 43 lb     **27.** 36.1 N at 300.8°     **28.** 74.6 mi/h at 135.9°     **29.** 3870 N at 308.2°
**30.** 325 N at 162.0°     **31.** 63.0 km at 145.4°     **32.** 630 km/h at 26° (64° east of north)
**33.** 67.7 mi at 98.6° (8.6° west of north)     **34.** 640 lb, 380 lb

## CHAPTER 12

### Exercises 12.1, Page 443

**1.** 2; $2\pi$

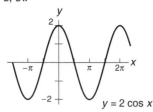

$y = 2 \cos x$

**3.** 3; $2\pi$

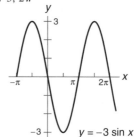

$y = -3 \sin x$

**5.** 1, $\dfrac{2\pi}{3}$

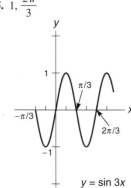

$y = \sin 3x$

**7.** 1, $\pi$

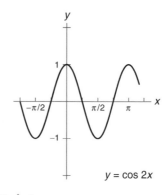

$y = \cos 2x$

**9.** 2, $\dfrac{\pi}{2}$

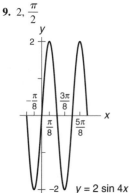

$y = 2 \sin 4x$

**11.** $\frac{5}{2}$, $4\pi$

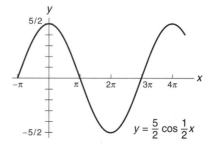

$y = \dfrac{5}{2} \cos \dfrac{1}{2}x$

**13.** $\frac{1}{2}$, $3\pi$

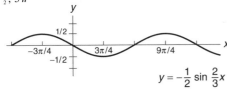

$$y = -\frac{1}{2}\sin\frac{2}{3}x$$

**15.** $2$; $\frac{2}{3}$

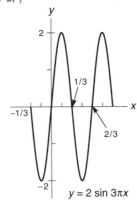

$y = 2\sin 3\pi x$

**17.** $3$; $2$

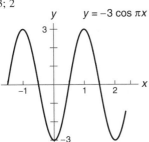

$y = -3\cos \pi x$

**19.** $6.5$; $\frac{1}{60}$

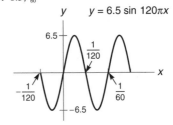

$y = 6.5\sin 120\pi x$

**21.** $40$; $\dfrac{\pi}{30}$

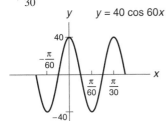

$y = 40\cos 60x$

**23.** $60$; $\frac{1}{40}$

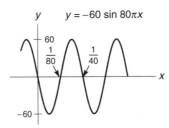

$y = -60\sin 80\pi x$

**Exercises 12.2, Page 446**

**1.** $1$, $2\pi$, $\dfrac{\pi}{3}$

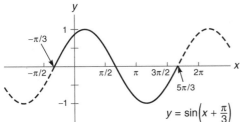

$$y = \sin\left(x + \frac{\pi}{3}\right)$$

**3.** $2$, $2\pi$, $-\dfrac{\pi}{6}$

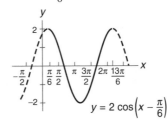

$$y = 2\cos\left(x - \frac{\pi}{6}\right)$$

**5.** $1, \dfrac{2\pi}{3}, -\dfrac{\pi}{3}$

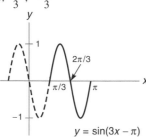

$y = \sin(3x - \pi)$

**7.** $1, \dfrac{\pi}{2}, \dfrac{\pi}{4}$

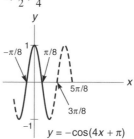

$y = -\cos(4x + \pi)$

**9.** $3, 4\pi, -\dfrac{\pi}{2}$

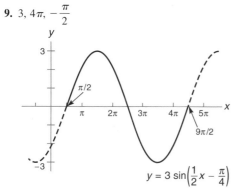

$y = 3\sin\left(\dfrac{1}{2}x - \dfrac{\pi}{4}\right)$

**11.** $2, \dfrac{3\pi}{2}, \dfrac{\pi}{4}$

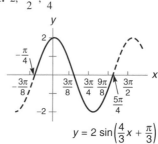

$y = 2\sin\left(\dfrac{4}{3}x + \dfrac{\pi}{3}\right)$

**13.** $3, 2, 1$

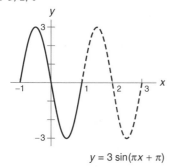

$y = 3\sin(\pi x + \pi)$

**15.** $2, 4, -1$

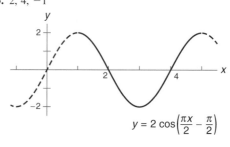

$y = 2\cos\left(\dfrac{\pi x}{2} - \dfrac{\pi}{2}\right)$

**17.** $40, \dfrac{\pi}{30}, \dfrac{\pi}{180}$

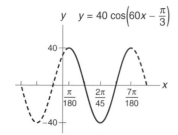

$y = 40\cos\left(60x - \dfrac{\pi}{3}\right)$

**19.** $120, \dfrac{1}{20}, \dfrac{1}{80}$

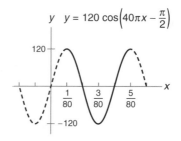

$y = 120\cos\left(40\pi x - \dfrac{\pi}{2}\right)$

**21.** $7.5, \dfrac{\pi}{110}, \dfrac{\pi}{220}$

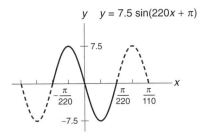

$y = 7.5 \sin(220x + \pi)$

**23.** $20, \dfrac{1}{60}, \dfrac{1}{30}$

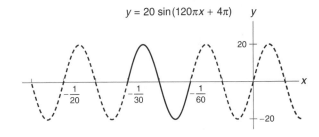

$y = 20 \sin(120\pi x + 4\pi)$

**Exercises 12.3, Pages 449–450**

**1.** $\dfrac{\pi}{3}$

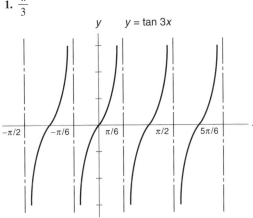

$y = \tan 3x$

**3.** $2\pi$

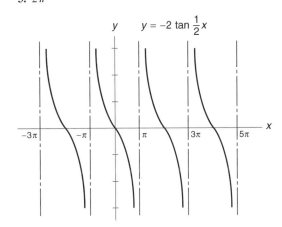

$y = -2 \tan \dfrac{1}{2}x$

**5.** $\dfrac{\pi}{6}$

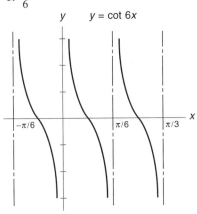

$y = \cot 6x$

**7.** $2\pi$

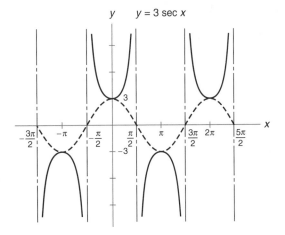

$y = 3 \sec x$

**9.** $2\pi$

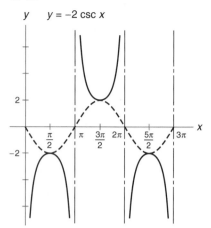

**11.** $\dfrac{\pi}{3}$

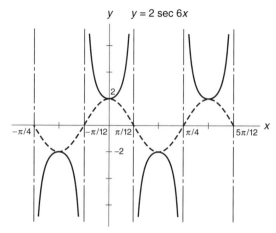

**13.** $4\pi$

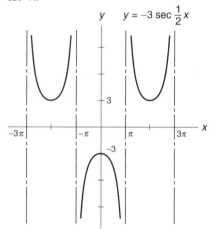

**15.** $\pi$

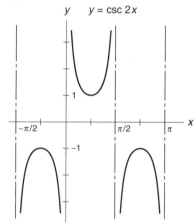

**17.**

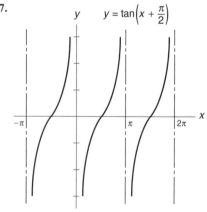

**19.**

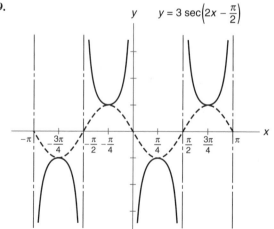

**Exercises 12.4, Pages 451–452**

**1.**

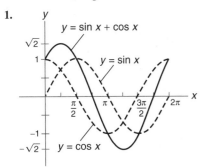

**3.**

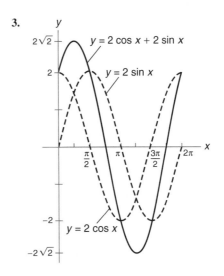

**5.**

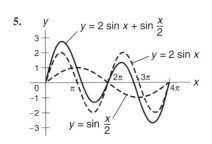

**7.**

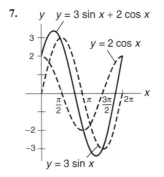

**9.**

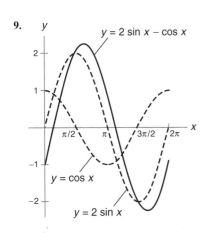

**11.**

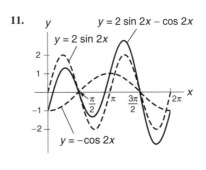

**13.**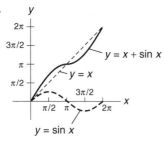

**15.**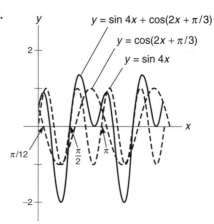

**Exercises 12.5, Pages 455–456**

**1.** $y = 8 \sin 10\pi t$

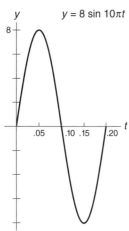

**3.** $i = 15 \sin\left(120\pi t + \dfrac{\pi}{4}\right)$

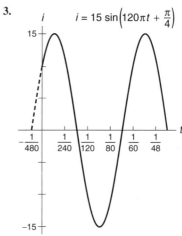

**5.** $e = 110 \cos 120\pi t$

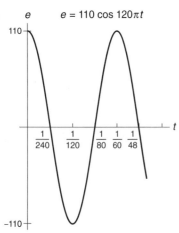

**7.** $y = 10 \sin\left(\dfrac{5\pi t}{2} - \dfrac{\pi}{2}\right)$ or $y = -10 \cos \dfrac{5\pi t}{2}$

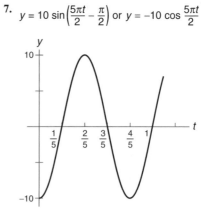

**9.**

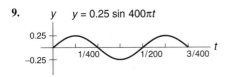

$y = 0.25 \sin 400\pi t$

**Chapter 12 Review, Page 457**

**1.** $4, \dfrac{\pi}{3}$

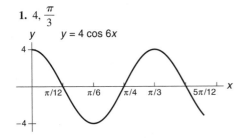

$y = 4 \cos 6x$

**2.** $2, 6\pi$

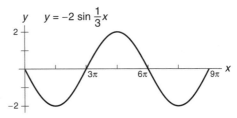

$y = -2 \sin \dfrac{1}{3}x$

**3.** $3, 1$

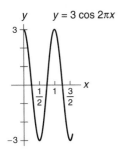

$y = 3 \cos 2\pi x$

**4.** $3, 2\pi, -\dfrac{\pi}{4}$

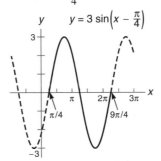

$y = 3 \sin\left(x - \dfrac{\pi}{4}\right)$

**5.** $1, \pi, \dfrac{\pi}{3}$

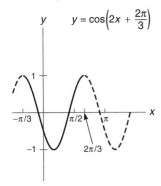

$y = \cos\left(2x + \dfrac{2\pi}{3}\right)$

**6.** $4, 2, \frac{1}{2}$

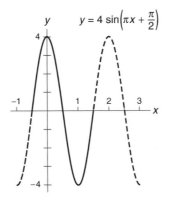

$y = 4 \sin\left(\pi x + \dfrac{\pi}{2}\right)$

**7.** $\dfrac{\pi}{5}$

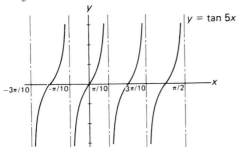

$y = \tan 5x$

**8.** $\dfrac{\pi}{3}$

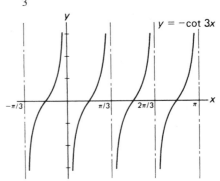

$y = -\cot 3x$

**9.** $\dfrac{\pi}{2}$

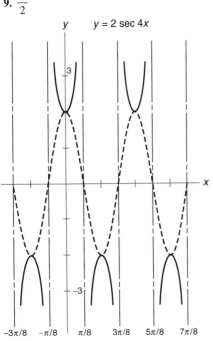

$y = 2 \sec 4x$

**10.**

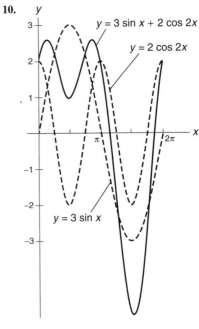

$y = 3 \sin x + 2 \cos 2x$

$y = 2 \cos 2x$

$y = 3 \sin x$

**11.**

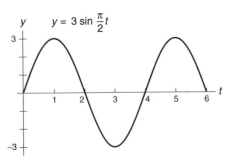

$y = 3 \sin \dfrac{\pi}{2}t$

# CHAPTER 13

## Exercises 13.2, Page 471

**21.** $\cos(M + N)$   **23.** $\sin 4\theta$   **25.** $\cos \theta$   **27.** $\tan 5\theta$   **29.** $2 \sin \theta \cos \phi$   **31.** $\sin(A + 2B)$

## Exercises 13.3, Pages 477–478

**1.** $\sin \dfrac{x}{2}$   **3.** $\cos 6x$   **5.** $\cos \dfrac{\theta}{8}$   **7.** $\cos \dfrac{x}{3}$   **9.** $10 \sin 8\theta$   **11.** $\cos 125°$   **13.** $4 \cos 2\theta$   **15.** $50 \sin 60t$

**17.** $\frac{24}{25}$   **19.** $\frac{120}{119}$   **21.** $-\sqrt{\dfrac{3 - \sqrt{5}}{6}}$

## Exercises 13.4, Page 481

**1.** $0°$   **3.** $45°, 225°$   **5.** $30°, 150°, 210°, 330°$   **7.** $45°, 225°$
**9.** $10°, 50°, 70°, 110°, 130°, 170°, 190°, 230°, 250°, 290°, 310°, 350°$   **11.** $90°, 270°, 210°, 330°$
**13.** $0°, 180°, 270°$   **15.** $21.5°, 158.5°$   **17.** $\dfrac{\pi}{2}, \dfrac{7\pi}{6}, \dfrac{11\pi}{6}$   **19.** $0, \pi$   **21.** $\dfrac{\pi}{4}, \dfrac{3\pi}{4}, \dfrac{5\pi}{4}, \dfrac{7\pi}{4}$
**23.** $\dfrac{\pi}{4}, \dfrac{5\pi}{4}, \dfrac{7\pi}{12}, \dfrac{11\pi}{12}, \dfrac{19\pi}{12}, \dfrac{23\pi}{12}$   **25.** $\dfrac{\pi}{3}, \dfrac{2\pi}{3}, \dfrac{4\pi}{3}, \dfrac{5\pi}{3}$   **27.** $\dfrac{\pi}{24}, \dfrac{5\pi}{24}, \dfrac{13\pi}{24}, \dfrac{17\pi}{24}, \dfrac{25\pi}{24}, \dfrac{29\pi}{24}, \dfrac{37\pi}{24}, \dfrac{41\pi}{24}$
**29.** $0, \dfrac{4\pi}{3}$   **31.** $\dfrac{2\pi}{3}, \pi$   **33.** $0$

## Exercises 13.5, Pages 485–486

**1.** $y$ is the angle whose sine is $x$.   **3.** $y$ is the angle whose cotangent is $4x$.
**5.** $y$ is three times the angle whose cosecant is $\frac{1}{2}x$.   **7.** $\dfrac{\pi}{3}, \dfrac{2\pi}{3}$   **9.** $\dfrac{\pi}{4}, \dfrac{5\pi}{4}$   **11.** $\dfrac{5\pi}{6}, \dfrac{7\pi}{6}$   **13.** $\dfrac{2\pi}{3}, \dfrac{4\pi}{3}$
**15.** $135°, 225°$   **17.** $225°, 315°$   **19.** $0°$   **21.** $63°, 243°$   **23.** $30°$   **25.** $120°$   **27.** $45°$   **29.** $150°$
**31.** $\dfrac{2\pi}{3}$   **33.** $\dfrac{5\pi}{6}$   **35.** $\dfrac{3\pi}{2}$   **37.** $\pi$   **39.** $45° + n \cdot 360°, 135° + n \cdot 360°$, for every integer $n$
**41.** $30° + n \cdot 360°, 210° + n \cdot 360°$, for every integer $n$   **43.** $30° + n \cdot 360°, 330° + n \cdot 360°$, for every integer $n$
**45.** $48° + n \cdot 360°, 132° + n \cdot 360°$, for every integer $n$   **47.** $\dfrac{\pi}{3} + 2n\pi, \dfrac{5\pi}{3} + 2n\pi$, for every integer $n$

**49.** $0 + 2n\pi$, for every integer $n$   **51.** $\dfrac{\pi}{3} + 2n\pi, \dfrac{5\pi}{3} + 2n\pi$, for every integer $n$

**53.** $4.17 + 2n\pi, 5.25 + 2n\pi$, for every integer $n$   **55.** $x = \frac{1}{3} \arcsin y$   **57.** $x = \arccos \dfrac{y}{4}$   **59.** $x = 2 \arctan \dfrac{y}{5}$

**61.** $x = 4 \operatorname{arccot} \dfrac{2y}{3}$   **63.** $x = 1 + \arcsin \dfrac{y}{3}$   **65.** $x = -\frac{1}{3} + \frac{1}{3} \arccos 2y$

## Exercises 13.6, Page 492

**1.** $\dfrac{\pi}{3}$   **3.** $-\dfrac{\pi}{6}$   **5.** $\dfrac{5\pi}{6}$   **7.** $\dfrac{2\pi}{3}$   **9.** $\dfrac{\pi}{4}$   **11.** $\dfrac{\pi}{3}$   **13.** $\dfrac{\pi}{4}$   **15.** $-\dfrac{\pi}{3}$   **17.** $1.234$   **19.** $2.554$

**21.** $1.889$   **23.** $-1.054$   **25.** $\frac{1}{2}$   **27.** $\dfrac{1}{\sqrt{2}}$   **29.** $0$   **31.** $\dfrac{\sqrt{3}}{2}$   **33.** $0.8$   **35.** $-0.1579$   **37.** $\sqrt{1 - x^2}$

**39.** $\dfrac{\sqrt{x^2 - 1}}{x}$   **41.** $\dfrac{1}{x}$   **43.** $x$   **45.** $\sqrt{1 - 4x^2}$   **47.** $2x\sqrt{1 - x^2}$

**49.**

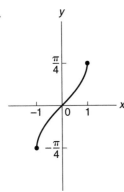

$y = \text{Arcsin } 2x$

**51.**

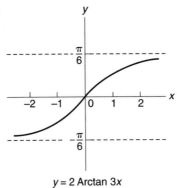

$y = 2 \text{ Arctan } 3x$

**Chapter 13 Review, Pages 494–495**

**11.** $\frac{1}{2}\sin 2\theta$    **12.** $\cos 6\theta$    **13.** $\cos^2 2\theta$    **14.** $\cos\dfrac{2\theta}{3}$    **15.** $\cos 5x$    **16.** $\sin x$    **17.** $-\frac{120}{169}$    **18.** $-\dfrac{\sqrt{5}}{5}$

**19.** $\dfrac{\pi}{2}, \dfrac{3\pi}{2}, \dfrac{\pi}{3}, \dfrac{5\pi}{3}$    **20.** $\dfrac{\pi}{6}, \dfrac{5\pi}{6}, \dfrac{7\pi}{6}, \dfrac{11\pi}{6}$    **21.** $\dfrac{\pi}{6}, \dfrac{5\pi}{6}, \dfrac{3\pi}{2}, \dfrac{7\pi}{18}, \dfrac{11\pi}{18}, \dfrac{19\pi}{18}, \dfrac{23\pi}{18}, \dfrac{31\pi}{18}, \dfrac{35\pi}{18}$    **22.** $0, \pi$

**23.** $\dfrac{3\pi}{2}$    **24.** $\dfrac{\pi}{8}, \dfrac{5\pi}{8}, \dfrac{9\pi}{8}, \dfrac{13\pi}{8}$    **25.** $\dfrac{\pi}{6}, \dfrac{5\pi}{6}$    **26.** $\dfrac{5\pi}{6}, \dfrac{11\pi}{6}$    **27.** $45°, 315°$    **28.** $120°, 240°$    **29.** $210°$

**30.** $\dfrac{2\pi}{3}$    **31.** $150° + n \cdot 360°, 330° + n \cdot 360°$, for every integer $n$    **32.** $\dfrac{3\pi}{2} + 2n\pi$, for every integer $n$

**33.** $x = \frac{4}{3}\arcsin 2y$    **34.** $\dfrac{\pi}{4}$    **35.** $-\dfrac{\pi}{6}$    **36.** $\pi$    **37.** $\dfrac{2\pi}{3}$    **38.** $\dfrac{\sqrt{3}}{2}$    **39.** $\sqrt{3}$    **40.** $\dfrac{\sqrt{x^2 + 1}}{x^2 + 1}$

**41.**

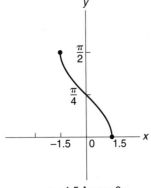

$y = 1.5 \text{ Arccos } 2x$

**CHAPTER 14**

**Exercises 14.1, Pages 501–502**

**1.** $7j$    **3.** $8j$    **5.** $2j\sqrt{3}$    **7.** $3j\sqrt{6}$    **9.** $-j$    **11.** $-1$    **13.** $j$    **15.** $-1$    **17.** $12 + 6j$    **19.** $2 - 8j$

**21.** $-j$    **23.** $19 + 2j$    **25.** $-22 + 7j$    **27.** $29$    **29.** $-7 - 24j$    **31.** $\dfrac{5 + 31j}{17}$    **33.** $-\dfrac{3j}{4}$    **35.** $\dfrac{-7 + 3j}{4}$

**37.** $2j, -2j$    **39.** $\dfrac{-2 \pm j\sqrt{23}}{3}$    **41.** $\dfrac{1 \pm 2j\sqrt{6}}{5}$    **43.** $\dfrac{1 \pm j\sqrt{2}}{3}$    **45.** $-1, \dfrac{1 \pm j\sqrt{3}}{2}$    **47.** $1, -1, j, -j$

**49.** $0, \pm4j\sqrt{5}$    **51.** $0, -3, \dfrac{3 \pm 3j\sqrt{3}}{2}$    **53.** $0, 1, \dfrac{-1 \pm j\sqrt{3}}{2}$    **55–59.**

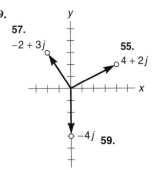

**57.** $-2 + 3j$    **55.** $4 + 2j$    **59.** $-4j$

**61.** $1 + 3j$    **63.** $4 - 8j$    **65.** $3 - 6j$    **67.** $7j$

**Exercises 14.2, Page 505**

**1.** $2\sqrt{2}(\cos 45° + j \sin 45°)$    **3.** $2(\cos 240° + j \sin 240°)$    **5.** $4(\cos 90° + j \sin 90°)$
**7.** $6\sqrt{2}(\cos 225° + j \sin 225°)$    **9.** $\sqrt{13}(\cos 124° + j \sin 124°)$    **11.** $2 + 2j\sqrt{3}$    **13.** $\sqrt{3} - j$    **15.** $-3 + 3j$
**17.** $-3j$    **19.** $7.00 - 2.01j$    **21.** $2e^{5.76j}$    **23.** $2e^{3.93j}$    **25.** $2\sqrt{13}e^{0.983j}$, or $7.21e^{0.983j}$
**27.** $3(\cos 77° + j \sin 77°) = 0.67 + 2.92j$    **29.** $4(\cos 330° + j \sin 330°) = 3.46 - 2j$
**31.** $2(\cos 57° + j \sin 57°) = 1.09 + 1.68j$

**Exercises 14.3, Pages 508–509**

**1.** $28e^{4j}$    **3.** $27e^{2j}$    **5.** $24e^{9.3j}$ or $24e^{3.0j}$    **7.** $-12 - 12\sqrt{3}j$    **9.** $-74.7 - 28.7j$    **11.** $-39.5 + 36.8j$

**13.** $12(\cos 113° + j \sin 113°)$    **15.** $9(\cos 300° + j \sin 300°)$    **17.** $7(\cos 135° + j \sin 135°)$    **19.** $\dfrac{25}{2} + \dfrac{25\sqrt{3}}{2}j$

**21.** $-12$    **23.** $54.1 + 14.5j$    **25.** $\dfrac{e^{4j}}{3}$    **27.** $4e^{-7j}$ or $4e^{5.57j}$    **29.** $\dfrac{e^{-2.2j}}{3}$ or $\dfrac{e^{4.1j}}{3}$    **31.** $-\sqrt{3} - j$

**33.** $-6.91 - 1.10j$    **35.** $3.94 - 3.08j$    **37.** $5(\cos 70° + j \sin 70°)$    **39.** $6(\cos 119° + j \sin 119°)$

**41.** $8(\cos 231° + j \sin 231°)$    **43.** $-\dfrac{9}{2} + \dfrac{9\sqrt{3}}{2}j$    **45.** $-\dfrac{\sqrt{3}}{10} + \dfrac{1}{10}j$    **47.** $3 - 3\sqrt{3}\,j$    **49.** $-8$    **51.** $-81j$

**Exercises 14.4, Page 514**

**1.** $243e^{7j}$ or $243e^{0.72j}$    **3.** $25e^{9.2j}$ or $25e^{2.9j}$    **5.** $81(\cos 80° + j \sin 80°)$    **7.** $32(\cos 30° + j \sin 30°)$
**9.** $\frac{1}{8}(\cos 0° + j \sin 0°)$, or $\frac{1}{8}$    **11.** $16$    **13.** $-64j$    **15.** $16 - 16\sqrt{3}\,j$    **17.** $-\frac{1}{324}$
**19.** $1, -\dfrac{1}{2} + \dfrac{\sqrt{3}}{2}j, -\dfrac{1}{2} - \dfrac{\sqrt{3}}{2}j$    **21.** $0.951 + 0.309j, j, -0.951 + 0.309j, -0.588 - 0.809j, 0.588 - 0.809j$
**23.** $3(\cos 135° + j \sin 135°) = -2.12 + 2.12j, 3(\cos 255° + j \sin 255°) = -0.776 - 2.90j,$
$3(\cos 15° + j \sin 15°) = 2.90 + 0.776j$
**25.** $\cos 45° + j \sin 45° = \dfrac{\sqrt{2}}{2} + \dfrac{\sqrt{2}}{2}j, \cos 225° + j \sin 225° = -\dfrac{\sqrt{2}}{2} - \dfrac{\sqrt{2}}{2}j$
**27.** $\cos 0° + j \sin 0° = 1, \cos 72° + j \sin 72° = 0.309 + 0.951j, \cos 144° + j \sin 144° = -0.809 + 0.588j,$
$\cos 216° + j \sin 216° = -0.809 - 0.588j, \cos 288° + j \sin 288° = 0.309 - 0.951j$
**29.** $\cos 36° + j \sin 36° = 0.809 + 0.588j, \cos 108° + j \sin 108° = -0.309 + 0.951j, \cos 180° + j \sin 180° = -1,$
$\cos 252° + j \sin 252° = -0.309 - 0.951\,j, \cos 324° + j \sin 324° = 0.809 - 0.588j$
**31.** $2(\cos 45° + j \sin 45°) = \sqrt{2} + \sqrt{2}\,j, 2(\cos 135° + j \sin 135°) = -\sqrt{2} + \sqrt{2}\,j,$
$2(\cos 225° + j \sin 225°) = -\sqrt{2} - \sqrt{2}\,j, 2(\cos 315° + j \sin 315°) = \sqrt{2} - \sqrt{2}\,j$
**33.** $\sqrt{5}(\cos 33° + j \sin 33°) = 1.88 + 1.22j, \sqrt{5}(\cos 153° + j \sin 153°) = -1.99 + 1.02j,$
$\sqrt{5}(\cos 273° + j \sin 273°) = 0.117 - 2.23j$

**Chapter 14 Review, Pages 516–517**

**1.** $9j$    **2.** $3\sqrt{2}\,j$    **3.** $-1$    **4.** $-j$    **5.** $1$    **6.** $j$    **7.** $5 + 10j$    **8.** $3 - 4j$    **9.** $44 - 23j$    **10.** $5 - 12j$
**11.** $\frac{2}{17} - \frac{9}{17}j$    **12.** $\frac{4}{29} + \frac{19}{29}j$    **13.** $6j, -6j$    **14.** $\dfrac{-3 \pm \sqrt{7}\,j}{4}$    **15.** $\sqrt{2}(\cos 135° + j \sin 135°), \sqrt{2}e^{2.36j}$

**16.** $2(\cos 300° + j \sin 300°)$, $2e^{5.24j}$ **17.** $3\sqrt{2} - 3\sqrt{2}\,j$, $6e^{5.50j}$ **18.** $-2\sqrt{3} - 2j$, $4e^{3.67j}$

**19.** $2(\cos 28° + j \sin 28°) = 1.77 + 0.94j$ **20.** $3(\cos 141° + j \sin 141°) = -2.33 + 1.89j$ **21.** $15e^{6j}$

**22.** $8(\cos 90° + j \sin 90°)$ **23.** $4e^{5j}$ **24.** $3(\cos 235° + j \sin 235°)$ **25.** $64e^{6j}$ **26.** $128(\cos 60° + j \sin 60°)$

**27.** $-64$ **28.** $64$ **29.** $-\frac{1}{4}$

**30.** $\cos 30° + j \sin 30° = \dfrac{\sqrt{3}}{2} + \tfrac{1}{2}j$, $\cos 150° + j \sin 150° = -\dfrac{\sqrt{3}}{2} + \tfrac{1}{2}j$, $\cos 270° + j \sin 270° = -j$

**31.** $\cos 45° + j \sin 45° = \dfrac{\sqrt{2}}{2} + \dfrac{\sqrt{2}}{2}j$, $\cos 135° + j \sin 135° = -\dfrac{\sqrt{2}}{2} + \dfrac{\sqrt{2}}{2}j$, $\cos 225° + j \sin 225° = -\dfrac{\sqrt{2}}{2} - \dfrac{\sqrt{2}}{2}j$,

$\cos 315° + j \sin 315° = \dfrac{\sqrt{2}}{2} - \dfrac{\sqrt{2}}{2}j$

**32.** $2(\cos 0° + j \sin 0°) = 2$, $2(\cos 90° + j \sin 90°) = 2j$, $2(\cos 180° + j \sin 180°) = -2$, $2(\cos 270° + j \sin 270°) = -2j$

**33.** $\sqrt{2}(\cos 63° + j \sin 63°) = 0.642 + 1.26j$, $\sqrt{2}(\cos 135° + j \sin 135°) = -1 + j$,

$\sqrt{2}(\cos 207° + j \sin 207°) = -1.26 - 0.642j$, $\sqrt{2}(\cos 279° + j \sin 279°) = 0.221 - 1.40j$,

$\sqrt{2}(\cos 351° + j \sin 351°) = 1.40 - 0.221j$

## CHAPTER 15

**Exercises 15.1, Page 523**

**1.**

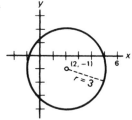

**3.**

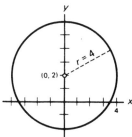

**5.** $(x - 1)^2 + (y + 1)^2 = 16$ **7.** $(x + 2)^2 + (y + 4)^2 = 34$ **9.** $x^2 + y^2 = 36$ **11.** $(0, 0)$; $r = 4$

**13.** $(-3, 4)$; $r = 8$ **15.** $(4, -6)$; $r = 2\sqrt{15}$ **17.** $(6, 1)$; $r = 7$ **19.** $(-\tfrac{7}{2}, -\tfrac{3}{2})$; $r = \sqrt{94}/2$

**21.** $x^2 + y^2 - 2y - 9 = 0$; $(0, 1)$; $r = \sqrt{10}$ **23.** $x^2 + y^2 + 10x - 40y = 0$; $(-5, 20)$; $r = 5\sqrt{17}$

**Exercises 15.2, Pages 531–532**

**1.**

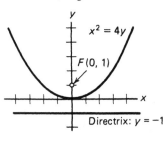

**3.**

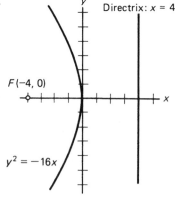

**5.**

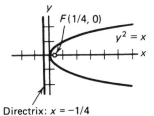

$F(1/4, 0)$

$y^2 = x$

Directrix: $x = -1/4$

**7.**

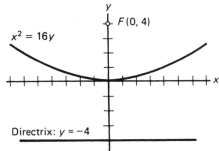

$x^2 = 16y$

$F(0, 4)$

Directrix: $y = -4$

**9.**

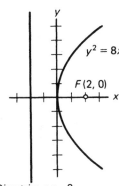

$y^2 = 8x$

$F(2, 0)$

Directrix: $x = -2$

**11.** $y^2 = 8x$    **13.** $y^2 = -32x$    **15.** $x^2 = 24y$    **17.** $y^2 = -16x$    **19.** $y^2 - 6y + 8x + 1 = 0$    **21.** 15 m, 7 m

**23.** $x^2 = 32y$

**25.**

$y = 2x^2 + 7x - 15$

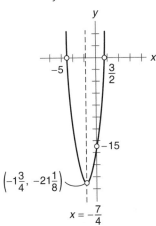

$-5$

$\dfrac{3}{2}$

$-15$

$\left(-1\dfrac{3}{4}, -21\dfrac{1}{8}\right)$

$x = -\dfrac{7}{4}$

**27.** $f(x) = -2x^2 + 4x + 16$

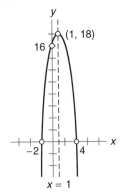

$(1, 18)$

$16$

$-2$    $4$

$x = 1$

**29. (a)** 1024 m    **(b)** 1024 m    **31.** 3600 m$^2$

**Exercises 15.3, Page 537**

| Vertices | Foci | Major axis | Minor axis |
|---|---|---|---|
| **1.** (5, 0) and (−5, 0) | (3, 0) and (−3, 0) | 10 | 8 |
| **3.** (4, 0) and (−4, 0) | ($\sqrt{7}$, 0) and (−$\sqrt{7}$, 0) | 8 | 6 |
| **5.** (0, 6) and (0, −6) | (0, $\sqrt{35}$) and (0, −$\sqrt{35}$) | 12 | 2 |
| **7.** (0, 4) and (0, −4) | (0, $\sqrt{7}$) and (0, −$\sqrt{7}$) | 8 | 6 |

**1.**

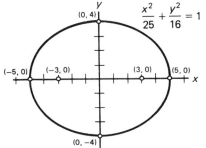

$(0, 4)$

$\dfrac{x^2}{25} + \dfrac{y^2}{16} = 1$

$(-5, 0)$    $(-3, 0)$    $(3, 0)$    $(5, 0)$

$(0, -4)$

**3.**

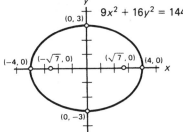

$9x^2 + 16y^2 = 144$

$(0, 3)$

$(-4, 0)$    $(-\sqrt{7}, 0)$    $(\sqrt{7}, 0)$    $(4, 0)$

$(0, -3)$

ANSWERS TO ODD-NUMBERED EXERCISES AND TO CHAPTER REVIEWS    **653**

**5.**

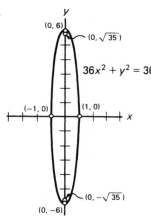

$36x^2 + y^2 = 36$

Labels: $(0, 6)$, $(0, \sqrt{35})$, $(-1, 0)$, $(1, 0)$, $(0, -\sqrt{35})$, $(0, -6)$

**7.**

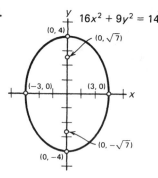

$16x^2 + 9y^2 = 144$

Labels: $(0, 4)$, $(0, \sqrt{7})$, $(-3, 0)$, $(3, 0)$, $(0, -\sqrt{7})$, $(0, -4)$

**9.** $\dfrac{x^2}{16} + \dfrac{y^2}{12} = 1$ or $3x^2 + 4y^2 = 48$  **11.** $\dfrac{x^2}{45} + \dfrac{y^2}{81} = 1$ or $9x^2 + 5y^2 = 405$

**13.** $\dfrac{x^2}{36} + \dfrac{y^2}{25} = 1$ or $25x^2 + 36y^2 = 900$  **15.** $\dfrac{x^2}{39} + \dfrac{y^2}{64} = 1$ or $64x^2 + 39y^2 = 2496$

**17.** $\dfrac{x^2}{5300^2} + \dfrac{y^2}{5292^2} = 1$

**Exercises 15.4, Page 543**

| | Vertices | Foci | Transverse axis | Conjugate axis | Asymptotes |
|---|---|---|---|---|---|
| **1.** | $(5, 0)$ and $(-5, 0)$ | $(13, 0)$ and $(-13, 0)$ | 10 | 24 | $y = \pm\frac{12}{5}x$ |
| **3.** | $(0, 3)$ and $(0, -3)$ | $(0, 5)$ and $(0, -5)$ | 6 | 8 | $y = \pm\frac{3}{4}x$ |
| **5.** | $(\sqrt{2}, 0)$ and $(-\sqrt{2}, 0)$ | $(\sqrt{7}, 0)$ and $(-\sqrt{7}, 0)$ | $2\sqrt{2}$ | $2\sqrt{5}$ | $y = \pm\sqrt{\frac{5}{2}}x$ |
| **7.** | $(0, 1)$ and $(0, -1)$ | $(0, \sqrt{5})$ and $(0, -\sqrt{5})$ | 2 | 4 | $y = \pm\frac{1}{2}x$ |

**1.**

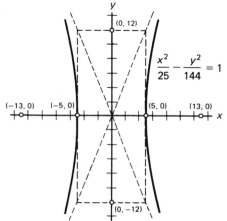

$\dfrac{x^2}{25} - \dfrac{y^2}{144} = 1$

Labels: $(0, 12)$, $(-13, 0)$, $(-5, 0)$, $(5, 0)$, $(13, 0)$, $(0, -12)$

**3.**

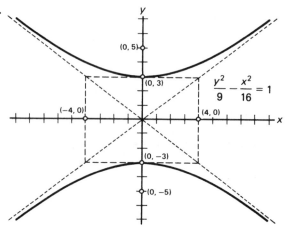

$\dfrac{y^2}{9} - \dfrac{x^2}{16} = 1$

Labels: $(0, 5)$, $(0, 3)$, $(-4, 0)$, $(4, 0)$, $(0, -3)$, $(0, -5)$

**5.**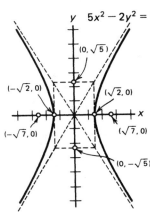

$5x^2 - 2y^2 = 10$

$(0, \sqrt{5})$
$(-\sqrt{2}, 0)$  $(\sqrt{2}, 0)$
$(-\sqrt{7}, 0)$  $(\sqrt{7}, 0)$
$(0, -\sqrt{5})$

**7.**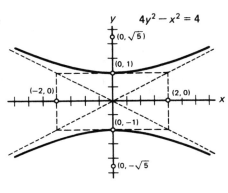

$4y^2 - x^2 = 4$

$(0, \sqrt{5})$
$(0, 1)$
$(-2, 0)$  $(2, 0)$
$(0, -1)$
$(0, -\sqrt{5})$

**9.** $\dfrac{x^2}{16} - \dfrac{y^2}{20} = 1$  or  $5x^2 - 4y^2 = 80$  **11.** $\dfrac{y^2}{36} - \dfrac{x^2}{28} = 1$  or  $7y^2 - 9x^2 = 252$

**13.** $\dfrac{x^2}{9} - \dfrac{y^2}{25} = 1$  or  $25x^2 - 9y^2 = 225$  **15.** $\dfrac{x^2}{25} - \dfrac{y^2}{11} = 1$  or  $11x^2 - 25y^2 = 275$

**17.**

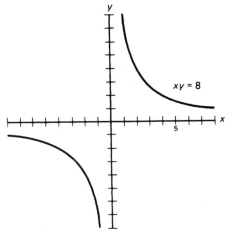

$xy = 8$

**Exercises 15.5, Pages 548–549**

**1.** $\dfrac{(x-1)^2}{16} + \dfrac{(y+1)^2}{12} = 1$  **3.** $\dfrac{(y-1)^2}{36} - \dfrac{(x-1)^2}{28} = 1$  **5.** $(y+1)^2 = 8(x-3)$

**7.** Parabola; vertex: $(2, -3)$

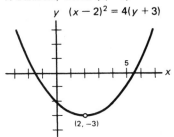

$y$  $(x-2)^2 = 4(y+3)$
$(2, -3)$

**9.** Hyperbola; center: $(-2, 0)$

$\dfrac{y^2}{9} - \dfrac{(x+2)^2}{16} = 1$

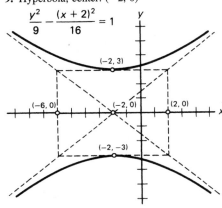

$(-2, 3)$
$(-6, 0)$  $(-2, 0)$  $(2, 0)$
$(-2, -3)$

**11.** Ellipse; center: (2, 0)

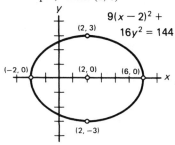

$9(x - 2)^2 + 16y^2 = 144$

**13.** Ellipse; center: (3, 1)

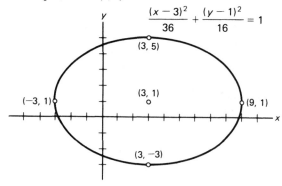

$$\frac{(x - 3)^2}{36} + \frac{(y - 1)^2}{16} = 1$$

**15.** Parabola; vertex: (1, −3)

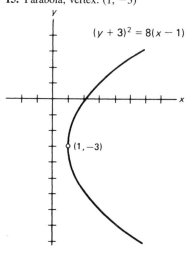

$(y + 3)^2 = 8(x - 1)$

**17.** Hyperbola; center: (−1, −1)

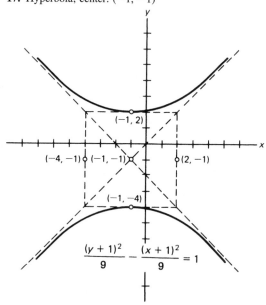

$$\frac{(y + 1)^2}{9} - \frac{(x + 1)^2}{9} = 1$$

**19.** Parabola; vertex: (2, −1)

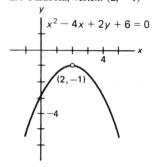

$x^2 - 4x + 2y + 6 = 0$

**21.** Ellipse; center: (−2, 1)

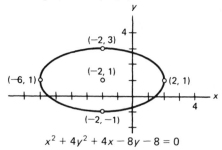

$x^2 + 4y^2 + 4x - 8y - 8 = 0$

**23.** Hyperbola; center: $(1, 1)$

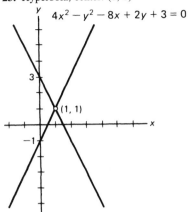

$$4x^2 - y^2 - 8x + 2y + 3 = 0$$

**25.** Hyperbola; center: $(-3, 3)$

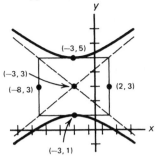

$$\frac{(y - 3)^2}{4} - \frac{(x + 3)^2}{25} = 1$$

**27.** Parabola; vertex: $(-8, -2)$

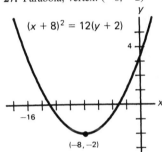

$$(x + 8)^2 = 12(y + 2)$$

**29.** Ellipse; center: $(-6, -2)$

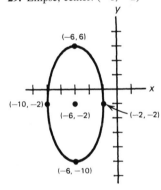

$$\frac{(x + 6)^2}{16} + \frac{(y + 2)^2}{64} = 1$$

**Exercises 15.6, Page 550**

**1.** Ellipse   **3.** Parabola   **5.** Hyperbola   **7.** Circle   **9.** Circle   **11.** Ellipse   **13.** Hyperbola
**15.** Parabola

**Exercises 15.7, Pages 553–554**

**1.** $(3, 3)$   **3.** $(1, \sqrt{3}), (1, -\sqrt{3})$   **5.** $(2\sqrt{6}, 6), (-2\sqrt{6}, 6)$   **7.** $(-2, 0), (2, 0)$   **9.** $(-6, 6), (6, 6)$
**11.** $(-2, 2), (-2, -2)$   **13.** $(2.3, 5.5), (-2.3, 5.5)$   **15.** $(5, 4), (5, -4), (-5, 4), (-5, -4)$
**17.** $(1, 4), (-1, -4), (4, 1), (-4, -1)$

**Exercises 15.8, Pages 561–562**

**1.**

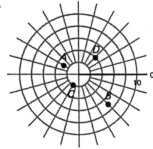

**3.**

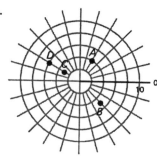

**5.** $(-3, 240°), (-3, -120°), (3, -300°)$   **7.** $(5, 135°), (-5, -45°), (5, -225°)$

**9.** $(-4, -315°), (-4, 45°), (4, 225°)$   **11.** $(-3, 7\pi/6), (-3, -5\pi/6), (3, -11\pi/6)$

**13.** $(9, 5\pi/3), (9, -\pi/3), (-9, -4\pi/3)$   **15.** $(4, -3\pi/4), (-4, \pi/4), (4, 5\pi/4)$

**17.**

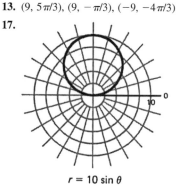

$$r = 10 \sin \theta$$

**19.**

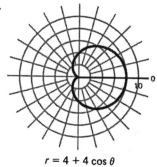

$$r = 4 + 4 \cos \theta$$

**21.**

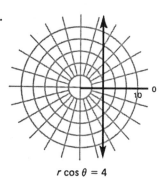

$$r \cos \theta = 4$$

**23.**

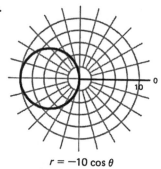

$$r = -10 \cos \theta$$

**25.**

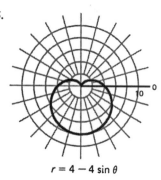

$$r = 4 - 4 \sin \theta$$

**27.**

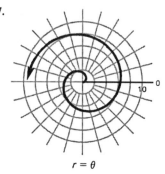

$$r = \theta$$

**29.** $(3\sqrt{3}/2, 3/2)$ **31.** $(1, \sqrt{3})$ **33.** $(2\sqrt{3}, -2)$ **35.** $(0, 6)$ **37.** $(2.5, -4.33)$ **39.** $(1.4, 1.4)$
**41.** $(7.1, 45°)$ **43.** $(4, 90°)$ **45.** $(4, 240°)$ **47.** $(4\sqrt{2}, 3\pi/4)$ **49.** $(2\sqrt{2}, 5\pi/6)$ **51.** $(4, 3\pi/2)$
**53.** $r \cos \theta = 3$ **55.** $r = 6$ **57.** $r + 2 \cos \theta + 5 \sin \theta = 0$ **59.** $r = 12/(4 \cos \theta - 3 \sin \theta)$
**61.** $r^2 = 36/(9 - 5 \sin^2 \theta)$ **63.** $r = 4 \sec \theta \tan^2 \theta$ **65.** $y = -3$ **67.** $x^2 + y^2 = 25$ **69.** $y = x$
**71.** $x^2 + y^2 - 5x = 0$ **73.** $x^2 + y^2 - 3x + 3\sqrt{3}y = 0$ **75.** $y^2 = 3x$ **77.** $xy = 1$
**79.** $x^4 + 2x^2y^2 + y^4 - 2xy = 0$ **81.** $y^2 = x^2(x^2 + y^2)$ **83.** $x^2 + 6y - 9 = 0$
**85.** $x^4 + 2x^2y^2 + y^4 + 4y^3 - 12x^2y = 0$ **87.** $x^4 + 2x^2y^2 + y^4 - 8x^2y - 8y^3 - 4x^2 + 12y^2 = 0$ **89.** $\sqrt{13}$
**91.** $d = \sqrt{r_1^2 + r_2^2 - 2r_1r_2 \cos(\theta_1 - \theta_2)}$

**Exercises 15.9, Page 569**

**1.**

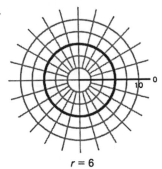

$r = 6$

**3.**

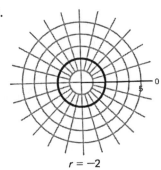

$r = -2$

**5.**

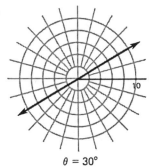

$\theta = 30°$

**7.**

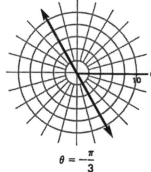

$\theta = -\dfrac{\pi}{3}$

**9.**

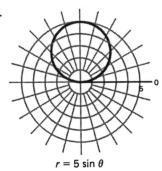

$r = 5 \sin \theta$

**11.**

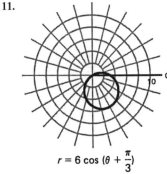

$r = 6 \cos \left(\theta + \dfrac{\pi}{3}\right)$

**13.**

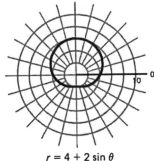

$r = 4 + 2 \sin \theta$

**15.**

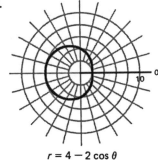

$r = 4 - 2 \cos \theta$

**17.**

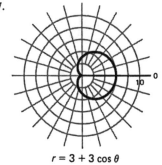

$r = 3 + 3 \cos \theta$

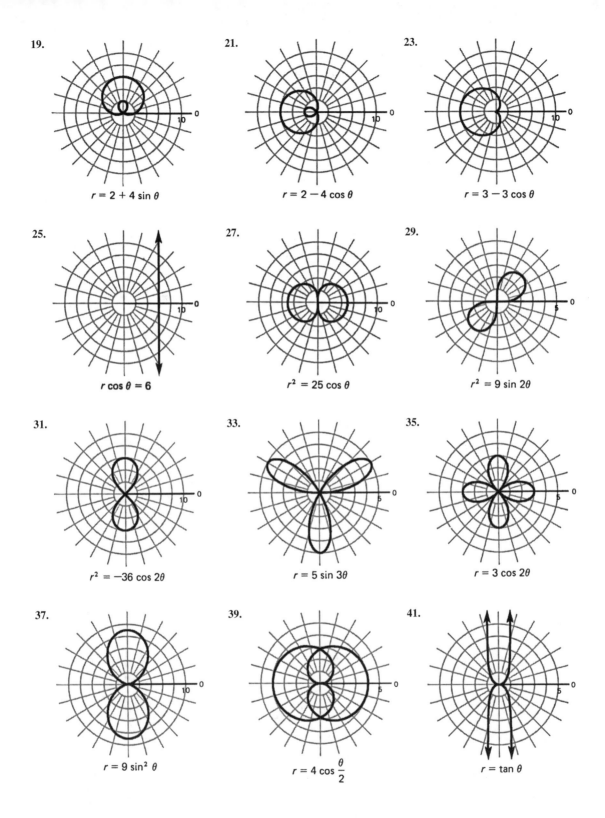

**19.**

$r = 2 + 4 \sin \theta$

**21.**

$r = 2 - 4 \cos \theta$

**23.**

$r = 3 - 3 \cos \theta$

**25.**

$r \cos \theta = 6$

**27.**

$r^2 = 25 \cos \theta$

**29.**

$r^2 = 9 \sin 2\theta$

**31.**

$r^2 = -36 \cos 2\theta$

**33.**

$r = 5 \sin 3\theta$

**35.**

$r = 3 \cos 2\theta$

**37.**

$r = 9 \sin^2 \theta$

**39.**

$r = 4 \cos \dfrac{\theta}{2}$

**41.**

$r = \tan \theta$

**43.**

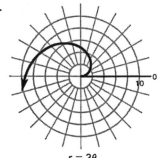

$r = 3\theta$

**45.**

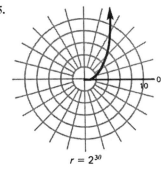

$r = 2^{3\theta}$

**47.**

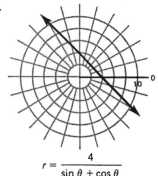

$r = \dfrac{4}{\sin\theta + \cos\theta}$

**49.**

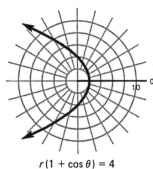

$r(1 + \cos\theta) = 4$

**Chapter 15 Review, Pages 572–573**

**1.** $(x - 5)^2 + (y + 7)^2 = 36$   or   $x^2 + y^2 - 10x + 14y + 38 = 0$     **2.** $(4, -3); 7$
**3.** $(0, \frac{3}{2}); y = -\frac{3}{2}$

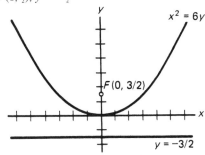

**4.** $y^2 = -16x$     **5.** $(y - 3)^2 = 8(x - 2)$   or   $y^2 - 6y - 8x + 25 = 0$
**6.** $V(7, 0), (-7, 0); F(3\sqrt{5}, 0), (-3\sqrt{5}, 0)$

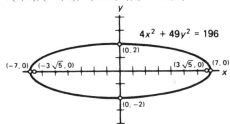

**7.** $\dfrac{x^2}{4} + \dfrac{y^2}{16} = 1$ or $4x^2 + y^2 = 16$

**8.** $V(6, 0), (-6, 0); F(2\sqrt{13}, 0), (-2\sqrt{13}, 0)$

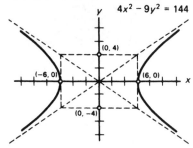

$4x^2 - 9y^2 = 144$

**9.** $\dfrac{y^2}{25} - \dfrac{x^2}{16} = 1$ or $16y^2 - 25x^2 = 400$  **10.** $\dfrac{(x-3)^2}{9} + \dfrac{(y+4)^2}{25} = 1$  **11.** $\dfrac{(x+7)^2}{81} - \dfrac{(y-4)^2}{9} = 1$

**12.** Hyperbola

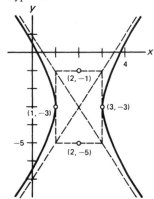

$16x^2 - 4y^2 - 64x - 24y + 12 = 0$

**13.** $(0, 0), (-12, -6)$  **14.** $(4, \sqrt{3}), (-4, \sqrt{3}), (4, -\sqrt{3}), (-4, -\sqrt{3})$

**15.**

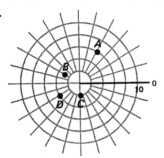

**16.**

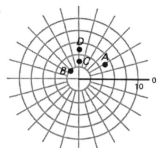

**17.** $(5, -225°), (-5, -45°), (-5, 315°)$  **18.** $(2, -11\pi/6), (-2, -5\pi/6), (2, \pi/6)$  **19.** $(-2.6, -1.5)$

**20.** $(-1, -1.7)$  **21.** $(-4.3, 2.5)$  **22.** $(0, 6)$  **23.** $(4.2, 135°)$  **24.** $(6, 270°)$  **25.** $(2, 120°)$  **26.** $(5, \pi)$

**27.** $(12, 5\pi/6)$  **28.** $(\sqrt{2}, 7\pi/4)$  **29.** $r = 7$  **30.** $r \sin^2 \theta = 9 \cos \theta$  **31.** $r = 8/(5 \cos \theta + 2 \sin \theta)$

**32.** $r^2 = 12/(1 - 5 \sin^2 \theta)$  **33.** $r = 6 \csc \theta \cot^2 \theta$  **34.** $r = \cos \theta \cot \theta$  **35.** $x = 12$  **36.** $x^2 + y^2 = 81$

**37.** $y = -\sqrt{3}x$  **38.** $x^2 + y^2 - 8x = 0$  **39.** $y^2 = 5x$  **40.** $xy = 4$  **41.** $x^4 + 2x^2y^2 + y^4 + 4y^2 - 4x^2 = 0$

**42.** $y = 1$  **43.** $x^4 + y^4 + 2x^2y^2 - 2x^2y - 2y^3 - x^2 = 0$  **44.** $x^2 = 4(y + 1)$

**45.**

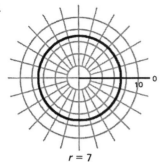

$r = 7$

**46.**

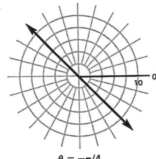

$\theta = -\pi/4$

**47.**

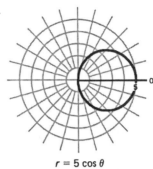

$r = 5 \cos \theta$

**48.**

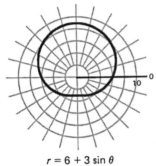

$r = 6 + 3 \sin \theta$

**49.**

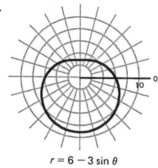

$r = 6 - 3 \sin \theta$

**50.**

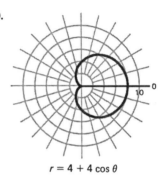

$r = 4 + 4 \cos \theta$

**51.**

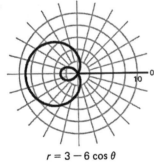

$r = 3 - 6 \cos \theta$

**52.**

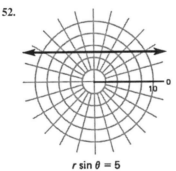

$r \sin \theta = 5$

**53.**

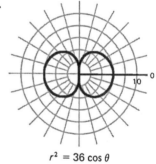

$r^2 = 36 \cos \theta$

**54.**

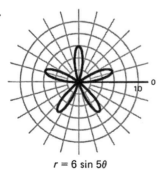

$r = 6 \sin 5\theta$

**55.**

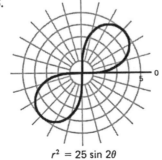

$r^2 = 25 \sin 2\theta$

**56.**

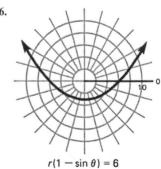

$r(1 - \sin \theta) = 6$

**Exercises C.1, Pages 586–587**

**1.** 14    **2.** −50    **3.** −144    **4.** −47    **5.** −65    **6.** 1    **7.** 0    **8.** −20    **9.** −122    **10.** −42
**11.** 430    **12.** 99

# Index

## Trigonometric Functions

1. $\sin \theta = \dfrac{y}{r}$

2. $\cos \theta = \dfrac{x}{r}$

3. $\tan \theta = \dfrac{y}{x}$

4. $\cot \theta = \dfrac{x}{y}$

5. $\sec \theta = \dfrac{r}{x}$

6. $\csc \theta = \dfrac{r}{y}$

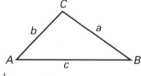

$P(x, y)$

$r = \sqrt{x^2 + y^2}$

## Geometric Formulas
### Area

Triangle: $A = \dfrac{1}{2}bh$

Rectangle: $A = \ell w$

Parallelogram: $A = bh$

Trapezoid: $A = \dfrac{1}{2}h(a + b)$

Circle: $A = \pi r^2$; (Circumference: $C = \pi d = 2\pi r$)

### Volume

Prism: $V = Bh$

Cylinder: $V = \pi r^2 h$

Pyramid: $V = \dfrac{1}{3}Bh$

Cone: $V = \dfrac{1}{3}\pi r^2 h$

Sphere: $V = \dfrac{4}{3}\pi r^3$

## Oblique Triangles

Law of Sines: $\dfrac{a}{\sin A} = \dfrac{b}{\sin B} = \dfrac{c}{\sin C}$

Law of Cosines: $a^2 = b^2 + c^2 - 2bc \cos A$

## Complex Numbers

$j = \sqrt{-1}, j^2 = -1, j^3 = -j, j^4 = 1, j^5 = j, \ldots$

| Rectangular Form | Trigonometric Form | Exponential Form |
|---|---|---|
| $a + bj$ | $= r(\cos \theta + j \sin \theta)$ | $= re^{j\theta}$ |

$r_1(\cos \theta_1 + j \sin \theta_1) \cdot r_2(\cos \theta_2 + j \sin \theta_2) = r_1 r_2[\cos(\theta_1 + \theta_2) + j \sin(\theta_1 + \theta_2)]$

$\dfrac{r_1(\cos \theta_1 + j \sin \theta_1)}{r_2(\cos \theta_2 + j \sin \theta_2)} = \dfrac{r_1}{r_2}[\cos(\theta_1 - \theta_2) + j \sin(\theta_1 - \theta_2)]$

## DeMoivre's Theorem

$[r(\cos \theta + j \sin \theta)]^n = r^n(\cos n\theta + j \sin n\theta)$